Lecture Notes in Artificial Intelligence 8992

Subseries of Lecture Notes in Computer Science

LNAI Series Editors

Randy Goebel
  University of Alberta, Edmonton, Canada
Yuzuru Tanaka
  Hokkaido University, Sapporo, Japan
Wolfgang Wahlster
  DFKI and Saarland University, Saarbrücken, Germany

LNAI Founding Series Editor

Joerg Siekmann
  DFKI and Saarland University, Saarbrücken, Germany

More information about this series at http://www.springer.com/series/1244

Reinaldo A.C. Bianchi · H. Levent Akin
Subramanian Ramamoorthy · Komei Sugiura (Eds.)

RoboCup 2014:
Robot World Cup XVIII

Springer

Editors

Reinaldo A.C. Bianchi
FEI University
São Bernardo do Campo
Brazil

H. Levent Akin
Boğaziçi University
Istanbul
Turkey

Subramanian Ramamoorthy
University of Edinburgh
Edinburgh
UK

Komei Sugiura
National Institute of Information
    and Communications Technology
Kyoto
Japan

ISSN 0302-9743   ISSN 1611-3349 (electronic)
Lecture Notes in Artificial Intelligence
ISBN 978-3-319-18614-6   ISBN 978-3-319-18615-3 (eBook)
DOI 10.1007/978-3-319-18615-3

Library of Congress Control Number: 2015938428

LNCS Sublibrary: SL7 – Artificial Intelligence

Springer Cham Heidelberg New York Dordrecht London

Printed on acid-free paper

Springer International Publishing AG Switzerland is part of Springer Science+Business Media
(www.springer.com)

Preface

The RoboCup – Robot World Cup – 2014 took place at the Convention Center "Poeta Ronaldo Cunha Lima" in João Pessoa, Brazil, from July 19–24. The RoboCup 2014 competition had a total of 2,631 participants from 45 countries in the Major and Junior leagues. If we consider also the exhibitors, staff, local organization, and volunteers, RoboCup 2014 had more than 2,900 participants. Approximately 100,000 visitors from different Brazilian states attended RoboCup 2014 during the five days that it was open to the general public.

The 18th annual RoboCup International Symposium was held in the same venue, on July 25th, and was attended by more than 800 members of the RoboCup Community. We also had 186 students from Brazilian universities, attending only the RoboCup Symposium.

For the 18th RoboCup International Symposium, we received a total of 66 regular paper submissions, which were reviewed carefully by the International Program Committee, consisting of 54 members. Each paper was reviewed by at least three reviewers. Overall, we accepted 36 of those (54%), of which 10 were chosen for oral presentation.

Besides the regular papers, this year we had three special tracks: Champion Teams, Open Source Development, and Advancement of the RoboCup Leagues Tracks.

The Champion Teams Papers Track contains 11 papers written by the teams who won the competition. This track is used to document and disseminate the state of the art in each League, and to introduce new ideas and promising technologies.

The Open Source Software and Hardware Development Track was introduced in 2013. This track encourages the sharing of technology, software implementations, and hardware projects among RoboCup researchers. This year we had three papers published in this track, including one that won the Best Paper Award for its engineering contribution, selected from among all papers submitted to the Symposium.

This year we introduced a new track to the Symposium: the Special Track on the Advancement of RoboCup Leagues. The aim of this track was to foster discussion on the future of RoboCup, league by league, with focus on changes in the next year and plans for future. Two panel discussions were held, one about the soccer leagues and another about the application leagues. The results of these discussions are reported in nine papers published in this book.

We also had two invited speakers: Rodney Brooks, Professor emeritus at MIT, robotics entrepreneur and Founder, Chairman and CTO of Rethink Robotics; and Sami Haddadin, Full professor and director of the Institute of Automatic Control (IRT) at Leibniz University Hanover. Professor Brooks' talk was about the Challenges for Robotics Research, which involve having robots that are able to support human society in the future, and making progress in three areas of research: Mobility, Manipulation, and Messiness. Professor Haddadin's talk was also focused on the interaction between robots and humans: he described his work on human-centered robot design, control, and planning, which may lead to robots that can safely act as human assistants and collaborators over a variety of application domains.

The award committee selected two papers, printed first in the book, as best papers:

- The Best Paper Award for its Scientific Contribution: Balanced Walking with Capture Steps, by Marcell Missura and Sven Behnke.
- The Best Paper Award for its Engineering Contribution: RoboCup@Home Spoken Corpus: Using Robotic Competitions for Gathering Datasets, by Emanuele Bastianelli, Luca Iocchi, Daniele Nardi, Giuseppe Castellucci, Danilo Croce, and Roberto Basili.

Finally, we would like to thank the chairs of RoboCup 2014, Flavio Tonidandel, Esther Luna Colombini, Alexandre da Silva Simões, and all the members of the Organizing Committee who brought together this splendid event; the Program Committee Members and additional reviewers, who ensured the quality of the contributions assembled in this book; and finally, all the participants and authors, for making this event a huge success.

March 2015

Reinaldo A.C. Bianchi
H. Levent Akin
Subramanian Ramamoorthy
Komei Sugiura

Organization

Symposium Co-chairs

Reinaldo A.C. Bianchi	FEI University, Brazil
H. Levent Akin	Boğaziçi University, Turkey
Subramanian Ramamoorthy	The University of Edinburgh, UK
Komei Sugiura	National Institute of Information and Communications Technology, Japan

Program Committee

Hidehisa Akiyama	Fukuoka University, Japan
Luís Almeida	University of Porto, Portugal
Jacky Baltes	University of Manitoba, Canada
Sven Behnke	University of Bonn, Germany
Ansgar Bredenfeld	Dr. Bredenfeld UG, Germany
Xiaoping Chen	University of Science and Technology of China, China
Esther Luna Colombini	FEI University, Brazil
Anna Helena Reali Costa	University of São Paulo, Brazil
Bernardo Cunha	University of Aveiro, Portugal
Klaus Dorer	Offenburg University of Applied Sciences, Germany
Amy Eguchi	Bloomfield College, USA
Bernhard Hengst	University of New South Wales, Australia
Dirk Holz	University of Bonn, Germany
Nobuhiro Ito	Aichi Institute of Technology, Japan
Jianmin Ji	University of Science and Technology of China, China
Gerhard Kraetzschmar	Bonn-Rhein-Sieg University of Applied Sciences, Germany
Tim Laue	DFKI Bremen, Germany
Pedro U. Lima	Instituto Superior Técnico, Portugal
Luis F. Lupian	La Salle University, México
Norbert Michael Mayer	National Chung Cheng University, Taiwan
Çetin Meriçli	Carnegie Mellon University, USA
Tekin Meriçli	Boğaziçi University, Turkey
Zhao Mingguo	Tsinghua University, China
Eduardo Morales	National Institute for Astrophysics, Optics and Electronics, México

Itsuki Noda   National Institute of Advanced Industrial Science
                             and Technology, Japan
Luis Paulo Reis   University of Minho, Portugal
Fernando Ribeiro   University of Minho, Portugal
Thomas Röfer   DFKI Bremen, Germany
Raul Rojas   Free University of Berlin, Germany
Javier Ruiz-del-Solar   University of Chile, Chile
Paulo E. Santos   FEI University, Brazil
Sanem Sariel-Talay   Istanbul Technical University, Turkey
Jesus Savage   National Autonomous University of Mexico,
                             México
Alexandre Simões   São Paulo State University UNESP, Brazil
Elizabeth Sklar   University of Liverpool, UK
Domenico G. Sorrenti   University of Milan-Bicocca, Italy
Mohan Sridharan   Texas Tech University, USA
Gerald Steinbauer   Graz University of Technology, Austria
Peter Stone   University of Texas at Austin, USA
Luis Enrique Sucar   National Institute for Astrophysics,
                             Optics and Electronics, México
Yasutake Takahashi   University of Fukui, Japan
Flavio Tonidandel   FEI University, Brazil
Manuela Veloso   Carnegie Mellon University, USA
Ubbo Visser   University of Miami, USA
Arnoud Visser   University of Amsterdam, The Netherlands
Oskar von Stryk   Technical University of Darmstadt, Germany
Alfredo Weitzenfeld   University of South Florida, USA
Feng Wu   University of Southampton, UK
Rong Xiong   Zhejiang University, China

Sponsoring Institutions

The National Council for Technological and Scientific Development – CNPq, Brazil.
Coordination for the Improvement of Higher Education Personnel – CAPES, Brazil.

Contents

Poster Presentations

Special Track on Open-Source Developments

Best Paper Award for its Scientific Contribution

Balanced Walking with Capture Steps

Marcell Missura[✉] and Sven Behnke

Autonomous Intelligent Systems, Computer Science,
University of Bonn, Bonn, Germany
{missura,behnke}@cs.uni-bonn.de
http://ais.uni-bonn.de

Abstract. Bipedal walking is one of the most essential skills required to play soccer with humanoid robots. Superior walking speed and stability often gives teams the winning edge when their robots are the first at the ball, maintain ball control, and drive the ball towards the opponent goal with sure feet. In this contribution, we present an implementation of our Capture Step Framework on a real soccer robot, and show robust omnidirectional walking. The robot not only manages to locomote on an even surface, but can also cope with various disturbances, such as pushes, collisions, and stepping on the feet of an opponent. The actuation is compliant and the robot walks with stretched knees.

1 Introduction

For the RoboCup initiative, which has the goal of defeating the human world champions in the game of soccer by the year of 2050, it is of particular interest to conceive a bipedal walk with human-like capabilities. However, the complexity of the walking motion, the formulation of sufficiently simple models that account for balance, and the difficulties that arise from controlling a humanoid body with a high number of degrees of freedom within a feedback loop, make this task particularly difficult. While a number of sophisticated approaches exist that promise some degree of robustness, the RoboCup experience shows that state of the art algorithms do not find their way into the dynamic world of low-cost robots competing on the soccer field. One of the reasons for this is simply that the required sensors, high-precision actuators, and computational power are not available on custom built prototypes and affordable standard platforms that are used in robotic soccer games. The amount of expertise required to successfully integrate a complex algorithm into already complex soccer software in a real robot environment is also not a negligible factor.

The Capture Step Framework [1] has been designed with the aforementioned limitations in mind. Using only postural information provided by motor encoders and an inertial measurement unit (IMU), it can produce a stable omnidirectional walk with push-recovery capabilities. Zero moment point control, foot placement, and step timing strategies are utilized simultaneously. The balance computations

This work is supported by Deutsche Forschungsgemeinschaft (German Research Foundation, DFG) under grants BE 2556/6 and BE 2556/10.

R.A.C. Bianchi et al. (Eds.): RoboCup 2014, LNAI 8992, pp. 3–15, 2015.
DOI: 10.1007/978-3-319-18615-3_1

are based on a two-dimensional linear inverted pendulum model, and can interface with already working open-loop gait pattern generators. The framework can operate on systems with high sensor noise, high latency, and imprecise actuation. Furthermore, it does not restrict the center of mass to a plane, or the feet to remain flat on the ground. In this paper, we demonstrate the capabilities of the framework on a real soccer robot in combination with compliant actuation and walking with stretched knees. We outline the theoretical concept of the Capture Step Framework, provide details about the implementation on a real robot, and show a video and experimental data as evidence of the most prominent features.

2   Related Work

Zero moment point (ZMP) tracking with preview control [2] is the most popular approach to bipedal walking to date. A number of footsteps planned ahead are used to define a future ZMP reference, e.g. by placing the ZMP in the center of the footsteps and allowing for a smooth transition from one foot to the other during the double support phase. A continuous center of mass (CoM) trajectory that minimizes the ZMP tracking error is then generated in a Model Predictive Control [3] setting. As long as the actual ZMP stays well inside the support polygon, stable walking is guaranteed. Using ZMP preview control, high quality hardware [4–6] can walk reliably on flat ground. Next generation walking controllers from the ZMP preview family [7–9] also consider foot placement in addition to ZMP control either by including the footstep locations in the optimization process, or by using a simplified model to compute a footstep plan online. The approaches that have matured beyond a theoretical state require either precise physical modeling, the estimation of an impact force, or the feedback of a measured ZMP location. All of these requirements are difficult to meet for soccer robots.

Recently, Urata et al. [10] presented an impressive foot placement-based controller on a real robot that is capable of recovering from strong pushes. Instead of optimizing the CoM trajectory for a single ZMP reference, a fast iterative method is used to sample a whole set of ZMP/CoM trajectory pairs for three steps into the future. Triggered by a disturbance, the algorithm selects the best available footstep plan according to given optimization criteria. Resampling during execution of the footstep plan is not possible. The robot has to be able to track a fixed motion trajectory for the duration of the recovery. Specialized hardware was used for meeting the precision requirements.

Englsberger et al. [11] proposed using a capture point trajectory as reference input for gait generation instead of the ZMP. As the capture point can easily be computed from the CoM state, it is more suitable for state feedback than the ZMP. We incorporated the core ZMP control equation as one of the building blocks of our framework. The capture point approach is potentially suitable for soccer robots, but it does not consider adaptive foot placement and step timing.

Focusing on the methods that are applied by leading teams of the Humanoid League, it is notable that the preferred algorithms are simple and light-weight.

Most of them are open-loop [12,13], sometimes with limited state feedback for posture control. Perhaps the most advanced closed-loop walk so far was presented for the Nao standard platform by Graf and Röfer [14], who proposed the online-adjustment of step parameters based on the solution of a system of linear pendulum equations. In the KidSize class, the DARwIn robot comes with a fast and reliable open-loop walk (Yi et al. [15]). Parameterized ZMP and CoM trajectories are generated analytically using simple linear inverted pendulum model equations. Zhao et al. [16] suggested an elegant open-loop gait generation technique inspired by passive walking down a shallow slope. This approach creates a virtual downwards slope by shortening the swing leg before support exchange and recharging energy by extending the leg again during the support phase.

The most closely related works published by the authors themselves are the Capture Step balance controller [1] and the open-loop central pattern generator [12] that were combined in this framework, and implemented on a real robot.

3   The Capture Step Framework

The Capture Step Framework is an omnidirectional bipedal gait generator with separable conceptual modules. The main software components are the state estimation, reference trajectory generation, balance control, and motion generation. The layout of the components is shown in Fig. 1. The framework simplifies the full-body dynamics of the robot to the trajectory of a single point mass, which is assumed to move like a linear inverted pendulum. In strong contrast to classic ZMP preview algorithms, motion

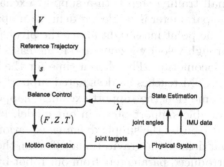

Fig. 1. Conceptual structure of our gait control architecture.

trajectories are expressed directly for the center of mass and adaptive footstep locations arise as the output of the trajectory generation process. In each iteration of the main control loop, an ideal CoM reference trajectory is computed that depends only on the desired walking velocity. The state estimation component maps the current pose of the robot to the position and velocity of a point mass. Then, the balance control module computes a zero moment point offset and an estimated time for the next support exchange to steer the current state towards the reference trajectory while preserving balance with an adequate step size. The timing and the step location encode a step motion that is generated on a lower level without further concern for balance.

3.1   Reference Trajectory Generation

It is common practice to compose a pendulum motion by superposing two uncoupled one-dimensional linear inverted pendulum models. One model describes the lateral motion of a point mass and the other the sagittal motion. The two models are synchronized at a shared moment of support exchange, where each model is reset to a post-step state. Figure 2 shows schematic trajectories for the lateral and the sagittal dimensions. There is an interesting conceptual difference between the two projections. In the sagittal dimension, the point mass crosses the pivot point in every step cycle. In the lateral dimension, however, the point mass oscillates between two supports and never crosses the pendulum pivot point.

We identify four parameters that characterize the walking motion. The lateral distance between the pivot point and the apex of the point mass trajectory is denoted as α. It is evident that the lateral component of the center of mass velocity equals zero in this point. As long as the apex distance is greater than zero, the point mass will return and is guaranteed to reach a support exchange location in a range bounded by δ and ω. When walking in place, we assume the support exchange to occur at the minimal distance δ. When walking with a nonzero lateral velocity, the walker first takes a long step with the leading leg and the support exchange occurs at a distance up to an upper bound ω, depending on the desired lateral walking velocity. The large leading step is followed by a small trailing step with a support exchange at δ. In the sagittal direction only one parameter is needed. σ defines an upper bound for the pass-through velocity of the point mass right above the pivot point. When walking in place, the pass-through velocity is zero, and it increases up to σ depending on the desired speed of locomotion. The chosen time for the support exchange is the moment when the CoM reaches its designated support exchange location between δ and ω. We assume that the support exchange happens instantaneously, and do not include a double support phase in our model. Finally, as the linear inverted pendulum is driven by the simple dynamic equation $\ddot{x} = C^2 x$, we regard C as a parameter that defines the gravitational effect on the point mass trajectory. We estimate the introduced parameters from our robot by having it walk in open-loop mode and averaging the trajectory apexes, the support exchange points, and the sagittal pass-through velocity.

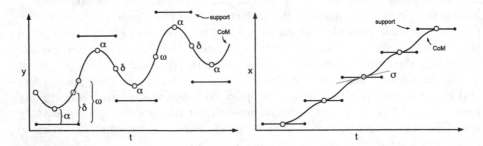

Fig. 2. The two-dimensional physical model is composed of a lateral motion (left) and a sagittal motion (right). The reference trajectory is described by four configuration parameters that define the lateral distance at the step apex (α), the minimal and maximal support exchange locations (δ and ω), and the maximum sagittal velocity at the step apex (σ).

We use this parameterized pendulum model to generate ideal reference trajectories for a permitted range of walking velocities. Since the reference trajectory follows the laws of the linear inverted pendulum model, a single end-of-step state s is sufficient to represent the entire trajectory. It defines a target state that the balance controller attempts to reach at the end of the step. The input into the gait control framework is a desired walking velocity vector $V = (V_x, V_y, V_\theta)$, $V \in [-\hat{V}_x, \hat{V}_x] \times [-\hat{V}_y, \hat{V}_y] \times [-\hat{V}_\theta, \hat{V}_\theta]$, with bounded components for the sagittal, lateral, and rotational directions. Let $\bar{V} = (\frac{V_x}{\hat{V}_x}, \frac{V_y}{\hat{V}_y}, \frac{V_\theta}{\hat{V}_\theta})$ be the componentwise normalized input velocity. Given the configuration parameters α, δ, ω, σ, and C, we compute the nominal support exchange state $s = (s_x, \dot{s}_x, s_y, \dot{s}_y)$:

$$s_x = \bar{V}_x \frac{\sigma}{C} \sinh(C\tau), \tag{1}$$

$$\dot{s}_x = \bar{V}_x \sigma \cosh(C\tau), \tag{2}$$

$$s_y = \begin{cases} \lambda\xi, & \text{if } \lambda = \text{sgn}(V_y) \\ \lambda\delta, & \text{else} \end{cases}, \tag{3}$$

$$\dot{s}_y = \lambda C \sqrt{s_y^2 - \alpha^2}, \tag{4}$$

$$\xi = \delta + |\bar{V}_y|(\omega - \delta), \tag{5}$$

$$\tau = \frac{1}{C} \ln\left(\frac{\xi}{\alpha} + \sqrt{\frac{\xi^2}{\alpha^2} - 1}\right), \tag{6}$$

where $\lambda \in \{-1, 1\}$ denotes the sign of the support leg. The nominal state s is expressed in coordinates relative to the current support foot. Please note that ξ and τ express meaningful quantities. ξ is the lateral support exchange location for the leading step, interpolated between the minimal support exchange location δ and the maximal support exchange location ω. τ is the "half step time" that the CoM travels, starting at the lateral apex α with a velocity of zero to the support exchange location ξ.

3.2 State Estimation

The state estimation module aggregates measurements from the physical robot to estimate the current pendulum state $c = (c_x, \dot{c}_x, c_y, \dot{c}_y)$ expressed in coordinates relative to the current support foot, and the sign $\lambda \in \{-1, 1\}$ of the support leg. More precisely, the joint angle information obtained from the motor encoders is used to update a kinematic model using a forward kinematics algorithm. Then, the entire model is rotated around the center of the current support foot such that the torso inclination matches the angle measured by the IMU. When the vertical coordinate of the swing foot has a lower value than the vertical coordinate of the support foot, the roles of the feet are switched, and a footstep frame is set to the ground projection of the new support foot preserving the global orientation that the new support had in this moment. The footstep frame remains fixed until the next support exchange occurs. With respect to the footstep frame, we measure the coordinates of the ground projection of the point in the center between the hip joints and present them to a Kalman filter to obtain a smoothed pendulum state c. The relocation of the footstep frame at the support exchange introduces an unavoidable discontinuity in the pendulum state trajectory. To compensate for undesired effects on the Kalman filter, we reinitialize the filter with the first coordinates measured after the support exchange and a velocity vector that is rotated into the new support frame, such that the continuity of the velocity in the global reference frame is preserved.

By tracking a fixed point on the robot frame instead of the true center of mass, we not only avoid having to provide masses and inertias to construct a physical model, but we also exclude noise due to moving body parts. We abstain from using quantities that are difficult to measure, such as torques, forces, and accelerations. Environmental disturbances and the dynamic influences of body parts that are strong enough to change the trajectory of this fixed point will still result in an immediate reaction of the balance controller. There is no need to estimate an impact force, or any other magnitude of a disturbance.

Without attempting to be overly precise, we used the same general humanoid kinematic model that we used previously in simulation [1], and adjusted the lengths of the body segments to the measured lengths from the real robot. In this general humanoid model, all degrees of freedom in each joint intersect at a point as a simplification. This is clearly not the case with the real hardware.

3.3   Balance Control

Given the current pendulum state c and the desired end-of-step state s, the balance control module computes a time T for the support exchange, a zero moment point offset Z, and the footstep location F where the swing foot is expected to touch down. The concept of the balance module is illustrated in Fig. 3. The zero moment point offset Z is expressed relative to the ankle joint. It steers the center of mass towards the target state s during the current step. However, as the zero moment point is physically bounded to remain inside the support polygon, the effect of the

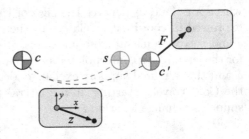

Fig. 3. Balance control computes a zero moment point offset Z that steers the center of mass c towards the nominal support exchange state s. The location of the next step F is computed with respect to the predicted achievable end-of-step state c'.

zero moment point is limited and the target state is not guaranteed to be reached. Based on the estimated support exchange time T and the zero moment point offset Z, we predict the achievable end-of-step state c' and use it to compute the step coordinates F expressed with respect to the predicted state c'. For the computation of all of these parameters, analytic formulae are derived from the linear inverted pendulum model in closed form. We compute the **lateral ZMP offset**

$$Z_y = \frac{s_y 2 C e^{C\check{T}} - c_y C(1 + e^{2C\check{T}}) + \dot{c}_y(1 - e^{2C\check{T}})}{C(e^{2C\check{T}} - 2e^{C\check{T}} + 1)} \qquad (7)$$

in a way that it attempts to reach the lateral support exchange location s_y (3) at the nominal step time $\check{T}$ of an ideal step and helps to maintain a desired step frequency. Z_y has to be bounded to a reasonable range, for example the width of the foot, and thus a fixed frequency cannot be guaranteed. We set $\check{T} = 2\tau$ (6) whenever a support exchange occurs and decrement it by the iteration time of the main control loop (in our case 12 ms) with every iteration. Please note that $\check{T}$ can have a negative value if the nominal step time is exceeded.

Due to an observed sensitivity of the lateral oscillation to disturbances [17], we attribute the computation of the predicted **step time** T entirely to the lateral direction. We want the support exchange to occur when the CoM reaches the nominal lateral support exchange location s_y. Taking the bounded lateral ZMP offset into account, T is given by

$$T = \frac{1}{C} \ln \left(\frac{s_y - Z_y}{c_y - Z_y + \frac{\dot{c}_y}{C}} + \sqrt{\frac{(s_y - Z_y)^2}{(c_y - Z_y + \frac{\dot{c}_y}{C})^2} - \frac{c_y - Z_y - \frac{\dot{c}_y}{C}}{c_y - Z_y + \frac{\dot{c}_y}{C}}} \right). \qquad (8)$$

However, there are two cases where the step time cannot be clearly determined. When after a strong disturbance the CoM is moving towards the pivot point in the lateral direction and is in danger of crossing it, the lateral orbital energy $E_y = \frac{1}{2}(\dot{c}_y^2 - C^2 c_y^2)$ is positive. In this case, we use a large constant time, e.g. 2 s, to "freeze" the robot and hope that it will return after all. The other case is when the support exchange location has already been crossed in the past, or will never be crossed due to a large disturbance. In this case it is advisable to step as soon as possible and the step time should be derived from the maximum allowed step frequency that the robot can handle. For the computation of the **sagittal ZMP offset**

$$Z_x = \frac{s_x + \frac{\dot{s}_x}{C} - e^{CT}(c_x + \frac{\dot{c}_x}{C})}{1 - e^{CT}}, \qquad (9)$$

we use the capture point based formula proposed by Englsberger et al. in [11]. It computes the sagittal ZMP such that if the CoM continues to move along an optimal linear inverted pendulum trajectory, the ZMP stays constant for the remainder of the step and the capture point of the CoM will match the capture point of the target state s by the time T of the support exchange. Since it is not possible to fulfill three constraints: the location, the velocity, and the time, with one constant ZMP offset per step, we opt for the simplicity of this good approximation. Finally, the sagittal ZMP offset also has to be bounded to a reasonable range.

Please note that both dimensions of the ZMP offset have been calculated without direct feedback of a measured zero moment point location. Only the position and the velocity of the center of mass have been used, which are easy to obtain. Given the bounded ZMP offset Z and the step time T, we can now compute the estimated achievable end-of-step state c'

$$c'_x = (c_x - Z_x)\cosh(CT) + \frac{\dot{c}_x}{C}\sinh(CT), \tag{10}$$

$$\dot{c}'_x = (c_x - Z_x)C\sinh(CT) + \dot{c}_x\cosh(CT), \tag{11}$$

$$c'_y = (c_y - Z_y)\cosh(CT) + \frac{\dot{c}_y}{C}\sinh(CT), \tag{12}$$

$$\dot{c}'_y = (c_y - Z_y)C\sinh(CT) + \dot{c}_y\cosh(CT). \tag{13}$$

The **footstep location** is then given by

$$F = (\frac{\dot{c}'_x}{C}\tanh(C\tau), \lambda\sqrt{\frac{\dot{c}'^2_y}{C^2} + \alpha^2}). \tag{14}$$

In the sagittal direction we compute the nominal step size that would result in the same end-of-step CoM velocity as the predicted one. In the lateral direction, the footstep location is computed with an extended capture point formula so that the CoM will pass the apex of the next step at distance α with a velocity of zero. Note that the footstep location F is expressed with respect to the future CoM state c'. It can be trivially converted to a foot-to-foot step size $S = F + (c'_x, c'_y)$.

3.4   Motion Generator

The hierarchical layout of the Capture Step Framework allows us to interface with virtually any walking motion generator that can exhibit control of stepping motions using step size and timing parameters. In this work, we use a central pattern generated gait (CPG) [12] that has been used in competition games with repeated success. The CPG can be combined with compliant actuation and provides a certain amount of open-loop stability out of the box. Instead of relying on inverse kinematics and end-effector trajectories in Cartesian space, the CPG operates in an abstract actuation space that makes it easy to produce stretched-knee walking.

For the integration of the motion generator, the output of the balance control module has to be transformed to gait control parameters in order to produce the desired step sizes and timings on the physical robot. The CPG expects a walking velocity control vector $V \in [-1, 1]^3$ with parameters for the sagittal, lateral, and rotational directions. Essentially, the velocity input parameters result in step amplitudes on the real robot. To map the output step size of the balance controller to the velocity input space of the CPG, we use the CPG in open-loop mode to generate data that describe the velocity control to step size mapping as measured by the sensors of the real robot and approximate it with a linear function. Then, we convert the balance control output step size S to a CPG velocity input V using the inverse of the linear approximation. The rotational component of the vector is ignored by the balance control layer. We simply pass the normalized rotational velocity input $\bar{V}_\theta$ through to the motion generator. To map the step time T to the motion phase of the CPG, we compute a phase increment such that the gait phase induces a support exchange at time T in the future. As T approaches zero when the support exchange is imminent, the computation of the phase increment becomes increasingly unstable. It is advisable to inhibit the timing adaptation near the support exchange and gait frequency bounds must be used to filter numerically unstable cases. The most significant innovation of the CPG integration is the fact that we did not link the inverted pendulum modeled CoM motion directly with the pelvis of the physical robot. This is typically done using inverse kinematics in conventional plane-restricted ZMP-preview walkers. This simplification was not only found to increase stability, but also allows for a non-level CoM motion simply by not forbidding it. Consequently, the ZMP offset is not explicitly transformed to a motion component, but since it is responsible for the step size variation, the physical system still reflects the commanded ZMP by increasing or decreasing the step size accordingly.

4   Real Robot Implementation

When dealing with real hardware, sensor noise and control loop latency are quite significant compared to simulation. An integral component of the real robot implementation is a predictive noise filter illustrated in Fig. 4. The filter smoothes the CoM trajectory using model-based assumptions, and predicts a short-term future state to overcome the latency. The first building block, denoted "rx", is the output of the state estimation, as it was described in Sect. 3.2. The second building block, denoted "mx", is the model

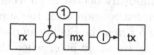

Fig. 4. A predictive noise filter is used to smooth the CoM state estimation and to predict a future state at the latency horizon.

state. In every iteration, the model state is forwarded by one time frame using the laws of the linear inverted pendulum model. The forwarded model is then linearly interpolated with the rx state using a blending factor $b \in [0, 1]$. The result is written into the new mx state, and forwarded in time by the latency l to compute a "tx" state at the latency horizon. The tx state is then presented to the balance controller for further processing. We have determined a latency $l = 65\,\text{ms}$ on our real hardware. This is quite significant considering that one step is approximately $420\,\text{ms}$ long. The unfortunate implication is that the support exchange has to be induced, and the first portion of the motion signal for "the other leg" has to be sent out, well before the change of the leg sign λ is detected by the kinematic model. This is achieved by forcing the tx model to step if the latency l exceeds the estimated step time T. Stepping is performed by setting the position of the tx state to $-F$ and resetting the step time to $T - 2\tau$. When the step time T reaches a negative value, the mx model itself is stepped, whether the real support exchange has been detected or not. This means that at times near the support exchange, the rx state and the mx state may assume different support leg signs and cannot be blended. In this case we set $b = 0$ and essentially switch to open-loop mode, where the mx model state does not

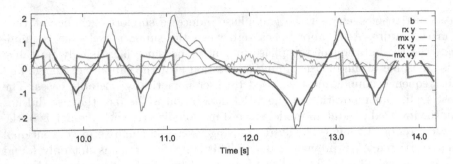

Fig. 5. Demonstration of the effects of the predictive filter.

receive any sensor feedback for a while and computes a linear inverted pendulum simulation from the last known state until the blending gate can be opened again. Independently of this, we utilize two additional mechanisms to adjust the value of the blending factor, and to control the behavior of the predictive filter. Firstly, we inhibit adaptation shortly before and after the support exchange by smoothly decreasing b when the step time T approaches zero and allowing it to increase again after the support exchange. And secondly, we decrease the value of b when the Euclidean distance between the rx and the mx model is small. This way we avoid jitter when the model state matches the measured input well, but do not sacrifice system response when the model state deviates significantly from the measured state. Figure 5 shows the lateral pendulum position and velocity recorded during an experiment. While the robot was walking on the spot, it was pushed from the side shortly after the time 11:0. The smoothing effect of the predictive filter can best be seen by comparing the rx and mx velocity data. The filter discards the high velocity peaks at the support exchange that differ strongly from the pendulum model and lead to bad predictions. The blending factor shows its highest peaks right after the support exchanges and after the push. The model adapts nicely to the new pendulum trajectory caused by the push, but eliminates the jittery noise shortly before the time 12:0. At the first step after the push, the rx and mx signals are slightly out of synchronization. The mx model steps earlier than the robot, but synchronization is quickly restored.

5   Experimental Results

We have implemented the Capture Step Framework on a bipedal robot with a weight of 7.5 kg and a height of 107 cm. TeenSize soccer robot Dynaped has demonstrated improved walking capabilities during a technology demonstration at the GermanOpen in Magdeburg in April 2014, as can be seen in the video [18], Fig. 7. Dynaped is equipped with a low cost, two-axis inertial measurement unit and Dynamixel EX-106 servo motors that we operate in a compliant mode. This makes the walk elastic and smooth, but also imprecise. We run the gait generation process with an update frequency of 83.3 Hz. Each iteration requires a computation time of 0.12 ms on a 1.3 GHz single core CPU. Figure 6 shows

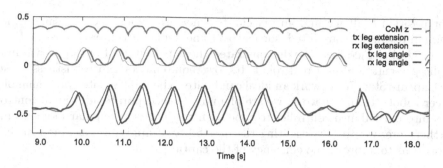

Fig. 6. CoM height, leg extension and leg angle during walking.

experimental data from the Dynaped robot. The experiment started with the robot walking in place. Then the robot accelerated and walked forward with its maximum velocity for approximately four seconds. When the robot came to a stop, it was immediately pushed from the back. The plot displays the CoM height on the top, and demonstrates the non-planar motion of the CoM. The data stream in the center shows the actuation signal (tx) for the leg extension as well as the signal received back from the robot (rx). The compliant actuation can be seen in the moments of the floor contact, where the swing leg automatically absorbs the impact force and the received signal deviates strongly from the actuation signal. The data stream in the bottom shows the motion signal of the leg angle. There is an evident delay between the activation signal and the received signal.

Fig. 7. Dynaped regaining balance after a push from the back by stepping forward [18].

6   Conclusions

We have contributed a robust omnidirectional gait generation method that is composed of an open-loop central pattern generator and a linear inverted pendulum based balance controller. Even though the balance controller is simplified to a point mass model, the controller is able to recover from disturbances that are strong enough to tilt the robot into an oblique pose using analytically computed step timing and foot placement adaptation. The direction or the magnitude of a disturbance does not need to be sensed. Our method can operate in a high latency environment with imprecise actuation using no more than an attitude sensor, joint position feedback and an inaccurate kinematic model. At the same time, common restrictions are lifted, such as bent knees, planar center of mass motion, and ground-aligned feet. We demonstrated the capabilities of our approach in a public presentation on a real robot. Its low requirements and robustness make the Capture Step Framework an ideal candidate to be implemented on humanoid soccer robots. In future work, we are planning to incorporate additional control laws into our balance controller to cope with the effects of angular momentum. Furthermore, we are investigating the possibility of applying machine learning algorithms to improve the efficiency of the capture steps.

References

1. Missura, M., Behnke, S.: Omnidirectional capture steps for bipedal walking. In: IEEE-RAS International Conference on Humanoid Robots (Humanoids) (2013)
2. Kajita, S., Kanehiro, F., Kaneko, K., Fujiwara, K., Harada, K., Yokoi, K.: Biped walking pattern generation by using preview control of zero-moment point. In: IEEE International Conference on Robotics and Automation (ICRA) (2003)
3. Wieber, P.-B.: Trajectory free linear model predictive control for stable walking in the presence of strong perturbations. In: Humanoids (2006)
4. Hirai, K., Hirose, M., Haikawa, Y., Takenaka, T.: The development of Honda humanoid robot. In: ICRA (1998)
5. Kajita, S., Morisawa, M., Miura, K., Nakaoka, S., Harada, K., Kaneko, K., Kanehiro, F., Yokoi, K.: Biped walking stabilization based on linear inverted pendulum tracking. In: IEEE/RSJ International Conference on Intelligent Robots and Systems (IROS) (2010)
6. Park, I.-W., Kim, J.-Y., Lee, J., Oh, J.-H.: Mechanical design of humanoid robot platform KHR-3 (KAIST humanoid robot 3: HUBO). In: Humanoids (2005)
7. Diedam, H., Dimitrov, D., Wieber, P.-B., Mombaur, K., Diehl, M.: Online walking gait generation with adaptive foot positioning through linear model predictive control. In: IEEE/RSJ International Conference on Intelligent Robots and Systems (IROS) (2008)
8. Morisawa, M., Kanehiro, F., Kaneko, K., Mansard, N., Sola, J., Yoshida, E., Yokoi, K., Laumond, J.-P.: Combining suppression of the disturbance and reactive stepping for recovering balance. In: IROS (2010)
9. Stephens, B.J., Atkeson, C.G.: Push recovery by stepping for humanoid robots with force controlled joints. In: Humanoids (2010)
10. Urata, J., Nishiwaki, K., Nakanishi, Y., Okada, K., Kagami, S., Inaba, M.: Online decision of foot placement using singular LQ preview regulation. In: IEEE-RAS International Conference on Humanoid Robots (Humanoids) (2011)
11. Englsberger, J., Ott, C., Roa, M.A., Albu-Schäffer, A., Hirzinger, G.: Bipedal walking control based on capture point dynamics. In: IROS (2011)
12. Missura, M., Behnke, S.: Self-stable omnidirectional walking with compliant joints. In: Workshop on Humanoid Soccer Robots, Atlanta, USA (2013)
13. Behnke, S.: Online trajectory generation for omnidirectional biped walking. In: IEEE International Conference on Robotics and Automation (ICRA) (2006)
14. Graf, C., Härtl, A., Röfer, T., Laue, T.: A robust closed-loop gait for the standard platform league humanoid. In: Workshop on Humanoid Soccer Robots (2009)
15. Yi, S.-J., Zhang, B.-T., Hong, D., Lee, D.D.: Online learning of a full body push recovery controller for omnidirectional walking. In: Humanoids (2011)

16. Dong, H., Zhao, M., Zhang, N.: High-speed and energy-efficient biped locomotion based on virtual slope walking. Auton. Robot. **30**(2), 199–216 (2011)
17. Missura, M., Behnke, S.: Dynaped demonstrates lateral capture steps. http://www.ais.uni-bonn.de/movies/DynapedLateralCaptureSteps.wmv
18. Missura, M., Behnke, S.: Walking with capture steps. http://www.ais.uni-bonn.de/movies/WalkingWithCaptureSteps.wmv

Best Paper Award for its Engineering Contribution

RoboCup@Home Spoken Corpus: Using Robotic Competitions for Gathering Datasets

Emanuele Bastianelli[2]([✉]), Luca Iocchi[1], Daniele Nardi[1], Giuseppe Castellucci[3], Danilo Croce[4], and Roberto Basili[4]

[1] DIAG, Sapienza University of Rome, Roma, Italy
{iocchi,nardi}@dis.uniroma1.it
[2] DICII, University of Rome, Tor Vergata, Roma, Italy
bastianelli@ing.uniroma2.it
[3] DIE, University of Rome, Tor Vergata, Roma, Italy
castellucci@ing.uniroma2.it
[4] DII, University of Rome, Tor Vergata, Roma, Italy
{croce,basili}@info.uniroma2.it

Abstract. The definition of high quality datasets for benchmarking single components and entire systems in intelligent robots is a fundamental task for developing, testing and comparing different technical solutions. In this paper, we describe the methodology adopted for the acquisition and the creation of a spoken corpus for domestic and service robots. The corpus has been inspired by and acquired in the RoboCup@Home setting, with the involvement of RoboCup@Home participants. The annotated data set is publicly available for developing, testing and comparing speech understanding functionalities of domestic and service robots, not only for teams involved in RoboCup@Home or in other competitions, but also for research groups active in the field. We regard the construction of the dataset as a first step towards a full benchmarking methodology for spoken language interaction in service robotics.

1 Introduction

The creation of data sets for benchmarking different components of an intelligent robot is an important task. Suitable and high-quality data sets allow for both developing and testing new solutions and to compare existing ones. However, creating high-quality data sets is not trivial, since: (1) a proper design of data collection must be performed, depending on the tasks to be measured with the data set; (2) a proper data acquisition campaign must be executed to ensure that data will meet the requirements defined in the design phase; (3) the generation of the ground truth needed to evaluate performance of the tested modules, typically time consuming and requiring a substantial human effort. Moreover, when the data set is related to human-robot interaction issues, the additional challenge is to collect a wide and diverse data set suitably representative of different users.

In this paper, we describe the design, the collection and the generation of the corresponding ground truth for a spoken corpus to be used for developing

© Springer International Publishing Switzerland 2015
R.A.C. Bianchi et al. (Eds.): RoboCup 2014, LNAI 8992, pp. 19–30, 2015.
DOI: 10.1007/978-3-319-18615-3_2

and testing speech recognition capabilities of domestic and service robots. In the definition of the scenario, we took inspiration from the RoboCup@Home environments and tasks. The corpus will be thus very relevant for RoboCup@Home teams. The resource is publicly available for researchers in the field at http:// sag.art.uniroma2.it/HuRIC.html, in particular for RoboCup@Home teams.

The main motivation of using RoboCup@Home (and in general robotic competitions) for acquiring data sets stems from the fact that competitions provide an ideal context for benchmarking functionalities. However, this kind of benchmarking is not actually performed during a competition, because the main focus is to evaluate and compare performance of entire systems. Conversely, the ability to benchmark individual system components is needed and efforts in this direction are ongoing [1]. RoboCup@Home provided the proper context for developing the benchmark, since at the competition venue many researchers that are addressing the problems to be benchmarked can provide feedback and suggestions and guarantee the quality and the significance of the acquired data.

Summarizing, by exploiting robotic competitions, it is possible to significantly improve the quality and the significance of data sets used for benchmarking important functionalities for intelligent robots. In this paper we present an instance of this method, applied to benchmarking speech understanding capabilities of a domestic and service robot through the RoboCup@Home competition.

The paper is organized as follows. Next section describes some related work in the development of linguistic resources for speech understanding. In Sect. 3 we describe the design and the implementation if the acquisition process. In Sect. 4 we provide details about the developed corpus, by defining the type of annotations adopted on the gathered data. Finally, a discussion about the use of competitions for collecting data sets for benchmarking is provided in the final section.

2   Related Works

Annotated resources have always been used in the Natural Language Processing field with the aim of learning language rules from observations. Semi-automatic methods to build grammars, as well as more advanced Machine Learning based system for *POS-tagging*, *Syntactic* and *Semantic Parsing* have been realized exploiting such resources. This brought to the development of large scale annotated corpora (e.g. FrameNet [2], Penn Treebank [9], PropBank [10]) inspired by sound linguistic theories that helped in the definition of many state-of-the-art Statistical Learning approaches for NLP tasks. Even though these resources are built to be as general as possible, they do not cover all the different cases and phenomena implied by human language. As a consequence, their reuse in heterogeneous domains is not straightforward. The generalization as attempted by ML algorithms is basically biased by the employed data. Large performance drops can be noticed in out-of-domain conditions, as reported in [6,11], where a Semantic Parsing system trained over a specific application-domain corpus shows a significant performance drop when applied to different domains.

For these reasons, in the recent years some corpora for the automatic understanding of robot commands in Natural Language have been produced. First of all, it is important to highlight the fact that NL Human Robot Interaction deals with different aspects of language processing. Spoken interaction implies a Speech Recognition stage, while understanding the meaning of a sentence representing a command needs some form of semantic parsing. Finally, also the translation of the meaning of a sentence in the final grounded representation can be learned, and thus a resource containing all the above information is interesting.

The resources available so far have taken into account only a subset of these different aspects. For example, the work by Bugmann *et al.* [3] focuses the attention on the analysis of the semantic primitives contained in the utterances pronounced by the user in a route instruction navigation task, providing utterances paired with the related recorded audio. Kuhlmann *et al.* [8] produced a corpus of commands for the Simulator League competition @Robocup. The meaning of the sentences representing the commands is here expressed using CLang (Coach Language), a specific language that can be compiled by the simulation environment of the competition in order to change the behavior of simulated soccer players. Other resources have been gathered using crowd-sourcing to produce data with a high degree of flexibility in term of language. For example, in Tellex *et al.* [12] a corpus of written commands for navigation and manipulation tasks have been realized and exploited. Here an analysis of how the spatial domain is modeled in such commands is carried out through *Spatial Description Clauses (SDCs)*. These are semantic structures composed by a figure, a verb, a spatial relation and landmark and represent a linguistic constituent that can be grounded in the real world. Similarly, in [4] Kais presents a corpus of natural language commands for a manipulator acting in a simulated discrete 3-dimensional board. The semantic information provided is modeled through the formal Robot Command Language, encoding both semantics about actions and spatial relations between objects.

However, these corpora are highly domain or system dependent. In this context, our main aim is to build a corpus containing information that are still specific of an application domain, e.g. the house service robotics, but at the same time based or inspired by general linguistic theories. By doing this, we want to offer a level of abstraction in our resource that is independent from the robotic platform, but yet motivated by largely supported theories. Multiple *semantic theories* can be applied to describe the aspects of the world that should be taken in account by a NL HRI system. We came to the point that, for our first investigation, two main features are required: first, the robots are supposed to execute actions, possibly corresponding to a user command; second, these actions take place in a physical environment. For the first issue, we pointed out *Frame Semantics* [5] as a possible solution to model the semantics of actions. For the second, we addressed the *Holistic Spatial Semantics* [13] to model the spatial referring expressions in spoken language.

Moreover, we wanted to offer information for each step of a possible NL processing chain (e.g. Speech Recognition, NL Understanding, etc.). For these reasons, each sentence in our corpus is paired with one or more audio files. We are also working on the possibility of providing the grounded version of the command, with respect to some environment (e.g. different house settings).

3 The Acquisition Methodology

The dataset described in this paper has been collected in two modalities: (i) by remote interaction with the Web system described in this section; (ii) by interviewing members of the teams participating at the RoboCup@Home 2013 competition. In both cases, the Web portal described in this section has been used for the acquisition.

The RoboCup@Home corpus is composed by a set of utterances representing commands in a home environment. Since our aim has been to produce a complete resource for NL HRI, we provide both audio and textual representation of each gathered command. Each recorded utterance is coupled with its correct transcription, that has been checked by an operator either controlling directly the user insertion or later, during a validation phase. Users have been also requested to pronounce sentences inserted by others, so that multiple spoken versions of the same sentence are included.

In the first phase of the acquisition process, users could access the Web portal showed in Fig. 1 to record the commands. General situations involved in an interaction were described in the portal by displaying text and images. Each user was asked to give a command inherent to the depicted situation. In order to provide data representing realistic conditions, a portion of the gathering took place in the competition venues and in a cafeteria, thus with different levels of background noise. Moreover, the users did not receive any constraint about what to command to the robot, except for the description of the situation. As a consequence, the uttered expressions exhibit large flexibility in lexical choices and syntactic structures, again reflecting a "realistic application" condition.

In a second phase, all sentences corresponding to the transcriptions have been annotated with different syntactic and semantic information. POS-tags and

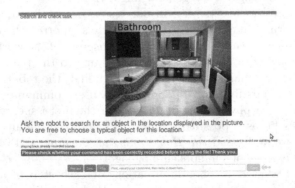

Fig. 1. The web portal used for the gathering through crowd-sourcing

syntactic dependency types have been automatically provided by the CoreNLP[1] system [7], and subsequently validated during the annotation process. Semantic information has been annotated according to *Frame Semantics* and *Holistic Spatial Semantics* by two expert annotators. In the last phase of the annotation process, all the tagged information has been validated by a third expert. A dedicated tool, the *Data Annotation Platform* (*DAP*) has been implemented and used in order to facilitate the annotation and validation process: its front-end is showed in Fig. 2. This tool provides the possibility to manage linguistic information at different levels as tagging the semantics, the syntax in term of dependency types, the POS-tag and allowing to change the lemma of each word. A specific functionality of DAP allows also the user to assign a quality score to each audio file, in order to reject the one that are too noisy. In a similar way, it is possible to mark syntactically wrong sentences that have been inserted by mistake.

Other information, as speakers' generalities (e.g. age, nationality, background experience in HRI) and the specific device used for the recordings are saved together with the annotations.

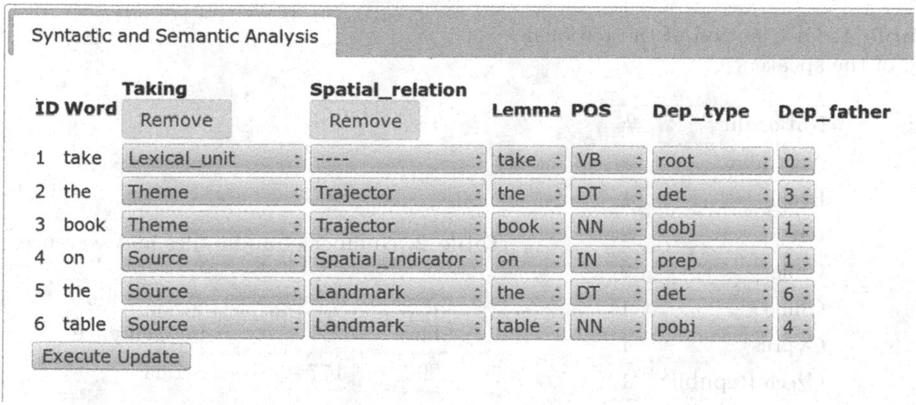

Syntactic and Semantic Analysis

ID	Word	Taking Remove	Spatial_relation Remove	Lemma	POS	Dep_type	Dep_father
1	take	Lexical_unit	----	take	VB	root	0
2	the	Theme	Trajector	the	DT	det	3
3	book	Theme	Trajector	book	NN	dobj	1
4	on	Source	Spatial_Indicator	on	IN	prep	1
5	the	Source	Landmark	the	DT	det	6
6	table	Source	Landmark	table	NN	pobj	4

Execute Update

Fig. 2. The Data Annotation Platform

4   Corpus Description

In this section an analysis of the corpus characteristics is carried out. General statistics about the composition of the corpus are here reported, as well as accurate measurements regarding the annotation process. More details are not provided here for lack of space, but are available on the official website of the resource (see Sect. 1).

---

[1] http://nlp.stanford.edu/software/corenlp.shtml.

4.1 Corpus Statistics

As previously stated, the RoboCup@Home corpus is composed by a set of audio files representing robot commands in a home environment. Each audio file is paired with its correct transcription. These are annotated with different linguistic information: lemmas, POS-tags, dependency trees, Frame Semantics and Spatial Semantics. Table 2 reports the number of audio files together with the number of sentences corresponding to their transcriptions. In order to provide training material for ASR engines, we also asked different speakers to pronounce the same command. The average number of sentences per audio file is reported in the aforementioned Table. The recordings took place during the Robocup 2013, so speakers with different nationalities have been interviewed. Involving nonnative English speakers has been a first step in the attempt of offering also training material for building nonnative accent acoustic models for ASR. Table 1 reports statistics about the nationality of the different speakers.

Each user has been required to insert and record 9 commands during the acquisition process. After removing the audio files considered too noisy, an average of about 8.1 audio files per speaker has been evaluated.

Table 1. Distribution of the nationality of the speakers

Nationality	#
Australia	3
Brazil	1
UK	2
Chile	2
China	1
Cyprus	1
Czech Republic	1
Holland	5
German	4
India	1
Indonesia	1
Italy	5
Japan	1
Mexico	1
Spain	2
Syria	1
USA	4
Total	36

Table 2. Number of audio files and sentences

#audio files	#sentences	#audio file per sentence
292	177	~1.64

Table 3. Distribution of utterance classes

Imperative	Descriptive	Definitional
150	14	13

Table 4. Fine-grain morpho-syntactic information

POS	#
CC	31
CD	16
DT	285
EX	11
FW	1
IN	134
JJ	43
JJS	1
MD	28
NN	365
NNS	22
POS	0
PRP	79

POS	#
PRP$	20
RB	16
RP	1
TO	66
UH	38
VB	165
VBD	6
VBG	2
VBN	1
VBP	11
VBZ	29
WDT	3
WRB	2

Table 5. Coarse-grain morpho-syntactic information

POS	#
CC	31
CD	16
DT	285
EX	11
FW	1
IN	134
J	44
MD	28

POS	#
N	387
P	99
RB	16
RP	1
TO	66
UH	38
V	214
W	5

The situations presented to the user through the Web portal belonged to three distinct categories, each corresponding to a different pragmatics of the command. In fact, each scene required the user to pronounce either a direct command as "*bring me the mug that is on the table*", or a description about the environment, e.g. "*there is a bottle on the table*", or the definition of a category of a referenced entity in the scene, e.g. "*this is the living room*". We then classified those sentences respectively as *imperative, descriptive* or *definitional*. Table 3 shows the number of sentences for each of these classes.

Statistics about the linguistic information are reported in Tables 4 and 5. Table 4 reports the number of fine-grain POS-tags annotated and validated in the whole corpus, while Table 5 shows the distribution of the general coarse-grain POS-tags, e.g. verbs or nouns.

4.2 Annotating Frame Semantics

One of the first purposes of the RoboCup@Home corpus was to provide linguistic information of different sort about natural language commands. The amount of information should enable a house service robot to completely understand their meaning.

In a house scenario, we expect mainly to have users giving commands to their "robotic butlers". Commands are then expressions of the expectation of a user to have a robot performing the desired action. For this reason we concluded that a way of representing how actions are modeled through language was necessary to fill the gap between the linguistic knowledge about the semantics of actions and the robotic actions. We pointed out that *Frame Semantics* fitted this case. This linguistic theory generalizes the notion of action by making reference to a situation, representing it as a *Semantic Frame* [5]. A frame is a micro-theory about a real world situation describing `actions`, such as *moving*, or more generally `events`, such as *natural phenomena* or `properties`. A set of semantic roles is associated to each frame, i.e. the descriptors of the different elements involved in the described situation (e.g. the *Goal* of a movement). Our hypothesis is that semantic frames represent a fundamental concept in NL HRI, as they can be straightforwardly linked to robot's actions. Moreover, linguistic resources providing Frame Semantics based information have been produced over the years, as FrameNet [2].

For the RoboCup@Home corpus a subset of FrameNet-inspired semantic frames have been selected, according to the most common actions that a house robot would perform. Table 6 reports statistics about the annotated frames, together with the relative frame elements. It is worth noting that some frames have been slightly adapted with respect to their definition in FrameNet, e.g. the frame *Scrutiny* has been called *Searching*. As an example, according to the defined set of frames, in the command *"go in front of the couch"* we annotated the *Motion* frame as evoked by the verb *go*. The phrase *in front of the couch* is labeled as the GOAL frame element, representing the destination of the motion action. The instantiated frame finally encodes all the information needed to the robot to understand what action to perform, together with the arguments involved in the command, i.e. in the example above the object near which to move.

The Frame Semantic annotation process usually follows three steps. First, all the expressed actions in a sentence must be recognized: this merely means finding all the possible words evoking a frame, and associating the correct semantic frame to each of them. This process is called *Frame Prediction (FP)*. Second, given a frame, the spans (in terms of words) of the different frame elements in the sentence must be identified. We refer to this task as the *Boundary Detection (BD)* process. Finally, the correct label representing the frame element name must be associated to each span identified during the BD task, e.g. GOAL. According to the practice in the generation of annotated resources, the *Inner-Annotator-Agreement* (IAA) between the two annotators has been evaluated as a measurement of the quality of the annotations. For each of the aforementioned steps, Precision, Recall and F-Measure have been measured, considering in turn one annotator as the gold standard and evaluating the other against him. The mean of the scores of the two annotators has been finally considered as the IAA. These results are reported in Table 7. For the BD and the AC subtasks, two different measures have been reported: the *exact match* and the *token match*. The first represents the percentage of roles that have been exactly tagged, meaning that a frame element has been correctly tagged only if its entire span perfectly matches the Gold Standard one. The second measure refers to the percentage of token correctly tagged inside the labeled spans. From this Table is possible to notice how difficult is tagging the Frame Semantics, especially the BD and AC steps. Different factors biased the scores of this two steps. First a slight misalignment in the FP phase reduces it, as tagging a wrong frame compromises the further processing. Second, in some cases the annotators disagreed on the span

Table 6. Distribution of Frames and related Frame elements

Attaching	2			
ITEM	2			
Being_in_category	14	*Giving*	2
CATEGORY	14	RECIPIENT	2
ITEM	14	THEME	2
Being_located	20	*Inspecting*	3
LOCATION	11	DESIRED_STATE	1
PLACE	6	GROUND	3
THEME	20	INSPECTOR	1
Bringing	37	*Motion*	39
AGENT	6	AREA	1
BENEFICIARY	13	GOAL	38
GOAL	24	MANNER	1
MANNER	1	PATH	1
SOURCE	9	THEME	8
THEME	37	*Placing*	10
Change_operational_state	3	AGENT	1
DEVICE	3	GOAL	10
OPERATIONAL_STATE	2	THEME	10
Closure	1	*Searching*	24
CONTAINER_PORTAL	1	COGNIZER	5
Entering	1	GROUND	7
GOAL	1	PHENOMENON	24
Following	30	PURPOSE	5
AREA	1	*Taking*	12
COTHEME	30	AGENT	4
GOAL	5	PURPOSE	2
MANNER	6	SOURCE	a
PATH	1	THEME	12
SPEED	1		
THEME	6		

of some frame elements, as the CATEGORY for the *Being_in_category* frame. For example, in the command *"this is a living room with a black table"*, one annotator tended to label only the phrase *a living room* as the CATEGORY, while the other used to annotate the whole span corresponding to *a living room with a black table*.

4.3   Annotating Spatial Semantics

After having found the way of linking the actions as they are represented through language and in the robot's world, it becomes fundamental for us to consider that these agents are supposed to act in a physical environment. We then focused on how the spatial domain is modeled through language, especially in human-robot standard interactions. Even though *Frame Semantics* is able to capture some

Table 7. *Frame semantics* Inter Annotators Agreement

	FP			BD			AC		
	P	R	F1	P	R	F1	P	R	F1
Exact Match	95.2	95.2	95.2	84.5	84.5	84.4	82.8	82.8	82.7
Token Match	-	-	-	89.9	89.9	89.8	85.0	85.0	85.0

of these aspects (e.g. some dynamic spatial references as the destination of a motion), we realized that in some cases the granularity level offered by this theory was not appropriate. Understanding the spatial relations holding between two or more entities can be crucial for HRI. If we consider the command "*move near the couch in the living room*", we find out that *Frame Semantics* is not able to capture the relation holding between *the couch* and *the living room.* as the whole sequence *near the couch in the living room* is considered as the destination of the motion trajectory, i.e. the GOAL frame element. Identifying such relation would allow a robot to understand which is the couch the user is referring to, among all the couches present in the world known by the robot.

We then looked at the *Holistic Spatial Semantics* [13] to model the static spatial relation expressed in the spoken commands. This theory defines the basic concepts in the domain of natural language spatial expressions. It helps to make reference to the location or the trajectory of a motion, usually involving one referent in a discourse. It defines the concept of *spatial relation*, as a composition of different *spatial roles* present in a sentence. These can be a TRAJECTOR, i.e. the entity whose location is of relevance, a LANDMARK, i.e. the reference entity by which the location of the trajectory of the motion is fully specified, or a SPATIAL_INDICATOR, i.e. the part of a sentence holding and characterizing the nature of the whole relation. For example, in the sentence "*go near the couch in the living room*", the preposition "*in*" is the SPATIAL_INDICATOR of the relation between "*table*" and "*kitchen*", respectively a TRAJECTOR and a LANDMARK. Even though Spatial Semantics defines also other spatial roles that model dynamic spatial relations, we decide to rely only on this restricted set in order to avoid an excessive overlap with the Frame Semantics. In fact the simple meaning representation structure of a spatial relation composed by three roles perfectly suits our needing. The LANDMARK and the SPATIAL_INDICATOR offer all the information needed to disambiguate the position of a referred entity (i.e. the TRAJECTOR), easily revealing which is respectively the reference point and the type of relation and the relation.

Spatial Semantics in term of these three roles have been annotated over the whole HuRIC. Table 8 reports the number of spatial relations annotated over the three datasets, together with the total number of spatial roles. It is worth noting that the number of LANDMARKs is different from the other two roles because sometimes it can be implicit, e.g. *go near* [*the table*]$_{\text{TRAJECTOR}}$ [*on the right*]$_{\text{SPATIAL_INDICATOR}}$. The average number of spatial relations and roles per sentence is also reported. The *Inter-Annotator-Agreement* has been evaluated

Table 8. Distribution of spatial relations and spatial roles

	#
Spatial_relation	47
TRAJECTOR	47
SPATIAL_INDICATOR	47
LANDMARK	41

Table 9. *Spatial semantics* Inter Annotators Agreement

	TRAJECTOR			SPATIAL IND.			LANDMARK		
	P	R	F1	P	R	F1	P	R	F1
Exact Match	85.8	88.6	85.7	81.4	81.4	81.3	84.7	84.7	84.6
Token Match	81.6	81.6	81.6	86.1	86.1	86.0	83.8	83.8	83.7

for each spatial role. It has been measured in the same way as for the *Frame Semantics*, and is reported in Table 9, considering both the *exact match* and the *token match* measures.

5   Discussion

Robotic competitions have an important role for testing integrated systems and compare performance of different teams in solving complex tasks, but are also very important settings for benchmarking specific functionalities of the robot. However, these benchmarking activities are rarely performed during a competition, usually because of time constraints and of the need to test entire systems. Nonetheless, the competition setting provides for an ideal context to acquire data that can be used for subsequent benchmarking.

We thus believe that robotic competitions, and RoboCup in particular, could gain if, in parallel with running the competitions, their set-up phases could be used to acquire data sets in typically more realistic scenarios than the ones each research group can recreate in its laboratory. Indeed, acquiring data during the competitions allows for reproducing similar characteristics, such as general environmental conditions, background noise, sensors, etc.

In this paper, we have described this approach applied to the speech understanding capability of a domestic and service robot, involved in RoboCup@Home competitions. Although the competition is focused on testing entire systems, the parallel acquisition of data for subsequent benchmarking of the speech understanding module in the same scenario of the actual competition is an important task for improving performance of this capability over time.

The publicly available RoboCup@Home spoken corpus described in this paper will thus help development, test and comparison of the speech understanding capabilities of domestic and service robots, not only for RoboCup@Home teams or teams participating to some competitions, but for any research group interested in the research field.

Acknowledgment. Authors are thankful to Cristina Giannone for her indispensable support in the development of the DAP system.

References

1. Amigoni, F., Bonarini, A., Fontana, G., Matteucci, M., Schiaffonati, V.: Benchmarking through competitions. Benchmarking, Technology Transfer, and Education, European Robotics Forum-Workshop on Robot Competitions (2013)
2. Baker, C.F., Fillmore, C.J., Lowe, J.B.: The berkeley framenet project. In: Proceedings of the ACL 1998, Association for Computational Linguistics, pp. 86–90. Stroudsburg, PA, USA (1998)
3. Bugmann, G., Klein, E., Lauria, S., Kyriacou, T.: Corpus-based robotics: a route instruction example. In: Proceedings of IAS-8, pp. 96–103 (2004)
4. Dukes, K.: Train robots: a dataset for natural language human-robot spatial interaction through verbal commands. In: ICSR. Embodied Communication of Goals and Intentions Workshop, October 2013
5. Fillmore, C.J.: Frames and the semantics of understanding. Quaderni di Semantica 6(2), 222–254 (1985)
6. Johansson, R., Nugues, P.: The effect of syntactic representation on semantic role labeling. In: Proceedings of COLING. Manchester, UK 18-22 August 2008
7. Klein, D., Manning, C.D.: Accurate unlexicalized parsing. In: Proceedings of ACL 2003, pp. 423–430. Stroudsburg, PA, USA (2003)
8. Kuhlmann, G., Stone, P., Mooney, R., Shavlik, J.: Guiding a reinforcement learner with natural language advice: initial results in RoboCup soccer. In: The AAAI-2004 Workshop on Supervisory Control of Learning and Adaptive Systems (2004)
9. Marcus, M.P., Santorini, B., Marcinkiewicz, M.A.: Building a large annotated corpus of English: The Penn Treebank. Comput. Linguist. 19(2), 313–330 (1993)
10. Palmer, M., Gildea, D., Xue, N.: Semantic Role Labeling. Synthesis Lectures on Human Language Technologies. Morgan & Claypool Publishers, San Rafael (2010)
11. Pradhan, S.S., Ward, W., Martin, J.H.: Towards robust semantic role labeling. Comput. Linguist. 34(2), 289–310 (2008)
12. Tellex, S., Kollar, T., Dickerson, S., Walter, M.R., Banerjee, A.G., Teller, S.J., Roy, N.: Understanding natural language commands for robotic navigation and mobile manipulation. In: Burgard, W., Roth, D. (eds.) Proceedings of the AAAI. AAAI Press, San Francisco (2011)
13. Zlatev, J.: Spatial Semantics. Handbook of Cognitive Linguistics pp. 318–350. Oxford University Press, New York (2007)

Champion Teams

UT Austin Villa: RoboCup 2014 3D Simulation League Competition and Technical Challenge Champions

Patrick MacAlpine[✉], Mike Depinet, Jason Liang, and Peter Stone

Department of Computer Science, The University of Texas at Austin, Austin, USA
{patmac,msd775,jliang,pstone}@cs.utexas.edu

Abstract. The UT Austin Villa team, from the University of Texas at Austin, won the 2014 RoboCup 3D Simulation League, finishing with an undefeated record. During the course of the competition the team scored 52 goals and conceded none. Additionally the team won the RoboCup 3D Simulation League technical challenge by accumulating the most points over a series of three league challenges: drop-in player, running, and free challenge. This paper describes the changes and improvements made to the team between 2013 and 2014 that allowed it to win both the main competition and the technical challenge.

1 Introduction

UT Austin Villa won the 2014 RoboCup 3D Simulation League for the third time in the past four years, having also won the competition in 2011 [1] and 2012 [2] while finishing second in 2013. During the course of the competition the team scored 52 goals and conceded none along the way to finishing with an undefeated record. Many of the components of the 2014 UT Austin Villa agent were reused from the team's successful previous years' entries in the competition. This paper is not an attempt at a complete description of the 2014 UT Austin Villa agent, the base foundation of which is the team's 2011 championship agent fully described in a team technical report [3], but instead focuses on changes made in 2014 that helped the team reclaim the championship.

In addition to winning the main RoboCup 3D Simulation League competition, UT Austin Villa also won the inaugural RoboCup 3D Simulation League technical challenge consisting of three league challenges: drop-in player, running, and free challenge. This paper also serves to document these challenges and the approaches used by the UT Austin Villa team when competing in the challenges.

The remainder of the paper is organized as follows. In Sect. 2 a description of the 3D simulation domain is given. Section 3 details changes and improvements to the 2014 UT Austin Villa team including those for kicking and passing, localization, and working with a new robot model having a toe, while Sect. 4 analyzes the contributions of these changes, and the use of heterogeneous robot types, in addition to the overall performance of the team at the competition. Section 5 describes and analyzes the league challenges that were used to determine the winner of the technical challenge, and Sect. 6 concludes.

© Springer International Publishing Switzerland 2015
R.A.C. Bianchi et al. (Eds.): RoboCup 2014, LNAI 8992, pp. 33–46, 2015.
DOI: 10.1007/978-3-319-18615-3_3

2   Domain Description

The RoboCup 3D simulation environment is based on SimSpark,[1] a generic physical multiagent system simulator. SimSpark uses the Open Dynamics Engine[2] (ODE) library for its realistic simulation of rigid body dynamics with collision detection and friction. ODE also provides support for the modeling of advanced motorized hinge joints used in the humanoid agents.

Games consist of 11 versus 11 agents playing on a 30 m in length by 20 m in width field. The robot agents in the simulation are modeled after the Aldebaran Nao robot,[3] which has a height of about 57 cm, and a mass of 4.5 kg. Each robot has 22 degrees of freedom: six in each leg, four in each arm, and two in the neck. In order to monitor and control its hinge joints, an agent is equipped with joint perceptors and effectors. Joint perceptors provide the agent with noise-free angular measurements every simulation cycle (20 ms), while joint effectors allow the agent to specify the torque and direction in which to move a joint.

Visual information about the environment is given to an agent every third simulation cycle (60 ms) through noisy measurements of the distance and angle to objects within a restricted vision cone (120°). Agents are also outfitted with noisy accelerometer and gyroscope perceptors, as well as force resistance perceptors on the sole of each foot. Additionally, agents can communicate with each other every other simulation cycle (40 ms) by sending 20 byte messages.

In addition to the standard Nao robot model, four additional variations of the standard model, known as heterogeneous types, are available for use. These variations, and rules regarding how they may be used, are described in Sect. 4.2.

3   Changes for 2014

While many components contributed to the success of the UT Austin Villa team, including an optimization framework used to learn low level behaviors for getting up, walking, and kicking via an overlapping layered learning approach [4], the following subsections focus only on those that are new for 2014. Analysis of the performance of these components is provided in Sect. 4.1.

3.1   Kicking

The 2014 UT Austin Villa agent includes sweeping changes to kicking from the 2013 agent. In addition to learning to kick further from a known starting point by mimicking another agent's existing kick as described in [5], the agent is also now able to reliably kick the ball after taking necessary steps to approach it. This latter improvement is achieved by learning a new kick approach walking parameter set for the team's omnidirectional walk engine, the purpose of which is to stop within a small bounding box of a target point while guaranteeing

---

[1] http://simspark.sourceforge.net/.

[2] http://www.ode.org/.

[3] http://www.aldebaran-robotics.com/eng/.

that the agent does not overshoot that target. This new parameter set is added to three existing walk engine parameter sets as described in [6]. With this new walk, the agent is able to successfully approach and kick a ball without thrashing around or running into the ball.

The kick approach parameter set updates target walk velocities in the X and Y directions based on the following equation:

$$desired[X,Y]Vel = \frac{\text{sqrt}(2 * \text{MAXDECEL}[X,Y]}{*(distToBall[X,Y] > 2 * \text{BUFFER} ?} \\ distToBall[X,Y] : distToBall[X,Y] - \text{BUFFER}))$$

The values for MAXDECEL[X,Y] and BUFFER are optimized using the CMA-ES [7] algorithm over a task where the robot walks up to the ball to a position which it can kick the ball from. The robot is given 12 s to reach a position where it can kick the ball and is given the following reward during optimization:

$$reward = \begin{array}{l} -timeTaken \ (\text{in seconds}) \\ +fFellOver ? \ -1 : 0 \\ +timeTaken > 12 \ \text{seconds} \ ? \ -0.7 : 0 \\ +fRanIntoBall ? \ -0.5 : 0 \\ +velocityWhenInPositionToKick > 0.005 \ \text{m/s} \ ? \ -0.5 : 0 \end{array}$$

The 2014 team also adds the ability to legally score on indirect kickoffs. This is the result of a multiagent optimization where one agent uses a static, accurate, long-distance kick and the other attempts to touch the ball while moving it as little as possible [5].

3.2 Passing with Kick Anticipation

When deciding where to kick the ball, the UT Austin Villa agent first checks to see if it can kick the ball and score from the ball's current location. If the agent thinks it can score then it tries to do so. If not, the agent then samples kicking the ball at targets in 10 degree direction increments and, for all viable kicking direction targets (those which don't kick the ball out of bounds or too far backwards), the agent assigns each a score based on Eq. 1. Equation 1 rewards kicks for moving the ball toward the opponent's goal, penalizes kicks that have the ball end up near opponents, and also rewards kicks for landing near a teammate. All distances in Eq. 1 are measured in meters. The kick direction chosen is the kick whose target ball location receives the highest score. When the agent is close (within 0.8 m) to the ball its chosen kick direction is fixed and held for 5 s to prevent thrashing between kick directions.

$$score(target) = \begin{array}{l} \|opponentGoal - target\| \\ \forall opp \in Opponents, - \max(25 - \|opp - target\|^2, 0) \\ + \max(10 - \|closestTeammateToTarget - target\|, 0) \end{array} \qquad (1)$$

Once an agent has decided on a target to kick the ball at it then broadcasts this target to its teammates. A couple agents then use "kick anticipation" where they run toward locations on the field that are good for receiving the ball based on the ball's anticipated location after it is kicked. The agents assigned to run to these anticipated positions are chosen by a dynamic role assignment system [8].

Kick anticipation[4] was first used by the team in 2012 [2] only when an agent was close to the ball and trying to position around the ball to kick it. New for 2014 kick anticipation has been extended such that an agent going to the ball will broadcast where it intends to kick the ball at any time, not just when close to the ball, as long as requirements for having time to kick the ball instead of dribbling it are met (no opponents are within two meters of the ball and no opponent is closer to the ball than the agent considering kicking it). By extending the amount of time that agents broadcast where they are going to kick the ball before doing so, teammates get more time to run to the anticipated location that the ball is to be kicked in order to receive a pass from the agent kicking the ball. Also new for 2014 teammates avoid getting in the way of the projected trajectory of the ball before it is kicked to prevent them from accidentally blocking the kick.

3.3 Localization with Line Data

In 2013 the UT Austin Villa team employed a particle filter for robot self localization that used only the observations of landmarks (four corner flags and two goal posts at each end of the field) along with odometry updates. It was noticed that sometimes robots would walk out of bounds near the middle of the field, where no landmarks are present, get lost and never return to the field of play. To try and prevent robots getting lost in this way, line information was added to the particle filter, based on previous work by Hester and Stone [9], to improve localization when landmarks are not present. In particular, the longest K observed lines were each compared to known positions of all the lines that exist on the field. Metrics such as the distance between endpoints, acute angle between the lines, and line length ratio were used to determine the similarity of an observed line with each actual line. For each observed line, the highest similarity value was expressed as a probability and used to update particles. Figure 1 shows how line information improves localization accuracy for various values of K.

As there are lines completely surrounding the field, assuming a robot is standing up it should always be able to see at least one line if it is currently on the field. if a robot doesn't see a line for a prolonged period of time (four seconds), the robot automatically assumes that it is now lost and off the field such that the robot just stops and turns in place until it sees a line to relocalize itself. Additionally, if a robot doesn't see any lines, it broadcasts to its teammates that it is not localized. If any teammates see a robot that reports itself as not being localized they will broadcast the current x, y position and (new for 2014) angle of orientation of the unlocalized robot so that it may use other robots' observations to localize itself. Empirically we have found that after incorporating line data into localization our agents no longer get lost when leaving the field.

---

[4] Videos of kick anticipation being used for passing can be found at http://www.cs. utexas.edu/~AustinVilla/sim/3dsimulation/AustinVilla3DSimulationFiles/2014/ html/kickanticipation.html.

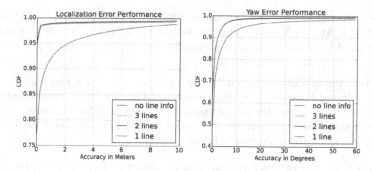

Fig. 1. CDF of localization error (left) and yaw error (right) for using K = 1,2,3 when incorporating line information. For comparison, not using line information (purple line) is shown as well (Color figure online).

3.4 Integration of Robot Toe Model

For the 2014 competition two new heterogeneous robot models were introduced including a robot model with a toe joint (known as a Type 4 model). Therefore, we modified the walk engine to use this added joint. The walk engine is described in depth in previous work [6]. The only modifications made to take advantage of the new joint was to add an offset to both the ankle pitch and to the new toe joint. We decided to alter the ankle pitch in addition to the toe joint as the ankle pitch can counteract the toe joint's effect on the robot's center of mass. This correction allows the remainder of the walk engine to perform as designed, resulting in a well-tuned walk. The offset to both joints takes the form of

$$\text{offset} = a\cos(t\pi + p) + c$$

where a is the amplitude of the movement, p controls the phase, and c is a constant offset. We chose to use a sinusoidal curve to maintain smooth movement that repeats once per step. The parameters for the ankle pitch and toe joint are not linked, resulting in an additional 6 parameters in the walk engine. These parameters were then optimized using CMA-ES as described in [6].

4 Main Competition Results and Analysis

In winning the 2014 RoboCup competition UT Austin Villa finished with an undefeated record of 13 wins and 2 ties.[5] During the competition the team scored 52 goals without conceding any. Despite finishing with an undefeated record, the relatively few number of games played at the competition, coupled with the complex and stochastic environment of the RoboCup 3D simulator, make it difficult to determine UT Austin Villa being better than other teams by a statistically significant margin. At the end of the competition, however,

---

[5] Full tournament results can be found at http://wiki.robocup.org/wiki/ Soccer_Simulation_League/RoboCup2014#3D.

all teams were required to release their binaries used during the competition. Results of UT Austin Villa playing 1000 games against each of the other 11 teams' released binaries from the competition are shown in Table 1.

Table 1. UT Austin Villa's released binary's performance when playing 1000 games against the released binaries of all other teams at RoboCup 2014. This includes place (the rank a team achieved at the competition), average goal difference (values in parentheses are the standard error), win-loss-tie record, goals for/against, and the percentage of own kickoffs which the team scored from.

Opponent	Place	Avg. goal diff	Record (W-L-T)	Goals (F/A)	KO score %
BahiaRT	5–8	2.075 (0.030)	990-0-10	2092/17	96.2
FCPortugal	4	2.642 (0.034)	986-0-14	2748/106	83.4
magmaOffenburg	3	2.855 (0.035)	990-0-10	2864/9	88.3
RoboCanes	2	3.081 (0.046)	974-0-26	3155/74	69.4
FUT-K	5–8	3.236 (0.039)	998-0-2	3240/4	96.3
SEU_Jolly	5–8	4.031 (0.062)	995-0-5	4034/3	87.6
KarachiKoalas	9–12	5.681 (0.046)	1000-0-0	5682/1	87.5
ODENS	9–12	7.933 (0.041)	1000-0-0	7933/0	92.1
HfutEngine	5–8	8.510 (0.050)	1000-0-0	8510/0	94.7
Mithras3D	9–12	8.897 (0.041)	1000-0-0	8897/0	90.4
L3M-SIM	9–12	9.304 (0.043)	1000-0-0	9304/0	93.7

UT Austin Villa finished with at least an average goal difference greater than two goals against every opponent. Additionally UT Austin Villa did not lose a single game out of the 11,000 that were played in Table 1. This shows that UT Austin Villa winning the 2014 competition was far from a chance occurrence. The following subsection analyzes some of the components described in Sect. 3 that contributed to the team's dominant performance.

4.1   Analysis of Components

Table 2 shows the average goal difference achieved by the following different versions of the UT Austin Villa team when playing 1000 games against top opponents at RoboCup 2014 as well as a version of itself that does not try and score on kickoffs (NoScoreKO).

UTAustinVilla. Released binary that does attempt to score on kickoffs.
NoScoreKo. Does not try and score on kickoffs but instead attempts to kick the ball as far as possible toward the opponent's goal posts without scoring.
NoKickAnt. Same as NoScoreKO but does not use kick anticipation.
Dribble. Same as NoScoreKO but always dribbles the ball and never kicks except for free kicks (e.g. goal kicks, corner kicks, kick-ins).
NoLines. Same as NoScoreKO but does not use any line observation data for localization.

Table 2. Average goal difference (standard error shown in parentheses) achieved by different versions of the UT Austin Villa team (rows) when playing 1000 games against both top opponents at the RoboCup 2014 competition and a version of the UT Austin Villa team which does not try and score on kickoffs (NoScoreKO).

Opponent	UTAustinVilla	Not scoring on kickoff			
		NoScoreKO	NoKickAnt	Dribble	NoLines
NoScoreKO	0.82 (0.03)	—	−0.20 (0.03)	−0.12 (0.03)	−0.32 (0.03)
BahiaRT	2.08 (0.03)	1.35 (0.03)	1.36 (0.03)	1.36 (0.03)	1.02 (0.03)
FCPortugal	2.64 (0.03)	2.13 (0.04)	1.98 (0.03)	1.69 (0.03)	1.76 (0.04)
magmaOffenburg	2.85 (0.03)	2.20 (0.04)	2.10 (0.03)	2.29 (0.04)	1.91 (0.03)
RoboCanes	3.08 (0.05)	2.54 (0.05)	1.77 (0.04)	1.53 (0.03)	2.09 (0.04)

When comparing the performance of UTAustinVilla, which tries to score on kickoffs, to that of NoScoreKO, which does not attempt to score on kickoffs, we see at least a half goal advantage for UTAustinVilla against all opponents. This is not surprising as Table 1 shows that the kickoff was able to score against almost all opponents over 80 % of the time. The only two opponents that the team had lower kickoff scoring percentages against were RoboCanes (69.4 %) and NoScoreKO (73.7 %). The ability to reliably score directly off the kickoff is a huge advantage. If it were possible to score 100 % of the time on kickoffs it would be almost impossible to lose — a team would quickly score on ensuing kickoffs right after their opponent scored in addition to scoring right at the beginning of a half when given the initial kickoff. Being able to score on kickoffs was a critical factor for the UT Austin Villa not to lose any games in Table 1 as the NoScoreKO team lost 10 games when playing 1000 games against all opponents (5 to RoboCanes, 4 to BahiaRT, and 1 to SEU_Jolly).

Advantages of using kick anticipation for passing described in Sect. 3.2 can be seen in Table 2 when evaluating the drop in performance from NoScoreKo to NoKickAnt, and gains in the ability to kick during game play by using the new walk approach detailed in Sect. 3.1 are evident in the performance drop between NoScoreKo and Dribble. Noticeable gains for using kicking and passing are seen against NoScoreKO, FCPortugal, and RoboCanes. This contrasts with the performance of previous year's teams when kicking was found to be detrimental [1] or negligible [2] to the team's performance. We believe the reason we do not see the same gains in performance against BahiaRT and magmaOffenburg is because they do not have as long kicks and kickoffs as the other opponents in Table 2, and thus do not kick the ball as much into open space where our kicking and passing components can best be utilized.

Using line data for localization as discussed in Sect. 3.3 is shown in the performance drop from NoScoreKO to NoLines. The incorporation of line data improved performance against all opponents as it helps to prevent agents from getting lost and wandering off the field.

4.2   Heterogeneous Types

At the RoboCup competition teams were given the option of using five different robot types with the requirement that at least three different types of robots must be used on a team, no more than seven of any one type, and no more than nine of any two types. The five types of robots available were the following:

Type 0: Standard Nao model
Type 1: Longer legs and arms
Type 2: Quicker moving feet
Type 3: Wider hips and longest legs and arms
Type 4: Added toes to foot

Table 3. Maximum speeds for general and sprint parameter sets, as well as median and maximum kick distances, for each of the different heterogeneous robot types used at the RoboCup 2014 competition.

	Type 0	Type 1	Type 2	Type 3	Type 4
Maximum general walk speed (m/s)	0.74	0.79	0.75	0.83	0.84
Maximum sprint walk speed (m/s)	0.78	0.94	0.78	0.97	0.89
Median kick length (m)	16	17	16	15	18
Maximum kick length (m)	19	20	19	17	20

Table 3 shows performance metrics after optimizing both walks [6] and kicks [5] for all robot types. The general walk speed is for moving to different target positions on the field while the sprint speed is for walking forward to targets within 15° of a robot's current heading. The median and maximum kick lengths are approximate values used by the robots to estimate where the ball will travel after it is kicked. Types 1 and 3 with longer legs have the fastest walking speeds. Type 4 with a toe is also relatively quick and has longer and more robust kicking than the other robot types.

Table 4 shows the results of playing teams consisting of all the same robot type against different opponents in order to isolate the performance of the different robot types. All the "not scoring on kickoff" teams do not try and score on the kickoff, but instead have the same behavior as NoScoreKO, so as to not have a robot type's ability to score on kickoffs overshadow the rest of its performance. From the data we see that Type 4 performs much better than the other types. For this reason we used seven Type 4 agents in the final round of the competition (the maximum number possible). We also used two Type 0 agents as they were the best at scoring on kickoffs. Our final two robots types used were a Type 3 because it could run the fastest, and a Type 1 robot as our goalie due to its larger body useful for blocking shots and good long kicks for goal kicks. The above number of robot types used are the same as those used in NoScoreKO.

To evaluate the incorporation of the Type 4 robot body's toe into our omnidirectional walk engine (detailed in Sect. 3.4), we optimized a walk for the type 4

Table 4. Average goal difference achieved by teams using different heterogeneous types (rows) when playing 1000 games against both top opponents at the RoboCup 2014 competition and a version of the UT Austin Villa team which does not try and score on kickoffs (NoScoreKO). All standard error values are in the range 0.03–0.05.

Opponent	Not scoring on kickoff					No kicking	
	Type 0	Type 1	Type 2	Type 3	Type 4	Type 4	Type 4 NoToe
NoScoreKO	−0.51	−0.57	−0.35	−0.56	0.32	−0.19	−0.53
BahiaRT	0.78	1.18	1.26	1.16	1.62	1.10	0.88
FCPortugal	1.30	1.67	1.89	1.58	2.50	1.37	1.06
magmaOffenburg	1.40	1.47	1.73	1.48	2.81	2.21	1.65
RoboCanes	1.13	2.01	1.90	1.98	3.13	1.38	1.11

robot type that kept the toe at a fixed default flat position as if the toe joint did not exist (Type 4 NoToe). Table 4 compares the results of Type 4 NoToe to that of Type 4 with the toe integrated into our omnidirectional walk engine. The "no kicking" teams using these walks for comparison were not allowed to kick, except for kicking the ball on kickoffs but not trying to score when doing so, so as to isolate the utility of walking and dribbling performed by the walk engine. Type 4 had better performance than Type 4 NoToe against all opponents revealing that integrating the toe joint into the omnidirectional walk engine was useful. Additionally Type 4 NoToe, with a maximum general walk speed of 0.80 m/s and a maximum sprint walk speed of 0.88 m/s, was slightly slower than Type 4.

5   Technical Challenges

New at RoboCup this year was an overall technical challenge consisting of three different league challenges: drop-in player, running, and free challenge. For each league challenge a team participated in points were awarded toward the overall technical challenge based on the following equation:

$$\texttt{points}(rank) = 25 - 20 * (rank - 1)/(number\ Of\ Participants - 1)$$

Table 5 shows the ranking and cumulative team point totals for the technical challenge as well as for each individual league challenge. UT Austin Villa earned the most points and won the technical challenge by taking first in the drop-in player and running challenges and second in the free challenge. The following subsections detail UT Austin Villa's participation in each league challenge.

5.1   Drop-in Player Challenge

The drop-in player challenge,[6] also known as an ad hoc teams challenge, is where agent teams consisting of different players randomly chosen from participants in

---

[6] Details of the drop-in player challenge at http://www.cs.utexas.edu/~AustinVilla/ sim/3dsimulation/2014_dropin_challenge/3D_DropInPlayerChallenge.pdf.

Table 5. Overall ranking and points totals for each team participating in the RoboCup 2014 3D Simulation League technical challenge as well as ranks and points awarded for each of the individual league challenges that make up the technical challenge.

Team	Overall		Drop-in player		Running		Free	
	Rank	Points	Rank	Points	Rank	Points	Rank	Points
UTAustinVilla	**1**	**71.67**	**1**	**25**	**1**	**25**	**2**	**21.67**
FCPortugal	2	60.5	4-5	18	4	17.5	1	25
magmaOffenburg	3	53.33	6	15	3	20	3	18.33
FUT-K	4	43.83	7	13	2	22.5	6	8.33
BahiaRT	5	34.5	10	7	6	12.5	4	15
L3M-SIM	6	30.67	9	9	7	10	5	11.67
SEU_Jolly	7	28.25	2-3	22	8-9	6.25	—	—
RoboCanes	8	22	2-3	22	—	—	—	—
KarachiKoalas	9	18	4-5	18	—	—	—	—
Mithras3D	10	16.25	11	5	8-9	6.25	7	5
ODENS	11	15	—	—	5	15	–	–
HfutEngine	12	11	8	11	—	—	—	—

the competition play against each other. Each participating team contributes two agents to one drop-in player team where drop-in player games are 10 vs 10 with no goalies. An important aspect of the challenge is for an agent to be able to adapt to the behaviors of its teammate. During the challenge agents are scored on their average goal differential across all games played.

Table 6 shows the results of the drop-in player challenge at RoboCup under the heading "At RoboCup 2014". The challenge was played across 5 games such that every agent played at least one game against every other agent participating in the challenge. UT Austin Villa used the same strategy employed in the 2013 drop-in player challenge [10], and in doing so was able to win this year's drop-in player challenge. The agent's performance was bolstered by longer kicks as discussed in Sect. 3.1, and also by using a Type 4 agent which was found to be the best performing type in Sect. 4.2.

Drop-in player games are inherently very noisy and it is hard to get statistically significant results when only playing 5 games. In order to get a better idea of each agents' true drop-in player performance we replayed the challenge with released binaries across all $\left(\binom{11}{5} * \binom{6}{5}\right)/2 = 1368$ possible team combinations of drop-in player games. Results in Table 6 of replaying the competition over many games show that UT Austin Villa has an average goal difference more than three times higher than any other team, thus validating UT Austin Villa winning the drop-in player challenge.

Table 6. Average goal differences for each team in the drop-in player challenge when playing all possible parings of drop-in player games (1386 games in total with each team playing 1260 games).

Team	Avg. goal diff	At RoboCup 2014	
		Rank	Avg. goal diff
UTAustinVilla	**1.782 (0.050)**	1	**1.75**
FCPortugal	0.574 (0.068)	4–5	0.80
SEU_Jolly	0.430 (0.068)	2–3	1.20
magmaOffenburg	0.170 (0.070)	6	−0.25
RoboCanes	0.129 (0.070)	2–3	1.20
BahiaRT	−0.178 (0.071)	10	−1.60
HfutEngine	−0.383 (0.070)	8	−0.75
KarachiKoalas	−0.525 (0.069)	4–5	0.80
L3M-SIM	−0.610 (0.069)	9	−1.50
FUT-K	−0.687 (0.068)	7	−0.40
Mithras3D	−0.703 (0.068)	11	−1.75

Table 7. Running challenge scores as well as speed and off ground values of optimized walks for each of the different robot types. Type X is a body type optimized for the challenge.

Robot model	Score	Speed	Off ground
Type 0	1.491	0.993	0.498
Type 1	1.741	1.241	0.500
Type 2	1.299	0.782	0.517
Type 3	1.839	1.339	0.500
Type 4	1.572	0.972	0.600
Type X	~1.75	~1.25	~0.5

5.2 Running Challenge

For the running challenge[7] robots were given 10 seconds to run forward as far as possible and then were given a score based on a combination of their average speed and the percentage of time both feet were off the ground. Teams were allowed to use any of the five robot types during the challenge. Teams were also allowed to submit their own custom robot types where the vertical offsets between the robot's hip, knee, and ankle joints could be changed within certain constraints as long as the the overall height of the robot remained the same.

---

[7] Details of the running challenge at http://www.cs.utexas.edu/~AustinVilla/sim/
3dsimulation/AustinVilla3DSimulationFiles/2014/files/RunChallenge.pdf.

Table 8. Scores as well as speed and off ground values for each of the participating teams in the running challenge.

Team	Score	Speed	Off Ground
UTAustinVilla	**1.845**	**1.343**	**0.502**
FUT-K	1.251	0.916	0.334
magmaOffenburg	1.152	0.986	0.166
FCPortugal	1.124	0.776	0.348
ODENS	0.632	0.632	0
BahiaRT	0.577	0.483	0.095
L3M-SIM	0.517	0.515	0.001
Mithras3D	0	0	0
SEU_Jolly	0	0	0

Table 7 shows the performance of various robot types where the walk engine parameters of the robot were optimized with CMA-ES [7] to maximize the running score. During optimization the average feet pressure of the robot was constrained within reasonable values to ensure that the robot did not learn strange running gaits (like running on its knees). For Type X, the body morphology and walk engine parameters were optimized simultaneously within the allowed constraints. Unfortunately, these constraints limited the top speed of the robot and its top running speed was not the fastest out of all the body types. The robots with longer legs (types 1 and 3) were able to achieve faster speeds and higher scores. UT Austin Villa used a Type 3 robot during the challenge.

Results of the running challenge are shown in Table 8. UT Austin Villa had the highest values for speed and off ground percentage, and won the challenge with a score almost 50 % higher than the next competitor.

5.3   Free Challenge

During the free challenge teams give a five minute presentation on a research topic related to their team. Each team in the league then ranks the top five presentations with the best receiving 5 votes and the 5th best receiving 1 vote. Additionally several respected research members of the RoboCup community outside the league vote as well with their votes being counted double. The winner of the free challenge is the team that receives the most votes. The top three teams were FCPortugal with 57 votes, UTAustinVilla with 53 votes, and magmaOffenburg with 44 votes.

UT Austin Villa's free challenge submission[8] focused on optimizing robot body types for the tasks of running and kicking. Running performance was evaluated on the same task as the running challenge, but additional body morphology

---

[8] Free challenge entry description at http://www.cs.utexas.edu/~AustinVilla/sim/3dsimulation/AustinVilla3DSimulationFiles/2014/files/FreeChallenge_Sim3D_UTAustinVilla.pdf.

parameters were optimized outside of the constraints of the running challenge. The final optimized body morphology allowed the robot to run at approximately 2.8 m/s running speed, and have its feet off the ground 55 % of the time, giving it a running challenge score of 3.35. A robot body type optimized for long distance kicking was able to kick the ball almost 27 m with the ball traveling over 17 m in the air (previous optimized kicks with fixed body morphologies only were not able to travel much farther than 22 m).[9]

6 Conclusion

UT Austin Villa won both the 2014 RoboCup 3D Simulation League main competition and technical challenge.[10] Data taken using released binaries from the competition show that UT Austin Villa winning the competition was statistically significant. The 2014 UT Austin Villa team improved dramatically from 2013 as it was able to beat the team's 2013 second place binary by an average of 1.525 (±0.034) goals and also beat the 2013 first place team (Apollo3D) by an average of 2.726 (±0.041) goals across 1000 games.

A large factor in UT Austin Villa's success in 2014 was due to improvements in kicking and passing where in previous years the team focused more on dribbling. This paradigm shift within the team is also reflected by the league as the other teams in the semifinals (RoboCanes, magmaOffenburg, and FCPortugal) all possess above average kicking and passing behaviors. In order to remain competitive, and challenge for the 2015 RoboCup championship, teams will likely need to improve multiagent team behaviors such as passing and marking.

Acknowledgements. Thanks to Samuel Barrett for implementing changes to UT Austin Villa's walk engine to support the toe joint of the Type 4 robot model. Also thanks to Klaus Dorer for putting together the running challenge. This work has taken place in the Learning Agents Research Group (LARG) at UT Austin. LARG research is supported in part by NSF (CNS-1330072, CNS-1305287) and ONR (21C184-01).

References

1. MacAlpine, P., Urieli, D., Barrett, S., Kalyanakrishnan, S., Barrera, F., Lopez-Mobilia, A., Ştiurcă, N., Vu, V., Stone, P.: UT Austin Villa 2011: a champion agent in the RoboCup 3D soccer simulation competition. In: Proceedings of 11th International Conference on Autonomous Agents and Multiagent Systems (AAMAS 2012) (2012)

---

[9] Videos of optimized robot body types running and kicking can be found at http://www.cs.utexas.edu/~AustinVilla/sim/3dsimulation/#2014challenges.

[10] More information about the UT Austin Villa team, as well as video highlights from the competition, can be found at the team's website: http://www.cs.utexas.edu/~AustinVilla/sim/3dsimulation/#2014.

2. MacAlpine, P., Collins, N., Lopez-Mobilia, A., Stone, P.: UT Austin Villa: RoboCup 2012 3D simulation league champion. In: Chen, X., Stone, P., Sucar, L.E., van der Zant, T. (eds.) RoboCup 2012. LNCS, vol. 7500, pp. 77–88. Springer, Heidelberg (2013)
3. MacAlpine, P., Urieli, D., Barrett, S., Kalyanakrishnan, S., Barrera, F., Lopez-Mobilia, A., Știurcă, N., Vu, V., Stone, P.: UT Austin Villa 2011 3D Simulation Team report. Technical report AI11-10, The University of Texas at Austin, Department of Computer Science, AI Laboratory (2011)
4. MacAlpine, P., Depinet, M., Stone, P.: UT Austin Villa 2014: RoboCup 3D simulation league champion via overlapping layered learning. In: Proceedings of the Twenty-Ninth AAAI Conference on Artificial Intelligence (AAAI-15) (2015)
5. Depinet, M., MacAlpine, P., Stone, P.: Keyframe sampling, optimization, and behavior integration: Towards long-distance kicking in the robocup 3d simulation league. In: Bianch, R.A.C., Akin, H.L (eds) RoboCup-2014: Robot Soccer World Cup XVIII. LNAI, vol. 8992, pp. 571-582. Springer, Berlin (2015)
6. MacAlpine, P., Barrett, S., Urieli, D., Vu, V., Stone, P.: Design and optimization of an omnidirectional humanoid walk: a winning approach at the RoboCup 2011 3D simulation competition. In: Proceedings of the Twenty-Sixth AAAI Conference on Artificial Intelligence (AAAI-12) (2012)
7. Hansen, N.: The CMA evolution strategy: a tutorial (2009). http://www.lri.fr/hansen/cmatutorial.pdf
8. MacAlpine, P., Price, E., Stone, P.: SCRAM: scalable collision-avoiding role assignment with minimal-makespan for formational positioning. In: Proceedings of the Twenty-Ninth AAAI Conference on Artificial Intelligence (AAAI-15) (2015)
9. Hester, T., Stone, P.: Negative information and line observations for monte carlo localization. In: IEEE Inernational Conference on Robotics and Automation (ICRA) (2008)
10. MacAlpine, P., Genter, K., Barrett, S., Stone, P.: The RoboCup 2013 drop-in player challenges: experiments in ad hoc teamwork. In: Proceedings of the IEEE/RSJ International Conference on Intelligent Robots and Systems (IROS) (2014)

ZJUNlict: RoboCup 2014 Small Size League Champion

Chuan Li, Rong Xiong$^{(\boxtimes)}$, Zeyu Ren, Wenjian Tang, and Yue Zhao

National Laboratory of Industrial Control Technology,
Zhejiang University, Hangzhou, People's Republic of China
rxiong@iipc.zju.edu.cn
http://www.nlict.zju.edu.cn/ssl/WelcomePage.html

Abstract. The Small Size League is one of the important events in
RoboCup Soccer. ZJUNlict got the first place in Robocup 2014. In this
paper, we introduce the improvement we have made in the past year.
We describe the overview of the mechanical design, show the design
of the protector for Infrared emission tube as well as the shield for
wheels. Simulation is given to show how our design works. Then the
lower level firmware architecture is illustrated. The dynamics analysis of
the robot is presented to help improving the robots' performance and
reducing motion deviation in y axis. Finally we present how we organize defense employing Close-Marking defense along with Zone defense
imitating human player.

1 Introduction

The Small Size League is one of the important events in RoboCup Soccer. It is
basically a game between two robot teams restricted to rules similar to human
soccer game. Each team consists of six robots and competes to goal more than
the opponent. The league is devoted to the advancement of mechanical design,
artificial intelligence and multi-agent cooperation of mobile robots with the distant expectation that robot will play a game with the FIFA champion in 2050.

ZJUNlict from Zhejiang University has participated in this League for ten years
since 2004. We received our first championship in RoboCup 2013, Netherland and
won the championship again in RoboCup 2014, Brazil. Although our robot is of one
of the best performance among the teams in Small Size League equipped with our
flexible and powerful strategies for both attack and defense, many problems and
defects still exist for us to deal with after Robocup 2013. Firstly, during the game
play, our robot is very likely to be damaged and needs carefully check and fixing
after each game, especially for the wheels. Secondly, the Soft Core architecture
needs to be reformed as the old design is difficult to maintain and introduces many
bugs caused by the nested reference. Thirdly, there is obvious movement deviation
in y axis of our robot, which influences movement accuracy. Finally, the defense
should be strengthened because the attack organized by other teams becomes more
threatening.

© Springer International Publishing Switzerland 2015
R.A.C. Blanchi et al. (Eds.): RoboCup 2014, LNAI 8992, pp. 47–59, 2015.
DOI: 10.1007/978-3-319-18615-3_4

The remainder of this paper is organized as follows. Section 2 introduces the mechanical design of our robot with detailed introduction of the protector and shield we design to make the robot more stable. Section 3 describes the firmware architecture. Section 4 analyzes the dynamics of the robot and addresses the cause of the deviation in y axis direction. Section 5 describes the defense strategy, which consists of Close-Marking defense and Zone defense. Section 6 concludes the paper.

2   Mechanical Design

2.1   Overview

As shown in Fig. 1, The robot is equipped with four omni-wheels, a dribbling device, a shooting device, and a chipping device.

Fig. 1. Mechanical design of the robot: (1) shooting device; (2) omni-directional wheel; (3) chipping device; (4) dribbling device

Each omni-wheel is composed of 16 distributed little wheel with o-ring rubber and an aluminous base wheel. The little wheel rotates freely along the direction perpendicular to the rotation axis of the base wheel so that the robot can carry out omnidirectional movement. Each base wheel is driven by a 50 watt brushless Maxon motors with a gearbox, of which the reduction ratio is 3.18:1.

The shooting device and the chipping device are driven by two solenoid respectively, of which the parameters are calculated accurately in advance. The electromagnet is charged by two big capacitors. The time and the force of kicking the ball can be controlled by specifying the capacitor voltage as well as the discharge time.

The dribbling device is designed to control the ball and not losing it. The core component is a stick swathed with a special pipe circum, which rotates the ball to keep the ball in the dribbling device. The material of the pipe has significant effects on the robot's ability to control the ball, so we have chosen a special kind of rubber after a series of experiment and long term tests.

2.2 Mechanical Structure Adjustment

Infrared Emission Tube Protector. In RoboCup 2013, we have found out that the infrared emission tube and the rubber rings on Omni-direction wheels of our robots easily got damaged due to mechanical collision with other robots in the game. We had to frequently replace these components, which are both cumbersome and not environmentally friendly. So this year in 2014, we have made some special structure adjustments to solve these problems and obtained remarkable results.

As shown in Fig. 2, the cylindrical inner wall which painted blue is the exactly place to install infrared emission tube. In order to ensure the ball detected once it runs into the dribbling device, the infrared emission tube is mounted in the front of the device within a small box. When crashed, the deformation of the box is likely to squeeze the tube. In our new design, a protector is being installed in front of the infrared emission tube. Impact force will disperse through the protector to the whole mechanical structure rather than concentrate in the front so as to reduce the deformation.

(a) Old mechanical design for infrared   (b) New mechanical design for infrared

Fig. 2. Mechanical parts (Color figure online)

We use the software Ansys to compare the vector displacement of the two mechanical structure [1]. The result is shown in Fig. 3. The vector displacement declines to $7.785 * 10^{-6}$ from $7.738 * 10^{-5}$.

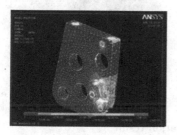   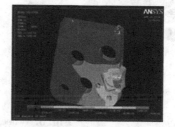

Fig. 3. The result of analysis about the protector

Omni-Wheel Shield. Another adjustment happens on the rubber rings of the omni-wheels. We increase the thickness of the big wheel flaps. The adjustment does not change the basic functions of the omni-wheel. Due to the increased thickness, the wheels will not be subjected to the direct mechanical impact. This adjustment received a remarkable effect in RoboCup 2014. The structures of the old version and the new version are shown in Fig. 4(a) and Fig. 4(b) respectively.

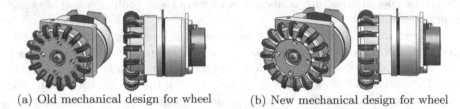

(a) Old mechanical design for wheel   (b) New mechanical design for wheel

Fig. 4. Wheel design

We have verified the effect of the proposed improvement using the SimulationExpress of Solidworks. Assume the collision speed between the robots is $3\,m/s$, the mass of the robot is $3\,kg$ and the collision time is $0.05\,s$, and after collision the robot will immediately stop. According to the theorem of momentum, the collision force is $180\,N$, which is the extreme case. Because of the protection of the big wheel flaps, the rubber wheels will not subject to the impacts any more. So we just analyze the situation that the impact happens on the big wheel flaps. Figure 5(a) shows the effect of the force applied in the simulation as well as the deformation of the wheel component. The material of the component is aluminum 7075, with the elastic modulus $7.2 * 10^{10}$ and the poissons ratio 0.33. From the Figure, we can conclude that the maximum deformation is $6.92 * 10^{-2}$ mm, which is so small that it can be neglected. Therefore the new design achieves a great effect in protecting the rubber rings of Omni-direction wheels.

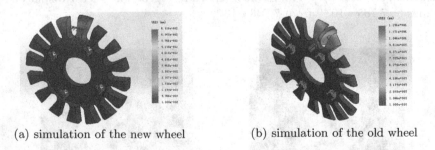

(a) simulation of the new wheel   (b) simulation of the old wheel

Fig. 5. Wheel simulation result

We have also made an analysis to the old mechanical structure. The little wheels are subjected to the direct mechanical impact so we add one to analyze.

And we have brought pressure to the rubber rings of omni-wheels and analysis under rated pressure what is the theoretically deformation of the rubber rings. Similarly Fig. 5(b) shows the constraints and force added to the part as well as the deformation of the rubber. The material is normal rubber with elastic modulus 10000 and the poissons ratio 0.45. The deformation of the rubber is $1.256 * 10^6$ mm, which means that the rubber will be definitely broken, and in real situation the deformation of the rubber can not be achieved at all. The theoretically result matches the actual phenomenon.

3 Firmware Design

3.1 Overview

We choose Altera Cyclone III as our central processor unit. The firmware in the chip is divided into Hard Core and Soft Core. The former defines the basic electronic resource(e.g. IOs, Timer, UART) available for the program written in Verilog and with SOPC Builder. The latter runs the main function written in C based on processor defined by Hard Core.

The Hard Core flowchart is shown in Fig. 6. The main part of the Hard Core consists of standard module, including Nios processor for Soft Core, Uart for chip communication, PIO for signal's input and output and so on. Other unique user modules featured with special communication protocols and control method are written in verilog, such as "motorcontrol" for BL motor control, which is of the greatest importance among our modules. The virtual pin of all this module is assigned to the physical pin of the Cyclone III chip.

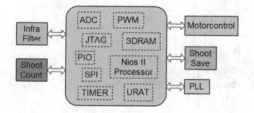
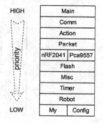

Fig. 6. Hard Core structure   **Fig. 7.** Soft Core flowchart

The function of the Soft Core is communication and motion execution [4]. Our previous Soft Core is not well designed and leaves with a lot of nested references among the c-files and h-files. As the robot's function extends, the code becomes more complicated and difficult to debug and maintain. This year, we reorganize the overall Soft Core architecture and specifies different priorities for different modules, which is shown as Fig. 7. As illustrated, modules on the same level has the same priority in execution. Each module has its specific function and priority. It is recognizable that Main has the highest priority and My has the lowest priority as well as Config. Functions in a module can only call functions in modules with the same or lower priority level.

4 Dynamics Analysis

Our robot is one with best performance in Robocup Small Size League. It can chip and shoot stably and execute smooth motion according to the visual feedback and the control strategy. However, when without vision feedback, it cannot go straight along y axis while go well along x axis. This is a common problem in Small Size League, so we decide to analyze the dynamics of the four-wheels driven robot to find out possible solution [2].

The robot's coordinate system (X, Y, θ) and the field coordinate system (X_L, Y_L, θ_L) are defined as shown in Fig. 8 and the positive rotation is counter-clockwise. So the coordinate-transformation matrix R_θ is shown as the Eq. (1). For each wheel, there is a similar coordinate system as Fig. 9 and a coordinate-transformation matrix as Eq. (2).

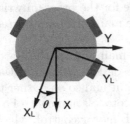

Fig. 8. Robot and field coordinate system

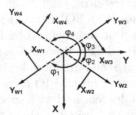

Fig. 9. Moment equilibrium

$$
\begin{bmatrix} \dot{y}_L \\ \dot{x}_L \\ \dot{\theta}_L \end{bmatrix} = \begin{bmatrix} \cos\theta & -\sin\theta & 0 \\ \sin\theta & \cos\theta & 0 \\ 0 & 0 & 1 \end{bmatrix} \begin{bmatrix} \dot{y} \\ \dot{x} \\ \dot{\theta} \end{bmatrix}
\tag{1}
$$

$$
\begin{bmatrix} F_{xi} \\ F_{yi} \\ T_i \end{bmatrix} = \begin{bmatrix} \cos\varphi_i & -\sin\varphi_i \\ \sin\varphi_i & \cos\varphi_i \\ R & 0 \end{bmatrix} \begin{bmatrix} F_{wi} \\ f_{wi} \end{bmatrix}
\tag{2}
$$

where

- F_{xi} is force that No.i wheel provides in the x direction,
- F_{yi} is force that No.i wheel provides in the y direction,
- T_i is torque that No.i wheel provides in the positive rotation,
- F_{wi} is force that No.i basic wheel provides,
- f_{wi} is resistance force of No.i little wheel.
- φ_i is the angle between the wheel and y axis, $\varphi_1 = -145°$, $\varphi_2 = -35°$, $\varphi_3 = 35°$, $\varphi_4 = 145°$
- R is radius of the robot.

The key to the analysis is to solve out F_{wi} and f_{wi}. During rotation, it is very likely that the onmi-wheels will dig into the carpet. In this condition, we assume that the force the wheel provide to drive the robot is proportional to their support force from the ground and the difference in speed between the wheel and the ground. The reaction force comes from two parts: one is the friction caused by pressure between the ground and the basic wheel, the other is the friction caused by pressure on the small wheels. F_{wi} and f_{wi} can be defined as

$$F_{wi} = KN_i(\omega_i r - \dot{x}_{wi}), \tag{3}$$

$$f_{wi} = -(abs(F_{wi}\mu_2 + N_i\mu_1))sgn((\dot{y}_{wi})). \tag{4}$$

K is the speed-force coefficient, μ_1 and μ_2 are coefficient of F_{wi} and N_i. These coefficient can be measured experimentally. r is the radius of the wheel. ω_i is the angular velocity of the No.i wheel. $\dot{x}_{wi}$ and $\dot{y}_{wi}$ can be derived as

$$\begin{bmatrix} \dot{x}_{w1} \\ \dot{x}_{w2} \\ \dot{x}_{w3} \\ \dot{x}_{w4} \end{bmatrix} = \begin{bmatrix} \cos\varphi_1 \sin\varphi_1 -R \\ \cos\varphi_2 \sin\varphi_2 -R \\ \cos\varphi_3 \sin\varphi_3 -R \\ \cos\varphi_4 \sin\varphi_4 -R \end{bmatrix} \begin{bmatrix} \dot{x} \\ \dot{y} \\ \dot{\theta} \end{bmatrix}, \begin{bmatrix} \dot{y}_{w1} \\ \dot{y}_{w2} \\ \dot{y}_{w3} \\ \dot{y}_{w4} \end{bmatrix} = \begin{bmatrix} -\sin\varphi_1 \cos\varphi_1 0 \\ -\sin\varphi_2 \cos\varphi_2 0 \\ -\sin\varphi_3 \cos\varphi_3 0 \\ -\sin\varphi_4 \cos\varphi_4 0 \end{bmatrix} \begin{bmatrix} \dot{x} \\ \dot{y} \\ \dot{\theta} \end{bmatrix} \tag{5}$$

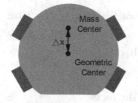

Fig. 10. Mass center deviation   **Fig. 11.** Moment equilibrium

When the robot is in still, the center of mass of the robot is not in its geometric center, the deviation, denoted by Δx, is shown in Fig. 10. There exist normal force N_{i0} to meet force equilibrium. During acceleration, there exists ΔN_i to meet the moment equilibrium, as shown in Fig. 11. The normal force N_i can be defined as

$$\begin{bmatrix} N_1 \\ N_2 \\ N_3 \\ N_4 \end{bmatrix} = \begin{bmatrix} N_{10} \\ N_{20} \\ N_{30} \\ N_{40} \end{bmatrix} + \begin{bmatrix} \frac{1}{2} & \frac{1}{2} \\ \frac{1}{2} & -\frac{1}{2} \\ -\frac{1}{2} & -\frac{1}{2} \\ \frac{1}{2} & \frac{1}{2} \end{bmatrix} \begin{bmatrix} \frac{H}{2R\cos\varphi} & 0 \\ 0 & \frac{H}{2R\cos\varphi} \end{bmatrix} \begin{bmatrix} F_x \\ F_y \end{bmatrix} \tag{6}$$

where

$$\begin{bmatrix} F_x \\ F_y \end{bmatrix} = \begin{bmatrix} m & 0 \\ 0 & m \end{bmatrix} R_\theta^{-1} \begin{bmatrix} \ddot{x}_L \\ \ddot{y}_L \end{bmatrix}, \tag{7}$$

$$N_{10} = N_{20} = \frac{mg}{4}(1 - \frac{\Delta x}{R}), N_{30} = N_{40} = \frac{mg}{4}(1 + \frac{\Delta x}{R}) \tag{8}$$

To describe all of the state variable, Eq. (6) can be transformed into the augmented matrix:

$$\begin{bmatrix} N_1 \\ N_2 \\ N_3 \\ N_4 \end{bmatrix} = \begin{bmatrix} N_{10} \\ N_{20} \\ N_{30} \\ N_{40} \end{bmatrix} + \begin{bmatrix} \frac{1}{2} & \frac{1}{2} & 0 \\ \frac{1}{2} & -\frac{1}{2} & 0 \\ -\frac{1}{2} & -\frac{1}{2} & 0 \\ \frac{1}{2} & \frac{1}{2} & 0 \end{bmatrix} \begin{bmatrix} \frac{H}{2R\cos\varphi} & 0 & 0 \\ 0 & \frac{H}{2R\cos\varphi} & 0 \\ 0 & 0 & 0 \end{bmatrix} \begin{bmatrix} m & 0 & 0 \\ 0 & m & 0 \\ 0 & 0 & J \end{bmatrix} R_\theta^{-1} \begin{bmatrix} \ddot{x}_L \\ \ddot{y}_L \\ \ddot{\theta}_L \end{bmatrix} \tag{9}$$

Substitute Eq. (9) into

$$\begin{bmatrix} m & 0 & 0 \\ 0 & m & 0 \\ 0 & 0 & J \end{bmatrix} R_\theta^{-1} \begin{bmatrix} \ddot{x}_L \\ \ddot{y}_L \\ \ddot{\theta}_L \end{bmatrix} = R_\theta \sum_{i=1}^{4} \begin{bmatrix} \cos\varphi_i & -\sin\varphi_i \\ \sin\varphi_i & \cos\varphi_i \\ R & 0 \end{bmatrix} \begin{bmatrix} F_{wi} \\ f_{wi} \end{bmatrix} \quad (10)$$

and we can run simulation using Matlab. The result is shown as Fig. 12.

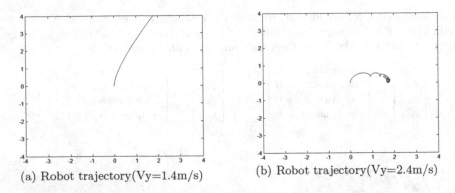

(a) Robot trajectory(Vy=1.4m/s)   (b) Robot trajectory(Vy=2.4m/s)

Fig. 12. Simulation result

The result simulated by Matlab is consistent with the facts.

When Robot is running in the direction of y axis, the f_{wi} and F_{wi} can be express specifically. It can be deduced from Eq. (1):

$$\begin{bmatrix} \ddot{x}_L \\ \ddot{y}_L \\ \ddot{\theta}_L \end{bmatrix} = \begin{bmatrix} \cos\theta & -\sin\theta & 0 \\ \sin\theta & \cos\theta & 0 \\ 0 & 0 & 1 \end{bmatrix} \begin{bmatrix} \ddot{x} \\ \ddot{y} \\ \ddot{\theta} \end{bmatrix} + \begin{bmatrix} -\sin\theta & -\cos\theta & 0 \\ \cos\theta & -\sin\theta & 0 \\ 0 & 0 & 0 \end{bmatrix} \begin{bmatrix} \dot{x}\dot{\theta} \\ \dot{y}\dot{\theta} \\ \dot{\theta} \end{bmatrix} \quad (11)$$

Substitute Eqs. (3), (4), (9), (10) into (11).

$$K \begin{bmatrix} mg\cos 35°(\cos 35° + \sin 35°\mu_2) & 0 & 0 \\ 0 & mg\sin 35°(\sin 35° - \cos 35°\mu_2) & \Delta xmg(-\sin 35° + \cos 35°\mu_2) \\ 0 & \Delta xmg\sin 35° & R^2 mg \end{bmatrix}$$

$$\left(\begin{bmatrix} \dot{x}_s \\ \dot{y}_s \\ \dot{\theta}_s \end{bmatrix} - \begin{bmatrix} \dot{x} \\ \dot{y} \\ \dot{\theta} \end{bmatrix} \right) - \mu_1 \begin{bmatrix} 0 \\ mg\cos 35° \\ 0 \end{bmatrix} = \begin{bmatrix} m & 0 & 0 \\ 0 & m & 0 \\ 0 & 0 & J \end{bmatrix} \left(\begin{bmatrix} \ddot{x} \\ \ddot{y} \\ \ddot{\theta} \end{bmatrix} + \begin{bmatrix} 0 & -1 & 0 \\ 1 & 0 & 0 \\ 0 & 0 & 0 \end{bmatrix} \begin{bmatrix} \dot{x}\dot{\theta} \\ \dot{y}\dot{\theta} \\ \dot{\theta} \end{bmatrix} \right)$$

$$(12)$$

It can be inferred that the rotation is coupled with the motion in the direction of y axis. When the robot is running in the positive direction of y axis, there exists angular velocity because of deviation of the center of mass as well as the deviation of the trajectory. It is possible to compensate the deviation by introducing velocity feedback to angular velocity which requires more sensor to detect the actual speed of the robot, for example, the visual system. In order to simplify the solution, we suggest to solve the issue fundamentally by designing the counterweight for the robot so as to make the mass center overlap with the geometrical center, and thus to reduce the deviation Δx.

It can also be inferred that the rotation is coupled with the acceleration of the robot. The deviation of the center of mass cannot be absolutely eliminated. So when the acceleration of the robot is large, there exists angular velocity leading to the deviation of the trajectory. So we introduce the velocity and acceleration limit into our program to reduce the deviation. For example, when the robot move fast, the velocity of the robot is high. High velocity and high acceleration will cause large deviation of the trajectory according to the analysis. We limit the acceleration in that case because the velocity can not reduce immediately.

5   Defense Strategy in AI System

The Intelligent Control System of ZJUNlict has been fully introduced in the Champion paper last year, which includes strategy selection and trajectory generation based on learning approach [5]. The play script written in Lua assigns tasks to each robot. In this section, we mainly introduce our defense which is a vital part for competition and a superiority of our team. We only lost one point in the RoboCup 2014 tournament, which verifies the effectiveness of our strategy. Our defense is divided into two parts just like real person game: One is Close-Marking Defense, the other is Zone Defense.

5.1   Close-Marking Defense

A flow chart of Close-Marking Defense is shown as Fig. 13. We get the information about all opponent robot and calculate feature value. According to the attribute value we will match the role for every opponent robot, such as leader, passer, etc. Then attack array is set up to describe the robot in order and design defense strategy at last.

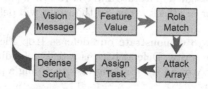

Fig. 13. Defend flow chart

Feature Value. We calculate all feature values for every opponent robot according to vision messages. Feature values, which is used to estimate the threatening level, reflects the state of opponent robots. We have more than a dozen of feature, some are simple value which can be calculated by vision message according to some obvious geometrical relationship, just like the distance between robot and ball, the distance between opponent robot and our goal, shoot angle and so on. The other are some complex feature value which can be calculated by simple feature value, just like Touch Ball value(the ability to receive ball and shoot), Chase Ball value(the ability to chase all and shoot), Pass Ball value(the ability to chase and shoot). For example, Touch Ball value of an opponent robot depends on the shoot angle of the opponent robot, whether other robots block in the pass line, the distance between the robot and our goal. All These features will be used in the next step opponent robots role matching.

Opponent Robots Role Match. Different roles have different priorities. A default role group includes:

- **receiver** is an offensive role which receives the ball and finishes shooting.
- **leader** is an offensive role which controls the ball.
- **attacker** is an offensive role which is always ready for the attack.
- **defender** is a defensive role which takes part in defense strategy.
- **goalie** is the goalie.

The role matched degree is calculated according to the features of robot. For example, when we judge whether the robot is a receiver, we will use three feature value: Touch Value(the ability to receive ball and shoot), Chase Value(the ability to chase all and shoot), Receive Angle(the angle between ball receiving and ball shooting). Matched degree will be calculated for each robot in priority decreasing order.

Attack Array. Attack array is a list of opponent robots. In attack array, We generate the attack array according to the match result. Now every opponent robot has their role and the matched degree for this role. We can compare the role priority and matched degree to set up attack array. But we don't need to mark for certain roles, such as goalie, opponent blocker in kick off area, which should be ignored.

Design Defend Script. We design our defence script in the form of a Finite State Machine [4]. The task is assigned in a state and can receive a parameter which is a number representing the order in attack array. So the robot will defend the corresponding opponent in attack array.

There is an example to demonstrate our defense [10]. Figure 14 shows the task assignment in the defense script. Defend Kick is a task defending the opponent robot which takes the free kick. We can see marking task receiving a parameter which is just an order in the attack array.

A simulation is shown as Fig. 15. Yellow team is the attack side. There are five attackers, four are of attacker role, and one is of leader role. Their defense superiority order is 4-3-1-5-2.

```
["DefendState"] = {
        Leader   = task.defendKick(),
        Special  = task.marking("First"),
        Middle   = task.marking("Second"),
        Defender = task.marking("Third"),
        Assister = task.marking("Fourth"),
        Goalie   = task.goalie(),
}
```

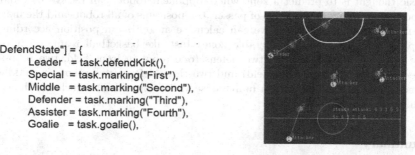

Fig. 14. A part of defense script

Fig. 15. Defense simulation (Color figure online)

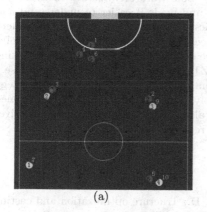

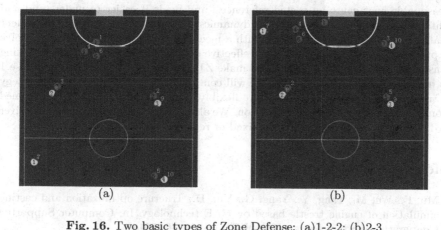

(a)   (b)

Fig. 16. Two basic types of Zone Defense: (a)1-2-2; (b)2-3

5.2 Zone Defense

Close-Marking Defense is our main defense strategy. But when competing with some strong teams, they may advance and get rid of our defense robots. We design Zone Defense which mainly takes effect in free kick defense to deal with this kind of situations.

A basic thought is to predict a zone where opponent robots can receive ball and finish shooting according to the angle of passer, the positions of all robots and the max marking range of our robot. Then we can calculate an optimum position according to the opponent robot in the predictable zone. Just like basketball team, we have two basic positioning type, one-two-two defense(one robot in front field, two robots in middle field, two robots in back field) and two-three defense(two robots in middle field, three robots in back field, it's a more conservative formation), which are shown as Fig. 16.

5.3 Summary

We design two kinds of defense to ensure we lose no points to the opponent. With the expansion of game field, Zone Defense will play a more and more important role. In 2015, how to mix Close-Marking Defense and Zone Defense together more flexibly will be a key focus of our team.

6 Conclusion

In this year, we focused on a more stable robot system and a more effective defense strategy. First, the protector for the infrared tube and the shield for the wheel was designed to reduce mechanical wastage, with simulations to analyze and verify the design effectiveness. Second, We reformed the Soft Core architecture to avoid the unreasonable bugs cause by nested references, and made it easier to understand and maintain. Third, we analyzed the robot dynamics to figure out the reason of movement deviation in y direction and come up with a mechanical counterweight design to solve this problem. Finally we design a more effective defense strategy using Close-Marking defense and Zone defense. These efforts make ZJUNlict score 43 goals and only lose 1 goal in RoboCup2014. In next year, we will continue to focus on the defense strategy and try mixing those two methods more flexibly. The counterweight will be designed to compensate the deviation in y direction. We also plan to come up with a mini-driver board for motor driving which can be fixed or replaced easily and flexibly.

References

1. Mu, F., Wu, M., Yang, Y., Yang, G., Yin, D.: Tructure optimization and casting simulation of engine trestle based on CAE technology. In: Computer Supported Cooperative Work in Design (CSCWD), pp. 41–46 (2014)
2. Byun, K.-S., Song, J.-B.: Design and construction of continuous alternate wheels for an omnidirectional mobile robot. J. Robot. Syst. 20(9), 569–579 (2003)
3. Ren, C., Ma, S.: Dynamic modeling and analysis of an omnidirectional mobile robot. In: Intelligent Robots and Systems (IROS), pp. 4860–4865 (2013)
4. Wu, Y., Zhao, Y., Xiong, R.: ZJUNlict Team Description Paper for RoboCup2013 (2013)

5. Zhao, Y., Xiong, R., Tong, H., Li, C., Fang, L.: ZJUNlict: RoboCup 2013 small size league champion. In: Behnke, S., Veloso, M., Visser, A., Xiong, R. (eds.) RoboCup 2013. LNCS, vol. 8371, pp. 92–103. Springer, Heidelberg (2014)
6. Wu, Y., Yin, P., Zhao, Y., Mao, Y., Xiong, R.: ZJUNlict Team Description Paper for RoboCup 2012 (2012)
7. Browning, B., Bruce, J., Bowling, M., Veloso, M.: STP: skills, tactics and plays for multi-robot control in adversarial environments. J. Syst. Control Eng. **219**(I1), 33–52 (2005)
8. Sheng, Y., Wu, Y.: Motion prediction in a high-speed, dynamic environment. In: Tools with Artificial Intelligence, pp. 703–705 (2005)
9. Thrun, S., Burgard, W., Fox, D.: Probabilistic robotics, vol. 1. MIT press, Cambridge (2005)
10. tolua++ Reference Manual. http://www.codenix.com/tolua/tolua++.html

Tech United Eindhoven, Winner RoboCup 2014 MSL

Middle Size League

Cesar Lopez Martinez, Ferry Schoenmakers, Gerrit Naus, Koen Meessen,
Yanick Douven, Harrie van de Loo, Dennis Bruijnen, Wouter Aangenent,
Joost Groenen, Bob van Ninhuijs, Matthias Briegel, Rob Hoogendijk,
Patrick van Brakel, Rob van den Berg, Okke Hendriks, René Arts, Frank
Botden, Wouter Houtman, Marjon van't Klooster, Jeroen van der Velden,
Camiel Beeren, Lotte de Koning(✉), Olaf Klooster, Robin Soetens,
and René van de Molengraft

Eindhoven University of Technology, Den Dolech 2, P.O. Box 513, 5600 MB
Eindhoven, The Netherlands
techunited@tue.nl
http://www.techunited.nl

Abstract. In this paper we discuss improvements in mechanical, electrical and software design, which we did to become RoboCup 2014 world champion. Regarding hardware and control our progress includes first steps towards improved passing accuracy via velocity feedback control on the shooting lever. In terms of intelligent gameplay we have worked on creating possibilities for in-game optimization of strategic decisions. Via qr-code detection we can pass coaching instructions to our robots and with a basic machine learning algorithm success and failure after free-kicks is taken into account. In the final part of this paper we briefly discuss progress we have made in designing a four-wheeled soccer robot with a suspension system.

Keywords: RoboCup soccer · Middle-size league · Multi-agent coordination · Mechatronic design · Motion control · Machine learning · Human-robot interaction · QR-codes

1 Introduction

Tech United Eindhoven is a RoboCup team of Eindhoven University of Technology. Our team consists of PhD, MSc and BSc students, supplemented with academic staff members from different departments. The team was founded in 2005, originally only participating in the Middle-Size League (MSL). Six years later service robot AMIGO was added to the team, which since also participates in the RoboCup@Home league. Knowledge acquired in designing our soccer robots proved to be an important resource in creating a service robot [3].

© Springer International Publishing Switzerland 2015
R.A.C. Bianchi et al. (Eds.): RoboCup 2014, LNAI 8992, pp. 60–69, 2015.
DOI: 10.1007/978-3-319-18615-3_5

This paper describes our major scientific improvements over the past year which helped us to become the winner of RoboCup 2014. First we introduce our current robot platform, followed by a description of the robot skills we have improved (we will focus on accurate shooting). Hereafter we describe our progress in strategy and human-robot interaction and lastly the advancements in a new four-wheeled soccer robot platform we designed in collaboration with an industrial partner.

Many of the points of improvement described in this paper are a direct result of rulechanges. In 2012 the mid-line passing rule was introduced, which was a large boost for the league in terms of stimulating smart team-play. Enforcing teams to make a pass before scoring provides an interesting academic challenge, but it also makes the matches more fun to watch for spectators.[1,2] Rule-changes for RoboCup 2014 limited continuous dribbling distance, allow robot coaching along channels that are natural to human beings and replace the mid-line passing rule by a more general 'pass before scoring' rule. The combination of these rule changes and introducing human-robot interaction to the middle-size league (Sect. 4.2) and moved the competition towards an even higher level of multi-agent coordination (Sect. 4).

2   Robot Platform

Our robots have been named TURTLEs (acronym for Tech United RoboCup Team: Limited Edition). Currently we are employing the fifth redesign of these robots, built in 2010, together with a goalkeeper robot which was built one year later (Fig. 1).

Three 12 V Maxon motors, driven by Elmec Violin 25/60 amplifiers and two Makita 24 V, 3.3 Ah batteries, are used to power our omnidirectional platform. Our solenoid shooting mechanism, powered by a 450 V, 4.7 mF capacitor, provides an adjustable, accurate and powerful shot [4]. Each robot, except for the goalkeeper, is equipped with an active ball handling mechanism, enabling it to control the ball when driving forwards, while turning, and even when driving backwards [1]. As said before, we aimed on improving passing abilities and conducted experiments on directly catching lob balls with our ball handling system (Sect. 3.1.1).

To acquire information about its surroundings, the robot uses an omnivision unit, consisting of a camera focussed on a parabolic mirror [2]. An electronic compass is implemented to differentiate between omnivision images on our own side versus on the opponent side of the field. We also added a kinect sensor to each robot. A detailed list of hardware specifications, along with CAD files of the base, upper-body, ball handling and shooting mechanism, has been published on a ROP wiki.[3]

---

[1] http://youtu.be/UagXSjp9nfk (Final Match RoboCup 2013 in Eindhoven).
[2] http://youtu.be/9XJc-jY90dE (Final Match RoboCup 2014 in Joao Pessoa).
[3] http://www.roboticopenplatform.org/wiki/TURTLE.

Fig. 1. Fifth generation TURTLE robots, with on the left the goalkeeper robot.

To facilitate data-acquisition and high-bandwidth motion control, the robots are equipped with EtherCAT devices provided by Beckhoff. These are connected to the onboard host computer via ethernet. Each robot is equipped with an industrial mini-pc running a preemptive Linux kernel. The software is automatically generated from Matlab/Simulink models via the RTW toolbox, recently renamed to 'Simulink Coder'. In order to allow asynchronous processing we have created a multitasking target for Simulinks code generation toolchain.[4]

Software for our robots is divided in three main executables: Vision, Worldmodel and Motion. On-board and robot-to-robot they communicate via a realtime database tool made by the CAMBADA team [5]. The vision module provides a localization of ball, obstacles and the robot itself. Hereafter the worldmodel combines this information with data acquired from other team members to get a unified representation of the world. While vision runs at 60 Hz and worldmodel at 20 Hz, motion contains the controllers for shooting, ball handling en driving. Therefore it samples at a much higher rate (1000 Hz). On top of the controllers, the motion executable also contains strategy and pathplanning, partly implemented as a subtask running at a much lower sample rate.

3   Improved Skills

Considering the rule changes in the middle size league, it is likely that passing and catching will become increasingly important compared to dribbling. During RoboCup 2014 in Brazil, this was indeed the case. In the section below, we will describe how we prepared for RoboCup 2014 by improving our accuracy for flat passing and by increasing our abilities to accurately catch and shoot a lob ball. The latter is not only beneficial for passing but also for shots at goal.

---

[4] http://www.techunited.nl/wiki/index.php?title=MultiTasking_Target_for_Linux.

3.1   Shooting

The electrical scheme of our kicker consists of a battery pack charging a capacitor via a DC-DC converter (Fig. 2). Once fully charged, in roughly 20 s, the capacitor can be discharged via an IGBT switch, creating a pulse-width modulated signal. The energy of the capacitor drives a solenoid actuator connected to a mechanical transmission (a shooting lever). The lever can be adjusted in height to allow for lob- and flat shots.

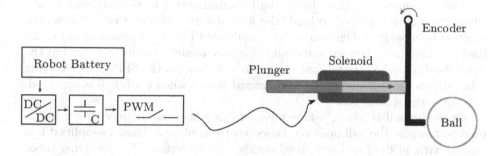

Fig. 2. Schematic overview of our shooting system. One half of the plunger is made of a non-magnetic material, the other half consists of a soft-magnetic material.

3.1.1   Shoot Lob Balls

To accurately shoot lob balls, the shooting system needs to be calibrated. Preferably we do this under conditions as close as possible to the conditions our robots face during the matches, i.e., on the official field with the same ball that will be used for competition. But during a tournament, testing time on the field is limited. Therefore our approach was to simply put the robot at the maximum distance it could take a lob shot from during a game, tune the PWM duty cycle until the ball lands exactly in the goal, and store the resulting duty cycle value. By linear interpolation between zero and the duty cycle we obtained during calibration, we could shoot from any spot within shooting range. The same calibration was used for all robots.

Although the above method is fast, it is also inaccurate. The relation between shooting distance and required duty cycle is non-linear, and since each robot has its own mechanical and electrical components, each robot has its own shooting characteristics. Therefore, calibration of each robot individually would be better.

For RoboCup 2014 we designed and implemented a tool to quickly do robot-dependent calibration. Furthermore, empirically we identified the relation between the shooting distance (x) and the required duty cycle (u) is exponential for a lob-shot (Eq. 1). Parameters a, b and c are robot-dependent parameters. They have to be obtained by measuring the travelled distance for multiple duty cycles. To make a correct fit at least four measurements are required, though more are preferred.

$$u = b^{-1}\ln(a^{-1}(c - x)) \qquad (1)$$

3.1.2   Catch Lob Balls

During the technical challenge of RoboCup 2013 we showed an initial attempt to shoot and catch lob passes. In terms of catching the ball, our approach there was to simply wait until the ball bounces were low enough to simply intercept it as if it were a flat pass. Building on these first tries, this year we worked on a much more challenging lob pass approach, where we use our current ball handling system to grab the ball exactly when it hits the ground after a flight-phase. We call this coordinate the point of intercept (POI).

The teammate shooting the lob ball communicates to the receiving robot, where the ball is expected to land (the feedforward position, FFP). When consecutive bounces are taken into account, multiple FFP's exist (example in Fig. 3). Each of them has a certain inaccuracy, for now modelled as a circle around the point itself. Based on the estimated time to reach each of the FFP's, the receiving robot drives towards one of the feedforward points when a lob ball is expected, but not actually shot yet.

Once the ball is in-air, a kinect camera mounted on the receiving robot is used to measure the ball position. Based on these observations, a simplified ball model, without drag and spin, predicts the ball trajectory. The receiving robot will respond to this ball-tracking based POI prediction, but only if it is located within the uncertainty circle. In case the estimated POI is located outside the circle, the robot will wait at the edge of the circle.

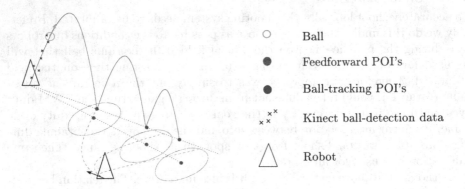

Fig. 3. Lob ball intercept strategy, the receiving robot chooses one of the points of intercept.

3.1.3   Shooting-Lever Velocity Feedback Control

Similar to what we described for lob shots, currently our control for flat shots and for flat passes is fully based on feedforward. As said before, many disturbances are robot-, ball- or field-dependent. Feedback control would allow to compensate for those.

We are using an encoder mounted on the rotational joint of the shooting lever (Fig. 2) as a feedback signal for velocity control. For full-power shots the end-effector of the shooting lever is pushed into the ball almost entirely before the ball itself even starts to move.[5] Using lever angular velocity as a feedback signal to control the resulting ball velocity would be hard in this case, because it is hard to exactly predict the dynamic behaviour of the deformed ball.

For slow shooting on the other hand, it is possible to make the lever and ball move as one body before the ball leaves the robot. Especially for passing, being able to accurately control ball-velocity would be of great help.

Fig. 4. Shooting lever end-effector for more accurate passing.

What was particularly challenging was the limited time one has available (a shot takes between 20 and 50 ms) and the limited spatial resolution of the encoder (130 ticks over the entire shooting lever stroke). Furthermore the solenoid actuator can only push in a single direction, therefore no overshoot is allowed (Fig. 4).

4   Improved Strategy

Our strategy takes into account the estimated positions of all peers and opponents, represented in a worldmodel. We developed a method to also use velocities and estimated game state to assess the feasibility of various tactical actions (plans). Instead of instantaneously seeking the free space on the field.

As a first step in moving to a more plan-based level of cognition, we have created a skill-selector, which we will describe in the upcoming section. Further we worked on in-game optimization of decision making in refbox tasks, either via human coaching (Sect. 4.2) or via machine learning (Sect. 4.3).

4.1   Skill Selector GUI

In our strategy, first we assign a unique role to each of the robots. Every role contains a number of actions/skills which can be executed during play. The main attacker for instance has five different skills to choose from: Flat shot, lob shot, pass, dribble and push-attack (i.e., bouncing the ball towards the goal with the side of the robot).

To decide on which skill to use at a certain moment in time, hard-coded conditional statements are evaluated. For the original system, these conditions were

---

[5] http://youtu.be/MF7mfItBriA (High-speed video of a full-power shot).

solemnly true/false evaluations (e.g., to shoot at goal, there must be a clear path to the goal). They are evaluated in the order they appear in programming and therefore immediately discard all other possible actions. This creates situations for which the TURTLEs do not take the optimal action. In order to solve this problem, a more generic framework for skill-selection has been developed.

In our improved skill-selector framework, for each of the skills the hard-coded conditions are complemented with normalized ranking functions (e.g., while turning towards the goal, the ranking for shooting at the goal will increase). After evaluating all ranking functions the skill selector chooses the skill with the highest overall ranking. In case multiple rankings are the same, the default skill 'dribble' will be selected. To make sure the chosen skill consistently ranks higher than the current skill, a hysteresis function has been added.

For debugging and tuning purposes we have created a graphical user interface which visualises skill-selector output for a given game state (Fig. 5).

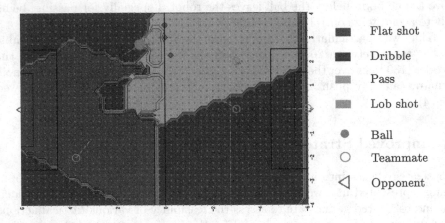

Fig. 5. Skill-selector visualization (Color figure online).

4.2   Human Coaching

For the world championships 2014 human-robot coaching in our league was allowed. Coaching instructions are intended to pass high-level instructions like 'shoot more often', as opposed to low-level commands like 'shoot now'. As a first step, this year we used qr-codes to tell our robots which predefined play to use, e.g., during a free-kick.

We use a freely available open source library to scan a video stream coming from our robots kinect sensor to scan for qr-codes.[6] With the maximum allowed qr-code size (i.e., 30×30 cm), containing three chars of encoded information, we experimentally searched for the maximum distance for which the code could be scanned. Averaged over 35 trails, using seven different char-combinations,

---

[6] http://zbar.sourceforge.net/ (ZBar, open source bar- and qr-code reader).

this distance turned out to be 5.1 m (with a standard deviation of 0.29). False positives within the code detection regularly occurred, especially against a non-plain background. But since none of these false positives matched any of the known strings, we could simply keep scanning until a combination of symbols was recognized that was actually grounded in the robots knowledge base.

In any trial of the experiment, if the code got detected, it was recognized within four seconds. Since in the current rules coaching is only allowed during 'dead time' between stop and start of a refbox task, we were interested how often a robot could actually get within five metres from the coaching spot and stay there for at least four seconds to receive a coaching instruction. Therefore we looked back at logged data of the final match during RoboCup 2013 in Eindhoven. In total this match involved 58 refbox tasks, 21 of them did not involve direct scoring risk (i.e., at least one of our robots was available to come to the side for coaching). Taking into account constraints on the robots acceleration and velocity, with our current qr-code detection system 17 coach moments would have succeeded.

4.3   Learning Refbox Play Decisions

In the previous section we described a way to do a hard, human-imposed, reset within our robots decision making. On top of these hard resets, we also worked on a basic reinforcement learning algorithm for a more subtle optimization of strategic play-choice during refbox tasks.

A reinforcement learning algorithm is built around actions, states and rewards [6]. Applying this framework to our free-kick strategy, we use six existing refbox plays as our action-space (single kick and shoot, double kick and pass etc.). Based on which opponent we face and the location of the free-kick (state), one play may result in slightly better scoring chances than the others. As a reward function we give high virtual reward for a scored goal, lower reward for a shot attempt, small punishment for loss of ball possession and severe punishment for a goal scored by the opponent (all weighted for time passed after the refbox task start signal).

Within this framework of rewards, states and actions we are able to store an expected reward for each state, based on past experience.

5   Four-Wheeled Platform with Suspension System

Already since our first generation of soccer robots, we have been using a robot base with three omniwheels positioned in a triangle. Such a three-wheeled design makes control easier because, regardless of field irregularities, all of the wheels will maintain in touch with the ground. But disadvantages also exist. Although driving straight forward is the most common direction of acceleration, it is also the direction for which our three-wheeled robot experiences the least traction during acceleration. The robot tends to tilt backwards, putting most of its pressure on the only wheel that cannot be used to transfer a torque to the ground when driving forward. For our current robot-design, traction is the limiting factor in achieving higher acceleration.

For a four-wheeled base accelerating forward, i.e., in the direction of the ball-handling mechanism, additional pressure is put on wheels that are actively used in acceleration. We worked with an industrial partner to realize a prototype of a four-wheeled robot.[7] On top of the RoboCup rulebook requirements with respect to weight and size, an additional requirement was created: Without any of the wheels losing contact with the floor, the robot should be able to take bumps of at least 10 mm in any direction while maintaining a ground clearance of 15 mm (Fig. 6).

To meet this latter requirement, a suspension system is needed. In the current prototype design, each of the wheels is equipped with an independent suspension system. Wheels and motor are still directly connected via a gearbox but the combination of the two is connected to the base via a passive spring-damper combination. The prototype of the four-wheeled base is being produced and during RoboCup 2014 the robot played several matches.

Fig. 6. Base structure.

6   Conclusions

In this paper we have discussed concrete steps towards more accurate shooting which, together with better ball tracking abilities, will enable passing via lob balls. Also we have presented proof of concept experiments for qr-code based human coaching and for learning algorithms in refbox strategy.

Altogether these improvements helped us to recapture the world title and this progress contributed to a higher level of dynamic an scientifically challenging robot soccer during RoboCup 2014. While at the same time maintaining the attractiveness of our competition for a general audience.

References

1. de Best, J.J.T.H., van de Molengraft, M.J.G., Steinbuch, M.: A novel ball handling mechanism for the RoboCup middle size league. Mechatronics **21**(2), 469–478 (2011)
2. Bruijnen, D., Aangenent, W., van Helvoort, J., van de Molengraft, R.: From vision to realtime motion control for the RoboCup domain. In: IEEE International Conference on Control Applications, pp. 545–550. Singapore (2007)
3. Lunenburg, J., Soetens, R., Schoenmakers, F., Metsemakers, P., van de Molengraft, R., Steinbuch, M.: Sharing open hardware through ROP, the robotic open platform. In: Behnke, S., Veloso, M., Visser, A., Xiong, R. (eds.) RoboCup 2013. LNCS, vol. 8371, pp. 584–591. Springer, Heidelberg (2014)

---

[7] Prodrive-Technologies, Science Park Eindhoven.

4. Meessen, K.J., Paulides, J.J.H., Lomonova, E.: A football kicking high speed actuator for a mobile robotic application. In: Proceedings of the 36th Annual Conference of the IEEE Industrial Electronics Society, pp. 1659–1664 (2010)
5. Neves, A.J.R., Azevedo, J.L., Cunha, B., Lau, N., Silva, J., Santos, F., Corrente, G., Martins, D.A., Figueiredo, N., Pereira, A., et al.: Cambada soccer team: from robot architecture to multiagent coordination. Robot Soccer pp. 19–45 (2010)
6. Sutton, R.S., Barto, A.G.: Introduction to Reinforcement Learning. MIT Press, Cambridge (1998)

RoboCup SPL 2014 Champion Team Paper

Jayen Ashar, Jaiden Ashmore, Brad Hall, Sean Harris, Bernhard Hengst[✉],
Roger Liu, Zijie Mei (Jacky), Maurice Pagnucco, Ritwik Roy, Claude Sammut,
Oleg Sushkov, Belinda Teh, and Luke Tsekouras

School of Computer Science and Engineering,
University of New South Wales, Sydney 2052, Australia
bernhardh@cse.unsw.edu.au
http://www.cse.unsw.edu.au

Abstract. Winning the Robocup SPL World Championship is not
accomplished in just one year, nor is it just a matter of writing effective software. Our success can also be attributed to the accumulation of
experience since 1999, strong institutional support, and dedicated collaborative teamwork. This paper summarises the key contributing innovations from the time the software was rewritten in 2010, and provides
some insight into team organisation. In this paper it is not possible to
cover all aspects and intricacies of the complex systems comprising the
rUNSWift software. We have therefore included an extensive list of references to our technical reports that provide detailed accounts of the
research, algorithms and results over the last 5 years. All the reports are
available on one website for easy access and make reference to many
external publications, including those from other teams in our league.

1 Introduction

Team *rUNSWift* from the University of New South Wales has been competing
in the Standard Platform League (SPL) since 1999. The league was formerly
called the Sony Four Legged League. We started as *UNSW United* and were
world champions three times in the years 2000 to 2003. It has taken eleven years
to regain the world title. In the interim we have seen major changes in both
the robots and the field. In 1999 each team played on a $6\,\mathrm{m}^2$ field with a border
using three first generation ERS-110 Sony AIBO robots. The field was uniformly
illuminated to 300 lux and localisation was aided by six beacons and coloured
goals. In contrast, the 2014 games were played on a $54\,\mathrm{m}^2$ borderless field using
standard venue lighting, uniform coloured goals, and without beacons. There are
now 6 Aldebaran V4 NAO bipedal robots per side including a coach robot.

While the SPL prohibits any hardware modification to the robots, fielding a
world champion team is not just a matter of good programming. Our experience
is that other requirements, including strong institutional support, a cohesive
team of motivated participants, a sound strategic plan, regular team meetings,
and many hours of testing and debugging are necessary. An additional UNSW
challenge is that this is largely a final year project for undergraduates. While

© Springer International Publishing Switzerland 2015
R.A.C. Bianchi et al. (Eds.): RoboCup 2014, LNAI 8992, pp. 70–81, 2015.
DOI: 10.1007/978-3-319-18615-3_6

Fig. 1. A subset of the 2014 rUNSWift Team. From left to right: Jaiden Ashmore, Zijie Mei (Jacky), Sean Harris, Bernhard Hengst, Brad Hall, Oleg Sushkov, Belinda Teh, Ritwik Roy.

we try to encourage participation over several years, the high student turnover means that we have more new recruits faced with a steep learning curve.

The current rUNSWift software architecture that led to the 2014 win had its genesis in 2010 when the code was rewritten from scratch. Although this report is authored by the 2014 team, their success is a culmination of all the developments from 1999, but particularly since 2010 [32]. The 2014 rUNSWift team members are Luke Tsekouras, Jaiden Ashmore, Zijie Mei (Jacky), Belinda Teh, Oleg Sushkov, Ritwik Roy, Roger Liu, Sean Harris, Jayen Ashar, and faculty members Brad Hall, Bernhard Hengst, Maurice Pagnucco, and Claude Sammut, several of whom are shown in the photo in Fig. 1.

The objective of this report is therefore to summarise both the technical approach and to describe the team organisation from 2010 to the world championship in 2014. This paper references many UNSW Computer Science and Engineering technical reports that provide detailed information and external references of rUNSWift developments throughout these years. Several of the reports have been published in international publications.[1]

The rest of this paper will cover our strategic planning process, the robot and software architectures, the major functional modules of perception, localisation, motion, and behaviour, our team organisation, and future work in the pipeline.

2 Strategic Planning and Development Methodology

Each year the new team plans developments using storyboards or just itemising objectives. For example, the 2010 version 1 storyboard [15] included innovations

---

[1] For completeness all the more than 40 reports are provided in one location at: http://cgi.cse.unsw.edu.au/~robocup/2014ChampionTeamPaperReports/.

such as a camera preprocessing stage we call the *saliency image* to reduce the vision processing load, and the RANSAC matching of outer field-edges with straight lines as additional field features.

Ensuing developments included visual identification of natural landmarks, foveated vision, iterative closest point combination of visual features, decision-tree learning of robot recognition, several iterations of multi-modal Kalman filters for localisation, and new motions for walking and kicking.

The 2014 strategy called for a major overhaul of the omni-directional locomotion, upgrade of the vision system to use higher definition camera images for field feature and goal detection, improved robot recognition, the reintroduction of a distributed multi-modal Kalman filter for localisation, and the rewriting of behaviours to improve robot team play on the large field.

Projects are selected by team participants based on interest and priority. New developments need to demonstrate improvement before being accepted into the code-base.

The development methodology mantra reinforced each year and adopted by the teams is "fail-fast, fail cheap". Our research strategy demands a complete integrated system at each stage, accepting poor performance initially, but quickly iterating through improved versions.

3   Robot Architecture

Figure 2 is a schematic of the rUNSWift robotic architecture showing the functional elements of sensor processing, world-modelling, and behaviour generation. This robotic architecture was first employed in 2000 [18] and has stood the test of time.

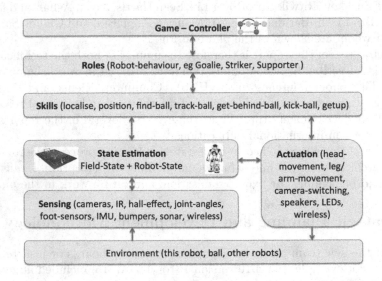

Fig. 2. The rUNSWift robotic architecture.

The rUNSWift robotic architecture can be envisaged as a *task-hierarchy* that consists of a set of finite state machines linked in a hierarchical lattice. The architecture is distributed in that each NAO robot tracks its own world-model rather than having one commander robot with a centralised view. This provides a level of redundancy in case individual robots are disqualified or stop working. Robots may have slightly different beliefs about the world formed from partially observable and noisy inputs. Robots share their world-model by communicating their position and that of the ball through wireless communication.

At the root-level, the game-controller changes the states of the game and player robots determine their roles implicitly from their world model.

4 Software Architecture

The Aldebaran NAO robot is equipped with a Linux operating system, as well as custom software from Aldebaran called *NaoQi* which allows interaction with the hardware. The *Device Communications Manager* (DCM) module of NaoQi actuates the joints and LEDs of the robot and reads sensors, including joint angle sensors, accelerometers, and sonars. We deploy two separate binary packages on the robot: *libagent* and *runswift*, which communicate using a block of shared memory and a semaphore.

The primary purpose of libagent is to provide an abstraction layer over the DCM that has the task of reading sensors and writing actuation requests. To facilitate in-game debugging, without the need to connect an external computer to the robot, a variety of features were added to libagent. These included a system of button-presses to perform various actions, such as releasing stiffness or running system commands. The libagent module also takes over the use of the LEDs for debugging purposes.

The *runswift* binary is a stand-alone linux executable, detached from NaoQi for safety and debugging. It reads frames from the two cameras, reads and writes to a shared memory block, synchronises with libagent to read sensor values and write actuation commands, and performs all the necessary processing to have the NAO robot play soccer. Because it is detached from NaoQi, it is easy to run runswift off-line, with the Motion thread disabled, allowing for vision processing and other testing to take place without physical access to a robot. It can be run using any of the standard linux debugging wrapper programs. The *runswift* executable is a multi-threaded process. The 6 threads are: Perception, Motion, Off-Nao Transmitter, NAO Transmitter, NAO Receiver and GameController Receiver.

The Perception thread is responsible for processing images, localising, and deciding actions using the behaviour module. The Motion thread is a near real-time thread, synchronised with libagent, and therefore the DCM, that computes appropriate joint values using the current action command and sensor values. The other threads are all networking related. A blackboard is used to share information internally between threads and externally between robots. We broadcast serialised blackboard information from each robot at 5 Hz to each of its teammates and Off-Nao.

Off-Nao is a desktop robot monitoring application which streams data from the NAO using a TCP/IP connection. Recordings can be reviewed in Off-Nao, to help determine the relationship between a sequence of observations and the resulting localisation status determined on the robot, as well as other correlations not determinable in real-time.

To facilitate the rapid development of behaviours, we chose to use Python. The Python interpreter was embedded into the *runswift* C++ executable. We monitor a directory on the robot containing Python code, and reload the interpreter whenever the Python code changes. Writing behaviours in a dynamic auto-reloadable language such as Python, means higher team productivity.

The software architecture developed in 2010 has largely been retained. More detailed information can be found in several reports [10,27] and the 2010 code release [9].

5 Perception

Vision is the primary sensor on the NAO and the one that consumes most of the computational resources. To stay within our processing budget, we sub-sample image pixels in a regular grid pattern to form a *saliency image* that is scanned for features of interest.

Unable to rely on colour alone due to poorer venue lighting we have added image edges as an additional modality to help identify objects. We still perform manual colour calibration, but supplement the colour-classified image with a gray-scale gradient image. Once points of interest have been identified, higher resolution rectangular foveas are used to focus processing resources while concurrently tracking several field features and ball hypotheses. In this way we can, for example, identify the ball from over half the field length away. Figure 3 shows the accurate detection of a distant poorly colour-classified ball. Related reports include [6,8,26,27].

RANSAC field-edge detection in 2010 was extended to field-line detection in 2011 [13]. The extension uses image colour and gradient to identify both straight lines and the centre circle with a novel RANSAC algorithm that concurrently

Fig. 3. Ball detection using colour and gradient in a high-resolution rectangular fovea (Color figure online)

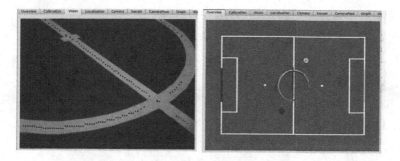

Fig. 4. Detecting field lines and the centre circle simultaneously using RANSAC (Color figure online)

Fig. 5. Robot Detection using a Naïve Bayes Classifier.

detects both features reliably (Fig. 4). Composite feature corners, T-intersections and parallel lines are passed to localisation via a unified sensor observation for the correction cycle of the Kalman filter.

Our vision robot detection algorithm has been subjected to several iterations over the years and together with sonar has been successful in detecting close robots. A region based approach in 2010 [27] was replaced by machine learning a decision tree from multi-modal vision features in 2011 [23] and improved using a naive Bayes classifier in 2014 [7] (Fig. 5). At close range we experimented with just foot detection using a Hough transform [20] but this development is not included in the current code.

The uniform coloured goals require methods for distinguishing the otherwise aliased ends of the field. A 2011 attempt was to use the zero-crossing points of the gradient image at the level of the horizon to map background features [12]. In 2012 a one dimensional version of the local feature point detector SURF, was found to be three orders of magnitude faster than the 2D version and viable

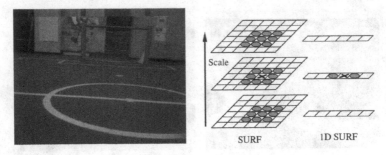

Fig. 6. Natural Landmarks. A strip of image pixels at robot eye level (between the red lines) is reproduced at the top of the image in gray-scale to find 1D SURF features (Colour figure onine).

on the NAO. With a distinguishing visual pattern at robot eye level behind the goals, robots were again able to tell the ends of the field apart (Fig. 6) [1,4,5]. The technique has been extended to a full 360 degree visual compass [2].

A unified field-feature sensor model utilises the extra information available from the specific combination of observed features. 2010 saw early incarnations of sensor models that use multiple goal-posts and field-edges [27]. In 2012 all the observed features were combined using a modified iterative closest point (ICP) algorithm that resulted in significant improvements in localisation accuracy [3,14].

6   Localisation

We used the combination of a particle filter (to solve the kidnapped robot problem) and a Kalman filter for localisation in 2010. In 2011 a multi-hypothesis linear Kalman filter with a mode selection algorithm using the combined field feature sensor model proved to be accurate and reliable [11]. In that year we also developed a separate Kalman filter model to track a moving ball [29].

It is often difficult to measure the ground truth adequately when evaluating localisation algorithms. An exemplary effort in 2011 found a method for tracking a NAO robot in real-time by un-distorting and stitching together two fish-eye images of the field from cameras mounted on our low ceiling in the lab [22].

From 2012 onwards goal colours were uniform. A combination of natural feature detection, the unified field-sensor model, using the ball as a disambiguation beacon and a multi-modal Kalman filter ensured accurate localisation [21]. No own-goals were scored throughout the entire competition that year.

In 2014 our approach to localisation combined the ICP unified field-sensor model with a distributed multi-modal Kalman filter, which is used for tracking the belief-state of all the robot positions on the field. One of the key features of this localisation system is distributed state tracking. Robots combine their observations via wireless communication to track the full global state of the team, including the ball, with the one filter. This Kalman filter algorithm is based on the successful AIBO 2006 rUNSWift localisation system [28].

7   Motion

The two new omni-directional walking motions developed in 2010 were *SlowWalk* and *FastWalk* [27]. SlowWalk is an open-loop walk that maintains balance by keeping the center-of-mass over the support polygon of the stance foot. This walk formed the basis of several variable strength kicking behaviours. Fastwalk is a closed-loop walk based on the inverted pendulum model with stabilisation feedback supplied via the foot sensors and accelerometers. Algebraic equations were ported from Matlab for an iterative inverse kinematic solution to position the feet while turning. For stability, walk parameters are adjusted slowly over time, making the walk style sluggish.

The 2010 walk was improved over the next three years [25,33]. Several approaches to sagittal and coronal stabilisation using reinforcement learning were tried [16,19,24] but the control policies were not deployed in competition as they were not smooth enough. The development of a directional kick in 2012 reduced the delay when approaching the ball and was used to good effect that year with rUNSWift scoring more goals than any other SPL team during the competition [30].

The walk engine was upgraded in 2014 to address several shortcomings including instability, slow side-stepping, robots overheating, and the poor method to change walk parameters. *Walk2014* [17] is based on a reinforcement learning policy for sagittal balance control. It integrates a stance that allows motor stiffness to be reduced to near-zero when not walking. The walk generator includes upgraded kicks, reacts to new action commands immediately on each support foot change and shifts the centre-of-gravity forward towards the centre of the support foot to reduce the incidence of backward falls. For the 2014 Open Challenge we experimented with a *heel-to-toe* walk in anticipation of a more natural walking style in the future [31].

8   Behaviour

Behaviours follow instructions from the game controller. For game-play an ongoing objective is to localise each robot on the field and to team-track the ball. Only then can robots dynamically allocate their roles for Striker and the three supporter roles (Defender, Midfielder, and Upfielder). The Goalie and Coach robots are dedicated. Role switching relies on hysteresis to avoid role vacillation. Striker allocation is based on kick-distance to the ball and supporter roles switch given their field position and that of the ball. More detailed descriptions of behaviours can be found for: 2010 Chap. 7 [27]; 2011 Goalie Chap. 4 [29], 2012 Striker Chap. 7 [30].

The 2014 striker uses a strategy to determine the appropriate action, namely whether to perform a straight kick, dribble the ball or dribble while turning. This strategy process was divided into two parts, a *high-level target* and *target adjustments*. The high-level target was an expression of intent without taking into consideration other opponents on the field. Target adjustments took into account the robot's immediate surroundings, and adapted the target in order to better achieve its goal. For example our *Ronaldo* adjustment would come into play if there is an opponent occluding the shot to the target. In this case we would adjust our aim to avoid the opponent and change a kick to a dribble. The 2014 Striker is described in Chap. 2 [31].

The joints of the NAO heat up following intensive use and fail safe by automatically reducing power to the motors or shutting down altogether – with devastating consequences. Our strategy for 2014 called for measures to reduce overheating. The integrated low-stiffness stand in the walk engine allowed us to rest the robots when they did not require to change their position, as is often the case for support roles and the goalie. Not only did this help reduce overheating, but it allowed the robots to stretch to maximum height and track the ball with a steady camera.

Behaviours are written in Python and generally rewritten each year. The perception thread reads sensor information, updates the world model, and calls a Python function in an embedded interpreter which returns actions for lighting LEDs and to physically move the robot. There are three core Python classes. *BehaviourTask* is the superclass for any task. *World* provides access to shared information about the world. It also sets a behaviour request in that world which is read at the end of the behaviour tick. *TaskState* is for tasks that are complex enough to warrant states, state transitions, and hysteresis.

Helper classes calculate geometries and check information about the team. For efficiency we have utility modules for constants, world model information, field geometry, team status, etc. These modules have the latest blackboard information available on each time-tick.

9   Team Organisation and the Competition

The number of undergraduates at UNSW that expressed interest in SPL participation was low in 2014. We therefore invited participants from previous years to join the 2014 team. This call was well received with some full-time employed alumni giving up their nights and weekends. The additional benefit to new team members was that past experience could be passed on more efficiently.

The team formally conducted 37 weekly meetings from October 2013 until just before the competition. At each meeting, progress was reviewed and objectives set for the following week. We use Google+ Hangouts for meetings and even had a participant call in while on a bus late for work. Participants would work together in small groups as time permitted for the rest of the week.

The rUNSWift code repository and wiki is hosted on GitHub. Before leaving for Brazil we took the precaution of transferring the code to a git repository on one of our laptops to ensure uninterrupted access in case of internet issues at the venue.

To help manage overheating we adopted a policy to rest robots at least two hours before key games to give robots a chance to cool down to ambient temperature. We tried several techniques to cool the NAOs below ambient temperature before and during play as the venue was not air-conditioned.

The robot gears progressively showed wear. Every robot was sent to the NAO clinic at least once during the competition. We reduce the aggressiveness of the walk when playing weaker teams to try to conserve the robots. Despite our best efforts we could not field all our robots for the duration of the final or whilst playing against the all-stars team in the drop-in final.

10   Concluding Discussion

After the competition the team members made a list of future developments to overcome weaknesses and add new functionality as a starting point for next year's team. Spectators enjoy seeing the robots fall down, but limiting the number of falls during a game is on our list of improvements. The team of robots rely heavily on wireless as was evident in the final when the communications failed and all the robots clustered around the ball. This is a perennial problem and could be addressed, not so much by improving wireless which has been unsuccessful, but by expecting the robots to play without it. To make this possible and improve team play, better algorithms to detect team member and opposition robots are needed to build and track the state of the whole game.

The progression of rUNSWift to SPL champions in 2014 started in 2010. After failing to reach the quarter finals in 2009, we were in the finals in 2010 with B-Human. In 2012 we were narrowly defeated (7-6) by the champions that year, UT Austin Villa. Our 2014 team included participants from both the 2010 and 2012 teams, and one individual from the 2006 Four Legged Sony League. The 2014 performance cannot be attributed to a single factor, or to a single year, but to a combination of integrated software development and team collaboration over several years. The 2014 code and accompanying wiki has been released: https:// github.com/UNSWComputing/rUNSWift-2014-release.

Acknowledgements. The 2014 team wish to acknowledge the legacy left by previous rUNSWift teams and the considerable financial and administrative support from the School of Computer Science and Engineering, University of New South Wales. We wish to pay tribute to other SPL teams that inspired our innovations in the spirit of friendly competition.

References identified as UNSW CSE Robocup reports and other Robocup related references in this paper are available in chronological order from: http://cgi.cse.unsw. edu.au/~robocup/2014ChampionTeamPaperReports/

References

1. Anderson, P.: New methods for improving perception in RoboCup SPL, UNSW CSE RoboCup Report 20120830-Peter.Anderson-ImprovingPerception.pdf (2012)
2. Anderson, P., Hengst, B.: Fast monocular visual compass for a computationally limited robot. In: Behnke, S., Veloso, M., Visser, A., Xiong, R. (eds.) RoboCup 2013. LNCS, vol. 8371, pp. 244–255. Springer, Heidelberg (2014)

3. Anderson, P., Hunter, Y., Hengst, B.: An ICP inspired inverse sensor model with unknown data association. In: Proceedings of the IEEE International Conference on Robotics and Automation (ICRA), pp. 2713–2718. IEEE (2013)
4. Anderson, P., Yusmanthia, Y.: Natural landmark localisation for RoboCup, UNSW CSE RoboCup Report 20111113-AndersonYusmanthia-NaturalLandmarks.pdf (2012)
5. Anderson, P., Yusmanthia, Y., Hengst, B., Sowmya, A.: Robot localisation using natural landmarks. In: Chen, X., Stone, P., Sucar, L.E., van der Zant, T. (eds.) RoboCup 2012. LNCS, vol. 7500, pp. 118–129. Springer, Heidelberg (2013)
6. Ashar, J., Claridge, D., Hall, B., Hengst, B., Nguyen, H., Pagnucco, M., Ratter, A., Robinson, S., Sammut, C., Vance, B., White, B., Zhu, Y.: RoboCup standard platform league - rUNSWift 2010. In: Proceedings of the Australasian Conference on Robotics and Automation (2010)
7. Ashmore, J.: Robot detection using bayesian machine learning, UNSW CSE RoboCup Report 20140831-Jaiden.Ashmore-RobotDetectionReport.pdf (2014)
8. Chatfield, C.: rUNSWift 2011 vision system: A foveated vision system for robotic soccer, UNSW CSE RoboCup Report 20110825-Carl.Chatfield-VisionFoveated.pdf (2011)
9. Claridge, D.: rUNSWift 2010 code release (2010). http://github.com/UNSWComputing/runswift
10. Claridge, D.: Generation of python interfaces for RoboCup SPL robots, UNSW CSE RoboCup Report 20110228-David.Claridge-Python.pdf (2011)
11. Claridge, D.: Multi-hypothesis localisation for the NAO humanoid robot in RoboCup SPL, UNSW CSE RoboCup Report 20110830-David.Claridge-MulitHypothesisLocalisation.pdf (2011)
12. Deng, Y.: Natural landmark localisation, UNSW CSE RoboCup Report 20110830-Yiming.Deng-NaturalLandmarksLocalisation.pdf (2011)
13. Harris, S.: Efficient feature detection using RANSAC, UNSW CSE RoboCup Report 20110824-Sean.Harris-RansacVision.pdf (2011)
14. Harris, S., Anderson, P., Teh, B., Hunter, Y., Liu, R., Hengst, B., Roy, R., Li, S., Chatfield, C.: Robocup standard platform league - rUNSWift 2012 innovations. In: Proceedings of the 2012 Australasian Conference on Robotics and Automation (ACRA 2012) (2012)
15. Hengst, B.: Robocup 2010 version 1 storyboard prototype, UNSW CSE RoboCup Report 20091013-Bernhard.Hengst-Robocup2010Storyboard.pdf (2009)
16. Hengst, B.: Reinforcement learning of bipedal lateral behaviour and stability control with ankle-roll activation. In: Proceedings of the 16th International Conference on Climbing and Walking Robots (CLAWAR2013), World Scientific Publishing Company, Sydney, Australia (2013)
17. Hengst, B.: rUNSWift Walk 2014 report, UNSW CSE RoboCup Report 20140930-Bernhard.Hengst-Walk2014Report.pdf (2014)
18. Hengst, B., Ibbotson, D., Pham, S.B., Sammut, C.: The UNSW United 2000 Sony legged robot software system, UNSW CSE RoboCup Report 20001114-HengstEtAl-RobocupReport.pdf (2000)
19. Hengst, B., Lange, M., White, B.: Learning ankle-tilt and foot-placement control for flat-footed bipedal balancing and walking. In: Proceedings of the 11th IEEE-RAS International Conference on Humanoid Robots (2011)
20. Hunter, Y.: Industrial training report (foot detection and ball stealing), UNSW CSE RoboCup Report 20110216-Youssef.Hunter-FootDetectionForBallStealing.pdf (2011)

21. Hunter, Y.: Humanoid robot localisation for the Robocup Standard Platform League, UNSW CSE RoboCup Report 20120824-Youssef.Hunter-Localisation.pdf (2012)
22. Jisarojito, J.: Tracking a robot using overhead cameras, UNSW CSE RoboCup Report 20110217-Jarupat.Jisarojito-OverheadCameras.pdf (2011)
23. Kurniawan, J.: Multi-modal machine-learned robot detection for RoboCup SPL, UNSW CSE RoboCup Report 20110824-Jimmy.Kurniawan-RobotDetection.pdf (2011)
24. Lange, M.: Developing a bipedal walk using a Cycloid II, UNSW CSE Robocup Report 20110731-Manuel.Lange-BipedalCycloidWalk.pdf (2011)
25. Liu, R.: Bipedal walk and goalie behaviour in Robocup SPL, UNSW CSE RoboCup Report 20130108-Roger.Liu-WalkAndGoalie.pdf (2013)
26. Ratter, A., Claridge, D., Ashar, J., Hengst, B.: Fast object detection with foveated imaging and virtual saccades on resource limited robots. In: Wang, D., Reynolds, M. (eds.) AI 2011: Advances in Artificial Intelligence. LNCS, vol. 7106, pp. 560–569. Springer, Heidelberg (2011)
27. Ratter, A., Hengst, B., Hall, B., White, B., Vance, B., Claridge, D., Nguyen, H., Ashar, J., Robinson, S., Zhu, Y.: rUNSWift team report 2010 robocup standard platform league, UNSW CSE RoboCup Report 20100930–2010rUNSWiftTeam Report.pdf (2010)
28. Sushkov, O.: Robot localisation using a distributed multi-modal kalman filter, and friends, UNSW CSE RoboCup Report 20060908-Oleg.Suchkov-Localisation.pdf (2006)
29. Teh, B.: Ball modelling and its application in robot goalie behaviours, UNSW CSE RoboCup Report 20110824-Belinda.Teh-BallModellingGoalieBehaviour.pdf (2011)
30. Teh, B.: Dynamic omnidirectional kicks on humanoid robots, UNSW CSE RoboCup Report 20120824-Belinda.Teh-OmniDirectionalKicks.pdf (2012)
31. Tsekouras, L.: A heel to toe gait for efficient bipedal walking and stricker behaviour for rUNSWift Robocup SPL 2014, UNSW CSE RoboCup Report 20141105-Luke.Tsekouras-AHeeltoToeGait.pdf (2014)
32. Tsekouras, L., Ashmore, J., Mei, Z., Teh, B., Sushkov, O., Roy, R., Liu, R., Harris, S., Ashar, J., Hall, B., Hengst, B., Pagnucco, M., Sammut, C.: Team rUNSWift University of NSW Australia, UNSW CSE RoboCup Report 20140630–2014Team-Robocup2014rUNSWiftTeamDescription.pdf
33. White, B.: Humanoid omni-directional locomotion, UNSW CSE RoboCup Report 20111010-Brock.White-OmniDirectionalLocomotion.pdf (2011)

CIT Brains KidSize Robot: RoboCup 2014 Best Humanoid Award Winner

Yasuo Hayashibara[1(✉)], Hideaki Minakata[1], Kiyoshi Irie[1],
Taiki Fukuda[1], Victor Tee Sin Loong[1], Daiki Maekawa[1], Yusuke Ito[1],
Takamasa Akiyama[1], Taiitiro Mashiko[1], Kohei Izumi[1],
Yohei Yamano[1], Masayuki Ando[1], Yu Kato[1], Ryu Yamamoto[1],
Takanari Kida[1], Shinya Takemura[1], Yuhdai Suzuki[1],
Nung Duk Yun[1], Shigechika Miki[2], Yoshitaka Nishizaki[3],
Kenji Kanemasu[4], and Hajime Sakamoto[5]

[1] Chiba Institute of Technology, 2-17-1 Tsudanuma, Narashino, Chiba, Japan
yasuo.hayashibara@it-chiba.ac.jp
[2] Miki Seisakusyo Co., Ltd., 1-7-28 Ohno, Nishiyodogawa, Osaka, Japan
[3] Nishizaki Co., Ltd., 1-7-27 Ohno, Nishiyodogawa, Osaka, Japan
[4] Yosinori Industry, Ltd., 1-1-7 Fukumachi, Nishiyodogawa, Osaka, Japan
[5] Hajime Research Institute, Ltd., 1-7-28 Ohno, Nishiyodogawa, Osaka, Japan
sakamoto@hajimerobot.co.jp

Abstract. In this paper, we describe the system design of the robots developed by our Team CIT Brains for the RoboCup soccer humanoid KidSize league. We have been participating in the Humanoid League for eight years. Two years ago, we redesigned the system to put a large weight on maintainability and usability. In RoboCup 2014, we won the first prizes of 4on4 soccer and technical challenge. Consequently, we were awarded the Louis Vuitton Humanoid Cup. The system we developed has high mobility, well-designed control system, position estimation by a monocular camera, user-friendly interface and a simulator. The robot can walk speedily and robustly. It also has a feedback system with a gyro sensor to prevent falls. It detects positions of landmarks by color-based image processing. A particle filter is employed to localize the robot in the soccer field fusing the motion model and landmark observation.

Keywords: Humanoid robots · Programming environment · Education robotics

1 Introduction

In this paper, we describe our system for the RoboCup soccer humanoid KidSize league. In RoboCup 2014 Brazil, we won the first prizes of 4on4 soccer and technical challenge. The results are indicated in Table 1. Consequently, we were awarded the Louis Vuitton Humanoid Cup. We started to develop the autonomous humanoid soccer robot in 2006 and have developed 22 robots in total. A history of the major KidSize robots in CIT

R.A.C. Bianchi et al. (Eds.): RoboCup 2014, LNAI 8992, pp. 82–93, 2015.
DOI: 10.1007/978-3-319-18615-3_7

Brains is shown in Fig. 1. From 2013, we use the fourth generation robots. Through the development, we have studied many issues related to our robots [1–13].

CIT Brains is a joint team consisting of Hajime Research Institute and Chiba Institute of Technology (CIT). Hajime Research Institute developed mechanisms and prototypes of control systems of the robots [14]. CIT developed computer systems and overall intelligence such as perception and planning. CIT also made contributions to improve the mechanisms and control systems. We would like to emphasize that 13 members out of 19 members from CIT are undergraduate students. Any students who want to join this development can join our team. Senior students teach new members from the basic knowledge of the robot system. We aim to make an educational and research platform of intelligent humanoid.

Table 1. Results of CIT Brains in RoboCup2014

Soccer 4 on 4	8 wins - 0 loss
	Total goals: 33 goals - 3 losses
Technical challenge	High-Kick Challenge: 10 points
	Obstacle Avoidance and Dribbling: 10 points
	Artificial Grass Challenge: 7 points
	Total: 27 points

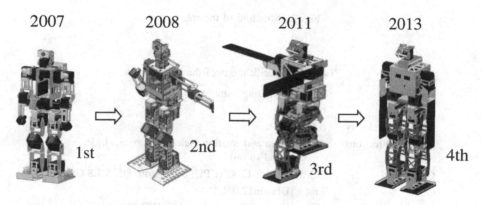

2007   2008   2011   2013

1st   2nd   3rd   4th

Fig. 1. A history of KidSize robots in CIT Brains

2 Overview of the System

A photograph of our robot is shown in Fig. 2. The specifications of the robot are summarized in Table 2. An overview of the system hardware is shown in Fig. 3. Our robot system consists of a USB camera, a computer board, an inertial measurement unit (IMU), 17 servomotors, a battery and several user interfaces such as switches.

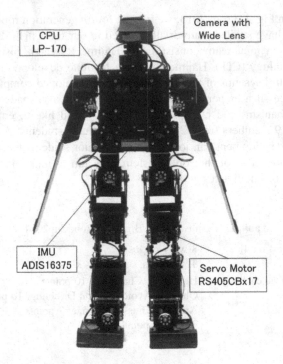

Fig. 2. Structure of the robot

Table 2. Specifications of the robot

Weight	3.5 kg (Including batteries)
Height	600 mm
Maximum velocity	0.4 m/s
Walking directions	All direction and rotation (select the angle, stride, period and so on)
CPU board	COMMEL LP-170C (CPU: Intel Atom D525 1.8 GHz)
OS	Linux (Ubuntu12.04LTS)
Interface	Ethernet × 1, USB × 1, Speaker, DIP switch × 4, Push switch × 1
Servomotor	Futaba RS405CB × 17
IMU	Analog Devices ADIS16375
USB camera	Shikino High-Tech KBCR-M05VU
Battery	3S (11.1 V, 5000 mAh)

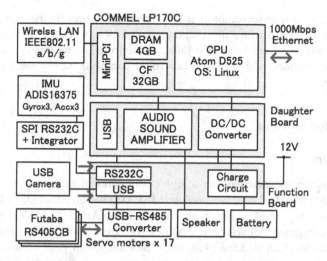

Fig. 3. Overview of the hardware system

3 Mobility

We achieved high-speed and stable mobility. The maximum speed of walk is approximately 0.4 m/s when carefully tuned. For playing soccer, it is also important to keep stable walking for long periods of time. Many soccer robots in the KidSize league tend to become unstable in the later part of games because of motor overheating. However, our robot can keep stable walk during full match. This is one of the key factors of our winning.

Gyro feedback is applied during walking to prevent falling down. Our robot does not usually fall down when it walks alone. However, in a soccer game, it often falls down when it is pushed by other robots. When our robot falls down, it detects the fall and stands up smoothly. The posture of the robot is estimated by fusing acceleration and angular velocity received from the IMU [15].

We designed a leg structure using parallel mechanism as shown in Fig. 4. This mechanism can keep pitch angles of the feet synchronized even if the motors are not synchronized completely. It contributes to the stable walking.

Fig. 4. Parallel mechanism of legs.

Choosing large torque servomotors is also important to prevent the overheating. We employ Futaba RS405CB with a maximum torque of 48 kg-cm. Active cooling fans are attached to motors for knee and ankle joints. Heat from the motors is suppressed by abovementioned factors, and therefore our robot can walk stably even after a long operation time.

4   Computer System

4.1   Hardware

The key advantages of the robot are a high computational capability and an ease of maintenance. A computer board we employ (LP-170C) has an Atom D525 CPU, which has substantially higher capability than DarwIn-OP's Atom Z530. We run Ubuntu Linux on the computer. All software modules we develop, including perception and control, are executed on it. Linux has many advantages in ease of installation and operation compared to other operating systems we have previously used (Windows and NetBSD). To improve the maintainability of electronic components, we designed a slot-in mechanism as shown in Fig. 5. The mechanism interfaces *function board* which supplies power and *daughter board* on which the main computer board is mounted. The pictures of the boards are shown in Fig. 6. By this mechanism we were able to eliminate a large number of cables compared with the previous model. The function board also has battery-charging functionality so that we can charge batteries without external chargers.

4.2   Development Environment

We have installed a development environment to each robot so that we can edit and compile source codes in the onboard computer. We directly operate the onboard computer by connecting a display and a USB keyboard (Fig. 7), or remotely operate it via VNC. The charging circuit significantly improved ease of development. While we are editing and compiling source codes, we plug the A/C adapter to the robot and charge the battery. Servomotors are automatically powered off when the A/C adapter is plugged. When we want to check software using the robot, we unplug the adapter then the power of the motors is automatically turned on and we only need to put the robot on the field. We achieved laptop-like usability.

5   Software System

5.1   Architecture

Figure 8 shows the architecture of the software system. All software modules including perception, planning, and control are executed on the main computer board. Two processes are executed on a single computer: one is for perception and planning and the other is for control. Images are captured by the USB camera, and processed on the

Fig. 5. Slot-in system

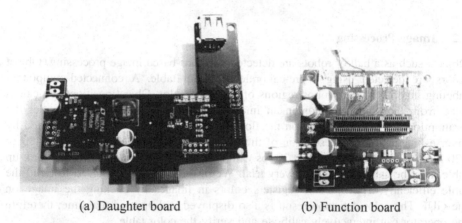

(a) Daughter board   (b) Function board

Fig. 6. Developed circuit boards

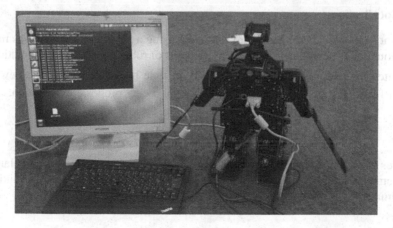

Fig. 7. Software development environment.

computer board to detect the position of the ball, other robots and landmarks. The robot continuously estimates self-position using the obtained information. Higher level of the robot behaviors such as following a ball are described in the soccer strategy programs. The soccer strategy programs are written in Python for ease of trial-and-error type of development while the rest of the software modules are written in C and C++.

The body control tasks are operated in a dedicated control process. It controls the posture of the robot according to commands sent from the main process. The status of the robot (e.g. posture) is periodically sent to the main process. The control process sends commands such as angles to servomotors. An IMU is used for gyro feedback and posture estimation.

We employ Internet Communication Engine (ICE) for communication among software components. ICE is a middleware for distributed computing including Remote Procedure Call (RPC). ICE is known to be computationally efficient compared to other middleware such as CORBA [16]. Our software modules running in different processes or computer hosts communicate with each other via ICE.

5.2   Image Processing

Objects such as a ball or robots are detected by color-based image processing. Object colors are detected using a pre-calibrated look-up table. A connected-component labeling algorithm then groups regions of the same color. Object positions are calculated from the object position in an image and the pose of the camera under the assumption that all objects are on the floor. The pose of the camera is calculated by inverse kinematics. The resolution of the images can be selected from 640×480 or 320×240. Our algorithm runs at 20 fps with the onboard computer. The color look-up table must be calibrated before every trial. We developed a GUI (Fig. 9) to build the table efficiently. The operator registers colors in images by clicking the image on the GUI. The color detection result is also displayed and updated real-time; therefore the operator can interactively calibrate and verify the color table.

5.3   Localization

The robot position and orientation are estimated using a particle filter that fuses motion estimations and observations [17]. The hypotheses of robot position and orientation are represented by a set of particles $\left\{ S^{(i)} \right\}_{i=1}^{N}$. Each time the robot moves, the new generation of particles are drawn according to the following proposal distribution:

$$S_t^{(i)} \sim P\left(x_t | S_{t-1}^{(i)}, \Delta x_t \right)$$

Here, Δx_t is the estimated robot motion from time $t = 1$ to t, which is calculated by the kinematics and integrating angular velocity from the IMU. The distribution is approximated by a normal distribution.

When landmarks such as white lines and goal posts are observed, the particles resampled according to the weight proportional to the observation likelihood

$$w^{(i)} \propto P\left(Z|s^{(i)}\right).$$

Errors of the landmark measurements are modeled by a normal distribution and correlation-based measurement model [18] is used for white line observations.

To deal with the localization ambiguity caused by the soccer field symmetry, the robot orientation is manually initialized when we put the robot on the field. The robot has an orientation reset button for this purpose and the robot handler pushes the button when the robot is manually repositioned during the game.

During soccer games, the localization system sometimes fails because of various reasons such as unexpected measurement errors and manual repositioning. To reduce the risk of own-goal, we integrated failure recovery algorithm proposed by Ueda et al. [19]. Our system keeps track of the measurement score and when the score falls below a certain threshold, the particles are diffused for re-localization.

5.4 Obstacle Avoidance

We use graph-based path planning for obstacle avoidance as illustrated in Fig. 10. Several control points are placed around the detected obstacles (i.e. other robots) and a complete graph is built by connecting all control points, start point, and destination point. Costs are set to all edges of the graph according to their length. Extra costs are added for edges that go across obstacles. Dijkstra algorithm is employed to find the path with minimal cost.

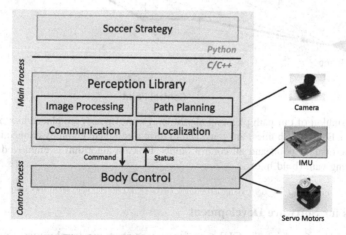

Fig. 8. Architecture of the software system.

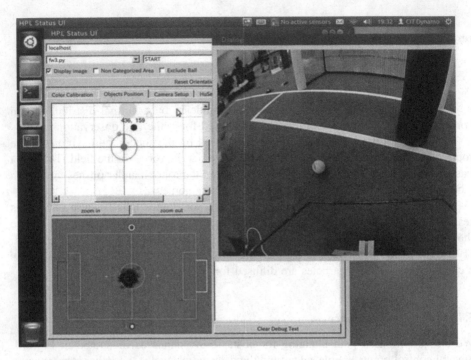

Fig. 9. Our graphical user interface. Positions of detected objects are visualized.

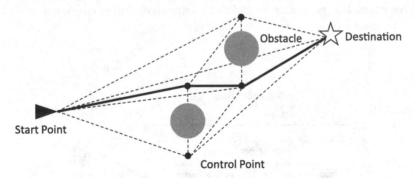

Fig. 10. Illustration of our path planning algorithm. Control points are placed around obstacles (small circles). Edges (dash lines) are generated to form a complete graph by connecting between start point, destination point and all control points. Dijkstra algorithm is employed to find the cost-minimizing path (bold line).

5.5   Tools for Software Development

We developed a user-friendly GUI tool for soccer strategy development environment (Fig. 11). Programmers can interactively check variety of items through this interface. The functionalities provided by the interface are as follows:

[User operation]

(1) Send commands to the control process
(2) Build a color look-up Table
(3) Execute strategy programs by its name

[Status monitoring]

(1) Image data (both raw and processed images)
(2) Detected objects and their positions
(3) Estimated robot position and particles
(4) Debugging messages
(5) Battery voltage and temperatures of servomotors

The GUI displays most of the significant status of the robot so that the programmer can check the algorithm and find problems easily.

We also developed a simulator based on V-REP. V-REP is an open source robot simulator made by COPPELIA ROBOTICS. Most of the behaviors of the robot are verified in the simulator. We check both the low-level walking control and high-level soccer strategy codes using the simulator and apply the verified codes to the real robot without any modification.

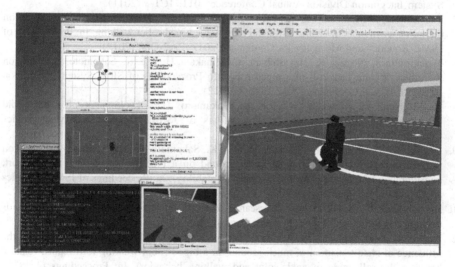

Fig. 11. Soccer strategy development environment and simulator software

6 Conclusion

In this paper we described our autonomous soccer humanoid system. Our system has high mobility, well-designed control system, position estimation by a monocular camera, user-friendly interface and simulator. High integrity of our humanoid system made it possible for us to win the Louis Vuitton Humanoid Cup at RoboCup 2014 Brazil.

References

1. Hayashibara, Y., Minakata, H., Sakamoto, H., Irie, K., Kaminaga, H., Fujita, S., Fukuta, M., Horiuchi, T.: Development of autonomous soccer robot system "CIT Brains and Hajime Robot" for RoboCup humanoid. In: Proceedings of JSME Robotics and Mechatronics Conference 2008, 2P1-J18 (2008)
2. Minakata, H., Hayashibara, Y., Ichizawa, K., Horiuchi, T., Fukuta, M., Fujita, S., Kaminaga, H., Irie, K., Sakamoto, H.: A method of single camera RoboCup humanoid robot localization using cooperation with walking control. In: The 10th International Workshop on Advanced Motion Control, TF-000787 (2008)
3. Hayashibara, Y., Minakata, H., Seike, Y., Ichizawa, K., Ogura, S., Irie, K., Sakamoto, H., Kaminaga, H., Kawakami, T., Fujita, S., Fukuta, M., Horiuchi, T.: Development of soccer robot system "CIT Brains". In: Proceedings of JSME Robotics and Mechatronics Conference 2009, 1P1-C07 (2009)
4. Hayashibara, Y., Minakata, H., Sakamoto, H.: Autonomous soccer humanoid system "CIT Brains". J. Jpn. Soc. Des. Eng. 44(6), 334–340 (2009)
5. Hayashibara, Y., Minakata, H., Irie, K.: Development of humanoid "Dynamo" for RoboCup. In: Proceedings of the 28th Annual Conference of the Robotics Society of Japan, 1A3-5 (2010)
6. Hayashibara, Y., Minakata, H., Irie, K., Ichizawa, K., Tsuchihashi, I., Nohira, K., Sakamoto, H.: Development of a humanoid "Dynamo2011" for RoboCup. In: Proceedings of SICE System Integration Division Annual Conference 2011, 1C1-3 (2011)
7. Tsuchihashi, I., Minakata, H., Irie, K., Nozaki, K., Sakamoto, H., Hayashibara, Y.: Study on the influence of state transition of touch ground on a biped robot walking. In: Proceedings of JSME Robotics and Mechatronics Conference 2012, 2A2-Q06 (2012)
8. Tsuchihashi, I., Nozaki, K., Hayashibara, Y., Minakata, H., Irie, K., Sakamoto, H.: Study on the effect of joint backlash of a biped robot. In Proceedings of SICE System Integration Division Annual Conference 2012, 3N2-2 (2012)
9. Hayashibara, Y., Minakata, H., Irie, K., Sakamoto, H.: Development of a humanoid "Dynamo2012" for RoboCup. In: Proceedings of JSME Robotics and Mechatronics Conference 2013, 1P1-A12 (2013)
10. Nozaki, K., Tsuchihashi, I., Minakata, H., Irie, K., Sakamoto, H., Hayashibara, Y.: Study on the influence of state transition of touch ground on a biped robot walking (2nd report: consideration of the sole angle and the walking stability). In: Proceedings of JSME Robotics and Mechatronics Conference 2013, 1P1-A13 (2013)
11. Hayashibara, Y., Minakata, H.: The research development of autonomous soccer humanoids through the RoboCup. J. Soc. Instrum. Control Eng. 52(6), 487–494 (2013)
12. Nozaki, K., Minakata, H., Irie, K., Sakamoto, H., Yoneda, K., Hayashibara, Y.: Study on the influence of state transition of touch ground on a biped robot walking (3rd report: relation between the roll angle of ankle joint and walking behavior). In: Proceedings of JSME Robotics and Mechatronics Conference 2014, 1P1-S03 (2014)
13. Fukuda, T., Minakata, H., Irie, K., Sakamoto, H., Hayashibara, Y.: Study on walking control of a humanoid robot considering motor temperatures (2nd report: relationship between walking patterns and motor temperatures). In: Proceedings of JSME Robotics and Mechatronics Conference 2014, 1P1-S01 (2014)
14. Sakamoto, H., Nakatsu, R.: Walking and running control of small size humanoid robot 'HAJIME ROBOT'. In: The Thirteenth International Symposium on Artificial Life and Robotics 2008, GS6-1 (2008)

15. Madgwick, S.O.H., An efficient orientation filter for inertial and inertial/magnetic sensor arrays. Technical report, University of Bristol University, UK (2010)
16. Khan, S., Qureshi, K., Rashid, H.: Performance comparison of ICEHORB, CORBA and Dot NET remoting middleware technologies. Int. J. Comput. Appl. 3(11), 15–18 (2010)
17. Dellaert, F., Fox, D., Burgard, W., Thrun, S.: Monte carlo localization for mobile robots. In: Proceedings of the IEEE International Conference on Robotics and Automation (ICRA), pp. 1322–1328 (1999)
18. Thrun, S., Burgard, W., Fox, D.: Probabilistic Robotics. MIT Press, Cambridge (2005)
19. Ueda, R., Arai, T., Sakamoto, K., Kikuchi, K. Kamiya, S.: Expansion resetting for recovery from fatal error in monte carlo localization - comparison with sensor resetting methods, In: Proceedings of the IEEE International Conference on Intelligent Robots and Systems (IROS), pp. 2481–2486 (2004)

RoboCup 2014 Humanoid AdultSize
League Winner

Seung-Joon Yi[1], Steve McGill[1(✉)], Qin He[1], Larry Vadakedathu[1], Hak Yi[2], Sanghyun Cho[2], Dennis Hong[2], and Daniel D. Lee[1]

[1] GRASP Lab, Engineering and Applied Science, University of Pennsylvania, Philadelphia, PA, USA
{yiseung,smcgill3,heqin,vlarry,ddlee}@seas.upenn.edu
http://www.seas.upenn.edu/~robocup
[2] University of California, Los Angeles, CA 90095, USA
{yihak,albe1022,dennishong}@ucla.edu

Abstract. The RoboCup Soccer Humanoid AdultSize League is a very challenging league that requires fast and reliable locomotion, a robust vision and localization system, and long term planning to score in a limited attack window. As a result, many adult-size robots are now highly optimized for the soccer task, with very lightweight construction that enables nimble locomotion with low power actuators. However, such a soccer optimized humanoid design excludes the possibility of using the humanoid robot for more general tasks that include, for instance, dextrous manipulation and navigating over rugged terrain. In this paper, we describe how we utilize the general purpose and full sized humanoid robot, THOR-OP (Tactical Hazardous Operations Robot - Open Platform), which was originally developed to compete at the DARPA Robotics Challenge (DRC) for the 2014 RoboCup AdultSize league robotic soccer competition.

1 Introduction

The University of Pennsylvania RoboCup Team has been competing in the Humanoid KidSize league since 2010 as Team DARwIn. During that time, we pushed forward to open source both the hardware and software platforms: the latest generation of DARwIn robots, the DARwIn-OP, is openly available in the form of blueprints and as a commercial product. The modular, portable and easy-to-use UPenn humanoid robotic software platform has also been released as open source. Now, a major portion of the KidSize teams are using a variant of the DARwIn-OP hardware; some with our open source software.

Playing soccer well requires excellence in many parts of robotics research, including dynamic locomotion, computer vision, world modeling, unsupervised autonomy and multi-robot coordination - all within constrained power and computation capability of individual robot, as well as the software and hardware robust to unknown environmental disturbances. The skill sets achieved in the framework of robot soccer push the boundaries for attaining human abilities

© Springer International Publishing Switzerland 2015
R.A.C. Bianchi et al. (Eds.): RoboCup 2014, LNAI 8992, pp. 94–105, 2015.
DOI: 10.1007/978-3-319-18615-3_8

Fig. 1. The modified THOR-OP robot playing soccer at RoboCup 2014.

in addition to just playing soccer [1]. And to solve more general problems - especially that includes manipulating objects - requires a humanoid robot that is both "full sized" – roughly with adult human characteristics – and "general purpose" – those with enough power and dexterity to handle various tasks. Until recently such humanoid robots have been accessible to only few researchers, as building and maintaining a heavy and powerful humanoid robot can take a large amount of time and effort Fig. 1.

However, with the introduction of new heavy-duty modular actuators, building and maintaining of full-size general purpose humanoid robot can now be as straightforward as those of smaller humanoid robots. In this paper, we describe the THOR-OP general purpose, full sized humanoid robot we use for RoboCup 2014, and how we handled additional difficulty that comes with using a heavy general purpose robot not specially designed for robot soccer task.

2   THOR-OP Humanoid Robot Platform

The THOR-OP (Tactical Hazardous Operations Robot - Open Platform) humanoid robot is developed by Robotis, Co. Ltd[1] as a general purpose disaster response robot to compete in the DARPA Robotics Challenge (DRC) [2]. The DRC requires a complete system with the mobility, dexterity, strength and

---

[1] http://www.robotis-shop-en.com/.

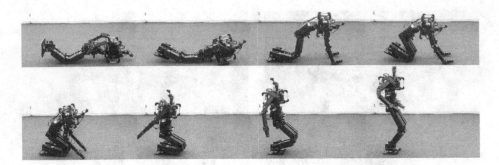

Fig. 2. THOR-OP robot performing getup motion from the prone posture.

endurance for a practical disaster response situation. In order to score in the various DRC tasks, the robot should be able to operate in unstructured environments that include rough terrain, ladders, doorways, piles of debris, and industrial valves. It is also crucial that the robot be capable of maneuvering unmodified power tools and vehicles designed for use by human beings. THOR-OP performed well during the DRC Trials and is among the finalists for the DRC Final in 2015, showing to be an advanced and comprehensive humanoid platform.

Table 1. Specs of AdultSize humanoid robots in RoboCup 2012-2014

Team	Robot	Height (cm)	Weight (kg)
EDROM	MABI	134	7
Team Sweaty	Sweaty	140	11.5
Team Charli	CHARLI-2	140	12.1
HuroEvolution AD	HuroEvolution	149	15
Tsinghua Hephaestus	THU_Strider	132	18.1
TU Eindhoven	TUlip	125	23
JoiTech	Tichno-RN	150	25
KW-1	KW-1	152	26
Robo-Erectus	Sr-2001	150	44
THORwIn	THOR-OP	147	43.6 / 51.6

As shown in Table 1, most of the AdultSize humanoid soccer robots are designed solely for soccer playing and extremely light-weight. They are relatively simple in terms of kinematics, lacking high DOF upper bodies necessary for manipulation tasks. In comparison to these single-purpose robots, THOR-OP has more DOF (waist pitch and yaw and 7DOF in arms) and a stronger mechanical structure for manipulating heavy tools and performing whole body movements such as get-up motion necessary on soccer fields and other challenging terrains (shown in Fig. 2). During the DRC Trials event, THOR-OP robot has successfully accomplished demanding tasks such as manipulating a drill, dragging the fire hose and turning tight valves.

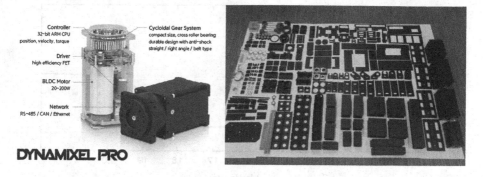

Fig. 3. The modular actuator and structural parts of THOR-OP robot.

2.1 Actuators and Structural Components

For the most high-load joints, the off-the-shelf Robotis Dynamixel Pro series actuators are used. Three different actuators types are used, rated at 20 W, 100 W and 200 W, which are fitted with either inline and parallel gearboxes. During the preparation of the DRC Trials, the actuators proved themselves strong enough for all the tasks we tried, which includes carrying a cordless power drill and cutting through the wall, rotating a tight valve, and stepping over uneven blocks. In addition to the actuators, the robot is mainly built with standardized structural components that are simply extruded aluminum tubings and brackets with regularly spaced bolt holes, and one can easily assemble them with hex bolts. Figure 3 shows the structural components required for a single THOR-OP robot; twenty four total man-hours is estimated to complete assembly from parts.

2.2 Modification for RobCup

THOR-OP originally had 7DOF arms with underactuated grippers, providing a large workspace and high dexterity for manipulation tasks. We have replaced the original arm due to a number of reasons: the arms are too long to conform with the RoboCup rules, and long arm with gripper gets in the way with vision and makes the robot posture wider, increasing the chance to touch the obstacles. A shorter and simpler arm design have therefore been introduced, which used acrylic rather than metal and has only 2DOF per arm, sparing 4 kg in weight from each arm. The new lighter arm still has enough range of motion and strength for picking up and holding a ball, which are needed for technical challenge.

We have also modified the feet of THOR-OP. The original feet are designed to be small and rigid to help the robot traverse uneven terrain better, but not ideal for the robot soccer where robot is supposed to walk over relatively flat surface. We have made a slightly wider feet in lightweight plastic, as well as putting a short wall at the toe section of each foot. Through testing and competition, we found that the increased dimensions of the feet proved to provide better support for the robot, especially when the robot is doing a strong kick, and helped the robot kick straight even with a slight misalignment.

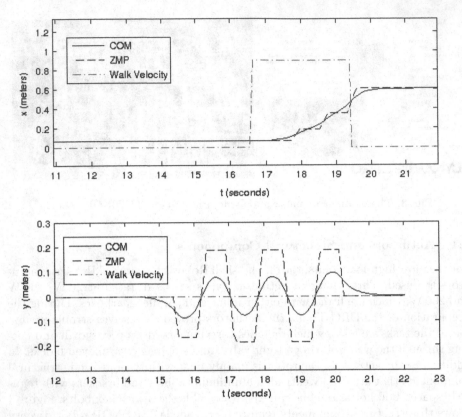

Fig. 4. X and Y axis COM, ZMP and commanded velocity during a walking sequence.

3   Walk and Kick Motion Generation

We have been using the analytic Zero Moment Point (ZMP) based controller [3] that can reactively change walk velocity for our previous KidSize robots. With the larger and heavier THOR-OP robot, however, we have found that we cannot apply the same approach, since the torso velocity discontinuity during transition between double and single support induces vibration that can destabilize the robot. For this reason, we have used the ZMP preview walk controller [4] during the DRC Trials competition, where the environment is static and the control lag does not matter much. However, ZMP preview controller is not preferable for robotic soccer application due to its control lag and high computational requirement.

3.1   Hybrid Walk Controller

In our previous work [5], we have suggested a hybrid walk controller that dynamically switches between the analytic ZMP based reactive controller and ZMP preview controller, and tested the controller on DARwIn-OP KidSize humanoid

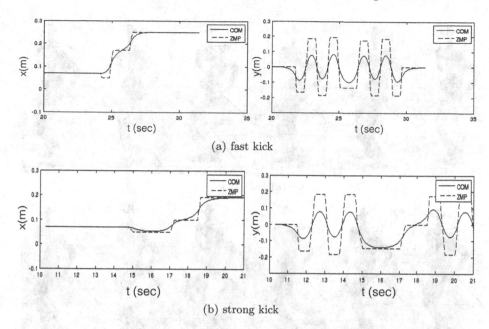

Fig. 5. X and Y axis COM and ZMP trajectories during two different kick motions.

robot. The robot mainly uses the reactive controller for locomotion; the ZMP preview controller is used only to perform more complex footstep motions such as dynamic kicking. For the THOR-OP robot, we use the ZMP preview controller also to generate the initial and final steps where the transitions between single and double support take place. We add the initial and final preview step entries into the step queue to initiate the initial and final steps. The initial step has zero foot displacement while the final step is set to make both feet parallel. To simplify designing the ZMP trajectory for the ZMP preview controller, we used the same trapezoidal ZMP trajectory as the one for our reactive controller [3]. Figure 4 shows the ZMP and Center of Mass (COM) trajectories of the robot where the robot starts walking, take a few step forward and stops.

3.2 Kick Motion Generation

We also use the hybrid walk controller formulation to generate the dynamic kick motion for the robot. A number of different dynamic kick motions are needed for the competition, because the attacker loses its penalty kick attempt if any of the following situations happen: direct scoring from the far side of the field, robot touching any of obstacles, or the kicked ball stopping inside the penalty box. Under those rules, the best strategy is to move the ball to cross the centerline first and do a strong kick to score.

We designed three different kicks. A fast kick with three to four meters of kicking range moves the ball from the defensive half to the offensive half of the

(a) Fast kick

(b) Strong kick

Fig. 6. Simulated THOR-OP robot performing a dynamic kick in middle of reactive walking.

field. A weak kick with a one meter kicking range moves the ball close to the middle of the offensive half. Last but not least, a slow and strong kick can directly score. Thanks to our hybrid walk controller formulation, we can use a single kick motion to initiate the kick in any phase of the walking. Figure 5 shows the COM and ZMP trajectories of two different kicks we have used for the competition, and Fig. 6 shows the kicking sequences in simulated environment.

4   Vision and Localization

4.1   Vision

Images from a Logitech C920 HD camera are captured and processed. We first use a look-up table to categorize pixels into one of eight color labels. The look-up table is generated beforehand using supervised learning approach with a Gaussian mixture model. Figure 7 shows the look-up table generating tool for a real robot camera frame, and how the generated lookup table performs in simulation. The labeled image is then down-sampled for faster processing and fed into high level blob detection routines and object classifiers. The robot is able to detect the ball, goal posts, and lines based on their color, size, and shape

Fig. 7. Look-up table can be made with a user interface for classifying colors (left). A MATLAB monitor displays the labeled image efficiently generated in real-time using the generated look-up table(right) (Color figure online).

information. Given the exact dimensions of objects, the distance and angle of detected objects are obtained by projection from the image frame to the robot's egocentric coordinates.

For AdultSize League in RoboCup 2014, there are two black cylindrical obstacles randomly placed between the robot and the goal. The similar blob detection techniques mentioned above could be used, however there is a large amount of black pixels in the labeled images which could affect the efficiency and reliability of the detection. We instead use a simpler detection and filtering approach for obstacles, with the knowledge that an obstacle is basically a chunk of black pixels with proper dimensions surrounded by green (field) pixels. The labeled image is scanned horizontally and the position of the lowest black pixel (*depth* of black) in each column is marked. The columns with black *depth* difference within a threshold are stitched together to be connected regions, and those are the candidates of obstacles which then go through a filtering process including height check, width check, on-the-ground check, and within-the-field check. A final filtering process is performed after transforming the positions of the detected obstacles from image frame to the robot's egocentric, polar coordinates. The potential obstacles are clustered into groups based on their angles to the robot, and statistical information is calculated for each group providing the certainty and weighed average position of each obstacle. The closest three obstacles (two obstacles and the opponent) are registered for later trajectory planning.

4.2   Localization

For previous years, we have used a particle filter to track the three-dimensional robot pose (x, y, θ) on the field, in which orientation is reset when the robot falls down. The particles are probabilistically updated using a motion model based on the odometry of the robot and yaw angle based on inertial measurement, as well as vision information that takes into account pre-specified positions of landmarks such as goal posts, lines and corners.

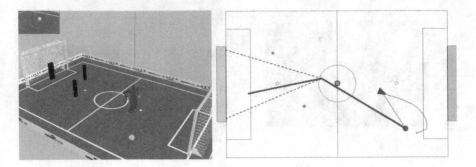

Fig. 8. Simulated THOR-OP robot follows the planned trajectory (red line) and generates an optimized attacking strategy (blue line) based on the perceived locations of obstacles (two red markers) (Color figure online).

This year we could further simplify this framework due to the uniqueness of AdultSize league: the robot always starts at exactly the same initial position. It is never supposed to fall down, and each trial has quite a short time limit of 90 s. This makes the inertial tracking of the robot heading angle θ very precise during the match, so we assume that the current estimate of heading direction is correct. Then we discard the distance information from goalpost observations, which can be quite noisy for distant goalposts, to handle the redundant information. This simplification greatly helped the localization during the whole competition, where the localization error was hardly visible.

5   Hierarchical Planning

As we have said above, the best tactic for the attacking robot is to move the ball to just cross the centerline and do a powerful kick to score. However the existence of obstacles and the high possibility of the ball rolling irregularly makes this not a trivial problem.

5.1   Kick Planning

At the highest level, we determine the optimal sequence of kick angles that maximizes the scoring chance while minimizing the failure chance. To reduce the computational load we use a discrete search tree, where each node represents a possible ball position. The root node is the current ball position, and each node branches into k different kick angles until it is possible to directly score from that position. Then every path is evaluated based on the scoring chance based on the goal angle at the final node, and the failure chance based on the minimum kick error that can result in out of bounds or colliding the obstacle. To prevent oscillating when traversing down the kick tree, we also penalize paths based on their difference from the best path chosen at the previous planning step.

5.2   Walk Trajectory Planning

Once the optimal kick sequence is planned, the robot has to move to the first kicking position in the shortest time possible. We use the parameterized curved trajectory [6] to better utilize the fast forward walking speed of our robot. Once the trajectory is generated, the maximum possible velocity within its stable region is selected as the commanded velocity.

5.3   Foot Step Planning

Our velocity based walking control scheme is based on the torso velocity and does not consider the foot stance, which is not preferable for the final approaching steps where we want the robot to stop with feet precisely aligned to the ball. For the KidSize league, we simply use progressively shorter steps when the ball is close. This approach can waste a lot of time with a big robot that has a much longer step duration and can be fatal when time is strictly limited. We use a greedy foot step planner to handle this problem, where each footstep is selected to be the closest position to the target foot position while satisfying the stride length constraint.

5.4   Testing in Simulated Environment

We use the Webots simulator to replicate the competition environment and tune various parameters. From a number of simulator runs, where the initial positions of the ball and obstacles varied a lot, we found that the stride length of 30 cm is enough to make the attacker robot score within the time limit of 90 s for all tested scenarios. Figure 8 shows the sequence of the THOR-OP robot performing two motions in simulated environment.

6   Results

The hybrid walk controller has been used successfully on the actual robot during the RoboCup competition, with very little parameter tuning required. We did not experience stability issues during the competition. Same as in simulation, the robot could kick the ball to the goal within the time limit almost every time. In the artificial grass challenge, the robot stably achieved a stride length of 40cm per step over cushy artificial grass terrain.

Figure 9 shows two kick sequences that our robot utilized during the RoboCup competition. Although our hybrid walk controller allows initiation of any dynamic kick motion during locomotion, we limited the strong kick to always start from double support. Initiation a big kick while walking led to inconsistent kick distances. With this limitation, the kick distances from each kick are quite consistent. The fast kick achieves around a four meter range, while the strong kick has more than a 10 m range.

Throughout the matches, the hierarchical planner reliably found optimal and safe kick sequences that lead to scores, even with frequent irregular rolling of

(a) Fast kick

(b) Strong kick

Fig. 9. THOR-OP performing a dynamic kick in the middle of reactive walking.

the kicked ball. During the last two games our robot succeeded in kicking to the goal, while avoiding all the obstacle. However, some of the kicks were blocked by the opposing goalie robot.

7   Conclusions

In this work, we described how we utilize a general purpose full sized humanoid robot for RoboCup robot soccer task. We utilize the hybrid walk controller to help the robot to initiate and end locomotion smoothly while keeping the reactive stepping capability required for dynamic environment. Additionally, it allows the robot to perform various dynamic kicks without stopping, which helped us greatly. We also present the vision and planning systems we developed for the RoboCup competition, which worked robustly throughout the competition.

Acknowledgments. We acknowledge the Defense Advanced Research Projects Agency (DARPA) through grant N65236-12-1-1002. We also acknowledge the support of the ONR SAFFIR program under contract N00014-11-1-0074.

References

1. McGill, S.G., Yi, S.-J., Zhang, Y., Lee, D.D.: Extensions of a robocup soccer software framework. In: Behnke, S., Veloso, M., Visser, A., Xiong, R. (eds.) RoboCup 2013. LNCS, vol. 8371, pp. 608–615. Springer, Heidelberg (2014). doi:10.1007/978-3-662-44468-9_56

2. Yi, S.-J., McGill, S., Vadakedathu, L., He, Q., Ha, I., Rouleau, M., Hong, D., Lee, D.D.: Modular low-cost humanoid platform for disaster response. In: 2014 IEEE/RSJ International Conference on Intelligent Robots and Systems (IROS) (2014)
3. Yi, S.-J., Zhang, B.-T., Hong, D., Lee, D.D.: Online learning of a full body push recovery controller for omnidirectional walking. In: IEEE-RAS International Conference on Humanoid Robots, pp. 1–6 (2011)
4. Kajita, S., Kanehiro, F., Kaneko, K., Fujiwara, K., Yokoi, K.H.K.: Biped walking pattern generation by using preview control of zero-moment point. In: Proceedings of the IEEE International Conference on Robotics and Automation, pp. 1620–1626 (2003)
5. Yi, S.-J., Hong, D., Lee, D.D.: A hybrid walk controller for resource-constrained humanoid robots. In: 2013 13th IEEE-RAS International Conference on Humanoid Robots (Humanoids), October 2013
6. Lee, D.D., et al.: Robocup 2011 humanoid league winners. In: Röfer, T., Mayer, N.M., Savage, J., Saranlı, U. (eds.) RoboCup 2011. LNCS, vol. 7416, pp. 37–50. Springer, Heidelberg (2012)

MRESim, a Multi-robot Exploration Simulator for the Rescue Simulation League

Victor Spirin[1], Julian de Hoog[2], Arnoud Visser[3(✉)], and Stephen Cameron[1]

[1] Oxford University, Oxford, UK
[2] University of Melbourne, Melbourne, VIC, Australia
[3] Universiteit van Amsterdam, Amsterdam, The Netherlands
A.Visser@uva.nl

Abstract. This paper describes MRESim, a multi-robot exploration simulator which aims to provide a middle ground between the RoboCup Agent and Virtual Robot competitions. A detailed description of this new infrastructure is provided, followed by examples and case studies of successful research outcomes arising from the use of MRESim. Our work on MRESim won the 2014 Infrastructure competition of the RoboCup Rescue Simulation League.

1 Introduction

The RoboCup Rescue competitions provide benchmarks for evaluating robot platforms' usability in disaster mitigation. Research groups should demonstrate their ability to deploy a team of robots that explore a devastated area and locate victims. RoboCup is moving towards longterm goals of sophisticated resource allocation in disaster situations, both at a large scale (the Agents competition) and at a small scale (the Virtual Robots competition). Each year the benchmarks are made more challenging to accommodate and encourage progress by all teams. The Infrastructure competition showcases innovations that have enabled extensions of these benchmarks.

The Infrastructure competition was started in 2004 with the aim of promoting the development of new tools to improve rescue simulation. The simulation of various disaster situations turns out to be complicated and difficult to validate. Therefore, the infrastructure competition was launched to promote the maintenance and development of simulation environments. For example, the fire simulator [24] was developed by the winner of the infrastructure competition in 2004. Another nice example of a component developed in the infrastructure competition is the flood simulator [27]. Recently, an extension towards flying robots has been proposed, both in the Virtual Robot [7] and Agent competitions [9].

The Amsterdam Oxford Joint Rescue Forces, and their predecessor the UvA Rescue Team, has participated several times in the Infrastructure competition. In 2010 the simulator of the Virtual Robot competition was extended with a realistic response of laser scanners to smoke; a circumstance which is quite common in disaster situations. The response of the laser scanners was validated in a number of experiments in a training center of the Dutch fire brigade [8]. In 2011

ⓒ Springer International Publishing Switzerland 2015
R.A.C. Bianchi et al. (Eds.): RoboCup 2014, LNAI 8992, pp. 106–117, 2015.
DOI: 10.1007/978-3-319-18615-3_9

a model of a humanoid robot was introduced in USARSim, which made it possible to model one of the robots in the RoboCup@Home League [31]. In 2012 a validated flying robot was introduced to USARSim and it was demonstrated that such a robot allows for fast exploration of disaster areas, while creating a visual map of the area [7]. This resulted in the UvA Rescue team winning the Infrastructure award in 2012.

What was still missing inside the Rescue Simulation League is an environment which is a common ground for both the Agent and Virtual Robot competitions. Inside the Agent competition the focus is on the coordination of rescue teams [20, 33]. The competition consists of a simulation environment which resembles a city after an earthquake. Teams of fire brigade, police and ambulance team agents try to extinguish fires and rescue victims in the collapsed buildings. In contrast, inside the Virtual Robot competition the focus is on sensor-data fusion on maps automatically generated by teams of robots which explore inside buildings. The multi-robot exploration simulator MRESim[1], described in this paper, is proposed as a middle ground between the Agent and Virtual Robot competitions.

In the following section the background behind this new infrastructure is given, followed by a section which gives an overview of the research performed with this multi-robot exploration simulator.

2   Simulator

2.1   Simulation Cycle

MRESim is a discrete-time simulator. Unlike the existing Virtual Robot competition simulator USARSim, which uses a three-dimensional representation, MRESim works from a two-dimensional grid representation. In each time step, the simulator moves the robots to their new positions, updates their sensor information, simulates communication between the robots and outputs the current state of the simulation. The ordering of the events that happen in each time step are outlined in Algorithm 1. This is equivalent with the discrete-time step procedure in the simulator environment of the Agent competition. All of the robots process their next steps simultaneously. Since this tends to be the most computationally intensive part of each time step, planning robot movements in parallel allows the simulator to take advantage of modern multi-core CPUs. Other events in each time step are processed sequentially.

2.2   User Interface and Environments

The user interface (Fig. 1) contains a panel on the right for each robot with toggle buttons to visualize the information which is the basis of each robot's decisions. Example of information that can be made visible is the following: location; path; free space; safe space; frontiers; communication range; skeleton and potential rendezvous points; exact rendezvous points; potential rendezvous points through obstacles; exact rendezvous points through obstacles. There are

---

[1] https://github.com/v-spirin/MRESim.

```
    Input: Set R of robots
    // Simulate movement, range data
 1  foreach r_i ∈ R do
    │   // described in Sect. 2.5
 2  │   while distance_moved(r_i) < max_speed(r_i) do
 3  │   │   nextStep = r_i.takeStep();
 4  │   │   if isValid(nextStep) then
 5  │   │   │   move(r_i, nextStep);
    │   │   │   // described in Sect. 2.4
 6  │   │   │   rangeData = findRangeData(r_i, nextStep);
 7  │   │   │   r_i.receiveRangeData(rangeData);
 8  │   │   else
 9  │   │   │   r_i.setError(true);
10  │   │   │   break;
11  │   │   end
12  │   end
13  end
    // Simulate communication, described in Sect. 2.6
14  foreach r_i ∈ R do
15  │   foreach r_j ∈ R, i ≠ j do
16  │   │   if isInRange(r_i, r_j) then
17  │   │   │   r_i.receiveMsg(r_j.createMessage());
18  │   │   │   r_j.receiveMsg(r_i.createMessage());
19  │   │   end
20  │   end
21  end
    // Complete cycle
22  logData();
23  updateGUI();
```

Algorithm 1. A single simulation cycle

further toggle buttons for the environment walls and the team hierarchy at the bottom right, and a run may at any time be paused, continued, or stopped using the green buttons at the bottom. This means that it is very easy to examine specific aspects of any supported exploration approach. In addition, simulation logs produced by MRESim can be replayed in the simulator; this means that each simulation run can be analyzed in detail after the run has completed.

Environments in MRESim can be uploaded from text-based or PNG files, as can be found on dataset repositories like Radish[2]. The environments are occupancy grid models, with each cell being either free or an obstacle.

2.3 Planning and Movement

In every time step, each robot has to decide where to move next. The next destination is decided by the robot by calling the *takeStep* method (line 3 of Algorithm 1).

---

[2] http://radish.sourceforge.net/.

Input: Robot r_i, simulator settings $simConfig$
// Replan if necessary, then retrieve the next path point
1 if $r_i.timeToReplan$ **or** $r_i.getPath().isEmpty()$ **then**
2 | $simConfig$.getExplorationStrategy().replan(r_i);
3 **end**
4 **return** $r_i.getPath().getNextPoint()$;

Algorithm 2. Description of the agent takeStep method

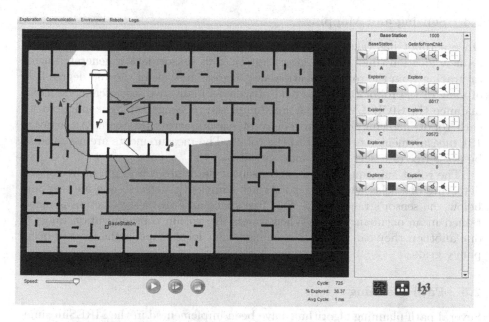

Fig. 1. Screenshot of MRESim. Obstacles are shown in black (━), unexplored area is shown in grey (■), area known at base station is shown in green (■). The blue arrows (▶) represent exploring robots, and the communication range of robot D is shown as the blue polygon (▭) (Color figure online).

If the robot has a path to its destination, and has replanned the path recently, then it simply continues on the current path by retrieving the next path point. Otherwise, it calls the corresponding *replan* method of the exploration strategy assigned to the robot team to generate a new path (line 2 of Algorithm 2). The exploration strategy then selects the best destination for the robot, and generates a path to the destination. The exploration strategy can use all the information known to the robot when making that decision — including the information about last known locations of the robot's teammates and agreements made in previous communication with other robots or the base station. Note that this information is not globally available, but communicated when in range. When no recent information is available, reasonable estimates are made.

Each robot r can move a maximum distance of d_r in each time step (line 5 of Algorithm 1). This is defined as the robot "speed", and can be set up to be different for each robot, allowing for heterogeneous teams of robots to be created. By default, this parameter is set for each robot to $DEFAULT\_SPEED = 3$, configured in the *Constants* class of the *config* package. In each time step, each robot moves a maximum distance of d_r along the path generated in the planning stage (see line 2 in Algorithm 2). If the robot reaches the planned destination before d_r is exhausted, it will not move any further in that timestep.

2.4 Sensing and Mapping

Once a robot has taken a step, the simulator provides it with sensor data at its new location (lines 6–7 of Algorithm 1). This sensor data is generated using raytracing from the robot's location at 1-degree intervals in the 180-degree field of view of the robot (the same field of view as many real laser scanners, as for instance the SICK LMS200). A maximum sensor range can be configured for each robot, allowing for heterogeneity of sensors across the team. An array of 181 measurements is returned to the robot. The measurements are assumed to be noise-free (see Sect. 2.7 for more detail).

The robot subsequently turns this sensor data into a polygon of free space, detecting obstacles where two points are sufficiently close to one another and below the sensor's range limit. This free space and obstacle detection is maintained in an occupancy grid. When robots are within communication range of one another, they can decide to exchange their local maps in the form of occupancy grids.

2.5 Path Planning

Several path planning algorithms have been implemented in the MRESim simulator. The most straightforward one is the A* algorithm [11], operating directly on the occupancy grid representation of the map. While A* generates optimal paths, it can be very slow with a sufficiently high resolution of the occupancy grid. In an attempt to overcome this problem, we have implemented several more efficient path planning algorithms. One of them is Jump Point Path search [10], which is a modification of the A* algorithm which significantly improves the performance in environments with large open areas. However, complex paths can still take several hundred milliseconds to compute.

We found that using A* search on a combination of a simple topological map that captures the connectivity between regions and the underlying occupancy grid map can significantly improve performance over the other implemented approaches. In each time step, if the map of the environment has been updated, we can generate a new topological map as follows. First, we perform thinning on the occupancy grid map in order to obtain a skeleton of the free space. Then, we uniformly select a set of nodes from the skeleton. Each point of the occupancy grid is then assigned to the nearest node. The set of nodes and edges connecting the nodes then represents a simple topological map of the environment. An illustration of the process is given in Fig. 2. We can then plan a path between any two points on the occupancy grid map by using the A* algorithm to find a path in the graph between the two corresponding nodes on the topological map, and using A* on the occupancy grid map to find paths from start and finish locations to their corresponding node locations. Generating paths then becomes very computationally efficient, which improves the speed of running the simulation and can increase the performance of exploration strategies by allowing them to calculate more accurate path lengths, instead of relying on crude approximations.

2.6 Communication

MRESim supports a variety of communication models. In principle messages from both robots are exchanged (as indicated in lines 17 and 18 of Algorithm 1), but only when the robots are *inRange* (line 16 of Algorithm 1). Simple models include a straight-line model (any robot within radius x is considered in range, regardless of obstacles) and a line-of-sight model (any two robots that can be connected by a line that doesn't hit an obstacle are in range). A more realistic communication model implemented in MRESim is a path loss model, originally proposed by Bahl and Padmanabhan [2]. This model is also used for simulating communication in the USARSim simulator used at RoboCup, and is considerably more realistic as it takes attenuation by walls into account. In this model, communication strength is calculated as follows:

$$S = P_{d_0} - 10 \times N \times log_{10} \frac{d_m}{d_0} \times \min(nW, C) \times WAF$$

where P_{d_0} is the signal strength at reference distance d_0, N is the rate of path loss, d_m is the distance, nW is the number of obstructing walls, WAF is the wall attenuation factor and C is the maximum number of walls where the attenuation factor needs to be considered.

Typical communication ranges using this model can be seen in Fig. 1.

2.7 Realism of Simulator Results

Clearly MRESim does not take into account a number of factors that any robot system in the real world would have to consider. The most significant ones are:

1. There is no sensor noise. In reality, wheel encoders provide inexact data, and laser range finders often have spurious measurements at either close or maximum range. Localisation remains a significant challenge in robotics, even if a number of techniques such as particle filters and scan matching show great promise.
2. Environments are two-dimensional and flat. In the real world this is almost never the case.
3. The simulator is discrete-time. Real robots would each run their own (multithread) processes, take different amounts of time to do the required processing, and would not always move the same distance in each time segment.
4. Communication in reality is highly variable and very difficult to predict. Nevertheless, for purposes of quick, simple, controlled, and repeatable comparison of exploration algorithms, MRESim remains a useful tool.

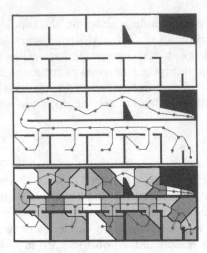

Fig. 2. Illustration of the process used by MRESim to generate topological maps from occupancy grid maps. On top, we see an occupancy grid of a partially explored environment, shown in white. In the middle, we apply "thinning" to obtain a skeleton, and sample uniformly points, shown in red (-•-), from the skeleton to act as nodes in the topological graph. In the bottom picture, we map each point of the occupancy grid map to the nearest node (Color figure online).

In the next section we give an overview of published work that used MRESim to evaluate exploration strategies. In some of the work, results are compared with those obtained from high-fidelity simulators such as USARSim, as well as with results obtained on real robots. Those experiments suggest that results obtained in MRESim have much in common with the results obtained in real life experiments.

3   MRESim as Benchmark

A number of exploration strategies have already been implemented in MRESim and are available "out-of-the-box". This makes it a potentially useful tool for benchmarking various exploration strategies against each other. According to [19], the performance of most algorithms is compared using the following three methods, from the best to the worst:

1. Using the same implementation that was used in other work.
2. Using a custom implementation, created from descriptions of the algorithm in published work.
3. Simply taking the results from those in other papers, without re-running the algorithm.

According to [1], most algorithm comparisons in robotics are currently conducted using the second approach. Publishing of the team's source-code, as done inside the Rescue Simulation League, made it possible to reimplement existing algorithms and to make a fair comparison (see for instance [1, 28]). Using MRESim, the first approach can be used more often:

1. Results of using different exploration strategies can be directly compared with each other.
2. Existing implementations of exploration strategies can be used.

3.1 Studies Based on MRESim

MRESim was first mentioned in [13]. In this study the effect of robots with a relay role was studied, when exploration was needed outside the direct communication range from the base station. This study inspired several other researchers [3, 5, 6] to base the coordination decisions on multiple criteria.

Important for the coordination of the explorer and relay role is the selection of an appropriate rendezvous point [15]. When those points are selected near gateways, the locations inside indoor environments become important junctions for a navigation algorithm. This work was followed up by several researchers [21, 23, 25] (although the concept of rendezvous points has already been known for a long time [26]). In addition, in [29] the possibility of communicating through obstacles, such as thin walls, was introduced into the planning.

Dividing the work into explorer and relay roles is already valuable, but sometimes the explorer finds a dead-end. This is an example of a situation where it is beneficial to switch roles dynamically using the *role swap rule*, as described in [14]. Other researchers have validated this results with real robots, such as [18].

In the study [35], the exploration indoors is extended to open areas which can be encountered in outdoor scenarios. The robots still use rendezvous points, but no longer near gateways. Instead, the robots divide the work into sectors and meet at the sector boundaries. This work has motivated several other studies [4, 17, 34].

The algorithms implemented in MRESim have been validated with Pioneer robots [16], although it is difficult to scale indoors to extensive environments with large robot teams. The MRESim environment was discussed in several dissertations [12, 22, 34] and a workshop contribution [32].

The latest extension are robot teams which can adjust their exploration strategy based on the information need of the base station [30]. When it is important that the base station gets timely updates more resources are allocated towards the relay role; when fast exploration is needed more resources are allocated towards the explorer role.

4 Discussion and Future Work

MRESim is a simulation environment with a well chosen level of detail, as discussed in 2.7. When the algorithms developed inside MRESim are applied in the real world issues will arise, which can be studied in separation inside the simulation. Precisely those issues could be identified as directions for future research.

In addition, the simulator itself can be extended in a number of ways to increase its realism and allow for using it to study a wider range of research problems:

1. **Variable Terrain.** In real applications, the terrain of the environment being explored can vary significantly, affecting the speed of robots navigating over such terrain. For example, robots are likely to be able to travel much faster just outside a partially-collapsed building, than inside where there is likely to be a lot of rubble. It may then be possible to improve the exploration speed by reducing the amount of travel over rough terrain, particularly for the relays.

2. **Additional Communication Channels.** It may be possible to use low-frequency, low-bandwidth, high range radio communication channels in disaster scenarios to transfer some control information between robots, as suggested in [29]. (We are about to test this hypothesis experimentally).

3. **Robustness Against Failing Robots.** Robots may fail for a number of reasons (and often do). In greedy approaches this does not affect the failed robot's teammates, which continue exploring as if nothing happened. In some other approaches, meetings between robots at pre-agreed times and locations can be planned for explicitly. In those cases, the failure of robots is a scenario that must be dealt with carefully. A useful extension to the simulator could therefore be the ability to specify scenarios that include robot failure, making it a useful tool to evaluate the robustness of exploration strategies to individual robot failure.

4. **Dynamic Environments.** In many robotics applications, the environment may change as the exploration effort progresses. A possible extension to the simulator would be to specify how the environment should change over time, or allow for random changes to the environment. This is precisely where the current Agent competition accelerates and it would be nice if for example part of the fire, earthquake or flood simulation could be incorporated in MRESim.

5. **Prior Knowledge.** In practice, some knowledge about the environment may be available to the team before the start of the mission, even though this information may not be accurate.

5   Conclusions

The original motivating questions for this research were: How can a team of robots be coordinated to explore a previously unknown and communication-limited environment as efficiently as possible; and how can new information obtained by this team be gathered at a single location as quickly and as reliably as possible?

These are precisely the research questions studied in both the Agent and the Virtual Robot competition of the RoboCup Rescue Simulation League. The studies performed with the simulation environment MRESim are well recognized, both inside the RoboCup community and outside. MRESim is applied to compare several coordination algorithms and could be easily extended with other coordination algorithms. In this paper some ideas for future research are given, but many other extensions are possible.

References

1. Amigoni, F., Schiaffonati, V.: Good experimental methodologies and simulation in autonomous mobile robotics. In: Magnani, L., Carnielli, W., Pizzi, C. (eds.) Model-Based Reasoning in Science and Technology. SCI, vol. 314, pp. 315–332. Springer, Heidelberg (2010)
2. Bahl, P., Padmanabhan, V.N.: Radar: An in-building rf-based user location and tracking system. In: Proceedings of the 19th Annual Joint Conference of the IEEE Computer and Communications Societies (INFOCOM). vol. 2, pp. 775–784 (2000)
3. Basilico, N., Amigoni, F.: Exploration strategies based on multi-criteria decision making for searching environments in rescue operations. Auton. Robots **31**(4), 401–417 (2011)
4. Bayram, H., Bozma, H.I.: Decentralized network topologies in multirobot systems. Adv. Robot. **28**(14), 967–982 (2014)
5. Capitan, J., Spaan, M.T., Merino, L., Ollero, A.: Decentralized multi-robot cooperation with auctioned pomdps. Int. J. Robot. Res. **32**(6), 650–671 (2013)
6. Cepeda, J.S., Chaimowicz, L., Soto, R., Gordillo, J.L., Alanís-Reyes, E.A., Carrillo-Arce, L.C.: A behavior-based strategy for single and multi-robot autonomous exploration. Sensors **12**(9), 12772–12797 (2012)
7. Dijkshoorn, N., Visser, A.: Urban Search and with micro aerial vehicles. In: Proc. CD of the 16th RoboCup International Symposium (June 2012)
8. Formsma, O., Dijkshoorn, N., van Noort, S., Visser, A.: Realistic simulation of laser range finder behavior in a smoky environment. In: Ruiz-del-Solar, J. (ed.) RoboCup 2010. LNCS, vol. 6556, pp. 336–349. Springer, Heidelberg (2010)
9. Gohardani, P.D., Ardestani, P., Mehrabi, S., Yousefi, M.A.: Flying Agent: An Improvement to Urban Disaster Mitigation in Robocup Rescue Simulation System. Mechatronics Research Laboratory, Qazvin (2013)
10. Harabor, D.D., Grastien, A.: Online graph pruning for pathfinding on grid maps. In: 25th AAAI Conference on Artificial Intelligence (2011)
11. Hart, P.E., Nilsson, N.J., Raphael, B.: A formal basis for the heuristic determination of minimum cost paths. IEEE Trans. Syst. Sci. Cybern. **4**(2), 100–107 (1968)
12. de Hoog, J.: Role-Based Multi-Robot Exploration. Ph.D. thesis, University of Oxford, May 2011
13. de Hoog, J., Cameron, S., Visser, A.: Role-based autonomous multi-robot exploration. In: Proceedings of the International Conference on Advanced Cognitive Technologies and Applications (Cognitive 2009), November 2009
14. de Hoog, J., Cameron, S., Visser, A.: Dynamic team hierarchies in communication-limited multi-robot exploration. In: Proceedings of the IEEE International Workshop on Safety, Security and Rescue Robotics (SSRR 2010), July 2010
15. de Hoog, J., Cameron, S., Visser, A.: Selection of rendezvous points for multi-robot exploration in dynamic environments. In: International Conference on Autonomous Agents and Multi-Agent Systems (AAMAS), May 2010
16. de Hoog, J., Jiménez-González, A., Cameron, S., Martínez de Dios, J.R., Ollero, A.: Using mobile relays in multi-robot exploration. In: Proceedings of the Australasian Conference on Robotics and Automation, December 2011
17. Hourani, H., Hauck, E., Jeschke, S.: Serendipity rendezvous as a mitigation of exploration's interruptibility for a team of robots. In: 2013 IEEE International Conference on Robotics and Automation (ICRA), pp. 2984–2991. IEEE (2013)
18. Jiménez-González, A., Martínez-de Dios, J.R., Ollero, A.: An integrated testbed for cooperative perception with heterogeneous mobile and static sensors. Sensors **11**(12), 11516–11543 (2011)

19. Johnson, D.S.: A theoretician's guide to the experimental analysis of algorithms. Data Structures, Near Neighbor Searches, and Methodology: Fifth and Sixth DIMACS Implementation Challenges, p. 215–250 (2002)
20. Kitano, H., Tadokoro, S., Noda, I., Matsubara, H., Takahashi, T., Shinjou, A., Shimada, S.: RoboCup Rescue: Search and rescue in large-scale disasters as a domain for autonomous agents research. In: IEEE Conference on Man, Systems, and Cybernetics (SMC-99) (1999)
21. Luo, C., Ward, P., Cameron, S., Parr, G., McClean, S.: Communication provision for a team of remotely searching uavs: A mobile relay approach. In: 2012 IEEE Globecom Workshops (GC Wkshps), pp. 1544–1549 (2012)
22. Martin, A.: A Framework for the Development of Scalable Heterogeneous Robot Teams with Dynamically Distributed Processing. Ph.D. thesis, University of Toronto (2013)
23. Meghjani, M., Dudek, G.: Combining multi-robot exploration and rendezvous. In: 2011 Canadian Conference on Computer and Robot Vision (CRV), pp. 80–85 (2011)
24. Nüssle, T.A., Kleiner, A., Brenner, M.: Approaching urban disaster reality: the resQ firesimulator. In: Nardi, D., Riedmiller, M., Sammut, C., Santos-Victor, J. (eds.) RoboCup 2004. LNCS (LNAI), vol. 3276, pp. 474–482. Springer, Heidelberg (2005)
25. Pham, V.C., Juang, J.C.: An improved active slam algorithm for multi-robot exploration. In: 2011 Proceedings of SICE Annual Conference (SICE), pp. 1660–1665. IEEE (2011)
26. Roy, N., Dudek, G.: Collaborative robot exploration and rendezvous: algorithms, performance bounds and observations. Auton. Robots 11(2), 117–136 (2001)
27. Shahbazi, H., Abdolmaleki, A., Salehi, S., Shahsavari, M., Movahedi, M.: Robocup rescue 2010 - rescue simulation league team description paper - brave circles - infrastructure competition. In: Proceedings of CD of the 14th RoboCup International Symposium, July 2010
28. Shayesteh, M., Salamati, M., Taleghani, S., Hashemi, A., Hashmi, S.: Mrl team description paper for virtual robots. Team Description Paper for the 2014 RoboCup competition, July 2014
29. Spirin, V., Cameron, S.: Rendezvous through obstacles in multi-agent exploration. In: Proceedings of the IEEE International Symposium on Safety, Security and Rescue Robotics (SSRR 2014), October 2014
30. Spirin, V., Cameron, S., de Hoog, J.: Time preference for information in multi-agent exploration with limited communication. In: Natraj, A., Cameron, S., Melhuish, C., Witkowski, M. (eds.) TAROS 2013. LNCS, vol. 8069, pp. 34–45. Springer, Heidelberg (2014)
31. van Noort, S., Visser, A.: Extending virtual robots towards robocup soccer simulation and @home. In: Chen, X., Stone, P., Sucar, L.E., van der Zant, T. (eds.) RoboCup 2012. LNCS, vol. 7500, pp. 332–343. Springer, Heidelberg (2013)
32. Visser, A., de Hoog, J., Jiménez-González, A., Martínez de Dios, J.R.: Discussion of multi-robot exploration in communication-limited environments. In: Workshop "Towards Fully Decentralized Multi-Robot Systems: Hardware, Software and Integration" at the ICRA Conference, May 2013
33. Visser, A., Ito, N., Kleiner, A.: Robocup rescue simulation innovation strategy. In: Bianchi, R.A.C., Akin, H.L., Ramamoorthy, S., Sugiura, K. (eds.) RoboCup 2015: Robot Soccer World Cup XIX. LNAI, vol. 8992, pp. 661–672. Springer, Heidelberg (2015)

34. Wellman, B.L.: Cooperation paradigms for overcoming communication limitations in multirobot wide area coverage. Ph.D. thesis, The University of Alabama (2011)
35. Wellman, B.L., de Hoog, J., Dawson, S., Anderson, M.: Using rendezvous to overcome communication limitations in multirobot exploration. In: Proceedings of the IEEE International Conference on Systems, Man, and Cybernetics, October 2011

Towards Highly Reliable Autonomy for Urban Search and Rescue Robots

Stefan Kohlbrecher$^{(\boxtimes)}$, Florian Kunz, Dorothea Koert, Christian Rose,
Paul Manns, Kevin Daun, Johannes Schubert, Alexander Stumpf,
and Oskar von Stryk

Department of Computer Science, Technische Universität Darmstadt,
Karolinenplatz 5, 64289 Darmstadt, Germany
rescue@sim.tu-darmstadt.de
http://www.gkmm.tu-darmstadt.de/rescue

Abstract. Participating in the RoboCup Rescue Real Robot League competition for approximately 5 years, the members of Team Hector Darmstadt have always focused on robot autonomy for Urban Search and Rescue (USAR). In 2014, the team won the RoboCup RRL competition. This marked the first time a team with a strong focus on autonomy won the championship. This paper describes both the underlying research and open source developoments that made this success possible as well as ongoing work focussed on increasing rescue robot performance.

The exploration of disaster environments poses a great challenge to both human rescuers and rescue robot systems (as well as canines). The RoboCup Rescue Robot League (RRL) competition benchmarks robot system performance in scenarios that are designed to represent many of the same challenges as real Urban Search and Rescue (USAR) situations in a systematic manner to enable reproducible benchmarking of robot capabilities.

In this work we present recent advances in software for USAR robots, provide a overview of performance at the RoboCup 2014 competition and present ongoing work towards further improved robot capabilities.

We use the open source Robot Operating System (ROS) as a middleware for our software. In the following sections, ROS package or stack names written in *italics* like *hector_slam* are available as open source software and can be found on the ROS wiki, e.g., www.ros.org/wiki/hector_slam.

The remainder of the paper is structured as follows. The next section presents related work on USAR robotics. It outlines the modularity concept of the hardware and stresses the importance of simulation capabilities for robots in the USAR context. The two subsequent sections show recent developments in navigation and perception software including victim search, navigation towards goal poses, path smoothing and victim localization. Afterwards, we assess the performance of the robots with respect to the benchmark that RoboCup Rescue League provides. Then, we give a short overview on ongoing work towards 3D mapping, planning and active gaze control. We close with a conclusion.

© Springer International Publishing Switzerland 2015
R.A.C. Bianchi et al. (Eds.): RoboCup 2014, LNAI 8992, pp. 118–129, 2015.
DOI: 10.1007/978-3-319-18615-3_10

1   Related Work

While capabilities for navigation or victim search using unmanned ground vehicles have been research topics for a long time, only few publications consider fully integrated autonomous UGV systems such as the one described in this work. For this reason, we focus on multiple aspects and modules that, in combination, enable autonomous operation in challenging environments. In previous work we provide a comprehensive description of components for world modeling (*hector_worldmodel*) [1] and SLAM (*hector_slam*) [2]. Our approach and open source software for navigation is described in [3]. Further detailed information about used hardware is available in the team description papers (TDPs) published yearly [4]. Multiple members of the team are also of member of the DARPA Robotics Challenge Team ViGIR. We are currently adapting their research in manipulation and 3D motion planning [5] for use with autonomous wheeled and tracked ground robots.

1.1   Hardware

We use two different types of mobility platforms for research in autonomous systems for USAR tasks. The Hector UGV (Fig. 1a) is a lightweight agile ground vehicle. The platform is based on a modified Kyosho Twin Force R/C model. The 4-wheel-drive with a differential gear for each axis allows the vehicle to move on slippery surfaces. The front and rear wheels can be steered independently, providing higher mobility than common 2-wheel-steering.

A major drawback of the Hector UGV platform is its limited mobility in rough terrain. To address this shortcoming, we introduced the tracked "Obelix" UGV platform, shown in Fig. 1b. This tracked robot with front and back flippers provides state of the art mobility, similar to other (response) robots used in challenging terrain.

A key feature of both platforms is their modularity. The main components are the mobility platform and the "autonomy box" featuring various sensors for navigation and victim detection as well as a state of the art (Intel Core i7) computer system. The "autonomy box" can be mounted on any mobile platform with minor modifications to high-level software. Only low-level interfaces (e.g. motion controllers) of the mobility platform have to be adapted to communicate with the ROS-based high-level software.

1.2   Simulation

To achieve high reliability of complex autonomous systems, the capability to perform comprehensive testing of the complete system is highly important. Setting up a real robot for such testing is both time and effort consuming, especially when (for instance large) scenarios are to be considered that cannot be provided physically. We simulate the complete system, including actuation, sensors and scenarios using *Gazebo*[1]. This enables testing of autonomous behaviors with

---

[1] http://gazebosim.org/.

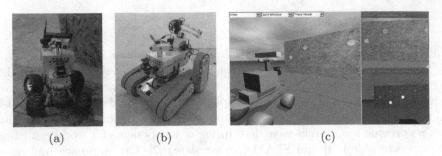

(a)   (b)   (c)

Fig. 1. Team Hector mobile platforms: (a) Hector UGV (b) Tracked vehicle Obelix (c) Hector UGV in Gazebo with simulation of thermal sensor data visualized.

negligible setup effort in multiple scenarios without the need for real hardware. As an example the simulation of a thermal camera can be seen in Fig. 1c. Simulated RoboCup Rescue arena scenarios serve to evaluate the robot in the USAR context. The *hector_nist_arena_designer* ROS package permits the fast and intuitive creation of user defined test scenarios.

2   Navigation

This chapter describes on software modules developed for autonomous navigation. We focus on recent research and refer the reader to prior publications for information about our approaches for SLAM [2] and navigation [3].

2.1   Exploration Planner

Based on the exploration transform approach [6] commonly used for frontier-based, risk-aware exploration, we developed multiple modifications in the *hector_exploration_planner* for improved performance in search and rescue missions.

Improved Victim Search. To detect objects of interest in disaster scenarios reliably, a thorough search has to be performed by (autonomous) robot systems tasked with the exploration of the environment. Exploration of unknown environments has been intensively studied in the past, but only few approaches consider different sensing modalities. Often, the environment that has been covered by a LIDAR sensor is considered explored. This approach is sufficient for tasks like mapping an unknown environment, but it does not take the limited range of sensors into account when searching for objects of interest like victims in a disaster environment. The reliable detection of victims trapped under rubble using thermal imaging requires the USAR robot system to come as close as possible to enable detection even if only very small portions of the victim are visible to sensors. To provide this capability, we modified the original exploration transform approach. In our modified version, we do not plan paths to map frontiers (grid cells that have a transition from known free space to unknown space),

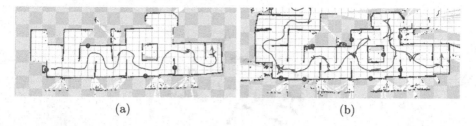

(a)   (b)

Fig. 2. Victim search at RoboCup 2014: (a) Use of standard exploration transform. (b) Exploration by approaching to target points close to the path.

but limit the generation of target poses to those that have a specified distance from the path that the robot has traveled so far. Using this approach, the system also explores tight spaces that would otherwise be ignored by LIDAR map based exploration. Figure 2 shows examples of standard exploration and our new approach.

Robust Navigation Towards Goal Poses. Both in autonomous and semi-autonomous operation, target poses for the robot system might be invalid due to obstacles preventing the robot from reaching it. Without handling of this, the planner will simply fail planning to such poses. In many practical cases, however, a pose close-by may be reachable, allowing continuation of the robot mission without interruption.

A common example we observed in the USAR scenario is approaching victims for inspection. To reach victims that are situated inside or close-by to obstacles, just forwarding their pose P_V to the planner as goal poses would naturally fail. Instead, a *observation pose* P_o has to be found that is sufficiently close to the victim to provide sensors with a view that allows to inspect and possibly confirm the victim hypothesis.

To achieve this task, the planner is augmented with an approach for generating valid observation poses. Based on an obstacle transform map, an area around the target pose is searched for collision free poses as shown in Fig. 3. For all candidates, the vector from the original target to the candidate is checked against the original target orientation. The candidate is only used if the orientation difference lies within a predefined threshold. This is motivated by the fact that the observation of the target pose is only possible if it lies in front of the robot.

2.2 Path Smoothing

Based on a discrete map representation, the exploration planner generates continuous piecewise linear paths. Two successive segments can be right- or even acute-angled with respect to each other. Such non-smooth paths cannot be reliably followed by motion controllers, so a smoothing approach is employed.

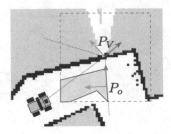

Fig. 3. Generating observation poses: P_V is the estimated victim pose. Constraints based on victim orientation (dashed red lines), distance from original pose (dashed blue lines) and distance from walls result in the mint green polygon being searched and P_o getting selected (Colour figure online).

The piecewise linear path is passed as the set of corner points. We introduce a variable t, which represents the integrated length along the path, and obtain a functional representation γ of the path as

$$\gamma(t) = \sum_{i=0}^{N-1} \chi_{[t_i, t_{i+1}]}(t) \left(x_i + \frac{t - t_i}{t_{i+1} - t_i}(x_{i+1} - x_i) \right) \tag{1}$$

where $x_i \in \mathbb{R}^3$ denotes the i-th one of the $N + 1$ corner points, t_i the integrated Euclidean length until x_i and χ_A the characteristic function for $A \subseteq \mathbb{R}$. We consider $\gamma(t)$ and x_i in $\mathbb{R}^3$, but the approach works in $\mathbb{R}^2$ as well.

This approach allows us to treat the path smoothing problem as a function smoothing problem. It is tackled by sampling γ with an intermediate discretization $D_M = \{\tilde{t}_0, \ldots, \tilde{t}_M\}$ and computing a Gaussian-weighted convex combination from the points $\gamma(\tilde{t}_i)$. The resulting smooth function $\tilde{\gamma}$ is given by

$$\tilde{\gamma}(t) = \sum_{m=0}^{M} \gamma(\tilde{t}_m) \frac{\exp\left(-\frac{(t-\tilde{t}_m)^2}{2\eta^2}\right)}{\sum_{m=0}^{M} \exp\left(-\frac{(t-\tilde{t}_m)^2}{2\eta^2}\right)} \tag{2}$$

The steepness of the Gaussian weight is controlled by the parameter η. It trades off accuracy versus smoothness of $\tilde{\gamma}$. After computing $\tilde{\gamma}$, it is sampled again, yielding another piecewise linear path $\hat{\gamma}$ which is fed into the base controller. Figure 4 shows an input path and its smoothed versions for different choices of η. One observes that a small value yields a path being very close to the original path while a high value implies big deviations and a centralization effect on the end points of $\hat{\gamma}$.

Compared to smooth function approximation approaches like Gaussian processes as described in [7] or curve smoothing approaches like B-splines as described in [8], our approach works more heuristically. However, it overcomes the need for solving optimization problems or linear equation systems for new input paths with complexities of $\mathcal{O}(n^2)$ or worse.

Complexity lies in $\mathcal{O}(n)$ which is beneficial given limited onboard computational resources and frequent path updates. Furthermore, the approach only

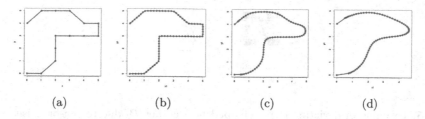

(a)   (b)   (c)   (d)

Fig. 4. Input path (a) and output path with different values for the temperature parameter η. (b) $\eta = 0.1$, (c) $\eta = 0.5$, (d) $\eta = 1.0$

requires to choose a single parameter, which easy to handle compared to other techniques that have more than three degrees of freedom and suffer from numerical instabilities if the prediction is computed without $\mathcal{O}(n3)$ techniques [9,10]. B-splines suffer from a very complex definition and a difficulty to control the trade-off between smoothness and overfitting [11]. To choose η, we propose to reduce big directional differences along the path while staying in an ε-tube around the original path.

3 Perception

The semantic world model backend of our perception system is described in [1] and details are omitted here for brevity.

Victim Localization. Reliable localization of human victims in unstructured post-disaster environments is a key issue for USAR robots. We tackle this problem by using complementary information provided by imagery cues such as RGB, thermal and depth images. Possible victims are identified by analyzing thermal images with a blob detection algorithm with configurable thresholds. While providing important cues for victim locations, thermal images can yield a high number of false-positives. They can be caused by other heat emitting spots, such as radiators or fire. We minimize the number of false positives by using the complementary information provided by visual perception and depth information provided by an RGB-D camera. Using this 3D information, false-positive victims can be identified by the shape of the environment as well as by missing depth information at the victim's location.

To identify the correct position of a victim after its detection by the robot's sensors and insert it into the map, we cast a 3D ray along the robot's gaze direction into the onboard octomap [12]. This map is created from the depth data of the RGB-D camera. Limited octomap resolution and state estimation noise might result in erroneous ray casts. Without consideration of this issue, a estimated victim location might be mapped onto a corner instead of the correct spot behind it. This case is illustrated in Fig. 5 where the victim would be mapped on the corner at p_0 instead of p_V behind it. To overcome this issue, we perform a spread fan ray cast of n rays which hit the octomap on the right and

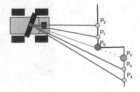

Fig. 5. Example of a victim at P_V mapped to a corner P_0 due to sensor noise. The additional rays (blue) allow detection of the false mapping (Colour figure online).

the left of the original gaze direction ray. The fan is scaled such that the points are equally spaced on a horizontal line around p_0. The ray casts return the wall points p_i. For all of them, we compute the Euclidean distances d_i to the camera origin c. If they differ significantly, we decide on a corner case and dismiss this object localization attempt. This provides the chance of a correct mapping when new sensor data arrive and the robot does not waste time during exploration of false-positives.

4   Results

The RoboCup Rescue competition represent a systematic and reproducible benchmark for testing and evaluation of teleoperated and autonomous USAR robots in a complex environment in direct competition with the approaches of other teams. We present results from both the RoboCup German Open 2014 and RoboCup 2014. In both competitions we won both the "Best in Class Autonomy" award as well as the overall competition.

Figure 7 shows the performance increase during the RoboCup 2014 competition with respect to the performance metrics *number of found victims* and *number of found QR-codes*. The number of found victims evaluates performance for the USAR task. The total number of victims was 12 in all missions. The number of found QR-codes correlates with the fraction of the arena that was explored by the robot during the mission. The performance increase is also visible when comparing maps generated in the preliminary round (Fig. 2) with those of the final missions at RoboCup 2014 (Fig. 6).

Intense training of the operator is required for the case that human control becomes desirable in difficult situations. As our robot allows flexible switching between full and semi-autonomous behavior during the mission, we leveraged this capabilities in multiple competition runs. In both RoboCup 2014 final missions, the robot successfully found 4 victims autonomously, after which the operator took over control to explore the rest of the arena. It should be noted that a operator mistake lead to the robot rolling over in the second final mission, ending the mission prematurely.

We provide combined external and operator control station video from multiple missions at the RoboCup German Open 2014 competition online[2].

---

[2] http://www.youtube.com/playlist?list=PLqdOEBv9QGrEiqlUklq0BI1QPU55
IhTO2.

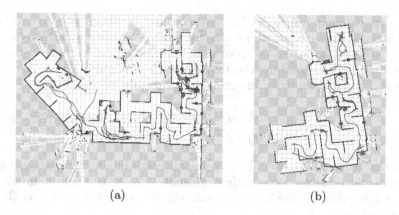

(a)   (b)

Fig. 6. RoboCup 2014 example maps: (a) Final mission 1 (b) Final mission 2. The starting pose of the robot is marked by yellow arrows. In both missions, the robot discovered 4 victims autonomously (right part of the arena) (Colour figure online).

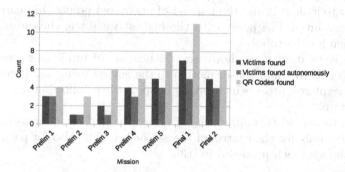

Fig. 7. Improvement of Team Hector's performance during the RoboCup 2014 competition.

5   Ongoing Work

Despite the good performance demonstrated in the previous section, there remain open issues that have to be tackled to increase robustness of the USAR robot system and allow deployment in real disasters in the future. In this section, directions and preliminary results of ongoing work are described.

5.1   Active Gaze Control

For victim detection, it is crucial to have reliable sensor coverage as to not miss any victims. It is crucial that the robot covers the scenario thoroughly in a minimal amount of time. Intelligent camera control thus can be a major factor for ensuring sufficient performance. Solutions for next best view point and coverage path planning on mobile robots must avoid high computational effort like computing all possible view points. We use a coverage mapping approach based on the octomap framework [12]. We extend it to include the viewing angle α_n that measures the deviation of a view ray from a surface normal. We consider a frontal view of a surface as the most useful in a victim detection scenario. The score of a cell being sensed by a certain sensor $S(c_i)$ is initialized with zero for all cells and updated for every incoming sensor reading according to

$$S(c_i)_{new} = \begin{cases} S(c_i)_{old} + \gamma(1 - (\frac{\alpha_n}{\pi})^2) & \text{if } \alpha_n < \frac{\pi}{4} \\ S(c_i)_{old} + \gamma & \text{otherwise} \end{cases} \tag{3}$$

such that views near a frontal viewing angle result in a better coverages score than views under a steeper angle. Here, γ denotes the update factor and all cells having a $S(c_i)$ value higher than a certain threshold are considered as already been covered.

The information provided by this coverage map can be used to compute the next gaze direction for the robot. Given the current pan and tilt of the camera, a randomized set of n possible next gaze points is sampled. Each point is weighted by the effort to look at it and the number of uncovered points that can be found in the surroundings. The point with the highest weight is chosen as the next gaze direction for the robot.

Tests in simulation have shown several benefits of intelligent gaze control. The robot is able to adapt sensor motion to the environment and therefore does not miss unexplored areas which might happen when following a fixed sensor motion pattern.

We plan to extend this approach for future use and consider using the coverage map not only for the control of the sensor head but also for path planning similar to the approach proposed in [13].

5.2   3D Mapping

We have been using the previously mentioned highly robust *hector_slam* approach in the past. With a 2D internal map representation, the capability for full 3D state estimation from LIDAR data is limited. Thus, although this system provides great reliability and robustness in various scenarios, it provides limited capability for use in multi-story scenarios.

For this reason, we are currently evaluating 3D SLAM approaches allowing deployment of tracked vehicles in challenging terrain, including significant roll and pitch angles and angular rates of the platform.

The state of the art *loam_back_and_forth* [14] and *ethzasl_icp_mappper* [15] ROS packages were selected for initial tests in simulation. These tests confirmed that major challenges are the tight spaces of USAR scenarios, high (roll/pitch) velocities of the platform and sensor occlusion when close to obstacles. We will investigate tackling these challenges in future work.

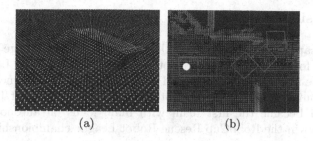

(a)   (b)

Fig. 8. Planning over a ramp (a) with low (green dots) and high (red dots) risk costs. The path (cyan) and the expanded states are projected onto the ground. (b) Checking a 2D path for stability, green poses stable, red pose instable (Colour figure online).

5.3   3D Path Planning

While reduction of 3D data to a 2D representation for planning can be computationally advantageous, it only allows avoiding harsh terrain, but not planning it's traversal.

The problem of planning a path from the robot position to a goal point can be reduced to finding a path in a graph. We base our planner on the SBPL library [16]. The set of nodes is a lattice of pose states $s(x, y, \theta)$ and the edges are the transitions connecting them. The available transitions are a discretized subset of possible robot movements. The Anytime Repairing A* (ARA*) algorithm [17] finds an initial solution and incrementally improves it. Therefore, it is well suited for real-time applications like ours.

We propose the following cost function that consists of a basic cost $c_b(s, s')$ for the movement and a factor $f_r(s') \geq 1$ for the risk associated with reaching the target position:

$$c(s, s') = f_r(s')c_b(s, s'). \tag{4}$$

f_r is computed with the Force Angle Stability Metric (FA) [18], which is based on the supporting polygon, i.e. the convex hull of the ground contact points.

Each of it's edges is considered a possible tip-over axis. FA is computed for them based on the gravity-induced moments. The value of the least stable axis serves as the stability estimate of the whole system.

To apply FA, the contact points need to be computed. To avoid the high computational cost associated with physics simulation, we developed an algorithm providing a sufficiently accurate estimation by imitating the physical behavior with geometric operations. It imitates dropping the robot onto the requested position and is based on a 3D octomap model generated from onboard sensors.

Evaluation shows that this planning approach enables the computation of safely traversable paths in 3D terrain (Fig. 8a). We also use FA for checking planned paths for their feasibility (Fig. 8b).

6   Conclusion

Autonomous capabilities for robots in USAR scenarios offer a huge potential of improvements for responders, but are not yet available in practice. In this paper, we describe the research approaches and technical achievements (including open source developments) underlying the autonomous USAR robots of Team Hector which in 2014 became the first team with main focus on autonomous robot capabilities to win the RoboCup Rescue Robot League championship.

Acknowledgements. This work has been supported in parts by the German Research Foundation (DFG) within the Research Training Group 1362 "Cooperative, Adaptive and Responsive Monitoring in Mixed-Mode Environments" of Technische Universit ät Darmstadt.

References

1. Meyer, J., et al.: A semantic world model for urban search and rescue based on heterogeneous sensors. In: Ruiz-del-Solar, J. (ed.) RoboCup 2010. LNCS, vol. 6556, pp. 180–193. Springer, Heidelberg (2010)
2. Kohlbrecher, S., Meyer, J., von Stryk, O., Klingauf, U.: A flexible and scalable SLAM system with full 3D motion estimation. In: Proceedings of IEEE International Symposium on Safety, Security and Rescue Robotics (SSRR), Kyoto, Japan, 1–5 November 2011, pp. 155–160. IEEE (2011)
3. Kohlbrecher, S., Meyer, J., Graber, T., Petersen, K., Klingauf, U., von Stryk, O.: Hector open source modules for autonomous mapping and navigation with rescue robots. In: Behnke, S., Veloso, M., Visser, A., Xiong, R. (eds.) RoboCup 2013. LNCS, vol. 8371, pp. 624–631. Springer, Heidelberg (2014)
4. Kohlbrecher, S., Meyer, J., Graber, T., Petersen, K., Klingauf, U., von Stryk, O.: RoboCupRescue 2014 - Robot League Team Hector Darmstadt (Germany). Technical report. Technische Universität Darmstadt (2014)
5. Kohlbrecher, S., Romay, A., Stumpf, A., Gupta, A., von Stryk, O., Bacim, F., Bowman, D.A., Goins, A., Balasubramanian, R., Conner, D.C.: Human-robot teaming for rescue missions: team ViGIR's approach to the 2013 DARPA robotics challenge trials. J. Field Robot. **32**, 352–377 (2014)
6. Wirth, S., Pellenz, J.: Exploration transform: a stable exploring algorithm for robots in rescue environments. In: IEEE International Workshop on Safety, Security and Rescue Robotics (SSRR), pp. 1–5 (2007)
7. Rasmussen, C.E.: Gaussian Processes for Machine Learning. MIT Press, Cambridge (2006)
8. Gallier, J.: Curves and Surfaces in Geometric Modeling: Theory and Algorithms. Morgan Kaufmann, San Francisco (1999)
9. Gärtner, T., Driessens, K., Ramon, J.: Graph kernels and gaussian processes for relational reinforcement learning. In: Horváth, T., Yamamoto, A. (eds.) ILP 2003. LNCS (LNAI), vol. 2835, pp. 146–163. Springer, Heidelberg (2003)
10. Foster, L., Waagen, A., Aijaz, N., Hurley, M., Luis, A., Rinsky, J., Satyavolu, C., Way, M.J., Gazis, P., Srivastava, A.: Stable and efficient gaussian process calculations. J. Mach. Learn. Res. **10**, 857–882 (2009)

11. Eilers, P.H.C., Marx, B.D.: Flexible smoothing with B-splines and penalties. Stat. Sci. **11**(2), 89–102 (1996)
12. Hornung, A., Wurm, K.M., Bennewitz, M., Stachniss, C., Burgard, W.: OctoMap: an efficient probabilistic 3D mapping framework based on octrees. Auton. Robot. **34**(3), 189–206 (2013). http://octomap.github.com
13. Dornhege, C., Kleiner, A., Kolling, A.: Coverage search in 3D. In: Proceedings of the Symposium on Safety, Security and Rescue Robotics (SSRR), pp. 1–8, October 2013
14. Zhang, J., Singh, S.: LOAM: lidar odometry and mapping in real-time. In: Robotics: Science and Systems Conference, July 2014
15. Pomerleau, F., Colas, F., Siegwart, R., Magnenat, S.: Comparing ICP variants on real-world data sets. Auton. Robots **34**(3), 133–148 (2013)
16. Likhachev, M.: Search-based planning with motion primitives (2009)
17. Likhachev, M., Gordon, G.J., Thrun, S.: ARA*: anytime A* with provablebounds on sub-optimality. In: Advances in Neural Information Processing Systems (2003)
18. Papadopoulos, E., Rey, D.A.: The force-angle measure of tipover stability margin for mobile manipulators. Veh. Syst. Dyn. **33**(1), 29–48 (2000)

The Intelligent Techniques in Robot
KeJia – The Champion of RoboCup@Home 2014

Kai Chen[✉], Dongcai Lu, Yingfeng Chen, Keke Tang, Ningyang Wang,
and Xiaoping Chen[✉]

Multi-Agent Systems Lab, Department of Computer Science and Technology,
University of Science and Technology of China, Hefei 230027, China
{chk0105,ludc,chyf,kktang,wny257}@mail.ustc.edu.cn,
xpchen@ustc.edu.cn
http://wrighteagle.org/en/robocup/atHome

Abstract. In this paper, we present the details of our team WrightEagle@Home's approaches. Our *KeJia* robot won the RoboCup@Home competition 2014 and accomplished two tests which have never been fully solved before. Our work covers research issues ranging from hardware, perception and high-level cognitive functions. All these techniques and the whole robot system have been exhaustively tested in the competition and have shown good robustness.

1 Introduction

The RoboCup@Home league was established in 2006 and aims at developing intelligent domestic robots which can perform tasks safely and freely in human daily lives. Our team WrightEagle@Home first participated in @Home league in 2009 and we got two 2nd places in 2011 and 2013. With continuous development in these years, we won the competition in 2014 with a clear advantage.

In this paper, we will first describe the setup of RoboCup@Home 2014. Then we will cover all the aspects of our *KeJia* robot and present our latest research progress. Section 3 gives an overview of *KeJia*'s hardware. The low-level functions for the robot are described in Sect. 4. Section 5 presents the techniques of complicated task planning while Sect. 6 introduces the grounding technique used in the open demonstrations. In Sect. 7 we report the performance of our robots at 2014 competition. Finally we conclude in Sect. 8.

2 RoboCup@Home 2014 Design and Setup

In RoboCup@Home competition, each team has to pass through two stages and a final demonstration to achieve the top places. In the 2014 competition, stage I has five tests while stage II has four. Each stage contains one open demonstration, fully open in stage I while scoped in stage II.

Each regular test in stage I focuses on some certain abilities of the robot. For example, the *Robo-Zoo* tests the robot's visual appearance and interaction with

R.A.C. Bianchi et al. (Eds.): RoboCup 2014, LNAI 8992, pp. 130–141, 2015.
DOI: 10.1007/978-3-319-18615-3_11

ordinary audience, *Emergency Situation* require the robot to detect abnormal status of human and take measures to prevent potential hazard. *Follow Me* test requires more abilities, including human detection, tracking and recognition, as well as dynamic obstacle avoidance in unknown environment.

While stage I tests focus on some certain but limited abilities, stage II tests expect the robot to handle difference situations and accomplish a complex task. Take the test *Restaurant* for example, during the limited time of this test, the robot is guided by one of the teammates through a completely unknown environment, which is or is like a restaurant settings. The guiding person takes the robot to three ordering places and two object locations (from where to retrieve a certain category of object, e.g. food or drinks), then return to the starting point. After the guiding phase, the robot is ordered to deliver three objects to two tables, simulating the robot being a waiter. So the robot is expected to drive to the object locations, look for and grasp the ordering objects, and take them to the correct tables. In such test, robots are requested to demonstrate more than one abilities, including human tracking, on-line SLAM, object recognition as well as grasping, making it a very challenging test.

After the two stages, the best five teams advance into the final, where each team is requested to perform an open demonstration and will be evaluated by both executive committee and an outer-league jury. More details of the competition are described in [9].

3 Hardware

The *KeJia* service robot is designed to manipulate common daily objects within an indoor environment. In order to move across narrow passages and avoid collisions with furniture, *KeJia* is equipped with a small sized chassis of 50 × 50 cm. The chassis is driven by two concentric wheels, each fixed with a motor of opposite direction. Another omnidirectional wheel is mounted to keep balance and makes the robot able to turn. A lifting system is mounted on the chassis, attached with the robot's upper body. The vertical range of the lifter is about 74 cm, allowing the robot to manipulate objects in different heights, ranging from the floor to adult chest height. When fully lifted, the *KeJia* robot is about 1.6 m high. Assembled with the upper body is a five degrees of freedom (DOF) arm. It is able

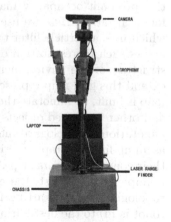

Fig. 1. Our robot KeJia

to reach objects over 83 cm far from mounting point and the maximum payload is about 500 g (fully stretched). This arm enables the robot to grasp common daily items such as bottles and mugs. The robot's power is supplied by a 20Ah battery, which guarantees the robot a continuous running of at least one hour.

As for perception demand, our robot is equipped with a Kinect and a high-resolution CCD camera. These two cameras are mounted on a Pan-Tilt Unit, which provides two degrees of freedom to ensure the cameras to acquire different views. Two 2D laser range finders are installed for self-localization and navigation. These two sensors are mounted in different height, one close to floor, another one about 15 cm above ground. This configuration makes our robot able to detect and avoid small obstacles, e.g. bottles on the floor. A working station laptop is used to meet the computation needs. An image of our *KeJia* robot is shown in Fig. 1.

4   Perception

4.1   Self-localization and Navigation

Precise indoor localization and navigation is an important prerequisite for domestic robots. We use the 2D laser scanners and the Kinect as main sensors for localization and navigation. First, a 2D occupancy grid map [5,7] is generated from the laser scanning results and odometry data. The scanning results are collected frame by frame and labeled with corresponding odometry information. In order to retrieve a full map of the working environment, our robot is driven by human control to visit all the rooms beforehand to get the full map. Each pixel in the grid map has a probability of occupancy, making it adaptive to potential changes in the environment. This feature is important as it allows the robot to deal with movable obstacles, such as chairs and walking people. With this pre-built occupancy map, a probabilistic matching technique is employed for localization. Here we use the adaptive Monte Carlo localization method [4], which uses a particle filter to track the pose of a robot against a known map.

As such representation does not provide any information of the topological structure of the environment, e.g. the rooms and their connective doors, we extend this grid map representation to contain such information. After the grid map is built, we annotate the different structures such as rooms, doors, furniture and other interested objects according to their semantic information. After the annotation, a topological map is automatically generated. This map is modeled as an undirected graph, vertices in the graph are rooms while the edges represent their connective doors or passages. With such topology, a layered path planning is implemented. First, the global path planner finds a connective path in the topological graph, starting from the start room (usually the current room the robot is in) to the destination room. This path consists of a series of key points. Between the key points within a room, a local path planner is used to search for the shortest path in the grid map. This is implemented via a heuristic A star algorithm. At last, during the navigation phase, VFH+ [12] is adopted for local obstacle avoidance, including both static and dynamic ones. A labeled map of our laboratory is shown in Fig. 2.

Beside the 2D localization and navigation, we also use a 3D depth camera, e.g. the Kinect, to avoid those obstacles not visible from the laser scanners. Here we use octo-tree [5] as the representation of 3D map. The acquired depth

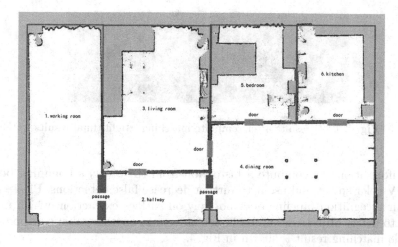

Fig. 2. A labeled map of our laboratory.

images are aligned with the 2D localization results and converted into octo-tree structure, and is used for avoiding local obstacles during navigation.

4.2 Object Recognition

In order to achieve fast yet robust object recognition, our recognition system is designed as a pipeline approach. Two cameras are used, a high resolution CCD camera and a Microsoft Kinect, to obtain aligned RGB-D images as well as high quality RGB images. Both cameras are calibrated so we can directly get the correspondence between the images. Figure 3 shows the pipeline of our approach.

As shown in Fig. 3, first we use the Line-MOD [8] template matching method to find possible candidates in the RGB-D image. Using the non-maximal suppression (NMS), we select a smaller set of candidates in different regions of the image. As the

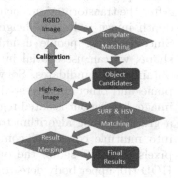

Fig. 3. The object recognition pipeline

Line-MOD method is characterized by a high recall low precision rates, this procedure can easily reject those regions where no possible targets are present. After this, all the regions are projected into the high quality RGB image, from where SURF [2] features are extracted and computed. These features are then matched against the database using a pre-built K-d tree. The matched features are grouped and filtered using RANSAC to check their geometry correspondence with the features extracted from known objects. If the number of inliers after RANSAC passes a certain threshold, then the object is treated as possible present. However, to overcome the weakness that SURF does not contain

Fig. 4. Left: results after template matching. Right: final results

color information, we compute a histogram from the object's bounding box in the HSV color space, and use it to further decrease false detections. Please note that our recognition pipeline does not rely on surface extraction, which makes it able to detect objects in human hands or even in another robot's gripper. An example matching result is shown in Fig. 4.

4.3 People Detection and Tracking

We developed a fast walking people detection method to efficiently detect standing or walking people. First the depth image is transformed into the robot's reference frame, where the origin is the center point of the two wheels projected to floor. After the transformation, each pixel's height in the depth image is its real height against the floor. Since standing people usually compose a fixed shape, we can use spatial information to quickly obtain such candidates. So we remove the floor plane and other uninterested areas in the depth image, leaving out isolated regions. Then we adopt a graph labeling algorithm to segment the image

Fig. 5. A segmentation result

into multiple connected components based on the relative distance between pixels. Figure 5 shows the output of segmentation. After that, a pre-trained HOD [10] upper body detector is used to classify each component human or not. As the segmentation procedure filters out many irrelevant regions, the classifier does not need to be run on the whole image, gaining a significant acceleration. As for sitting people, we use HAAR [13] face detector to detect and localize human faces. If present, the *VeriLook SDK* is used to identify each face.

4.4 Speech Recognition

For speech synthesis and recognition, we use a software from iFlyTek[1]. It is able to synthesis different languages including Chinese, English, Spanish etc. As for recognition, a configuration represented by BNF grammar is required. Since

---

[1] http://www.iflytek.com/en/index.html.

each test has its own set of possible speech commands, we pre-build several configurations to include all the possible commands for each test.

5   Integrated Decision-Making

One of the most challenging tests in the RoboCup@Home competition is the *General Purpose Service Robot*, where a robot is asked to fulfill multiple requests from an open-ended set of user tasks. In the *KeJia* robot, the integrated decision-making module is implemented using Answer Set Programming (ASP), a logic programming language with Prolog-like syntax under stable model semantics originally proposed by Gelfond and Lifschitz (1988). We pre-define and implement a set of atomic actions, where each action is implemented by a parameterized pre-defined program. All these actions are designed as primitives for the task planning module. With the specification of atomic actions, the division between the task and motion planning modules is clearly defined. Some of the atomic actions are listed in Table 1.

Table 1. Samples of atomic actions

Action	Function
Goto a location	Drive to the target location
Pick up an item	Pick up the assigned item
Put down at a position	Put down the item in hand at the assigned position
Search for an object	Search for the assigned object

An important feature of *KeJia*'s atomic actions is underspecifiedness, which provides flexibility of representation. In fact, for example, all phrases semantically equivalent to "search for an object" are identified by *KeJia* as instances of *findobj* action. Generally, people cannot afford to explicitly spell out every detail of every course of action in every conceivable situation. For example, the user may express a query like "bring me the drink on a table". If *KeJia* knows more than one table or drink in the environment, it will ask the ordering person for further information. To realize these features, *KeJia*'s task planning module should be able to *plan under underspecification* to generate underspecified high-level plans, and provides *KeJia* with the possibility of acquiring more information when necessary. Once new information is received, the task planning module should update its world model and re-plan if needed. For this purpose, non-monotonic inference is required. This is one of the main reasons that we choose ASP as the underlying inference tool. Also, ASP provides a unified mechanism of handling commonsense reasoning and planning with a solution to the frame problem. There are many ASP program solver tools which can produce a solution to an ASP program. The solver *KeJia* uses is *iclingo* [6]. A more detailed explanation of the decision-making module could be found in [3].

6 Grounding

In this section, we present our grounding system. The goal of this system is to map natural language queries to their referents in a physical environment, we decompose it into two sub-problems: *semantic parsing*, i.e. learning the semantics of natural language query, and *spatial relations processing*, i.e. learning to recognize objects and extracting a knowledge base containing nouns and spatial relations. Both kinds of knowledge are necessary to achieve one query's grounding. For example, Fig. 6(a) and (b) show the input and results of the two processing modules. Through *spatial relations processing* module, the visual input of Fig. 6(a) is processed, the objects labeled and referred as numerical IDs. A knowledge base is also extracted which shown in Table 2. The query sentence in Fig. 6(b) is parsed by the *semantic parsing* module, which transforms the natural language query into logical form (semantic in Fig. 6(b)). Finally, both pieces of information are combined in the grounding system through logical rules which only consist of conjunctions and existential quantifiers to extract the grounding result (Fig. 6(c)). In case of ambiguity, the robot would ask the querying person to provide more information.

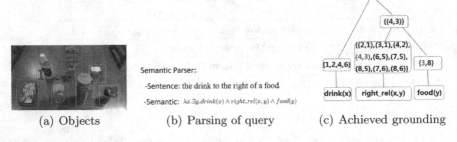

(a) Objects   (b) Parsing of query   (c) Achieved grounding

Fig. 6. An example process of the grounding system

6.1 Semantic Parser

The Semantic Parser is used to transform natural language query to internal logical form that the planning module can handle. The queries in our training data are annotated with expressions in a typed lambda-calculus language [1]. Our Semantic Parser is based on CCG [11]. A CCG specifies one or more logical forms for each sentence that can be parsed by the grammar. The core of any CCG is the semantic lexicon. In any lexicon, a word with its syntactic category and semantic form is defined. For example, the lexicon for our Semantic Parser would be as follows:

$$box := N : \lambda x.box(x)$$
$$food := N : \lambda x.food(x)$$
$$right := N/PP : \lambda f.\lambda x.\exists y.right\_rel(x,y) \wedge f(y)$$

Table 2. Extracted knowledge base

Object	Categories	Relations
1(amsa)	drink, bottle	behind:{5 ,6}, left:{2, 3, 4}
2(ice tea)	drink, box	behind:{6, 7}, left:{4}, top:{3}, right:{1}
3(porridge)	food, can	behind:{6, 7}, left:{4}, under:{2}, right:{1}
4(acid milk)	drink, box	behind:{7, 8}, right:{2, 3, 1}
5(pretz)	snack, box	before:{1}, left:{6, 7, 8}
...	...	...

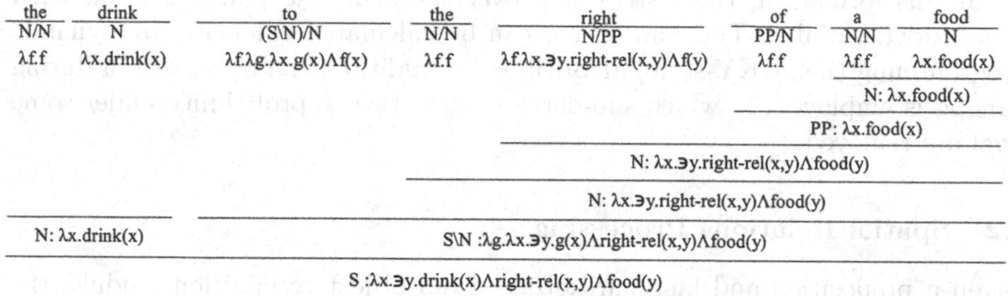

the	drink	to	the	right	of	a	food
N/N	N	(S\N)/N	N/N	N/PP	PP/N	N/N	N
λf.f	λx.drink(x)	λf.λg.λx.g(x)∧f(x)	λf.f	λf.λx.∃y.right-rel(x,y)∧f(y)	λf.f	λf.f	λx.food(x)

N: λx.food(x)

PP: λx.food(x)

N: λx.∃y.right-rel(x,y)∧food(y)

N: λx.∃y.right-rel(x,y)∧food(y)

N: λx.drink(x)　　　　　S\N :λg.λx.∃y.g(x)∧right-rel(x,y)∧food(y)

S :λx.∃y.drink(x)∧right-rel(x,y)∧food(y)

Fig. 7. An example parse of "the drink to the right of a food". The first row of the derivation retrieves lexical categories from the lexicon, while the remaining rows represent applications of CCG combinators.

In the above lexicon, *box* has the syntactic category N which stands for the linguistic notions of noun, and the logical form denotes the set of all entities x such that *box* is true. In addition to the lexicon, a CCG also has a set of *combinatory rules* to combine syntactic categories and semantic forms which are adjacent in a string. The basic combinatory rules are the functional application rules:

$$X/Y : f \ \ Y : g \Rightarrow X : f@g \quad (>)$$
$$Y : g \ \ X\backslash Y : f \Rightarrow X : f@g \quad (<)$$

The first rule is the forward ($>$) application indicating that a syntactic category X/Y with a semantic form f can be combined with a syntactic category Y with a semantic form g to generate a new syntactic category X whose semantic form is formed by λ-applying f to g. Symmetrically in the second rule, backward ($<$) application, generates a new syntactic category and semantic form by applying the right one to its left. Figure 7 illustrates how CCG parsing produces a syntactic tree t and a logical form l using combinatory rules.

When defining a CCG parser, we make use of a conditional log-linear model to select the best scored parses. It defines the joint probability of a logical form Z constructed with a syntactic parse Y, given a sentence X:

$$P(Y, Z|X; \theta, \Lambda) = \frac{exp[\theta \cdot f(X, Y, Z)]}{\sum_{Y', Z'} exp[\theta \cdot f(X, Y', Z')]} \tag{1}$$

Where Λ is the semantic lexicon, $f(X, Y, Z)$ is a feature vector evaluating on the sub-structures within (X, Y, Z). In this paper we make use of the lexical features. Each lexical feature counts the number of times that a lexical entry is used in y. With the probabilistic model, the semantic parsing then turn to compute:

$$\underset{Z}{argmax}\, P(Z|X; \theta, \Lambda) = \underset{Z}{argmax} \sum_{y} P(Y, Z|X; \theta, \Lambda) \tag{2}$$

In this formalism, the distribution over the syntactic parses y is modeled as a hidden variable. The sum over y can be calculated efficiently by dynamic programming (i.e., CKY-style) algorithms. In addition, the beam-search during parsing is employed, in which sub-derivations with low probability under some level are removed.

6.2 Spatial Relations Processing

Given a preposition and landmark object from object recognition module, the *spatial relations processing* module first calculates the probability of each target object, and extract the 3D point cloud segment which is located at the given preposition in relation to the landmark object. Then a logical knowledge Base τ containing object instances and spatial relations is constructed.

Here six spatial relations are used: {above, below, in_front_of, behind, to_the_left_of, to_the_right_of}, where the positions are described in the robot's perspective, e.g. object A *in_front_of* object B means A is closer to the robot. We model these spatial relations using a probabilistic distribution that predicts the identity of a target object (or 3D point) T conditioned on a preposition W, and landmark object L. To obtain the spatial relations between a target object and a landmark object (except itself), our module computes the maximum probability on each relation: $argmax_W\, P(T|W; L)$. A 3D point cloud segment is then extracted and its center position calculated as the pose of landmark object, which is denoted as (x, y, z). So the probability of this spatial relation is:

$$\underset{W}{argmax}\, P(T|W; L) = \underset{W}{argmax}\, P((x, y, z)|W; (x', y', z')) \tag{3}$$

In the Eq. 3, (x, y, z) is the 3D pose of object T, (x', y', z') is the 3D pose of object L. we assume that V is a six-vector represent {in_front_of, behind, to_the_left_of, to_the_right_of, below, above}. For example, $V = (1, 0, 0, 0, 0, 0)$ is represent the $W = $ *in_front_of*, we also assume that $\overrightarrow{delta} = ((x' - x), (x - x'), (y' - y), (y - y'), (z' - z), (z - z'))$. Now we can calculate the right side of the Eq. 3:

$$\underset{W}{argmax}\, P((x, y, z)|W; (x', y', z')) = \underset{W}{argmax}\, V * \overrightarrow{delta} \tag{4}$$

After the module obtains all spatial relations between landmark objects, a logical knowledge base τ is constructed. The knowledge base produced by perception module is a collection of ground predicate instances and spatial relations (see Table 2).

6.3 Demo in the Open Challenge

In the *Open Challenge* test, the grounding system was used to fulfill a task where ordering person did not know the names of the objects. The person used descriptive expressions to describe his intention, e.g. "The food to the left of a drink." If such denotation was not clear, the robot would ask more questions to eliminate ambiguity. Finally, the robot understood human's intention and retrieved the correct object for him. Here the logical rules described in Sect. 6 are as follows:

$$if \ \ l = \lambda x.c(x) \ \ then \ \ g = g^c$$
$$if \ \ l = \lambda x.\lambda y.r(x,y) \ \ then \ \ g = g^r$$
$$if \ l = \lambda x.l_1(x) \wedge l_2(x), \ then \ \ g(e) = 1 \ iff \ g_1(e) = 1 \wedge g_2(e) = 1$$
$$if \ l = \lambda x.\exists y.l_1(x,y), \ then \ \ g(e_1) = 1 \ iff \ \exists e_2.g_1(e_1, e(2)) = 1$$

7   Competition Results at RoboCup@Home 2014

At last we discuss our team's performance at the competition. The scores of most tests are shown in Table 3. In the *Robo-Zoo* test, *KeJia* failed to attract most audience due to a not-that-friendly appearance, thus resulting in a very low score. As for *Basic Functionalities Test*, our robot met some problems in entering the door, but finished the other two sub-tasks afterwards. In the *Emergency* test, almost all teams failed to enter the door, making a nearly blank score column. But starting from the test *Follow Me*, *KeJia* began to show better performance. In the *Follow Me* test, *KeJia* succeeded in following the guiding person till the very end. It passed around intercepting human, entered and left elevator, and find the person again after she sneaked through a group of unknown people. This was the first time this test has been fully solved by participating team in RoboCup@Home. Team Nimbro from Germany also completed this test, but us finishing in shorter time resulted in a higher score. In the *Open Challenge* in stage I, our robot performed the grounding demo described in Sect. 6.3, showing it able to not only recognize objects, but also understand location relationships and reason with such knowledge. During the *General Purpose Service Robot* test in Stage II, *KeJia* successfully completed one command in which important information was left out, but failed to recover from audience noise when executing the second command. Then in the *Cocktail Party*, *KeJia* almost finished the whole test, showing the abilities of robust speech recognition, human detection and recognition, object recognition and grasping, as well as navigation through crowded rooms. *KeJia* achieved 1750 points from a total of 2000 points in this test, making another record of best single test score since *Cocktail Party*

Table 3. Scores of most tests (except for *Emergency*)

	Robo-Zoo	Follow-Me	BFT	Open	GPSR	Cocktail-Party	Restaurant	Demo
Our score	87	**811**	600	**1507**	**750**	**1750**	**2600**	450
Average	211.2	368.8	415	763.4	195	410	442.5	225
Best	500	811	800	1507	750	1750	2600	750

was set. In the last most challenging regular test *Restaurant* which has never been even half-solved, our robot was able to carry out the whole test smoothly, taking a big lead against all the other teams. This was the third record that *KeJia* broke during the competition. Finally, we got a total of 9305 points in the first two stages, with a clear advantage over the former champion team Nimbro (Germany, 5701 points) and the third team TU/e (Netherlands, 5656 points).

At last in the *Final*, two *KeJia* robots collaborated to serve the ordering person a cup of coffee with sugar, in which two robots executed parts of the task in parallel and cooperated with each other when opening the cap of a sugar can. This demonstrated *KeJia*'s ability of multi-agent task planning, motion planning and precise motion control. In the end we achieved a total normalized score of 97 points, followed by TU/e (79 points) and Nimbro (74 points).

8   Conclusion

In this paper we present our contributions of our team WrightEagle@Home to the RoboCup@Home competition 2014. Our robot *KeJia* is designed to accomplish tasks involved in environment perception, human robot interaction, speech understanding and reasoning. In the competition, *KeJia* has shown great abilities and good robustness thus winning the competition with a big lead. Furthermore, we have demonstrated our robot's high level understanding and reasoning functions, which means a good integration of AI techniques into robotics. In future work, we plan to increase the generality of *KeJia*'s object recognition and grasping abilities. We also aim at further increase *KeJia*'s cognition functions by utilizing internet open knowledge.

Acknowledgement. This work is supported by the National Hi-Tech Project of China under grant 2008AA01Z150, the Natural Science Foundations of China under grant 60745002, 61175057, and USTC 985 project. Team members beside the authors are Zhe Zhao, Wei Shuai and Jiangchuan Liu. We also thank the anonymous reviewers for their valuable comments and suggestions on the manuscript of this work.

References

1. Carpenter, B.: Type-Logical Semantics. MIT Press, Cambridge (1997)
2. Bay, H., Tuytelaars, T., Van Gool, L.: SURF: speeded up robust features. In: Leonardis, A., Bischof, H., Pinz, A. (eds.) ECCV 2006. LNCS, vol. 3951, pp. 404–417. Springer, Heidelberg (2006)

3. Chen, X., Xie, J., Ji, J., Sui, Z.: Toward open knowledge enabling for human-robot interaction. J. Hum.-Rob. Interact. **1**(2), 100–117 (2012)
4. Dellaert, F., Fox, D., Burgard, W., Thrun, S.: Monte carlo localization for mobile robots. In: Proceedings of IEEE International Conference on Robotics and Automation, vol. 2, pp. 1322–1328 (1999)
5. Elfes, A.: Using occupancy grids for mobile robot perception and navigation. Computer **22**(6), 46–57 (1989)
6. Gebser, M., Kaminski, R., Kaufmann, B., Ostrowski, M., Schaub, T., Thiele, S.: Engineering an incremental ASP solver. In: Garcia de la Banda, M., Pontelli, E. (eds.) Logic Programming. LNCS, vol. 5366, pp. 190–205. Springer, Heidelberg (2008)
7. Grisetti, G., Stachniss, C., Burgard, W.: Improved techniques for grid mapping with rao-blackwellized particle filters. IEEE Trans. Robot. **23**(1), 34–46 (2007)
8. Hinterstoisser, S., Holzer, S., Cagniart,C., Ilic, S., Konolige, K., Navab, N., Lepetit, V.: Multimodal templates for real-time detection of texture-less objects in heavily cluttered scenes. In: IEEE International Conference on Computer Vision (2011)
9. Holz, D., del Solar, J. R., Sugiura, K., Wachsmuth, S.: On robocup@home - past, present and future of a scientic competition for service robots. In: Proceedings of the RoboCup International Symposium (2014)
10. Spinello, L., Arras,K. O.: People detection in RGB-D data. In: Proceedings of the 2011 IEEE/RSJ International Conference on Intelligent Robots and Systems, pp. 3838–3843. IEEE (2011)
11. Steedman, M.: Surface Structure and Interpretation. MIT Press, Cambridge (1996)
12. Ulrich, I., Borenstein, J.: Vfh+: reliable obstacle avoidance for fast mobile robots. In: IEEE International Conference on Robotics and Automation, vol. 2, pp. 1572–1577 (1998)
13. Viola, P., Jones, M.: Rapid object detection using a boosted cascade of simple features. In: Proceedings of the 2001 IEEE Computer Society Conference on Computer Vision and Pattern Recognition, vol. 1, pp. 511–518 (2001)

Winning the RoboCup@Work 2014 Competition: The smARTLab Approach

Bastian Broecker, Daniel Claes[✉], Joscha Fossel, and Karl Tuyls

smARTLab, Department of Computer Science,
University of Liverpool, Liverpool, UK
{bastian.broecker,daniel.claes,j.fossel,k.tuyls}@liverpool.ac.uk

Abstract. In this paper we summarise the approach of the smART-Lab@Work team that has won the 2014 RoboCup@Work competition. smARTLab (swarms, multi-agent, and robot technologies and learning lab) is a robotics research lab that is part of the Agent ART group at the computer science department of the University of Liverpool. This team has won the competition for the second year in a row. The team previously competed as swarmlab@Work and changed name after the move of professor Tuyls and his research group from Maastricht University to the University of Liverpool, UK. The various techniques that have been combined to win the competition come from different computer science domains, including machine learning, (simultaneous) localisation and mapping, navigation, object recognition and object manipulation. While the RoboCup@Work league is not a standard platform league, all participants use a (customised) KUKA youBot. The stock youBot is a ground based platform, capable of omnidirectional movement and equipped with a five degree of freedom arm featuring a parallel gripper. We present our adaptations to the robot, in which the replacement of the gripper was the most important upgrade comparing to the version of the robot that was used last year.

1 Introduction

Although fixed industrial robotics has achieved unrivalled productivity in mass production lines executing predefined repetitive tasks, such automation is still unreachable in customised flexible production or assembly situations. Such situations typically involve a high mixture of different types of products or objects that need to be transferred from one configuration into another. For example this could involve a task in a car garage in which robots need to collect specific parts from a bin or cupboard from one side of the workshop and bring these to another part or even put them back in a specific configuration, requiring planning, learning and reasoning capabilities as well as multi-robot and human-robot cooperation.

Recent years have seen new initiatives in this direction by the initiation of new robot competitions that focus on future industrial robot settings [11]. The most famous competition of its type is RoboCup@Work [10], which is specifically targeted to mobile industrial robotics in futuristic settings (the factory of

© Springer International Publishing Switzerland 2015
R.A.C. Bianchi et al. (Eds.): RoboCup 2014, LNAI 8992, pp. 142–154, 2015.
DOI: 10.1007/978-3-319-18615-3_12

the future). The competition, in particular, focuses on the use of robots in various work-related or industrial scenarios, such as navigating to specific locations and moving and manipulating objects. It aims to foster the research and development of innovative industrial robots for current and future industrial applications, where robots cooperate with humans on complex tasks ranging from manufacturing and automation to general logistics. Currently this competition is limited to one single mobile industrial robot (Kuka youBot) that has to achieve a number of predefined challenges. The current single-robot setting still poses important scientific and engineering challenges i.e. navigation, object detection, planning, learning and manipulation from the robotic platform in several industrial settings.

The competition has been founded in 2012 and has matured in the past two years. Some teams (including ours) have made their source code from the last competition publicly available[1].

2   smARTLab@Work

smARTLab is the research laboratory of the Department of Computer Science at University of Liverpool that focuses on designing Autonomous Systems. The general research goal and mission of the lab is to create adaptive systems and robots that are able to autonomously operate in complex and diverse environments. The smARTLab@Work team has been established in the beginning of 2014, consisting of 3 PhD candidates and 2 senior faculty members. The team has won the @Work competitions of the 2013 RoboCup German Open, and the 2013 RoboCup world championship under the name swarmlab@Work. After migrating to the University of Liverpool smARTLab@Work also placed first in the 2014 German Open and the 2014 RoboCup world championship.

The team's mission is (a) to apply smARTLab research achievements in the area of industrial robotics and (b) identify new research challenges that connect to the core smARTLab research areas: autonomous systems, reinforcement learning and multi-robot coordination.

In the remainder of this section we introduce the smARTLab@Work robot platform that was used for the 2014 RoboCup@Work world championship. Especially, we describe the necessary hardware modifications that were made to adapt this standard platform to the winning configuration.

2.1   YouBot Platform

The youBot is an omni-directional platform that has four mecanum [8] wheels, a 5 DoF manipulator and a two finger gripper. The platform is manufactured by KUKA[2], and is commercially available at the youBot-store[3]. It has been

---

[1] https://github.com/swarmlab/swarmlabatwork.
[2] http://kuka.com.
[3] http://youbot-store.com/.

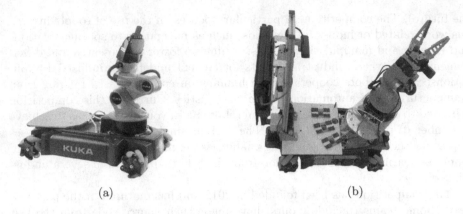

Fig. 1. (a) Stock youBot. (b) smARTLab@Work modified youBot for world championships 2014.

designed to work in industrial like environments and to perform various industrial tasks. With this open-source robot, KUKA is targeting educational and research markets [2]. Figure 1a shows a model of the stock youBot.

The youBot comes with a 5-degree-of-freedom arm that is made from casted magnesium, and has a 2-degree-of-freedom gripper. The standard gripper has two detachable fingers that can be remounted in different configurations. The gripper has a stroke of 20 mm and a reach of 50 mm, it opens and closes with an approximate speed of 1 cm/s. All hardware modules of the robot communicate internally over real-time EtherCat [9].

2.2   Hardware Modifications

In order to meet the requirements we demand from the youBot platform to tackle the challenges, we made a number of modifications to the robot. In this paragraph we describe which parts are modified and why these modifications are a necessity for our approach. Figure 1b shows the modified smARTLab@Work youBot setup that was used in the world championships in Brazil 2014.

The main adaptations with respect to last years version of the youBot are the replacement of the standard gripper with a custom made gripper and the design of a backplate to allow transportation of multiple objects. Since the new custom gripper is longer than the standard gripper, the arm was elevated by an additional 10 cm to still be able pick objects in a top-down fashion.

The standard gripper is replaced by two FESTO FinGripper fingers[4] mounted on two Dynamixel AX-12A[5] servo motors. This increases the stroke to more than 20 cm and the speed of the gripper to up to 10 cm/s. Also the fingers passively adapt to the shape of the objects as illustrated in Fig. 2. Through the custom

---

[4] http://www.festo.com/rep/en_corp/assets/pdf/Tripod_en.pdf.
[5] http://support.robotis.com/en/product/dynamixel/ax_series/dxl_ax_actuator.htm.

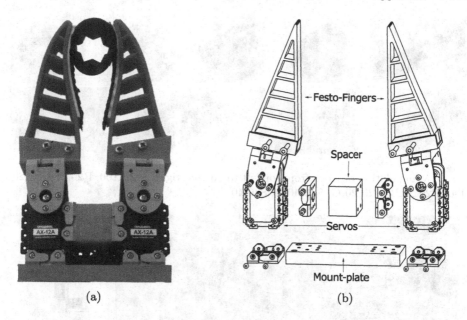

(a)   (b)

Fig. 2. Figure (a) illustrates the adaptation of the gripper to the round shape of the object. Figure (b) shows a drawing of the separate parts of the new designed gripper in more detail.

gripper is the youBot is now able to grasp every competition object, where the previous standard gripper was limited to the thinner objects due to his maximal opening width of 2 cm.

The new backplate (see Fig. 1b) offers three high friction pockets to place the objects on the robot. These pockets are designed to hold the items in place during transportation.

In order to detect and recognise manipulation objects, an ASUS Xtion PRO LIVE RGBD camera is attached to the last arm joint. This camera is mounted, so that it faces away from the manipulator, as can be seen in Fig. 7a. The main idea behind this mounting position originates from the minimal distance of the RGB-D camera, which is approximately ∼0.5 m. Using this special pre-grip pose, we can scan the area and perceive objects from a top-down perspective. An additional camera is mounted on the back of the robot. This camera is used to get a second view of certain objects, i.e. an object is picked up and it is moved by the arm such that it can be seen from a different angle. This helps to resolve confusions that are inevitable when only looking from a top-down fashion with the ASUS RGBD camera.

In this section we described the new hardware modifications needed to optimally execute the required RoboCup@Work tasks, in the next section these tasks will be describes in more detail, as well as the methods that we used to solve the specific tasks.

Fig. 3. Manipulation objects, from top left to bottom right: M20_100, R20, V20, M20, M30, F20_20_B, F20_20_G, S40_40_B, S40_40_G.

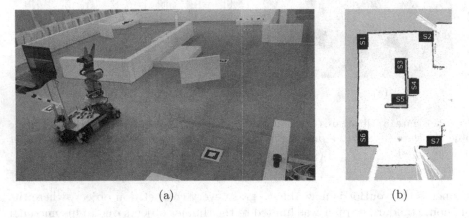

(a)   (b)

Fig. 4. (a) 2014 arena during final run with extra obstacles. (b) Map of the arena annotated with the service areas.

3 RoboCup@Work

In this section, we introduce the various tests of the 2014 RoboCup@Work world championship. Note that while the tests are slightly different for various levels of difficulty, we only give a description for the difficulty chosen by smARTLab@Work. For a more detailed description we refer the reader to the RoboCup@Work rule book[6]. A list of manipulation objects used in 2014 is given in Fig. 3. A picture of the 2014 arena is shown in Fig. 4a, in which several extra static obstacles are placed. In the arena, there are seven service areas as annotated in Fig. 4b. The term service area refers to a small table from which objects have to be manipulated. Figure 5a shows our robot performing a grasp in one of the service areas.

Basic Navigation Test. The purpose of the Basic Navigation Test (BNT) is testing mapping, localisation, path planning, obstacle avoidance and motion

---

[6] http://robocupatwork.org/resources.html.

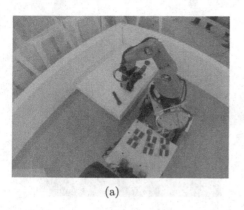
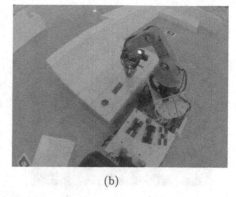

(a)   (b)

Fig. 5. (a) Grasping the S40_40_B in a service area. (b) A PPT placement into the M20 hole.

control. The robot has to autonomously enter the (known) arena, visit multiple way-points, and exit the arena again. Each way-point has to be accurately covered by the robot for a given duration in a given orientation. Additionally, obstacles may be spawned randomly in the arena, where a penalty is given if the robot collides with an obstacle.

Basic Manipulation Test. In the Basic Manipulation Test (BMT), the robot has to demonstrate basic manipulation capabilities such as grasping and placing an object. For this purpose three target and two decoy objects are placed on a service area, and the robot has to detect the three actual target objects and move them to another service area that is close by. All three objects are allowed to be carried at the same time. Implicitly this also tests whether the object recognition works sufficiently well. Since navigation is not a part of this test, the robot may be placed at any location before the test starts and is allowed to end anywhere in the arena.

Basic Transportation Test. The Basic Transportation Test (BTT) is a combination of the BNT and BMT. Here, six objects have to be picked up from various service areas and transported to specific destination areas. Similar to BNT and BMT both decoy objects and obstacles are placed on the service areas and in the arena, respectively. The robot has to start and end the BTT outside of the arena. In addition to the challenges individually posed by BNT and BMT, the robot now has to approach multiple service and destination areas within a certain time window, hence for optimal performance determining the optimal route and payload configuration is necessary. Only three objects are allowed to be carried at any given point in time.

Precision Placement Test. The Precision Placement Test (PPT) consists of grasping objects (see BMT), followed by placing them very accurately in object-specific cavities. Every object has one specific cavity for vertical, and one specific

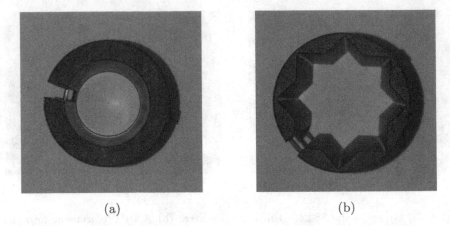

(a)   (b)

Fig. 6. (a) R20 object with detected circle. (b) V20 object with no detected circle.

cavity for horizontal placement. Each cavity has the same outline as the object plus 10 percent tolerance. Figure 5b shows the PPT destination area with some sample cavities.

Final Test. The final tests consists of a combination of all the tests described above. More specifically, in the arena various unknown static obstacles are present. There are six objects distributed over three service platforms with other decoy objects present. Of these objects two have to be placed precisely into the corresponding holes as in the PPT and the rest has to be placed on different service platforms as in the BTT.

4   Approach

In this section, the different techniques used to tackle the above mentioned tests are explained. We developed different modules for many different capabilities, e.g. basic global navigation, fine-positioning, object recognition, inverse kinematics of the arm, etc. We also explain our various recovery methods. By combining these capabilities in state-machines we are able to solve the tasks specified above.

The main adaptation with respect to last year, is the addition of the possibility to transport objects on the backplate of the robot. This allows us to transport more than one object at a time and actually planning the task becomes an important factor in the execution of the test. Furthermore, the object recognition was improved by combining the depth information together with the RGB information for the features. Also, the additional camera on the back was used to distinguish the R20 and V20 object, since the object shapes are very similar as shown in Fig. 3.

Thus for the descriptions of the approaches for mapping, localisation, inverse-kinematics and fine-positioning, we refer to [1].

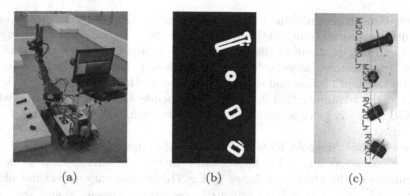

Fig. 7. (a) Pre-grip scan position. (b) Pre-processing of the image. (c) Detected objects, classification and grasp position of the closest object.

Navigation. A necessary capability of the robot is to navigate in the known environment without colliding with obstacles. The map is created with gmapping [6] and AMCL [4] is used for the localisation in the map. The global path is computed by an A* algorithm and is then executed using a dynamic window approach [5] trajectory rollout for the local path planning. This planner samples different velocities and ranks them according to the distance to the goal and the distance to the path, while velocities that collide with the environment are excluded. We tuned the trajectory rollout in such a way that it incorporates the omni-directional capabilities of the robot and also included scores for the correct destination heading, when being in close distance to the goal. The ROS implementation of the planner only takes the goal position into account and disregards the goal orientation until the robot has reached the goal position. The robot then performs a turn in place in order to achieve the correct orientation. Additionally, the robot tends to drive diagonally instead of straight, since the magnitude of the velocity can be higher and velocities with non-negative y values are sampled first. To tackle these problems, two additional scoring functions were added to the evaluation for each velocity sample:

$$c_p = |y| * w_p \tag{1}$$

$$c_h = |\theta_r - \theta_g| * w_h \tag{2}$$

where c_p is the penalty for considering velocities that contain a y component. c_h is the costs for the difference of the current heading θ_r in comparison to the goal heading θ_g. w_p and w_h are the corresponding weights for the two cost values. Equation 1 ensures that the robot prefers driving straight instead of diagonally. Equation 2 is applied if the robot is in proximity to the goal position, i.e. 1.5 m. Thus, the robot rotates towards the target heading while driving and it does not need to turn in place.

Object/Hole Recognition. Besides the navigation tasks, object detection and recognition is crucial to be able to interact with the environment, i.e. picking up objects and placing them in the correct target locations. We use the openCV-library[7] to detect the objects. We use two different detection pipelines, one is based on the depth image and the other is based on the RGB image. The depth image has the advantage that it is mostly independent of light conditions, while the RGB image is more accurate and allows to distinguish between black and silver objects.

An adaptive threshold filter is applied to the input image. Afterwards the image is converted into a black and white image and this is used to detect the contours of the objects as shown in Fig. 7b. We use various features of the detected objects, e.g., length of principal axis, average intensity and area, and use a labeled data-set that we created on the location to train a J4.8 decision tree in weka [7] for the recognition of the objects. This decision tree is then used for the online classification of the objects. Figure 7c shows the detection in a service area.

In order to distinguish between the R20 and the V20 object, we use an additional camera on the back. The objects have the same shape from the outside, but inside there different, i.e. round or star-shaped as shown in Fig. 6. This means that we have to pick the object up and hold in such that the additional camera can look inside the object. We use the Hough circle detection method [12] in order to determine if the inside of the object is round or star-shaped.

Recovery Methods. Of course, in robotics many things can go wrong, so we include various recovery methods. The navigation can easily get stuck, since the robot is tuned to drive very close to obstacles. Therefore, we implemented a force field [3] recovery. More specifically, the robot moves away from every obstacle that is picked up by the laser scan for a couple of seconds and the navigation is restarted. The servos in the arm tend to overshoot sometimes, especially when they get hot. Since the new fingers are compliant, it is not too problematic to slightly hit the ground, but in some cases it can lead to an overload of the arm. Thus, we measure the voltages in the arm, and as soon as a certain voltage is exceeded, we recover the arm to an upwards position and back up a couple of states in the state machine to be able to try again.

4.1   State Machine

For the different tests, we combine the various capabilities explained in the previous section in different state machines to complete the tasks. For the BNT, we do not need any of the manipulation techniques, so we have a very small state machine that combines the navigation and the fine-positioning. More specifically, we use the navigation to get close to the position and then the fine-positioning

---

[7] http://opencv.org.

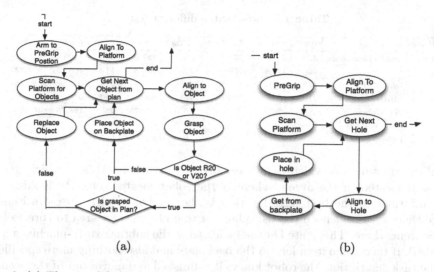

Fig. 8. (a) The simplified subtask-state-machine that is used for grasping objects at a service area. (b) The simplified subtask-state-machine that is used for dropping objects into the holes for PPT.

moves the robot exactly to the asked position. If the navigation fails, the force-field recovery is activated. Afterwards it retries to move to the current location. When it fails more than a fixed number of times, the current goal is skipped.

In principle the BMT, BTT, PPT and the final tasks do not differ very much. For instance, the BMT is basically a BTT, in which only one source and one target location are used and the PPT is a BMT in which the objects have to be placed in a specific location instead of anywhere on the platform. Thus for all these other tests we created so-called subtask state-machines that are shown in Fig. 8. For sake of simplicity, the recovery connections are left out. The subtasks are for example an behavior sequence for aligning to a platform a grasping all objects that need to be taken according to the task specification and placing them on the backplate of the robot as shown in Fig. 8a. Before any object is taken, a full scan of the service platform is performed. This enables us to reason about the objects, if there are any classifications that are uncertain. For instance, if there is only one R20 or V20 found on the platform, and the specification says that we have to get one R20 from that service area, we know that the object should be a R20 and the extra check does not need to be performed. Another subtask is placing all objects that need to be precisely placed at this location, shown in Fig. 8b. This task is very similar to the previously described behavior, since it is the reversed idea. The last subtask is the normal place of objects at target location. This is also similar to the PPT place, but since the objects can be placed anywhere on the platform, we only need to move once to the center of the platform and use the inverse kinematics of the arm to place the objects.

These subtask state machines are combined by a decision state that plans where to go and what to do next. This state gets all the necessary global and

Table 1. Scores for the different tests.

Place	Team	BNT	BMT#1	BMT#2	BTT#1	BTT#2	PPT	Final	Total
1	smARTLab@Work	420	0	250	400	187.5	500	1100	2857.5
2	b-it-bots	420	187.5	150	275	196.875	112.5	468.75	1810.625
3	Lyon CPE	187.5	93.75	112.5	275	84.375	225	0	978.125
4	LUHbots	210	50	100	75	300	150	0	885
5	WF Wolves	200	37.5	75	0	84.375	75	168.75	640.625
6	Robo-Erectus	0	0	0	0	0	0	0	0
7	Robotica UBM	0	0	0	0	0	0	0	0

local information, i.e. which items are currently on the backplate, which items are still located in the arena, where is the robot located, etc. By looking at this information the best next location is chosen, i.e. to pick up more items, since there is space left on the backplate, or to go to a target area to (precisely) place items there. This state is called each time the subtask-state-machines are finished. If there is no item left on the backplate and also nothing more specified in the task description, the robot knows it is finished and moves out of the arena.

5   Competition

Table 1 shows the final scores for the 2014 RoboCup championships.

The team scored full points for almost every test, except for BMT#1 and BTT#2. During the first run of the BMT tests, the robot tried to align to the target service area using the ASUS RGBD camera. Since the arm was in a fully tucked position before the tests, and the joint to which the camera is attached tends to overshoot heavily, no line was found in the image to which the robot could align to. The behaviour that was implemented as solution, was to "blindly" drive in the assumed direction of the platforms for a couple of centimeters. However, since the robot was already very close, it crashed into the platform and the run had to be aborted.

This issue relates to the problem that we had in the second BTT run. There we used part of the program to initialise the arm to a different position, but by doing that we disabled a part of the state machine that we forgot to enable again. Thus the state-machine stopped prematurely, and the run was stopped after grasping all necessary objects from one platform and placing them at another platform.

The rest of the tests went flawlessly, even in the final test, the combination of all of the above, we got the maximum points that were achievable in the difficulty level that was chosen. A video of the final run can be found here: http://youtu.be/FzVgDqrWiOY.

6   Future Work

The smARTLab@Work code and hardware designs will be released on Github[8]. For future competitions we plan on further improving the object recognition by

---

[8] http://github.com/smartlab-liv.

adding state of the art techniques, such as deep learning algorithms in order to be even more robust against changing light conditions and also to be able to quickly adapt to new objects. Furthermore, we are currently working on a learning by demonstration approach to allow for learning new arm trajectories on the fly. This might enable grasping various 'more difficult to grasp objects', e.g. glasses. It will also help to cope with future RoboCup@Work tests, which are yet to be introduced, e.g. assembly tasks or picking up moving objects.

Additionally, this year we were almost able to compete on the highest difficulty level for the tasks, which includes the detection and grasping of upright standing objects. However, as the object recognition currently did not perform sufficiently reliable for this case, we opted to not tackle grasping upright standing objects in the competition.

Hardware-wise, we plan on mounting the arm on a rotating extension plate, so that the arm can be turned towards the optimal grasp position. Additionally, the backplate can be optimized, maybe even by including servos to hold different kinds of objects in place during transportation.

Lastly, we are looking into an open demonstration that involves the cooperation of many robots, e.g. some Turtlebots[9] for transportation and the youBot to grasp and place the objects on the Turtlebots.

References

1. Alers, S., Claes, D., Fossel, J., Hennes, D., Tuyls, K., Weiss, G.: How to win robocup@work? In: Behnke, S., Veloso, M., Visser, A., Xiong, R. (eds.) RoboCup 2013. LNCS, vol. 8371, pp. 147–158. Springer, Heidelberg (2014)
2. Bischoff, R., Huggenberger, U., Prassler, E.: Kuka youbot - a mobile manipulator for research and education. In: 2011 IEEE International Conference on Robotics and Automation (ICRA), pp. 1–4, May 2011
3. Chong, N. Y., Kotoku, T., Ohba, K., Tanie, K.: Virtual repulsive force field guided coordination for multi-telerobot collaboration. In: Proceedings of the 2001 IEEE International Conference on Robotics and Automation (ICRA 2001) (2001)
4. Fox, D., Burgard, W., Frank D., and Thrun, S.: Monte carlo localization: efficient position estimation for mobile robots. In: Proceedings of the Sixteenth National Conference on Artificial Intelligence (AAAI 1999) (1999)
5. Fox, D., Burgard, W., Thrun, S.: The dynamic window approach to collision avoidance. IEEE Robot. Autom. Mag. **4**, 23–33 (1997)
6. Grisetti, G., Stachniss, C., Burgard, W.: Improved techniques for grid mapping with rao-blackwellized particle filters. IEEE Trans. Robot. **23**, 34–46 (2007)
7. Hall, M., Eibe, F., Holmes, G., Pfahringer, B., Reutemann, P., Witten, I.H.: The weka data mining software: an update. SIGKDD Explor. Newsl. **11**(1), 10–18 (2009)
8. Hon, B.E.: Wheels for a course stable selfpropelling vehicle movable in any desired direction on the ground or some other base. US Patent 3,876,255 (1975)
9. Jansen, D., Buttner, H.: Real-time ethernet: the ethercat solution. Comput. Control Eng. **15**(1), 16–21 (2004)

---

[9] http://www.turtlebot.com.

10. Kraetzschmar, G.K., Hochgeschwender, N., Nowak, W., Hegger, F., Schneider, S., Dwiputra, J., Berghofer, J., Bischof, R.: Robocup@work: competing for the factory of the future. In: RoboCup Symposium (2014)
11. TAPAS Project. Robotics-enabled Logistics and Assistive Services for the Transformable Factory of the Future (TAPAS) (2013). http://tapas-project.eu/
12. Xu, L., Oja, E., Kultanen, P.: A new curve detection method: Randomized Hough transform (RHT). Pattern Recogn. Lett. **11**(5), 331–338 (1990)

Decisive Factors for the Success of the Carologistics RoboCup Team in the RoboCup Logistics League 2014

Tim Niemueller[1]($\boxtimes$), Sebastian Reuter[2], Daniel Ewert[2], Alexander Ferrein[3], Sabina Jeschke[2], and Gerhard Lakemeyer[1]

[1] Knowledge-based Systems Group, RWTH Aachen University, Aachen, Germany
niemueller@kbsg.rwth-aachen.de
[2] Institute Cluster IMA/ZLW and IfU, RWTH Aachen University, Aachen, Germany
[3] Electrical Engineering Department, FH Aachen, Aachen, Germany

Abstract. The RoboCup Logistics League is one of the youngest application- and industry-oriented leagues. Even so, the complexity and level of difficulty has increased over the years. We describe decisive technical and organizational aspects of our hardware and software systems and (human) team structure that made winning the RoboCup and German Open competitions possible in 2014.

1 Introduction

The Carologistics RoboCup Team is a cooperation of the Knowledge-based Systems Group, the IMA/ZLW & IfU Institute Cluster (both RWTH Aachen University), and the FH Aachen University of Applied Sciences initiated in 2012. Doctoral, master, and bachelor students of all three partners participate in the project and bring in their specific strengths tackling the various aspects of the RoboCup Logistics League sponsored by Festo (LLSF): designing hardware modifications, developing functional software components, system integration, and high-level control of a group of mobile robots.

Our approach to the league's challenges is based on a distributed system where robots are individual autonomous agents that coordinate themselves by communicating information about the environment as well as their intended actions. In this paper we outline decisive factors for our successes, like building on and extending proven methods and components and hosting events to attract new students to cope with the challenge of students leaving the team.

Our team has participated in RoboCup 2012, 2013, and 2014 and the RoboCup German Open (GO) 2013 and 2014. We were able to win the GO 2014 (cf. Fig. 1) as well as the RoboCup 2014 in particular demonstrating a flexible task coordination scheme, and robust collision avoidance and self-localization.

In the following Sect. 2 we give an overview of the LLSF. Then we describe our team's robots, software components (Sect. 3), and aspects of our task coordination and simulation (Sect. 4). We detail our involvement in the league's organization and outreach events in Sect. 5, before concluding in Sect. 6.

R.A.C. Bianchi et al. (Eds.): RoboCup 2014, LNAI 8992, pp. 155–167, 2015.
DOI: 10.1007/978-3-319-18615-3_13

Fig. 1. Carologistics (three Robotino 2 with laptops on top) and BavarianBendingUnits (two larger Robotino 3) during the LLSF finals at RoboCup 2014 (Color figure online)

2   The RoboCup Logistics League

RoboCup [1] is an international initiative to foster research in the field of robotics and artificial intelligence. The basic idea of RoboCup is to set a common testbed for comparing research results in the robotics field. RoboCup is particularly well-known for its various soccer leagues. In the past few years focus application-oriented leagues such as urban search and rescue or domestic service robotics received more and more attention. In 2012, the new industry-oriented Logistics League Sponsored by Festo (LLSF) was founded to tackle the problem of production logistics. Groups of up to three robots have to plan, execute, and optimize the material flow in a factory automation scenario and deliver products according to dynamic orders. Therefore, the challenge consists of creating and adjusting a production plan and coordinate the group of robots [2]. The LLSF has attracted an increasing number of teams since established (8 teams in 2014).

The LLSF competition takes place on a field of $11.2\,m \times 5.6\,m$ surrounded by walls (Fig. 1). Two teams are playing at the same time competing for points, (travel) space and time. Each team has an exclusive input storage (blue areas) and delivery zone (green area). Machines are represented by RFID-readers with signal lights on top. The lights indicate the current status of a machine, such as "ready", "producing" and "out-of-order". There are three delivery gates, one recycling machine, and twelve production machines per team. Material is represented by orange pucks with an RFID tag. At the beginning all pucks have the raw material state S_0 and are in the input storage, and can be refined through several stages to final products using the production machines. These machines are assigned a type randomly at the start of a match which determines what inputs are required and what output will be produced, and how long this conversion will take [3]. Finished products must then be taken to the active gate in the delivery zone.

The game is controlled by the referee box (refbox), a software component which keeps track of puck states, instructs the light signals, and posts orders to the teams [4]. After the game is started, no manual interference is allowed, robots receive instructions only from the refbox. Teams are awarded with points for delivering ordered products, producing complex products, and recycling. The refbox can be seen as a higher-level production planning entity as used in industry, e.g. ERP or MES-Systems.

3   The Carologistics Platform

The standard robot platform of this league is the Robotino by Festo Didactic [5]. The Robotino was developed for research and education and features omni-directional locomotion, a gyroscope and webcam, infrared distance sensors, and bumpers. It is also equipped with a static puck holder to move pucks. The teams may equip the robot with additional sensors and computation devices.

3.1   Hardware System

The robot system currently in use is still based on the Robotino 2 which is now replaced by the new version 3. The modified Robotino depicted in Fig. 2(a) used by the Carologistics RoboCup team features two additional webcams and a Sick laser range finder. One of the webcams is used for recognizing the signal lights of the production machines, the other to detect pucks in front of the robot. The former omni-directional camera is no longer used as it was prone to distortion and time-intensive calibration. Tracking pucks especially during rotational movement presented another challenge while the benefit of detecting pucks anywhere next to the robot was minimal. The webcams are mounted with serrated locking plates for a firm adjustment to defined angles. The Sick TiM551 laser scanner is used for collision avoidance and self-localization. In comparison to the Hokuyo laser scanner with a scanning range of 4 m we used last year, the Sick TiM551 has a maximal scanning of 10 m. An additional laptop on the robot increases the computation power and allows for more elaborate methods for self-localization, computer vision, and navigation. A custom-made passive guidance device is mounted to the front of the robots to allow for proper control of the pucks. Optical sensors mounted to the guidance device are used to measure the longitudinal distance for approaching the signal lights.

Next year we intend to migrate to the new Robotino 3. This is a requirement to cope with the new field stations coming in 2015 [4]. Preliminary experiments indicated a smooth migration of our control system to the new platform.

3.2   Architecture and Middleware

The software system of the Carologistics robots combines two different middlewares, Fawkes [6] and ROS [7]. This allows us to use software components from both systems. The overall system, however, is integrated using Fawkes.

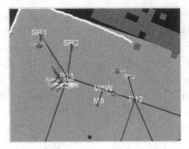

(a) Carologistics Robotino 2014   (b) Visualization of a scene in rviz

Fig. 2. Carologistics Robotino, sensor processing, and visualization

Adapter plugins connect the systems, for example to use ROS' 3D visualization capabilities (cf. Fig. 2(b)). The overall software structure is inspired by the three-layer architecture paradigm [8]. It consists of a deliberative layer for high-level reasoning, a reactive execution layer for breaking down high-level commands and monitoring their execution, and a feedback control layer for hardware access and functional components. The lowest layer is described in Sect. 3.3. The upper two layers are detailed in Sect. 4. The communication between single components – implemented as *plugins* – is realized by a hybrid blackboard and messaging approach [6]. This allows for information exchange between arbitrary components. As shown in Fig. 3, information is written to or read from *interfaces*, each carrying certain information, e.g. sensor data or motor control, but also more abstract information like the position of an object. The information flow is somewhat restricted – by design – in so far as only one component can write to an interface. Reading, however, is possible for an arbitrary number of components. This approach has proven to avoid race conditions when for example different components try to instruct another component at the same time. The principle is that the interface is used by a component to provide state information. Instructions and commands are sent as messages. Then, multiple conflicting commands can be detected or they can be executed in sequence or in parallel, depending on the nature of the commands.

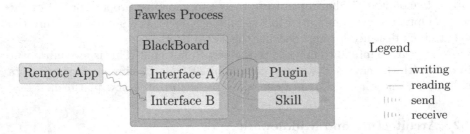

Fig. 3. Components communicate state data via interfaces stored in the blackboard. Commands and instructions are send as messages. Communication is universally shared among functional plugins and behavioral components.

3.3 Functional Software Components

A plethora of different software components is required for a multi-robot system. Here, we discuss the lowest layer in our architecture which contains functional modules and hardware drivers. All functional components are implemented in Fawkes. Drivers have been implemented based on publicly available protocol documentation, e.g. for our laser range finders or webcams. To access the Robotino base platform hardware we make use of a minimal subset of OpenRobotino, a software system provided by the manufacturer.

Localization is based on Adaptive Monte Carlo Localization which was ported from ROS and then extended. For locomotion, we integrated the collision avoidance module [9] which is also used by the AllemaniACs[1] RoboCup@Home robot.

A computer vision component for robust detection of the light signal state on the field has been developed specifically for this domain. A new such component we developed is a vision-based machine detection module. It will allow to detect and approach the machines more precisely as it yields a 3D pose. Figure 4 shows the visualization of the extracted features.

Fig. 4. Vision-based machine detection providing 3D pose information to approach a machine.

4 High-level Decision Making and Task Coordination

Task coordination is performed using an incremental reasoning approach [10]. In the following we describe the behavior components, and the reasoning process in two particular situations from the rules in 2014. For computational and energy efficiency, the behavior components need also to coordinate activation and deactivation of the lower level components to solve computing resource conflicts.

4.1 Behavior Components for the LLSF

Tasks that the high-level reasoning component of the robot must fulfill in the LLSF are:

Exploration: Gather information about the machine types by sensing and reasoning to gain more knowledge, e.g. the signal lights' response to certain types of pucks

Production: Complete the production chains as often as possible dealing with incomplete knowledge.

Execution Monitoring: Instruct and monitor the reactive mid-level Lua-based Behavior Engine.

---

[1] See the AllemaniACs website at http://robocup.rwth-aachen.de.

A group of three robots perform these steps cooperatively, that is, they communicate information about their current intentions, acquire exclusive control over resources like machines, and share their beliefs about the current state of the environment. This continuous updating of information suggests an incremental reasoning approach. As facts become known, the robot needs to adjust its plan.

4.2   Lua-based Behavior Engine

In previous work we have developed the Lua-based Behavior Engine (BE) [11]. It mandates a separation of the behavior in three layers, as depicted in Fig. 5: the low-level processing for perception and actuation, a mid-level reactive layer, and a high-level reasoning layer. The layers are combined following an adapted hybrid deliberative-reactive coordination paradigm with the BE serving as the reactive layer to interface between the low- and high-level systems.

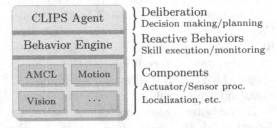

Fig. 5. Behavior Layer Separation

The BE is based on hybrid state machines (HSM). They can be depicted as a directed graph with nodes representing states for action execution, and/or monitoring of actuation, perception, and internal state. Edges denote jump conditions implemented as Boolean functions. For the active state of a state machine, all outgoing conditions are evaluated, typically at about 15 Hz. If a condition fires, the active state is changed to the target node of the edge. A table of variables holds information like the world model, for example storing numeric values for object positions. It remedies typical problems of state machines like fast growing number of states or variable data passing from one state to another. Skills are implemented using the light-weight, extensible scripting language Lua.

4.3   Incremental Reasoning Agent

The problem at hand with its intertwined world model updating and execution naturally lends itself to a representation as a fact base with update rules for triggering behavior for certain beliefs. We have chosen the CLIPS rules engine [12], because using incremental reasoning the robot can take the next best action at any point in time whenever the robot is idle. This avoids costly re-planning (as with approaches using classical planners) and it allows us to cope with incomplete knowledge about the world. Additionally, it is computationally inexpensive.

The CLIPS rules are roughly structured using a fact to denote the current overall state that determines which subset of the rules is applicable at any given time. For example, the robot can be idle and ready to start a new sub-task, or it may be busy moving to another location. Rules involved with physical interaction typically depend on this state, while world model belief updates often do not.

The state is also required to commit to a certain action and avoid switching to another one if new information, e.g., contributed by other robots on the field, becomes available. While it may be better in the current situation to pursue another goal, aborting or changing an action usually incurs much higher costs.

The rules explained in the following demonstrate what we mean by incremental reasoning. The robot does not create a full-edged plan at a certain point in time and then executes it until this fails. Rather, when idle it commits to the 'then-best' action. As soon as the action is completed and based on its knowledge, the next best action is chosen. The rule base is structured in six areas: exploration, production step decision, coordination with other robots, process execution, world modeling, and utilities.

```
(defrule load-T5-with-S0
  (declare (salience ?*PRIORITY-LOAD-T5-WITH-S0*))
  (phase PRODUCTION)
  ?s <- (state IDLE)
  (holding NONE|S0)
  (team-color ?team-color)
  (machine (mtype T5) (loaded-with $?l&~:(member$ S0 ?l))
    (incoming $?i&~:(member$ BRING_S0 ?i)) (name ?name)
    (produced-puck NONE) (team ?team-color))
  =>
  (printout t "PROD: Loading T5 " ?name " with S0" crlf)
  (assert (proposed-task (name load-with-S0) (args create$ ?name))))
  (retract ?s)
  (assert (state TASK-PROPOSED))
)
```

Fig. 6. CLIPS production process rule

In Fig. 6 we show a simplified rule for the production process. The game is in the production phase, the robot is currently idle and holds a raw material puck S_0 or no puck: (phase PRODUCTION) (state IDLE)(holding NONE|S0). Furthermore there is a $T5$-machine, whose team-color matches the team-color of the robot, which has no produced puck, is not already loaded with an S_0, and no other robot is currently bringing an S_0. If these conditions are satisfied and *PRIORITY-LOAD-T5-WITH-S0* is the highest priority of currently active rules, the rule fires proposing to load the machine with the name ?name with an puck in state S_0. It also switches the state.

There is a set of such production rules with their conditions and priorities determining what the robot does in a certain situation, or – in other terms – based on a certain belief about the world in the fact base. This simplifies adding new decision rules. The decisions can be made more granular by adding rules with more restrictive conditions and a higher priority.

After a proposed task was chosen, the coordination rules of the agent cause communication with the other robots to announce the intention and ensure that there are no conflicts. If the coordination rules accept the proposed task, process execution rules perform the steps of the task (e.g. getting an S_0 from the input storage and bringing it to the machine). Here, the agent calls the Behavior Engine to execute the actual skills like driving to the input storage and loading a puck.

The *world model* holds facts about the machines and their state, what kind of puck the robot is currently holding (if any) and the state of the robot. A simplified examples for a world model update is shown in Fig. 7. The world model update rule is invoked after a task or sub-task from the production rule presented above was successfully completed, i.e. an S_0 puck was taken to a machine of the type T_5. The rule shows the inference of the output puck type given a machine's reaction. The conditions (`state GOTO-FINAL`) (`goto-target ?name`) denote that the robot finished locomotion and production at the target machine `?name`, Furthermore, the robot sees only a green light at the machine, which indicates that the machine successfully finished the production. If all these conditions hold, the rule updates the world model about what kind of puck the robot is holding. Additionally it assumes all pucks removed that were loaded in the machine and increases the amount of consumed pucks. The world model is synchronized with other robots with another set of rules.

In comparison to 2013, the agent evolved to enable a tighter cooperation of the three agents. This required smaller atomic tasks, which are performed by the agents, a coordination mechanism to ensure the robots perform no redundant actions, more fine-grained production rules, and synchronization of the world model. The latter allows for dynamically adding or removing robots without interference to the overall production process. Furthermore, the agent became more robust against failure of behavior execution and wrong perception by adding a set of more distinctive world model update rules.

4.4   Multi-robot Simulation in Gazebo

The character of the LLSF game emphasizes research and application of methods for efficient planning, scheduling, and reasoning on the optimal work order of

```
(defrule goto-proc-complete
  (declare (salience ?*PRIORITY-WM*))
  (state GOTO-FINAL)
  (goto-target ?name)
  ?h <- (holding ?)
  (lights GREEN-ON YELLOW-OFF RED-OFF)
  (machine (name ?name) (output ?output) (loaded-with $?lw) (junk ?jn))
  =>
  (printout t "Production completed at " ?name "|" ?mtype crlf)
  (retract ?h)
  (assert (holding ?output))
  (foreach ?puck ?lw
    (assert (worldmodel-change (machine ?name)
             (change REMOVE_LOADED_WITH) (value ?puck)))
  )
  (assert (worldmodel-change (machine ?name) (change SET_NUM_CO)
                              (amount (+ ?jn (length$ ?lw)))))
)
```

Fig. 7. CLIPS world model update rules

production processes handled by a group of robots. An aspect that distinctly separates this league from others is that the environment itself acts as an agent by posting orders and controlling the machines' reactions. This is what we call *environment agency*. Naturally, dynamic scenarios for autonomous mobile robots are complex challenges in general, and in particular if multiple competing agents are involved. In the LLSF, the large playing field and material costs are prohibitive for teams to set up a complete scenario for testing, let alone to have two teams of robots. Additionally, members of related communities like planning and reasoning might not want to deal with the full software and system complexity. Still they often welcome relevant scenarios to test and present their research. Therefore, we have created an *open simulation environment* [3] to support research and development. There are three core aspects in this context: (1) The simulation should be a turn-key solution with simple interfaces, (2) the world must react as close to the real world as possible, including in particular the machine responses and signals, and (3) various levels of abstraction are desirable depending on the focus of the user, e.g. whether to simulate laser data to run a self-localization component or to simply provide the position (possibly with some noise) Fig. 8.

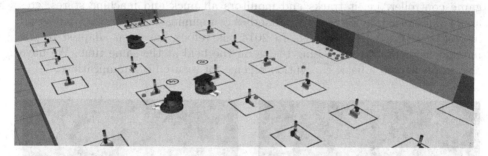

Fig. 8. The simulation of the LLSF in Gazebo. The circles above the robots indicate their localization and robot number.

In recent work [3], we provide such an environment. It is based on the well-known Gazebo simulator addressing these issues: (1) its wide-spread use and open interfaces already adapted to several software frameworks in combination with our models and adapters provides an easy to use solution; (2) we have connected the simulation directly to the referee box, the semi-autonomous game controller of the LLSF, so that it provides precisely the reactions and *environment agency* of a real-world game; (3) we have implemented *multi-level abstraction* that allows to run full-system tests including self-localization and perception or to focus on high-level control reducing uncertainties by replacing some lower-level components using simulator ground truth data. This allows to develop an idealized strategy first, and only then increase uncertainty and enforce robustness by failure detection and recovery. More information, media, and the software itself are available at http://www.fawkesrobotics.org/projects/llsf-sim/.

In the LLSF, the large playing field and material costs are prohibitive for teams to set up a complete scenario, let alone to have two teams of robots.

Additionally, members of related communities like planning and reasoning might not want to deal with the full software and system complexity. Still they often welcome relevant scenarios to test and present their research. Therefore, we propose a new simulation sub-league for the LLSF based on our simulation [3].

5   Continued Involvement and Outreach

We have been active members of the Technical and Organizational Committees and proposed various ground-breaking changes for the league like merging the two playing fields or using physical processing machines in 2015 [2,4]. Additionally we introduced and currently maintain the autonomous referee box for the competition as explained in the next section.

5.1   LLSF Referee Box

The Carologistics team has developed the autonomous referee box (refbox) for the LLSF which was deployed in 2013 [2]. It strives for full autonomy on the game controller, i.e. it tracks and monitors all puck and machine states, creates (randomized) game scenarios, handles communication with the robots, and interacts with a human referee. In 2014 the refbox has been adapted to the merged fields and two opposing teams on the field at the same time. We have also implemented a basic encryption scheme for secured communications.

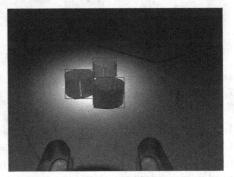

(a) Field with illuminated obstacles where the robot has to find objects in the dark with just a headlight.

(b) View from the robot's front camera with detected pucks in the light cone of the headlight mounted on the robot.

Fig. 9. The 2014 Carologistics Hackathon challenge

5.2   Organizational and Didactic Aspects

One of the primary goals of the RoboCup Logistics League lies in providing a test bed for cyber-physical systems [13]. The competition also serves as an excellent educational tool to give students a hands-on experience in dealing with robotics in industrial applications and future challenges for industry and research.

To improve the outreach of the league and involve a larger group of students the team and its associated institutes host *Hackathons*: students are invited to delve into the teams robotic system and develop a solution for a specific simplified challenge in a night's time. This year, the task was to explore an arena and find color-coded objects. A particular complication was that the arena was dark, i.e. without external lighting (see Fig. 9). The robots were equipped with headlights to enable perception with a considerably smaller field of view.

Hackathons also serves as recruiting platform for new team members, which are always needed due to the natural fluctuation, e.g. due to students leaving the university. Until now, Hackathons have been held each year since 2012, attracting up to 50 attendees for each event. The majority of the current student team members entered through one of the Hackathons.

The LLSF is also a formidable teaching platform. The KBSG offers regular lab courses where students are introduced to the robot platform and have to work on a specific problem. Example topics are inter-robot communication[2] or a new task coordination component[3] based on Procedural Reasoning Systems [14].

5.3   Research

The LLSF provides an excellent domain for research, in particular for a focus on task coordination for multi-robot systems. Compared to other RoboCup leagues, the problem is less dynamic (compared to soccer) and less cluttered and unorganized (compared to urban search and rescue or domestic service robots). That does not mean that it is easy to compete in the league but it does provide an interesting balance for researchers from related fields like planning or knowledge representation and reasoning to apply their results in a robotic environment.

In our context, creating a new central planning component is slated for inclusion as a part for the next phase of an ongoing research project on hybrid reasoning[4]. There, we want to explore the possibility to have a globally optimizing reasoner that offers suggestions for the robot group to maximize the overall achievable score with the given resources. This inclusion in larger scale research projects is crucial to advance the state of the art in a team or a league.

6   Conclusion

In this paper we have outlined several decisive factors for winning all competitions and technical challenges in 2014 without loosing a single game. It is important to have a joint team from different areas to cope with the large variety of challenges, from hardware modification to software integration. Our robust and proven base system that has been developed and used for and in other RoboCup

---

[2] http://kbsg.rwth-aachen.de/teaching/WS2012/LabRoCoCo.

[3] http://kbsg.rwth-aachen.de/teaching/WS2014/LabPRoGrAMR.

[4] http://www.hybrid-reasoning.org.

leagues allowed us to focus on the domain-specific challenges like flexible and efficient task coordination. In particular our focus on behavioral and multi-robot coordination components and the availability of our Gazebo-based simulation environment were crucial advantages. Even though the more elaborate approach meant a disadvantage through higher complexity for the simpler rules in 2012 and 2013, it was worthwhile since we were able to cope with the increasing level of difficulty of the league more quickly. Finally, our outreach program organizing large yearly Hackathons to attract and recruit new students is crucial to keep the team vivid and compensate leaving team members.

The website of the Carologistics RoboCup Team with further information and media can be found at http://www.carologistics.org.

Acknowledgments. The team members in 2014 are: Daniel Ewert, Alexander Ferrein, Sabina Jeschke, Nicolas Limpert, Gerhard Lakemeyer, Matthias Löbach, Randolph Maaßen, Victor Mataré, Tobias Neumann, Tim Niemueller, Florian Nolden, Sebastian Reuter, Johannes Rothe, and Frederik Zwilling.

We gratefully acknowledge the financial support of RWTH Aachen University and FH Aachen for participation at RoboCup 2014 in João Pessoa, Brazil. We thank the Bonding student organization for co-organizing and providing food, caffeinated drinks, and support for the Hackathons 2013 and 2014.

F. Zwilling and T. Niemueller were supported by the German National Science Foundation (DFG) research unit *FOR 1513* on Hybrid Reasoning for Intelligent Systems (http://www.hybrid-reasoning.org).

References

1. Kitano, H., Asada, M., Kuniyoshi, Y., Noda, I., Osawa, E.: RoboCup: The Robot World Cup Initiative. In: Proceedings of the First International Conference on Autonomous Agents (1997)
2. Niemueller, T., Ewert, D., Reuter, S., Ferrein, A., Jeschke, S., Lakemeyer, G.: RoboCup logistics league sponsored by Festo: a competitive factory automation testbed. In: Behnke, S., Veloso, M., Visser, A., Xiong, R. (eds.) RoboCup 2013. LNCS, vol. 8371, pp. 336–347. Springer, Heidelberg (2014)
3. Zwilling, F., Niemueller, T., Lakemeyer, G.: Simulation for the RoboCup logistics league with real-world environment agency and multi-level abstraction. In: RoboCup Symposium (2014)
4. Niemueller, T., Lakemeyer, G., Ferrein, A., Reuter, S., Ewert, D., Jeschke, S., Pensky, D., Karras, U.: Proposal for advancements to the LLSF in 2014 and beyond. In: ICAR - 1st Workshop on Developments in RoboCup Leagues (2013)
5. Karras, U., Pensky, D., Rojas, O.: Mobile robotics in education and research of logistics. In: IROS 2011 - Workshop on Metrics and Methodologies for Autonomous Robot Teams in Logistics (2011)
6. Niemueller, T., Ferrein, A., Beck, D., Lakemeyer, G.: Design principles of the component-based robot software framework fawkes. In: Ando, N., Balakirsky, S., Hemker, T., Reggiani, M., von Stryk, O. (eds.) SIMPAR 2010. LNCS, vol. 6472, pp. 300–311. Springer, Heidelberg (2010)

7. Quigley, M., Conley, K., Gerkey, B.P., Faust, J., Foote, T., Leibs, J., Wheeler, R., Ng, A.Y.: ROS: an open-source Robot Operating System. In: ICRA Workshop on Open Source Software (2009)
8. Gat, E.: Three-layer Architectures. In: Kortenkamp, D., Bonasso, R.P., Murphy, R. (eds.) . Artificial Intelligence and Mobile Robots, pp. 195–210. MIT Press, Cambridge (1998)
9. Jacobs, S., Ferrein, A., Schiffer, S., Beck, D., Lakemeyer, G.: Robust collision avoidance in unknown domestic environments. In: Baltes, J., Lagoudakis, M.G., Naruse, T., Ghidary, S.S. (eds.) RoboCup 2009. LNCS, vol. 5949, pp. 116–127. Springer, Heidelberg (2010)
10. Niemueller, T., Lakemeyer, G., Ferrein, A.: Incremental task-level reasoning in a competitive factory automation scenario. In: Proceedings of AAAI Spring Symposium 2013 - Designing Intelligent Robots: Reintegrating AI (2013)
11. Niemüller, T., Ferrein, A., Lakemeyer, G.: A Lua-based behavior engine for controlling the humanoid robot nao. In: Baltes, J., Lagoudakis, M.G., Naruse, T., Ghidary, S.S. (eds.) RoboCup 2009. LNCS, vol. 5949, pp. 240–251. Springer, Heidelberg (2010)
12. Wygant, R.M.: A powerful development and delivery expert system tool. Comput. Ind. Eng. **17**(1–4), 546–549 (1989)
13. Niemueller, T., Ewert, D., Reuter, S., Karras, U., Ferrein, A., Jeschke, S., Lakemeyer, G.: Towards benchmarking cyber-physical systems in factory automation scenarios. In: Timm, I.J., Thimm, M. (eds.) KI 2013. LNCS, vol. 8077, pp. 296–299. Springer, Heidelberg (2013)
14. Alami, R., Chatila, R., Fleury, S., Ghallab, M., Ingrand, F.: An architecture for autonomy. Int. J. Robot. Res. **17**(4), 315–337 (1998)

Oral Presentations

RoboCup@Work: Competing for the Factory of the Future

Gerhard K. Kraetzschmar[1], Nico Hochgeschwender[1(✉)], Walter Nowak[2],
Frederik Hegger[1], Sven Schneider[1], Rhama Dwiputra[1], Jakob Berghofer[3],
and Rainer Bischoff[3]

[1] Bonn-Rhein-Sieg University, Sankt Augustin, Germany
`nico.hochgeschwender@h-brs.de`
[2] Locomotec GmbH, Augsburg, Germany
[3] KUKA Laboratories GmbH, Augsburg, Germany

Abstract. Mobile manipulators are viewed as an essential component
for making the factory of the future become a reality. RoboCup@Work
is a competition designed by a group of researchers from the RoboCup
community and focuses on the use of mobile manipulators and their integration with automation equipment for performing industrially-relevant
tasks. The paper describes the design and implementation of the competition and the experiences made so far.

1 Introduction

For a long time, research on mobile robots and on industrial manipulators have
been performed in separate communities. Recently, the robotics and automation
industry is shifting its attention towards robotics scenarios involving the integration of mobility and manipulation, larger-scale integration of service robots
and industrial robots, cohabitation of robots and humans, and cooperation of
multiple robots and/or humans. All of these ideas are part of the factory of the
future (FoF) (see also [10]), which is considered a foundation for conserving or
reviving industrial competitiveness in the future.

An essential component in FoF scenarios are *mobile manipulators*, the combination of a mobile robot with one or more manipulators. Recent development efforts of robots manufacturers led to research platforms now available for
research and the market. Examples include the youBot [2] and the omniRob
platforms by KUKA, and the rob@work platform by Fraunhofer IPA depicted
in Fig. 1.

Although industry views this technology as an essential component for the
factory of the future, real applications are still rare. It became quickly obvious
that the control concepts and algorithms developed independently for robotic
arms and for mobile robots could not easily be combined and integrated. More
research is necessary to exploit the capabilities of mobile manipulators in innovative applications.

In early 2012, a group of researchers concluded that the progress in mobile
manipulation research could benefit significantly by organizing a competition

© Springer International Publishing Switzerland 2015
R.A.C. Bianchi et al. (Eds.): RoboCup 2014, LNAI 8992, pp. 171–182, 2015.
DOI: 10.1007/978-3-319-18615-3_14

Fig. 1. Mobile manipulators from KUKA Laboratories GmbH (left) and Fraunhofer IPA (right). (Image courtesy of KUKA Laboratories GmbH and Fraunhofer IPA.)

targeted towards the factory of the future: RoboCup@Work [7]. The new competition should address research issues and appeal to both new and existing RoboCup teams who want to demonstrate their competences in a more industry-related setting. The competition itself should be designed to combine the public appeal of robotic soccer, the competition structure of RoboCup@Home [4], and the benchmarking technology from RoboCup Rescue [5,9]. This paper describes the motivation and context of RoboCup@Work, the competition design and implementation, and the experiences made so far.

2   Research Issues and Challenges

RoboCup@Work targets the use of robots in work-related scenarios and tackles open research challenges in industrial and service robotics. Examples for the work-related scenarios targeted by RoboCup@Work include

- loading and unloading containers with objects of the same or different size,
- pickup or delivery of parts from/to structured or unstructured storage/heaps,
- operation of machines, including pressing buttons, opening/closing doors and drawers, and similar operations with underspecified or unknown kinematics,
- flexible planning and dynamic scheduling of production processes involving multiple agents (humans, robots, machines, tools, and parts),
- cooperative assembly of non-trivial objects with other robots or humans,
- cooperative collection of objects in large-scale spatial areas, and
- cooperative transportation of objects (robot with robots or humans).

The RoboCup@Work scenarios target difficult, still mostly unsolved problems in robotics, artificial intelligence, and advanced computer science, in particular in perception, path planning and motion planning, mobile manipulation, planning and scheduling, learning and adaptivity, and probabilistic modeling, to name just a few. Solutions to the problems posed by RoboCup@Work require sophisticated and innovative approaches and methods and their effective integration. The scenarios are defined such that the problems are sufficiently general and independent of particular industrial applications, but also sufficiently close to real application problems such that the solutions can be adapted to particular application problems with reasonable effort.

3   Competition Design

The design of the RoboCup@Work competition utilizes proven ideas and concepts from other RoboCup competitions to tackle open research challenges in industrial and service robotics.

3.1   Overview and General Ideas

As stated previously, the new competition should combine the public appeal of robotic soccer, the competition structure of RoboCup@Home, and the benchmarking technology from RoboCup Rescue.

Being appealing to the public is not easy for a competition with industrial relevance. Using small robots instead of large industrial manipulators and keeping the competition arena small helps. Two essential ingredients for public appeal are using easy to understand (*what is the robot supposed to do?*) and easy to observe (*is the robot doing what it is supposed to do?*) tasks and realistic objects, which humans can associate with these tasks.

The competition structure of RoboCup@Home served as a model for structuring the RoboCup@Work competition. The basic element are well-defined tests, which focus on a specific task. The task can be executed by a team within a limited period of time (typically 10 min). Tests may be executed repeatedly, typically on another competition day, which gives teams some room for applying bug fixes and improvements to their code or remedy hardware problems.

For scoring, RoboCup Rescue served as a role model for RoboCup@Work. We define clear objectives and performance criteria for each test and tasks the robots must perform. Scores achieved in all tests are accumulated to determine the overall ranking of teams in the competition. Furthermore, awards are given for special achievements or excellent performance on specific functionalities and tasks (*best-in-class* awards).

3.2   The RoboCup@Work Competition Arena

The environment *initially* designed for RoboCup@Work is depicted in Fig. 2. This design is meant only as a guiding example, as the environment should be *adaptable* within reasonable margins. The size of the arena is a rectangular area no less than 2 m × 4 m and no more than 8 m × 10 m. The competition arena is *partially* surrounded by walls. One or more gates may be foreseen, where robots can enter or leave the arena.

The arena contains designated service areas (tagged D_i or S_i in the figure). Examples include loading and unloading areas, conveyor belts, rotators, storage areas, etc. Service areas may contain specific environment objects, such as racks, shelves, etc. A set of designated places is defined, which are locations in the arena that can be referred to by a unique, symbolic identifier, which are used in task specifications. The arena may foresee the use of obstacles. Obstacles may by passive or active (other robots or humans).

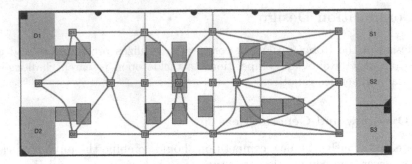

Fig. 2. Conceptual design of the RoboCup@Work competition arena.

3.3 Manipulation Objects and Cavities Used by RoboCup@Work

The manipulation objects used in RoboCup@Work must be objects relevant in industrial applications. An excerpt of the initial set of objects includes screws, nuts, and aluminium profiles as shown in Fig. 3. The objects are standardized DIN and ISO objects used in industry and are available worldwide. The objects pose challenging problems in terms of detection (e.g. relatively small screws) and recognition (e.g. distinguishing between nuts which differ only on the surface structure). In Fig. 4 the corresponding cavities for depositing and fitting are shown.

3.4 BNT: Basic Navigation Test

The first and simplest test in RoboCup@Work is the Basic Navigation Test (BNT). Teams must demonstrate their ability to safely navigate well in their environment, i.e. in a goal-oriented, autonomous, robust, and safe manner, by efficiently visiting a sequence of locations specified in a symbolic manner.

(a) M20 bolt   (b) M20 nut   (c) alu profile   (d) R20 bushing   (e) M30 nut

Fig. 3. Examples of manipulation objects

(a) M20 bolt   (b) M20 nut   (c) alu profile   (d) R20 bushing   (e) M30 nut

Fig. 4. Examples of shapes for depositing/fitting manipulation objects.

Task Description: A single robot is used. The robot is given a sequence of locations it is supposed to visit in exactly the order given by the sequence. When the test starts, the robot must request a task description from the referee box using a given protocol. The task description sent by the referee box is a string containing a sequence of triples. Each triple specifies a place, an orientation, and pause duration. Examples include

BNT<(S6,N,3), (S2,N,3), (D1,S,3), (S5,W,3), (D3,E,3), (D4,S,3)>

and

BNT<(S6,N,3), (S3,W,3), (S7,W,3), (S2,W,3), (D3,E,3)>

The robot has to move to the places specified in the task string, in the order as specified by the string, orient itself according to the orientation given, cover a place marker, pause its movement for the time in seconds as specified by the pause length, and finally leave the arena through the gate.

Scoring: The robot obtains 50 points for each successfully achieved subtask and an additional 150 points for completing the complete task specification. 20 negative points are awarded as penalty every time the robot touches an obstacle or an environment object. If two or more teams achieve the same score, time needed to achieve the equivalent result is used for tie-breaking.

3.5 BMT: Basic Manipulation Test

The second test is the Basic Manipulation Test (BMT). Teams must demonstrate their ability to perceive objects and perform pick and place operations with them. The focus is on the manipulation and on demonstrating safe and robust grasping and placing of objects of different size and shape.

Task Description: A single robot is used. The objective is to move a set of objects from one service area into another. The task specification consists of (i) an initial place, (ii) a source location where the objects will be found, (iii) a (nearby) destination location where the objects will have to put, (iv) a desired spatial configuration of the manipulation objects, to be produced by the robot, and (v) a final place for the robot. Manipulation objects are referenced by symbolic descriptors as defined in the rule book. Only one spatial configuration has been used so far: a line of objects. Two examples for full task specifications are:

BMT<S6, S6, S7, line(V20,R20, F20_20_B), S7>
BMT<S6, S6, S7, line(F20_20_G, F20_20_B, M20_100), S7>

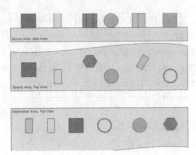

(a) Spatial configurations of objects   (b) Example scenario for BMT

Scoring: The robot obtains 100 points for each successfully picked-up object and another 50 points for putting it in its correct place. An additional 150 points are awarded for completing the complete task specification. If two or more teams achieve the same score, time needed to achieve the equivalent result is used for tie-breaking.

3.6 BTT: Basic Transportation Test

The third test is the Basic Transportation Test (BTT). Teams must demonstrate their ability to combine navigation and manipulation abilities to perform simple transportation tasks.

Task Description: A single robot is used, which is initially positioned outside of the arena near a gate to the arena. The task is to get several objects from the source service areas and to deliver them to the destination service areas. Robots may carry up to three objects simultaneously. The task specification consists of two lists: The first contains for each service area a list of manipulation objects. The second list contains for each destination service area a configuration of objects the robot is supposed to produce. Two example task specifications are:

```
BTT<init(<S6,line(M20_100,F20_20_B)> <S7,line(V20)>);
    goal(<S1,line(M20_100,F20_20_B)> <S2,line(V20)>)>
BTT<init(<S6,line(S40_40_B,F20_20_B)> <S7,line(M20_100)>);
    goal(<S1,line(F20_20_B,F20_20_B)><S2,line(M20_100)>)>
```

The robots can fetch objects from any source area and deliver them to destination areas in any order; they may also decide themselves whether to carry a single object or several ones simultaneously.

Scoring: The robot obtains 100 points for each successfully picked-up object and another 50 points for putting it in its correct place. An additional 150 points are awarded for completing the complete task specification. If two or more teams achieve the same score, time needed to achieve the equivalent result is used for tie-breaking.

3.7   PPT: Precision Placement Test

The Precision Placement Test (PPT) was introduced in the second year of RoboCup@Work. Its purpose is to assess the robots ability to grasp and place objects into object-specific cavities. This demands advanced perception and reasoning abilities, and the capability for precise manipulation of objects.

Fig. 5. Examples of cavities arrangement and environment setup.

Task Description: A single robot is placed by the team in front of the service area which stores the objects to be manipulated. The objective is to pick each object placed in the service area and place it in the corresponding cavity in the special PPT service area (see Fig. 5), which contains five cavity tiles. A PPT task specification consists of a triple which specifies the source and destination locations and a list of objects. An example is: PPT<S1,(M20, S_40_40_G),S2>, where S1 is the location where to pick up the objects M20 and S_40_40_G, and S2 designates the location where the objects need to be placed then in the corresponding cavities.

Scoring: The robot obtains 100 points for each successfully picked-up object and another 100 points for putting it in its correct cavity. A robot gets a penalty of 50 points for putting an object in the wrong cavity. An additional 200 points are awarded for completing the complete task specification. If two or more teams achieve the same score, time needed to achieve the equivalent result is used for tie-breaking.

3.8   CBT: Conveyor Belt Test

The Conveyor Belt Test (CBT) is another challenging test for mobile manipulators. Teams must demonstrate their ability to deal with predictable dynamics in the environment by picking or placing objects from/onto a moving conveyor belt. The test demands fast perception and manipulation skills in order to pick up objects from a operating conveyor belt.

Task Description: A single robot is used, which is placed in some starting position outside of the arena. Initially the conveyor is switched off and three objects are already placed — with a clear distance between 5 cm and 10 cm to each other, at one end of the belt by the referee(s). The task of the robot is to navigate to the location of the conveyor belt and to grasp all three objects from the moving belt before they fall off at the other end of the belt. The robot is supposed to place the grasped objects on a suitable tray on the robot itself.

A CBT task specification simply consists of the specification of the service area where the conveyor belt is found: CBT<srv_area>.

Scoring: The robot obtains 200 points for each successfully picked-up object and another 200 points for grasping all three objects. A robot gets a penalty of 50 points for dropping an object. If two or more teams achieve the same score, time needed to achieve the equivalent result is used for tie-breaking (Fig. 6).

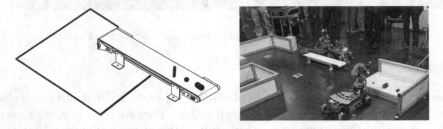

Fig. 6. Conveyor belt design (left) and example environment setup (right).

4   Implementation: Getting the League Off the Ground

Designing a competition is not sufficient to start a new league. It is also important to disseminate the idea and concept, to build a community of interested teams, and to develop a suitable infrastructure for organizing competition events.

Community Building: The idea of RoboCup@Work was initially developed in the context of the EU project BRICS [1], and the people driving the league development effort were mainly Ph.D. students and senior researchers working in BRICS. Two committees were created: a Technical Committee, responsible for designing the competition and developing the rule book, and an Organizing Committee for preparing demonstration events and recruiting teams. The BRICS project organized also a series of research camps between 2010 and 2012, each of which attracted about 20–25 Ph.D. students from around the world. The camps provided a great opportunity to discuss and test the interest of students and to try out some ideas that were later reused in the tests.

In May 2012, the First International RoboCup@Work Camp[1] was organized at Bonn-Rhein-Sieg University (BRSU). The camp attracted 22 participants

---

[1] The camp received financial support from the RoboCup Federation via a Promoting RoboCup project. This is gratefully acknowledged.

from institutes in Turkey, Germany, Austria, and Italy. Furthermore, we demonstrated the concept and organized demonstration competitions at RoboCup German Open in 2012, 2013 and 2014 (Magdeburg), RoboCup World Championships in 2012 (Mexico City) and 2013 (Eindhoven), and at IROS 2012 (Vilamoura).

League Infrastructure Development: Simultaneously with building the community, we developed the tangible league infrastructure: the competition arena, a set of parts suitable for the manipulation tasks, and a referee box for performing and controlling tests.

For 2012, we developed a low-cost, custom-built version of the initial environment as depicted in Fig. 2. In order to increase flexibility, the group at BRSU designed and built a new competition arena for 2013, based on an industrial-quality aluminium profile system (see Fig. 7). This environment allows to quickly change the layout of the arena and has been used in all events since 2013. The cavities used for the Precision Placement Test were designed and 3D-printed at the mechanical workshop of the BRSU Engineering Department.

Fig. 7. The 2013 RoboCup@Work competition arena in the BRSU lab.

Selecting the set of objects for manipulation tasks was more difficult than expected. The objects should be real industrial objects, large enough to be perceivable with current perception technology, sufficiently small and light-weight so that they can be grasped and manipulated by the target robot platforms, like the KUKA youBot.

The actual execution of tests in RoboCup@Work requires a software tool that allows to define task specifications and send them to the robots. The tool also performs time-keeping operations and sends start and stop signals to the robots. An initial version of the referee box has been developed in a project course of the Master of Autonomous Systems course at BRSU, and was subsequently maintained and enhanced by the community.

5   Participation, Performance, and Lessons Learned

Participation and Performance: During the last two years we attracted a solid amount of teams (between 4 and 6) which regularly participate at RoboCup@Work competitions. So far, teams are coming from the following home countries: Netherlands, Germany, UK, Turkey, and Singapore. We are working on attracting more and more teams to make the league more sustainable. In particular, we intend to attract (see the RoboCup@Work event at IROS-2012) teams which have never participated in other RoboCup leagues, but are researching on mobile manipulation. In general, the performance of the teams increased significantly which is reported in the following.

- *Basic Navigation Test:* In 2012 teams moved mainly in differential-mode even though the youBot is an omni-platform and every team used available open-source software to implement navigation capabilities. Apparently it was a good idea to introduce the BNT in order to stress the importance of robust navigation for mobile manipulation. In 2014 the BNT can be considered to be solved also in the presence of dynamic obstacles.
- *Basic Manipulation Test:* From the very beginning we have seen diverse hardware setups effecting manipulation. Even though, so far every team participates with a KUKA youBot the mounting and placement of the manipulator on the base differs. Some teams use the default position and others placed the arm at the front. Each configuration has pros and cons concerning object reachability, field of view, and more. Further, from a perception point of view the setup differs. Some teams work solely with RGB-D cameras mounted on a sensor tower whereas others mount cameras on the arm itself. In 2012 imprecise object grasping and collisions with the environment where the normal case. In 2013 and 2014 we can report significant improvement. In particular, collisions with the environment and false-positive grasping decreased. However, some objects remain difficult to manipulate in certain object positions and teams started to replace the youBot gripper with custom-made grippers in order to grasp also bigger objects.
- *Basic Transportation Test:* In 2012 navigation problems obviously effected also the BTT. Every team scored very low. In 2013 and 2014 test performance improved with very few wrong placements and overall quite fast task execution.
- *Precision Placement Test:* We introduced the test in 2013 and almost no team prepared the test in advance. However, in 2014 we can report an improved performance. Even though, some particular objects remain advanced to be placed precisely (e.g., in the vertical configuration).
- *Conveyor Belt Test:* As for the PPT the CBT remains quite challenging and no teams showed real mobile manipulation (moving base and arm while the object is moving on the belt). However, as object perception and manipulation is getting faster and more reliable as seen in the BMT and BTT we strongly believe that the test will be solved soon.

Lessons Learned. Mobile manipulation remains challenging. The systematic integration of perception, planning, and control as required for the tests introduced in RoboCup@Work calls for research on various levels. In particular, we believe that for a new team the sheer amount of capabilities and competences required to solve the tasks can be overstraining. Hence, we recently started to share code among teams in order to learn from other solutions and also to identify *best practices*. As in other leagues this code releasing will be further established. Further, we would like to link software releases with educational activities such as camps as this will also help new teams to enter the league.

6 Future Directions

The RoboCup@Work competition has developed well so far and is considered in a stable state. Participation is increasing slowly but steadily. Some ideas for the future are expected to further increase interest and participation.

Scenario Evolution. As some of the initial tests are performed more and more robustly by the teams, the league is looking into designing more advanced tests. More interaction with automation equipment (like with the conveyor belt in the CBT test) is on the agenda. Under discussion are scenarios involving multiple robots and human-robot cooperation. Furthermore, we need to make every effort to address industrial relevance of the scenarios. Therefore, we plan to include industrial stakeholders more actively in the scenario design.

Festo Logistics League. Almost simultaneously with RoboCup@Work, Festo Didactics undertook its own initiative for establishing a league with industrial relevance. Festo adopted the concept of a manufacturing logistics scenario and, in several annual iterations, gradually tuned its competition, initially designed for a different purpose, to this idea [6]. We have been seeking cooperation with Festo from the very beginning, but due to strategic and marketing consideration Festo did not want to have their platform directly compete against other robot platforms. Nevertheless, we continue to work towards cooperation between the leagues, possibly by connecting the arenas somehow and developing scenarios where teams must control both RoboTinos, youBots and other robots.

RoCKIn Liaison. A very interesting future direction is the liaison with the EU project RoCKIn [8], where both BRSU and KUKA Labs are partners, together with three other long-time RoboCup researchers. The RoCKIn project has developed its own versions of the RoboCup@Home and RoboCup@Work competitions, with special attention to benchmarking and networked robotics. It is a clear objective that the RoCKIn@Work ideas and concepts will be assimilated by RoboCup@Work in the near future.

EuROC Liaison. Last but not least, we would like to point to the EuROC European Robotics Challenge [3], an EU-funded project started in 2014. EuROC defines three challenges, two of which have picked up ideas proposed by RoboCup@Work: the Reconfigurable Interactive Manufacturing Cell (RIMC) challenge and the Shop Floor Logistics and Manipulation (SFLM) challenge. To some extent, this effort can already be viewed as an impact of the RoboCup@Work effort.

7   Conclusion

This paper presents the ideas and concepts behind the RoboCup@Work Demonstration League. We have discussed, motivation, research challenges, competition design and implementation in detail and reported about the development of the league, lessons learned, and future directions. The conclusion is that RoboCup@Work is a timely and well-defined competition with high industrial relevance, and a very interesting extension of the spectrum of RoboCup leagues.

References

1. Bischoff, R., Guhl, T., Prassler, E., Nowak, W., Kraetzschmar, G., Bruyninckx, H., Soetens, P., Haegele, M., Pott, A., Breedveld, P., Broenink, J., Brugali, D., Tomatis, N.: BRICS - best practice in robotics. In: 41st International Symposium on Robotics (ISR) and 6th German Conference on Robotics (ROBOTIK), pp. 1–8, June 2010
2. Bischoff, R., Huggenberger, U., Prassler, E.: Kuka youbot - a mobile manipulator for research and education. In: IEEE International Conference on Robotics and Automation (2011)
3. EuRoC Project: European robotics challenges (2014). http://www.euroc-project.eu/
4. Holz, D., Iocchi, L., van der Zant, T.: Benchmarking intelligent service robots through scientific competitions: the RoboCup@Home approach. In: Proceedings of the AAAI Spring Symposium Designing Intelligent Robots: Reintegrating AI II (2013)
5. Kitano, H., Tadokoro, S.: RoboCup rescue - a grand challenge for multiagent and intelligent systems. AI Mag. **22**, 39–52 (2001)
6. Niemueller, T., Ewert, D., Reuter, S., Ferrein, A., Jeschke, S., Lakemeyer, G.: RoboCup logistics league sponsored by Festo: a competitive factory automation testbed. In: Behnke, S., Veloso, M., Visser, A., Xiong, R. (eds.) RoboCup 2013. LNCS, vol. 8371, pp. 336–347. Springer, Heidelberg (2014)
7. RoboCup@Work: RoboCup@Work (2014). http://www.robocupatwork.org
8. RoCKIn: Rockin (2014). http://www.rockinrobotchallenge.eu/
9. Sheh, R., Kimura, T., Mihankhah, E., Pellenz, J., Schwertfeger, S., Suthakorn, J.: The RoboCupRescue robot league: guiding robots towards fieldable capabilities. In: IEEE Workshop on Advanced Robotics and its Social Impacts (ARSO), pp. 31–34, Oct 2011
10. TAPAS Project: Robotics-enabled logistics and assistive services for the transformable factory of the future (tapas) (2013). http://www.tapas-project.eu/

Simulation Leagues: Analysis of Competition Formats

David Budden[1]([⊠]), Peter Wang[2], Oliver Obst[2], and Mikhail Prokopenko[2]

[1] The University of Melbourne, Parkville, VIC 3010, Australia
david.budden@unimelb.edu.au
[2] Statistical Learning, CSIRO Computational Informatics,
Epping, NSW 1710, Australia
{peter.wang,oliver.obst,mikhail.prokopenko}@csiro.au

Abstract. The selection of an appropriate competition format is critical for both the success and credibility of any competition, both real and simulated. In this paper, the automated parallelism offered by the RoboCupSoccer 2D simulation league is leveraged to conduct a 28,000 game round-robin between the top 8 teams from RoboCup 2012 and 2013. A proposed new competition format is found to reduce variation from the resultant statistically significant team performance rankings by 75 % and 67 %, when compared to the actual competition results from RoboCup 2012 and 2013 respectively. These results are statistically validated by generating 10,000 random tournaments for each of the three considered formats and comparing the respective distributions of ranking discrepancy.

1 Introduction

1.1 The RoboCup Humanoid Challenge

RoboCup (the "World Cup"of robot soccer) was first proposed in 1997 as a standard problem for the evaluation of theories, algorithms and architectures in areas including artificial intelligence (AI), robotics and computer vision [1]. This proposal followed the observation that traditional AI problems were increasingly unable to meet these requirements and that a new challenge was necessary to initiate the development of next-generation technologies.

The overarching RoboCup goal of developing a team of humanoid robots capable of defeating the FIFA World Cup champion team, coined the "Millennium Challenge", has proven a major factor in driving research in AI and related areas for over a decade, with a search for the term "RoboCup"in a major literature database yielding over 23,000 results. Since 1997, researchers and competitors have decomposed this ambitious pursuit into two complementary categories [2]:

- **Physical robot league:** Using physical robots to play soccer games. This category now contains many different leagues for both wheeled robots (small-sized [3] and mid-sized leagues [4]) and humanoids (standard platform league [5]

R.A.C. Bianchi et al. (Eds.): RoboCup 2014, LNAI 8992, pp. 183–194, 2015.
DOI: 10.1007/978-3-319-18615-3_15

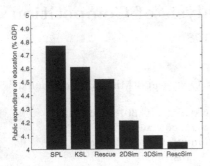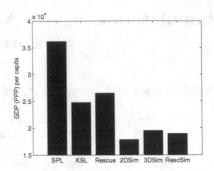

Fig. 1. PEoE (public expenditure on education as a percentage of GDP [23]) and GDP/cap (gross domestic product at purchasing power parity per capita [24]) for the home country of each participating RoboCup 2013 team, averaged over each of the six largest RoboCup leagues. Each of the three major simulation leagues (2D, 3D and rescue) exhibit significantly lower values than those requiring the purchase or development of physical robots.

and humanoid league [6]), with each focusing on different aspects of physical robot design [7], motor control and bipedal locomotion [8,9], real-time localisation [10,11] and computer vision [12–14].
- **Software agent league:** Using software or synthetic agents to play soccer games on an official soccer server over a network. This category contains both 2D [15–17] and 3D [18] simulation leagues.

The annual RoboCup competition, which attracted 2,500 participants and 40,000 spectators from 40 countries in 2013 [19], now exhibits a number of non-soccer competitions. The oldest and largest of these, RoboCupRescue, is also separated into physical and simulation leagues [20,21].

1.2   Significance of Simulation Leagues

The RoboCupSoccer simulation leagues traditionally involve the largest number of international participating teams, reaching 40 in 2013 [22]. The ability to simulate soccer matches without physical robots abstracts away low-level issues such as image processing and motor breakages, allowing teams to focus on the development of complex team behaviours and strategies for a larger number of autonomous agents. The remainder of this section expands upon some specific contributions of the RoboCupSoccer simulation leagues toward the "Millennium Challenge".

Financial Inclusiveness of Competing Nations.   The physical robots required by non-simulation leagues remain particularly expensive. By removing these costs and those associated with robot repairs and transportation, the simulation leagues allow institutes with access to less funding to actively contribute to and participate in the RoboCup initiative. To quantify this claim, Fig. 1 presents the PEoE (public expenditure on education as a percentage of GDP [23]) and GDP/cap (gross domestic product at purchasing power parity per capita [24]) for the home country of each participating RoboCup 2013 team, averaged over each of the six largest RoboCup leagues. The countries participating in the standard platform league, which requires teams to field five Aldebaran Nao humanoid robots, have the highest average PEoE and GDP/cap of any league considered. The kid-sized humanoid and rescue leagues, each of which require the purchase or construction of physical robots, also involve countries with a high average PEoE and GDP/cap. Each of the three major simulation leagues (2D, 3D and rescue) exhibit significantly lower values, suggesting that the inclusion of simulation leagues supports financial inclusiveness within the competition.

Statistically Significant Analyses by Automated Competition Parallelism. The automation of multiple parallel games makes RoboCupSoccer simulation leagues ideal platforms for analysing the complexities of complex team behaviours. Most team games and sports (both real and virtual) are characterised by rich, dynamic interactions that influence the contest outcome. As described by Vilar *et al.*, "quantitative analysis is increasingly being used in team sports to better understand performance in these stylized, delineated, complex social systems" [25]. Early examples of such quantitative analysis include *sabermetrics*, which attempts to "search for objective knowledge about baseball" by considering statistics of in-game activity [26]. A recent study by Fewell *et al.* involved the analysis of basketball games as networks, with properties including degree centrality, clustering, entropy and flow centrality calculating from measurements of ball position throughout the game [27]. This idea was extended by Vilar *et al.*, who considered the local dynamics of collective team behaviour to quantify how teams occupy sub-areas of the field as a function of ball position [25]. Recently, Cliff *et al.* presented several information-theoretic methods of quantifying dynamic interactions in football games, using the RoboCupSoccer 2D simulation league as an experimental platform [28].

The ability to automate thousands of simulation league games allows for the analysis of competition formats to determine which best approximate the true performance rankings of competing teams. The selection of an appropriate competition format is critical for both the success and credibility of any competition. Unfortunately, this choice is not straightforward: The format must minimise randomness relative to the true performance ranking of teams while keeping the number of games to a minimum, both to satisfy time constraints and retain the interest of participants and spectators alike. Furthermore, maintaining competition interest introduces a number of constraints to competition formats: As an example, multiple games between the same two opponents (the obvious method of achieving a statistically significant ranking) should be avoided.

The remainder of this paper quantifies the appropriateness of different tournament formats (a major consideration in many human sports) by determining the statistically significant performance rankings of 2012 and 2013 RoboCup-Soccer 2D simulation teams. A new competition format is then proposed and verified by leveraging the automated parallelism facilitated by the 2D simulation league platform. In addition to demonstrating the utility of simulation leagues for statistical analysis of team sport outcomes given some system perturbation, it is anticipated that the adoption of the proposed format would improve the success and credibility of the RoboCupSoccer simulation leagues in future years.

2 Previous Competition Formats

The following two competition formats were adopted by the RoboCupSoccer 2D simulation league in 2012 and 2013:

- In 2012, a total of 20 games were played to determine the final rank of the top 8 teams. Specifically, the top 4 teams played 6 games each (3 quarterfinal round-robin, 2 semifinal and 1 final/third place playoff), and the bottom 4 teams player 4 games each.
- In 2013, a *double-elimination* system was adopted, where a team ceases to be eligible to place first upon having lost 2 games [29,30]. A total of 16 games were played to determine the final rank of the top 8 teams. Specifically, 14 games were played in the double-elimination format (i.e. $2n - 2, n = 8$) in addition to 2 classification games.

Previously, it has been unclear whether this change in competition format improves the fairness and reproducibility of the final team rankings. In general, lack of reproducibility is due to non-transitivity of team performance (a well-known phenomena that occurs frequently in actual human team sports). This may be addressed by a round-robin competition (where all 28 possible pairs of teams play against one another), yet it is also unclear whether this increase in the number of games is guaranteed to improve ranking stability.

3 Methods of Ranking Team Performance

Before evaluating different competition formats, it is necessary to establish a fair (i.e. statistically significant) ranking of the top 8 RoboCupSoccer 2D simulation league teams for 2012 and 2013. This was accomplished by conducting an 8-team round-robin for both years, where all 28 pairs of teams play approximately 1000 games against one another. In addition, two different schemes were considered for point calculation:

- **Continuous scheme:** Teams are ranked by sum of average points obtained against each opponent across all 1000 games.
- **Discrete scheme:** Firstly, the average score between each pair of teams (across all 1000 games) is rounded to the nearest integer (e.g. "1.9 : 1.2" is rounded to "2 : 1"). Next, points are allocated for each pairing based on these rounded results: 3 for a win, 1 for a draw and 0 for a loss. Teams are then ranked by sum of these points received against each opponent.

The final rankings generated for 2012 and 2013 RoboCupSoccer 2D simulation league teams under these two schemes are presented in Sect. 5.1. Finally, in order to formally capture the overall difference between two rankings $\mathbf{r^a}$ and $\mathbf{r^b}$, the L_1 distance is utilised:

$$d_1(\mathbf{r^a}, \mathbf{r^b}) = \|\mathbf{r^a} - \mathbf{r^b}\|_1 = \sum_{i=1}^{n} |r_i^a - r_i^b|, \tag{1}$$

where i is the index of the i-th team in each ranking, $1 \leq i \leq 8$. The difference between rankings for different competition formats are presented in Sect. 5.2.

4   Proposed Competition Format

Section 3 describes two schemes under which statistically significant rankings of RoboCupSoccer 2D simulation league teams can be achieved. However, it remains unclear whether the previously adopted competition formats are able to replicate these rankings with minimal noise for considerably fewer games, or whether a new format may achieve improved results in this regard. One possible format involves the following two steps:

– Firstly, a preliminary round-robin is conducted where 1 game is played for all 28 pairs of teams.
– Following the rankings obtained in the previous step, 4 classification games are played: The final between the top 2 teams and playoffs between third and fourth, fifth and sixth, and seventh and eighth places. It is possibly to use the best-of-three format for each of these classification games.

The 32 games required involved in this competition format could still fit readily in a 1–2 day time frame, particularly with 2 games running simultaneously as per RoboCup 2013.

5   Results

5.1   Statistically Significant Rankings Versus Previous Competition Formats

Following iterated round-robin and two point calculation schemes described in Sect. 3, statistically significant rankings were generated for the top 8 RoboCupSoccer 2D simulation league teams for 2012 and 2013. These results are presented below.

RoboCup 2012 Results. The final round-robin results of the top 8 teams for RoboCup 2012 are presented in Tables 1 and 2, for the continuous and discrete scoring schemes described in Sect. 3 respectively. Results are ordered according to actual performance at RoboCup 2012, $\mathbf{r^a}$.

Table 1. Round-robin results (average goals scored) for the top 8 teams from RoboCup 2012, ordered according to their final competition rank, $\mathbf{r^a}$. The final points for each team were determined by summing the average points scored against each opponent over approximately 1000 games, resulting in the round-robin with continuous point allocation scheme ranking, $\mathbf{r^c}$.

$\mathbf{r^a}$	Team	Helios	Wright	Marlik	Gliders	GDUT	AUT	Yushan	RobOTTO	Points	Goal Diff	Rank, $\mathbf{r^c}$
1	Helios		1.397	2.442	2.517	2.948	2.970	2.880	2.998	18.152	+ 26.0	2
2	Wright	1.406		2.792	2.835	2.900	2.998	2.970	2.998	18.899	+ 38.7	1
3	Marlik	0.309	0.129		1.147	2.121	2.804	0.874	2.615	9.999	+ 0.3	5
4	Gliders	0.261	0.102	1.396		1.809	2.957	0.903	2.863	10.291	+ 3.4	4
5	GDUT	0.029	0.074	0.633	0.960		2.955	0.552	2.597	7.800	- 6.0	6
6	AUT	0.007	0.001	0.107	0.026	0.024		0.003	0.209	0.377	- 39.3	8
7	Yushan	0.084	0.021	1.822	1.875	2.316	2.994		2.993	12.105	+ 6.5	3
8	RobOTTO	0.001	0.001	0.233	0.087	0.228	2.418	0.005		2.973	- 29.6	7

Table 2. Round-robin results (average goals scored and discretised points allocated) for the top 8 teams from RoboCup 2012, ordered according to their final competition rank, $\mathbf{r^a}$. Discretised points are determined by calculating the average number of goals scored over approximately 1000 games rounded to the nearest integer, then awarding 3 points for a win, 1 point for a draw and 0 points for a loss. The resultant round-robin with discrete point allocation scheme ranking, $\mathbf{r^d}$, is equivalent to that generated under the continuous scheme.

	Helios	Wright	Marlik	Gliders	GDUT	AUT	Yushan	RobOTTO	Goals	Points	$\mathbf{r^d}$
Helios		2.3 : 2.3	1.4 : 0.1	1.6 : 0.1	4.4 : 0.2	7.7 : 0.0	4.5 : 0.7	7.6 : 0.1	29.5 : 3.5	19	2
Wright	2.3 : 2.3		3.2 : 0.3	3.3 : 0.2	5.8 : 1.2	12.1 : 0.1	7.2 : 1.0	10.1 : 0.2	44.0 : 5.3	19	1
Marlik	0.1 : 1.4	0.3 : 3.2		0.46 : 0.56	1.4 : 0.4	2.3 : 0.1	0.7 : 1.2	2.1 : 0.2	7.4 : 7.1	10	5
Gliders	0.1 : 1.6	0.2 : 3.3	0.56 : 0.46		1.9 : 1.2	4.3 : 0.1	1.4 : 2.2	4.6 : 0.8	13.1 : 9.7	12	4
GDUT	0.2 : 4.4	1.2 : 5.8	0.4 : 1.4	1.2 : 1.9		4.0 : 0.2	2.0 : 3.9	3.4 : 0.8	12.4 : 18.4	6	6
AUT	0.0 : 7.7	0.1 : 12.1	0.1 : 2.3	0.1 : 4.3	0.2 : 4.0		0.1 : 7.1	0.7 : 3.1	1.3 : 40.6	0	8
Yushan	0.7 : 4.5	1.0 : 7.2	1.2 : 0.7	2.2 : 1.4	3.9 : 2.0	7.1 : 0.1		6.7 : 0.4	22.8 : 16.3	13	3
RobOTTO	0.1 : 7.6	0.2 : 10.1	0.2 : 2.1	0.8 : 4.6	0.8 : 3.4	3.1 : 0.7	0.4 : 6.7		5.6 : 35.2	3	7

Table 2 presents the continuous (non-rounded) scores averaged across the approximately 1000 games for each pair in the round-robin, in addition to the points allocated according to the discretisation scheme (3 for a win, 1 for a draw and 0 for a loss). The tie-breaker is the rounded goal difference (not shown), which was used only to separate first place (WrightEagle, +39 points) from second (Helios, +26 points). The final ranking corresponds exactly with that generated under the continuous scheme, as presented in Table 1.

Despite the agreement between continuous and discrete scoring schemes, it is obvious that this ranking (generated from the results of approximately 28,000 games) disagrees significantly from the actual RoboCup 2012 results. This can be quantified using the distance metric defined in (1):

$$d_1(\mathbf{r^a}, \mathbf{r^c})_{2012} = |1-2| + |2-1| + |3-5| + |4-4| + |5-6| + |6-8| + |7-3| + |8-7| = 12,$$

Table 3. Round-robin results (average goals scored) for the top 8 teams from RoboCup 2013, ordered according to their final competition rank, r^a. The final points for each team were determined by summing the average points scored against each opponent over approximately 1000 games, resulting in the round-robin with continuous point allocation scheme ranking, r^c.

r^a	Team	Wright	Helios	Yushan	Axiom	Gliders	Oxsy	AUT	Cyrus	Points	Goal Diff	Rank, r^c
1	Wright		1.877	2.470	2.880	2.397	2.901	2.991	2.792	18.308	+ 22.5	1
2	Helios	0.883		2.841	2.940	2.194	2.343	2.969	2.767	16.937	+ 14.9	2
3	Yushan	0.406	0.093		2.506	1.892	1.557	2.059	0.921	9.434	- 1.3	4
4	Axiom	0.072	0.042	0.367		0.590	0.395	1.224	1.023	3.713	- 14.5	8
5	Gliders	0.437	0.490	0.884	2.249		1.612	1.871	0.828	8.371	- 2.0	6
6	Oxsy	0.065	0.385	1.159	2.437	1.105		2.225	2.167	9.543	- 2.2	3
7	AUT	0.006	0.017	0.718	1.491	0.878	0.575		0.731	4.416	- 14.0	7
8	Cyrus	0.137	0.136	1.791	1.740	1.926	0.632	2.046		8.408	- 3.4	5

Table 4. Round-robin results (average goals scored and discretised points allocated) for the top 8 teams from RoboCup 2012, ordered according to their final competition rank, r^a. Discretised points are determined by calculating the average number of goals scored over approximately 1000 games rounded to the nearest integer, then awarding 3 points for a win, 1 point for a draw and 0 points for a loss. The tie-breaker is the total of rounded goal differences (not shown). The resultant round-robin with discrete point allocation scheme ranking, r^d, is slightly different to that generated under the continuous scheme.

	Wright	Helios	Yushan	Axiom	Gliders	Oxsy	AUT	Cyrus	Goals	Points	r^d
Wright		1.9 : 1.2	2.8 : 0.9	4.9 : 0.3	2.5 : 0.7	5.4 : 0.8	6.4 : 0.3	3.4 : 0.6	27.3 : 4.8	21	1
Helios	1.2 : 1.9		2.8 : 0.2	4.1 : 0.2	1.2 : 0.2	2.2 : 0.4	4.1 : 0.1	2.5 : 0.2	18.1 : 3.2	18	2
Yushan	0.9 : 2.8	0.2 : 2.8		2.7 : 0.8	1.8 : 1.1	1.4 : 1.2	1.7 : 0.8	0.9 : 1.4	9.6 : 10.9	11	3
Axiom	0.3 : 4.9	0.2 : 4.1	0.8 : 2.7		1.0 : 2.3	0.7 : 2.7	0.9 : 1.1	1.4 : 2.0	5.3 : 19.8	1	8
Gliders	0.7 : 2.5	0.2 : 1.2	1.1 : 1.8	2.3 : 1.0		1.3 : 1.0	1.8 : 1.1	0.9 : 1.7	8.3 : 10.3	7	6
Oxsy	0.8 : 5.4	0.4 : 2.2	1.2 : 1.4	2.7 : 0.7	1.0 : 1.3		2.3 : 0.8	2.2 : 1.0	10.6 : 12.8	11	4
AUT	0.3 : 6.4	0.1 : 4.1	0.8 : 1.7	1.1 : 0.9	1.1 : 1.8	0.8 : 2.3		0.8 : 1.8	5.0 : 19.0	1	7
Cyrus	0.6 : 3.4	0.2 : 2.5	1.4 : 0.9	2.0 : 1.4	1.7 : 0.9	1.0 : 2.2	1.8 : 0.8		8.7 : 12.1	10	5

where r^a represents the actual RoboCup 2012 rankings and r^c represents the ranking generated under continuous scoring scheme round-robin. This large difference suggests that the 2012 competition format did not succeed in capturing the true team performance ranking.

RoboCup 2013 Results. The final round-robin results of the top 8 teams for RoboCup 2013 are presented in Tables 3 and 4, for the continuous and discrete scoring schemes described in Sect. 3 respectively. Results are ordered according to actual performance at RoboCup 2013, r^a, and presented in the same format as Tables 1 and 2 for RoboCup 2012.

Unlike RoboCup 2012, there is a slight disagreement between the rankings generated using continuous and discrete scoring schemes, with a swap between third

Table 5. Combined actual and average results for the top 8 teams from RoboCup 2012, ordered according to their final competition rank. Each goal difference represents the actual (integer) game results from RoboCup 2012 where possible. As this previous format does not necessarily require all pairs of teams to play against one another, some of these results are not available: In these cases, the average (continuous-valued) scores from Table 2 were utilised. Using these results, it is possible to infer the final ranking, r^p, for RoboCup 2012 under the competition format proposed in Sect. 4.

	Helios	Wright	Marlik	Gliders	GDUT	AUT	Yushan	RobOTTO	Points	Rank	r^P
Helios		4 : 1	4 : 0	1 : 0	4.4 : 0.2	1 : 0	4.5 : 0.7	2 : 0	21	1	1
Wright	1 : 4		2 : 1	2 : 0	5 : 1	12.1 : 0.1	6 : 1	10.1 : 0.2	18	2	2
Marlik	0 : 4	1 : 2		1 : 0	1 : 0	2.3 : 0.1	1 : 1	2.1 : 0.2	13	3	4
Gliders	0 : 1	0 : 2	0 : 1		1.9 : 1.2	2 : 0	1.4 : 2.2	3 : 0	9	5	5
GDUT	0.2 : 4.4	1 : 5	0 : 1	1.2 : 1.9		1 : 0	3 : 2	3.4 : 0.8	9	6	6
AUT	0 : 1	0.1 : 12.1	0.1 : 2.3	0 : 2	0 : 1		0.1 : 7.1	1 : 0	3	7	8
Yushan	0.7 : 4.5	1 : 6	1 : 1	2.2 : 1.4	2 : 3	7.1 : 0.1		3 : 1	10	4	3
RobOTTO	0 : 2	0.2 : 10.1	0.2 : 2.1	0 : 3	0.8 : 3.4	0 : 1	1 : 3		0	8	7

and fourth teams. Again using the distance metric defined in (1), the difference between these rankings and the actual RoboCup 2013 results can be quantified:

$$d_1(\mathbf{r^a}, \mathbf{r^c})_{2013} = |1-1|+|2-2|+|3-4|+|4-8|+|5-6|+|6-3|+|7-7|+|8-5| = 12,$$

$$d_1(\mathbf{r^a}, \mathbf{r^d})_{2013} = |1-1|+|2-2|+|3-3|+|4-8|+|5-6|+|6-4|+|7-7|+|8-5| = 10,$$

where $\mathbf{r^a}$ represents the actual RoboCup 2013 rankings, while $\mathbf{r^c}$ and $\mathbf{r^d}$ represent the ranking generated under continuous and discrete scoring schemes of round-robins respectively. It is evident that the 2013 double-elimination format yielded as much overall divergence as the 2012 single-elimination format, but with slightly fewer individual discrepancies. It is also clear that, given very small points differences between adjacent teams, it may be necessary to play classification games even after a statistically significant round-robin. It is therefore proposed that the format described in Sect. 4 should improve reliability of the competition outcomes.

5.2 Evaluation of Proposed Competition Formats

In order to evaluate the proposed competition format described in Sect. 4, the actual game results from RoboCup 2012 and 2013 were used where possible. As these previous formats do not necessarily require all pairs of teams to play against one another, some of these results are not available: In these cases, the average scores from Tables 2 and 4 were utilised for RoboCup 2012 and 2013 respectively.

Using these results, it is possible to infer final rankings for RoboCup 2012 and 2013 under the proposed competition format. These results are presented below.

Table 6. Combined actual and average results for the top 8 teams from RoboCup 2013, ordered according to their final competition rank. Each goal difference represents the actual (integer) game results from RoboCup 2013 where possible. As this previous format does not necessarily require all pairs of teams to play against one another, some of these results are not available: In these cases, the average (continuous-valued) scores from Table 4 were utilised. Using these results, it is possible to infer the final ranking, $\mathbf{r^P}$, for RoboCup 2013 under the competition format proposed in Sect. 4.

	Wright	Helios	Yushan	Axiom	Gliders	Oxsy	AUT	Cyrus	Points	Rank	$\mathbf{r^P}$
Wright		3 : 1	2.8 : 0.9	4.9 : 0.3	2.5 : 0.7	5 : 3	6.4 : 0.3	7 : 0	21	1	1
Helios	1 : 3		2 : 0	4.1 : 0.2	2 : 0	2.2 : 0.4	4.1 : 0.1	4 : 0	18	2	2
Yushan	0.9 : 2.8	0 : 2		4 : 1	0 : 1	3 : 0	1.7 : 0.8	2 : 0	12	3	3
Axiom	0.3 : 4.9	0.2 : 4.1	1 : 4		3 : 3	1 : 6	2 : 1	1.4 : 2.0	4	7	8
Gliders	0.7 : 2.5	0 : 2	1 : 0	3 : 3		4 : 0	1.8 : 1.1	0.9 : 1.7	10	4	4
Oxsy	3 : 5	0.4 : 2.2	0 : 3	6 : 1	0 : 4		2.3 : 0.8	2.2 : 1.0	9	5	5
AUT	0 : 7	0.1 : 4.1	0.8 : 1.7	1 : 2	1.1 : 1.8	0.8 : 2.3		3 : 1	3	8	7
Cyrus	0.6 : 3.4	0 : 4	0 : 2	2.0 : 1.4	1.7 : 0.9	1.0 : 2.2	1 : 3		6	6	6

RoboCup 2012 Results. The combined actual and average results of top 8 teams from RoboCup 2012 are presented in Table 5, in addition to the inferred final ranking, $\mathbf{r^P}$, for RoboCup 2012 under the competition format proposed in Sect. 4. Using the distance metric defined in (1), the difference between $\mathbf{r^P}$ and the ranking generated from the 28,000 game round-robin, $\mathbf{r^c}$, can be quantified:

$$d_1(\mathbf{r^P}, \mathbf{r^c})_{2012} = |1-2|+|2-1|+|4-5|+|5-4|+|6-6|+|8-8|+|3-3|+|7-7| = 4.$$

This is a considerably smaller difference than the 12 produced under the actual RoboCup 2012 format, suggesting that the proposed format better captures the true team performance ranking. Furthermore, this result is achieved using a majority of actual game results (i.e. 18 from 28 pairs, with only 10 using the averages from Table 2).

RoboCup 2013 Results. The combined actual and average results of top 8 teams from RoboCup 2013 are presented in Table 6, in addition to the inferred final ranking, $\mathbf{r^P}$, for RoboCup 2013 under the competition format proposed in Sect. 4. Again using the distance metric defined in (1), the difference between $\mathbf{r^P}$ and the ranking generated from the 28,000 game round-robin, $\mathbf{r^c}$ or $\mathbf{r^d}$, can be quantified:

$$d_1(\mathbf{r^P}, \mathbf{r^c})_{2013} = |1-1|+|2-2|+|3-4|+|8-8|+|4-6|+|5-3|+|7-7|+|6-5| = 6$$

$$d_1(\mathbf{r^P}, \mathbf{r^d})_{2013} = |1-1|+|2-2|+|3-3|+|8-8|+|4-0|+|5-4|+|7-7|+|6-5| = 4$$

Similarly to the results for 2012, these are considerably smaller differences than the 12 (or 10) produced under the actual RoboCup 2013 format, providing further evidence that the proposed format better captures the true team performance ranking. Again, this result is achieved using a majority of actual game results (i.e. 15 from 28 pairs, with only 13 using the averages from Table 4).

Statistical Validation. To statistically validate that the proposed competition format is significantly more appropriate than those adopted at RoboCup 2012 and 2013, 10,000 tournaments were generated for each format by randomly sampling game results from the 28,000 game round-robin. For each tournament, the L_1 distance, $d_1(\mathbf{r^a}, \mathbf{r^b})$ (see Eq. (1)), was calculated to capture the discrepancy between the tournament and true team rankings. These results are presented in Fig. 2 for the top 8 teams from RoboCup 2012 (a) and 2013 (b). In both cases, it is evident that the proposed format yields more statistically robust rankings (i.e. smaller L_1 distance) than the formats adopted in RoboCup 2012 (second best) and 2013 (worst).

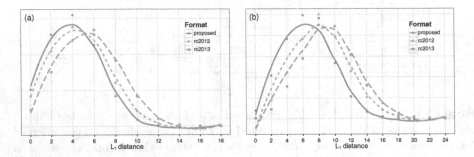

Fig. 2. Discrepancy between tournament and true team rankings, captured as an L_1 distance (see Eq. (1)), for 10,000 randomly-generated tournaments structured according to the three considered formats. It is evident that the proposed format (red) yields more statistically robust rankings (i.e. smaller L_1 distance) than the formats adopted in RoboCup 2012 (second best, green) and 2013 (worst, blue), considering the top 8 teams from both RoboCup 2012 (a) and 2013 (b) (Color figure online).

6   Conclusions

The selection of an appropriate competition format is critical for both the success and credibility of any competition. This is particularly true in the RoboCupSoccer 2D simulation league, which provides an ideal computational platform for examining different formats by facilitating automated parallel execution of a statistically significant number of games.

A 28,000 game round-robin competition was conducted between the top 8 2D simulation league teams from both RoboCup 2012 and 2013. The difference between the resultant rankings was calculated relative to the actual results of RoboCup 2012 and 2013 (12 and 12 respectively) and compared to those that would have resulted under a proposed new format (4 and 6 respectively). This suggests a significant reduction in randomness relative to true team performance rankings while only requiring the number of games to be increased to 32; a number that would still fit readily in a 1–2 day time frame, particularly utilising the round-robin parallelism enabled by the stable 2D simulation platform.

The RoboCup "Millennium Challenge" requires robots to exhibit both physical and strategic prowess, necessitating the decomposition of the larger problem into both physical robot and simulation leagues. Although often overlooked, the simulation leagues contribute significantly to this goal, both through improving financial inclusiveness of competing nations and providing a stable platform for statistically significant analysis of team behaviour and competition format. In addition to highlighting the latter of these contributions, it is anticipated that the introduction of the proposed new format will improve the reliability of final competition rankings and consequently success and credibility of the RoboCupSoccer simulation leagues in future years.

References

1. Kitano, H., Asada, M., Kuniyoshi, Y., Noda, I., Osawa, E.: RoboCup: the robot world cup initiative. In: Proceedings of the First International Conference On Autonomous Agents, pp. 340–347. ACM (1997)
2. Kitano, H., Asada, M.: The robocup humanoid challenge as the millennium challenge for advanced robotics. Adv. Robot. **13**(8), 723–736 (1998)
3. RoboCup Technical Committee: Laws of the RoboCup Small Size League (2013). http://robocupssl.cpe.ku.ac.th/_media/rules:ssl-rules-2013-2.pdf
4. RoboCup Technical Committee: Middle size robot league rules and regulations for (2013). http://wiki.robocup.org/images/9/98/Msl_rules_2013.pdf
5. RoboCup Technical Committee: RoboCup Standard Platform League (NAO) rule book (2013). http://www.tzi.de/spl/pub/Website/Downloads/Rules2013.pdf
6. RoboCup Technical Committee: RoboCup soccer humanoid league rules and setup (2013). http://www.tzi.de/humanoid/pub/Website/Downloads/Humanoid LeagueRules2013-05-28.pdf
7. Ha, I., Tamura, Y., Asama, H., Han, J., Hong, D.W.: Development of open humanoid platform DARwIn-OP. In: Proceedings of SICE Annual Conference (SICE) 2011, pp. 2178–2181. IEEE (2011)
8. Fountain, J., Walker, J., Budden, D., Mendes, A., Chalup, S.K.: Motivated reinforcement learning for improved head actuation of humanoid robots. In: Behnke, S., Veloso, M., Visser, A., Xiong, R. (eds.) RoboCup 2013. LNCS, vol. 8371, pp. 268–279. Springer, Heidelberg (2014)
9. Budden, D., Walker, J., Flannery, M., Mendes, A.: Probabilistic gradient ascent with applications to bipedal robot locomotion. In: Australasian Conference on Robotics and Automation (ACRA). (2013)
10. Annable, B., Budden, D., Mendes, A.: NUbugger: A visual real-time robot debugging system. In: Behnke, S., Veloso, M., Visser, A., Xiong, R. (eds.) RoboCup 2013. LNCS, vol. 8371, pp. 544–551. Springer, Heidelberg (2014)
11. Budden, D., Prokopenko, M.: Improved particle filtering for pseudo-uniform belief distributions in robot localisation. In: Behnke, S., Veloso, M., Visser, A., Xiong, R. (eds.) RoboCup 2013. LNCS, vol. 8371, pp. 385–395. Springer, Heidelberg (2014)
12. Budden, D., Fenn, S., Walker, J., Mendes, A.: A novel approach to ball detection for humanoid robot soccer. In: Thielscher, M., Zhang, D. (eds.) AI 2012. LNCS, vol. 7691, pp. 827–838. Springer, Heidelberg (2012)
13. Budden, D., Fenn, S., Mendes, A., Chalup, S.: Evaluation of colour models for computer vision using cluster validation techniques. In: Chen, X., Stone, P., Sucar, L.E.,

van der Zant, T. (eds.) RoboCup 2012. LNCS, vol. 7500, pp. 261–272. Springer, Heidelberg (2013)

14. Budden, D., Mendes, A.: Unsupervised recognition of salient colour for real-time image processing. In: Behnke, S., Veloso, M., Visser, A., Xiong, R. (eds.) RoboCup 2013. LNCS, vol. 8371, pp. 373–384. Springer, Heidelberg (2014)

15. Chen, M., Foroughi, E., Heintz, F., Huang, Z., Kapetanakis, S., Kostiadis, K., Kummeneje, J., Noda, I., Obst, O., Riley, P., Steffens, T., Wang, Y., Yin, X.: RoboCup Soccer Server. http://wwfc.cs.virginia.edu/documentation/manual.pdf

16. Prokopenko, M., Obst, O., Wang, P., Budden, D., Cliff, O.: Gliders 2013: Tactical analysis with information dynamics. In: RoboCup 2013 Symposium and Competitions: Team Description Papers, Eindhoven, The Netherlands, June 2013. (2013)

17. Prokopenko, M., Wang, P., Obst, O.: Gliders 2014: Dynamic tactics with voronoi diagrams. In: Robocup 2014 Symposium And Competitions: Team Description Papers, Joo Pessoa, Brazil, July 2014. (2014)

18. RoboCup Technical Committee: RoboCup Soccer Simulation League 3D competition rules and setup (2013). http://homepages.herts.ac.uk/sv08aav/RCSoccerSim3DRules2013.1.pdf

19. Butler, K.: RoboCup 2013: Humanoid robots play soccer for world title (2013). http://www.upi.com/Science_News/Blog/2013/07/01/RoboCup-2013-Humanoid-robots-play-soccer-for-world-title/9671372691642/

20. Kitano, H., et al.: The RoboCup synthetic agent challenge 97. In: Kitano, Hiroaki (ed.) RoboCup 1997. LNCS, vol. 1395, pp. 62–73. Springer, Heidelberg (1998)

21. Kitano, H., Tadokoro, S.: Robocup rescue: A grand challenge for multiagent and intelligent systems. AI Mag. 22(1), 39 (2001)

22. Bai, A., Chen, X., MacAlpine, P., Urieli, D., Barrett, S., Stone, P.: Wrighteagle and ut austin villa: robocup 2011 simulation league champions. In: Röfer, T., Mayer, N.M., Savage, J., Saranlı, U. (eds.) RoboCup 2011. LNCS, vol. 7416, pp. 1–12. Springer, Heidelberg (2012)

23. World Bank: World Development Indicators 2012. http://data.worldbank.org/data-catalog/world-development-indicators

24. World Bank: GDP per capita, PPP (current international $) (2012). http://data.worldbank.org/indicator/NY.GDP.PCAP.PP.CD

25. Vilar, L., Araújo, D., Davids, K., Bar-Yam, Y.: Science of winning soccer: emergent pattern-forming dynamics in association football. J. Syst. Sci. Complex. 26(1), 73–84 (2013)

26. Grabiner, D.: The sabermetrics manifesto. Three Rivers Press, New York (2004). http://seanlahman.com/baseball-archive/sabermetrics/sabermetric-manifesto/

27. Fewell, J., Armbruster, D., Ingraham, J., Petersen, A., Waters, J.: Basketball teams as strategic networks. PloS one 7(11), e47445 (2012)

28. Cliff, O.M., Lizier, J.T., Wang, X.R., Wang, P., Obst, O., Prokopenko, M.: Towards quantifying interaction networks in a football match. In: Behnke, S., Veloso, M., Visser, A., Xiong, R. (eds.) RoboCup 2013. LNCS, vol. 8371, pp. 1–12. Springer, Heidelberg (2014)

29. David, H.A.: The method of paired comparisons. DTIC Document, vol. 12 (1963)

30. Edwards, C.T.: Double-elimination tournaments: counting and calculating. Am. Stat. 50(1), 27–33 (1996)

Cosero, Find My Keys! Object Localization and Retrieval Using Bluetooth Low Energy Tags

David Schwarz, Max Schwarz[✉], Jörg Stückler, and Sven Behnke

Computer Science Institute VI: Autonomous Intelligent Systems,
Rheinische Friedrich-Wilhelms-Universität Bonn, Friedrich-Ebert-Allee 144,
53113 Bonn, Germany
max.schwarz@uni-bonn.de

Abstract. Personal robots will contribute mobile manipulation capabilities to our future smart homes. In this paper, we propose a low-cost object localization system that uses static devices with Bluetooth capabilities, which are distributed in an environment, to detect and localize active Bluetooth beacons and mobile devices. This system can be used by a robot to coarsely localize objects in retrieval tasks. We attach small Bluetooth low energy tags to objects and require at least four static Bluetooth receivers. While commodity Bluetooth devices could be used, we have built low-cost receivers from Raspberry Pi computers. The location of a tag is estimated by lateration of its received signal strengths. In experiments, we evaluate accuracy and timing of our approach, and report on the successful demonstration at the RoboCup German Open 2014 competition in Magdeburg.

1 Introduction

Intelligent devices continuously find their way into our homes. As such, we envision also personal service robots to solve mobile manipulation tasks within our future smart homes. In this paper, we consider the problem of localizing and retrieving items with a personal robot. Instead of tedious visual search, we propose to attach low-cost Bluetooth Low Energy tags to objects. The robot operates in an environment that is instrumented with static wireless devices to estimate the position of the tags. The coarse position estimate hints to typical placement locations in the environment, from which the robot can retrieve the objects.

Imagine a scenario in which you may have displaced your wallet. You command your personal service robot to find and fetch it. A Bluetooth tag included in the wallet acts as a beacon which is detected by static Bluetooth receiving devices in the surrounding. The receivers measure the signal strength of the Bluetooth tag. Using knowledge about the position of the receivers in the environment, the position of the tag can be estimated through lateration. The robot assesses the signal strengths and estimates the coarse position of the tag.

R.A.C. Bianchi et al. (Eds.): RoboCup 2014, LNAI 8992, pp. 195–206, 2015.
DOI: 10.1007/978-3-319-18615-3_16

It searches for the wallet close to the detected position, grasps it, and happily brings it to you.

Technically, our low-cost object localization system consists of small, cheap Bluetooth Low Energy (BLE) tags and at least four receivers. The receivers are built from Raspberry Pi computers. We propose to estimate the position of the tags through lateration of received signal strengths. We assess the localization performance of our system in terms of accuracy and timing. We also report on the successful demonstration of our approach at the RoboCup German Open 2014 competition in Magdeburg.

2   Related Work

Since GPS is not available indoors, other approaches need to be taken for localizing objects in indoor environments. In recent years, some approaches have been developed that find wireless devices.

A simple approach is to use proximity detection [1], e.g. to detect RFID tags close to one of several RFID readers that have been placed in an environment. The location of the object is then simply associated with the position of the reader that best detected the tag. Approaches that utilize less instrumentation of the environment are methods based on propagation principles [2–4] and fingerprints [5].

Propagation-based approaches use distance and angle information estimated from the electrodynamical properties of the wireless signals. Most common is the use of received signal strength (RSS, e.g., [2]) which is related to the distance between the signal emitter and the receiver. Other approaches are received signal phase methods (RSP, e.g., [3]), which measure the phase-shift of signals, or time of arrival methods (TOA, e.g. [4]) that directly measure the propagation time of the signal. The common principle behind these methods is that they physically explain measured properties of the signal, and determine a continuous estimate of the position of the wireless emitter. All propagation-based methods make model assumptions that need to be fit to actual data acquired for the specific wireless setup and environment.

Fingerprinting methods (e.g., [5]), on the other hand, do not make such model assumptions, but directly represent a map of typical signal measurements, when the emitter is at specific locations. To find the emitter for new measurements, the location with the best matching example measurements is retrieved. The advantage of fingerprinting methods is that they do not make strong model assumptions and can work even in difficult environments with e.g. shadowing or multi-path effects. However, they are tedious to train—signal strengths need to be measured from the emitter placed at a dense grid of locations in the environment—and require adaptation to every change of wireless setup and environment.

Using Bluetooth to localize objects [5,6] has a number of advantages over using other wireless modalities. Bluetooth is a low-cost, low-energy technology. Many mobile devices and computers are nowadays equipped with Bluetooth,

Fig. 1. The cognitive service robot *Cosero*. Left: Cosero opens a bottle during the final of the RoboCup@Home German Open competition 2014 in Magdeburg. Right: Cosero pushes a chair to its location during the Demo Challenge of RoboCup 2013, Eindhoven.

which makes Bluetooth solutions widely applicable without a special augmentation of the environment with e.g. RFID tags [7]. In our work, we deploy Bluetooth devices within a smart environment. For instance, small active Bluetooth tags can be attached to objects such as wallets, keys, or other items that users frequently displace. A cognitive service robot can then localize objects of interest, and retrieve them. While a robot could search visually for the object [8], using wireless sensing simplifies search significantly, due to the large field-of-view and the unique identification of Bluetooth devices.

3   System Description

3.1   Cognitive Service Robot Cosero

We designed our cognitive service robot Cosero, shown in Fig. 1, to cover a wide range of tasks in domestic indoor environments [9,10]. It has been equipped with a flexible torso and two anthropomorphic arms that provide human-like reach. A linear actuator moves the whole upper body up and down, allowing the robot to grasp objects from a wide range of heights—even from the floor. Cosero's anthropomorphic upper body is mounted on a mobile base with narrow footprint and omnidirectional driving capabilities. By this, the robot can maneuver through narrow passages that are typically found in indoor environments, and it is not limited in its mobile manipulation capabilities by holonomic constraints. The human-like appearance of our robot also supports intuitive interaction with human users.

For perceiving its environment, we equipped the robot with diverse sensors. Multiple laser scanners on the ground, on top of the mobile base, and in the torso measure objects, persons, or obstacles for navigation purposes. We use a

Microsoft Kinect RGB-D camera in the head to perceive tabletop objects and persons.

Cosero navigates in indoor environments on horizontal surfaces. Hence, we use the 2D laser scanner on the mobile base as the main sensor for mapping and localization. It navigates to goal poses by planning obstacle-free paths in the environment map, extracting waypoints, and following them. Obstacle-free local driving commands are derived from paths that are planned towards the next waypoint in a local collision map that includes 3D measurements from the lasers in the robot.

After approaching the estimated coarse location of the object, the robot detects objects in its vicinity either by segmenting 2D scans of its torso laser or the depth images of its head-mounted RGB-D camera. It selects a grasp on the detected object and executes a parametrized motion primitive to grasp it [11].

3.2   Bluetooth Localization

Bluetooth Low Energy (BLE) was chosen as the technology for wireless communication because it offers several advantages. Firstly, its low transmit power allows for very small tags with long lifetimes, which is crucial in our application. If the tag batteries would have to be replaced every week, this would be an additional burden on the user, which we want to avoid.

Secondly, BLE allows direct signal strength measurement in a standardized way. Bluetooth adapters report an RSSI (Received Signal Strength Indication) for each scan answer they receive from a Bluetooth tag. This RSSI value is directly related to the signal strength and, hence, to the distance of the tag from the receiver.

Thirdly, Bluetooth components are available off-the-shelf at a very low price and in small form factors, e.g. in coin size (see Fig. 2). Finally, the system can also be used to track other Bluetooth devices, such as most smart phones, tablets and other mobile equipment.

Our system is set up as follows: We employ a number of receivers, which are mounted at fixed locations in the apartment. The locations are chosen such that the distance between receivers is large. The receivers consist of a small single board computer (Raspberry Pi) and a Bluetooth USB receiver. The Raspberry Pi already has an Ethernet connector for relaying the data. If wireless communication is desired, a USB WiFi adapter can be added. All in all, we can buy a receiver set for under 50 US$.

We simplify the problem with the assumption that receivers and tag lie in a horizontal plane. Otherwise, we would have to consider the optimization problem in 3D. Because our achieved accuracy lies in the range of 1 m, this simplification does not impede successful application of the method in reasonable room sizes.

The receivers send the detected signals and their strength to a central *location estimation unit (LEU)*. This LEU could be, for instance, the robot, or a designated receiver. The LEU then estimates tag positions using the lateration approach described below and forwards the result to the robot or other data consumers.

Fig. 2. Used Bluetooth tags (left: Estimote Inc., right: StickNFind) with coin for size comparison.

3.3   Handheld Teleoperation Interface

To extend the possibilities of human-robot interaction, we also developed a handheld teleoperation interface on a tablet computer, which can be used to remotely interact with the robot. The general system, which is targeted at immobile users, is described in [12].

The interface offers three levels of teleoperation: On the body level, the user can directly operate parts of the robot such as the omnidirectional drive, gaze direction and the end-effectors. The skill level can be used to parametrize and trigger robot skills, e.g. by setting navigation goals or commanding objects to be grasped. Finally, the operator can configure autonomous high-level behaviors that sequence skills on the task level.

Situational awareness is gained through several visualizations of the robot perceptions. We also integrated the localization system into the user interface. The tag locations are always displayed to the user within a map of the environment (see Fig. 3). Appropriate actions (e.g. combining a navigation and grasping command to retrieve the object) can then be triggered using the already present user interface components for command input.

4   From Signal Strength to Relative Distance

In the LEU, the RSSI readings of all BLE receivers are processed. In a first step, we filter the measured signal strength by finding the highest three signal strengths observed in a time window of 10 s. We average those three highest measurements to obtain a robust signal strength estimate. The rationale behind this approach is that while signal strength is easily underestimated (e.g., because of temporal occlusions or multi-path echoes), it is difficult to overestimate it.

In our observations, the measured signal strength does not directly correlate with distance. There are other factors, such as occlusions of the line-of-sight between tag and receiver, receiver angle dependencies, and so forth. Because of this, we estimate a relative distance from the signal strength in a first step.

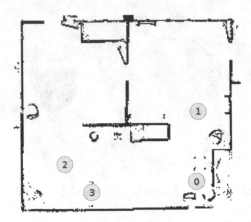

Fig. 3. Tablet user interface showing the tag locations on a map of the RoboCup@Home arena at German Open 2014 in Magdeburg.

This relative distance is assumed to be proportional to the real distance, with the same proportionality factor over all receivers. We make the assumption here that a loss of signal strength is always isotropic, i.e. experienced by all receivers at the same degree. This assumption is correct in many cases, such as objects inside a container.

For an isotropic sender of electromagnetic waves, the output power P perceived at a distance r follows the proportionality:

$$P \sim \frac{1}{r^2} \Leftrightarrow r \sim \frac{1}{\sqrt{P}}, \tag{1}$$

which we can turn into an equality for the estimated relative distance D:

$$D := r \cdot k = \frac{1}{\sqrt{P}}. \tag{2}$$

The linear relationship between distance r and estimated relative distance D was experimentally verified in a lab experiment (see Fig. 4). A Bluetooth tag was placed sequentially on a grid with 1 m resolution in our lab room (7×5 m). Four receivers mounted in the corners of the room were used to obtain signal strength data for each tag location. Furthermore, ground truth distance of the tag to the receivers was measured.

5   Position Estimation

For position estimation, the over-constrained lateration problem needs to be solved. The receivers report relative distances D_i to the tag. With perfect data, a single coefficient k would exist, which would make the circles around each receiver with radius $\frac{D_i}{k}$ intersect in a single point (see Fig. 5(a)). Because the data is not

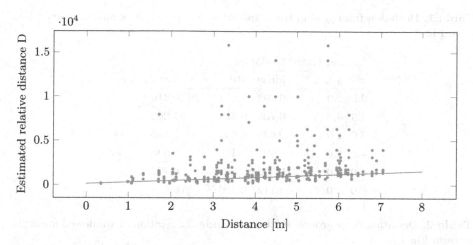

Fig. 4. Relationship of estimated relative distance D to real distance. The tag was placed at varying positions and ground truth distance to the receivers was measured. The red line shows the linear trend observed (Color figure online).

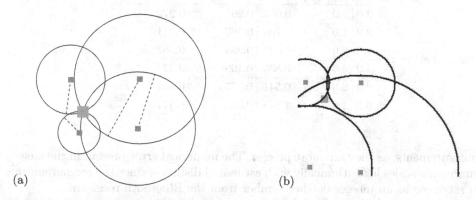

Fig. 5. Solutions for the lateration problem found by the optimization method. Green rectangles are the Bluetooth receivers, the cyan rectangle denotes the estimated tag position. The black circles show the relative distance D scaled by the found k parameter. (a) solution for an ideal problem (single intersection point); (b) solution for real-world data (Color figure online).

perfect, we treat the problem as an optimization problem. We simultaneously optimize for the tag position x and coefficient k.

The minimized cost function is specified as:

$$f(x,k) := \sqrt{\sum_{i=1}^{n} \left(\frac{||x - p_i||_2 - k \cdot D_i}{\ln (1 + D_i)} \right)^2}, \tag{3}$$

where D_i denotes the relative distance estimated by receiver i, and p_i is the position of receiver i. We divide by $\ln(1 + D_i)$ to give more weight to short-range

Table 1. Deviation from ground truth and standard deviation of measurements in m from Fig. 6

Ground truth		Deviation		
x	y	Mean	Std. dev. x	Std. dev. y
2.0	2.0	0.702	0.264	0.416
2.0	4.0	0.703	0.678	1.262
4.0	2.0	0.635	0.655	1.165
4.0	4.0	0.895	0.473	0.538
6.0	2.0	0.503	0.598	0.708
6.0	4.0	1.212	0.580	0.672

Table 2. Deviation from ground truth and standard deviation of windowed means in m from Fig. 7

Ground truth		Deviation		
x	y	Mean	Std. dev. x	Std. dev. y
2.0	2.0	0.697	0.265	0.272
2.0	4.0	0.705	0.367	0.711
4.0	2.0	0.621	0.233	0.563
4.0	4.0	0.836	0.199	0.322
6.0	2.0	0.541	0.377	0.451
6.0	4.0	1.215	0.303	0.411

measurements, as they are more precise. The numerical error present in the measurements scales logarithmically with estimated distance, since the measurement is retrieved as an integer decibel number from the Bluetooth receivers.

For minimization, we use the implementation of the Simplex algorithm of Nelder and Mead [13] available in the GNU scientific library. We initialize the simplex to the center of the environment, in order to find a reasonable minimum of the cost function. The resulting position estimate is still noisy, so we smooth it with a windowed mean over 30 s.

6 Results

6.1 Evaluation

We evaluated the accuracy and timing of our localization method in a lab experiment. The tag was sequentially placed on six positions spaced evenly in a 5×7 m room. The raw position estimates obtained without temporal smoothing are shown in Fig. 6 and their accuracy can be seen in Table 1. The estimates with 30 s windowed mean smoothing can be seen in Fig. 7 and Table 2. In almost all cases, the deviation between the ground truth tag position and the estimated

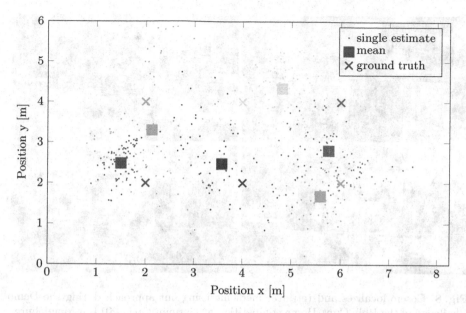

Fig. 6. Evaluation of position estimation (raw optimization results). Items with same color belong to one tag position (Color figure online).

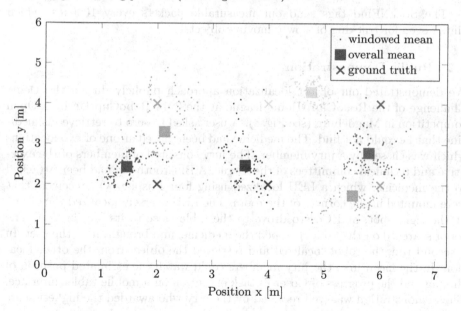

Fig. 7. Evaluation of position estimation (windowed means). Items with same color belong to one tag position (Color figure online).

position is below 1 m. This is more than sufficient for our application, since our robot can locate objects in close vicinity using its other sensors.

Fig. 8. Cosero localizes and retrieves medicine using our approach during the Demo Challenge of the RoboCup@Home competition at German Open 2014 in Magdeburg.

The StickNFind tags send out measurable packets every 1000 ms, which allows even for tracking of slowly moving objects.

6.2   Public Demonstration

We demonstrated our object localization approach publicly during the Demo Challenge of the RoboCup@Home league at the 2014 RoboCup German Open competition in Magdeburg (see Fig. 8). A user asked Cosero to retrieve his medicine that he could not find. The medicine had been placed at one of two locations which was chosen by a jury member. The jury consisted of members of the executive and technical committees of the league. A Bluetooth tag had been attached to the medicine, which a LEU localized using four RaspberryPi receivers that were mounted in the corners of the room. The LEU coarsely localized the object at the right spot, and Cosero drove to the table close to its localization. The robot searched on the table, grasped the medicine, and brought it to the user. In a second run, the robot localized and retrieved the object from the other location. In the meantime, the jury members could watch the estimated position of the tag and the progress of Cosero's task execution on a mobile tablet interface. The demonstration was well received by the jury who awarded the highest score in this test to our team.

7   Conclusions

We proposed a simple and affordable system for object localization and retrieval with a personal service robot using Bluetooth Low Energy devices. Our method places static receiver devices in a smart home environment to measure the signal strength of active Bluetooth tags that can be attached to objects. The signal strength is interpreted as relative distance, which we use to laterate the position of the tag. In addition to the location of the tag, we also estimate the proportionality factor between the signal strength and the distance to the tag. In experiments, we evaluate accuracy and timing of our localization approach. We also report on the successful demonstration of our integrated object localization and retrieval approach at RoboCup German Open 2014.

Ideas for future work include the investigation of static reference tags for system calibration. It might also be possible to combine lateration with fingerprints or local models in order to improve localization accuracy. Finally, it could be worth to equip the robot itself with Bluetooth receivers, which would make signal strength-based taxis towards tagged objects in its vicinity possible.

References

1. Ni, L.M., Liu, Y., Lau, Y.C., Patil, A.P.: Landmarc: Indoor location sensing using active RFID. Wireless Netw. **10**(6), 701–710 (2004)
2. Li, X.: Ratio-based zero-profiling indoor localization. In: IEEE 6th International Conference on Mobile Adhoc and Sensor Systems, 2009. MASS '09, pp. 40–49, Oct 2009
3. Sallai, J., Lédeczi, Á., Amundson, I., Koutsoukos, X., Maróti, M.: Using RF received phase for indoor tracking. In: Proceedings of the 6th Workshop on Hot Topics in Embedded Networked Sensors, HotEmNets '10, pp. 13:1–13:5 (2010)
4. Guvenc, I., Chong, C.-C.: A survey on toa based wireless localization and nlos mitigation techniques. Commun. Surv. Tutorial IEEE **11**(3), 107–124 (2009)
5. Pei, L., Chen, R., Liu, J., Tenhunen, T., Kuusniemi, H., Chen, Y.: Inquiry-based bluetooth indoor positioning via rssi probability distributions. In: 2010 Second International Conference on Advances in Satellite and Space Communications (SPACOMM), pp. 151–156, June 2010
6. Subhan, F., Hasbullah, H., Rozyyev, A., Bakhsh, S.T.: Indoor positioning in bluetooth networks using fingerprinting and lateration approach. In: 2011 International Conference on Information Science and Applications (ICISA), pp. 1–9, April 2011
7. Nickels, J., Knierim, P., Könings, B., Schaub, F., Wiedersheim, B., Musiol, S., Weber, M.: Find my stuff: supporting physical objects search with relative positioning. In: Mattern, F., Santini, S., Canny, J.F., Langheinrich, M., Rekimoto, J. (eds.) ACM International Joint Conference on Pervasive and Ubiquitous Computing, pp. 325–334 (2013)
8. Shubina, K., Tsotsos, J.K.: Visual search for an object in a 3D environment using a mobile robot. Comput. Vis. Image Underst. **114**(5), 535–547 (2010). Special Issue on Intelligent Vision Systems
9. Stückler, J., Holz, D., Behnke, S.: RoboCup@Home: demonstrating everyday manipulation skills in RoboCup@Home. Robot. Autom. Mag. **19**(2), 34–42 (2012)
10. Stückler, J., Badami, I., Droeschel, D., Gräve, K., Holz, D., McElhone, M., Nieuwenhuisen, M., Schreiber, M., Schwarz, M., Behnke, S.: NimbRo@Home: winning team of the RoboCup@Home competition 2012. In: Chen, X., Stone, P., Sucar, L.E., van der Zant, T. (eds.) RoboCup 2012. LNCS, vol. 7500, pp. 94–105. Springer, Heidelberg (2013)

11. Stückler, J., Steffens, R., Holz, D., Behnke, S.: Efficient 3d object perception and grasp planning for mobile manipulation in domestic environments. Robot. Auton. Syst. **61**(10), 1106–1115 (2013)
12. Schwarz, M., Stückler, J., Behnke, S.: Mobile teleoperation interfaces with adjustable autonomy for personal service robots. In: 9th ACM/IEEE International Conference on Human-Robot Interaction (HRI), pp. 288–289 (2014)
13. Nelder, J.A., Mead, R.: A simplex method for function minimization. Comput. J. **7**(4), 308–313 (1965)

Object Recognition for Manipulation Tasks in Real Domestic Settings: A Comparative Study

Luz Martínez, Patricio Loncomilla, and Javier Ruiz-del-Solar[✉]

Department of Electrical Engineering and Advanced Mining Technology Center,
Universidad de Chile, Santiago, Chile
{ploncomi, jruizd}@ing.uchile.cl

Abstract. The recognition of objects is a relevant ability in service robotics, especially in manipulation tasks. There are many different approaches to object recognition, but they have not been properly analyzed and compared by considering real conditions of manipulation tasks in domestic setups. The main goal of this paper is to analyze some popular object recognition methods and to compare their performance in realistic manipulation setups. Object recognition methods based on SIFT, SURF, VFH, OUR-CVFH and color histogram descriptors are considered in this study. The results of this comparison can be of interest for researchers working in the development of similar systems (e.g. RoboCup @home teams).

Keywords: Object recognition · RoboCup @home · RGB-D images · Benchmark

1 Introduction

The recognition of objects is of paramount importance in service robotics, especially in manipulation tasks. The main differences with standard computer vision applications are the requirements of real-time operation with limited on-board computational resources, and the constrained observational conditions derived from the robot geometry, limited camera resolution and sensor/object relative pose. An enormous amount of articles addresses the recognition of objects in computer and robot vision. Recent approaches used in service and/or domestic robots (e.g., the ones used in RoboCup @home) are mainly based on a pipeline that first detects horizontal surfaces (e.g., a table or the floor) for restricting the search area of the possible object's positions, and then it computes features in order to recognize the objects. Popular features include the use of visual, appearance-based local interest points (keypoints) and descriptors (e.g. SIFT [7] and SURF [8]) and/or the use of 3D feature descriptors such as feature histograms obtained from range images (e.g. PFH [4] and VFH [5]). In most of the cases, the robotic platforms are equipped with RGB and/or RGB-D cameras for the data acquisition.

In this context, we believe that it is important to analyze the performance of these different approaches in domestic setups by considering real conditions. Those conditions

© Springer International Publishing Switzerland 2015
R.A.C. Bianchi et al. (Eds.): RoboCup 2014, LNAI 8992, pp. 207–219, 2015.
DOI: 10.1007/978-3-319-18615-3_17

must include variability on the typical objects to be manipulated in domestic contexts, variable illumination, dynamic backgrounds, more than one object in the robot's field of view, search trials with more than 25 object types in the database, occlusions, typical sensors used in domestic robots (e.g. kinect sensors and low-resolution RGB cameras), and typical sensor-object pose conditions. These last two aspects are very important, because very often one of the main difficulties for the recognition task is the low resolution of the images. For instance, in Fig. 2 it is shown the Bender robot as an example of a typical domestic setup. It can be observed that for the cases of objects placed on a table and on the floor, the distance between the sensors and the objects are 104 cm and 177 cm, respectively, and the view angle of 56° respect to the horizon. Under these conditions, and considering a typical image resolution of 640 × 480 or even 1280 × 720 pixels, the objects are observed at low resolution and in a non-frontal view.

Although in several articles the performance of object recognition methods has been analyzed and compared, these comparisons are focused on computer vision applications or they do not consider the mentioned real-world conditions, but only situations with centered high-resolution object's images.

Thus, the main goal of this paper is to analyze some popular object recognition methods and to compare their performance in realistic manipulation setups. Object recognition methods based on SIFT, SURF, VFH, OUR-CVFH and color histogram descriptors are considered in this study. The results of this comparison can be of interest for researchers working in the development of similar systems (e.g. RoboCup @home teams).

This paper is organized as follows. In Sect. 2 some related work is presented. In Sect. 3, the object recognition methods under comparison are outlined. In Sect. 4 the evaluation of the different methods is described. Finally, conclusions are given in Sect. 5.

2 Related Work

In recent years, several methodologies addressing object recognition have been developed, by both using RGB images and range images, i.e. point clouds. The development of repeatable local descriptors computed from RGB images like SIFT [7] and SURF [8] made possible the creation of a broad family of powerful methods for robust object recognition of textured objects in cluttered backgrounds. Object recognition is based on the matching between local descriptors from two different images. Coherent sets of descriptor-to-descriptor matches are found by using Hough transform clustering [7] or RANSAC [26], and they indicate possible object poses on the image. The pose of an object can be recovered by matching a test image against a set of training images captured from different viewpoints [7, 1, 2, 3] or by reconstructing a 3D descriptor cloud by using structure from motion techniques [15, 20]. However, a better recognition accuracy is obtained by generating several 2D keyframes (views) from the 3D descriptor cloud [21, 22], i.e., transforming the recognition problem back from 3D to 2D, and then using 3D descriptors for retrieving an accurate pose. Also stereo images can be used for recovering the object's pose with a better accuracy [18, 19]. The recognition of non-textured objects from RGB images is harder to achieve, and it requires recognizing the object's boundaries [16], although global approaches like ensemble of exemplar HoG SVMs detectors [17] that use one positive example against multiple negative ones are an interesting alternative for solving this problem.

The availability of low cost 3D capture devices like Kinect and ASUS Xtion gave rise to the development of new kind of approaches for recognizing objects; however, several of them are adaptations of 2D recognition concepts into the 3D space by using normal-based methods instead of gradient-based ones. Some methods are based on the matching of 3D local shape descriptors [23–25] by using variants of RANSAC [26] for pose estimation. They are able to handle clutter and occlusion, but they do not work on objects having simple shapes as they do not enable the generation of distinctive local descriptors. Also, they are sensible to both noise and variable sampling step between points. Other approaches use global or semi-global descriptors [5, 6] that represent the full point cloud at once by using histograms of normals of the object, but they require to previously segment the object from the background. This is easy to enforce if the object is placed on a planar surface. An innovative approach is the use of point pair features [27] computed from pairs of points with its respective normals. The object is described globally by a set of four-dimensional training features constructed from several random pairs of oriented points, and they are stored in a two-dimensional hash table. When a test frame arrives, four-dimensional features on the point cloud are computed, and matched against the training features. Matched features vote for an object pose that is stored in a 3D accumulator, then peaks indicate possible object poses. These methods have been extended to use information from the RGB image [28, 29]. This family of methods enables recognition of objects under occlusion or cluttered backgrounds.

3 Object Recognition Methods Under Analysis

The task addressed in this work is the recognition of objects for manipulation purposes. The objects are placed on a planar surface, and they are recognized by using RGB and depth images. The height of the surface (a table or the floor) can be used for segmenting zones in the image that are upper than the surface, enabling focused object recognition. There are two pipelines for object recognition: the first one is used for visual recognition, and the second one is used for point cloud based recognition.

The pipeline for visual object recognition (Fig. 1 (a)) uses a RGB image and a depth image as inputs. The depth image is used only for selecting pixels that are upper than the plane, and it is an optional stage. The RGB image is used for extracting descriptors, which are matched against descriptors stored in a training database. Extra verifications can be used for discarding unfeasible transformations. The pipeline for point cloud based object recognition (Fig. 1 (b)) uses a depth image as input. Pixels upper than the surface are selected, and a global/semi-global point cloud based descriptor is extracted and matched against descriptors stored in a training database. In this case, extra verifications are not performed.

In this work, seven visual recognition methods, two point-cloud based recognition methods and a hybrid method are compared. The visual recognition methods are: *L&R*

SIFT [1, 2], *L&R SIFT segm* (L&R SIFT plus object segmentation), *obj_rec_surf* [2], *obj_rec_surf segm* (obj_rec_surf plus object segmentation), *L&R SURF*, *L&R SURF segm* (L&R SURF plus object segmentation), and color histograms. The point cloud based recognition methods are VFH [5] and OUR-CVFH [6]. The hybrid method is SIFT-VFH. These methods are described in the following paragraphs.

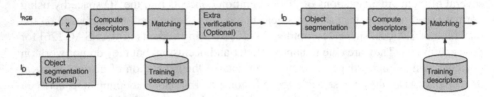

Fig. 1. Block diagrams of (a) the visual object recognition pipeline, and (b) the point cloud recognition pipeline. I_{RGB}: RGB image. I_D: Depth image.

- *SIFT based methods.* Methods based on matching local descriptors for visual object recognition are very popular. SIFT [7] is a methodology that enables computation of local descriptors by selecting interest points in the scale-space of an image, and then computing oriented, gradient-based feature vectors on image patches centered on the interest points. Descriptors from training images are stored in a kd-tree structure. Descriptors from a test image are matched against the training descriptors, and the descriptor-to-descriptor matches are clustered by using a Hough transform defined over the similarity transformation space. The *L&R SIFT* system [1] performs additional verification stages over the Hough transform bins. The stages are designed for discarding ill-defined transformations that have strong distortions, are undefined on some spatial direction or map pixels with different intensities on both images. The L&R SIFT system includes a non-maximal suppression test, a linear correlation test, a fast probability test, an affine distortion test, Lowe's probability test, a RANSAC test, a transformation fusing test, a semi-local constraints test and a pixel correlation test. Two variants of the L&R SIFT system are tested: the first variant, *L&R SIFT*, applies the object detection method over the whole image, while the second variant, *L&R SIFT segm*, applies it over a subset of the image (segmented image). The subset is obtained by detecting a planar surface in the depth image and then selecting only the regions that are upper than the plane.
- *SURF based methods.* The SURF method [8] is similar to SIFT, but rectangular functions are used instead of Gaussians for computing the scale-space. Four variants of the SURF matching algorithm are tested: *obj_rec_surf* [2] is a methodology that matches descriptors from the test image against training descriptors by using a Hough transform based clustering, using a simple probability test for accepting or rejecting the transformations. This method was the winner of the RoboCup@Home technical challenge on 2012. It is based on the pan-o-matic [11] SURF descriptor generation implementation, which is used by Hugin [12], a panorama photo stitcher. This descriptor generator was shown to outperform other open source SURF implementations as shown by the team homer@UniKoblenz [13], and become parallelized (parallel_surf [30]) and integrated on their object recognition pipeline

obj_rec_surf. The second variant, *obj_rec_surf segm*, is similar to *obj_rec_surf*, but SURF descriptors are computed by using the segmented image. The third variant, *L&R SURF*, is similar to *L&R SIFT*, but instead of using SIFT it uses SURF, specifically OpenSURF [14]. The fourth variant, *L&R SURF segm*, is similar *to L&R SURF*, but the descriptors are computed by using the segmented image.

– *VFH method*. The method [5] computes a global descriptor, which is formed by a histogram of the normal components of the object's surface. The histogram captures the shape of the object, and the viewpoint from which the point cloud is taken. In the first place, the angles α, ϕ and θ are computed for each point based on its normal and the normal of the point cloud's centroid c_i. The viewpoint-dependent component of the descriptor is a histogram of the angles between the vector p_c- p_v and each normal point. The other component is a SPFH estimated for the centroid of the point cloud, and an additional histogram of the distances of the points in the cloud to the cloud's centroid. The VFH descriptor is a compound histogram representing four different angular distributions of surface normals. In this work, the PCL implementation [9] is used, where each of those four histograms have 45 bins and the viewpoint-dependent component has 128 bins, totaling 308 bins.

– *OUR-CVFH method*. The method [6] computes a semi-global descriptor, and it is based on semi-global unique reference frames (SGURFs) [6] and the CVFH [10] descriptors. It exploits the orientation provided by the SGURFs to efficiently encode the geometrical properties of the object's surface. Given a surface S, the method computes the first three components of CVFH and the viewpoint component as presented in [10]. The viewpoint component is however encoded using 64 bins instead of the original 128, as normals are always pointing towards the sensor position. For each cluster c_i, a reference frame RF_i is created, then a transformation is applied to the surface S so that the points can be easily divided into the eight octants defined by the signed axes (x−, y−, z−) … (x+, y−, z−) … (x+, y+, z+). Each of the octants has an associated histogram, all of them are used for representing the surface S.

– *Color Histograms*. Color histograms are computed by transforming the images from the RGB space into the HSV one, and then filling an histogram matrix defined over the HS space. The matrix has size 30 in the H dimension, and size 32 in the S dimension. Matching is done by using Hellinger distance, which is related to Bhattacharyya coefficient.

$$d(H_1, H_2) = \sqrt{1 - \frac{1}{\sqrt{\overline{H_1}\,\overline{H_2}}N^2} \sum_I \sqrt{H_1(I)H_2(I)}} \qquad (1)$$

Color histograms are computed only on segmented images, because when the full image is used the background clutter generates a large impact on the resultant histograms.

– *SIFT-VFH*. In this hybrid method, the depth image is obtained for generating blobs, and the SIFT algorithm is applied on the image. If the blobs remain not identified, the VFH algorithm is applied on them. This algorithm is able to use both visual and shape information. The SIFT algorithm is applied before VFH because of its higher precision.

4 Evaluation

4.1 Setup and Methodology

The robot Bender from the Uchile Homebreakers team was used as platform for testing the different object recognition approaches. The robot has a Kinect camera and a RGB camera mounted over its head. Both are placed at a height of approximately 1.6 [mt], pointing downwards with an angle of 56° respect to the horizon. In the reported experiments, two kinds of object placements are used: objects are placed on a table or they are placed on the floor (see Fig. 2). The Kinect has a resolution of 640 × 480 and an angular field of view of 57° horizontally and 43° vertically. The RGB camera has a resolution of 1280 × 720 pixels, and has an angular field of view of 60° horizontally and 45° vertically. The mean distance between the robot's cameras and objects on the table is 104.1 [cm], while the mean distance between the robot's cameras and objects on the floor is 176.8 [cm]. As the objects are far from the camera, the area they cover on the images is small.

A set of 40 objects was selected for performing the tests; the objects are typical in a home environment (see Fig. 3). From the set of objects, 20 objects have visual textures, and the other 20 objects have uniform surfaces (no textures). For each object, 12 different views are captured by rotating the objects 30° between two consecutive frames. For each view, both an RGB and a depth image are captured. Therefore, a total of 480 RGB images and 480 depth images are used as gallery (database).

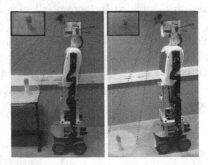

Fig. 2. Bender robot observing an object on a table (see main text for details).

Different setups are used for evaluating the performance of the different object recognition methodologies under comparison. The possible setups differ in the following conditions: (a) number of objects in each image: 1 object/6 objects; (b) image background: white/brown/different backgrounds; (c) Illumination: normal/low; (d) occlusion: no occlusion/50 % occlusion; and (e) surface : table/floor. The selected testing setups are the following: S1: One object on a table, white background, normal illumination, no

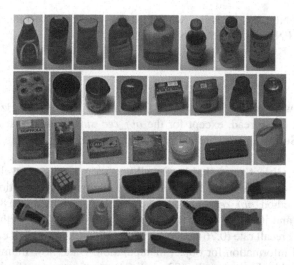

Fig. 3. Set of 40 objects used for evaluating the object recognition algorithms.

occlusion; S2: One object on a table, white background, low illumination, no occlusion; S3: One object on a table, brown background, normal illumination, no occlusion; S4: One object on a table, different backgrounds, normal illumination, no occlusion; S5: One object on a table, white background, normal illumination, 50 % occlusion; S6: One object on the floor, white background, normal illumination, no occlusion; S7: One object on the floor, white background, low illumination, no occlusion; S8: One object on the floor, brown background, normal illumination, no occlusion; S9: One object on the floor, different backgrounds, normal illumination, no occlusion; S10: One object on the floor, white background, normal illumination, 50 % occlusion; S11: Six objects on the table, white background, normal illumination, no occlusion; S12: Six objects on the table, white background, low illumination, no occlusion; S13: Six objects on the table, brown background, normal illumination, no occlusion; and S14: Six objects on the table, different backgrounds, normal illumination, no occlusion.

For each of these 14 setups, 160 experiments were carried out by selecting each object 4 times; each time the object's view is chosen randomly. The random view is selected by putting the object inside the field of view of the cameras, and then selecting a random number between 0° and 360° for setting the object's orientation. The recognition is considered successful if the correct object is identified, independently of the recovered viewpoint; i.e., the current object must be matched correctly against the 480 images (40 objects) in the database. In the case that six objects are being matched, results (success or failure) from individual matches are added. From the 160 experiments per setup, two measures, precision and recall, are used for describing the accuracy of the recognition system. When an object is put on the table or the floor, and the object recognizer is executed there are three possible outcomes: the object is successfully recognized (true positive), the object is detected but mislabeled and confused with another object (false positive) or it is not detected (false negative). From the detection statistics, precision and recall are computed as TPR/(TPR+FPR) and TPR/(TPR+FNR), respectively, with TPR the true positive rate, FPR the false positive rate and FNR the false negative rate.

4.2 Results

The experiments were carried in a ultrabook with Intel Core i7 @ 2 GHz and 2048 MB RAM using only one thread, except for the *obj_rec_surf* method that uses all of the available cores. Surface planes are detected using PCL.

Results for One Object on a Table. Recall/Precision results, as well as execution times are shown in Tables 1 and 2. When normal illumination conditions are considered (S1), the method *obj_rec_surf* achieves a good recall rate (0.84), but a low precision (0.72), and it works better than *obj_rec_surf segm*. The method L&R SIFT gets an acceptable recall rate (0.76) and a good precision (0.96), then it can be used as a reliable source of information for object manipulation. Under low illumination (S2) or when using variable backgrounds (S3 and S4), performance of visual detection methods falls down, and then point cloud based recognizers VFH and OUR-CVFH perform better with an acceptable recall (between 0.67 and 0.72) but with a limited precision (between 0.58 and 0.75). Under these unfavorable conditions, the best visual recognition method is L&R SIFT; it has a bad recall (between 0.22 and 0.36) but a good precision (between 0.92 and 0.97). When occlusions are present (S5), precision of point cloud based methods falls down, then the method with best recall is *obj_rec_surf seg* (0.65) but it has a bad precision (0.43) that makes it unusable for manipulation purposes. In that case (S5), the most recommendable method is L&R SIFT, it achieves a lesser recall (0.58) but a good precision (0.98). The mean of the results of recognition of one object shows that SIFT-VFH is the method with the highest recall, and it outperforms both VFH and OUR_CVFH. L&R SIFT is the best visual method by having a good recall and an excellent precision. Color histograms have a poor performance, and they are overcome by the other visual and shape based methods. OUR-CVFH is much faster than VFH. As both have similar recall and precision, the first one is recommended. *L&R SIFT segm* is much faster than L&R SIFT, but the first one has a lesser recall, and then there exists a tradeoff as no one of the methods outperforms the other. Methods *obj_rec_surf* and *obj_rec_surf seg* require a considerable processing time and have a limited precision, then the L&R SIFT variants perform better. Then, the recommended method for object recognition of objects on tables is SIFT-VFH.

Table 1. Recall/Precision for recognition of one object on a table

Setup	L&R SIFT	L&R SURF	L&R SIFT seg	L&R SURF seg	obj_rec_ surf	obj_rec_ surf seg	VFH	OUR- CVFH	Hist.	SIFT-VFH
S1	0.76/0.96	0.39/**1**	0.53/0.98	0.24/**1**	0.84/0.72	**0.82**/0.51	0.63/0.58	0.64/0.65	0.45/0.56	0.78/0.91
S2	0.36/0.97	0.09/**1**	0.21/**1**	0.06/**1**	0.29/0.70	0.27/0.41	0.67/0.58	**0.69**/0.65	0.29/0.53	0.66/0.88
S3	0.31/0.96	0.12/0.91	0.21/**1**	0.04/**1**	0.19/0.57	0.25/0.40	**0.72**/0.75	0.67/0.69	0.02/0.27	0.61/0.78
S4	0.22/0.92	0.03/**1**	0.21/0.94	0.05/**1**	0.17/0.75	0.29/0.53	**0.67**/0.72	**0.67**/0.70	0/**1**	0.66/0.88
S5	0.58/0.98	0.31/**1**	0.44/0.99	0.21/**1**	0.57/0.77	**0.65**/0.43	0.29/0.25	0.22/0.21	0.22/0.31	0.58/0.85
Mean	0.45/0.96	0.19/0.98	0.32/0.98	0.12/**1**	0.42/0.70	0.46/0.46	0.6/0.58	0.58/0.58	0.20/0.53	**0.66**/0.86

Table 2. Execution time (ms) for recognition of one object on a table.

Setup	L&R SIFT	L&R SURF	L&R SIFT seg	L&R SURF seg	obj_rec_ surf	obj_rec_ surf seg	VFH	OUR-CVFH	Hist.	SIFT-VFH
S1	1278	229	245	124	1735	2155	2180	432	**152**	464
S2	1103	325	262	**120**	1595	1422	2041	423	134	393
S3	1031	242	376	**103**	1604	1221	1882	426	115	412
S4	1253	**231**	643	525	1430	1157	3418	650	448	1485
S5	1241	372	230	203	1681	1941	2243	491	**148**	427
Mean	1181	280	351	215	1609	1579	2353	485	**199**	636

Results for One Object on the Floor. Recall/Precision results, as well as execution times are shown in Tables 3 and 4. As it can be observed, the point cloud based methods VFH and OUR-CVFH have a very small recall that makes them non useful for manipulation purposes, and SIFT-VFH is overcome by obj_rec_surf, then visual detection methods must be used. The methods with higher recall *are obj_rec_surf* and *obj_rec_surf segm*, but they have a limited precision. L&R SIFT has a good precision but a very low recall. Then, the analysis of the data shows that there is not a reliable way to detect objects on the floor by using a camera on the robot's head; main reason is the low resolution of the objects in the images. The use of high-resolution cameras could revert this situation.

Table 3. Recall/Precision for recognition of one object on the floor

Setup	L&R SIFT	L&R SURF	L&R SIFT seg	L&R SURF seg	obj_rec_ surf	obj_rec_ surf seg	VFH	OUR-CVFH	Hist.	SIFT-VFH
S6	0.49/0.97	0.06/1	0.26/1	0.02/1	**0.77**/0.67	0.54/0.20	0.19/0.07	0.01/0.01	0.07/0.02	0.44/0.42
S7	0.19/1	0.01/1	0.12/1	0.01/1	0.47/0.56	**0.48**/0.48	0.19/0.09	0/1	0.05/0.02	0.2/0.29
S8	0.19/1	0.02/1	0.08/1	0.01/1	0.26/0.52	**0.47**/0.28	0.15/0.07	0.01/0.01	0/0	0.11/0.23
S9	0.14/0.92	0.01/1	0.08/0.93	0.01/1	0.22/0.72	**0.45**/0.38	0.07/0.05	0.02/0.01	0/0	0.09/0.15
S10	0.28/1	0.03/1	0.17/1	0.02/1	0.39/0.57	**0.56**/0.22	0.07/0.03	0/0	0.08/0.03	0.25/0.37
Mean	0.26/0.98	0.02/1	0.14/0.99	0.01/1	0.42/0.61	**0.50**/0.31	0.13/0.06	0.01/0.01	0.04/0.21	0.22/0.28

Table 4. Execution time (ms) for recognition of one object on the floor

Setup	L&R SIFT	L&R SURF	L&R SIFT seg	L&R SURF seg	obj_rec_ surf	obj_rec_ surf seg	VFH	OUR-CVFH	Hist.	SIFT-VFH
S6	1127	270	542	**379**	1725	7469	974	617	648	2390
S7	1281	**238**	514	457	1753	5048	892	563	618	2414
S8	989	**250**	561	377	1692	5778	902	594	484	2609
S9	1125	**252**	320	275	1839	9214	725	418	441	1307
S10	992	**233**	566	440	1813	6339	1041	609	718	1619
Mean	1103	**249**	501	386	1765	6770	907	560	582	2067

Results for Six Objects on a Table. In these experiments, six random objects on a table must be recognized. 160 tests are performed as follows: 40 times using only textured objects, 40 times using only objects without texture, and 80 times using objects with and without texture. Recall/Precision results, as well as execution times are shown in Tables 5 and 6. As it can be observed, the method with the highest recall is OUR-CVFH; however, its precision is limited (between 0.67 and 0.69). OUR-CVFH outperformes VFH. Visual methods have a poor performance in normal conditions (S11) and perform even worse in the other cases (S12, S13, S14). *L&R SIFT* outperforms *obj_rec_surf* in recall, precision and speed. *L&R SIFT* and *L&R SIFT segm* have similar accuracies, but the second one is faster. The use of segmented images in *obj_rec_surf* increases its recall in a 50 %, but decreases its precision in a 40 %, making it unreliable for object recognition from a single frame. It is noticeable that in general the recall of the methods decreases but their precision increments as the number of objects becomes higher. Altogether, by considering the performance in terms of precision and recall, as well as the execution time, the best method is SIFT-VFH.

Table 5. Recall/Precision for recognition of six objects on a table

Setup	L&R SIFT	L&R SURF	L&R SIFT seg	L&R SURF seg	obj_rec_surf	obj_rec_surf seg	VFH	OUR-CVFH	Hist.	SIFT-VFH
S11	0.36/0.97	0.18/1	0.36/0.97	0.13/1	0.36/0.93	0.50/0.52	0.47/0.60	0.5/0.69	0.25/0.65	**0.60/0.88**
S12	0.1/0.99	0.04/1	0.08/0.96	0.02/1	0.07/0.90	0.12/0.45	0.36/0.58	**0.43/0.68**	0.10/0.42	0.41/0.8
S13	0.15/0.99	0.05/1	0.14/0.98	0.02/1	0.09/0.85	0.15/0.47	0.45/0.63	**0.50/0.69**	0.04/0.39	0.48/0.8
S14	0.10/0.94	0.03/1	0.09/0.99	0.01/0.12	0.09/0.96	0.15/0.69	0.44/0.63	**0.53/0.69**	0.01/1	0.36/0.77
Mean	0.18/0.97	0.07/1	0.17/0.97	0.04/0.78	0.15/0.91	0.23/0.53	0.43/0.61	**0.49/0.69**	0.10/0.61	0.47/0.81

Table 6. Execution time (ms) for recognition of six objects on a table

Setup	L&R SIFT	L&R SURF	L&R SIFT seg	L&R SURF seg	obj_rec_surf	obj_rec_surf seg	VFH	OUR-CVFH	Hist.	SIFT-VFH
S11	1800	586	493	313	1851	4601	2982	912	**185**	884
S12	1385	456	553	400	1779	4964	3381	1041	**284**	1101
S13	1411	309	377	**170**	1760	4544	3035	924	185	856
S14	1839	586	996	527	1849	2077	4880	1192	**522**	2025
Mean	1608	484	605	352	1810	4047	3569	1017	**294**	1217

5 Discussion and Conclusions

Tradeoffs between recall, precision and execution time are present in the object perception task. Several different object recognition methodologies could be applied in parallel in multicore systems, and their results fused. In the case of the existence of only one core, the execution of different object recognition methodologies will slow down the object detection frame rate. The L&R SIFT detections have a high precision, then they could be used immediately for object manipulation purposes. The other algorithms give not enough precision for being used independently, then several instances of them must be run in a serial or parallel way, and then a global decision using the multiple detection results must be done. L&R SIFT outperforms obj_rec_surf in precision, accuracy and speed in all of the cases, except when the object is very far. The addition of the L&R verifications to the *obj_rec_surf* pipeline could create a system with both very high recall and precision for both near and far objects and it could be addressed in a future work.

The L&R SURF implementation have a poor performance when compared to both L&R SIFT and *obj_rec_surf*. This is caused because the OpenSURF library [14] that is used in L&R SURF generates less descriptors than the *obj_rec_surf* descriptor generator [2, 11, 13] when used with its default parameters. It is also reflected in the algorithm runtimes, as L&R SURF is much faster than obj_rec_surf in all of the tests. OpenSURF starts with half of the scale than obj_rec_surf, then the last generates more keypoints and is better for detecting objects far from the camera. Also, descriptors generated by *obj_rec_surf* have a better repeatability [13].

Respect to execution time, the three fastest methods are L&R SURF, L&R SURF segm and color histograms; however, their recall and precision are poor and they cannot be considered as feasible alternatives. OUR-CVFH and SIFT-VFH outperform VFH in both accuracy and speed. L&R SIFT segm is faster than L&R SIFT, but its recall is lower. The obj_rec_surf method is slower than the L&R variants, and obj_-rec_surf segm is even slower.

Altogether, L&R SIFT seems to be the best visual method and SIFT-VFH performs better than the only shape-based methods by considering all time, recall and precision. L&R SIFT obtains the best performance among all SIFT and SURF methods. It is also interesting to note that the factor affecting most the performance of the methods is the low resolution of the objects in the images. Important improvement in the results of the recognition process could be obtained by using cameras having higher resolutions.

Acknowledgments. This work was partially funded by FONDECYT grant 1130153.

References

1. Ruiz-del-Solar, J., Loncomilla, P., Devia, C.: Fingerprint verification using local interest points and descriptors. In: Ruiz-Shulcloper, J., Kropatsch, W.G. (eds.) CIARP 2008. LNCS, vol. 5194, pp. 519 –526. Springer, Heidelberg (2008)
2. Ruiz-del-Solar, J., Loncomilla, P.: Robot head pose detection and gaze direction determination using local invariant features. Adv. Robot. **23**(2009), 305–328 (2009)
3. Seib, V., Kusenbach, M., Thierfelder, S., Paulus, D.: Object recognition using hough-transform clustering of SURF features. RoboCup@home Technical Challenge 2012 paper (2012)
4. Rusu, R.B., Blodow, N., Marton, Z.C., Beetz, M.: Aligning point cloud views using persistent feature histograms. In: Proceedings of 21st IEEE/RSJ International Conference on Intelligent Robots and Systems (IROS), pp. 3384-3391, Nice, France, 22–26 September 2008
5. Rusu, R.B., Bradski, G., Thibaux, R., Hsu, J.: Fast 3d recognition and pose using the viewpoint feature histogram. In: Proceedings of the 2010 IEEE/RSJ International Conference on Intelligent Robots and Systems (2010)

6. A, Aldoma., Tombari, F., Rusu, R., Vincze, M.: OUR-CVFH – oriented, unique and repeatable clustered viewpoint feature histogram for object recognition and 6DOF pose estimation. In: Pinz, A., Pock, T., Bischof, H., Leberl, F. (eds.) DAGM and OAGM 2012. LNCS, vol. 7476, pp. 113–122. Springer, Heidelberg (2012)

7. Lowe, D.G.: Distinctive image features from scale-invariant keypoints. Int. J. Comput. Vis. **60**(2), 11–91 (2004)

8. Bay, H., Ess, A., Tuytelaars, T., Van Gool, L.: SURF: speeded up robust features. Comput. Vis. Image Underst. (CVIU) **110**(3), 346–359 (2008)

9. PCL–Point Cloud Library 1.7. http://www.pointclouds.org

10. Aldoma, A., Blodow, N., Gossow, D., Gedikli, S., Rusu, R.B., Vincze, M., Bradski, G.: CAD-model recognition and 6d of pose estimation using 3d cues. In: 3DRR Workshop, ICCV (2011)

11. Pan-o-matic. http://aorlinsk2.free.fr/panomatic/

12. Hugin-Panorama photo stitcher. http://hugin.sourceforge.net/

13. Gossow, D., Decker, P., Paulus, D.: An evaluation of open source surf implementations. In: Ruiz-del-Solar, J. (ed.) RoboCup 2010. LNCS, vol. 6556, pp. 169–179. Springer, Heidelberg (2010)

14. OpenSURF library. http://www.chrisevansdev.com/computer-vision-opensurf.html

15. Zillich, M., Prankl, J., Morwald, T., Vincze. M.: Knowing your limits-self-evaluation and prediction in object recognition. In: IROS 2011, pp. 813–820 (2011)

16. Liu, M.Y., Tuzel, O., Veeraraghavan, A., Chellappa, R.: Fast directional chamfer matching. In: Proceedings of the IEEE International Conference on Robotics and Automation (ICRA 2010), Anchorage, Alaska, May 2010

17. Malisiewicz, T., Gupta, A., Efros, A.A.: Ensemble of exemplar-SVMs for object detection and beyond. In: ICCV 2011 (2011)

18. Azad, P., Asfour, T., Dillmann, R.: Stereo-based vs. monocular 6-DoF pose estimation using point features: a quantitative comparison. In: Dillmann, R., Beyerer, J., Stiller, C., Zöllner, M., Gindele, T. (eds.) Autonome Mobile Systeme (AMS), Springer, Heidelberg (2009)

19. Grundmann, T., Eidenberger, R., Schneider M., Fiegert M., Wichert G.V.: Robust high precision 6D pose determination in complex environments for robotic manipulation. In: Workshop of Best Practice in 3D Perception and Modeling for Mobile Manipulation at the International Conference on Robotics and Automation ICRA 2010 (2010)

20. Martinez, M., Collet, A., Srinivasa S.S.: MOPED: a scalable and low latency object recognition and pose estimation system. In: ICRA 2010 (2010)

21. Kim, K., Lepetit, V., Woo, W.: Keyframe-based modeling and tracking of multiple 3D objects. In: ISMAR, 2010, pp. 193–198 (2010)

22. Kim, K., Lepetit, V., Woo, W.: Real-time interactive modeling and scalable multiple object tracking for AR. Comput. Graph. **36**(8), 945–954 (2012)

23. Steder, B., Rusu, R.B., Konolige, K., Burgard, W.: Point feature extraction on 3d range scans taking into account object boundaries. In: Proceedings of the IEEE International Conference on Robotics and Automation (ICRA 2011) (2011)

24. Johnson, A.E., Hebert, M.: Using spin images for efficient object recognition in cluttered 3d scenes. Pattern Anal. Mach. Intell. **21**(5), 433–449 (1999)

25. Mian, A., Bennamoun, M., Owens, R.: On the repeatability and quality of keypoints for local feature-based 3d object retrieval from cluttered scenes. Int. J. Comput. Vis. **89**(23), 348–361 (2010)

26. Fischler, M.A., Bolles, R.C.: Random sample consensus: a paradigm for model fitting with applications to image analysis and automated cartography. Comm. ACM **24**(6), 381–395 (1981)

27. Drost, B., Ulrich, M., Navab, N., Ilic, S.: Model globally, match locally: efficient and robust 3D object recognition. In: IEEE Computer Society Conference on Computer Vision and Pattern Recognition (CVPR) (2010)
28. Drost, B., Ilic, S.: 3D Object detection and localization using multimodal point pair features. In: Second Joint 3DIM/3DPVT Conference on 3D Imaging, Modeling, Processing, Visualization & Transmission (3DIMPVT), Zurich, Switzerland (2012)
29. Choi, C., Christensen, H.I.: 3D pose estimation of daily objects using an RGB-D camera. In: IROS 2012, pp. 3342–3349. IEEE (2012)
30. Parallel SURF. http://sourceforge.net/apps/mediawiki/parallelsurf

Simulation for the RoboCup Logistics League with Real-World Environment Agency and Multi-level Abstraction

Frederik Zwilling, Tim Niemueller[✉], and Gerhard Lakemeyer

Knowledge-Based Systems Group, RWTH Aachen University, Aachen, Germany
{zwilling,niemueller,gerhard}@kbsg.rwth-aachen.de

Abstract. RoboCup is particularly well-known for its soccer leagues, but there are an increasing number of application leagues. The newest one is the Logistics League where groups of robots take on the task of in-factory production logistics. It has two unique aspects: a game environment which itself acts as an agent and a focus on planning and scheduling in robotics. We propose a simulation based on Gazebo that takes these into account. It uses the exact same referee box to simulate the environment reactions similar to the real game and it supports multiple levels of abstraction that allow to focus on the planning with a high level of abstraction, or to run the full system on simulated sensor data on a lower level for rapid integration testing. We envision that this simulation could be a basis for a simulation sub-league for the LLSF to attract a wider range of participants and ease entering the robot competition.

1 Introduction

Research on autonomous mobile robots in industrial applications has significantly increased during the last years. In industry, mobile robots are the most complex variant of cyber-physical systems, embedded devices that combine computational resources with physical interaction. The context is the Industry 4.0 movement [1] whose goals are adaptive production capabilities, where material flows and production processes are dynamic and factories can output a variety of product types. In this scenario, a group of mobile robots can be used to move material and handle machines, eventually delivering the resulting products.

The RoboCup Logistics League Sponsored by Festo (LLSF) strives to address these issues in a competitive scenario to foster research, system integration, and robotics education relevant to industry. While aspects like competition and scoring points are clearly influenced by RoboCup, other aspects like the referee box as an order issuing system and time as a critical factor are driven by demands from industrial applications. Many basic robotics problems like self-localization, collision avoidance and perception must also be tackled in the LLSF. The character of the game emphasizes research and application of methods for efficient planning, scheduling, and reasoning on the optimal work order of production processes handled by a group of robots. An aspect that distinctly separates this

© Springer International Publishing Switzerland 2015
R.A.C. Bianchi et al. (Eds.): RoboCup 2014, LNAI 8992, pp. 220–232, 2015.
DOI: 10.1007/978-3-319-18615-3_18

league from others is that the environment itself acts as an agent by posting orders and controlling the machines' reactions. This is what we call *environment agency*. Naturally, dynamic scenarios for autonomous mobile robots are complex challenges in general, and in particular if multiple competing agents are involved. In the LLSF, the large playing field and material costs are prohibitive for teams to set up a complete scenario for testing, let alone to have two teams of robots. Additionally, members of related communities like planning and reasoning might not want to deal with the full software and system complexity. Still they often welcome relevant scenarios to test and present their research.

Therefore, we have created an *open simulation environment* to support research and development. There are three core aspects in this context:

1. The simulation should be a turn-key solution with simple interfaces,
2. the world must react as close to the real world as possible, including in particular the machine responses and signals, and
3. various levels of abstraction are desirable depending on the focus of the user, e.g. whether to simulate laser data to run a self-localization component or to simply provide the position (possibly with some noise).

With this work, we provide such an environment. It is based on the well-known Gazebo simulator addressing these issues: (1) its wide-spread use and open interfaces in combination with our models and adapters provides an easy to use solution; (2) we have connected the simulation directly to the referee box, the semi-autonomous game controller of the LLSF, so that it provides precisely the reactions and *environment agency* of a real-world game; (3) we have implemented *multi-level abstraction* that allows to run full-system tests including self-localization and perception or to focus on high-level control reducing uncertainties by replacing some lower-level components using simulator ground truth data. This allows to develop an idealized strategy first, and only then increase uncertainty and enforce robustness by failure detection and recovery.

We propose a new simulation sub-league for the LLSF based on the Gazebo simulator at different levels of difficulty using the multi-level abstraction, to attract more teams and ease entering the LLSF robotics competition.

In Sect. 2 we give a brief introduction to the LLSF, followed by related work in Sect. 3, before we get into the details of the simulation in Sect. 4. We describe several applications of the simulation in game strategy evaluation in Sect. 5. We propose a simulation sub-league and conclude in Sect. 6.

2   RoboCup Logistics League

RoboCup is an international robotics competition particularly well-known for its various soccer leagues. The Logistics League Sponsored by Festo (LLSF) is an application-oriented major league since 2012 regarding simplified production logistics. Groups of up to three robots have to plan, execute, and optimize the material flow in a factory automation scenario and deliver products according to dynamic orders. Therefore, the challenge consists of creating and adjusting a production plan and coordinate the group of robots [2].

Fig. 1. Carologistics (three robots with omni-vision tower) and TUMBendingUnits (robots on the left and right) during the LLSF finale at the German Open 2014 (Color figure online)

Since 2014, the two formerly separate playing fields have been merged into one 11.2 m × 5.6 m in size (cf. Fig. 1). Two teams are playing at the same time competing for points, (travel) space and time. Each team has an exclusive input storage (blue areas) and delivery gates (green area). Machines are represented by the RFID-readers with signal-lights on top. The signal-lights indicate the current status of a machine, such as "ready", "producing" and "out-of-order". There are three delivery gates, one recycling machine, and twelve production machines per team. Material is represented by orange pucks with an RFID tag. At the beginning all pucks have the raw material state S_0, and can be refined through several stages to final products using the production machines. These machines are assigned a type which determines what inputs are required and what output will be produced, and how long this conversion will take. Figure 2 shows the production trees for two final products, P_1 and P_3. The latter is rather simple, it requires only a single step, but it scores only a small amount of points. For P_1, four refinement steps are required. Only raw material pucks are available at the beginning, all others must be produced by the robot using the appropriate machines. The machines are distributed randomly across the field (mirrored at the narrow middle axis so both teams have the same travel

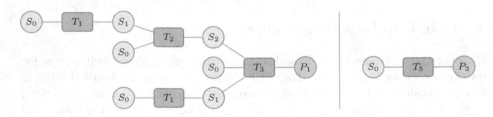

Fig. 2. Production Chains for two types of products.

distances). The rulebook [3] describes the game in more detail. For 2015, the league has decided to introduce real production steps through the use of the Festo Modular Production System [4] to make the game easier to understand.

The game is controlled by the referee box (refbox), a software component which keeps track of puck states, instructs the light signals, and posts orders to the teams [4]. After the game is started, no manual interference is allowed, robots receive instructions only from the refbox. Teams are awarded with points for delivering ordered products, producing complex products, and recycling.

The standard robot platform of this league is the Robotino by Festo Didactic [5]. The Robotino was developed for research and education and features omni-directional locomotion, a webcam, infrared distance sensors and bumpers. It is also equipped with a static puck holder to move pucks. The teams may equip the robot with additional sensors and computation devices. For example, the robots of the Carologistics team in Fig. 1 are equipped with a laptop, a omni-directional camera and a laser range finder. In 2014, Festo released the new version 3 of the robot featuring better driving, computing unit, and extensibility.

3   Related Work

There is a plethora of existing robotic simulations. Here, we mention two particular examples used in other RoboCup leagues and the Robotino simulator provided by Festo as a comparison. We also mention the Fawkes robot software framework and related systems, as it is used for agent strategy evaluation and as a prototyping development environment for the simulation.

3D Soccer Simulation League (3DSSL). The 3DSSL is based on the Open Source multi-robot simulator SimSpark and currently uses the Nao as model for the players. Software agents communicate with a soccer server to announce their actions but also to communicate with other players. SimSpark features multi-platform support, is scriptable via Ruby and supports automatically enforced rules to realize soccer rules. Today, solutions for body control and perception must be developed similar to as if a real robot were used. The agent has to take limited skills of the simulated robot and possible problems in their execution into account [6]. SimSpark is mostly specialized on the 3DSSL, thus Gazebo brings many more existing models for, e.g., sensors we could reuse. Especially due to the extensibility of the robots in the LLSF, we therefore prefer Gazebo.

Rescue Simulation League (RSL). The RSL aims to benchmark software agents and robots in a disaster scenario. A part of the RSL is the Virtual Robot Competition (VRC) which has the task to find victims in a disaster environment with a team of robots and a human operator. Important research issues of this league are victim detection, utility-based mapping of the environment, autonomous navigation and multi-robot coordination. To foster transfer of the research results to a real application, the simulation features graphical and physical realism [7]. The VRC is based on the multi-robot simulator USARSim. Initially focused on urban search and rescue simulations it has evolved into a general purpose simulator which is also used to simulate RoboCup@Home scenarios [8]. USARSim uses the Unreal graphics and PhysX physics engine. These allow a graphically and physically realistic environment. The simulator can be interfaced with ROS what allows having the same interface as on a real robot [9,10]. We preferred Gazebo instead of USARSim because of the plethora of available models and because USARSim relies on non-free components like the Unreal engine.

Robotino Sim Professional. Robotino Sim Professional[1] is a simulator for the Robotino developed by its manufacturer Festo. It is used for the Robotino in general. An environment resembling the LLSF is *not* provided. Its closed nature make it rather hard to extend or modify and it does not run on the Linux operating system. Therefore it is unsuitable as an open simulation.

3.1   Fawkes

Robot software systems are increasingly complex in the number of components as well as in their interactions. Therefore, a robot software framework which provides a suitable middleware, basic modules and auxiliary libraries is typically used. An often used candidate is the *Robot Operating System* [11] (ROS). In this work, we use the component-based framework Fawkes[2] [12]. It uses a hybrid blackboard and messaging middleware for inter- and intra-component communication. Compared to ROS, it supports a closer integration of the various components. These are implemented as run-time loadable plugins consisting of one or more threads. These threads can be invoked coordinated in a common main loop or run concurrently.

We use Fawkes for two primary functions. First, it is used to implement and integrate the software components that drive our robot, including self-localization, navigation and collision avoidance, and behavior control. Fawkes serves as one particular example how to connect to the simulation. Similarly other frameworks like ROS could be used. Second, we have implemented simulated inter-robot communication and the connection between the referee box and the simulation as Fawkes plugins. We intend to integrate these two aspects into Gazebo plugins for general use at a later time. We will give more details on this in the following section.

4   Simulation

A robotic simulation is a tool to ease testing and debugging of robotic applications, and in the context of the LLSF we also want to lower the entry barrier to participate. We have developed a simulation of the LLSF based on Gazebo [13]. An example scene is shown in Fig. 3. The simulation is designed to achieve

---

[1] http://www.festo-didactic.com/int-en/learning-systems/software-e-learning/robotino-sim-view/robotino-sim-professional.htm.

[2] Fawkes is Open Source software and available at http://www.fawkesrobotics.org.

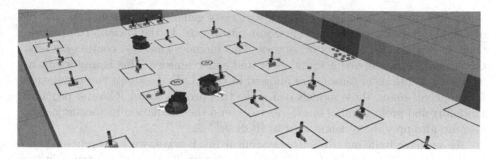

Fig. 3. The simulation of the LLSF in Gazebo. The circles above the robots indicate their localization and robot number.

the five goals realism, multi-level abstraction, compatibility with the real robot, expandability and allowing multi-robot strategy evaluation. The realism determines how similar a robotic system behaves in the simulation and the real world. That includes sensor data, physical and logical behavior in the simulation that is similar to their real-world counterpart, e.g. with similar noise and precision. Compatibility with the real robot allows the robot software to operate in the simulation with the same interfaces as on the real robot. This allows an easy transfer from a robot system that works in the simulation to the real world. The LLSF is a multi-robot scenario. Therefore we need to simulate multiple robots at the same time efficiently. The LLSF allows additions of arbitrary sensors and the rules are steadily evolving [4]. Therefore the simulation must be expandable.

4.1 Gazebo

As a basis for our simulation, we use the Open Source robot simulator Gazebo [14] (http://gazebosim.org). It can simulate various robots and their environment in a 3D world and is used in many applications, for example in the *DARPA Virtual Robotics Challenge* which is about solving challenging tasks in a disaster scenario and requires a high realism of the simulator. Gazebo uses the *Ogre* rendering engine for graphically realistic environments with reflections, shadows and detailed textures. This is important for the simulation of camera sensors as reflections of light sources and inhomogeneous lighting can cause problems. Gazebo can use both physics engines *Open Dynamics Engine (ODE)* or *Bullet*.

A Gazebo simulation environment is described as a *world* which contains certain objects according to models, the light sources, and parameters e.g. for the physics simulation or rendering. *Models* describe physical objects such as a table or a robot. The model is built out of *links* of various geometries and *joints* which connect two links and define possible relative motions. Beside the physical and visual description of objects, there are also *plugins*. These consist of executable code that interacts with the simulation at run-time. They typically model the behavior of objects. For example, there are plugins for getting sensor data, applying motor commands and world-plugins for spawning new models. We will also use plugins later to connect the simulation to the LLSF referee box.

Out of the box, Gazebo includes a variety of generalized sensors and models of common robots, sensors and environment-objects. In the Gazebo framework, there is support for a variety of sensors. This includes cameras, contact-sensors, GPS, inertial measurement units, laser and sonar sensors. This is important as arbitrary additions of sensors are allowed in the LLSF. To adapt the simulation for a specific team, these models can make this process faster. Gazebo provides modeling and programming interfaces to extend the simulation to specific needs. Gazebo also provides a connection to ROS.

As we have built our Robotino system using Fawkes, we have replaced the hardware accessing plugins by simulation adapters. The interfaces towards the other components remain unchanged, such that sensor processing, path planning, or behavior control continue to function without modifications.

Protocol Buffers (protobuf) are a data interchange format developed by Google[3]. Given a message specification it generates native (de-)serialization code for various programming languages. Protocol buffers are used for Gazebo's internal communication and as an external interface by means of a publisher/subscriber model (orange connections in Fig. 4). The refbox uses protobuf messages to communicate with robots (UDP broadcast) and the simulation (multiplexed stream protocol). Additionally we have implemented a module to simulate typical wifi communication problems like latency or packet loss.

4.2 Architecture

The basic parts for the simulation environment are referenced in the world file, which contains the walls, ground floor, the machines, pucks, and markings. In Fig. 4, the middle box represents the Gazebo process. Based on the models (red) Gazebo can run and visualize the environment. It provides an application programming interface (API, green). The protobuf-based messaging middleware is used for internal communication as well as an external interface. Gazebo also hosts a number of plugins (blue). These plugins are active components (executable code) that typically maintain, provide, and process simulation data. For example, the LLSF environment plugin communicates with the refbox to provide the same field reactions as in the real world. If a puck is moved under a machine, a message is sent to announce this to the refbox, which in return sends instructions how the light signals change. The robot actually consist of a number of plugins, typically one per sensor or actuator. We use, for example, the built-in plugin for the laser data, and have a custom plugin for the robot movements.

The robot software itself is connected again using protobuf. It gathers data from and sends instructions to the robot-specific plugins. For now, we are using Fawkes for this connection. But the existing Gazebo-ROS integration would make the adaptation simple. Also, the refbox connection is currently proxied through a Fawkes plugin, which we will change before the public release such that the simulation itself can be accessed from any framework.

---

[3] https://developers.google.com/protocol-buffers.

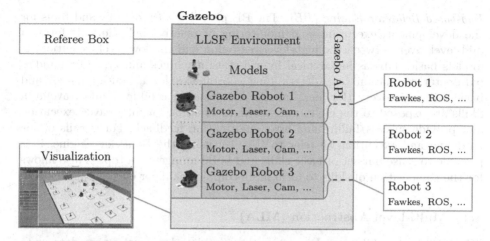

Fig. 4. Architecture overview of the simulation; blue means Gazebo plugins, red models, green Gazebo API and middleware, orange are components connected via protobuf middleware. The robots are driven by a system like Fawkes or ROS, referee box and visualization operate and visualize the environment (Color figure online).

To handle the simulation speed, which can be smaller than real-time on a slow computer or greater to speed up automated test runs, we publish the simulation time and the current simulation speed, which can be used to estimate the simulation time between two synchronization messages. We extended the refbox to use the simulation time. In Fawkes, the simulation time is provided instead of the system time.

4.3 Simulation Interfaces

We provide three different ways to interact with the simulation: direct interaction using native Gazebo messaging, accessing the simulation through Fawkes, or using our Behavior Engine instruction interface. With this variety of interfaces it is possible to use existing own software, or re-use (parts of) our components.

Direct Interaction. Using Gazebo's protobuf-based middleware most aspects of the simulation are directly accessible via topics. For example, models can be directly manipulated, e.g. to implement specific robot modifications. It also allows direct access to internal data, e.g. from additional sensors. The direct interaction is also accessible from the ROS connection provided by Gazebo.

Fawkes. Our system already comes with many components for interaction with the simulation. Communication with these components happens over the Fawkes middleware. For example, Fawkes provides access to generated camera images to run a perception module, or directly access the puck positions over the blackboard for a higher abstraction level. The interfaces are unified across simulation and the real robots, allowing a seamless migration between the two.

Lua-based Behavior Engine (BE). The BE provides a framework and tools for the development, execution, and monitoring of reactive behaviors. It forms a mid-level layer between the high-level reasoning and the low-level execution. It models basic skills as hierarchical hybrid state machines and typically handles parameter estimation (e.g. determining coordinates for an entity name) and basic failure recovery (for local failures like losing a product while traveling). Skills are exposed to the high level system as function for interleaved execution and provide success/failure and error information feedback. For details of the modeling and execution we refer to [15]. By using the Behavior Engine, it is possible to reuse our set of basic skills and build an agent on top. It also allows for the composition of skills to form more complex skills or even an agent itself.

4.4   Multi-level Abstraction (MLA)

MLA [16] is the ability to choose to either simulate low-level sensor data or to directly extract higher-level information. In terms of simulation interfaces, the abstraction level is defined based on the Gazebo topic chosen as input data. In Fawkes, this manifests in different components which are used to access simulation data. For example, as depicted in Fig. 5, the simulation (red boxes) provides both, laser data and pose information in different topics. In Fawkes, either an accessor plugin is used (higher abstraction), or the simulated laser data is fed to the localization component (lower abstraction). Additionally, for actuation MLA, we provide the BE as a basis for high-level control software. Note that Fawkes and the BE serve as one particular example, but they are not a requirement. Similar structures can be implemented for example using ROS.

While this is not an aspect unique to our simulation, the explicit specification is necessary to clarify the capabilities of the simulation, in particular to attract new users from the planning community. MLA is also useful during development, as it allows to perform full integration tests with all software components (including sensor processing), or to focus on high-level evaluation.

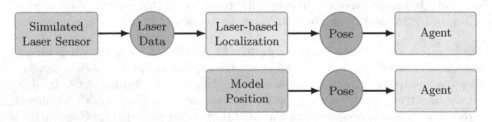

Fig. 5. Multi-level abstraction example: the simulation can either provide the data to run a laser-based self-localization component or directly provide the position (Color figure online).

5 Applications

We have used the simulation for rapid testing, performance evaluation and as a development environment. We have also conducted experiments to verify the similarity between simulation and real world robot behavior.

5.1 Agent Strategy Evaluation

With a focus on the planning and scheduling aspects in the LLSF, efficient testing of such systems becomes an important concern. Our first application of the simulation is aimed at the evaluation of different game strategies. We have developed tools to run several games unattended over night and present the results later for comparison, e.g. in terms of achieved score. Our reasoning component, based on the CLIPS rules engine, implements incremental task-level reasoning [17]. We evaluated different numbers of robots on the field or static and dynamic strategies, where robots would either pursue the same goal (like producing a P_1 product) the whole time or where they could change their role at run-time [13]. This allowed us to steer our development efforts more informed to improve the system's overall performance.

The CPU usage of a simulation game on a desktop with an Intel Core i7-3770 at 3.40 GHz with 16 GB of RAM is shown in Fig. 6. The system was utilized at about 60 % of capacity including the simulation and 3 robots running all necessary components except visual perception which was given by the simulation.

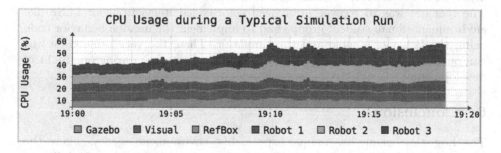

Fig. 6. CPU usage during a simulation game; X axis shows the system time during the experiment; Y axis areas show percent of CPU time used stacked on top of each other.

5.2 Simulation vs. Real World

There is typically a gap between the simulation and the real robot. For example, the lighting of the scene tends to be more uniform and stable in a simulation. Still some gaps can be bridged, for example using accurate friction parameters for the pucks or to have similar noise and precision in the laser data. We have compared simulation runs to our performance at the German Open 2014. We ran the same software versions, the only differences being that hardware modules were replaced by simulation modules and machine-signal perception is given by the simulation. In ten simulated games the system achieved an average score of 75.3 points with a standard deviation of 10.1. In the last five games of the competition we achieved 61.6 points with a standard deviation of 11.6. Overall teams scored up to 71 points with average 19.1 ± 23.3. While this is certainly not an exhaustive comparison, we believe that the data provides strong evidence that the simulation behaves accurately compared to the real world.

We have further corroborated this by comparing trajectory repeatability in simulation
and the real world. Figure 7 shows the concave
hull around position information (robot's self-
localization) recorded at 1 Hz during ten back
and forth drive operations between machines
M1 and M12. As we can see the overall trajectory execution is similar with only small differences at larger free areas where higher speeds
can be achieved. Perception performance on
simulated images for the camera perspective
depends on the methods used as producing the
bright spotlights on active lights is problematic.
Example images are shown on the website.

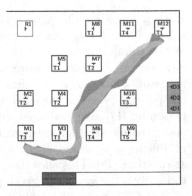

Fig. 7. Trajectory comparison of real world moves (blue) and simulated moves (red) on a half field (Color figure online).

Student Hackathon 2013. We conducted a hackathon with about 30 participants in cooperation with the student organization Bonding.
The scenario was to recover prioritized color-coded items from an LLSF-like environment. Small student groups had to implement the mission behavior code using the Behavior Engine and the simulator. Then, the very same code was run on the real robot. Several teams accomplished the task during one night of development.

6   Conclusion

The Logistics League has unique aspects like environment agency, where the environment acts like an agent itself, and a focus on planning and scheduling in robotics. We have developed a simulation based on Gazebo which takes these into account. It connects to the referee box of the LLSF to provide the same reactions as the real environment. Multi-level abstraction allows to choose whether to process simulated sensor data like images or directly use information like signal light states. This way, the simulation can provide multiple levels of difficulty. This allows to run full integration tests or to focus on the development of the planning and scheduling system. As Gazebo is widely used and provides a messaging middleware, it is open for systems of other teams. We provide adapters for the Open Source robot software framework Fawkes, but others can be added easily.

Having to develop and maintain a team of robots can be prohibitively costly in terms of maintaining a local playing field. Other communities we want to reach might not even be interested in low-level robotics software. Therefore, we propose to establish a simulation league directly associated with the LLSF. It would serve as an entry to the LLSF, which is simplified by multi-level abstraction and the components we provide, and as a catalyst for new planning strategies, which could be developed and tested in simulation before being ported to the real robots. At RoboCup 2014, we seek a discussion with interested parties within and outside the LLSF to form an interest group to pursue this goal.

Our software components, instructions how to setup the simulation, all evaluation data, and further information is available on the project website at http://www.fawkesrobotics.org/projects/llsf-sim/.

Acknowledgments. We thank the Carologistics RoboCup Team for their tremendous effort to develop a system which served as a basis for the presented work.

F. Zwilling and T. Niemueller were supported by the German National Science Foundation (DFG) research unit *FOR 1513* on Hybrid Reasoning for Intelligent Systems (http://www.hybrid-reasoning.org). We thank the anonymous reviewers.

References

1. Kagermann, H., Wahlster, W., Helbig, J.: Recommendations for implementing the strategic initiative INDUSTRIE 4.0. Final Report, Platform Industrie 4.0 (2013)
2. Niemueller, T., Ewert, D., Reuter, S., Ferrein, A., Jeschke, S., Lakemeyer, G.: RoboCup logistics league sponsored by festo: a competitive factory automation testbed. In: Behnke, S., Veloso, M., Visser, A., Xiong, R. (eds.) RoboCup 2013. LNCS, vol. 8371, pp. 336–347. Springer, Heidelberg (2014)
3. LLSF Technical Committee: RoboCup Logistic League sponsored by Festo - Rules and Regulations 2014 (2014). http://www.robocup-logistics.org/rules
4. Niemueller, T., Lakemeyer, G., Ferrein, A., Reuter, S., Ewert, D., Jeschke, S., Pensky, D., Karras, U.: Proposal for advancements to the LLSF in 2014 and beyond. In: ICAR - 1st Workshop on Developments in RoboCup Leagues (2013)
5. Karras, U., Pensky, D., Rojas, O.: Mobile robotics in education and research of logistics. In: IROS 2011 - Workshop on Metrics and Methodologies for Autonomous Robot Teams in Logistics (2011)
6. Boedecker, J., Asada, M.: SimSpark-concepts and application in the RoboCup 3D soccer simulation league. In: SIMPAR - WS on RoboCup Simulators (2008)
7. Akin, H.L., Ito, N., Jacoff, A., Kleiner, A., Pellenz, J., Visser, A.: RoboCup rescue robot and simulation leagues. AI Mag. **34**, 78–86 (2013)
8. van Noort, S., Visser, A.: Extending virtual robots towards RoboCup soccer simulation and @Home. In: Chen, X., Stone, P., Sucar, L.E., van der Zant, T. (eds.) RoboCup 2012. LNCS, vol. 7500, pp. 332–343. Springer, Heidelberg (2013)
9. Carpin, S., Lewis, M., Wang, J., Balakirsky, S., Scrapper, C.: USARSim: a robot simulator for research and education. In: IEEE International Conference on Robotics and Automation (ICRA) (2007)
10. Kootbally, Z., Balakirsky, S., Visser, A.: Enabling codesharing in rescue simulation with USARSim/ROS. In: Behnke, S., Veloso, M., Visser, A., Xiong, R. (eds.) RoboCup 2013. LNCS, vol. 8371, pp. 592–599. Springer, Heidelberg (2014)
11. Quigley, M., Conley, K., Gerkey, B.P., Faust, J., Foote, T., Leibs, J., Wheeler, R., Ng, A.Y.: ROS: an open-source Robot Operating System. In: ICRA Workshop on Open Source Software (2009)

12. Niemueller, T., Ferrein, A., Beck, D., Lakemeyer, G.: Design principles of the component-based robot software framework Fawkes. In: Ando, N., Balakirsky, S., Hemker, T., Reggiani, M., von Stryk, O. (eds.) SIMPAR 2010. LNCS, vol. 6472, pp. 300–311. Springer, Heidelberg (2010)
13. Zwilling, F.: Simulation of the RoboCup logistic league with Fawkes and Gazebo for multi-robot coordination evaluation. Bachelor's thesis, RWTH Aachen University, Knowledge-Based Systems Group, December 2013
14. Koenig, N., Howard, A.: Design and use paradigms for gazebo, an open-source multi-robot simulator. In: International Conference on Intelligent Robots and Systems (2004)
15. Niemüller, T., Ferrein, A., Lakemeyer, G.: A lua-based behavior engine for controlling the humanoid robot Nao. In: Baltes, J., Lagoudakis, M.G., Naruse, T., Ghidary, S.S. (eds.) RoboCup 2009. LNCS, vol. 5949, pp. 240–251. Springer, Heidelberg (2010)
16. Beck, D., Ferrein, A., Lakemeyer, G.: A simulation environment for middle-size robots with multi-level abstraction. In: Visser, U., Ribeiro, F., Ohashi, T., Dellaert, F. (eds.) RoboCup 2007. LNCS (LNAI), vol. 5001, pp. 136–147. Springer, Heidelberg (2008)
17. Niemueller, T., Lakemeyer, G., Ferrein, A.: Incremental task-level reasoning in a competitive factory automation scenario. In: AAAI Spring Symposium 2013 - Designing Intelligent Robots: Reintegrating AI (2013)

Automatic Robot Calibration for the NAO

Tobias Kastner[1]($^{\boxtimes}$), Thomas Röfer[2], and Tim Laue[1]

[1] Fachbereich 3 – Mathematik und Informatik, Universität Bremen,
Postfach 330 440, 28334 Bremen, Germany
{dyeah,tlaue}@informatik.uni-bremen.de

[2] Deutsches Forschungszentrum Für Künstliche Intelligenz, Cyber-Physical Systems,
Enrique-Schmidt-Str. 5, 28359 Bremen, Germany
thomas.roefer@dfki.de

Abstract. In this paper, we present an automatic approach for the kinematic calibration of the humanoid robot NAO. The kinematic calibration has a deep impact on the performance of a robot playing soccer, which is walking and kicking, and therefore it is a crucial step prior to a match. So far, the existing calibration methods are time-consuming and error-prone, since they rely on the assistance of humans. The automatic calibration procedure instead consists of a self-acting measurement phase, in which two checkerboards, that are attached to the robot's feet, are visually observed by a camera under several different kinematic configurations, and a final optimization phase, in which the calibration is formulated as a non-linear least squares problem, that is finally solved utilizing the *Levenberg-Marquardt* algorithm.

1 Introduction

Calibration is the process of determining the relevant parameters of a robotic system by comparing the prediction of the system's mathematical model with the measurement of a known feature, which is considered as the *standard*. If the difference between the prediction and the measurement exceeds a certain tolerance, it is inevitable to compensate for this mismatch to allow the robot to operate as desired. The automatic calibration method presented in this paper is customized for the humanoid robot NAO [4] from *Aldebaran Robotics*. The NAO has 21 degrees of freedom and is equipped with two cameras. The images are taken at a frequency of 30 Hz by each camera while other sensor data (such as joint angle measurements) are updated at 100 Hz. The operating system is an embedded Linux that is powered by the *Intel-Atom* processor at 1.6 GHz. Since 2008, the NAO is the official platform of the RoboCup Standard Platform League (SPL). To play soccer, a NAO has to be able to walk fast and to score goals. The overall performance of these two essential tasks strongly depends on prior precise calibration, which, so far, is a manual procedure that requires the assistance of human users. It is time-consuming and error-prone, since the errors in the kinematics are estimated by eye and hand. These errors arise from the imprecise assembly of the robot itself and damage and wearout of the motors

© Springer International Publishing Switzerland 2015
R.A.C. Bianchi et al. (Eds.): RoboCup 2014, LNAI 8992, pp. 233–244, 2015.
DOI: 10.1007/978-3-319-18615-3_19

during its operation. Consequently, it is often necessary to recalibrate a NAO after a match. The goal of the automatic calibration that is presented in this paper is to reduce the workload of human users, while providing comparable appropriate calibration parameters.

Markowsky [8] developed a method to calibrate the NAO's leg kinematics by using a *hill climbing* algorithm and a cost function that computes the distribution of the current and weight loads of the two legs. The assumption was that the NAO is calibrated if the loads of each leg are similar. Due to poor sensor readings, this method turned out to be inapplicable. This paper is rather inspired by the works of Birbach *et al.* [2], Pradeep *et al.* [11], and Nickels [10]. These methods are based on the *hand-eye calibration*, which is explained by Strobl *et al.* [13]. The goal of these calibration procedures is to find the poses of the cameras in the robot's head as well as the joint offsets that affect the proper positioning of the robot's end effectors. The robot's end effectors are driven to pre-recorded positions and markers attached to these end effectors are visually detected by the robot's cameras. Each position of the marker's features in pixel coordinates, as well as the respective joint angle measurements of the kinematics are gathered and finally used to minimize the deviation between the actual visual measurements and the predicted ones based on the mathematical model of the robot, i. e. forward kinematics and extrinsic and intrinsic camera parameters.

In this paper, the NAO is calibrated by moving two checkerboards, which are attached to the robot's feet, in front of the lower camera of the NAO's head. Similar to the related work presented, the image position of each vertex (point of contact of two isochromatic tiles) on the checkerboard is measured together with the current joint angles. To formulate the calibration as a problem of non-linear least squares, the forward kinematics of the NAO are computed with homogeneous coordinate transformation and the projection function of the cameras is described as a pinhole camera model. Using an a-priori intrinsic camera calibration and the *Levenberg-Marquardt* algorithm, the 21 parameters involved (12 joint offsets, 3 camera rotation errors and 6 correction parameters for imprecisely mounted checkerboards) are estimated by minimizing the sum of squared residuals of each measured vertex in pixels and the projection of the corresponding vertex from robot-relative coordinates to 2D image points, utilizing the joint angles measured.

The remainder of this paper is structured as follows: in the next section, the properties of the checkerboard and foot-assembly as well as the formulation of the calibration as a problem of non-linear least squares are discussed. Section 3 presents the results achieved, followed by Sect. 4, which concludes the paper and gives an outlook on possible future work and improvements to the calibration procedure presented.

2   Automatic Robot Calibration

First of all, the properties of the checkerboard sandals are discussed, followed by some implementation details and the definition of the calibration as a non-linear least squares problem.

2.1   Checkerboard Sandals

The checkerboard pattern with 35 tiles (7 columns and 5 rows) is printed on a customized aluminum composite panel (see Fig. 1b) with a square size of 2.5 cm. Each board is mounted with four pins that exactly fit into the recesses located at the force-sensing resistors (FSR) on the bottom of the NAO's feet (soles) and a further fixture with a hook and loop fastener tape. Since the positions of the FSRs (see Fig. 1a) are known relative to the projection of the ankle joint on the sole, the positions of all 24 checkerboard vertices can be computed easily, considering that the center of the checkerboard is situated 17 cm away from the joint projection mentioned. These positions are used to predict the image coordinates of each vertex using the mathematical model of the NAO that still has to be defined.

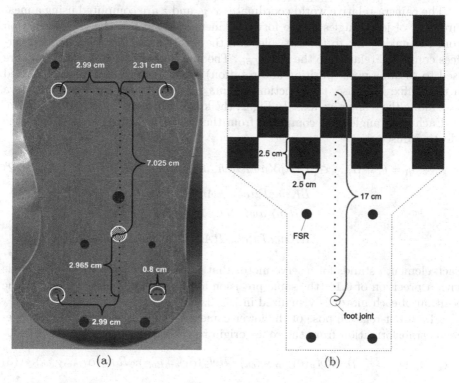

(a)   (b)

Fig. 1. The positions of the FSRs on the sole (a). The checkerboard sandal with some dimension information (b).

2.2   Implementation

The calibration is implemented using the $C++$ framework [12] for modular development of robot control programs, published by the SPL team *B-Human*. In the context of this paper, a motion control engine, a checkerboard recognizer, an optimization algorithm, and a calibration control module were developed. Any more in-depth information on those software components can be found in the corresponding thesis [5].

As proposed by Wiest [15] and Elatta *et al.* [3], the calibration is subdivided into four phases *modeling, measurement, identification,* and *compensation.*

Modeling. The conversion from world to image coordinates is modeled with the pinhole-camera model P:

$$P(x, y, z) = O_C - (y, z)^T \frac{f_l}{x}. \tag{1}$$

The optical center O_C and the focal length f_l were determined in a prior intrinsic camera calibration, as proposed by Zhang [16].

The camera-relative world coordinates x, y, and z are computed using a measured set of joint angles q, the forward kinematics $T_{LS}^O(q)$, the transformation from the ankle joint to the left sole, and the positions of the checkerboard vertices computed relative to the feet $p_{LS\alpha}$. The functions Rot_x, Rot_y, and Rot_z are used to rotate a pose (position and rotation) around the respective axis, named in the suffix. Likewise, the functions $Trans_x$, $Trans_y$, and $Trans_z$ translate a pose along the respective axis. In Fig. 2, the kinematic tree of the NAO is shown.

Each joint angle q_i is composed from the actual measured angle θ_i and the affecting offset α_i:

$$q_i = \theta_i + \alpha_i, \ i \in \Big\{ LHipYawPitch, \ LHipRoll, \ LHipPitch, \tag{2}$$
$$LKneePitch, \ LAnklePitch, \ LAnkleRoll,$$
$$RHipYawPitch, \ RHipRoll, \ RHipPitch,$$
$$RKneePitch, \ RAnklePitch, \ RAnkleRoll \Big\}.$$

Each element i stands for a servo motor that is able to measure its current angle with a precision of $0.1°$ (the same precision applies to commanded angles). The position of each motor is visualized in Fig. 2.

The robot-relative pose of the lower camera is computed with T_C^O (homogeneous transformation from the robot origin to the camera):

$$T_C^O(q, \alpha) = T_C^O(q) \cdot Rot_y(\alpha_{CamPitch}) \cdot Rot_x(\alpha_{CamRoll}) \cdot Rot_z(\alpha_{CamYaw}). \tag{3}$$

It is assumed that the actual camera pose $T_C^O(q)$ is affected by three rotational offsets $\alpha_{CamPitch}$, $\alpha_{CamRoll}$, and α_{CamYaw}.

With the information on the checkerboard given in Sect. 2.1, the coordinates of all vertices p_{LS}, relative to the soles, can be computed.

$$p_{LS\alpha} = Rot_z(\alpha_{LRotZ}) \cdot Trans_x(\alpha_{LTransX}) \cdot Trans_y(\alpha_{LTransY}) \cdot p_{LS}. \tag{4}$$

Since the pins on the boards are glued, it is possible that the assembly on the soles is imprecise. In addition to the joint and camera correction offsets, three

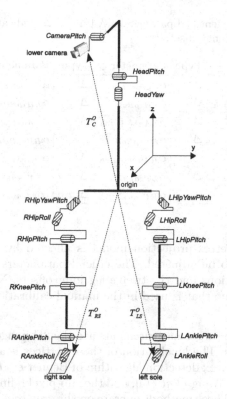

Fig. 2. The kinematic tree of the NAO. The origin is located between the two hip motors. The *CameraPitch* is a fixed assembly angle, taken from the official NAO documentation. Any other motor angles are modifiable.

further parameters for each board are considered. It is assumed that the boards have a rotational offset α_{LRotZ} around the z-axis and a translational error in x and y direction ($\alpha_{LTransX}$ and $\alpha_{LTransY}$).

The final model is defined with $proj(p_{LS}, \theta, \alpha)$:

$$proj(p_{LS}, \theta, \alpha) = P(T_C^O(q, \alpha)^{-1} \cdot T_{LS}^O(q) \cdot p_{LS\alpha}). \qquad (5)$$

A given sole-relative vertex coordinate is transformed into origin coordinates, which gets further converted into camera-relative coordinates and is finally projected onto the NAO's camera plane, resulting in a prediction, where a vertex should be located in the image, considering the current joint data and the calibration parameters.

The definitions given above are for the left leg kinematics. The missing projection function for the right leg is computed analogously, using the right leg's joint data and the vertex positions from the right checkerboard. Also, only the leg's kinematics are adjusted, since, so far, there are no actions that require precisely calibrated arms.

Table 1. Listing of all identified parameters. A type of $\Delta°$ indicates a rotational offset and a Δmm a translational offset.

Parameter	Type	Parameter	Type	Parameter	Type
α_{CamYaw}	$\Delta°$	$\alpha_{CamPitch}$	$\Delta°$	$\alpha_{CamRoll}$	$\Delta°$
$\alpha_{LHipYawPitch}$	$\Delta°$	$\alpha_{LHipRoll}$	$\Delta°$	$\alpha_{LHipPitch}$	$\Delta°$
$\alpha_{LKneePitch}$	$\Delta°$	$\alpha_{LAnklePitch}$	$\Delta°$	$\alpha_{LAnkleRoll}$	$\Delta°$
$\alpha_{RHipYawPitch}$	$\Delta°$	$\alpha_{RHipRoll}$	$\Delta°$	$\alpha_{RHipPitch}$	$\Delta°$
$\alpha_{RKneePitch}$	$\Delta°$	$\alpha_{RAnklePitch}$	$\Delta°$	$\alpha_{RAnkleRoll}$	$\Delta°$
α_{LRotZ}	$\Delta°$	$\alpha_{LTransX}$	Δmm	$\alpha_{LTransY}$	Δmm
α_{RRotZ}	$\Delta°$	$\alpha_{RTransX}$	Δmm	$\alpha_{RTransY}$	Δmm

Altogether, the vertex-projection model is affected by 21 parameters (see Table 1), that need to be adjusted. The crucial parameters are the three rotational camera corrections and the twelve leg joint offsets. Note that this set of parameters is the same that is used in the manual calibration procedures.

Measurement. The target measurements are the 2D image coordinates of the checkerboard vertices. The localization of these vertices is accomplished with a combination of the vertex detection algorithm of Bennet *et al.* [1], the chessboard extraction method of Wang *et al.* [14], and the sub-pixel refinement implemented by Birbach *et al.* [2]. These methods operate on the grey-scale images of the lower camera with VGA resolution.

To assure that the calibration can be performed secure and unsupervised, the NAO is lying on its back, while driving to 24 different leg stances (for both legs). For each configuration, the checkerboard is observed incorporating three different head postures. The configurations are chosen heuristically, considering that the checkerboard is fully visible, while having different rotations and distances to the camera. Note that the checkerboard vertices are only detected after the robot reached its desired stance, since there is a lag between the arrival of a new camera image and the sensor readings. The complete self-acting measurement phase is visualized in Fig. 3.

For each vertex observed, a new measurement m is collected and added to a set M:

$$M = \Big\{ m_0, m_1, \ldots, m_n \Big\}. \tag{6}$$

A measurement is a triple, consisting of the image position p_{img}, the corresponding sole-relative 3D coordinate of the vertex p_{cb}, and the measured joint angles θ:

$$m = (p_{img}, p_{cb}, \theta). \tag{7}$$

Identification. After the completion of the measurement phase, the optimization procedure is initialized. First of all, the set of residuals R is built, calculating

Fig. 3. Some snapshots taken during a calibration experiment. The upper images show different joint angle configurations, whereas the lower figures depict the corresponding points of view of the lower camera.

the deviation of each measurement's image coordinate $m[p_{img}]$ and the corresponding prediction, using the projection model (see Eq. 5) and the measurement's joint data $m[\theta]$, the sole-relative vertex position $m[p_{cb}]$, and the current parameter set α (see Table 1):

$$R = \left\{ r \middle| m \in M \land r = \left\| m[p_{img}] - proj(m[p_{cb}], m[\theta], \alpha) \right\| \right\}. \tag{8}$$

By summing up the squared residuals, the calibration can be formulated as a problem of non-linear least squares:

$$\arg \min_{\alpha} \sum_{r \in R} r^2. \tag{9}$$

To find a parameter set α that minimizes the sum of squared residuals, the algorithm of Levenberg [7] and Marquardt [9] is used, in which the initial parameter set α is set to zero.

Compensation. To compensate for the incorrect servo motor's sensor readings, the optimized joint offsets are added to the requested joint angles and subtracted from the measured joint values. The rotational camera offsets are considered in the calculation of the NAO's camera matrix (see Röfer et al. [12]).

3 Results

This section outlines the results of the automatic robot calibration, beginning with a simulated calibration and a concluding calibration experiment executed with a real NAO.

3.1   Simulation

With the help of *SimRobot* [6], the calibration's feasibility was tested. A simulated NAO was modeled with random erroneous joint offsets and with imprecise checkerboard assembly. As Fig. 4a and b imply, all errors were correctly compensated. Since the errors for each leg chain are different, the residual distribution for the left and right leg exhibit different appearances (see Fig. 4a). Apparent from Fig. 4b, the adjusted parameters create a normal distribution with zero mean and a standard deviation of 0.0486 pixels in x and 0.0403 pixels in y direction.

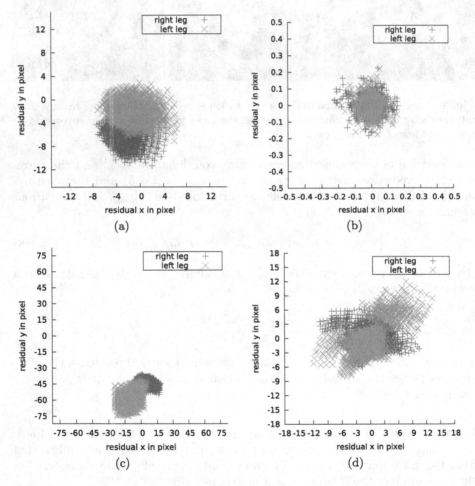

Fig. 4. Residual distribution of a simulated calibration (a) without and (b) with optimized parameters. The lower figures visualize the residual distribution of the first out of three calibration experiments on a real NAO: distribution (c) without and (d) with adjusted parameters. Note that the axes have different ranges.

Table 2. Resulting parameters of three consecutive calibration runs with a NAO.

		1	2	3	∅
α_{CamYaw}	[°]	0.4764	0.4751	0.4466	0.4660 ± 0.0168
$\alpha_{CamPitch}$	[°]	−4.5627	−4.5430	−4.4956	$−4.5338 \pm 0.0345$
$\alpha_{CamRoll}$	[°]	−1.7991	−1.8891	−1.8180	$−1.8354 \pm 0.0475$
$\alpha_{LHipYawPitch}$	[°]	−2.2938	−2.7094	−2.5055	$−2.5029 \pm 0.2078$
$\alpha_{LHipRoll}$	[°]	0.9692	0.9620	0.8863	0.9392 ± 0.0459
$\alpha_{LHipPitch}$	[°]	0.2299	1.3066	0.8783	0.8049 ± 0.5421
$\alpha_{LKneePitch}$	[°]	−1.1442	−1.9709	−1.6455	$−1.5869 \pm 0.4165$
$\alpha_{LAnklePitch}$	[°]	1.6708	1.5798	1.6349	1.6285 ± 0.0458
$\alpha_{LAnkleRoll}$	[°]	0.0267	−0.2078	−0.2618	$−0.1476 \pm 0.1534$
$\alpha_{RHipYawPitch}$	[°]	−0.9800	−0.9297	−0.9599	$−0.9565 \pm 0.0253$
$\alpha_{RHipRoll}$	[°]	0.1893	0.3001	0.1881	0.2258 ± 0.0643
$\alpha_{RHipPitch}$	[°]	1.4916	1.3355	1.3064	1.3778 ± 0.0996
$\alpha_{RKneePitch}$	[°]	−0.0156	−0.1881	0.1841	$−0.0065 \pm 0.1863$
$\alpha_{RAnklePitch}$	[°]	0.6081	0.9223	0.6810	0.7371 ± 0.1644
$\alpha_{RAnkleRoll}$	[°]	−0.5310	−1.0132	−0.7909	$−0.7784 \pm 0.2413$
α_{LRotZ}	[°]	−0.6010	−0.2357	−0.4890	$−0.4419 \pm 0.1871$
$\alpha_{LTransX}$	[mm]	−0.3878	−1.3591	−0.8144	$−0.8538 \pm 0.4868$
$\alpha_{LTransY}$	[mm]	0.5075	0.7773	0.8117	0.6988 ± 0.1666
α_{RRotZ}	[°]	0.8058	0.9085	0.7725	0.8289 ± 0.0709
$\alpha_{RTransX}$	[mm]	−2.0173	−1.0657	−1.4765	$−1.5198 \pm 0.4773$
$\alpha_{RTransY}$	[mm]	2.2007	1.7893	1.5274	1.8391 ± 0.3394
Iterations		41	47	35	41
Duration	[s]	591	600	579	590
rms x	[px]	3.22829	3.17493	3.07256	-
rms y	[px]	2.60487	2.53476	2.49331	-

3.2 NAO

On a real NAO, three consecutive automatic calibrations were executed[1]. Beforehand, an intrinsic camera calibration was done, resulting in a focal length of 562.5 pixels and an optical center of $(324, 189)^T$ pixels. The resulting parameters, as well as further information, such as the calibration duration, are shown in Table 2. One calibration round took 590 s on average and the Levenberg-Marquardt algorithm converged after 41 iterations on average. The root mean squared error was reduced to roughly three pixels in x and two and a half pixels in y direction. It is noticeable, regarding the optimized parameters, that the left

---

[1] The gathered data of the three experiments can be found as CSV files on: https://sibylle.informatik.uni-bremen.de/public/calibration/.

(a)   (b)

Fig. 5. The checkerboard projection without calibration (a). The resulting projection with optimized parameters α (b).

leg's pitch motor deviations are rather big, with a value of $0.5421°$ at a max. This implies that there are several different configurations that minimize the sum of squared residuals. In contrast, the right leg's parameters have a maximal deviation of $0.2413°$. The vast majority of the parameters were similar after the three calibration runs, in particular the three camera rotation offsets. The checkerboard assembly correction parameters are adequately small.

It is obvious that the resulting parameters are not a perfect solution to the minimization problem, since there are many residuals bigger than the threefold standard deviation (see Fig. 4d). Unlike the depiction of Fig. 5, where (a) is showing the projection of the checkerboard before and (b) after a calibration, there are countless kinematic configurations resulting in an imprecise projection. A standard test for a good calibration of the camera's pose is the projection of the modeled lines of an official SPL field back into the image from a known position on the field and comparing them with the lines that are actually seen. However, using the optimized parameters, this test showed unsatisfactory results. The most probable cause is backlash that has a different impact on a robot lying on its back than on a standing robot.

Using the parameters of the third calibration run, the NAO was able to play soccer for the duration of a half (10 min), while falling down 6 times. Without a calibration, the NAO fell down 11 times in the same amount of time. Further experiments with four other NAOs exhibited a similar result: a calibration never negatively influenced the performance of a NAO playing soccer. In one case, a robot, that was initially not able to walk half a meter, was able to play a complete half (still being very shaky and unstable).

4 Conclusion

In this paper, we presented a method to define an automatic robot calibration as a problem of non-linear least squares, customized for the humanoid robot NAO.

The NAO was modeled with the help of homogeneous coordinate transformation and the pinhole camera model. The resulting least squares problem was solved with the Levenberg-Marquardt algorithm.

The simulated calibrations resulted in perfect estimates for the compensation of the identified errors and therefore prove the plausibility of the presented approach. Performing the calibration with real NAOs turned out to be less satisfying. Apart from the fast overall operation time with roughly 10 min, the deviations of some parameters after consecutive calibrations vary strongly, the projection of field lines was rather imprecise, and there are many joint configurations that resulted in an insufficient re-projection of the checkerboard. However, the kinematic parameters can be used as a better initial guess for a manual calibration.

We think that the extension of the projection model with non-geometric errors, such as joint elasticities, could improve the results. Birbach *et al.* [2] and Wiest [15] modeled elasticities with a spring, using the torques that affect each motor. Since the NAO lacks of torque sensors, only static torques can be used. It is also conceivable that the lengths of the limbs may vary from the values in the official NAO documentation. The most difficult unregarded errors are those that result from backlash, which, according to Gouaillier *et al.* [4], might have a range of $\pm 5°$. In addition, the offsets of the two head joints might also impair the calibration result, because they are not adjusted during the optimization, since they might have a linear dependency to the camera rotation offsets.

The improvement of this calibration approach will be further investigated by considering the possible problems mentioned.

Acknowledgement. We would like to thank the members of the team B-Human for providing the software framework for this work. This work has been partially funded by DFG through SFB/TR 8 "Spatial Cognition".

References

1. Bennett, S., Lasenby, J.: ChESS - quick and robust detection of chess-board features. Comput. Vis. Image Underst. **118**, 197–210 (2014)
2. Birbach, O., Bäuml, B., Frese, U.: Automatic and self-contained calibration of a multi-sensorial humanoid's upper body. In: Proceedings of the IEEE International Conference on Robotics and Automation (ICRA), St. Paul, MN, USA, pp. 3103–3108 (2012)
3. Elatta, A.Y., Gen, L.P., Zhi, F.L., Daoyuan, Y., Fei, L.: An overview of robot calibration. IEEE J. Robot. Autom. **3**, 74–78 (2004)
4. Gouaillier, D., Hugel, V., Blazevic, P., Kilner, C., Monceaux, J., Lafourcade, P., Marnier, B., Serre, J., Maisonnier, B.: The NAO humanoid: a combination of performance and affordability. CoRR abs/0807.3223 (2008)
5. Kastner, T.: Automatische Roboterkalibrierung für den humanoiden Roboter NAO. Diploma thesis, Universität Bremen (2014)
6. Laue, T., Spiess, K., Röfer, T.: SimRobot – a general physical robot simulator and its application in RoboCup. In: Bredenfeld, A., Jacoff, A., Noda, I., Takahashi, Y.

(eds.) RoboCup 2005. LNCS (LNAI), vol. 4020, pp. 173–183. Springer, Heidelberg (2006)

7. Levenberg, K.: A method for the solution of certain problems in least squares. Q. Appl. Math. **2**, 164–168 (1944)

8. Markowsky, B.: Semiautomatische Kalibrierung von Naogelenken. Bachelor thesis, Universität Bremen (2011)

9. Marquardt, D.W.: An algorithm for least-squares estimation of nonlinear parameters. SIAM J. Appl. Math. **11**(2), 431–441 (1963)

10. Nickels, K.M., Baker, K.: Hand-eye calibration for Robonaut. Technical report, NASA Summer Faculty Fellowship Program Final Report, Johnson Space Center (2003)

11. Pradeep, V., Konolige, K., Berger, E.: Calibrating a multi-arm multi-sensor robot: a bundle adjustment approach. In: Khatib, O., Kumar, V., Sukhatme, G. (eds.) Experimental Robotics. STAR, vol. 79, pp. 211–225. Springer, Heidelberg (2012)

12. Röfer, T., Laue, T., Müller, J., Bartsch, M., Batram, M.J., Böckmann, A., Böschen, M., Kroker, M., Maaß, F., Münder, T., Steinbeck, M., Stolpmann, A., Taddiken, S., Tsogias, A., Wenk, F.: B-Human team report and code release 2013 (2013). http://www.b-human.de/downloads/publications/2013/CodeRelease2013.pdf

13. Strobl, K.H., Hirzinger, G.: Optimal hand-eye calibration. In: Proceedings of the 2006 IEEE/RSJ International Conference on Intelligent Robots and Systems (IROS 2006), pp. 4647–4653. IEEE, Beijing (2006)

14. Wang, Z., Wu, W., Xu, X.: Auto-recognition and auto-location of the internal corners of planar checkerboard image. In: International Conference on Intelligent Computing, Heifei, China, pp. 473–479 (2005)

15. Wiest, U.: Kinematische Kalibrierung von Industrierobotern. Berichte aus der Automatisierungstechnik. Shaker Verlag, Aachen (2001)

16. Zhang, Z.: A flexible new technique for camera calibration. IEEE Trans. Pattern Anal. Mach. Intell. **22**, 1330–1334 (1998)

Single- and Multi-channel Whistle Recognition with NAO Robots

Kyle Poore[(✉)], Saminda Abeyruwan, Andreas Seekircher, and Ubbo Visser

Department of Computer Science, University of Miami,
1365 Memorial Drive, Coral Gables, FL 33146, USA
{kyle,saminda,aseek,visser}@cs.miami.edu

Abstract. We propose two real-time sound recognition approaches that are able to distinguish a predefined whistle sound on a NAO robot in various noisy environments. The approaches use one, two, and four microphone channels of a NAO robot. The first approach is based on a frequency/band-pass filter whereas the second approach is based on logistic regression. We conducted experiments in six different settings varying the noise level of both the surrounding environment and the robot itself. The results show that the robot will be able to identify the whistle reliability even in very noisy environments.

1 Introduction

While much attention in autonomous robotics is focused on behavior, robots must also be able to interact and sense trigger-events in a human environment. Specifically, apart from direct interaction between humans and robots, it is appropriate for robots to sense audio signals in the surrounding environment such as whistles and alarms. Digital Audio Signal Processing (DASP) techniques are well established in the consumer electronics industry. Applications range from real-time signal processing to room simulation.

Roboticists also develop DASP techniques, only tailored for their needs on specific kind of robots. Literature shows a whole spectrum of techniques, starting with techniques that aim for the recognition of specific signals with one microphone on one end, e.g. [13], to complete systems that combine the entire bandwidth between single signals to microphones arrays combined with speech recognition and other tasks such as localization on the other end, e.g. [8]. The available literature reveals that there are many cases of audio processing/ recognition situations as there are different robots and environments, including real-time processing, combining human speech and other audio signals etc.

A lot of research has been devoted to audio signals featuring humanoid robots, especially in the past decade. Audio signals can be important sensor information as they can be used for various purposes, e.g. for the improvement of the robot's self-localization, the communication between multiple robots, or using the audio signals as the only source for self-localization when an existing Wi-Fi network might be down. A demonstration within the SPL in 2013 in

© Springer International Publishing Switzerland 2015
R.A.C. Bianchi et al. (Eds.): RoboCup 2014, LNAI 8992, pp. 245–257, 2015.
DOI: 10.1007/978-3-319-18615-3_20

Eindhoven by the team RoboEireann revealed how difficult it is to communicate between NAOs on the soccer field in a noisy environment.

The technical committee of SPL announced a challenge where the robots have to recognize predefined static signals emitted by a global sound system. Similar to the horn-like audible alarms in ice hockey, where half-time starts and ends are signaled using the horn, future RoboCup tournaments could rely on this mechanism to signal GameController or referee messages. Teams are also required to bring one whistle that has to be recognized by the teams robots. This part of the challenge brings in a real soccer aspect to the SPL. In this paper, we focus on recognizing the sound of a whistle utilizing several NAO robots. We present two approaches, one general idea of a naive one-channel approach and one using a multi-channel learning approach.

The paper is organized as follows: we discuss relevant work in the next section and describe our approach in Sect. 3. Our experimental setup and the conducted robot tests is explained in Sect. 4. We discuss the pros and cons of our results in Sect. 5 and conclude and outline future work in the remaining Sect. 6.

2   Related Work

When consulting the literature one finds a number of research papers that relate to our work. Saxena and Ng [13] present a learning approach for the problem of estimating the incident angle of a sound using just one microphone, not connected to a mobile robot. The experimental results show that their approach is able to accurately localize a wide range of sounds, such as human speech, dog barking, or a waterfall. Sound source localization is an important function in robot audition. Most existing research investigates sound source localization using static microphone arrays. Hu et al. [4] propose a method that is able to simultaneously localize a mobile robot and in addition an unknown number of multiple sound sources in the vicinity. The method is based on a combinational algorithm of difference of arrival (DOA) estimation and bearing-only SLAM. Experimental results with an eight-channel microphone array on a wheeled robot show the effectiveness of the proposed method. Navigation is part of another study where the authors developed an audio-based robot navigation system for a rescue robot. It is developed using tetrahedral microphone array to guide a robot finding the target shouting for help in a rescue scenario [14]. The approach uses speech recognition technology and using a time DOA method (TDOA). The authors claim that the system meets the desired outcome.

ASIMO, the remarkable humanoid developed by HONDA also uses the auditory system for its tasks. An early paper from 2002 introduces the use of a commercial speech recognition and synthesis system on that robot. The authors state that the audio quality and intonation of voice need more work and that they are not yet satisfactory for use on the robot [12]. Okuno et al. [11] present a later version of ASIMO's ability to use the auditory system for tasks at hand. They use the HARK open-source robot audition software [9] and made experiments with speech and music. The authors claim that the active audition improves the localization of the robot with regard to the periphery.

Speech/dialogue based approaches for the NAO also exist. Kruijff-Korbayová et al. [5], e.g., present a conversational system using an event-based approach for integrating a conversational Human-Robot-Interaction (HRI) system. The approach has been instantiated on a NAO robot and is used as a testbed for investigating child-robot interaction. The authors come to the conclusion that the fully autonomous system is not yet mature enough for end-to-end usability evaluation. Latest research such as the paper by Jayagopi et al. [15] suggest that significant background noise presented in a real HRI setting makes auditory tasks challenging. The authors introduced a conversational HRI dataset with a real-behaving robot inducing interactive behavior with and between humans. The paper however does not discuss the auditory methods used in detail. We assume that the authors use the standard auditory recognition that comes with the NAO.

Athanasopoulos et al. [1] present a TDOA-based sound source localization method that successfully addresses the influence of a robot's shape on the sound source localization. The evaluation is made with the humanoid robot NAO. The authors state that this approach allows to achieve reliable sound source location.

All mentioned approaches differ from our approach (a) in the method used, (b) in the purpose of the audio recognition, and (c) in us using the RoboCanes framework. Here, all audio modules have been implemented from scratch and run within the robot's system loop. We are synchronizing the audio signals with the update of the vision system of our NAO robots.

3 Approach

The recognition of whistle sounds will provide information that can be used by the behavior control of the robot to react to signals, which, for example, may be given by a referee. The behavior is mostly based on information gained from the camera images. Therefore, most behavior modules are running in a control loop synchronized with the camera (in our experiments 30 fps). To minimize the delay in reacting to whistle signals, we need to run the audio processing with the same rate. In every cycle of the decision making, the whistle detection needs to check the most recent audio data from the microphones. However, integrated in the behavior control the time between two executions of the audio processing module can vary slightly. Processing all audio data since the last cycle would result in a slightly varying amount of recorded audio samples to be processed, since the microphones of NAO provide a constant stream of audio samples with 48 kHz. To be independent of the exact execution frequency of the audio processing, we select the block of audio sample to process using a moving window. Every cycle we use the most recent 2,048 audio samples. The time between two executions of the whistle detection will be approximately 33 ms (30 fps), thus a window length of 42.67 ms on the audio data (2,048 samples at 48 kHz) is a sufficient size to not skip any samples. When multiple microphones are available, this process is done for each channel independently, such that we obtain new microphone measurements in the form of equally sized blocks of audio samples. The audio

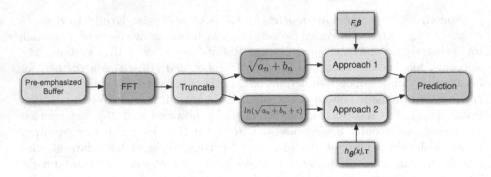

Fig. 1. The whistle identification framework for Sect. 3.

signal in the time domain can then be transformed to the frequency domain by using a Fast Fourier Transformation (FFT) on those blocks of 2,048 audio samples (Fig. 1).

If the size of the input to the FFT is N, then the output contains $\frac{N}{2}+1$ coefficients [7]. We have used these coefficients to generate the energy or log-energy profiles for each block of audio samples. These energy or log-energy profiles will be the input data for the whistle detection approaches. In the following, we will call a set of those coefficients a sample (as in sample input data or training sample, not audio sample). In our case, the output of the FFT is 1,025 coefficients. The preliminary analysis has shown that the majority of the energies or log-energies resides within the first 400 frequency components. Therefore, our samples contain feature vectors with 400 components, such that each feature consists of $\sqrt{a_n^2 + b_n^2}$ or $\ln(\sqrt{a_n^2 + b_n^2}+\epsilon)$, where $n = \{1,\ldots,400\}$, a_n represents the real coefficients, b_n represents imaginary coefficients, and $\epsilon = 2.2204e^{-16}$ is a small positive number. We would also add a bias term to provide more expressivity to our learning models. We have collected positive and negative samples, and have annotated the target of each sample indicating the presence of the whistle. It is to be noted that we have collected our samples at the rate the system outputs the coefficients, which would amount to approximately 40–50 ms. For datasets containing multiple channels, for each sampling point, we have collected multiple samples proportional to the number of channels. We have tested two approaches, a simple approach using a frequency/band-pass filter to isolate the wanted frequency from the audio signal and another approach using logistic regression with l^2-norm regularization.

3.1   Frequency/Band-Pass Filter

In frequency/band-pass filter approach, we investigate the recognition of a whistle given the energies of the frequency spectrum. The fundamental frequency is usually the dominant frequency in the sample. For this reason, an attempt was made to exploit this correlation to provide a fast, memory efficient algorithm for whistle recognition. The algorithm takes as input a sample **x** (the energy profile), the known frequency of the whistle F, and a frequency error parameter β.

We iterate over the elements in the sample and record the index of the element with the highest amplitude. The index of the maximum element is translated to a frequency value by multiplying by the sample rate to frame ratio, where the sample rate is the number of samples taken per second in the original audio signal and the number of frames is the number of time-domain samples used to compute the FFT. If the computed frequency is within the bounds defined by $F \pm \beta$, the sample is assumed to have been taken in the presence of a whistle.

The frequency F may be selected by analyzing several positive data samples and computing the average fundamental frequency across these samples. β may be selected by trial and error, although there are fundamental limits to its potential values; since the frequency granularity of the output of the FFT is the $\frac{S}{f}$, where S is the sample rate and f is the number of frames used to compute the FFT, β cannot be chosen to be less than half of $\frac{S}{f}$, as this will prevent any recognition at all. In practice, it is desirable for β to be much larger than $\frac{S}{f}$; as β increases, the recall of the set should increase to 1.0, while the precision may decrease due to the inclusion of an increased number of false-positives. The value of β should also not be chosen to be high either, as while this will ensure excellent recall, it will include far too many false positives to be a useful recognition system. This algorithm may be improved by averaging the calculations of the fundamental frequency across multiple channels of input before testing the frequency's inclusion in $F \pm \beta$.

3.2 Logistic Regression with l^2-norm Regularization

Our datasets contain log-energy profiles as well as indications of the availability of the whistle. Therefore, we can formulate our original goal mentioned in Sect. 1 as a binary classification problem using logistic regression [2]. The outcome or the target of the methods such as logistic regression is quite suitable for robotic hardware, as it consumes minimal computational and memory resources. We represent our training examples by the set $\{(\mathbf{x}_i, y_i)\}_{i=1}^{M}$, where, the feature vector $\mathbf{x}_i \in \mathbb{R}^{N+1}$ with bias term, $y_i \in \{0, 1\}$, $M \gg N$, and $M, N \in \mathbb{Z}_{>0}$. Hence, we define the design matrix $\mathbf{X}$ to be a $M \times (N+1)$ matrix that contains training samples in its rows. We also define a target vector $\mathbf{y} \in \mathbb{R}^M$ that contains all the binary target values from the training set. Our hypotheses space consist of vector-to-scalar sigmoid functions, $h_{\boldsymbol{\theta}}(\mathbf{x}) = \frac{1}{1+e^{-\boldsymbol{\theta}^T \mathbf{x}}}$, with adjustable weights $\boldsymbol{\theta} \in \mathbb{R}^{N+1}$. Similarly, we define the matrix-to-vector function, $\mathbf{h}_{\boldsymbol{\theta}}(\mathbf{X})$, which results in a column vector with i^{th} element $h_{\boldsymbol{\theta}}(\mathbf{X}_i)$, where, $\mathbf{X}_i$ is the ith row of the design matrix $\mathbf{X}$. We use a cross-entropy cost function, $J(\boldsymbol{\theta}) = -\frac{1}{M}(\mathbf{y}^T \ln(\mathbf{h}_{\boldsymbol{\theta}}(\mathbf{X})) + (\mathbf{1} - \mathbf{y})^T \ln(\mathbf{1} - \mathbf{h}_{\boldsymbol{\theta}}(\mathbf{X}))) + \frac{\lambda}{2M} \boldsymbol{\theta}^T \boldsymbol{\theta}$, with l^2-norm regularization. Here, the natural logarithmic function, $\ln(.)$, is applied element wise and $\mathbf{1}$ is an M-dimensional column vector with all elements equal to one. It is a common practice to avoid regularizing of the bias parameter. We have regularized the bias weight in the cost function, in order to present the equations without too much clutter. In practice, we normally do not regularize the weight associated with bias term. Taking the gradient of the cost function with respect to $\boldsymbol{\theta}$, we obtain $\nabla_{\boldsymbol{\theta}} J = \frac{1}{M} \mathbf{X}^T (\mathbf{h}_{\boldsymbol{\theta}}(\mathbf{X}) - \mathbf{y}) + \frac{\lambda}{M} \boldsymbol{\theta}$.

Table 1. Dataset description for different environments, and robot activities. Table shows the number of positive and negative samples collected for each environment and robot combinations, and the number of channels active on the robot.

Set	Description	Positive	Negative	Channels
1	Indoor silent environment; silent robot	1005	1000	1
2	Indoor silent environment; active robot	2328	2032	1 – 2
3	Outdoor mildly noisy environment; silent robot	2112	2154	1 – 2
4	Outdoor noisy environment; silent robot	2030	2010	1 – 2
5	Indoor noisy environment (1); silent robot	4022	4170	1 – 2
6	Indoor noisy environment (2); silent robot	8024	8000	1 – 2
7	Indoor noisy environment (3); silent robot	8324	7996	1 – 4
8	Indoor noisy environment (3); active robot	8272	8000	1 – 4

We have trained our logistic regression classifiers in batch mode with the state of art L-BFGS quasi-Newton method [10] to find the best θ. We predict the availability of the whistle if and only if $h_\theta(\mathbf{x}) \geq \tau$, where, $0 < \tau \leq 1$. Therefore, λ and τ would be the hyper-parameters that we need to modify to find the best solution. We have used standard parameter sweeping techniques to find the λ that provides the best trade off between the bias and the variance, while precision, recall, and F_1-score have been used to obtain the suitable τ value. As a preprocessing step, the features, except the bias, have been subjected to feature standardization. We have independently set each dimension of the sample to have zero-mean and unit-variance. We achieved this by first computing the mean of each dimension across the dataset and subtracting this from each dimension. Then each dimension is divided by its standard deviation.

4   Experiments and Results

We have conducted all experiments using audio data recorded on NAO robots. We have used several different setups to evaluate the performance of the different approaches on a range of recorded data with different characteristics and different amounts of noise. Each recorded sample contains the log-energy profile of a the captured audio signal of one time step. During the recording, the samples were manually marked as positive, $y = 1$, or negative, $y = 0$, samples.

The whistle identification methods, that we will be describing in this paper, have used the datasets shown in Table 1 and Fig. 2. The samples in the datasets 1, 2, 5, 6, and 7 were collected from indoor environments, while the samples in the datasets 3, and 4 were collected from outdoor environments. The datasets 5, 6, and 7 contain samples from noisy RoboCup environments simulated though speakers. We have simulated three different noisy environments with a combinations of silent and active robots to collect samples. The datasets 2–6 have used channels 1 and 2 to collect samples, the datasets 7 and 8 have used all four channels, and the first dataset have used only the first channel.

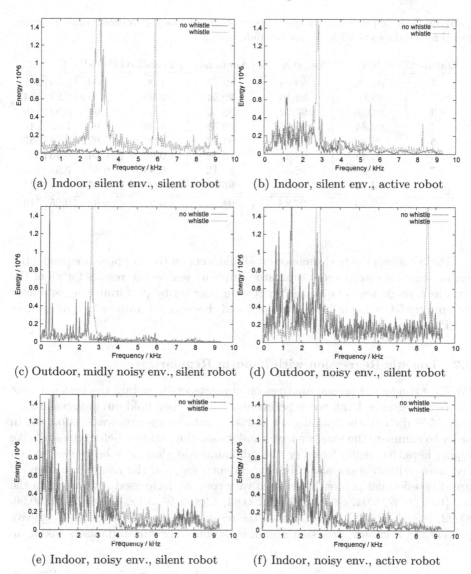

Fig. 2. Example frequencies for the different setups. Each figure shows one example for a positive sample (green) and one example for a negative sample (red) (Color figure online).

4.1 Frequency/Band-Pass Filter

We have analyzed the data using the maximum frequency technique, and for each dataset, we found best values for β such that the F_1-score was maximized. For each tuning of β and for each dataset, a random 70 % of the data was chosen as a training set, while the remaining 30 % served as a cross-validation set.

Table 2. Positive percentage, negative percentage, accuracy, precision, recall, F_1-score, and β for all datasets with all samples independently.

Dataset	Positive %	Negative %	Accuracy	Precision	Recall	F_1	β
1	100.00	99.66	99.83	1.00	1.00	1.00	271
2	99.86	99.18	99.54	0.99	1.00	0.99	154
3	100.00	97.82	98.90	0.98	1.00	0.99	130
4	99.84	99.67	99.75	1.00	1.00	1.00	247
5	89.28	98.10	93.82	0.98	0.89	0.93	457
6	94.46	98.32	96.38	0.98	0.94	0.96	226
7	84.49	98.18	91.12	0.98	0.84	0.91	154
8	93.72	98.85	96.19	0.99	0.94	0.96	247
1-8	92.00	97.89	94.96	0.98	0.92	0.95	319

Table 2 shows the performance on all datasets on the samples independently; each channel is considered a separate sample as well as the results for all of the data as a single set. The values for β were selected by performing a parameter sweep from 50 to 800 in increments of 1 and choosing the value which maximizes the F_1-score.

4.2   Logistic Regression with l^2-norm Regularization

We have conducted several analyses on our datasets to obtain the best outcome on the predictions. In all our experiments, we have used hold-out cross validation with 70 % data on the training set and 30 % data on the cross-validation set. In order to eliminate the bias, we have randomized the datasets before the split. We report here the results based on the minimum cost that have been observed on the cross-validation set after 30 independent runs, and the results are rounded up to two decimal points. In order to vary cost, we have used λ values from the set {0.01, 0.02, 0.04, 0.08, 0.16, 0.32, 0.64, 1.28, 2.56, 5.12, 10.24, 20.48, 40.96, 81.92, 163.84, 327.68}. Table 3 shows the results for positive percentage, negative percentage, accuracy, precision, recall, F_1-score, and τ for all datasets taking all

Table 3. Positive percentage, negative percentage, accuracy, precision, recall, F_1-score, and τ for all datasets with all samples independently.

Dataset	Positive %	Negative %	Accuracy	Precision	Recall	F_1	τ
1	100.00	100.00	100.00	1.00	1.00	1.00	0.5
2	99.86	100.00	99.92	1.00	0.99	0.99	0.5
3	99.53	100.00	99.77	1.00	0.99	0.99	0.5
4	99.51	99.67	99.59	0.99	0.99	0.99	0.4
5	94.70	94.97	94.84	0.95	0.95	0.95	0.5
6	97.88	98.38	98.13	0.98	0.98	0.98	0.5
7	96.68	97.08	96.88	0.97	0.97	0.97	0.5
8	95.57	97.92	96.72	0.98	0.96	0.97	0.7

Table 4. Positive percentage, negative percentage, accuracy, precision, recall, F_1-score, and τ for all datasets dependently (averaging).

Dataset	Positive %	Negative %	Accuracy	Precision	Recall	F_1	τ
2	100.00	100.00	100.00	1.00	1.00	1.00	0.5
3	100.00	100.00	100.00	1.00	1.00	1.00	0.5
4	99.67	100.00	99.84	1.00	0.99	0.92	0.5
5	96.69	95.85	96.26	0.96	0.97	0.96	0.5
6	98.67	98.83	98.75	0.99	0.99	0.99	0.5
7	98.56	98.50	98.53	0.99	0.99	0.99	0.5
8	96.30	99.33	97.79	0.99	0.96	0.98	0.6

Table 5. Overall performance on the combined dataset. The datasets 2–8 have 1–2 channels in common. The combined dataset have been tested on 400 + 1 features independently and averaging. We have performed analysis on combining the adjacent two channels to generate 800 + 1 features and tested the performance independently. Finally, we have analyzed the performance independently and dependently on all channels for datasets 2–8 on 400 + 1 features.

Dataset	Ch.	Method	Feat.	Pos.	Neg.	Acc.	Prec.	Recall	F_1	τ
2–8	1–2	Independently	400 + 1	94.26	96.78	95.51	0.97	0.94	0.96	0.5
2–8	**1–2**	**Dependently**	**400 + 1**	**96.47**	**97.60**	**97.03**	**0.98**	**0.97**	**0.97**	**0.5**
2–8	1–2	Independently	800 + 1	95.68	97.95	96.80	0.98	0.96	0.97	0.6
2–8	1–4	Independently	400 + 1	94.58	96.90	95.73	0.97	0.95	0.96	0.5
2–8	**1–4**	**Dependently**	**400 + 1**	**96.57**	**98.21**	**97.38**	**0.98**	**0.97**	**0.97**	**0.5**

samples independently, i.e., we have assumed that the samples from each channel is independent in the cross-validation set. Therefore, a sample is predicted positive if and only if $h_\theta(\mathbf{x}) \geq \tau$. We have conducted a parameter sweep for τ from the set $\{0.3, 0.4, 0.5, 0.6, 0.7, 0.8\}$ and selected the value with the highest F_1-score.

Table 4 shows the results for all datasets dependently, i.e., during cross validation, we select hyper-parameters based on the average values of the channels active while the samples were collected. For example, when we collect samples for dataset eight, every time we would collect a sample, there are four active channels. During the cross validation phase, in order to determine the number of correctly classified positive samples, we have summed-up the probabilities of the samples of the adjacent four channels above the given threshold and divided by four. Therefore, when there are k-channels ($k \in \{1, 2, 3, 4\}$) active in a dataset, at every sampling point, we would collect k samples. Therefore, when we calculate the scores in Table 4 for cross-validation set, we have used the averaging formula (hence, dependently), $f(\mathbf{x}_1, \ldots, \mathbf{x}_k) = \frac{1}{k} \sum_{i=1}^{k} h_\theta(\mathbf{x}_i) \geq \tau$, where $\{\mathbf{x}\}_1^k$ are features of the adjacent samples, and $f(\mathbf{x}_1, \ldots, \mathbf{x}_k) : \{\mathbb{R}^{N+1}\}_1^k \mapsto [0, 1]$, to predict a positive sample. When $k = 1$, independent and dependent scores will be similar. It is clearly evident from the Table 4 that the averaging has improved the prediction capabilities. We have not used the first dataset in Table 4 as it contains samples only from channel 1.

Table 6. Performance of the channels 1, 2, 3, and 4 separately and independently for the combined datasets (1–8).

Channel	Positive	Negative	Accuracy	Precision	Recall	F_1	τ
1	91.31	91.09	91.20	0.91	0.91	0.91	0.6
2	91.78	86.38	89.10	0.87	0.91	0.89	0.6
3	94.94	94.33	94.64	0.95	0.95	0.95	0.5
4	95.34	96.75	96.03	0.97	0.95	0.96	0.7

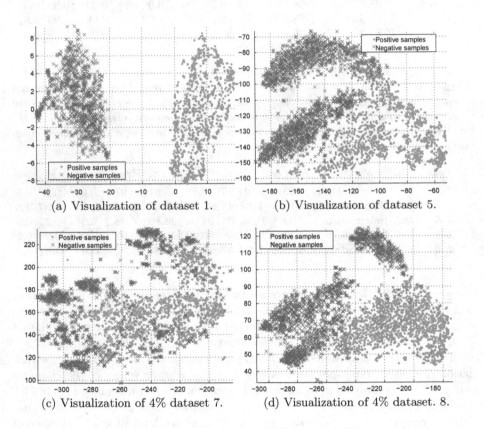

(a) Visualization of dataset 1.   (b) Visualization of dataset 5.

(c) Visualization of 4% dataset 7.   (d) Visualization of 4% dataset. 8.

Fig. 3. Visualization of datasets 1, 5, 7 (4 %), and 8 (4 %) using t-SNE.

Table 5 shows the overall performance on the combined dataset. We have conducted several analyses on the combined dataset. Firstly, Table 1 shows that the channels 1–2 are common to all datasets. Therefore, we have extracted all samples from channels 1–2 and analyzed the performance on $400 + 1$ features independently and dependently (averaging). Secondly, we have expanded the adjacent two channels to create a feature vector of $800 + 1$ features and analyzed the performance independently. Finally, we have analyzed the performance independently and dependently for all channels from the combined datasets 2–8 on $400 + 1$ features. Table 5 concludes that for both robots with only two active channels (1–2) and for robots with all active channels (1–4) it is best to use weights learned from averaging for $400 + 1$ features. Finally, we have observed the performance of the channels 1, 2, 3, and 4 separately and independently for the combined datasets (1–8), which is given in Table 6.

We have concluded from our findings that performance on averaging provides best results for our datasets. Once we have decided the hyper-parameters, we have learned the weights from the complete datasets. On the robot, we have used the average history of 16 decision points to decide the availability of a whistle. We have used a threshold of 0.8 for these averaging, and the robot has detected a whistle 100 % on a separate test set.

5 Discussion

When working with audio signals, it is a common practice to use Mel-frequency cepstral coefficients (MFCCs) [3] as features. In our work, we have used a truncated power or log-power spectrum of the signal as features. The main reason behind this choice is motivated by (1) the shape of the distribution of the samples in the high-dimensional space; and (2) the detection of a whistle signal at every sampling point. If we were to change the problem to identify particular patterns of whistle signals, then MFCCs would have been our primary choice as the feature extractor. Figure 3 shows the distribution of the samples in 2D for datasets 1, 5, 7, and 8 using t-Distributed Stochastic Neighbor Embedding (t-SNE) [6]. The distribution of the samples in dataset 1 (Fig. 3a) is clearly linearly separable, therefore, we have obtained 100 % accuracy in the first row of Tables 2, 3, and 4. Figure 3b shows the distribution of the samples of the dataset 5. The approach 2 has found solutions with 94.84 % and 96.26 % (Tables 3 and 4) accuracies, but the frequency/band-pass filter approach has shown slightly inferior (Table 2 fifth row) performance. The main reason behind the drop of performance for this approach is that it uses the frequency of the highest magnitude. When we collected samples for the dataset 5, we had explicitly whistled with less strength. Therefore, the energies of the whistle signal may not have enough strength to overcome the energies of the ambient sounds. Our second approach has managed to learn a statistically significant classifier for dataset 5. Figure 3c and d show the distribution of 4 % (approximately 4000) of the samples in the datasets 7 and 8. These were the hardest datasets that we had collected. Tables 3 and 4 show that approach 2 has found better solutions than approach 1 (Table 2 last row). Both approaches are fast enough to be executed in real-time on the NAO (Intel Atom Z530 1.6 GHz). The audio capture and FFT takes 2.4 ms. The whistle detection using approach 1 adds 0.1 ms, approach 2 adds 0.27 ms. Overall, our findings conclude that approach 2 has outperformed approach 1, and is suitable for practical usage.

For approach 1, as a future work, we have considered attempts to learn the frequency profile of the noise in the signal. The method takes the ordering of the samples into account, and rather than computing the frequency of maximum amplitude, computes the frequency with the highest impulse; a characteristic of most whistles is that they usually cause a large difference in a particular frequency in a short period of time. This method accomplishes this by computing a normalization vector $\mathbf{v}$ such that $\mathbf{v}_t \odot \mathbf{x}_t = 1$. The frequency impulse is then obtained by computing $\mathbf{v}_{t-1} \odot \mathbf{x}_t = \mathbf{w}$. The vector $\mathbf{v}$ is then adjusted such that $\mathbf{v}_t = \alpha \mathbf{v}_{t-1} + (1 - \alpha) \frac{1}{\mathbf{x}_{t-1}}$, where α is a resistance factor that determines how easily $\mathbf{v}$ conforms to the new environment. We can then determine if frequencies within the range $F \pm \beta$ have experienced a sufficient impulse between $\mathbf{x}_{t-1}$ and $\mathbf{x}_t$.

6   Conclusion

We have presented two approaches to identify an existence of a whistle sound on a NAO robot in various noisy environments using one, two, and four microphone channels. The first approach is based on a frequency/band-pass filter, whereas the second approach is based on logistic regression. The results show that the robot will be able to identify the whistle reliably even in very noisy environments. Even though both approaches are real-time compatible on predictions, the second approach has outperformed the first approach in all datasets and combined datasets and it is the most suitable method for practical usage. In future, we are planning to conduct classification using a multi-layer perceptron and support vector machines [2], and to extend our work to recognize different whistle patterns. We also plan to use the approach to improve robot localization.

References

1. Athanasopoulos, G., Brouckxon, H., Verhelst, W.: Sound source localization for real-world humanoid robots. In: Proceedings of the SIP, vol. 12, pp. 131–136 (2012)
2. Bishop, C.M.: Pattern Recognition and Machine Learning (Information Science and Statistics). Springer-Verlag New York Inc., Secaucus (2006)
3. Davis, S., Mermelstein, P.: Comparison of parametric representations for mono-syllabic word recognition in continuously spoken sentences. IEEE Trans. Acoust. Speech Signal Proc. **28**(4), 357–366 (1980)
4. Hu, J.S., Chan, C.Y., Wang, C.K., Lee, M.T., Kuo, C.Y.: Simultaneous localization of a mobile robot and multiple sound sources using a microphone array. Adv. Robot. **25**(1–2), 135–152 (2011)
5. Kruijff-Korbayová, I., Athanasopoulos, G., Beck, A., Cosi, P., Cuayáhuitl, H., Dekens, T., Enescu, V., Hiolle, A., Kiefer, B., Sahli, H., et al.: An event-based conversational system for the NAO robot. In: Proceedings of the Paralinguistic Information and its Integration in Spoken Dialogue Systems Workshop, pp. 125–132. Springer (2011)
6. Van der Maaten, L., Hinton, G.: Visualizing data using t-SNE. J. Mach. Learn. Res. **9**(11), 2579–2605 (2008)
7. Mitra, S.: Digital Signal Processing: A Computer-based Approach. McGraw-Hill Companies, New York (2010)
8. Mokhov, S.A., Sinclair, S., Clement, I., Nicolacopoulos, D.: Modular Audio Recognition Framework and its Applications. The MARF Research and Development Group, Montréal, Québec, Canada, v. 0.3.0.6 edn., December 2007

9. Nakadai, K., Takahashi, T., Okuno, H.G., Nakajima, H., Hasegawa, Y., Tsujino, H.: Design and implementation of robot audition system 'hark'–open source software for listening to three simultaneous speakers. Adv. Robot. **24**(5–6), 739–761 (2010)
10. Nocedal, J., Wright, S.: Numerical Optimization. Springer Series in Operations Research and Financial Engineering, 2nd edn. Springer, New York (2006)
11. Okuno, H.G., Nakadai, K., Kim, H.-D.: Robot audition: missing feature theory approach and active audition. In: Pradalier, C., Siegwart, R., Hirzinger, G. (eds.) Robotics Research. STAR, vol. 70, pp. 227–244. Springer, Heidelberg (2011)
12. Sakagami, Y., Watanabe, R., Aoyama, C., Matsunaga, S., Higaki, N., Fujimura, K.: The intelligent asimo: system overview and integration. In: IEEE/RSJ International Conference on Intelligent Robots and Systems, vol. 3, pp. 2478–2483. IEEE (2002)
13. Saxena, A., Ng, A.Y.: Learning sound location from a single microphone. In: IEEE International Conference on Robotics and Automation, ICRA 2009, pp. 1737–1742. IEEE (2009)
14. Sun, H., Yang, P., Liu, Z., Zu, L., Xu, Q.: Microphone array based auditory localization for rescue robot. In: 2011 Chinese Control and Decision Conference (CCDC), pp. 606–609. IEEE (2011)
15. Wrede, S., Klotz, D., Sheikhi, S., Jayagopi, D.B., Khalidov, V., Wrede, B., Odobez, J.M., Wienke, J., Nguyen, L.S., Gatica-Perez, D.: The vernissage corpus: a conversational human-robot-interaction dataset. In: Proceedings of the 8th ACM/IEEE International Conference on Human-Robot Interaction. No. EPFL-CONF-192462 (2013)

Towards Spatio-Temporally Consistent Semantic Mapping

Zhe Zhao[✉] and Xiaoping Chen

University of Science and Technology of China, Hefei, China
zhaozhe@mail.ustc.edu.cn, xpchen@ustc.edu.cn

Abstract. Intelligent robots require a semantic map of the surroundings for applications such as navigation and object localization. In order to generate the semantic map, previous works mainly focus on the semantic segmentations on the single RGB-D images and fuse the results by a simple majority vote. However, single image based semantic segmentation algorithms are prone to producing inconsistent segments. Little attentions are paid to the consistency over the semantic map. We present a spatio-temporally consistent semantic mapping approach which can generate the temporal consistent segmentations and enforce the spatial consistency by Dense CRF model. We compare our temporal consistent segment algorithm with the state-of-art approach and generate our semantic map on the NYU v2 dataset.

Keywords: Spatio-temporal consistency · Semantic mapping · Indoor scene understanding

1 Introduction

Semantic understanding of the environment from images plays an important role in robotic tasks, such as human-robot interaction, object manipulation and autonomous navigation. The release of Microsoft Kinect, and the wide availability of RGB-D sensors, changes the landscape of indoor scene analysis. Many works have investigated single RGB-D image-based indoor scene understanding using different techniques such as SVM [6], CRF [13], CPMC [2]. Although, many advancements have been reported on the indoor image dataset NYU, single image based semantic segmentation algorithms are prone to producing inconsistent segments, such as Fig. 1. Many reasons cause this inconsistent segments including the motion of the camera, lighting, occlusion and even the sensor noises. These inconsistent predictions can have a significant impact on robotic tasks in practice. For example, obstacles may suddenly appear in front of the robot in one frame and vanish in the next.

In order to constrain the semantic segmentations in images to be temporally consistent, we need to find the correspondences between frames. For RGB images, optical flow algorithm is often used: for any pixel $z_{n'}^{t+1}$, find the optical flow m_t from z_n^t to $z_{n'}^{t+1}$ ($n' = n + m_t$), and let $z_{n'}^{t+1}$ takes the same label as z_n^t.

© Springer International Publishing Switzerland 2015
R.A.C. Bianchi et al. (Eds.): RoboCup 2014, LNAI 8992, pp. 258–269, 2015.
DOI: 10.1007/978-3-319-18615-3_21

Fig. 1. An example of the inconsistent semantic labels of the indoor scene. Independent segmentations are prone to producing inconsistent superpixels and further resulting in inconsistent semantic precisions. (These images are best seen in color) (Colour figure online).

In this case, the temporal consistent constraints can be easily formulated as a binary relation between one pixel in frame t and one in frame $t+1$, in the form of $f(z_{n'}^{t+1}, z_n^t)$. However, the biggest problem of the exact pixel motion assumption is the optical flow algorithms are often not as reliable as we expect it to be. For example, holes and dragging effects often occur as the results of inaccurate optical flows [3]. In order to improve the performance, complex appearance models or motion-appearance combined approaches [1,4] are investigated. However, depth information and the RGB-D images are less being researched to improve the performance.

In this paper, we want to build a spatio-temporally consistent semantic map for indoor scenes. Typically, incrementally building a semantic map for RGB-D images involve two steps: (1) generate the semantic segmentation for RGB-D images. (2) fuse the semantic segmentation using the estimated camera poses and get the global semantic map. However, in step 1, simply applying the semantic segmentation algorithms to each image is not sufficient as it does not enforce temporal consistent. Different from previous works that transfer the label through 2D optical flow or complex appearance models, we build the correspondence with the help of depth information. As we observed, the depth information can help build a more easily and more accurate consistent segmentation algorithm to constrain the temporal consistency. In order to explain its effective, we compare our segmentation algorithms with the state-of-art consistent segmentation algorithm in experiments. In step 2, since many points will be fused over time, simply taking the last label as the 3D point label is suboptimal. A straightforward approach is to keep track of the different pixels, that are fused in a

3D point. Then the label is selected by a simple majority vote. Although, this visually improves the results, however it will not lead to a spatial consistency. For this purpose, we use a Dense CRF (Conditional Random Field) model to distribute the information over neighboring points depending on the distance and color similarity. In summary, our algorithm consists of two main contributions:

- We use depth information to help building the temporal consistent segmentations over RGB-D images and probability transfer the semantic label according to the correspondence. In the experiment, we compare our consistent segmentations with the state-of-art algorithm and show that our method is more effective and accurate.
- We use a Dense CRF to enforce the spatial consistency and generate the spatio-temporally consistent semantic maps for NYU v2 dataset. The detailed results will be reported in experiment.

In the rest of the paper, we first briefly describe the related work in Sect. 2. In Sect. 3, we introduce the details of the temporal consistent segmentation algorithm. Section 4 discusses the spatial consistency method. Experimental results and statistics are described in Sect. 5 and the conclusion with discussions about future work in Sect. 6.

2 Related Work

Robots need a semantic map of the surrounding environment for many applications such as navigation and object localization. In RoboCup@Home, a set of benchmark tests is used to evaluate the robots abilities and performance in a realistic non-standardized home environment setting. The ability of building a consistent semantic map is useful to some tests, such as the Clean UP, General Purpose and Restaurant. For example, in the Restaurant test, a robot needs not to be guided by a human if it can build a semantic map by itself. However, the previous works mainly focus on the single RGB-D image based indoor scene understanding [2,6,11,13,14] which will cause inconsistent result for image sequences. Reference [8] labels the indoor scene over the global point cloud, but it is unclear how long the feature computation takes. For robots, it is typically suitable to build the semantic map incrementally, rather than doing all of the work on the global 3D scene. In this paper, we build a spatio-temporally consistent semantic map by the temporal consistent segmentation and Dense CRF.

3 Temporal Consistent Segmentation

Independently segment images into superpixels is prone to producing inconsistent segments. Given a segmentation S_t of an RGB-D image I_t at time t, we wish to compute a segmentation S_{t+1} of the image I_{t+1} at the time $t+1$ which is consistent with the segments S_t. An example of the independent segmentations and temporal consistent segmentations are shown in Fig. 2. Compare to independent segmentations, the temporal consistent segmentations can capture the

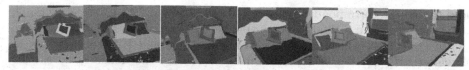

independent segmentations

temporal consistent segmentations

Fig. 2. A comparison of the independent segmentations and temporal consistent segmentations. (The images are best seen in color) (Colour figure online).

correspondence for superpixels between frames. Each corresponding superpixel presences in every frame and will not vanish in one frame and suddenly appear in the next. These temporal consistent segmentations will lead to more consistent semantic precisions than independent segmentations.

3.1 3D Transformation of the Camera

In the 2D image, researchers often use optical flow to find the motion of pixels. However, the accuracy of the optical flow can not be guaranteed. For RGB-D images, if the scene is static which means only the camera moves and objects in the scene keep static, the 3D transformation can be computed effectively with the help of the depth information. If the 3D transformation T is known, we back-project the image I_{t+1} into image I_t. In this way, the binary relation between one pixel in frame t and one in frame $t + 1$ is founded. These binary relations are more accurate than optical flow method as they use the 3D information.

In order to compute the 3D transformation of the camera, SURF and corner points are employed. We detect Shi-Tomasi [12] corner points and adopt the Lucas-Kanade optimal flow method to track the corner points between frames. For SURF keypoints, a KD-tree based feature matching approach is used to find the correspondence. Then RANSAC (Random Sample Consensus) procedure is running to find a subset of feature pairs corresponding to a consistent rigid transformation. At last, a rigid transformation is estimated by SVD from the subset of feature pairs. If the RANSAC algorithm fails to find enough inliers, an ICP algorithm is used which is initialized by the transformation between the previous two frames.

3.2 Graph Matching

Before generating the consistent segmentations S_{t+1} for image I_{t+1}, we first independently segment the image I_{t+1} into superpixels S'_{t+1} using the 2D segmentation algorithm [5]. Once the image I_{t+1} is independently segmented, we need to find the correspondence between segments S_t and S'_{t+1}. We use a graph matching procedure to find the correspondence. The vertices of G include two sets of vertices: V_t which are the set of regions of S_t and V'_{t+1} that are the set of regions of S'_{t+1}. We back-project image I_{t+1} into the image I_t by the 3D transformation and make edges if there exist an overlap between the segment v_i and the back-projected segment v'_j. The edge weights are defined by the size of overlap region:

$$w_{ij} = |r_i \cap r_j| \tag{1}$$

where r_i is the number of pixels of region r_i, $|r_i \cap r_j|$ is the number of pixels in the overlap region.

We find the correspondence using the forward flow and reverse flow together. That means for each segment of S'_{t+1}, we find its best corresponding segment in S_t by maximizing the w_{ij}. Symmetrically, for each segment of S_t, its best corresponding segment in S'_{t+1} is identified. These two corresponding sets in forward flow and reverse flow can be used for generating the final consistent segmentation S_{t+1}.

3.3   Final Segmentation

According to the correspondences in the forward flow and reverse flow, we define that segment $f(s)$ in S'_{t+1} is the best correspondence of the segment s in S_t, the segment $r(s')$ in S_t is the best correspondence of the segment s' in S'_{t+1}. Four cases may appear:

- $f(r(s')) = s' \wedge r(f(s)) = s$, that means s' should have the same segment label as $r(s')$ in S_t.
- $r(s') = NULL$ or $f(r(s')) \neq s'$, that means s' has no matching region or s' is a part of a larger segment $r(s')$ in S_t. In this case, s' need have a new segment label.
- $r(f(s)) \neq s$, that means the segment s in S_t disappears in S'_{t+1}. In this case, the segment s should be propagated in S'_{t+1} by the 3D transformation and keep its segment label.

3.4   Semantic Transfer

To classify the superpixels, we compute the geometry and appearance features as in [6] and train SVM (Support Vector Machine) classifiers based on them. The scores obtained by the classifiers are used as the semantic prediction for superpixels. In order to obtain temporal consistent semantic predictions over images, we transfer the semantic predictions based on the consistent segmentations.

For each segment s in S_{t+1}, if the segment label of s comes from the previous segmentation S_t, we transfer the semantic prediction by the following equation:

$$y_i^{t+1} = y_i^{t+1} + c y_j^t \tag{2}$$

where the semantic prediction y_j^t of s_j^t in S_t is transferred into the segment s_i^{t+1} in S_{t+1}, c is the weight that controls the transfer ratio.

4 Build the Semantic Map

We generate and voxelize the 3D global point cloud using the estimated 3D transformation. In order to fuse the temporal consistent semantic segmentation results into the global point cloud, a straightforward approach is a simple majority vote from the different pixels that are fused in the same 3D point. However, this will not lead to a spatial consistent result. For this purpose, we want to use the CRF to model the local interactions between neighbor 3D points. However, for 3D point cloud, the computation cost for finding point neighbors is expensive. So, we adopt the full connected CRF model in this work. This means the graph with node set V is defined by $G = \{V, V \times V\}$, each node can directly influence every other node by a pairwise potential. If we consider a fully connected CRF over the 3D global point cloud, this will result in billions of edges in a graph. Here the traditional inference algorithms are no longer feasible. In order to render this feasible, [9] provides a Dense CRF that limits the pairwise potentials to a linear combination of Gaussian kernels. This approach can produce good labeling performance within 0.2-0.7 s in our experiments.

Formally, we denote object category $y_i \in \{1, \ldots, C\}$ where C is the number of the object category. X is the global point cloud and the Dense CRF graph $G = \{V, V \times V\}$ contains vertices v_i corresponding to the random variables y_i. The energy of the Dense CRF model is defined as the sum of unary and pairwise potentials:

$$E(Y|X) = \sum_{i \in V} \psi_u(y_i|X) + \sum_{(i,j) \in V \times V} \psi_p(y_i, y_j|X). \tag{3}$$

For the unary potential $\psi_u(y_i|X)$, we multiply the likelihoods of pixels which are fused in the same 3D point and normalize it using the geometric mean:

$$\psi_u(y_i|X) = -log(\{\prod_{p_i \in v_i} p(y_i|p_i)\}^{\frac{1}{N}}) \tag{4}$$

where p_i is the image pixel that is fused in the 3D point v_i, N is the number of pixels that are fused in the 3D point v_i, the probability of $p(y_i|p_i)$ is obtained through SVM classifiers.

For the pairwise potential $\psi_p(y_i, y_j|X)$, we compute it by a linear combination of Gaussian kernels:

$$\psi_p(y_i, y_j|X) = \mu(y_i, y_j) \sum_{m=1}^{K} w^{(m)} k^{(m)}(f_i, f_j) \tag{5}$$

where $\mu(y_i, y_j)$ is a label compatibility function, the vector f_i, f_j are feature vectors and $k^{(m)}$ are Gaussian kernel $k^{(m)}(f_i, f_j) = exp(-\frac{1}{2}(f_i - f_j)^T \Lambda^{(m)}(f_i - f_j))$. Each kernel $k^{(m)}$ is characterized by a symmetric, positive-definite precision matrix $\Lambda^{(m)}$, which defines its shape.

We define two Gaussian kernels for the pairwise potential: appearance kernel and smoothness kernel. The appearance kernel is computed as follow:

$$w^{(1)} \exp(-\frac{|p_i - p_j|^2}{2\theta_\alpha^2} - \frac{|I_i - I_j|^2}{2\theta_\beta^2}) \tag{6}$$

where p is the 3D location of the point and I is the color vector. This kernel is used to model the interactions between points with a similar appearance. The smooth kernel is defined as follow:

$$w^{(2)} \exp(-\frac{|p_i - p_j|^2}{2\theta_\gamma^2})$$ (7)

This kernel is used to removes small isolated regions.

Instead of computing the exact distribution $P(Y)$, the mean field approximation method is used to compute a distribution $Q(Y)$ that minimizes the KL-divergence $D(Q\|P)$ among all distributions Q that can be expressed as a product of independent marginal, $Q(Y) = \prod_i Q_i(Y_i)$. The final point label is obtained by $y_i = argmax_l Q_i(l)$.

5 Experiment

5.1 Dataset

We test the system on the NYU v2 RGB-D dataset. The dataset is recorded using Microsoft Kinect cameras at 640*480 RGB and depth image resolutions. The NYU dataset contains 464 RGB-D scenes taken from 3 cities and 407024 unlabeled frames. It comes with 1449 images with ground-truth labeling of object classes. And it has been split into 795 training images and 654 test images. Furthermore, the dataset contains a large number of raw images, which we use for our semantic mapping. We select 4 RGB-D scenes to generate their semantic map and train 12 core classes on the 795 training images. Firstly, we evaluate our temporal consistent segmentation algorithm with the state-of-art approach StreamGBH for the 4 RGB-D scenes. Secondly, we evaluate our semantic mapping method on the 4 RGB-D scenes.

5.2 Evaluate the Temporal Consistent Segmentation

We compare our consistent segmentation algorithm with the state-of-art approach StreamGBH using the video segmentation performance metrics from [15]. The StreamGBH is a hierarchical video segmentation framework which is available as part of LIBSVX software library, which can be download at http://www.cse.buffalo.edu/~jcorso/r/supervoxels/. The 4 RGB-D scenes are densely labeled with semantic category by ourselves. We evaluate consistent segmentation using the undersegmentation error, boundary recall, segmentation accuracy and the explained variation. Instead of evaluating singe image metrics, these evaluation metrics are used for 3D supervoxel in the 3D space-time which is mentioned in [15].

2D and 3D Undersegmentation Error: It measures what fraction of voxels goes beyond the volume boundary of the ground-truth segment when mapping the segmentation onto it.

$$UE(g_i) = \frac{\{\sum_{\{s_j | s_j \cap g_i \neq \emptyset\}} Vol(s_j)\} - Vol(g_i)}{Vol(g_i)}$$ (8)

where g_i is the ground-truth segment, Vol is the segment volume. 2D underseg-mentation error is defined in the same way.

2D and 3D Segmentation Accuracy: It measures what fraction of a ground-truth segmentation is correctly classified by the supervoxel in 3D space-time. For the ground-truth segment g_i, a binary label is assigned to each supervoxel s_j according to the majority part of s_j that resides inside or outside of g_i.

$$ACCU(g_i) = \frac{\sum_{j=1}^{k} Vol(\bar{s}_j \cap g_i)}{Vol(g_i)} \qquad (9)$$

where $\bar{s}_j$ is the set of correctly labeled supervoxels.

2D and 3D Boundary Recall: It measures the spatio-temporal boundary detection. For each segment in the ground-truth and supervoxel segmentations, we extract the within-frame and between-frame boundaries and measure recall using the standard formula as in [15].

Explained Variation: It considers the segmentation as a compression method of a video which is proposed in [10]:

$$R^2 = \frac{\sum_i (\mu_i - \mu)^2}{\sum_i (x_i - \mu)^2} \qquad (10)$$

where x_i is the actual voxel value, μ is the global voxel mean and μ_i is the mean value of the voxels assigned to the segment that contains x_i.

The detailed comparison is shown in Table 1. Our approach is obviously bet-ter than the StreamGBH method. Sometimes StreamGBH's performance is not ideal, for example the 3D segmentation accuracy is only 0.3942 for scene 1. However the 3D accuracy in the well-known xiph.org videos for the StreamGBH is about 0.77. The main reasons of the bad performance for StreamGBH are twofold: (1) In the 2D videos, the difference or motion between neighboring frames is small. However, in the semantic mapping procedure, the neighboring frames are keyframes [7] which have a larger difference and motion. StreamGBH can not handle the large motion for neighboring frames. (2) StreamGBH only uses the color cue which is difficult to find the exact correspondence. 3D motion is more helpful and accurate for semantic mapping procedure which only camera moves and objects remain static. The visualizations for the StreamGBH and our algorithm are shown in Fig. 3.

The limit of our method is that we assume the objects in the scene remain static. If the objects move, we can not compute 3D transformation for neighbor-ing frames, the global 3D point cloud can not be obtained, either.

5.3   Evaluate the Spatio-Temporally Consistent Semantic Map

For semantic labeling, we train 12 core classes on the 795 training images. We compute the geometry and appearance features using the algorithm in [6] and

Table 1. The comparison between StreamGBH and our approach.

Method	2D UE	3D UE	2D accuracy	3D accuracy	2D boundary	3D boundary	EV
Scene 1							
StreamGBH	12.5185	15.695	0.4214	0.3942	0.0993	0.5365	0.2823
Our	4.3192	7.339	0.8292	0.7793	0.2	0.9043	0.7835
Scene 2							
StreamGBH	4.5693	6.7917	0.652	0.5955	0.1705	0.8163	0.7166
Our	2.1626	2.41	0.8332	0.8109	0.2531	0.9411	0.8581
Scene 3							
StreamGBH	7.8218	17.068	0.5722	0.3454	0.1427	0.5873	0.4433
Our	1.6341	3.7838	0.8862	0.7072	0.2653	0.9606	0.7647
Scene 4							
StreamGBH	3.3975	10.6291	0.8161	0.7657	0.2047	0.8493	0.7875
Our	1.6685	4.6865	0.8627	0.7706	0.3149	0.954	0.8326

StreamGBH

Our temporal consistent segmentation

Fig. 3. Some examples of the StreamGBH and our temporal consistent segmentation. From the examples, we can see that the green and brown regions often change their locations. This situation happens as the StreamGBH can not handle large motion and difference between neighboring frames. (The images are best seen in color) (Colour figure online).

train SVM classifiers based on these features. We test the classifiers' performance on the 654 test images. The average pixel precision and class precision are listed in Table 2. We compute the class precision as follow:

$$\frac{TruePositives}{TruePositives + FalsePositives + FalseNegatives} \tag{11}$$

We get an average pixel accuracy in 73.45 % and average class accuracy in 45.42 %.

Table 2. Performance of the labeling on the object category.

wall	floor	cabinet	bed	chair	table	window
73.69	81.98	44.34	49.74	46.81	31.78	35.15
bookshelf	picture	blinds	ceiling	tv	per-pixel	per-class
30.57	37.91	37.91	53.08	22.13	73.45	45.42

Table 3. Semantic mapping results

	Baseline	Our
Scene 1	88.61 %	91.73 %
Scene 2	84.50 %	86.93 %
Scene 3	86.32 %	88.82 %
Scene 4	78.03 %	82.92 %
Average	84.36 %	87.60 %

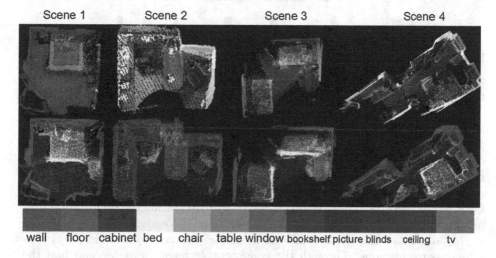

wall   floor cabinet bed   chair   table window bookshelf picture blinds   ceiling   tv

Fig. 4. The visualization of the semantic maps. (The images are best seen in color) (Colour figure online).

In order to get the global 3D point cloud for sequence RGB-D scene, we use the 3D transformation between frames to get the estimated camera poses and register the RGB-D sequences. As a baseline, we independently segment each RGB-D image and classify the superpixels, the 3D point is labeled by a simple majority vote for the labeled 2D pixels which are fused in the same 3D point. In our approach, the temporal consistent segmentations and consistent semantic labels are obtained for each RGB-D image through the method mentioned in Sect. 3. Then a Dense CRF model is used to inference the global 3D point labels. The comparisons for 4 RGB-D scenes are given in Table 3.

We get an average precision of 87.60 % with an improvement of 3.24 %. Compared to the temporal consistent segmentation, independent segmentations often lead to inconsistent semantic precisions. For example, one superpixel which is labeled as window in frame t may vanish in frame $t+1$. Through the consistent segmentation, the consistent semantic precisions over superpixels are enforced. Compared to the majority voting strategy, Dense CRF smooths and distributes the information over neighboring points and the spatial consistency is guaranteed. The visualizations of the semantic maps are shown in Fig. 4, which the global point cloud and semantic labels are shown.

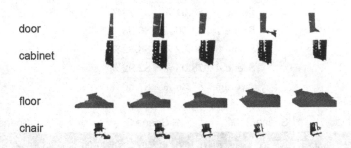

Fig. 5. Sample corresponding objects discovered by temporal consistent segmentations.

6   Conclusion

In this paper, we propose a spatio-temporally consistent semantic mapping approach for indoor scene. By using 3D transformation between frames, we can compute more accurate correspondence for superpixels than the optical flow algorithm. In this way, the performance of our temporal consistent segmentation approach is superior than the state-of-art StreamGBH approach. Also, Dense CRF model enforces the spatial consistency by distributing the information over neighboring points depending on the distance and color similarity.

In fact, more consistent constraints can be enforced by the temporal consistent segmentations. Through the consistent segmentations, we can find the corresponding objects on the different images, as shown in Fig. 5. These corresponding objects on different images should be labeled by the same semantic class. This cue can be enforced by CRF or high-order CRF model. Also, the temporal consistent segmentation can be used for unsupervised object discovery. In this further, we will explore them both.

References

1. Bai, X., Wang, J., Simons, D., Sapiro, G.: Video snapcut: robust video object cutout using localized classifiers. ACM Trans. Graph. (TOG) 28, 70 (2009)
2. Banica, D., Sminchisescu, C.: CPMC-3D-O2P: Semantic segmentation of RGB-D images using CPMC and Second Order Pooling. CoRR, abs/1312.7715 (2013)

3. Chen, A.Y., Corso, J.J.: Propagating multi-class pixel labels throughout video frames. In: 2010 Image Processing Workshop (WNYIPW), Western New York, pp. 14–17. IEEE (2010)

4. Criminisi, A., Cross, G., Blake, A., Kolmogorov, V.: Bilayer segmentation of live video. In: 2006 IEEE Computer Society Conference on Computer Vision and Pattern Recognition, vol. 1, pp. 53–60. IEEE (2006)

5. Felzenszwalb, P.F., Huttenlocher, D.P.: Efficient graph-based image segmentation. Int. J. Comput. Vis. **59**(2), 167–181 (2004)

6. Gupta, S., Arbelaez, P., Malik, J.: Perceptual organization and recognition of indoor scenes from RGB-D images. In: 2013 IEEE Conference on Computer Vision and Pattern Recognition (CVPR), pp. 564–571. IEEE (2013)

7. Henry, P., Krainin, M., Herbst, E., Ren, X., Fox, D.: RGB-D mapping: using kinect-style depth cameras for dense 3D modeling of indoor environments. I.J. Robotic Res. **31**(5), 647–663 (2012)

8. Koppula, H.S., Anand, A., Joachims, T., Saxena, A.: Semantic labeling of 3D point clouds for indoor scenes. In: NIPS, pp. 244–252 (2011)

9. Krähenbühl, P., Koltun, V.: Efficient inference in fully connected crfs with gaussian edge potentials. CoRR, abs/1210.5644 (2012)

10. Moore, A.P., Prince, S., Warrell, J., Mohammed, U., Jones, G.: Superpixel lattices. In: 2008 IEEE Conference on Computer Vision and Pattern Recognition, pp. 1–8. IEEE (2008)

11. Ren, X., Bo, L., Fox, D.: RGB-(D) scene labeling: features and algorithms. In: CVPR, pp. 2759–2766 (2012)

12. Shi, J., Tomasi, C.: Good features to track. In: 1994 IEEE Computer Society Conference on Computer Vision and Pattern Recognition, pp. 593–600. IEEE (1994)

13. Silberman, N., Fergus, R.: Indoor scene segmentation using a structured light sensor. In: ICCV Workshops, pp. 601–608 (2011)

14. Silberman, N., Hoiem, D., Kohli, P., Fergus, R.: Indoor segmentation and support inference from RGBD images. In: Fitzgibbon, A., Lazebnik, S., Perona, P., Sato, Y., Schmid, C. (eds.) ECCV 2012, Part V. LNCS, vol. 7576, pp. 746–760. Springer, Heidelberg (2012)

15. Xu, C., Corso, J.J.: Evaluation of super-voxel methods for early video processing. In: 2012 IEEE Conference on Computer Vision and Pattern Recognition (CVPR), pp. 1202–1209. IEEE (2012)

A New Real-Time Algorithm to Extend DL Assertional Formalism to Represent and Deduce Entities in Robotic Soccer

Saminda Abeyruwan[(⊠)] and Ubbo Visser

Department of Computer Science, University of Miami, 1365 Memorial Drive,
Coral Gables, FL 33146, USA
{saminda,visser}@cs.miami.edu

Abstract. Creating, maintaining, and deducing accurate world knowledge in a dynamic, complex, adversarial, and stochastic environment such as the RoboCup environment is a demanding task. Knowledge should be represented in real-time (i.e., within ms) and deduction from knowledge should be inferred within the same time constraints. We propose an extended assertional formalism for an expressive $\mathcal{SROIQ}(\mathcal{D})$ Description Logic to represent asserted entities in a lattice structure. This structure can represent temporal-like information. Since the computational complexity of the classes of description logic increases with its expressivity, the problem demands either a restriction in the expressivity or an empirical upper bound on the maximum number of axioms in the knowledge base. We assume that the terminological/relational knowledge changes significantly slower than the assertional knowledge. Henceforth, using a fixed terminological and relational formalisms and the proposed lattice structure, we empirically bound the size of the knowledge bases to find the best trade-off in order to achieve deduction capabilities of an existing description logic reasoner in real-time. The queries deduce instances using the equivalent class expressions defined in the terminology. We have conducted all our experiments in the RoboCup 3D Soccer Simulation League environment and provide justifications of the usefulness of the proposed assertional extension. We show the feasibility of our new approach under real-time constraints and conclude that a modified FaCT++ reasoner empirically outperforms other reasoners within the given class of complexity.

Keywords: $\mathcal{SROIQ}(\mathcal{D})$ Logic · Symbol grounding · RoboCup agents

1 Introduction

The OWL 2 Web Ontology Language[1], recommended by the World Wide Web Consortium (W3C) as part of the existing "Semantic Web" technologies, provides an explicit specification of a conceptualization that allows computers to

---

[1] http://www.w3.org/TR/owl2-overview/.

© Springer International Publishing Switzerland 2015
R.A.C. Bianchi et al. (Eds.): RoboCup 2014, LNAI 8992, pp. 270–282, 2015.
DOI: 10.1007/978-3-319-18615-3_22

intelligently search, combine, and process "data" (e.g., visual percepts, sensori-motor streams) on the basis of its meaning, i.e., the semantics. Therefore, similar to humans, computers can interpret data and deduce conclusion from them in its day-to-day operations. The conceptualization provides an abstract, simplified view of the world being modeled [5], the ability to use reasoning subsystems to draw meaningful conclusions from these models, and to exchange complex information or conclusions among multi-agent systems unambiguously. Complex robotic systems such as soccer playing robots in RoboCup [8] environments or self-driving cars (e.g., [18]) need substantial awareness of its surroundings. There is usually a substantial gap between the information that a robotic system actually collects via its modeling mechanisms and the high-level knowledge that could be used in order to obtain the appropriate decisions. Generally, high-level knowledge varies at a slower time scale than the modeling information, and most of the systems are bound to a faster duty cycle. We have investigated an ontological methodology to reduce this gap, ground information with respect to a domain of discourse, and reason in real-time.

The OWL 2 specification is based on $\mathcal{SROIQ(D)}$ Description Logic [7]. It is a less expressive, but highly structured, decidable fragment of the first-order predicate logic. Efficient reasoning engines exist that use DL constructs to infer about the domain of discourse. Though DL is decidable, its worst-case complexity is exponential, which demands upper bounds on the size of the knowledge base. It is common that OWL 2 ontologies are reasoned off-line and use the deduced axioms later in the process to scale for practical problems [11]. Even though DL has exponential worst case complexity, it provides constructs that can be used in real-time robotic systems to model data, derive conclusions from them, and exchange the now semantically grounded data among similar robotic systems. Here, we empirically investigate the ability to use DL in a real-time system setting. The proposed method uses a fixed terminological (TBox) and a fixed relational (RBox) formalism, and provides the justification of using "an extended assertional formalism (ABox)" to represent modeling information in a lattice structure, that has temporal-like structures, without explicitly adding new constructs to $\mathcal{SROIQ(D)}$ DL. This gives us the opportunity to use existing DL reasoners in a real-time setting. Since the general reasoning problem is N2ExpTime-complete, we maintain upper bounds on the number of axioms in each formalism, and empirically study the behavior of the extended ABox.

The RoboCup 3D Soccer Simulation environment provides a dynamic, real-time, complex, adversarial, and stochastic multi-agent environment for simulated agents. The agents formalize their goals in two layers: (1) the physical layers – controls related to walking, kicking, and so forth are conducted; and (2) the decision layers – high-level actions are taken to emerge behaviors. Our proposed method resides in the decision layer to assert modelling information and deduce soccer domain specifications. We have conducted all our experiments in the RoboCup 3D Soccer Simulation League Server[2]. Since, every simulation cycle is limited to 20 ms, we consider the upper-bound of the real-time reasoning

---

[2] http://simspark.sourceforge.net/wiki/.

within 5 ms, 10 ms, or 15 ms. We also consider and discuss situation in which multiple duty cycles, e.g., five cycles amounts to 100 ms, can be used with a threading architecture to harness the idle processing time of the CPU. The fixed TBox contains class expressions to deduce individuals. An example would be the definition of a pass between two players or intercept a moving ball and so forth. We can compose queries to the system and use several heuristics to control the axiom count. The heuristics are activated based on pre-defined criteria such as active region surrounding the ball.

DL provides an appropriate trade-off between expressivity and scalability in practice (the reader is referred to [2,3] a comprehensive discussion on DL syntax[3], semantics, and model construction). In the proposed extension, the TBox and RBox is stationary, while the ABox changes over time. Therefore, complexity is dominated by the data complexity, which is NP-hard for $\mathcal{SROIQ(D)}$ DL ABoxes and N2ExpTime-complete for the combined TBox, RBox, and ABox. Thus, the real-time systems need an upper bound for the size of the ABox, while retaining as much as logical consequences as possible. Modern $\mathcal{SROIQ(D)}$ DL reasoners such as the (1) tableau-based FaCT++ [17] and Pellet [15] reasoners; and the (2) hyper-tableau HermiT [14] reasoner, use intelligent heuristics and optimization methods to perform inferencing as efficiently as possible. We investigate the real-time performances of tableau-based reasoners with respect to the proposed ABox extension.

2   Related Work

In AI, an ontology defines a formal specification of a conceptualization [5]. The conceptualization is defined using concepts, individuals, and relations among them. The formal specification allows agents in a multi-agent system to share information, and it provides a base to agents to act rationally to achieve common goals. The knowledge an agent possesses has the distinct feature of time dependence. But instead of committing to a temporal architecture, we are extending an ABox to a variable lattice structure that captures temporal-like information within the constructs given in DL. Therefore, we explicitly fixed the conceptualization encoded in the TBox and RBox, and change the ABox conceptualization. Similar to our approach, the $\mathcal{TL}\text{-}\mathcal{ALCF}$ DL extends static $\mathcal{ALCF}$ to represent interval-based temporal networks using Allen's interval-based temporal logic [9]. Our approach differs from this work in that we use $\mathcal{SROIQ(D)}$ DL and we encode the temporal-like information (only in the ABox) in a lattice structure that captures the dynamics of the changing knowledge. Therefore, we can directly use existing $\mathcal{SROIQ(D)}$ DL reasoners without substantial modifications. OWL-Time [6] allows representing temporal concepts and temporal relations in $\mathcal{SHOIN(D)}$ DL to represent new languages such as tOWL [10] to conceptualize concrete-domains. Our work significantly differs from these

---

[3] As a convention, we indicate entities in a conceptualization using sans serif letters (e.g., ∀hasParticipant.Thing, ∃hasID.nonNegativeInteger[>0], PassBall, HoldBall).

approaches as we directly represent the temporal-like assertions in a lattice structure and constrain the size of the ABox to support real-time requirements.

Allen's temporal interval algebra [1] captures the ability to represent intervals and temporal properties, and their evolution over those intervals. There are many instances where these constructs are presented in OWL DL ontologies (c.f., [12]) and we use an approach similar to that of Open Biological and Biomedical Ontologies Relation Ontology (OBO RO) [16] to represent temporal-like constructs within the ABox lattice structure.

OWL DL provides resources to represent entities in Semantic Web ontologies. These ontologies are large in nature (T/R/A/Box) and the main focus of many of the research approaches is to investigate: (1) the inference characteristics in expressivity, correctness, worst-case computational complexities of DL languages, incremental classification, rules, justification abilities, and large ABox reasoning; and (2) empirical performance indicators with respect to classification, satisfiability, subsumption, consistency, performance, and heap space and time [4]. These ontologies generally require minutes or hours to finish the reasoning tasks, while we consider the tasks that finish within a few milliseconds (e.g., $\sim$10 ms), yet using all of the functionalities of the reasoner. This is a conflicting objective that needs compromises in different degrees.

A perdurantist (four-dimensionalist) approach has been introduced in [20] to represent entities that change information over time. Instead of depending on time directly, we have used the concepts of continuants and occurrents to represent entities on our domain of discourse. A continuant represents an entity that exists in whole at any time in which it exists at all, and persists through time while maintaining its identity. It has no temporal parts. An example would be the team of an agent. An occurrent is an entity that has temporal parts, and if it occurs, unfolds or develops through time [16]. An example would be the orientation, and two-dimensional location of an agent. Our method uses a combination of continuant and occurrent concepts to create assertions in the extended ABox.

An approach presented in [13] recognizes and predicts spatio-temporal patterns within games of the RoboCup 3D Soccer Simulation League. The method recognizes situations in real-time, and has the the ability to learn and predict the opponent behavior. Recognition, learning, and prediction is performed using Bayesian Networks, and the method requires on average $\sim$40 ms to compute inferences. The work most closely related to our work is presented in [19]. This method introduces a knowledge processing pipeline to detect complex events and action sequences as a spatio-temporal pattern sequence generated from qualitative scene descriptions. The method has been tested under tournament conditions with 5 Hz resulting in precise and also incomplete perceptions.

3   DL Assertional Formalism Extension

In this section, we present the syntax and semantics of the assertional formalism (ABox) extension to $\mathcal{SROIQ}(\mathcal{D})$ DL to represent entities in RoboCup 3D soccer simulation league. Firstly, we provide the definition, secondly, we describe the extension with respect an illustrative examples, thirdly, we describe a few real world examples from our knowledge base, and finally, we describe the extended ABox algorithm. Our extended ABox definition goes as follows:

Definition 1 *(ABox Extension). Given a fixed TBox and an RBox as defined in Sect. 1, the extended ABox is defined as follows*[4]:

(1) There exists sampling points, $t_i \in \mathbb{Z}_{>0}$, such that, when pre-defined criteria are matched, a set of individual assertions are generated.

(2) These assertions are of the form $[C(a_{t_i})]_{t_j}$ for class expressions, and

(3) $[R(a_{t_i}, b_{t_j})]_{t_j}$, $t_i \leq t_j$ for relations with given individuals a_{t_i} and b_{t_j} at the sampling point t_j.

(4) The individual assertions are realized with a timeToLive $\in \mathbb{Z}_{\geq 0}$ data type property, and they will be active for timeToLive >0.

(5) The assertions are active for maximum sampling points of latticeLength $\in \mathbb{Z}_{>0}$, and they are first created with timeToLive $=$ latticeLength.

(6) At each sampling points, the timeToLive data value of all individuals except the individuals with timeToLive $\neq$ latticeLength is decremented by one, and the assertions are purged when timeToLive $= 0$.

(7) Lattice structure query expression $[C(\text{refinement})]$ for an equivalent class expression C and an optional user defined refinement for which the individuals of C should bind to.

The semantics of the Definition 1 is given by same constructs used in Sect. 1. The extended ABox does not include additional constructs, yet provides an efficient framework to manage the number of asserted axioms. Each individual in the extended ABox is annotated with timeToLive data property. The individuals are generated with timeToLive $=$ latticeLength, and they are purged when timeToLive $= 0$.

3.1   An Illustrative Example

Figure 1 (a) shows an illustrative example of an extended ABox with the lattice structure with latticeLength four. In this example, time increases from left-to-right. The sampling points are t_1, t_2, t_3, and t_4 such that, $t_1 < t_2 < t_3 < t_4$, and $t_i \in \mathbb{Z}_{\geq 0}, \forall i \in \mathbb{Z}_{\geq 0}$. The symbol "•" shows an individual in the extended ABox (we will call the extended ABox as ABox at this point forward, and distinguishing the difference, if ambiguity occurs), and the Internationalized Resource Identifier (IRI) is shown to the right. Let's assume that before the sampling point t_1, the ABox is empty. Let's assume that at t_1 four individuals, **1**, **2**, **3**, and **4**, are added to the ABox. Therefore, ABox$_{t_1}$ contains these four individuals. If they are asserted with types, they will be of the form $C(1)_{t_1}, \ldots$ for some class expressions in TBox. These individuals are also asserted with the timeToLive

---

[4] $\mathbb{Z} = \{\ldots, -2, -1, 0, 1, 2, \ldots\}$, $\mathbb{Z}_{>0} = \{n \in \mathbb{Z}; n > 0\}$, and $\mathbb{Z}_{\geq 0} = \{n \in \mathbb{Z}; n \geq 0\}$.

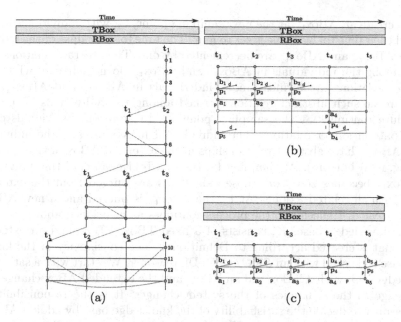

Fig. 1. (a) An extended ABox with the lattice structure, (b) An extended ABox that matches an instance of the equivalent class expression PassBall, and (c) An extended ABox that matches an instance of the equivalent class expression HoldBall.

concrete property with value four. It means that the individuals, that are created at this sampling point, will be lasted for latticeLength - 1 sampling points in the future. In this example, they will last for three more sampling points. All individuals in ABox_{t_1} will have the same timeToLive value. In addition, we also add other abstract and concrete properties to the individuals in ABox_{t_1} that match any pre-defined criteria. All these assertions are represented in a vertical line at t_1.

3.2   Real World Examples

In the next sampling point, t_2, we create ABox_{t_2} and add this to the extended ABox. At this point, all the timeToLive concrete properties in the ABox_{t_1} individuals are decremented by one. Let's assume that the individuals **5**, **6**, and **7** belong to ABox_{t_2}. The timeToLive value is set to latticeLength. The horizontal lines show all existing abstract relations between the individuals from ABox_{t_1} to ABox_{t_2}. At sampling point t_2, a situation could occur that there exists some individuals that may not have a corresponding individual from the previous ABox. e.g., individual **2** does not have a corresponding individual from ABox_{t_2}. The abstract properties are from the RBox, and they could be of the form atomic roles or generalized role inclusion axioms. In addition, individuals in ABox_{t_1} and ABox_{t_2} may add additional abstract roles as shown from the dashed line in Fig. 1 (a), hence, initiating a lattice structure.

We create an $ABox_{t_3}$ at the next sampling point, t_3, with individuals t_8 and t_9, and their timeToLive value is set to four. The timeToLive values of the individuals in $ABox_{t_1}$ and $ABox_{t_2}$ are decremented by one. The abstract relations that exist among the individuals in $ABox_{t_1}$ and $ABox_{t_2}$ do not change, while new abstract relations are formed among individuals in $ABox_{t_2}$ and $ABox_{t_3}$. Say there are no such abstract relations formed among the individuals. The same procedure continues at the sampling point t_4. In addition, individuals could participate in longer relations. The individual 4 in $ABox_{t_1}$ and the individual 13 in $ABox_{t_4}$ have abstract relationships in the extended ABox in this example (cf. Fig. 1 (a) bottom). At sampling t_5, the timeToLive value of the individuals in $ABox_{t_1}$ becomes zero, and those individuals are purged from the extended ABox with all related axioms. At t_5, the $ABox_{t_1}$ is purged and a new $ABox_{t_5}$ will be created. Henceforth, the process continues as mentioned above.

Our knowledge base, K, consists of a fixed TBox, RBox, and an extended ABox that is created according to Definition 1. The consistency of the knowledge base is checked with an $\mathcal{SROIQ(D)}$ DL reasoner. We start with a satisfiable knowledge base with a TBox and an RBox, and the extended ABox changes the knowledge as the dynamics of the system changes. It is the responsibility of the reasoner to decide the satisfiability of the knowledge base by adding $ABox_{t_i}$, $i = 1, \ldots$, to the extended ABox. If the knowledge base is unsatisfiable, then the $ABox_{t_i}$ will be removed from the knowledge base. We query for assertions using equivalent class expressions and user defined refinements.

In this subsection we provide a few examples from our knowledge base to understand the process. We have developed an ontology to represent entities in the RoboCup 3D Soccer Simulation environment based on prior knowledge.

(1) The equivalent class, Object $\equiv$ $\exists$timeToLive.nonNegativeInteger, defines an object in our domain of discourse as any entity that has a positive time-to-live value.

(2) We define an agent using equivalent class expression; Agent $\equiv$ Object $\sqcap$ $\exists$hasID.nonNegativeInteger$[> 0]$, and hasID $\in \mathbb{Z}_{>0}$, as any object in the domain of discourse that has a strictly positive identification number.

(3) Therefore, we define a home agent and an opponent agent; HomeAgent $\sqsubseteq$ Agent and OpponentAgent $\sqsubseteq$ Agent. The agents are disjoint: HomeAgent $\sqcap$ OpponentAgent $\sqsubseteq \bot$.

(4) Most entities in the RoboCup 3D soccer simulation have poses (rotation and two dimensional position on the field). We define a pose: Pose2D $\equiv$ ($\exists$rotation.int $\sqcap$ $\exists$xcoord.int $\sqcap$ $\exists$ycoord.int), with rotation, xcoord, ycoord $\in \mathbb{Z}$. Any *thing* in the domain of discourse which has an orientation and (x, y) coordinates in a two-dimensional plane. The distances are annotated with millimeters (mm), while the angles in radians are subjected to the mapping function $f : [-\pi, \pi] \mapsto [0, 2048]$.

(5) Ball GCI axioms are: Ball $\sqsubseteq$ $\exists$locatedIn.Pose2D and Ball $\sqsubseteq$ Object.

(6) Using axioms 1, 2, 3, 4, and 5, we can query for all objects potentially close to the ball from the extended ABox as ObjectsWithBallContact $\equiv$ Object $\sqcap$

∃hasParticipant.Ball ⊓ ∃hasParticipant.(Participant ⊓ ∃distance.int[<500]), and distance ∈ ℤ, any object in the domain of discourse which has a ball participant and the ball participant is *close* to the object.

hasParticipant is a transitive object property. We use N-ary relationship representations[5] to state the connection between agents, participants, and soccer ball representations. We use a distance threshold, which is given as prior knowledge from the domain experts. This class expression uses a 500 mm distance threshold to quantify the closeness property.

(7) Let's define a class expression for the HoldBall skill, which queries for agents that control the ball. We define a refinement such that the individuals should have different timeToLive values and there must exists at at least five individuals in the class expression. We have defined the equivalent class expression: HoldBall ≡ Agent ⊓ ∃hasParticipant.(BallParticipant ⊓ ∃distance.int[<150]), demands classification of agents that have some ball participants within *a close proximity*. [HoldBall(refinement)] query expression uses several ABox parameter choices and the prior knowledge of duration in which an agent should be in close proximity to the ball. An instance of the ABox that matches the query expression is given in Fig. 1 (c). The refinements are executed after the deduction process is finished.

Our assumptions are: (A1) latticeLength is five; (A2) user define refinements; and (A3) given sampling points t_i, $i = 1, \ldots, 5$. The individuals are: (I1) (I2) b_j, $j = 1, \ldots, 5$, represents an instance of class expression Ball; (I3) p_j, $j = 1, \ldots, 5$, represents an instance of class expression Participant; and (I3) a_j, $j = 1, \ldots, 5$, represents an instance of class expression Agent, and it is a realization of the same agent over the extended ABox sampling points. The relations are: (R1) p and l represent the transitive abstract properties hasParticipant and locatedIn respectively; (R2) d represents the concrete property distance; and (R3) there exists diverse variety of relations among individuals that does not influence the given outcome. Using these criteria, and if the ball is within a close proximity (e.g., ≤500 mm), a DL reasoner can deduce that a_1 is the only instance of the class HoldBall within the parameters of the given ABox. Using our TBox, we can determine the type of the agent, and using RBox and ABox we can determine the identification and other assertions. If the definition of the class expression HoldBall needs to be more specific, such as whether the agent needs to remain stationary or stay as far away from the opponent as possible, these conditions should be explicitly stated in the class expression. These additional constraints increase the number of axioms, and there will be a trade-off between computational complexity and the number of queries we can define.

The interpretation of the statement "in close proximity" is based on prior knowledge and design parameters. We can define multiple subclasses of HoldBall with different refinements that meets our criteria. According to the query expression, the result set contains either home agents or opponent agents. In addition, a DL reasoner deduces the fact that HoldBall ⊑ WithBallContact.

---

[5] http://www.w3.org/TR/swbp-n-aryRelations/.

(8) PassBall equivalent class expression queries for home agents that pass the ball to its teammates. It is defined as:

PassBall ≡ HomeAgent ⊓ ∃hasParticipant.(BallParticipant ⊓ ∃distance.int[<100]) ⊓ ∃hasParticipant.(HomeAgent ⊓ ∃hasParticipant.(BallParticipant ⊓ ∃distance.int[<150]) ⊓ ∃ hasParticipant.(BallParticipant ⊓ ∃distance.int[>1000]).

PassBall class subsumes individuals in the extended ABox close to the ball, and within the same ABox, locate another agent of the same team that is close to the ball. In order to conduct a pass, the ball is required to be relatively close to an agent (e.g., <150 mm) and the ball should travel some distance (e.g., >1000 mm), and the receiving ball should be also close to an agent.

There are some limitations in the given definition. First, we neither can write the requirement that the passing agent and the receiving agent should be different in the class expression nor a refinement for PassBall using DL expressivity. Second, there could be situations where external forces or disturbances from another agent or environment could cause the ball to move from the close-by-agent to another agent in the same team. PassBall definition does not capture these special cases. In order to capture the degree of PassBall confidence, extra systems with probabilistic interpretation must be used.

In RoboCup 3D Soccer Simulation, there are no constructs to define a *kick* directly. Therefore, we have defined the pass without explicitly committing to a notion of a kick. Similarly, we define a pass ball behavior to opponent agents. Hence, we generalize PassBall class expression to deduce either a home or an opponent agent as the passing agent using logical union conjunction. An instance of the ABox that matches the class expression is given in Fig. 1 (b). The symbol ∘ in Fig. 1 (b) represents a *black node* that connects relevant individuals. We have used the same assumption and parameters that are defined in Example 7. We have made a slight modification to the individuals such that, the sets $\{a_1, a_2, a_3\}$ and $\{a_4\}$ are disjoint realizations of physically different agents (we have used the agent identification number to make this distinction). A DL reasoner can deduce that a_1 is an instance of the class PassBall for this particular example.

4   Experiments

In order to establish a baseline, we have compared the distributions of deduction times of four reasoners[6]. We have modified the FaCT++ implementation to use hash tables instead of binary tree implementations, when it is necessary, and slightly changed the caching mechanisms. This modification has positive effects on TBox and RBox reasoning. We expect improvements in ABox reasoning, when there are individuals with many data type axioms. We have used

---

[6] (1) HermiT (1.3.8): http://hermit-reasoner.com/; (2) Pellet (2.3.1): http://clarkparsia.com/pellet/; and (3) FaCT++ (1.6.2): http://code.google.com/p/factplusplus/.

Table 1. Average reasoning times (95 % confidence).

Reasoner	Time in milliseconds
FaCT++ (Modified)	*1.092 ± 0.002*
FaCT++ (1.6.2)	1.403 ± 0.009
HermiT (1.3.8)	2.289 ± 2.654
Pellet (2.3.1)	11.634 ± 2.465

the modified version of FaCT++ in baseline establishment and it is labeled as FaCT++ (Modified). Table 1 shows the average reasoning times and the 95 % confidence intervals. We have used 80−logical axioms (62−entities) from the ontology (TBox and RBox) for this experiment. This calculation uses 500 independent trials from each reasoner. We observe that the modified FaCT++ DL reasoner shows a statistically better performance over the other reasoners[7]. Our modifications to FaCT++ DL reasoner have improved 22 % over the original FaCT++ implementation. Henceforth, we have selected the modified FaCT++ DL reasoner to be used with the simulated agents, and we have used the reasoners in non-incremental mode. The baseline establishment has set the empirical lower bound to zero individuals and ∼150 axioms. This corresponds to an empty ABox. In order to estimate the empirical upper bound, we have conducted the following experiment: Firstly, we have added the examples mentioned in subsect. 3.2 to the ontology. Secondly, we have created a hypothetical world model for an agent. This world model changes its believes randomly about teammates, opponents, and ball poses. We have used a uniform distribution to sample entities in the belief model. All poses are randomly generated inside the simulated soccer field. Thirdly, we have chosen a value from $[2, 20]$ for latticeLength to generate axioms. Finally, we ran 10 sets of 100 sampling points for every latticeLength setting. It corresponds to a set of 19, 000 data points. Figures 2 and 3 show the concluding results (the error bars show one-standard deviation of bins of ten).

In Fig. 2, the KernelConsistentTime plot shows that the consistency checking scales linearly with the number of axioms. The classification (KernelClassifyTime plot) produces an exponential growth. Its contribution immensely affects the reasoning time and the empirical upper bound. It also affects the variability of timing. After 1, 000 axioms, there is significant variance, which is undesirable for real-time systems. The realization (KernelRealizeTime plot) and expression (ExpressionTime plot) times are relatively negligible. The expression time exhibits our operations for refinements and to track the evolution of the extended ABox. It shows linear time complexity and it is justifiable for real time operations.

Figure 3 shows the number of individuals (active and cached) with respect to total deduction time. The extended ABox algorithm uses the caching mechanism to reuse individuals, which improves the expression time. We can conclude

---

[7] We have used 2.2 GHz Core−i7 (4 GB) laptop for all experiments. Authors can provide the modified FaCT++ implementation and the ontology upon request.

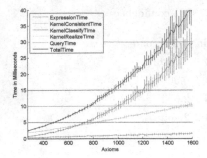

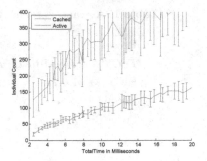

Fig. 2. Establishment of empirical upper bound.

Fig. 3. Relationship between reasoning time and number of individuals.

from these figures that in order to operate within 5 ms, we can keep ~500 axioms ~50 individuals. Similarly, we can use ~800 axioms and ~100 individuals for 10 ms, and ~1000 axioms and ~150 individuals for 15 ms. This suggests that with a given latticeLength, as long as we bound the size of the axioms and individuals, our system can operate in real-time. We have investigated the requirement whether DL-based constructs are suitable for real-time operations. Figures 2 and 3 emphasize the facts that: (1) there is an upper bound where DL boxes are effective to deduce conclusions in real-time; and (2) the total deduction time exponentially increase with the number of axioms. Therefore, it suggests that we can use multiple cycles (e.g., 100 ms corresponds to five cycles in our domain) to execute the reasoning process. This requires an agent equipped with a low-priority thread that uses extra clock cycles of the processing units.

In order to bound the size of the axioms and individuals, we have developed explicit heuristics for which the agents should react to. Agents keep assertions about Ball in all sampling points to produce concise decisions. All agents maintain assertions about themselves and w.r.t. the Ball. We have considered only two players close to the ball from each team to participate with Ball individuals. Each agent keeps track of two close players from each team. The agents participate with other agents through Participant objects. This provides a clean and simple mechanism in which an agent could include Allen's temporal constructs to be used within the ABox. In our on-line setting, we ran 11 vs 11 games with the given heuristics. We set the latticeLength to five for this experiment. An agent tracks 49.18 ± 4.86 individuals and 529.44 ± 105.31 axioms in 6.57 ± 2.81 ms. Therefore, we can justify that our algorithm is real-time compatible on a RoboCup 3D simulated robot.

5   Conclusion

We presented a new approach of using an extended ABox structure to represent temporal-like information and deducing conclusions in real-time. Our approach has extended the $\mathcal{SROIQ}(\mathcal{D})$ DL ABox with a lattice structure and it provides flexibility to use existing DL reasoners. We have tested and validated our approach in an off-line and on-line settings for the RoboCup 3D soccer domain. The approach enables autonomous agents to successfully interpret its believes about the world. We have showed that the deduction complexity and the computation complexity produce a conflicting objective. Therefore, our approach has empirically bounded the size of DL boxes and modified the FaCT++ DL reasoner to be compatible with real-time operations. We intend to use our approach with incremental reasoning on a physical robot to model believes and interpret entities in uncertain environments in the near future.

References

1. Allen, J.F., Ferguson, G.: Actions and events in interval temporal logic. J. Log. Comput. **4**(5), 531–579 (1994)
2. Baader, F., Nutt, W.: Basic description logics. In: Baader, F., Calvanese, D., McGuinness, D.L., Nardi, D., Patel-Schneider, P.F. (eds.) Description Logic Handbook, pp. 43–95. Cambridge University Press, Cambridge (2003)
3. Brown, D., Chumakina, M., Corbett, G.G. (eds.): Canonical Morphology and Syntax. Oxford University Press, New York (2013)
4. Dentler, K., Cornet, R., ten Teije, A., de Keizer, N.: Comparison of reasoners for large ontologies in the OWL 2 EL profile. Semant. Web **2**(2), 71–87 (2011)
5. Gruber, T.R.: Towards principles for the design of ontologies used for knowledge sharing. In: Guarino, N., Poli, R. (eds.) Formal Ontology in Conceptual Analysis and Knowledge Representation. Kluwer Academic Publishers, Deventer (1993)
6. Hobbs, J.R., Pan, F.: An ontology of time for the semantic web. ACM Trans. Asian Lang. Process. (TALIP) **3**(1), 66–85 (2004). Special issue on Temporal Information Processing
7. Horrocks, I., Kutz, O., Sattler, U.: The even more irresistible SROIQ. In: Proceedings of the 10th International Conference on Principles of Knowledge Representation and Reasoning (KR2006), pp. 57–67. AAAI Press, June 2006
8. Kitano, H., Asada, M., Kuniyoshi, Y., Noda, I., Osawa, E.: RoboCup: the robot world cup initiative. In: Proceedings of the First International Conference on Autonomous Agents, pp. 340–347. AGENTS 1997. ACM, New York (1997)
9. Lutz, C., Wolter, F., Zakharyaschev, M.: Temporal description logics: a survey. In: Demri, S., Jensen, C.S. (eds.) TIME, pp. 3–14. IEEE Computer Society (2008)
10. Milea, V., Frasincar, F., Kaymak, U.: tOWL: a temporal web ontology language. IEEE Trans. Syst. Man Cybern. Part B **42**(1), 268–281 (2012)
11. Motik, B., Horrocks, I., Kim, S.M.: Delta-reasoner: a semantic web reasoner for an intelligent mobile platform. In: Mille, A., Gandon, F.L., Misselis, J., Rabinovich, M., Staab, S. (eds.) WWW (Companion Volume), pp. 63–72. ACM (2012)
12. Petnga, L., Austin, M.: Ontologies of time and time-based reasoning for MBSE of cyber-physical systems. In: Paredis, C.J.J., Bishop, C., Bodner, D.A. (eds.) CSER. Procedia Computer Science, vol. 16, pp. 403–412. Elsevier (2013)
13. Rachuy, C., Visser, U.: Behavior-analysis and -prediction for agents in real-time and dynamic adversarial environments. In: Proceedings of the 2010 conference on ECAI 2010: 19th European Conference on Artificial Intelligence, pp. 979–980. IOS Press, Amsterdam, The Netherlands (2010)

14. Shearer, R., Motik, B., Horrocks, I.: HermiT: a highly-efficient OWL reasoner. In: Ruttenberg, A., Sattler, U., Dolbear, C. (eds.) Proceedings of the 5th International Workshop on OWL: Experiences and Directions, pp. 26–27. Karlsruhe, Germany, October 2008

15. Sirin, E., Parsia, B., Grau, B.C., Kalyanpur, A., Katz, Y.: Pellet: a practical OWL-DL reasoner. Web Semant.: Sci. Serv. Agents World Wide Web 5(2), 51–53 (2007)

16. Smith, B., Ceusters, W., Klagges, B., Kohler, J., Kumar, A., Lomax, J., Mungall, C., Neuhaus, F., Rector, A., Rosse, C.: Relations in biomedical ontologies. Genome Biol. 6(5), R46+ (2005)

17. Tsarkov, D., Horrocks, I.: FaCT++ description logic reasoner: system description. In: Furbach, U., Shankar, N. (eds.) IJCAR 2006. LNCS (LNAI), vol. 4130, pp. 292–297. Springer, Heidelberg (2006)

18. Wang, C.C., Thorpe, C., Thrun, S.: Online simultaneous localization and mapping with detection and tracking of moving objects: theory and results from a ground vehicle in crowded urban areas. In: Proceedings of the IEEE International Conference on Robotics and Automation (ICRA). Taipei, Taiwan, September 2003

19. Warden, T., Lattner, A.D., Visser, U.: Real-time spatio-temporal analysis of dynamic scenes in 3D soccer simulation. In: Iocchi, L., Matsubara, H., Weitzenfeld, A., Zhou, C. (eds.) RoboCup 2008. LNCS, vol. 5399, pp. 366–378. Springer, Heidelberg (2009)

20. Welty, C.A., Fikes, R.: A reusable ontology for fluents in OWL. In: Bennett, B., Fellbaum, C. (eds.) Formal Ontology in Information Systems. Frontiers in Artificial Intelligence and Applications, vol. 150, pp. 226–236. IOS (2006)

Poster Presentations

An Event-Driven Operating System
for Servomotor Control

Geoff Nagy[✉], Andrew Winton, Jacky Baltes, and John Anderson

Autonomous Agents Lab, University of Manitoba, Winnipeg, MB R3T 2N2, Canada
{geoffn,jacky,andersj}@cs.umanitoba.ca, umwintoa@cc.umanitoba.ca

Abstract. Control of a servomotor is a challenging real-time problem. The embedded microcontroller is responsible for fast and precise actuation of the motor shaft, and must handle communication with a master controller as well. If additional tasks such as temperature monitoring are desirable, they must take place often enough to be useful, but not so frequently that they interfere with the operation of the servo. Since microcontrollers have limited multi-tasking capabilities, it becomes difficult to perform all of these tasks at once. It was our goal to create servo firmware with high communication speeds for humanoid robots, and our solution is generalizable to non-humanoid motor control as well. In this paper, we present an event-driven operating system for the Robotis AX-12 servomotor. By using interrupts to drive functionality that would otherwise require polling, our operating system meets the real-time constraints associated with controlling a servomotor.

1 Introduction

Servomotors are used extensively in robotics to give rise to motion—arms, legs, torsos, and other appendages contain servos which, in response to commands from a master controller, must move in a coordinated, timely fashion to walk, run, or perform other tasks. Control of a servomotor is a task with many real-time constraints. Of these, the most crucial is timely response to positional feedback data, in order to actuate the servo to a target position. Failing to meet this real-time constraint could mean anything from a crashed remote-controlled airplane to the loss of more expensive equipment. Actuation commands are given by a master controller, and thus a servo is also responsible for listening to these commands and responding as quickly as possible: a servo moving in response to an actuation command that was given two seconds ago is virtually useless.

The ability of a servo to monitor its status (temperature, torque load, etc.) is also a desirable feature. Overheating, for example, is a situation that should be avoided to prevent damage. Servos that are able to sense their own status can shut down in dangerous conditions, or provide relevant sensor data to a master controller to produce a more intelligent response and avoid the same equipment losses associated with failure to meet real time constraints.

The main challenge in meeting these requirements stems from a microcontroller's limited ability to multi-task. Analog-to-digital (A2D) conversions for

R.A.C. Bianchi et al. (Eds.): RoboCup 2014, LNAI 8992, pp. 285–294, 2015.
DOI: 10.1007/978-3-319-18615-3_23

reading onboard sensors take time away from the servo's primary function, which is to actuate to a specific position with minimal latency or jitter. Although temperature reads, for example, do not need to occur frequently, a servo must continually check the potentiometer connected to the motor shaft to make sure it is rotating to the correct position. Doing so requires constant A2D conversions, which take time and can cause latency in the servo response if implemented inefficiently. This becomes a challenging real-time problem, and it is the responsibility of a servo firmware designer to ensure that these constraints are met.

It was our goal to create a servo firmware implementation that would allow us to communicate with multiple servos at extremely high speeds. Our intention was to use this firmware to control the motors in humanoid robots such as the DARwIn-OP or Bioloid, but our solution has applications in other areas as well. High communication speeds with motors is desirable because latency in complex motor tasks can lead to equipment failure. (For example, a bipedal robot might fall over if it loses its balance due to servo communication latency.) As the number of motors increases, this problem is compounded since addressing more servos requires more time.

Our target motor platform was the Robotis AX-12 servo, as shown in Fig. 1. Our approach applies to other programmable servos as well. In order to meet our goal of faster communication times, we needed to create firmware that would be capable of handling the servo tasks described above. To support this, we endeavoured to develop an operating system for the AX-12's embedded Atmel ATmega8 microcontroller.

It was immediately apparent to us that polling techniques for serial communications or actuation would not suffice. Simply blocking and waiting for A2D reads or bytes via serial would not be practical, since spending too much time on one task would take time away from others. Our solution was to develop an interrupt-driven operating system, and was motivated by several factors. In general, interrupt-based systems eliminate the need for polling, and are ideal for low-power microcontrollers [1,2]. They are also ideal for memory-constrained applications [1,3,4]. In our case, embracing an event-driven approach allowed us to update the motor control pulse-width modulation (PWM) signals as often as A2D conversions could take place, and enabled the use of interrupts to handle serial communications and timer functions. The following section describes related work, and the section afterwards describes our operating system, Dorkeus, in detail.

2   Related Work

Baltes et al. [5] developed a multi-threaded, real-time kernel for the Atmel ATmega128 microprocessor called *Freezer OS*. It handled scheduling of multiple tasks pre-emptively in a round robin fashion, and included support for interrupt-based A2D conversions and serial communications. The goal in its creation was to ease application development of robotic firmware. In contrast, our approach is entirely interrupt-based and was specifically designed for servomotor control applications.

Fig. 1. A Robotis AX-12 servo.

Node.js is a Javascript framework that firmly embraces the event-driven paradigm [6,7]. Using this framework, Web developers specify events that must be responded to, and provide callbacks that should execute to handle the events requiring response. This interrupt-based approach is similar to our own, whereby events are handled without the need for polling. Our work differs from Node.js in that our approach has been tailored to meet various real-time constraints associated with servo control. Additionally, our work applies the event-driven paradigm to a control application, rather than a web-based server-side framework.

TinyOS is an event-driven framework that supports the development of real-time embedded operating systems [1]. Intended for use in sensor networks, its small size and event-based design make it well-suited for running on low power, low cost microcontrollers. TinyOS programs can be written modularly without the need for a global understanding of the entire system [1], and this is made possible by the self-contained nature of the interrupt handlers that respond to various events. In this regard, our approach is quite similar. The TinyOS system, however, was designed specifically for large sensor networks, while our application is meant to address the real-time constraints associated with motor control. A similar system is *Contiki* [8], a C-implemented operating system developed for use in large sensor networks. It runs an event-driven kernel that also supports pre-emptive multi-threading. Our approach differs from Contiki in that Dorkeus is entirely event-driven, and is meant for servo control applications.

Åarzén [9] described an event-based proportional-integral-derivative (PID) controller that only generates output responses when the input signal goes beyond a certain threshold. Improvements were made by Durand and Marchand [10], in which the controller avoids re-computation of the output signal in cases where the input signal remains unchanged. While our own approach is event-driven, our PID control is not triggered by the measured input signal, but rather by the completion of an A2D conversion. In addition to PID control, our operating system is responsible for handling serial communications and timer facilities as well.

OpenServo [11] is an online community project dedicated to developing firmware for various servomotors on different microcontrollers. It supports a large number of serial commands for servo control and is partially driven by interrupts. In contrast, our approach is entirely event-driven, and all of our processing takes place in the interrupt context. The *ActuatedCharacter* [12] project is another effort whose goals involve improving communication speeds with servos. This approach specifically addresses the AX-12, and features a small, fast communications protocol. Unlike our approach, it is not entirely interrupt-driven.

3   Dorkeus Operating System

The Dorkeus OS is a fully event-driven servo operating system—it relies entirely on interrupts to drive its functionality. Our implementation is such that virtually all of our processing occurs within the interrupt service routine (ISR) context. While simple computations taking place in an interrupt context do not pose a problem, performing expensive computation is often not desirable for a number of reasons. These include (a) the possibility that an interrupt of the same type can be missed without taking extra precautions, and (b) the fact that ISR code might need to disable interrupts for longer periods of time in order to avoid race conditions, possibly resulting in lost events. One solution to both problems is to offload time-consuming computations to the main task context. Fortunately, the computation performed in our operating system is intentionally minimal. Thus, the processing associated with each event does not require it to be handled outside of an ISR context, resulting in simpler, event-driven code that is easier to maintain and debug.

Our operating system is composed of three subsystems as shown in Fig. 2: the control loop, the serial communication system, and the timer queue. The following subsections describe these systems in detail.

3.1   Control Loop

The control loop in Dorkeus is responsible for actuating the motor to a specific position—this is the highest-priority task of any servomotor firmware. In our approach, the main control algorithm is a PID controller. Positional feedback is obtained by reading the value of the motor shaft potentiometer using the embedded ATmega8's analog-to-digital converter (ADC). An A2D conversion on an ATmega8 can take anywhere between $13\,\mu s$ to $260\,\mu s$ [13], which means that pausing to wait for a positional read would result in a delay that could otherwise be used to perform useful work. We used an interrupt-based approach to mitigate this problem.

In our approach, we trigger an A2D conversion inside an infinite loop. Whenever an A2D conversion completes, the corresponding interrupt is executed. Inside this interrupt, the PID controller is used to compute the response of the motor. Then, this response is used to adjust the output PWM signal. The hardware configuration of the AX-12 dictates that this signal is output using the ATmega8's only 16-bit timer. Once this process is complete, another A2D conversion begins.

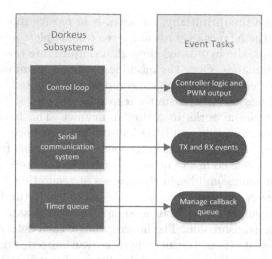

Fig. 2. The three subsystems in Dorkeus and the processing associated with each of their events.

This method has the advantage of not continually polling and waiting for the ADC result: when the conversion result is ready, it is fed into the controller logic. This ensures that the servo responds with minimal latency, since it responds to positional errors as quickly as the A2D conversions can finish. Using interrupts in this fashion means that while an A2D conversion is taking place, the ATmega8 is free to perform other useful work.

We chose to use a PID controller in our approach due to its generalizability and ease of implementation. It is also used in the original AX-12 firmware provided by Robotis [14]. Our firmware implementation does not depend on any particular control algorithm, and so it would be a simple matter to substitute our PID controller for another control algorithm.

3.2 Serial Communication System

The serial communication system is responsible for listening to serial commands from the master controller and responding as necessary. We chose to embrace an event-driven approach for our serial communications. This eliminated the need to poll for serial data.

Supported commands include actuation instructions, pings, and servo position read requests. A key element in our implementation is the communication protocol's support of a *synchronous read* command that does not exist in the original Robotis firmware. This command enables each servo to send positional data one after the other, after an initial synchronous read request, without further management from the master controller. This was inspired by the *chain reply* that is used by the ActuatedCharacter firmware [12].

The AX-12 hardware is configured to use a half-duplex communication protocol with a main controller (i.e., it can either send or receive data, but it cannot do both at the same time). Initially, the servo is in receive mode. Every time a byte is received by the ATmega8 hardware, a *receive-complete* interrupt executes. Within the interrupt, the byte is fed through a simple state machine that interprets the serial commands. Once the machine reaches a terminal state (i.e., the whole message is received), the complete command is executed. If the command requires it, the ATmega8 will transmit a response.

Message transmission works in a similar manner. The master-slave architecture of the communications and the half-duplex hardware dictate that the servo should only transmit messages in response to commands from a main controller. Therefore, once a complete message is received, Dorkeus can switch from receive to transmit mode, and begin the process of generating a response. With *transmission-complete* interrupts enabled, the servo sends the first byte of the response. The transmission-complete interrupt then fires as a consequence of the completed byte transmission. The interrupt logic then determines if another byte needs to be sent. If so, the next byte is transmitted, triggering another interrupt, and so on. This cascading effect allows all response bytes to be sent from the ATmega8 as quickly as possible. Eventually, when all bytes have been sent, the last transmission-complete interrupt finishes without sending anything, thus ending the chain of byte transmissions. The AX-12 then switches back to receive mode to await the next serial command.

3.3   Timer Queue

Certain low-priority events, such as measuring slow-changing sensor data (e.g., temperature), do not have to occur frequently, but must happen regularly enough to be useful. For example, it is only necessary to check once approximately every second if a servo is overheating. To facilitate this, we have developed an interrupt-based timer system with which events can be scheduled for execution in the future. Events are registered by the user specifying (a) the time delay in milliseconds, and (b) the callback function that should be executed when that time expires. Listing 1 shows the pseudocode of our implementation.

The timer system is implemented as a priority queue, sorted by time in non-descending (i.e., increasingly distant) order. When an event is added to the front of the timer queue, the ATmega8's 8-bit timer should be set to expire exactly when the first event should occur. Unfortunately, the maximum amount of time allowed by an 8-bit timer in our implementation is 16.32 ms, which is generally too small an amount to be useful. This limitation is due to the speed of the external clock used (16 Mhz), and even with a timer prescaler of 1024, the maximum timer value is rather short. We are unable to use the 16-bit timer on the ATmega8 since it is already in use by the PWM output logic. Our calculations are shown below in Eq. 1.

$$maxtime = \frac{255}{\left(\dfrac{16\,\text{MHz}}{1024}\right)} = 16.32\,\text{ms} \tag{1}$$

```
void timerExpire ()
{
  TimerEvent e = queue.getFront ()
  e.elapseTime (getTimerElapsed ())
  if e.isTimeExpired
    queue.dequeue.fire ()
    foreach i in queue
      i.elapseTime (front.getFullDelay ())
      if i.isTimeExpired
        queue.dequeue ().fire ()
  else setTimer (e)
  if !queue.isEmpty () setTimer (queue.getFront ())
}
void setTimer (TimerEvent t)
{
  if t.getDelay () > MAX_TIME
    setTimerInterrupt (MAX_TIME)
  else
    setTimerInterrupt (t.getDelay ())
}
void registerTimerEvent (Callback c, int ms)
{
  TimerEvent e = TimerEvent (c, ms)
  Time t = getTimerElapsed ()
  if queue.isEmpty () setTimer (e)
  else
    int delay = queue.getFront ().getDelay ()
    if ms < (delay - t)
      setTimer (e)
      queue.getFront ().elapseTime (t)
  queue.orderedEnqueue (c)
}
```

Listing 1. Pseudocode for our interrupt-based timer implementation.

To bypass this limitation of the 8-bit timer, events that are scheduled to occur sufficiently far into the future (i.e., later than 16 ms from when the event was scheduled) set the 8-bit timer to expire in 16 ms. When the expiration occurs, the timer logic examines the event at the head of the queue and determines if there is any time remaining before the event fires. If there is time remaining, the timer is reset again for either (a) the remaining amount, if the delay is less than 16 ms, or (b) the maximum time if there are still 16 ms or more remaining.

If the timer expires and an event is ready to fire, the function associated with that event is invoked. Figure 3 illustrates the execution of two events. The total

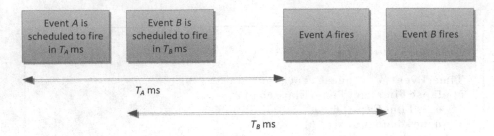

Fig. 3. Shows the scheduling and execution of two timer events.

amount of time in milliseconds that it took the front event to fire is subtracted from the respective times of the remaining events to allow for their consideration of the passage of time. The timer queue is then checked for any additional events whose time may have expired. Any events that are ready to execute then do so in the proper order.

It should be noted that there is some imprecision associated with our timer implementation. Our timer delays are expressed in whole milliseconds, and the maximum delay that can be implemented using an 8-bit timer in our approach is 16.32 ms. When setting the 8-bit timer counter register, we multiply the delay in milliseconds by 15.625 (since $\frac{16\,\text{MHz}}{1024} = 15.625$ ticks per millisecond) and then round the resulting number to the nearest integer. We were not concerned with sub-millisecond accuracy, so imprecisions due to rounding were not an issue. If these imprecisions were problematic, we could have solved this problem by using a different prescaler (such as 64, which would give $\frac{16\,\text{MHz}}{64} = 250$ ticks per millisecond) that would remove the need to round. An alternative method would be to use a prescaler of 256, which would give $\frac{16\,\text{MHz}}{256} = 62.5$ ticks per millisecond, and would allow us to specify delays in terms of half-milliseconds. A third option would be to simply alter our timer queue to work in microseconds instead of milliseconds, but this would necessitate the use of **unsigned longs** and would use more memory. We chose to use the highest prescaler available to ensure that the system spent as little time handling the timer queue as possible.

4   Performance

Our operating system achieves our goal of faster communication speeds with a master controller. Using our interrupt-based approach and minimal communication protocol, we have achieved highly reliable communication at speeds as high as 1 Mbps with up to 10 servos. Both the original Robotis firmware and Dorkeus support the ability to send and receive positional data to and from servos, and additional tests have revealed that our protocol enables us to continually send and receive positional data at a rate nearly ten times faster than the original firmware on the AX-12 for an equal number of servos. The use of our synchronous read command contributes greatly to these reduced communication times when compared to the original Robotis firmware, which requires positional data to be requested individually from each servo, one at a time. Since our communication protocol can address up to 32 unique IDs, the system as it stands should be able to perform with similar success with a larger number of servos. This would enable much faster control of the servos in a robot, allowing faster motion updates and helping to reduce the probability of falling or becoming unbalanced. We intend to discuss our protocol in more depth in a future paper.

Our AX-12 operating system is entirely event-based; all actions that occur do so as a result of either an internal or external event. This has certain implications and presents challenges to which traditional time-based serial implementations are not susceptible. Consider the Watchdog Timer (WDT) present in the ATmega8.

The WDT's primary function is to detect nonprogress in the operation of a microcontroller. At regular intervals, onboard firmware should reset the WDT counter to indicate that progress is being made. Should the WDT expire (perhaps due to an infinite loop), a system reset command will be issued if the proper configurations are made. This ensures that the system is restarted if it fails to make progress. While this method works with conventional approaches, it will not succeed in an event-driven environment where there is no guarantee that progress will be made in finite time. It is conceivable that the WDT could expire in between events, undesirably resetting the system.

While disabling the WDT reset functionality (by, for example, enabling the WDT interrupt) can prevent undesirable resets, this defeats the main purpose of the WDT, which is to detect nonprogress. Since inter-event progress is impossible to determine in an interrupt-based implementation, progress must be considered at the event level itself. In other words, if a single event fails to make progress on its own, the system should reset. A detailed examination of possible approaches is beyond the scope of this paper, but one solution might be to implement an idle task that continually runs in the background, resetting the WDT counter at regular intervals. Should progress be halted in an event handler, the idle task will not run, causing the WDT to expire, which will in turn effect a system reset.

5   Future Work

The minimal nature of the computations performed by the ATmega8 in our implementation means that all of our events can be handled in an ISR context. More complex events and processing—such as inverse kinematic computations— would require much more intense calculations and should be run in the main task (i.e., non-interrupt) context. We intend to investigate the possibility of including such processing into our operating system to better support the inclusion of more advanced computational tasks.

Our minimal serial communication protocol, while beyond the scope of this paper, has helped contribute to the fast response times we have observed. We intend to investigate possible extensions to it in the future. Another one of our goals is to more strongly evaluate the performance of our operating system and its communication protocol by using it to control the servos on a humanoid robot—such as the Bioloid—while walking or performing other complex tasks.

6   Conclusion

In this paper, we have presented an event-driven operating system for the Robotis AX-12 servomotor for use in humanoid robotics. This approach is valuable because it allows efficient management of the many tasks that a servo must perform in real time, and is not limited to any particular servo application. Although our implementation utilizes an AX-12 motor, our approach is applicable to virtually any servo with a microcontroller. Combined with an efficient communication protocol, our operating system greatly reduces a servo's response time. In this case, polling techniques simply cannot offer the efficiency granted by the use of interrupt-based approaches. Our firmware embraces this paradigm and offers a more modern alternative in the face of multi-tasking and real-time constraints.

References

1. Levis, P., Madden, S., Polastre, J., Szewczyk, R., Whitehouse, K., Woo, A., Gay, D., Hill, J., Welsh, M., Brewer, E., et al.: Tinyos: an operating system for sensor networks. In: Weber, W., Rabaey, J.M., Aarts, E. (eds.) Ambient Intelligence, pp. 115–148. Springer, Heidelberg (2005)
2. Shih, E., Bahl, P., Sinclair, M.J.: Wake on wireless: an event driven energy saving strategy for battery operated devices. In: Proceedings of the 8th Annual International Conference on Mobile Computing and Networking, pp. 160–171. ACM (2002)
3. Hill, J., Szewczyk, R., Woo, A., Hollar, S., Culler, D., Pister, K.: System architecture directions for networked sensors. In: ACM SIGOPS Operating Systems Review, vol. 34, pp. 93–104. ACM (2000)
4. Dunkels, A., Schmidt, O., Voigt, T., Ali, M.: Protothreads: simplifying event-driven programming of memory-constrained embedded systems. In: Proceedings of the 4th International Conference on Embedded Networked Sensor Systems, pp. 29–42. ACM (2006)
5. Baltes, J., Iverach-Brereton, C., Cheng, C.T., Anderson, J.: Threaded C and freezer OS. In: Li, T.-H.S., et al. (eds.) FIRA 2011. CCIS, vol. 212, pp. 170–177. Springer, Heidelberg (2011)
6. Deitcher, A.: Simplicity and performance: Javascript on the server. Linux J. **2011**(204), 3 (2011)
7. Tilkov, S., Vinoski, S.: Node. js: using javascript to build high-performance network programs. IEEE Internet Comput. **14**(6), 80–83 (2010)
8. Dunkels, A., Gronvall, B., Voigt, T.: Contiki-a lightweight and flexible operating system for tiny networked sensors. In: 29th Annual IEEE International Conference on Local Computer Networks, pp. 455–462. IEEE (2004)
9. Årzén, K.E.: A simple event-based PID controller. In: Proceedings of the 14th IFAC World Congress, vol. 18, pp. 423–428 (1999)
10. Durand, S., Marchand, N., et al.: Further results on event-based PID controller. In: Proceedings of the European Control Conference 2009, pp. 1979–1984 (2009)
11. OpenServo Community: OpenServo. http://www.openservo.com. Accessed 10 june 2014
12. ActuatedCharacter: AX12 Firmware | Actuatedcharacter's Blog. http://actuated.wordpress.com/ax12firmware/. Accessed 13 june 2014
13. Atmel Corporation: ATmega8/L Datasheet, February 2013
14. Robotis: Dynamixel AX-12, June 2006

How Can the RoboCup Rescue Simulation Contribute to Emergency Preparedness in Real-World Disaster Situations?

Tomoichi Takahashi[1][(✉)] and Masaru Shimizu[2]

[1] Meijo University, Shiogamaguchi, Tempaku-ku, Nagoya 468-8502, Japan
ttaka@meijo-u.ac.jp
[2] Chukyo University, YagotoHonmachi, Showa-ku, Nagoya 466-8666, Japan
shimizu@sist.chukyo-u.ac.jp

Abstract. The RoboCup Rescue project is based on the situations that occurred during the Great Hanshin Earthquake in 1995. Various types of disasters occurred after the Hanshin-Awaji Earthquake, and the experiences from managing such disasters has shown that evacuation with prompt and appropriate information at the initial period of the disaster is as important as the rescue operations during the disasters. In this paper, we discuss validation and verification of the agent based systems that can simulate the behaviors of individuals and collective evacuation behavior during emergency situations. Collective behavior is difficult to verify by executing evacuation drills in the real world. The effectiveness of evacuation simulations is shown and a plan of experiments at RoboCup venues is proposed to challenge the validation and verification of social simulation and to prove the usefulness of as real-world applications.

1 Introduction

The RoboCup Rescue project was initiated in 1999 to promote research and development related to search-and-rescue operations. The project was initially designed on the basis of the situations encountered during the Great Hanshin Earthquake of 1995 in Kobe, Japan. Since then, there have been many major disasters such as the September 11 attacks (U.S.A., 2001), earthquake off the West Coast of Northern Sumatra (Indonesia, 2004), Hurricane Katrina (U.S.A., 2005), Eastern Sichuan Earthquake (China, 2008), Great East Japan Earthquake (Japan, 2011), and Typhoon Haiyan (Philippines, 2013). The types and causes of these disasters were different; however, saving human lives is a common challenge during and after such disasters.

The potential of robots designed in the Rescue Robot League (RRL) was revealed during the search-and-rescue operations carried out at the World Trade Center [8]. Robots entered areas where human rescuers cannot enter because of adverse environments and transmitted photos from inside the debris formed by the collapse of the buildings. In 2011, robots including QUINCE developed in the RRL were used for search operations inside the Fukushima Daiichi Nuclear Plant that was damaged by a tsunami [1].

© Springer International Publishing Switzerland 2015
R.A.C. Bianchi et al. (Eds.): RoboCup 2014, LNAI 8992, pp. 295–305, 2015.
DOI: 10.1007/978-3-319-18615-3_24

The Rescue Simulation League (RSL) proposed a new application field to facilitate research on issues regarding the agent community [6]. Supporting search-and-rescue operations after a disaster is one of the major purposes of robots that were designed in the initial days of the RSL. During the one and half decades that have passed after the RSL started, the expectations regarding rescue simulations have changed. Experiences from managing the disasters have showed that having an evacuation support system at the initial period of a disaster is as important as having the rescue operation support system. Some projects applied rescue simulation systems to real-world domains including training systems [14].

In this paper, we discuss validation and verification of results that are simulated for situations of future emergencies. New application areas of the agent based rescue systems are demonstrated and the effectiveness of the applications are shown in making prevention plans to save human lives for possible accidents. The remainder of this paper is organized as follows. In Sect. 2, related works are introduced, and it is shown that individual behaviors are modeled in an agent system. Section 3 discusses the verification methods and applications of rescue simulations. The effectiveness of the virtual evacuation drill is discussed as an example in Sect. 4. Finally, a summary is provided in Sect. 5.

2   Background and Applications of Simulation

A disaster can happen at any time and any place. Table 1 lists some major disasters during which many lives were lost and massive rescue operations were launched at the sites. The causes of disasters can be natural or man-made. Fires, building collapses, and tsunami/flood result in the loss of human lives and properties. In the case of earthquakes, there is less time for evacuation than in the case of other types of disasters, even though emergency earthquake alerts are available.

Table 1. Disasters that occurred after the Hanshin-Awaji Earthquake

		Man made	Nature		Cause		ASET
			E. (B. C)	H. (F.)	Fire	Tunami	
1995.1	Great Hanshin E.		√		√		-
2001.9	September 11 attacks	√			√		H
2004.12	W. C. of Northern Sumatra		√			√	-
2005.8	Hurricane Katrina			√		√	D
2008.5	Eastern Sichuan		√		√		-
2011.3	Great East Japan E		√			√	H (for tsunami)
2013.11	Typhoon Haiyan			√		√	D

W.C.of Northern Sumatra: Off The West Coast of Northern Sumatra
E.: Earthquake, H.: Hurricane, B.C.:Building Collapse, F.:Flood
ASET (Available Safe-Escape Time): -, H and D represent no time, an hourly basis time and a daily basis time to evacuate, respectively.

The International Organization for Standardization (ISO) published a technical report that specifies the parameters related to the assessment of the conditions of a building's occupants with respect to time [5]. The assessment related to occupants includes the numbers, locations, characteristics, conditions, and so on. The parameters that focus primarily on the evacuation of occupants are the available safe-escape time (ASET) and the required safe-escape time (RSET). For instance, to ensure compliance with the fire-safety guidelines of a building, the ASET should be greater than the RSET. Although making RSET short is an approach different from allocating rescue agents effectively - which the RSL targets -, simulating evacuation times using agent-based simulation (ABS) could play an important role in disaster preparedness and help significantly in saving human lives.

During emergencies, humans behave differently than during usual times. People's mental condition at these times greatly affects their behavior. For example, when people fear for their physical safety, they tend to think of only themselves and would flee from a building without consideration for anything else. Documents released by the National Institute of Standards and Technology (NIST) related to egress from the World Trade Center buildings on September 11, 2001, and reports from the cabinet office of Japan on the manner in which people evacuated during the Earthquake and the resulting tsunami on March 11, 2011, revealed several evacuation behavior patterns: some people evacuate immediately, while others do not evacuate despite hearing announcements made by authorities [3,9]. It is interesting to note that individuals' behaviors were similar to the behaviors of individuals during a flood in Denver, U.S.A. on June 16, 1965, even though communication methods have changed during this intervening fifty years [4]. Approximately 3,700 families were suddenly evacuated from homes. The family behaviors during the flood that occurred following the provision of warnings were categorized as follows: (1) some families evacuated immediately, (2) other families attempted to confirm the threat of disaster, and (3) some families ignored the initial warning and continued with routine activities.

Perry et al. used their study's empirical findings to summarize the influence of these human relationship factors on the decision-making process [13]. Pelechano et al. illustrated that communication among people improved evacuation rates by using an ABS [12]. They devised a scenario that focused on two types of agents: leaders who help others and explore new routes; and agents who might panic during emergencies that occur in unknown environments. Tsai et al. developed ESCAPES, a multi-agent evacuation simulation system, by incorporating four key features: different types of agents, emotional interactions, informational interactions, and behavioral interactions [15]. Okaya et al. proposed a model of information dissemination among people during evacuation and presented simulation results using a large number of people [11].

Human behavior depends on available evacuation times and the number of evacuees involved during emergencies. Evacuation simulations that consider the human factors are assumed to be useful in analyzing the egress behaviors at emergencies and estimating evacuations processes such as ASET and RSET. They are assumed to be useful in design evacuation processes and preparing prevention plans for emergencies. The plans can be checked more properly in estimating the loss of human lives than using simulations without considering the factors.

3   Problems in Estimating Simulations Involving Human Factors

3.1   Estimating Effects of Human Factors on Simulation Results

Reports on the past disasters indicate that human factors; mental status and the amount of guidance what they heard, play important roles at emergencies. It is hoped to execute evacuation drills by changing the conditions that affect human behaviors, however, from the participants safety it is difficult to execute such drills. Evacuation simulation systems provide useful perform to replace the evacuation drills, however, it is hard to verify the results of simulation.

The validation and verification (V & V) problem has been one of the most important issues in simulations [7]. Michel et.al represented the V & V problem in the form of the following questions:

Q1. Does the model accurately represent the source system?
Q2. Does the model accommodate the experimental frame?
Q3. Is the simulator correct?

It is ideal that simulations involving human behaviors are based on theoretical models and the results of simulations are tested against empirical data. At present, we cannot compare the results with real data, but we can improve prevention planning for disasters by using simulations in which many significant factors are represented.

In the RoboCup RSL, teams compete to improve the performance of their rescue agents in various disaster environments. The performance of the rescue agents is evaluated on the basis of how they decrease the damages after the disasters. The parameters of disaster scenarios include the places where earthquakes occur, fire ignition locations, number and locations of agents, and so on. Sub-simulations are programmed based on some models of disaster. The parameters of disaster simulations are such as the strength of earthquakes or the spreading speed of fires. The simulation results have not been tested by other methods and the situations are the same as in other evacuation simulations.

In the NIST report, evacuation times were calculated by several commercial evacuations systems. The calculations were crosschecked with the results of each other. The results were obtained under conditions in which all occupants started their egress at the same time. Actually, the simultaneous start of egress is not different from the situations that were presented in the NIST report. ABS provides a method to model such human behavior and can answer the first V & V question (Q1). Qualitative analysis of the simulation results by experts will help to make effective prevention plans without executing real evacuation drills.

3.2   Rescue Simulation to Improve Prevention Planning

The following questions show cases that simulations can be used to save lives for future emergencies.

Information Guidance by New Media. It is reported that the users of the microblogging site Twitter exchanged information during the Great East Japan Earthquake, 2011. Our first question is if Twitter had been available in 2001, would it have been useful in evacuating more people safely from the buildings or how many lives could have been saved at the time of the September 11 attacks.

To answer such questions, a multi-agent system provides a good platform on which a simulation system can be designed. Indeed, it is hard to carry out experiments in real-world situations and show that the simulation results have good V & V support. For instance, simulations can show the difference in the results without/with information diffusion by Twitter. The results can be used to improve the quality of prevention planning during future emergencies.

Variety in Evacuee's Behaviors and Guidance at Unfamiliar Buildings. Figure 1 shows two snapshots of Junior League venues at RoboCup 2010 (Singapore) and 2011 (Istanbul). Many families participate in the Junior League. Their evacuation and communication behaviors at the venue would be different from that of people in major leagues with no parent-child human relationships. It is hard to plan safety measures for emergencies against various possible cases even though other conditions such as the number of people are similar ones. In the case of evacuation planning from the venue of events, people are not familiar with the layout of the venue. This also makes it hard to egress smoothly. Our second question is whether RSL can provide an effective tool to save people when emergency occur at the venue of RoboCup or not.

Fig. 1. Snapshots of RoboCup Junior Competition venues (Singapore and Istanbul)

4   Possible Applications and Practical Answers

4.1   Virtual Evacuations Drills for Occupants Egress and Rescue Operations

Evacuation drills are conducted with the goal of ensuring smooth evacuations from buildings and improving the speed of rescue operations at emergency sites. Consider the following scenario: A fire breaks out in a five-story building and

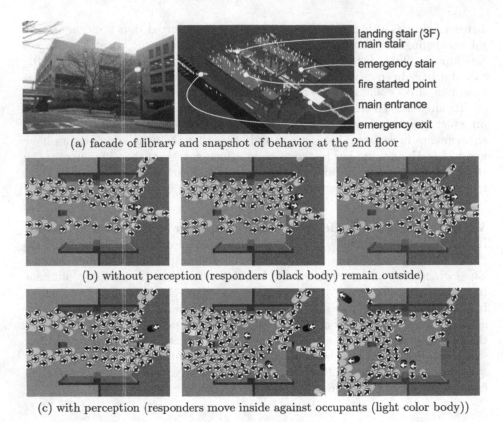

(a) facade of library and snapshot of behavior at the 2nd floor

landing stair (3F)
main stair

emergency stair

fire started point

main entrance

emergency exit

(b) without perception (responders (black body) remain outside)

(c) with perception (responders move inside against occupants (light color body))

Fig. 2. The simulation of evacuation from a building. (Rescue responders (blue) enter from the right.) (Colour figure online)

evacuation guidance is announced inside the building. One thousand occupants (200 occupants on each floor) evacuate from the building, and a rescue team enters the building to extinguish the fire. Does the rescue team accomplish its mission?

Figure 2 (a) shows the facade of a five-story library building and a snapshot of the agents' behaviors on the second floor. TENDENKO, RoboCup Rescue simulation system based evacuation simulator, was use to simulate the behaviors [10]. Figure 2(b) and (c) show the counterflow of occupants and fire responders at the main entrance. The occupants (light color body with black arrow) egress from left to right and the responders (dark color body with white arrow) enter the building from right. The arrows on the occupants' heads indicate the direction of their movements. The time-sequence is ordered from left to right and the simulation time steps are 40, 45, and 50, respectively. Table 2 shows the number of occupants who evacuated from the building and the number of responders who entered. The difference in simulations (b) and (c) is whether agents have the ability of perception or not.

Table 2. Number of evacuated occupants and entering responders at the main entrance.

Time step	Number of			
	Evacuted occupants		Entering responders	
	without perception	with perception	without perception	with perception
35	66 (21)	83 (27)	0	0
40	89 (23)	106 (23)	0	0
45	109 (20)	113 (7)	0	3 (3)
50	129 (20)	115 (2)	0	5 (2)
55	153 (24)	129 (14)	0	5 (0)
60	169 (16)	148 (19)	0	5 (0)

Numbers in the parentheses are different from the previous step.

In emergencies, people move aside to let the rescue team in. The perception of the other agent's social roles is an important feature required in heterogeneous MAS as well as the functions of agents. We categorize the roles as follows:

G_g (agents without priority): an agent gives no special considerations to them and the agent expects that no considerations would be made for itself.
G_h (agents with high priority): the agent gives special consideration to them.
G_l (agents with low priority): the agent expects that special considerations are expected from them.

A normal agent gives consideration to the rescuers and the disabled, who are categorized as G_h. For occupants, rescue personnel are G_h agents, whereas the rescuers are categorized as G_g by their colleagues. More occupants evacuate when they do not have the perception than when have the perception. In the case of occupants without perception, the rescue team cannot enter the building against the flow of evacuating occupants. In the case of occupants with perception, the occupants recognize the rescue agents as G_h and make way for the responders to move through the building. The rescue team can therefore enter the building and move to the appointed position in the building.

4.2   Evacuation Guidance in Commercial and Public Facilities

At emergencies, how can we egress from huge buildings where we are unfamiliar with the layout? Smartphones have spread rapidly, and applications (apps) for mobile phones for navigating routes to destinations or guiding people to downtown area are used. An app at the AAMAS2013 conference was prepared for participants by the organizers [2]. Figure 3 shows the displays of the application: the location of the hotel, the floor plan, and the program of the conference from left to right. It was useful for people who were unfamiliar with the layout of the conference hotel to find the rooms they wanted to join.

Figure 4 (a) is a guidance board showing a floor map including the locations of the stairs and fire extinguishers that are of use during emergencies. The board

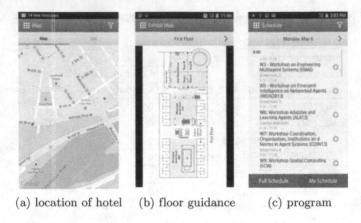

(a) location of hotel   (b) floor guidance   (c) program

Fig. 3. Application delivered at AAMAS2013

assists people with alarms or announcements recommending use of the nearest stairs and exit to the outside of the building. The places in which the board is displayed are limited, and it is difficult to inform many people simultaneously of proper guidance during emergencies.

Figure 4 (b) shows one solution to the problem.

1. Authorities simulate the evacuation behaviors during various cases of emergencies. The images at the top row of Fig. 4 (b) are photographs of the behaviors observed during evacuation from a building to the outside. Authorities plan routes for smooth evacuations and prepare guidance menus for the emergencies.
2. The evacuation routes are saved in an XML file and transferred to smartphones with other menus that are used for everyday services.
3. When an emergency occurs, people switch the app to emergency mode. The app displays the exit routes in a same fashion as the event menus. Circles in the image represent people, and the circle movements show efficient evacuation routes.

4.3 Evacuation Drills in Real Lives and by Simulations

Evacuation drills are carried out regularly in real lives. The purposes of the drill are to check the effectiveness of the desk plan and to familiarize the operations required during emergencies. The drills are carried out under the following conditions:

1. Participants in the drills know when the drill will start and how they are expected to behave during the drill.
2. The behaviors of people and the process of rescue operations are scheduled according to manuals that were planned beforehand.

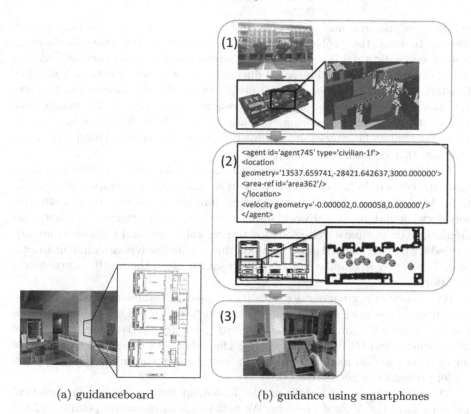

(a) guidanceboard   (b) guidance using smartphones

Fig. 4. Evacuation guidance board showing the floor layout including the locations of the stairs and fire extinguishers.

To verify the effectiveness of a given evacuation drill, it is necessary to simulate and analyze movement from the viewpoints of building occupants and the first responders. The concern of occupants is how quickly and safely they can evacuate a building, whereas the first responders are concerned with the ease and efficiency necessary to reach target points and start rescue operations.

The conditions of drills such as the number of people involved, time and weather when disasters occur are different from that of real cases. To verify simulations, the drill data obtained under one specific condition can be used as the experimental data. This limited data can be used as a qualitative characteristic, but not a quantitative characteristic, to verify the simulation results. The drills are useful for verifying the suitability of manuals and the level of emergency preparedness.

5   Discussions and Summary

After RSL started, many types of disasters occurred and new technologies and systems assumed to be useful in emergencies were announced. We expect such technologies and systems to be efficient, and try to utilize them during past disaster to solve the problems that were encountered. For example, through prompt evacuation guidance using new communication media such as Twitters and so on, more lives could be saved during the September 11 attacks and other disasters by guiding reuse teams and people promptly. We showed the counter-flow movements between occupants and rescue teams (Fig. 2). The counterflow movements that occurred during evacuation of the WTC buildings at the time of the September 11 attacks are assumed to occur during evacuation from large buildings that contain many occupants.

Agent-based simulations that include human factors such as emotions, human relations, types of behaviors and so on, can simulate the behaviors of individuals more closely to that of real at emergencies; thus the simulation results can support decision making in real-world disaster situations. At present, the evacuation simulations do not have well-grounded theoretical bases, and the results are not verified with experimental data. The experiments themselves are difficult to execute. The V & V of simulations are important for using the results in real-world situations.

RSL provides a great service with its open-source simulation platform. The simulations presented in this paper were executed through a system that is similar to the RSL simulator. The modified parts are the traffic-simulator, the three-dimensional GIS[1], and the agents. The agents are developed to behave as one of three reaction models presented in Sect. 2 and according to the perception model presented in Sect. 4.1.

The next step toward realizing the RoboCup Rescue project for practical use is to solve the V & V problem. We believe that evacuation planning of the RoboCup venue will provide an excellent test bed for the problem, and propose the following timeline.

Approximately two years prior to the RoboCup competition: When the future RoboCup site is established, the three-dimensional model is presented to the community.

Approximately one-half year before the RoboCup competition: Participants prepare evacuations plans using their simulation systems and present their plans to the RoboCup organizations. The organizations select some of the plans with cooperation from local governments.

During the RoboCup competition: The organizations prepare an app that includes the evacuation guide at emergencies, as well as the usual information, such as schedules and a layout of the venue, if possible, the participants rate the guides after experiencing drills at the venue.

Challenging this test bed will answer positively to the questions presented in Sect. 3 and demonstrated the effectiveness of social simulations. We hope participants will join the RoboCup competitions safely.

This work was supported by JSPS KAKEN Grant Number 24500186.

---

[1]   RSL targets a decision support for rescue operation at a wide area. Two-dimensional GIS represent the area.

References

1. Mid-and-long-term roadmap towards the decommissioning of fukushima daiichi nuclear power units 1–4. http://www.tepco.co.jp/en/nu/fukushima-np/roadmap/conference-e.html. (Accessed 26 June 2013)
2. http://aamas2013.cs.umn.edu/node/103. Accessed 9 September 2013
3. Averill, J.D., Mileti, D.S., Peacock, R.D., Kuligowski, E.D., Groner, N.E.: Occupant behavior, egress, and emergency communications (NIST NCSTAR 1–7). Technical report. National Institute of Standards and Technology, Gaitherburg (2005)
4. Drabek, T.E.: Social progress in disaster: family evacuation. Soc. Probl. 6(3), 336–349 (1968)
5. ISO:TR16738:2009. Fire-safety engineering - technical information on methods for evaluating behaviour and movement of people
6. Kitano, H., Tadokoro, S., Noda, I., Matsubara, H., Takahashi, T., Shinjou, A., Shimada, S.: Robocup rescue: search and rescue in large-scale disasters as a domain for autonomous agents research. In: IEEE International Conference on System, Man, and Cybernetics (1999)
7. Michel, F., Ferber, J., Drogoul, A.: Multi-agent systems and simulations: a survey from the agent community's perspective. In: Uhrmacher, A.M., Weyns, D. (eds.) Multi-Agent Systems Simulation and Applications, pp. 3–52. CRC Press, Boca Raton (2009)
8. Murphy, R.: Rescue robots at the world trade center. J. Jpn. Soc. Mech. Eng. Spec. Issue Disaster Robot. 102, 794–802 (2003)
9. C. O. G. of Japan: Prevention Disaster Conference, the Great West Japan Earthquake and Tsunami. Report on evacuation behavior of people, (in Japanese). http://www.bousai.go.jp/kaigirep/chousakai/tohokukyokun/7/index.html. (Accessed 21 January 2014)
10. Okaya, M., Niwa, T., Takahashi, T.: TENDENKO: agent-based evacuation drill and emergency planning system (demonstration). In: The Autonomous Agents and MultiAgent Systems (AAMAS), pp. 169–1670 (2014)
11. Okaya, M., Southern, M., Takahashi, T.: Dynamic information transfer and sharing model in agent based evacuation simulations. In: International Conference on Autonomous Agents and Multiagent Systems, AAMAS 2013, pp. 1295–1296 (2013)
12. Pelechano, N., Malkawi, A.: Comparison of crowd simulaton for building evacuationand an alternateive approach. In: Building Simulation 2007, pp. 1514–1521 (2007)
13. Perry, R.W., Mushkatel, A. (eds.): Disaster Management: Warning Response and Cummunity Relocation. Quorum Books, Westport (1984)
14. Schurr, N., Marecki, J., Kasinadhuni, N., Tambe, M., Lewis, J.P., Scerri, P.: The defacto system for human omnipresence to coordinate agent teams: The future of disaster response. In: AAMAS 2005, pp. 1229–1230 (2005)
15. Tsai, J., Fridman, N., Bowring, E., Brown, M., Epstein, S., Kaminka, G., Marsella, S., Ogden, A., Rika, I., Sheel, A., Taylor, M.E., Wang, X., Zilka, A., Tambe, M.: Escapes: evacuation simulation with children, authorities, parents, emotions, and social comparison. In: The 10th International Conference on Autonomous Agents and Multiagent Systems, AAMAS 2011, Richland, SC, vol. 2, pp. 457–464. International Foundation for Autonomous Agents and Multiagent Systems (2011)

RoboGrams: A Lightweight Message Passing Architecture for RoboCup Soccer

Elizabeth Mamantov[1]([✉]), William Silver[2], William Dawson[3], and Eric Chown[2]

[1] University of Michigan, Ann Arbor, USA
mamantov@umich.edu
[2] Bowdoin College, Brunswick, USA
{wsilver,echown}@bowdoin.edu
[3] Okta, San Francisco, USA
wdawson@okta.com

Abstract. RoboGrams is a lightweight and efficient message passing architecture that we designed for the RoboCup domain and that has been successfully used by the Northern Bites SPL team. This unique architecture provides a framework for separating code into strongly decoupled modules, which are combined into configurable dataflow graphs. We present several different architecture types and preexisting message passing implementations, but among all of these, we contend that RoboGrams' features make it particularly well suited for use in RoboCup. As a success story, we describe the Northern Bites' use of RoboGrams and the benefits it has provided to a single team, but we also suggest that it could help SPL teams collaborate in the future.

Keywords: Message passing · Software architecture · RoboCup · SPL

1 Introduction

The choice of architecture can make or break a software development project. A coherent and well-designed architecture facilitates progress on a system that may have many interacting parts. Conversely, a confusing or unwieldy architecture can hinder improvements to the system. Fitting pieces together awkwardly or forcing code into patterns demanded by a bad architecture can cause frustration for developers and can waste a lot of time.

These statements are particularly true in robotic systems because many different subsystems play a role both in the robot's online processing and in the offline tools needed for development. For example, a typical Standard Platform League (SPL) team will have specialized systems that perform some or all of the following crucial tasks: sensor acquisition, vision, localization, ball location modeling, behavior decisions, motion control, network communication, and data recording. Each of these requires data from one or more of the others, but their methods are not interrelated; it should be possible to develop these modules separately and fit them together in the context of the team's architecture.

© Springer International Publishing Switzerland 2015
R.A.C. Bianchi et al. (Eds.): RoboCup 2014, LNAI 8992, pp. 306–317, 2015.
DOI: 10.1007/978-3-319-18615-3_25

This paper presents RoboGrams, a unique message passing software architecture that has been specifically designed to meet these requirements of a robotic system. Its development was inspired by the needs of the Northern Bites, a team that competes in the SPL. After RoboCup 2012, the team decided to convert its code base to a new framework, which presented an opportunity to develop a new software architecture that would be used and evaluated in the RoboCup domain. Since this was a chance to design a unique new framework, we settled on a message passing architecture, which, to our knowledge, is an approach that no other SPL team has attempted. Although there are many existing message passing solutions, we built RoboGrams from the ground up and produced an extremely lightweight and efficient framework with a small dependency footprint. RoboGrams was first used in competition in the 2013 US Open and has been shown to serve the needs of a RoboCup team well. As such, we have released it for the community to use, and the source code can be found online.[1]

2 Background

In an end-to-end robot system, such as a RoboCup soccer player, the agent program must accomplish many different tasks that all rely on the same input data from the robot's sensors and interact in a complex web of data dependencies. As an example of the complexity of data communication in a robot control program, the Northern Bites' system overview is presented in Fig. 1. Ideally, each of the tasks will be a black box with known inputs and outputs to any other task that needs its data, and the data-sharing transactions will occur according to clear rules defined by the architecture. Thus, the architecture core needs to provide the "boxes" for processing modules to fill and data "pipes" that connect the modules in a standardized way.

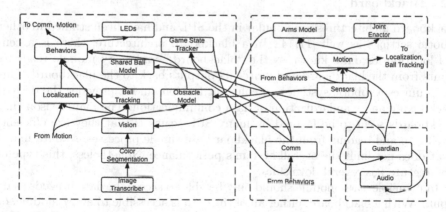

Fig. 1. A dataflow graph of the Northern Bites' online agent program. Separate threads are indicated with dashed boxes.

---

[1] http://github.com/northern-bites/robograms.

The different approaches to solving the problem of providing modularity and data communication result in several common architecture styles seen in the SPL today, which are discussed in the following sections.

2.1 Monolithic

When a team, by default or by decision, does *not* explicitly modularize its code, a monolithic code base results. For many years, the Northern Bites worked within this "non-architecture" framework, in which modules are not black-boxed and interact by calling methods on each other [8]. While this may seem like simply avoiding the architectural design question, it is a valid choice; the main functional systems are still largely separate from each other and data is passed between them. It has the advantages of being both very fast, with no architectural machinery, and very flexible, with no predefined rules.

Unfortunately, these strengths are outweighed by the weaknesses of this approach in the long run. First, it makes adding new functionality difficult because every new system requires a new interface, and with no explicit model of data flow, making a new module interact with existing ones can be tricky. In particular, in a multithreaded environment, setting up each interface individually to avoid data races is an error-prone and difficult process. Second, the lack of explicit black-boxing means that once such an interface is set up, changing one side often requires changing the other; that is, it is hard to update a single processing module without understanding how others work as well. Third, it is impossible to compile and test or use a single component from a monolithic system since everything is tightly coupled to everything else. All in all, a monolithic design may be functionally adequate, but, from the Northern Bites' experience, this setup will limit future team growth and development.

2.2 Blackboard

Blackboard architectures are popular in the SPL and have been studied in other robotic settings as well [11,14]. In a blackboard architecture, there is a central location for data, known as the "blackboard," and every module reads its inputs from the blackboard and writes its outputs back to the blackboard. This setup unites a robotic system in a coherent fashion, and it provides the desired blackboxing and standardization of data communication. The main reason that blackboards seem to be favored by many SPL teams is that they are efficient; writing to and reading from the blackboard are simple processes, so data movement does not limit the overall system's performance. Nonetheless, this type of architecture has several downsides.

In principle, each module should only be able to access the data it needs as its inputs. With a blackboard, either all of the data is accessible to every module, or bookkeeping must be done to keep track of which modules access which data [3]. That is, a straightforward blackboard architecture does not enforce explicit rules about the visibility of data. Another complication is that a blackboard could be accessed by several threads of execution, so locking mechanisms are necessary for data to be safe. A poorly implemented blackboard will be a hotbed for data races, and a well designed blackboard may require relatively complex locking mechanisms. This issue can be solved by having one blackboard per thread of execution, as in the B-Human framework [11], but then another communication method must be devised and incorporated into the same architecture to transfer information between processes. This design results in a less coherent overall approach. Obviously, many teams have overcome the issues described above, but the key point is that, in practice, a blackboard approach is less elegant than it sounds in theory.

2.3   Message Passing

Our chosen approach is a message passing architecture, in which each black-boxed module takes its inputs and produces its outputs in the form of "messages." Rather than keeping all of the messages in a centralized location, messages are passed directly from module to module via input-output connectors. By standardizing the format of a module and specifying how data flow occurs between inputs and outputs, a message passing architecture provides modularity and a standard data communication system, just like a blackboard architecture.

The main argument against message passing in the SPL setting is that it is generally considered less efficient than a blackboard. This is a fair criticism, since heavyweight mechanisms are often employed to move data around a system, making communication costly. However, a message passing architecture lends a robotic system enhanced clarity and structure over blackboards and particularly monolithic systems. With message passing, the setup ensures that the flow of data through the system is made explicit through input-output connections. A message passing system is the perfect software analogy for a dataflow diagram such as Fig. 1. It also provides the data-access rules that are lacking in a simple blackboard implementation—modules can only access data for which they explicitly have inputs—and allows for a single, simple solution to the data race problem that is perpetuated by default across all input-output connections. With these features in mind, message passing is a natural and elegant solution to the architecture design problem, provided that the architecture core is implemented efficiently.

3   System Requirements

There were several requirements we adhered to when developing the RoboGrams message passing architecture: speed, modularity, configurability, intra-process organization, and minimal dependencies. Each of those is discussed below.

3.1   Speed

The Nao's camera is capable of capturing images at a rate of 30 per sec. [1], and our goal was to be able to process each image and run the agent program in real time. A real-time RoboCup system should be capable of handling the full 30 frames per sec. (fps); any overhead from the software architecture needs to be small enough to not cause a decrease from the maximum frame rate. Given that soccer-playing robots need to function in a dynamic environment in real time, we were not willing to compromise on the speed of the architecture.

3.2 Modularity

The biggest problem with a monolithic and tightly coupled system is that, without black-boxing, modules often have to control or access each others' functionality. Therefore, no single system can be improved or replaced without also editing other modules. For the new architecture, we decided that each module should know nothing about other modules' functionality; the only agreements between two interacting modules would be the format of the data they exchange. Furthermore, no module should need to know where its input data comes from or where its output is sent to because the configuration of the rest of the platform should have no influence on the internal workings of a module.

3.3 Configurability

With the modules completely black-boxed, a related requirement was that the network of modules should be able to be "wired" together in many different ways. Any module that produces output of type T should be able to provide that data to another module which is set up to receive input of type T. As an illustration of the need for this requirement, consider offline vision development, depicted in Fig. 2. On a workstation, we need to populate the vision module's inputs with prerecorded images from a file and pass the module's output to a user interface that displays it. The same vision module should be able to run on the robot, except that its inputs would come from a camera module and its outputs would feed into localization and any other system that uses vision data. In a configurable system, the difference between competition, debug, and offline setups is simply a rewiring of most of the same component modules.

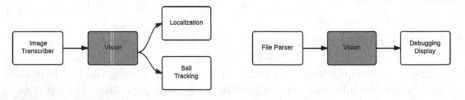

(a) Use of a vision module online.   (b) Use of the same vision module offline.

Fig. 2. Demonstration of how the same module may be used in two different system configurations.

3.4   Intra-process Organization

The Northern Bites and other SPL teams run a robot's agent program as a single, multithreaded process loaded into the Nao's built-in control program, naoqi [2]. Alternatively, the agent program can be broken into multiple processes. Message passing is often used in inter-process communication (IPC), and some SPL teams employ it in this way [11]. However, our goal was *intra*-process communication and organization; it should not be necessary to run every module as its own process to be able to enforce a message passing relationship between them. We thus required that a single process could contain any number of modules, with black-boxing and data communication policies enforced by the architecture core rather than operating system process divisions and IPC.

3.5   Minimal Dependencies

This idea applies in two related ways. First, we required that our core architecture depend on as few outside libraries or tools as possible. Not only are outside dependencies often difficult to manage from an administration and setup perspective, but we did not want to make the core of our system dependent on tools we could not completely control. Furthermore, many publicly available libraries have many more features and capabilities than we actually needed, which would have resulted in useless bloat in our platform. The second way this requirement works is within our own platform: each module should have as few dependencies as possible, so the architecture core should be absolutely minimal. In the monolith version of the Northern Bites' platform, compiling any given module necessitated compiling every other related module, so in practice, compiling and running one module required also including the entire platform. With the new architecture, we wanted each module to be dependent only on a small shared core with no extraneous features.

4   Alternative Architectures

As there are many existing message passing architectures, one might wonder if we are reinventing the wheel by implementing our own from scratch. In this section we give an overview of some alternatives and describe why they do not meet our requirements. One obvious option for a robotic message-passing architecture is Robot Operating System (ROS). ROS is an open-source framework that provides the "middleware" to connect various processing "nodes," or modules, via a publish/subscribe communication system [13]. ROS has been tested on the Nao and was proposed for use in the SPL [5]. However, ROS is a large external system that must be painstakingly installed on the Nao and thus does not fulfill our requirement of minimal dependencies. More importantly, it does not provide intra-process organization since every ROS node is a process and the middleware essentially provides IPC. Its performance on the Nao is also uncertain.

One less obvious option that we considered was the Qt framework, which offers signal/slot event-driven connections that could be used to implement data exchange between modules. Qt was developed for use in configurable graphical user interfaces (GUIs) and provides large libraries geared toward GUI development [9]. The core signal/slot functionality, however, can be used for any application, and the Qt core comes pre-installed on the Nao. Nonetheless, our code base would still be dependent on the bulky Qt libraries, so Qt fails our requirement of minimal dependencies. More importantly, without considerable amounts of boilerplate code to enforce system-wide rules, Qt also fails the requirement of modularity. Signals and slots can be linked between modules in such a way that the two modules actually control the timing or flow of each others' processing. We needed our architecture to provide stronger black-boxing.

It is also worth discussing the architectures used by other SPL teams, all of which use blackboards. Among SPL teams, B-Human provides one of the most clearly articulated architectures, and their framework has been adapted for use by several other SPL teams [7,11]. This architecture was built from the earlier German Team architecture, which successfully integrated the work of several different university teams, a testament to its modularity and the strong organization it provided [10]. The B-Human framework organizes processing into several threads of execution that pass inter-thread messages, but each thread also contains a blackboard. rUNSWift provides a very different overall architecture design, with hierarchical modules that are run in a multithreaded executable [2]. These modules write to and read from a blackboard in order to share information. With yet another alternative blackboard design, Austin Villa separates all robot memory into a "Memory Module" that provides particularly well-structured blackboard functionality for the rest of the team's code [3].

5   The RoboGrams Architecture

With our requirements clearly stated and the need for a novel architecture established, we present RoboGrams, our lightweight message passing architecture. A high-level overview of the different components of RoboGrams is given in Fig. 3. It is implemented in C++ and has been tested on Windows and Linux (x86).

5.1   Modules

Modules are the building blocks of a RoboGrams system; each one is a black-boxed unit of processing. To keep the architecture's structure coherent, the abstract class Module must be a base of any class that functions as a processing module. To create a derived class of Module, one pure virtual function, run_(), must be overridden. This function contains the main processing loop of the module. The decision of when to run the module is handled by a higher level of the architecture, so the developer's only concern is providing a module's functionality within the black box of the Module interface.

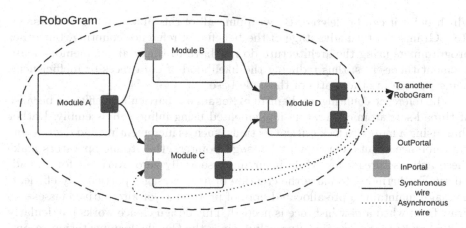

Fig. 3. An overview of the different components of the RoboGrams architecture, with a hypothetical RoboGram composed of several modules.

5.2   Messages

Modules exchange data in the form of the RoboGrams `Message` object. A `Message` is a templated wrapper class that can hold any C++ class that one module needs to pass to another. In the context of the Northern Bites' platform, the contents of a `Message` is typically a Google Protocol Buffer ("protobuf") [6]. Google designed protobufs to be lightweight data containers that also provide serialization and deserialization functionality. Protobufs were already incorporated into the Northern Bites platform before the message passing overhaul, and we decided to continue using them due to their efficiency and simple interface. Each message type is defined in a `.proto` file, and a provided compiler automatically creates C++ classes to hold and manipulate the specified data types.

The RoboGrams architecture itself, however, is message-type agnostic. Any C++ class can be the contents of a message; for example, protobufs are not ideal for working with images, so we have implemented specialized classes that handle images efficiently. Since RoboGrams does not require messages to be serialized for transfer, that particular feature of protobufs is a benefit for logging data but not a necessity from the architecture point of view. Although using a wrapper class like `Message` rather than simply passing a protobuf or other class directly may be an unintuitive choice, `Message` actually provides a core piece of the architecture's functionality: reference counting.

Reference counting is the mechanism that affords RoboGrams its reliability and efficiency. The main goal of having `Message` count references is so that a copy of a protobuf or other inter-module message class can be made in constant time for each of the modules that needs it. That is, rather than performing a deep copy to ensure that data is safe while it is being accessed by a particular module, all modules simply reference a single existing instance of the data. The number of these references is tracked in `Message` so that the data remains in existence only as long as it is needed—until there are zero remaining references—at which point it can be deleted. By wrapping all of this functionality in `Message`, RoboGrams can take advantage of the benefits of reference counting, but other programmers using the architecture do not have to worry about memory management themselves, thus reducing the likelihood of data races or inefficiencies being written in other parts of the code base.

The reference counting mechanism in `Message` is particularly efficient because of thread-safe atomic operators implemented using inline x86 assembly. Rather than using a mutex to lock critical sections, such as incrementing and decrementing reference counts and copying `Message` pointers, the atomic operators make these actions thread-safe while incurring essentially zero overhead for thread-safety. Furthermore, to make the creation of new messages particularly efficient, `Message` maintains a pre-allocated pool of protobufs—or alternative classes—to draw from when a new instance is needed. This design choice works particularly well when `Message` contains a protobuf, since the Google documentation recommends reusing protobufs in this way [6], but it can be beneficial for other classes as well. Having a pool means that time is not wasted reconstructing objects. All in all, `Message` contains several sophisticated mechanisms that make RoboGrams uniquely efficient, while keeping them wrapped behind a simple interface so that other programmers can use them reliably.

5.3 Portals

Portals are the RoboGrams objects that accomplish the main work of passing messages between modules; a module's data interface is specified by its portals. RoboGrams provides `InPortal` and `OutPortal` objects for modules' inputs and outputs, respectively. Portals must be declared within a `Module`, and their type cannot be changed at runtime. Both `InPortal` and `OutPortal` are template classes that must be provided with a message type, such as a particular protobuf class. `InPortal` provides two key methods, the first of which is `wireTo(OutPortal, bool)`, which allows an `InPortal` to be linked to an `OutPortal` of the same data type. The boolean argument specifies whether this link will be *synchronous* or *asynchronous*; the difference between these options is discussed later in this section. Note that `wireTo` can be called again at runtime, meaning that the wiring configuration can be changed dynamically.

The second `InPortal` method, `latch()`, is used to fetch the message from a connected `OutPortal`. When a message is accessed via `latch()`, its reference count is incremented. The count will automatically decrease when the `latch`ed message goes out of scope, so there is no way for a programmer to forget to release a message back to the pool. The other main functionality that `latch()` provides is that it orders the running of modules so that inputs for a module are produced before that module runs. When a module calls `latch()`, `run()` is called on the connected module if it has not already been run during the current processing cycle so that it can produce updated data.

There is a caveat to this setup, however, since it only works as expected if modules are linked in a directed acyclic graph (DAG) running in a single thread. In fact, in a real RoboCup system, backedges, or links that make loops in the graph, are common, as are links to modules in other threads. We do not want such connections to cause a module to run since backedges would cause infinite run loops and requests from another thread would cause data races. Thus, these two types of edges must be marked asynchronous when `wireTo` is used to make the connections. In sum, `latch()` on synchronous edges causes input modules to run, while on asynchronous edges, it simply accesses existing data.

5.4   RoboGrams

RoboGrams provides an additional, higher organizational level than most message passing systems: the `RoboGram`, a shortening of "RoboCup Wiring Diagram." As its name suggests, `RoboGram` is a collection of `Modules` that are typically wired together in specifying C++ code and run at the same rate. `RoboGram` simply holds a list of `Modules` and calls `run()` on each of them. Due to the way that `latch()` works (see previous section), the modules are automatically processed in topological order; data dependencies are produced before running the module that requires them. Each module is run just once per diagram cycle.

Each `RoboGram` structure is its own thread of execution that can run asynchronously with any other `RoboGrams` in the same system. That is, each diagram maps naturally onto a thread in a multithreaded process. In the Northern Bites' system, for example, there are four `RoboGrams`, each running in its own pthread thread at different rates. For example, the motion diagram runs at 100 Hz to synchronize with the Nao's DCM, whereas the cognition diagram, which contains vision, world modeling, and behaviors, runs at 30 Hz to synchronize with the Nao's camera.

6   Impact

The RoboGrams architecture has proven to be an overwhelming success based on the Northern Bites' experience using it for RoboCup development. The team's code base has been decluttered and streamlined, and no sacrifices were made in terms of runtime. The most obvious difference facilitated by the new architecture is a dramatic improvement in code quality and readability, accompanied by a drop in overall lines of code in the code base. That is, from our experience, RoboGrams can improve a team's development environment.

Legacy code is a significant challenge for any RoboCup team with a long history, so one major benefit of the modular nature of RoboGrams is that is leads to a more forgiving learning curve for new team members. In a truly modular code base, a new member can focus on one module and make valuable contributions to the team relatively quickly. This is a particularly important consideration for teams composed mostly or only of undergraduates, including the Northern Bites, where new team members may have little programming experience and membership cycles quickly as students graduate.

Beyond the striking simplification, RoboGrams can also facilitate the addition of new functionality and support an improved testing environment. Because

RoboGrams modules can be tested individually, the Northern Bites were able to build and make use of a continuous integration framework to replace our previous, less rigorous testing methods [4]. Also, the barrier to adding new modules in a RoboGrams-based system is extremely low; the team was able to add a module for obstacle modeling in just a few days, which resulted in noticeable improvements to our robots' competition play.

While all of the above improvements are valuable, RoboGrams would have been unusable if it added significant time overhead to a system; for use in RoboCup, the architecture absolutely needed to be capable of handling a real-time agent program. Fortunately, the pains taken to make our message passing implementation extremely efficient paid off. Based on a recent set of five time trials, the Northern Bites' current average frame rate is 29.31 fps, and there was no significant difference in frame rate between this code, which actually includes several new processing modules, and pre-RoboGrams code. This relationship strongly suggests that the bottleneck on our processing is the Nao's camera frame rate, not the RoboGrams architecture.

7   Conclusion

We have presented RoboGrams, an elegant and lightweight architecture for robotic systems that fills a unique niche among existing software architecture styles and has been very successfully applied in the SPL domain. Like all of the software developed by the Northern Bites, RoboGrams is open-source and is freely available online. While the new architecture's positive impact on the Northern Bites alone has been dramatic, we believe that sharing it among other SPL teams could result in increased collaboration and league-wide progress.

Since all SPL teams are attacking the same research problem and using the same hardware, in some sense, the SPL is one big software development project. Ideally, any team could borrow another's vision module, for example, and swap it into their own system for testing or try to improve the original. To keep the barrier to entry reasonably low for new teams, this sort of sharing—particularly of walking engines—is crucial. However, in practice, it is usually a struggle to separate one component of a team's code base, and teams often end up using more of another team's code than they need.

Due to its flexibility and simplicity, RoboGrams can enable this type of true collaboration. Each RoboGrams module is strongly decoupled from the rest of a given team's system and can be easily adopted for use by another team—even if that team's code is not based on RoboGrams. To make use of a module is simple: the correct data format must be supplied to its inputs. Thus a team can take advantage of the higher diagram organization, or not; in either case, RoboGrams facilitates sharing. Furthermore, since the architecture can run in a process with a single thread, a multithreaded process, or even be extended to work in multiple processes via IPC, any team's setup will be RoboGrams-compatible.

An initial set of RoboGrams modules are provided open-source in the Northern Bites' code base.[2] These include a module version of B-Human's 2011 walking

---

[2] http://github.com/northern-bites/nbites.

engine [12]. While RoboGrams has been invaluable to the Northern Bites and is clearly already a success, we hope that it can also help spark greater SPL-wide collaboration in the future. We thus share RoboGrams first as an innovative and elegant take on a robotic software architecture and second as a tool that the rest of the SPL and RoboCup as a whole is invited to utilize.

References

1. Aldebaran: video camera. https://community.aldebaran-robotics.com/doc/1-14/family/robots/video_robot.html#robot-video. Accessed February 2014
2. Ashar, J., Claridge, D., Hall, B., Hengst, B., Nguyen, H., Pagnucco, M., Ratter, A., Robinson, S., Sammut, C., Vance, B., et al.: Robocup standard platform league–rUNSWift 2010. In: Australasian Conference on Robotics and Automation (2010)
3. Barrett, S., Genter, K., He, Y., Hester, T., Khandelwal, P., Menashe, J., Stone, P.: UT austin villa 2012: standard platform league world champions. In: Chen, X., Stone, P., Sucar, L.E., van der Zant, T. (eds.) RoboCup 2012. LNCS, vol. 7500, pp. 36–47. Springer, Heidelberg (2013)
4. Dawson, W.J.: Extensible continuous integration framework. Undergraduate honors thesis, Bowdoin College, Brunswick, ME (2013)
5. Forero, L.L., Yáñez, J.M., Ruiz-del-Solar, J.: Integration of the ROS framework in soccer robotics: the NAO case. In: Behnke, S., Veloso, M., Visser, A., Xiong, R. (eds.) RoboCup 2013. LNCS, vol. 8371, pp. 664–671. Springer, Heidelberg (2014)
6. Google: protocol buffer developer guide. https://developers.google.com/protocol-buffers/docs/overview. Accessed February 2014
7. Hofmann, M., Schwarz, I., Urbann, O.: Nao devils team report (2013)
8. Neamtu, O., Dawson, W., Googins, E., Jacobel, B., Mamantov, E., McAvoy, D., Mende, B., Merritt, N., Ratner, E., Terman, N., et al.: Northern bites code release (2012)
9. Qt: Qt project. http://qt-project.org/. Accessed February 2014
10. Röfer, T.: An architecture for a national robocup team. In: Kaminka, G.A., Lima, P.U., Rojas, R. (eds.) RoboCup 2002. LNCS (LNAI), vol. 2752, pp. 417–425. Springer, Heidelberg (2003)
11. Röfer, T., Laue, T.: On B-human's code releases in the standard platform league – software architecture and impact. In: Behnke, S., Veloso, M., Visser, A., Xiong, R. (eds.) RoboCup 2013. LNCS, vol. 8371, pp. 648–655. Springer, Heidelberg (2014)
12. Röfer, T., Laue, T., Müller, J., Fabisch, A., Feldpausch, F., Gillmann, K., Graf, C., de Haas, T.J., Härtl, A., Humann, A., Honsel, D., Kastner, P., Kastner, T., Könemann, C., Markowsky, B., Riemann, O.J.L., Wenk, F.: B-human team report and code release (2011). Available online: http://www.b-human.de/downloads/bhuman11_coderelease.pdf
13. ROS: communications infrastructure. http://www.ros.org/#communications_infrastructure. Accessed February 2014
14. Tzafestas, S., Tzafestas, E.: The blackboard architecture in knowledge based robotic systems. In: Jordanides, T., Torby, B. (eds.) Expert Systems and Robotics, NATO ASI Series, vol. 71, pp. 285–317. Springer, Berlin Heidelberg (1991)

Towards Optimal Robot Navigation in Domestic Spaces

Rodrigo Ventura and Aamir Ahmad[✉]

Institute for Systems and Robotics (ISR), Instituto Superior Técnico (IST),
Av. Rovisco Pais 1, 1049-001 Lisbon, Portugal
{yoda,aahmad}@isr.ist.utl.pt

Abstract. The work presented in this paper is motivated by the goal of dependable autonomous navigation of mobile robots. This goal is a fundamental requirement for having autonomous robots in spaces such as domestic spaces and public establishments, left unattended by technical staff. In this paper we tackle this problem by taking an optimization approach: on one hand, we use a Fast Marching Approach for path planning, resulting in optimal paths in the absence of unmapped obstacles, and on the other hand we use a Dynamic Window Approach for guidance. To the best of our knowledge, the combination of these two methods is novel. We evaluate the approach on a real mobile robot, capable of moving at high speed. The evaluation makes use of an external ground truth system. We report controlled experiments that we performed, including the presence of people moving randomly nearby the robot. In our long term experiments we report a total distance of 18 km traveled during 11 h of movement time.

1 Introduction

Domestic spaces are significantly different from laboratory environments and office floors. Most of the robot navigation and path planning algorithms developed so far assumes the latter and are inherently designed for such controlled environments. People's homes vary not only with the cultural aspects of a country but also depending on individual choices. Consequently, it is quite difficult to generalize home environments. However, if the robot navigation algorithms accounted for certain recurring features in people's homes, e.g., presence of randomly moving humans, pets, unknown object, displaced pieces of small furniture as well as hard-to-perceive surfaces, robot motion would be perceived more natural in people's homes. In this context, it becomes necessary for such robots to not only execute an optimal path from a given start to goal pose in the environment but to also avoid previously-unmapped obstacles and randomly-moving people and pets.

In this paper we present a novel design for the motion of such robots in home-like environments by coupling an optimal path planning strategy with a navigation

Work supported by FCT projects PEst-OE/EEI/LA0009/2013 and FP7-ICT-9-2011-601033 (MOnarCH).

R.A.C. Bianchi et al. (Eds.): RoboCup 2014, LNAI 8992, pp. 318–331, 2015.
DOI: 10.1007/978-3-319-18615-3_26

algorithm that inherently avoids previously-unmapped obstacles and randomly moving people. To this effect, we integrate Fast Marching Method [1] for path planning and dynamic window approach [2] to compute the motion velocity commands for the robot in real time that automatically avoids previously-unmapped static and dynamic obstacles in the environment. The novelty of this paper consists in the way these two methods are integrated and implemented on a real robot. In addition, we performed experiments where the robot was expected to execute safe motion within human-occupied spaces. In order to evaluate the performance of our technique, we also implemented a ground truth system that estimates the robot's actual path during the experiments. Comparing the actual executed path by the robot with the optimal planned path will provide an insight into how well our proposed method behaved in real-time application.

The rest of the paper is structured as follows. Sect. 2 describes related work with this paper, followed by Sect. 3 where we describe the path planning, guidance and obstacle avoidance as well as their integration. This is followed by the robot description, details of the ground truth evaluation system and results of real robot experiments in Sect. 4. We conclude with a remark on future work in Sect. 5

2   Related Work

Domestic service robots operating in domestic spaces require various challenging functionalities, e.g., navigation, perception and manipulation, to accomplish a variety of tasks [3]. The human factor, in the context of developing such functionalities, is the most essential one. Consequently, new methods for robot navigation in the presence of humans are being studied extensively. Authors in [4] present a thorough survey of such methods where they identify comfort, naturalness and scalability as the three key issues addressed by the existing human-aware robot navigation methods so far. However, in most such methods, time and energy efficiencies of the robotic systems often get suppressed.

In [5] a method for human-centered navigation is presented which is based on various heuristics, such as, to maintain a given distance to the robot, to keep humans within certain visibility cone and to alter the robot velocity w.r.t. the human motion in the environment. In a similar direction of work, authors in [6] explore the possibility of robots learning the paths traversed by the humans, which is subsequently used by the robot's own path planner and obstacle avoidance system which predicts human trajectories. Such methods not only tend to become computationally heavy but also their reliance on human-centric heuristics might not hold true universally. To circumvent the need for heuristics some works have employed visual perception for real time feedback and continuous re-planning. Such a method has been used in [7] to solve complex human-robot cooperative tasks that includes navigation during the accomplishment of the whole task. However, visual perception in itself is a challenge that can seriously affect the optimality and robustness of the whole solution.

Some works, e.g., [8,9], have explored planning in 3D representations of environment maps leading to much robust navigation that could inherently account for the robot's full height and complete traversability in the environment.

More recently, some very successful approaches, e.g., [10,11], have made significant efforts in integrating efficient path planning and obstacle avoidance methods for navigation of domestic service robots in home environments. Apart from being robust and reliable, one very interesting property of these integrated methods is that they automatically account for human presence in the environment and do not depend on any heuristics. In this paper we not only subscribe to the aforementioned property but also focus on the optimality of the overall navigation of the robot in a home-like environment, in addition to being robust and reliable.

3   Navigation

Navigation in this context is understood as the capability of a robot to move autonomously in the environment with the goal of reaching a pre-specified final pose. The time taken by the robot to execute this task should be minimal, while avoiding collisions with obstacles as well as maintaining a certain clearance to them.

In this paper we take the classical approach of dividing navigation into self-localization and guidance, assuming knowledge of a map of the environment. We also assume that unmapped static or moving obstacles may appear in the environment, while the robot is expected to deal with them in an appropriate way. We further assume an existing self-localization system, possibly (but not necessarily) based on data fusion of odometry and range sensor matching with the map.

The guidance problem is approached as a two step process. First, given a goal location, the robot plans its path from the current pose to the goal pose. And second, the plan is executed by the robot, in real time, while avoiding unmapped obstacles. These two steps are described in the following two sections.

3.1   Optimal Path Planning

The path planning problem consists in determining a path for the robot to traverse the environment, given a map and a goal pose. Rather than explicitly planning for a path, we take a potential field approach. This potential field should have the property that, for any given robot location, the path resulting from following the gradient descent is the optimal path to the goal, while maintaining a certain clearance to the obstacles in the map. Such fields can be obtained using a Fast Marching Method approach applied to optimal path planning [1], which we closely follow. The process is explained next.

This potential field is obtained by considering, for each point x within the free region $\Omega \subset \mathbb{R}^2$ of the map, the minimal time it takes for a wave to propagate from the goal location x_g to the current robot position. The computation of this time for each point x in the free region Ω results in a field $u(x)$. Thus, the set of points that satisfy $u(x) = T$ corresponds to the set of points belonging to the wave front at time T. This representation of a set is also known as *level set* [12]. It is well known that the path resulting from solving the ODE $\dot{x} = -\nabla u(x)$ from an initial $x(0) = x_0$ results in the optimal path from x_0 to the initial wave front [1]. We set this wave front to an arbitrarily small ball $\Gamma \subset \Omega$ around the goal location x_g. The relation between the goal, the field, and its gradient at the robot location is shown in Fig. 1(a).

Given an initial wave front $\Gamma \subset \Omega$, the field $u(x)$ is the solution of the Eikonal equation

$$|\nabla u(x)|F(x) = 1$$
$$u(\Gamma) = 0 \tag{1}$$

where $x \in \Omega$ is the free space of robot position, $\Gamma \subset \Omega$ the initial wave front, and $F(x)$ is the wave propagation speed at point x. The specification of this speed allows the resulting path to maintain a certain clearance to the mapped obstacles, since the optimal path tends to keep away from areas with lower propagation speeds. Note that this speed is anisotropical, that is, independent from the propagation direction of the wave front. It should also be noted that this speed is completely unrelated with the actual robot speed, which is specified elsewhere. The wave propagation speed is only used to promote clearance from mapped obstacles.

The field $u(x)$ that solves the Eikonal equation has two key properties: (1) it shows no local minima, and (2) the gradient descent path is optimal, given a wave propagation speed function $F(x)$, *i.e.*, it is the smooth path $\gamma(\tau)$ that minimizes the integral

$$\int_{x_0=\gamma(0)}^{x_g=\gamma(L)} [F(\gamma(\tau))]^{-1}d\tau \tag{2}$$

where τ is the arclength parameterization of γ, that is $||\dot{\gamma}|| = 1$, and L is the total length of the path [1].

To solve numerically the Eikonal equation we employ a standard Fast Marching Method: given a discretization of the map in an occupancy grid, we supply to this method the region of free space Γ, the propagation speed function $F(x)$, and the goal point, and in return we obtain a numerical approximation to the solution of the Eikonal equation on the grid points. The speed function $F(x)$ is obtained by reducing the speed for the points near the mapped obstacles. Let $d(x)$ denote the euclidean distance to the nearest mapped obstacle point. Then,

$$F(x) = \begin{cases} -\frac{1}{C}d(x)^2 + 2d(x), & d(x) < C \\ C, & \text{otherwise.} \end{cases} \tag{3}$$

where C stands for a threshold distance beyond which the wave propagation speed is constant (see Fig. 1(b)). Therefore, the smaller the C, the lower the clearance of the resulting path from mapped obstacles will be. This approach is

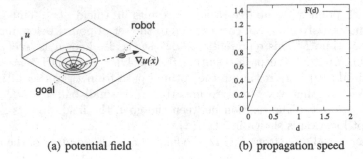

(a) potential field   (b) propagation speed

Fig. 1. Aspects of the path planning method: (a) illustration of the field $u(x)$ resulting from FMM, together with the gradient over that field starting at an arbitrary position; (b) plot of the wave propagation speed F as a function of the distance d to the nearest obstacle (for a $C = 1$).

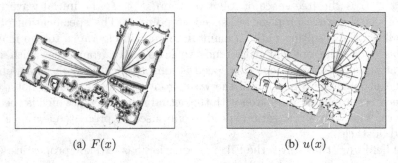

(a) $F(x)$ (b) $u(x)$

Fig. 2. Example of the propagation speed function $F(x)$ and the field $u(x)$ for the real scenario used for experimentation, as contour plots, together with a set of gradient descents from poses laid on a grid (within the scenario).

quite similar to [13], however, we allow for a plateau in $F(x)$ beyond a clearance distance C. This allows the robot to get closer to mapped obstacles, and thus resulting in shorter paths than in [13].

The resulting field $u(x)$, which is obtained once after a goal position is given, will then be used for the actual navigation. This navigation will aim at following a gradient descend of $u(x)$, while deviating from obstacles. Examples of several gradient descents over such a field, for the real scenario used for experimentation (Sect. 4), can be found in Fig. 2.

3.2   Guidance and Obstacle Avoidance

The goal of guidance is to compute in real time the robot actuation, in terms of motion velocity, given a FMM field $u(x)$ embedding the optimal path to the goal. We solve this problem by taking a Dynamic Window Approach (DWA) [2,14]. That is, given the robot's current velocity, pose and available sensor data, DWA computes the next motion velocity command. It is done by formulating a constrained optimization problem over a discrete set of candidate velocity commands.

The outline of the algorithm is the following:

1. generate a set of candidate linear velocity commands
2. discard the velocity values beyond a specified maximum absolute value
3. discard the velocity values which could lead to a collision, that is, the robot is unable to stop, at the maximum de-acceleration, in time before hitting an obstacle
4. compute an evaluation value for each candidate by weighting three contributions: (i) progress towards the goal, (ii) clearance from obstacles, and (iii) absolute speed
5. select candidate maximizing the evaluation value
6. compute angular velocity based on the direction of the selected linear velocity, such that the robot front tends to be aligned with the motion direction.

This algorithm follows closely the DWA as initially proposed in [2], except for novel methods for both computing the clearance, taking into consideration the robot shape, and the progress, based on the potential field obtained from FMM. Next, we will describe each one of the algorithm steps in detail.

Let the initial set of candidate velocities for (discrete) time instant t be $\mathcal{C}_0(t) = \{v^i(t)\}$, for $i = 1, \ldots, N$, expressed in the body frame. These candidates are assumed admissible, that is, they must comply with the kinematic constraints of the robot. In the case of an omnidirectional robot, as in the case of the targeted robot, we can independently control the motion velocity along these two directions. Thus, each candidate has the form $v^i(t) = (v_x^i(t), v_y^i(t))$ representing the tangent and normal linear velocities. Otherwise, the actuation space has to be appropriately parametrized, e.g., in the of a differential drive, a possible parametrization is a linear and angular velocities pair. These velocities are chosen in a grid of values around the current robot velocity, within the acceleration limits of the robot. That is, $||v^i(t) - v^i(t-1)|| \leq A_{max}T$, where A_{max} and T are the maximum linear acceleration and the period of the control loop (from here on, we will drop the dependence on (t) for the sake of clarity). The next candidate set $\mathcal{C}_1$ contains the velocity candidates within the maximum linear velocity, that is

$$\mathcal{C}_1 = \{v \in \mathcal{C}_0 : ||v|| < V_{max}\} \tag{4}$$

where V_{max} is the maximum linear velocity.

Computation of both the clearance and the collisions, for each candidate, makes use of a 2-D point cloud obtained from the range sensors, e.g., a laser range finder. These points are here represented with respect to the body frame, being the union of the points perceived from all sensors: $\mathcal{L} = \{p^j\}$, for $j = 1, \ldots, M$, where $p^j = (p_x^j, p_y^j)$ are the point coordinates in robot body frame. Given a velocity candidate $v^i = (v_x^i, v_y^i)$, the point cloud is projected into a reference frame aligned with the candidate direction, that is, the tangent and normal unit vectors e_t^i and e_n^i, computed using

$$e_t^i = \frac{v^i}{||v^i||} \qquad e_n^i = \begin{pmatrix} 0 & -1 \\ 1 & 0 \end{pmatrix} e_t^i \tag{5}$$

Thus, we get the projected points into this frame

$$p_t^{ij} = \langle p^j, e_t^i \rangle \qquad p_n^{ij} = \langle p^j, e_n^i \rangle \tag{6}$$

where $\langle \cdot, \cdot \rangle$ stands for the standard Euclidean inner product.

To compute the possibility of collision, we need to consider the physical space occupied by the robot, i.e., its shape. Rather than considering C-obstacles generated by the point cloud, which would require a computationally expensive convolution operation, we project the robot shape along its motion direction, and compare it with the obstacle point cloud. Let $\mathcal{S}$ be the region of the space, on the robot body coordinate frame, occupied by the robot. Depending on the motion direction, the robot body will span a certain longitudinal and a certain lateral space. We approximate this span with a rectangular bounding box, aligned with the motion direction. That is, given a motion direction e_t^i associated with a velocity candidate i, a rectangular bounding box aligned to the (e_t^i, e_n^i) axes is determined. This bounding box is delimited by the points[1] (b_F^i, b_L^i), $(-b_B^i, b_L^i)$, (b_B^i, b_R^i), and (b_F^i, b_R^i), expressed in the (e_t^i, e_n^i) frame. The relation between this bounding box, the robot shape, and a candidate velocity is shown in Fig. 3(b).

Assuming that the robot will be moving at speed $||v^i||$ along direction e_t^i, given a maximum (de)acceleration A_{max}, the minimum stop time is $T_{min}^i = ||v^i||/A_{max}$ and the corresponding minimum stop distance is $D_{min}^i = ||v^i||^2/(2A_{max})$. Thus, any obstacle within a distance of D_{min}^i of the robot front, across its extent, is here considered to result in a collision. Any candidate with obstacles meeting this criterion are discarded at this point. It should be noted that this criteria is an approximation, as we are approximating the spatial span of the robot by a bounding box. Formally, the new candidate set $\mathcal{C}_2$ is obtained using

$$\mathcal{C}_2 = \mathcal{C}_1 \setminus \left\{ v^i \in \mathcal{C}_1 \mid \exists_j - b_R^i \le p_n^{ij} \le b_L^i \wedge 0 \le p_t^{ij} \le b_F^i + D_{min}^i \right\} \tag{7}$$

The selection of these candidates, which take into account both the maximum robot speed and possibility of collision, is illustrated in Fig. 3(a).

Now, the next step is the computation of the evaluation value for each candidate. This evaluation function is a weighted sum of three contributions:

$$V_i = a\, P_i + b\, C_i + c\, S_i \tag{8}$$

where a, b, and c are positive weighting coefficients, and P_i, C_i, and S_i quantify the progress to goal, the clearance, and the speed of each candidate, respectively.

Progress to goal is computed from the inner product between the FMM field gradient and the velocity candidate, both expressed in the world frame. The field gradient is numerically computed from the field values on a neighbor of a cell (x, y) as

$$Du = \begin{bmatrix} \frac{u(x+1,y)-u(x-1,y)}{2h} \\ \frac{u(x,y+1)-u(x,y-1)}{2h} \end{bmatrix}, \tag{9}$$

---

[1] Subscripts F, B, L, and R stand for front, back, left, and right.

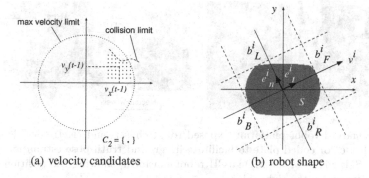

(a) velocity candidates   (b) robot shape

Fig. 3. Illustration of some aspects of the guidance method: (a) the velocity candidates form a grid (dynamic window) around the current velocity value; candidates are excluded when either the robot maximum speed or found to lead to a collision with a detected obstacle, resulting on a candidate set C_2; (b) how the bounding box relates to the robot shape S and the candidate velocity v^i, both expressed on the robot body reference frame.

where h is the grid size and x and y are the robot position coordinates, scaled and discretized to grid indices. The candidate velocity is normalized to maximum speed V_{max}, so that its relative contribution to the evaluation function is independent from the maximum speed scale. The gradient is normalized into a unit vector, so that its contribution to the progress to goal is independent from its absolute value (which, following (1), is $1/F(x)$). Thus, the progress to goal contribution is

$$P_i = \left\langle \frac{Du}{||DU||}, \frac{v^i}{V_{max}} \right\rangle \tag{10}$$

Clearance is computed, for each candidate, from the closest distance to an obstacle after moving along the candidate direction:

$$C_i = \min ||(p_t^{ij} - d_i, p_n^{ij})|| \tag{11}$$

where $d_i = ||v_i||T$ is the distance travelled along the candidate direction during the control loop period.

Finally, the absolute speed contribution is simply the absolute value of the candidate velocity

$$S_i = ||v^i|| \tag{12}$$

The optimal candidate results simply from the maximization of the evaluation values, that is

$$v^* = \arg \max_{v^i \in C_2} V_i \tag{13}$$

Since the target platform is omnidirectional, we have complete freedom on the choice of the heading of the robot, provided that the direction of movement is covered by range sensors. This extra degree of freedom can be used, for instance, to convey expressiveness to the robot motion. Currently we are not

Fig. 4. MOnarCH robot platform: Exposed robot chassis and electronics (left-most). Robot with a color coded plate to facilitate its ground truth pose estimation (second from left). Side views of the MOnarCH robot platform, showing the position of each LRF (two images on the right side)

exploiting this, and thus we use a simple heading controller, based on a feedback proportional gain with saturation

$$\omega = \max\left\{-\omega_{max}, \min\left\{Ke_\theta, \omega_{max}\right\}\right\} \tag{14}$$

where K is the proportional gain, $e_\theta = \arctan 2(v_y^*, v_x^*)$ is the angular deviation of the candidate velocity with respect to the robot heading, and ω_{max} the specified maximum angular speed.

4   Experiments and Results

4.1   Robot Details

The path planning and navigation algorithms described in this paper were implemented on a 4-wheeled omni-directional robot platform (Fig. 4). This robot has been specifically developed for an ongoing European FP7 project: MOnarCH[2]. In addition to various other sensors and actuators as described in [15], it is equipped with two laser range finders (LRFs) which are used for mapping, navigation, and obstacle avoidance (as shown in Fig. 4). In particular, we use two LRF together for the guidance method: the point cloud $\mathcal{L}$ consists of the union between the latest scan of each LRF, with the appropriate coordinate transformation to the common robot body reference frame.

The platform runs Linux Ubuntu and the integration middleware is the widely used ROS framework[3]. For mapping we use the gmapping package[4] and for localization we use the AMCL package[5], both available out of the box from the base ROS installation. The former is an implementation of a Rao-Blackwellized particle filter [16], while the latter implements an Augmented Monte Carlo Localization (AMCL) method [17].

---

[2] Project reference: FP7-ICT-2011-9-601033.

[3] http://www.ros.org.

[4] http://wiki.ros.org/gmapping.

[5] http://wiki.ros.org/amcl.

Fig. 5. Testbed in the 8th floor of ISR with the MOnarCH robot during the experiment from the left and right cameras of the GT system. The blue/yellow poles were knwon obstacles included in the map, while all the other objects on the field were previously unmapped (Color figure online).

4.2 Ground Truth Evaluation

In order to evaluate the performance of our navigation algorithm, i.e., to infer how well the DWA-based navigation performs w.r.t. the path planned using the FMM (gradient descent along the potential field, as explained in Sect. 3.1), it is necessary to compare the FMM-planned path with the actual path taken by the robot during the experiments. For the actual path taken by the robot, one cannot use the self-localization information as estimated by the robot itself because this information is one of the inputs to the DWA-based navigation algorithm. Therefore, re-using the self-localization information to evaluate the performance of navigation will be incorrect. Consequently, it is imperative that one must obtain the ground truth (GT) poses (position and orientation) of the robot during the experiment from a totally external GT system. To this end, we developed such a GT system as described further.

The hardware of the GT system to evaluate the pose of MOnarCH robot is the same as the one described in [18]. It consists of two gigabit ethernet cameras in a stereo baseline approximately 13 m apart. They are connected to a machine with Quad Core Intel(R) Core(TM) i5 CPU 750 @ 2.67 GHz, 8 GB RAM, running a Linux operating system. The model of the camera is Basler acA1300-30gc with a maximum acquisition frame rate of ∼25 frames per second (fps) and a resolution of 1294×964 pixels (1.2 megapixels). The robot and the GT system were time-synchronized using the network time protocol (NTP). In [18], this hardware was used to evaluate the 3D-position GT of a spherical-shaped object of known color and size therefore making the use of the stereo cameras. However, in this work we use this hardware to construct a GT system to estimate GT poses of the MOnarCH robot. For this purpose, we placed a bi-colored plate (stuck with two adjacent, colored, A4-sized markers) on top of the robot such that the center of the plate was directly above the robot's center of mass and the line joining the center of each colored markers on the plate was aligned with the robot's heading. This arrangement can be visualized in Fig. 4 (second from left). As this robot moves on a fixed plane, we could safely assume that the height of the

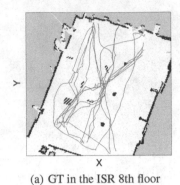
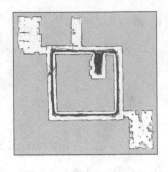

(a) GT in the ISR 8th floor   (b) self-localization in the ISR 6th floor

Fig. 6. Robot actual trajectories during experiments: (a) GT experiments, in the 8th floor of ISR which is also shown in the GT camera images of Fig. 5, and (b) a long 3 Km test, according to the robot self-localization. The ragged lines on the left corridor correspond to a situation where the robot got lost and required manual re-localization.

robot's plate above the ground plane was constant throughout the experiment. This height was pre-measured and used further in the GT pose evaluation of the robot.

Images from both the GT cameras (Fig. 5) were captured during the experiment and post-processed. The process pipeline then consisted of color-based image segmentation (for the colors of the robot's plate) and blob detection. The centers of the blobs, which denoted the centers of the colored markers on the robot's plate, were transformed from image coordinates to the world coordinates using the camera parameters under the assumption that these centers were at a constant height above the ground plane (the ground plane is denoted by $Z = 0$ plane in our world coordinate system). The camera parameters (intrinsic and extrinsic) were obtained prior to the experiments. Using simple geometry on the center of the colored markers, the position and orientation GT of the robot was calculated. Recalling that we had 2 cameras present in our GT system, we used one camera for either sides of the environment in which the experiment was carried out. As the height of the markers, and therefore the robot, was known a priori, only one camera would be required to estimate the robot's pose GT. Nevertheless, we assume that between the two cameras, placed on either sides of the environment, the one closer to the robot could produce more trustworthy GT of the robot's pose.

4.3   Experimental Results

We performed two main experiments, one in a controlled environment using the GT system, and another one consisting of a longer duration run (1 h and 50 min, about 3 km traveled distance). Both took place at ISR, in separate floors of the North Tower (IST). In these experiments, $V_{max} = 0.75\,\text{m/s}$, $A_{max} = 0.6\,\text{m/s}^2$, and $T = 50\,\text{ms}$. The trajectories for each of these experiments are shown in Fig. 6.

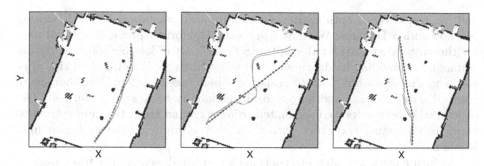

Fig. 7. Three autonomous navigation tasks, showing the path planned using FMM in square-dotted blue and the actual trajectory, according to self-localization, in dotted green, and according to the ground truth system, in continuous magenta. From left to right: a trajectory without unmapped obstacles (1:14), a trajectory including unmapped obstacles (0:50), and a trajectory with moving people (2:41). We include in parenthesis the approximate time (in minutes) each trajectory begins in the accompanying video (Colour figure online).

In the accompanying video (available at http://youtu.be/QW40yepKtuY), we show all the trajectories of the first experiment, overlaying one of the GT camera's image stream with the 2D re-projection of the world coordinate estimates (robot's actual (GT), self-localization and FMM-based poses). This video shows several actual trajectories of the robot, in an environment where only some of the obstacles are on the map, namely the three yellow/blue poles (shown with dark dots in the robot's map in Figs. 6(a) and 7). The previously-unknown obstacles for the robot (all objects on the field in Fig. 5 other than the poles) are shown in Figs. 6(a) and 7 as dashed regions. In the latter part of the run, randomly walking people, totaling 5 persons, were introduced in the scenario. In Fig. 7 we show three of those trajectories, with different obstacle types.

To access whether the proposed approach is getting us closer to our goal of dependable autonomous navigation of mobile robots, we have been running long-term tests with the platform in the environment shown in Fig. 6(b). These tests consists in assigning random goal locations within the free space of the environment. At the time of writing, the platform has traveled a total of **18 km** during **11 h** of movement. During this time, the platform lost localization a handful of times, mostly due to inaccuracies in odometry together with the range of the LRF being lower than half of the corridor length. However, we have no significant collisions to report. The only exceptions are light touching of persons' foot, explicable by the fact that typical feet lie entirely below the LRF scanning plane. That is, it only perceives the persons legs.

5 Conclusions and Future Work

This paper presented a novel design for dependable autonomous robot navigation, targeting domestic spaces, as well as populated public spaces. To this end, we employed an optimization based approach, combining a Fast Marching Method with a Dynamic Window Approach. The former provides optimal plan for the path, whereas the latter optimizes over a set of feasible robot actuation commands, providing guidance and obstacle avoidance. We evaluated this approach on a mobile robot platform, capable of high speed motion. Both short-term controlled experiments and long-term dependability tests were performed. The controlled experiments were evaluated using a ground truth tracking system to compare the actual trajectory with the one resulting from the Fast Marching Method.

As future work, we intend to continue long-term dependability tests, both in controlled environments as well as in populated indoor public spaces.

References

1. Sethian, J.A.: Fast marching methods. SIAM Rev. **41**(2), 199–235 (1999)
2. Fox, D., Burgard, W., Thrun, S.: The dynamic window approach to collision avoidance. IEEE Robot. Autom. Mag. **4**(1), 23–33 (1997)
3. Cakmak, M., Takayama, L.: Towards a comprehensive chore list for domestic robots. In: 2013 8th ACM/IEEE International Conference on Human-Robot Interaction (HRI), pp. 93-94 March 2013
4. Kruse, T., Pandey, A.K., Alami, R., Kirsch, A.: Human-aware robot navigation: a survey. Robot. Auton. Syst. **61**(12), 1726–1743 (2013)
5. Kruse, T., Kirsch, A., Sisbot, E., Alami, R.: Exploiting human cooperation in human-centered robot navigation. In: RO-MAN, pp. 192–197. IEEE September 2010
6. Yuan, F., Twardon, L., Hanheide, M.: Dynamic path planning adopting human navigation strategies for a domestic mobile robot. In: 2010 IEEE/RSJ International Conference on Intelligent Robots and Systems (IROS), pp. 3275–3281, October 2010
7. Stuckler, J., Behnke, S.: Following human guidance to cooperatively carry a large object. In: 2011 11th IEEE-RAS International Conference on Humanoid Robots (Humanoids), pp. 218–223, October 2011
8. Klaess, J., Stueckler, J., Behnke, S.: Efficient mobile robot navigation using 3D surfel grid maps. In: 7th German Conference on Robotics; Proceedings of ROBOTIK 2012, pp. 1–4, May 2012
9. Stückler, J., Droeschel, D., Gräve, K., Holz, D., Kläß, J., Schreiber, M., Steffens, R., Behnke, S.: Towards robust mobility, flexible object manipulation, and intuitive multimodal interaction for domestic service robots. In: Röfer, T., Mayer, N.M., Savage, J., Saranlı, U. (eds.) RoboCup 2011. LNCS, vol. 7416, pp. 51–62. Springer, Heidelberg (2012)
10. Holz, D., Kraetzschmar, G.K., Rome, E.: Robust and computationally efficient navigation in domestic environments. In: Baltes, J., Lagoudakis, M.G., Naruse, T., Ghidary, S.S. (eds.) RoboCup 2009. LNCS, vol. 5949, pp. 104–115. Springer, Heidelberg (2010)
11. Jacobs, S., Ferrein, A., Schiffer, S., Beck, D., Lakemeyer, G.: Robust collision avoidance in unknown domestic environments. In: Baltes, J., Lagoudakis, M.G., Naruse, T., Ghidary, S.S. (eds.) RoboCup 2009. LNCS, vol. 5949, pp. 116–127. Springer, Heidelberg (2010)

12. Sethian, J.A.: Level Set Methods and Fast Marching Methods Evolving Interfaces in Computational Geometry, Fluid Mechanics, Computer Vision, and Materials Science. Cambridge Monograph on Applied and Computational Mathematics. Cambridge Press, Cambridge (1999)
13. Garrido, S., Moreno, L., Abderrahim, M., Blanco, D.: FM2: a real-time sensor-based feedback controller for mobile robots. Int. J. Robot. Autom. **24**(1), 48–65 (2009)
14. Brock, O., Khatib, O.: High-speed navigation using the global dynamic window approach. In: Proceedings of the IEEE International Conference on Robotics and Automation, vol. 1, pp. 341–346. IEEE (1999)
15. Messias, J., Ventura, R., Lima, P., Sequeira, J., Alvito, P., Marques, C., Carrico, P.: A robotic platform for edutainment activities in a pediatric hospital. In: Proceedings of the IEEE International Conference on Autonomous Robot Systems and Competitions (2014) (accepted)
16. Grisetti, G., Stachniss, C., Burgard, W.: Improved techniques for grid mapping with Rao-Blackwellized particle filters. IEEE Trans. Robot. **23**(1), 34–46 (2007)
17. Thrun, S., Burgard, W., Fox, D.: Probabilistic Robotics. MIT Press, Cambridge (2005)
18. Ahmad, A., Xavier, J., Santos-Victor, J., Lima, P.: 3D to 2D bijection for spherical objects under equidistant fisheye projection. Comput. Vis. Image Underst. **125**, 172–183 (2014)

Safely Grasping with Complex Dexterous Hands by Tactile Feedback

Jose Sanchez[✉], Sven Schneider, and Paul Plöger

Department of Computer Science, Bonn-Rhein-Sieg University of Applied Science,
Sankt Augustin, Germany
jose.sanchez@smail.inf.h-brs.de,
{sven.schneider,paul.ploeger}@h-brs.de

Abstract. Robots capable of assisting elderly people in their homes will become indispensable, since the world population is aging at an alarming rate. A crucial requirement for these robotic caregivers will be the ability to safely interact with humans, such as firmly grasping a human arm without applying excessive force. Minding this concern, we developed a reactive grasp that, using tactile sensors, monitors the pressure it exerts during manipulation. Our approach, inspired by human manipulation, employs an architecture based on different grasping phases that represent particular stages in a manipulation task. Within these phases, we implemented and composed simple components to interpret and react to the information obtained by the tactile sensors. Empirical results, using a Care-O-bot 3® with a Schunk Dexterous Hand (SDH-2), show that considering tactile information can reduce the force exerted on the objects significantly.

Keywords: Robot grasping · Domestic robot · Tactile feedback

1 Introduction

The percentage of people over 65 years in the world is projected to increase from eight percent to twelve percent by the year 2030 [1], which will increment further the already high demand of elderly care. To address this issue, global leaders carried out a UN World Assembly on Aging in 2002 which had, as one of its main topics, the objective of providing enabling and supportive environments for the elderly [2].

As a solution to this workforce shortage of elderly care, the governments of nations such as Japan, US and Germany have been encouraging the introduction of robots in nursing homes. Recent efforts towards this goal can be seen in the work of Nagai et al., which provides an analysis of the challenges to introduce robots into these environments in [3]. Pineau et al. introduced a robot assistant that autonomously guided the elderly and also reminded them of their schedules [4]. In Germany, the Fraunhofer Institute for Manufacturing Engineering and Automation (IPA) developed the Care-O-bot 3® with the objective of assisting

© Springer International Publishing Switzerland 2015
R.A.C. Bianchi et al. (Eds.): RoboCup 2014, LNAI 8992, pp. 332–344, 2015.
DOI: 10.1007/978-3-319-18615-3_27

elderly people in domestic environments [5]. The need for such domestic service robots can also be seen in the RoboCup@Home competition [6], especially the "Emergency situation" scenario, where robots deal with an accident in a home environment.

Despite the introduction of robots in domestic environments, their physical interaction with humans remains limited, a skill that is crucial for robots to eventually become reliable caregivers. The robots have to operate safely in these highly dynamic and uncertain environments. Manipulation under these conditions, while extremely complicated for robots, is performed effortlessly by humans. This proficiency achieved by humans, highly depends on their tactile sensing abilities while executing manipulation tasks [7]. Based on this insight, together with the improvement of tactile sensing technology, robotic researchers have produced algorithms inspired by human tactile sensing [8], and used tactile feedback to reactively adjust grasps [9,10]. Although most of these approaches allow robots to interact physically in a domestic environment, their main concern is manipulation of objects.

With the long-term goal of enabling a robot, namely a Care-O-bot 3®, to safely interact with humans (e.g. guiding people with vision impairment in a nursing home), we develop a grasping approach that considers the pressure exerted during manipulation to prevent the application of excessive grasping forces. The pressure information provided by the tactile sensors of the SDH-2 hand is used as a feedback signal to control the fingers' motion and react to contacts. Aside from the tactile information, force-torque sensors of the manipulator are used to enable the detection of contacts between the robot's arm and its environment. Furthermore, the high-level control of our implementation is based on the phases observed in human manipulation.

To validate our work, we recorded empirical data of grasps on a set of objects[1] with distinct features such as hardness, shape, and size. Our approach effectively reduced the exerted force on the grasped objects, by at least, half of the original force. In a particular case, the applied force was reduced by a factor of 20, while still successfully executing the grasp. We also analyze the limitations of this approach and compare its performance to an open-loop grasp approach. An early version of this work has been demonstrated during the competitions of RoboCup@Home German Open 2013 in Magdeburg and RoboCup@Home World Championship 2013 in Eindhoven.

The remainder of the paper is organized as follows. Section 2 provides a brief description on human grasp and the involved tactile information, as well as current applications of tactile sensors in robotics. Section 3 describes our approach and the hardware it uses. In Sect. 4 the evaluation method is detailed and the results obtained are reported. A summary of the paper is presented in Sect. 5.

---

[1] Although the motivation of this work is to ultimately grasp humans, due to safety reasons, we evaluated our approach on objects.

2   Related Work

2.1   Human Manipulation

Johansson and Flanagan noticed the importance of tactile signals during manipulation by humans [7]. These tactile signals are denoted as tactile afferents[2] by Johansson. They can end at skin level (type I) or, deeper, at the dermis (type II); and they can have fast or slow frequency responses. Thus, the tactile afferents used by the hand are: fast-adapting type I (FA-I), slow-adapting type I (SA-I), fast-adapting type II (FA-II), and slow-adapting type II (SA-II). Besides studying these tactile signals they analyzed the phases involved during a manipulation task. The phases of a simple pick and place task, as described by Johansson in [7], are:

1. *Reach:* Fingers make contact with the object and FA-I afferents are activated.
2. *Load:* Enough force is applied to the object to obtain a firm grip. During this phase the SA-I and SA-II afferents are triggered.
3. *Lift:* The object is lifted off the support surface and the FA-II afferents are activated.
4. *Hold:* Forces are applied to the object to prevent its slippage. SA-I and SA-II afferents are activated in this phase.
5. *Replace:* The object makes contact with the support surface and the FA-II afferents are triggered in this phase.
6. *Unload:* The fingers release the object and FA-I afferents are activated.

2.2   Tactile Sensing in Robotics

Robots with tactile sensors have recently been used in object recognition [11], evaluation of grasp stability [12], and grasp adjustment. Our review of related work focuses on the latter application.

Hsiao et al. [9] apply corrective actions, using the tactile information of a PR2 gripper, to improve the location of the contacts. They define corrective actions to open the PR2 gripper when a contact is sensed, and moving the wrist in the direction of the sensed contact. This approach is able to compensate for position errors to yield better stability of the grasp. Prats [10] also improved the performance of a robotic control system by adding tactile information as feedback. Their previous approach only considered visual and force signals for feedback. The tactile feedback drives a controller that moves three degrees of freedom of a robotic arm to open a slide door. Romano et al. [8] developed an approach, inspired also on human manipulation, that uses tactile sensors to design low-level signals and control loops that mimic the tactile afferents FA-I, SA-I, FA-II. However, this implementation is specific to the PR2 gripper, a parallel jaw gripper with only one actuator. We therefore seek to extend their work to control a gripper with more than one degree of freedom, e.g. a SDH-2. Compliant grasps have also been achieved without the use of tactile sensing [13].

---

[2] A tactile afferent is a conduct that conveys signals to the brain.

3   Approach

3.1   Hardware

The SDH-2 is a servo-electric 3-finger gripping hand with seven degrees of free-dom[3] (DoF). The three fingers are actuated by two joints each, one is rooted to the hand's palm and the other is in the middle of the finger. Both of these joints have a range of motion of -90° to +90°, and enable the extension and flexion of the fingers. The seventh actuator allows two fingers to rotate simultaneously in opposite directions and generates an abduction or adduction movement. The range of motion of this actuator reaches 0° to +90°. Figure 1 depicts how these motions are executed by both a human finger and an SDH finger. Moreover, each finger has two phalanges: a *proximal phalanx* which is closer to the palm, and a *distal phalanx* which is further away from the palm. Each phalanx is equipped with a tactile sensor matrix.

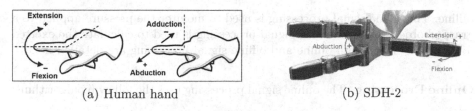

| (a) Human hand | (b) SDH-2 |

Fig. 1. Flexion/extension and abduction/adduction motions of (a) the human hand [14] and (b) the SDH-2.

The six tactile sensors from Weiss Robotics [15] provide the contact information. This information is represented as a matrix that either contains 6×14 tactile elements (tactels), for the proximal phalanges, or 6×13 tactels for the distal phalanges. Each tactel produces an integer value between 0, when there is no pressure, and 4095, the maximum pressure value that represents $250\,KPa$. The tactels in the proximal phalanges have identical sizes of $3.4 \times 3.4\,mm$. However, the sizes of the tactels in the distal phalanges vary slightly, because the tactile arrays are curved. For simplicity of the calculations, the size of all tactels is assumed to be the same. Figure 2 shows a diagram of a tactile sensor together with a visualization of a contact sample.

The SDH-2 is mounted to a KUKA Lightweight Robot with seven DoF that provides torque signals in each joint [16].

3.2   Tactile Signal Processing

Following the idea of considering the information produced by the tactile sensors as grayscale images, as proposed in [12], we process the tactile data online and

---

[3] http://www.schunk.com/schunk_files/attachments/SDH_DE_EN.pdf.

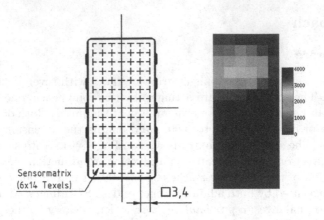

Fig. 2. Left: A diagram of a tactile sensor, right: the pressure profile of a contact.

offline. The online signal processing is used to monitor the pressure applied to a grasped object, and the offline signal processing is used to calculate the exerted force of the grasp. Both online and offline signal processing are detailed next.

Online Processing. The online signal processing uses the following algorithms:

- **detect_contacts:** Given a number of tactile arrays, with their respective threshold values, it returns a Boolean array that indicates if a tactel is exceeding the contact threshold value. A 0 is assigned for no contact, and a 1 represents that a tactile array has a contact.
- **detect_thresholds:** This inverts the result of *detect_contacts*. I.e. a 0 in the returned array, indicates a contact, while a 1 represents no contact.

Each element in the Boolean arrays controls the motion of a single phalanx. The output of *detect_contacts* selects which phalanges to move, this is to only move those phalanges that are in contact with the object (i.e. is used while the phalanges are not moving). The output of *detect_thresholds* is used to stop the movement of the phalanges that have reached the desired contact value.

Offline Processing. The offline processing is used after the hand has stopped moving. We applied the steps applied by Li et al. [17], namely:

- **Threshold:** For each tactile array, and their respective pressure threshold, it sets the tactels below this threshold to zero. The application of this thresholding is optional to remove low contact values, which may be caused by pressure applied to an adjacent tactel. Due to the rubber layer covering the sensor, pressure applied to a single tactel also activates its neighbors [18].
- **Label:** Using the connected-component labeling algorithm with a 4-connectivity criteria [19], it labels the contact regions in each tactile array.

The purpose of this step is to segment areas of contact for further classification (e.g. determine the largest contact area or the strongest contact area).

- **Extract:** This step differs from the one described in [17], by extracting the *strongest contact* region instead of the largest contact region. The strongest region is defined by its normal force. The normal force of each contact region is calculated with the equation $F = P * A$ (where P represents a normalized pressure of a contact region, and A is the area of the region). The region with the highest normal force is selected as the strongest. The normalized pressure P is calculated as the ratio of the maximum pressure range (250 KPa) to the maximum displayed value (4095 bits) times the average value of the active tactels (i.e. tactels with a contact value greater than zero). The area A is calculated by multiplying the individual area of a tactel times the number of active tactels. As noted in Sect. 3.1, the size of the tactels on the distal phalanges is assumed to be the same as on the proximal phalanges.

- **Locate:** For each tactile array, this step calculates the centroid of the contact regions. The centroids are calculated, as suggested in [18], using the raw moment formula:

$$M_{pq} = \sum_x \sum_y x^p y^q I(x, y) \tag{1}$$

where x and y represent the coordinates of a tactel in a tactile array and $I(x, y)$ is the intensity (i.e. pressure value) in tactel x, y. The order of x and y is determined by p and q, respectively. A centroid can be then calculated using:

$$\begin{bmatrix} x_0 \\ y_0 \end{bmatrix} = \frac{1}{M_{00}} \begin{bmatrix} M_{10} \\ M_{01} \end{bmatrix} \tag{2}$$

3.3 Architecture

Our architecture follows the human manipulation phases, as described by Johansson [7]. Figure 3 illustrates our architecture. Note that this architecture is based on a pick-and-place task. When grasping a human, the `lift/hold` and `place` phases will be different.

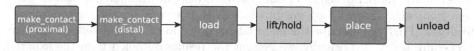

Fig. 3. Phase-based architecture, inspired by human manipulation. The colored boxes indicate the phases implemented in this work.

The `make_contact` phases move the phalanges (first the proximal phalanges, followed by the distal phalanges) from their initial, open configuration, to a desired closed configuration. Each phalanx is controlled by the loop shown in Fig. 4, where P is the pressure of the highest-valued tactel in the tactile sensor, P_{ref} is the pressure threshold and $\dot{\theta}$ is the desired joint velocity. The controller is a simple bang-bang controller that sets $\dot{\theta} = 0$ when $P_{err} \leq 0$. Once all joint velocities have been set to zero the make_contact phase ends and the load phase is started. During the load phase each joint, except for the one generating abduction/adduction movements, is also actuated using the pressure control loop shown in Fig. 4. The make_contact phase shapes the hand according to the object, while the load phase regulates the pressure to achieve a stable grasp.

When the load phase has finished, the object is grasped and the lift/hold phase raises the arm to lift the object from a surface and it holds the object during the transportation to the placement pose. Next, the place phase moves the arm downward, while using a force monitor to detect an abrupt change of force exerted on the hand, that indicates a contact between the object an a surface. This allows to safely place the object on the surface. Finally, the unload phase opens the hand to release the object.

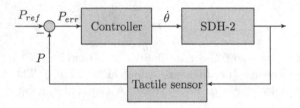

Fig. 4. Control loop using tactile feedback.

Each of these phases is composed by simpler components, which can be replaced without modifying the overall structure of a phase, thus separating concerns as described in [20]. These components implement algorithms that perform computations and communicate their outputs publishing messages through ROS topics [21].

4   Experimental Evaluation

To evaluate the performance of our *reactive* grasp we compared it to the current approach, an *open-loop* grasp, which does not consider the grasp force as feedback. First, we describe the materials involved in the experiments, then the procedure is detailed. Finally, we present the results obtained from the experimentation.

4.1   Materials

The platform used to carry out the experimental evaluation was the Care-O-bot 3 [5], with a KUKA Ligthweight Robot (LWR4) [16]. The end-effector located at the end of the LWR4 is a SDH-2. Furthermore, 18 objects were selected to represent the following three features:

- **Hardness:** The objects were regarded as *deformable* (D) when the open-loop grasp would either leave a mark on the object, or change its shape or size. If no mark or modification was observed, the object was labeled as *non-deformable* (N).
- **Shape:** The shape of an object was considered to be one of the following: *prismatic* (Pr), *spherical* (Sp) and *cylindrical* (Cy).
- **Size:** An object was classified as *small* (S), *medium* (M), or *large* (L).

A sample of the selected objects can be seen in Fig. 5, and their classification is shown in Fig. 6.

	Hardness	Shape	Size
Chocolate milk	D	Pr	S
Bathroom cleaner	D	Pr	M
Milk carton	D	Pr	L
Small ball	D	Sp	S
Medium ball	D	Sp	M
Large ball	D	Sp	L
Soda can (empty)	D	Cy	S
Chips can	D	Cy	M
Soup	D	Cy	L
Alloy profile	N	Pr	S
Dictionary	N	Pr	M
Dried coffee	N	Pr	L
Candle	N	Sp	S
Orange	N	Sp	M
Melon	N	Sp	L
Soda can (full)	N	Cy	S
Coffee bottle	N	Cy	M
Noodles	N	Cy	L

Fig. 5. A sample of the test objects. Missing objects in the figure are: dictionary, orange, melon and soda can (full).

Fig. 6. Categorization of the test objects, according to their hardness (D/N), shape (Pr/Sp/Cy) and size (S/M/L).

4.2   Procedure

For each test object, the robot's arm started in a predefined pose (see Fig. 7a). Both approaches were executed three times[4], for each of the six locations shown

---

[4] The reactive approach was partially executed. The phases executed were: *make contact*(proximal/distal), *load* and *lift* phase.

in Fig. 7b. Spherical and cylindrical objects were centered on the marked locations, while prismatic objects were placed on the marked locations along their edges. These locations were chosen to cover a range of positions within the grasp, e.g., *close/away* from the wrist and close to the *fingers/thumb*.

(a) Arm position used throughout the evaluation.

(b) Grasp locations.

Fig. 7. Experimental procedure.

4.3   Results

The results obtained by the experiments conducted on 18 different objects, using both the *open-loop* grasp and the *reactive* grasp, are summarized next. Table 1 shows the success rate of both approaches. Based on the performance of the grasps, two aspects were further analyzed: the grasp force applied by each grasp, and the cause of each failed grasp. The grasp forces of the objects that had a 100 % success rate with both approaches are displayed in Table 2. The force on each object is the average of all trials (i.e. 18 trials). This average represents the sum of the forces on each tactile sensor.

To conclude this section, the 84 failed grasps along with their causes are presented in Table 3. The majority of the reactive approach failures (i.e. no grasp) were caused by the phalanges pushing the objects out of the grasp (27 failed grasps), and by the phalanges not receiving their required stop commands, either because the specified joint limits could not be reached or the desired grasp force could not be reached (15 failed grasps). The overall success rate for the reactive grasp approach was 78.64 %, and of 94.17 % for the open-loop grasp approach.

Table 1. Success rate of both approaches, the *open-loop* grasp (OLG) and the *reactive* grasp (RG).

	Trials	OLG		RG	
		Success	Rate	Success	Rate
Chocolate milk	18	18	100 %	18	100 %
Bathroom cleaner	18	18	100 %	15	83.3 %
Milk carton	18	18	100 %	18	100 %
Small ball	18	18	100 %	17	94.4 %
Medium ball	18	18	100 %	16	88.9 %
Large ball	18	18	100 %	18	100 %
Soda can (empty)	18	18	100 %	18	100 %
Chips can	18	18	100 %	17	94.4 %
Soup	18	18	100 %	15	83.3 %
Alloy profile[a]	3	3	100 %	0	0 %
Dictionary	18	18	100 %	18	100 %
Dried coffee	18	18	100 %	5	27.8 %
Candle	18	5	27.8 %	5	27.8 %
Orange	18	13	72.2 %	10	55.6 %
Melon	18	18	100 %	3	16.7 %
Soda can (full)	18	18	100 %	16	88.9 %
Coffee bottle	18	18	100 %	16	88.9 %
Noodles	18	18	100 %	18	100 %

[a]The alloy profile was tested only three times, in position 1, for each grasp approach. The other positions were not tested, because in those positions the alloy profile was damaging the tactile sensors.

Table 2. Grasp forces applied by both approaches, in Newtons.

	OLG	RG	Force reduction
Chocolate milk	12.3	6.8	44.9 %
Milk carton	10.5	0.5	95.4 %
Dictionary	23.5	6.6	71.7 %
Noodles	32.1	8.9	72.3 %
Soda (empty)	18.6	5.5	70.7 %
Large ball	43.4	7.1	83.6 %

Table 3. Categorization of failures.

	OLG	RG
No grasp	0 (0 %)	42(63.6 %)
No lift	6 (33.3 %)	4 (6.1 %)
Rotated	0 (0 %)	5 (7.6 %)
Slip	12 (66.7 %)	15 (22.7 %)
Total	**18**	**66**

5   Conclusions and Future Work

This paper presented a software architecture that emulates the human manipulation phases together with an approach that significantly reduces the grasp force through the use of tactile feedback. The approach was specifically tuned for the SDH-2. The pressure information of all experiments, was recorded and made available at https://github.com/jsanch2s/tactile_info. However, the success rate of our reactive grasp approach was not as high as the open-loop grasp approach (78.64 % vs 94.17 %), mainly due to the following limitations:

– The tactile sensors do *not completely cover* the fingers, causing the reactive grasp to not reach the desired contact values.
– *Low sensitivity* of the tactile sensors hinders the ability to detect light contacts (this accounts for 40 % of the failures). Integrating the signals of a force-torque sensor signals, as demonstrated in [22] could improve contact detection.
– *Insufficient force* on the grasp caused objects to slip or rotate within the grasp, caused by low values of the contact thresholds.

Future work will be focused on the implementation of the offline signal processing and the improvement of individual components to detect contacts that tactile sensors cannot (e.g. using force-torque sensors), improve the location of contact points using arm motions as Hsiao et al. demonstrated in [9], and detect slippage by analyzing temporal readings from the tactile sensors. A video showing the capabilities of our reactive grasp is available at https://www.youtube.com/watch?v=fJoSDVKSdm0. The video shows a slower version due to safety reasons.

Acknowledgments. We gratefully acknowledge the support by the b-it Bonn-Aachen International Center for Information Technology.

References

1. Dobriansky, P.J., Suzman, R.M., Hodes, R.J.: Why population aging matters: a global perspective, U.S. Department of State, U.S. Department of Health and Human Services, National Institute on Aging, NIH, Washington, DC, p. 132 (2007)

2. Kinsella, K.G., Phillips, D.R.: Global aging: the challenge of success. Popul. Bull. **60**(1), 1–44 (2005)
3. Nagai, Y., Tanioka, T., Fuji, S., Yasuhara, Y., Sakamaki, S., Taoka, N., Locsin, R.C., Ren, F., Matsumoto, K.: Needs and challenges of care robots in nursing care setting: a literature review.In: Proceedings of IEEE International Conference on Natural Language Processing and Knowledge Engineering (NLP-KE), pp. 1–4 (2010)
4. Pineau, J., Montemerlo, M., Pollack, M., Roy, N., Thrun, S.: Towards robotic assistants in nursing homes: challenges and results. Robot. Auton. Syst. **42**(3–4), 271–281 (2003)
5. Graf, B., Reiser, U., Hagele, M., Mauz, K., Klein, P.: Robotic home assistant Care-O-bot$^\circledR$ 3 - product vision and innovation platform. In: Advanced Robotics and its Social Impacts (ARSO), pp. 139–144 (2009)
6. Chen, K., Holz, D., Rascon, C., Ruiz del Solar, J., Shantia, A., Sugiura, K., Stückler, J., Wachsmuth, S.: RoboCup@Home 2014: Rules and Regulations (2014). http://www.robocupathome.org/rules/2014_rulebook.pdf
7. Johansson, R.S., Flanagan, J.R.: Coding and use of tactile signals from the fingertips in object manipulation tasks. Nat. Rev. Neurosci. **10**(5), 345–359 (2009)
8. Romano, J.M., Hsiao, K., Niemeyer, G., Chitta, S., Kuchenbecker, K.J.: Human-inspired robotic grasp control with tactile sensing. IEEE Trans. Robot. **27**(6), 1067–1079 (2011)
9. Hsiao, K., Chitta, S., Ciocarlie, M., Jones, E.G.: Contact-reactive grasping of objects with partial shape information. In: IEEE/RSJ International Conference on Intelligent Robots and Systems, pp. 1228–1235 (2010)
10. Prats, M., Sanz, P.J., Del Pobil, A.P.: Vision-tactile-force integration and robot physical interaction. In: Proceedings of IEEE International Conference on Robotics and Automation, pp. 3975–3980 (2009)
11. Montano, A., Suarez, R.: Object shape reconstruction based on the object manipulation. In: Proceedings of International Conference on Advanced Robotics (ICAR), pp. 1–6 (2013)
12. Bekiroglu, Y., Huebner, K., Kragic, D.: Integrating grasp planning with online stability assessment using tactile sensing. In: Proceedings of IEEE International Conference on Robotics and Automation (ICRA), pp. 4750–4755 (2011)
13. Rosales, C., Surez, R., Gabiccini, M., Bicchi, A.: On the synthesis of feasible and prehensile robotic grasps. In: Proceedings of IEEE International Conference on Robotics and Automation (ICRA), pp. 550–556 (2012)
14. Li, K., Chen, I.M., Yeo, S.H., Lim, C.K.: Development of finger-motion capturing device based on optical linear encoder. J. Rehabil. Res. Dev. **48**(1), 69–82 (2011)
15. Weißand, K., Wörn, H.: The working principle of resistive tactile sensor cells. In: Proceedings of IEEE International Conference on Mechatronics and Automation, vol. 1, pp. 471–476 (2005)
16. Bischoff, R., Kurth, J., Schreiber, G., Koeppe, R., Albu-Schäffer, A., Beyer, A., Eiberger, O.: The KUKA-DLR lightweight robot arm - a new reference platform for robotics research and manufacturing. In: Proceedings of International Symposium on Robotics (ISR), pp. 1–8 (2010)
17. Li, Q., Elbrechter, C., Haschke, R., Ritter, H.: Integrating vision, haptics and proprioception into a feedback controller for in-hand manipulation of unknown objects. In: Proceedings of IEEE/RSJ International Conference on Intelligent Robots and Systems (IROS), pp. 2466–2471 (2013)

18. Nagatani, T., Noda, A., Hirai, S.: What can be inferred from a tactile arrayed sensor in autonomous in-hand manipulation? In: Proceedings of IEEE International Conference on Automation Science and Engineering (CASE), pp. 461–468 (2012)
19. Di Stefano, L., Bulgarelli, A.: A simple and efficient connected components labeling algorithm. In: Proceedings of International Conference on Image Analysis and Processing, pp. 322–327 (1999)
20. Bruyninckx, H., Klotzbücher, M., Hochgeschwender, N., Kraetzschmar, G., Gherardi, L., Brugali, D.: The BRICS component model: a model-based development paradigm for complex robotics software systems. In: Proceedings of the 28th Annual ACM Symposium on Applied Computing, pp. 1758–1764 (2013)
21. Quigley, M., Conley, K., Gerkey, B., Faust, J., Foote, T., Leibs, J., Wheeler, R., Ng, A.Y.: ROS: an open-source robot operating system. In: ICRA Workshop on Open Source Software, vol. 3(3.2) (2009)
22. Pastor, P., Righetti, L., Kalakrishnan, M., Schaal, S.: Online movement adaptation based on previous sensor experiences. In: Proceedings of IEEE/RSJ International Conference on Intelligent Robots and Systems, pp. 365–371 (2011)

A Formalization of the Coach Problem

G.Y.R. Schropp[1]([✉]), J-J. Ch. Meyer[1], and S. Ramamoorthy[2]

[1] Utrecht University (UU), Utrecht, The Netherlands
gwendolijn.schropp@phil.uu.nl
[2] University of Edinburgh (UoE), Edinburgh, UK

Abstract. Coordination is an important aspect of multi-agent team-work. In the context of robot soccer in the RoboCup Standard Platform League, our focus is on the *coach* as an external observer of the team, aiming to provide his teammates with effective tactical advice during matches. The coach problem can be approached from different angles: in order to adapt the behaviour of his teammates, he should at first be able to perform *plan recognition* on their observable actions. Furthermore, in providing them with appropriate advice, he should still adhere to the norms and regulations of the match to prevent penalties for his team. Also, when teammates' profiles and attributes are unknown or the system is only partially observable, coordination should be more 'ad hoc' to ensure robustness of the Multi-Agent System (MAS). In this work, we present a formalization of the problem of designing a coach in robot soccer, employing a temporal deontic logical framework. The framework is based on *agent organizations*[10], in which social coordination and norms play an important part.

Keywords: Agent organization · Multi-agent system · Teamwork · Coordination · Logic · Plan recognition

1 Introduction

RoboCup's Standard Platform League (SPL) now allows for a coach robot, whose main role is to provide tactical advice to his team. Besides that, he could help the team in disambiguating signals (e.g. in localization) and give them high level suggestions to improve their play. As the coach is a novel addition to the SPL, there is currently little guidance on what a good coach should do. In contrast, there is a lot of work on for example path planning and localization. As such, the need for a formal understanding of the coach's role arises. This formalization is the goal of this paper. In the simulation and middle size leagues however, coaching has been the topic of research since 2001 [37].

For the formalization of the robot soccer system we use the concept of agent organizations: a set of entities or agents that are regulated by social order in achieving common goals. A logic-based framework enables a precise and formal but abstract specification, which could guide many different kinds of detailed future implementations via agent programming languages [8,10]. Because agents

© Springer International Publishing Switzerland 2015
R.A.C. Bianchi et al. (Eds.): RoboCup 2014, LNAI 8992, pp. 345–357, 2015.
DOI: 10.1007/978-3-319-18615-3_28

are left unspecified in our framework, they can be developed according to the needs of the designer, yielding a flexible and adaptable system [10]. The contribution of this paper is to present a formalization using deontic temporal logic, capturing the specification of the coach within the RoboCup SPL, and describing the rules and temporal aspects of such a system. Once a type of agents have been chosen, this framework is ready to be implemented. Suggestions for implementation are also given in this work.

2   Related Research

There are few formal MAS for robot soccer based on agent organizations. However, aspects of our framework are related to ideas that have appeared previously in the literature on multiagent systems and robotics. Agent organizations and developing methods have been presented in [10,12,13,26,39]. Esteva et al. [12] and Dignum [10] introduce the notion of *norms* in combination with the agent *roles* already used by Odell et al. [26], Ferber and Gutknecht [13] and Wooldridge [39]. Roles can be used to decompose the tasks to be performed by the MAS into sub-objectives to increase efficiency [14,34,36]. The role-based approach to ad hoc teamwork by Genter et al. [14] determines role selection on the team's utility. In robot soccer, roles have been based on absolute position in the field [25], position relative to the ball [1,24] and robot trajectories, sometimes also considering positions of other players [40]. Other work on team coordination through roles and dynamic positioning can be found for example in [9,21]. Desheng et al. [9] achieve collaboration between simulated agents via roles in situation calculus. One of the characteristics of the human organizations framework used in this work is communication or negotiation. However, coordination without negotiation [17] is more appropriate if an ad hoc approach is required. Tracking multiple soccer players' trajectories can for example be done using the camera feed, in combination with an analyzing system that segments motions into classified ball actions [1]. Extracting tactic events from human soccer video feeds has also been done using spatio-temporal interaction among players and the ball [40].

In order to handle the team's knowledge representation, it is important that the robots share the same definitions of concepts in their environment. Moreover, these concepts should be *grounded* in order to link them to the robot's percepts of the real world, for example via *ontologies* [16,28] and pattern recognition [19]. Also, a communication language for the coach should be defined, for example the COACH UNILANG language, enabling both high-level and low-level coaching through coach instructions [30].

As a first step towards integrating the framework in a coach robot, we pay special attention to the problem of *plan recognition*. There's related work on the use of *roles* to infer a robot's plans or intentions [2,4,14]. 'Plans' can be interpreted in various ways, ranging from a sequence of actions currently performed according to some behaviour [4,32] to a robot's entire internal state (e.g. belief base, goals). The latter approach is sometimes called 'opponent modelling' and used to learn an optimal strategy against the modelled type of agent [5,31]. Although plan recognition was first introduced in the field of logical inference by [20], the current state of the art is mainly based on probabilistic methods like Bayesian Networks [3,4,6] and Markov Models [3,38]. For example in [31], simulated robot soccer players adapt their positions strategically in adaptive response to the opponent's behaviour throughout the game. As for logic-based approaches, the work by Sindlar et al. [35] seems to fit our framework best: the highly structured and regulated character of robot soccer seems suitable for their abductive reasoning approach to intention recognition (more in Sect. 5). Besides abductive reasoning, intention recognition has also been modelled formally, in a language based on situation calculus [15,22]. In this approach, observed behaviour is incrementally matched to an annotated library of plans, which makes it more restricted than Sindlar's approach.

3   Formal Framework

For the formalization of the coach problem in the context of RoboCup SPL, we used Dignum's OperA framework development methodology [10], as it is formal and elaborate enough to precisely describe the system while still being flexible, reusable and adaptable to specific agent designs and future innovations.

In OperA, structures are described in a formal logic called 'logic for contract representation' (LCR), which is a combination of CTL* (computation tree logic, a temporal logic), STIT ('sees to it that') and Deontic expressions [10]. Deontic logic is the logic of norms (obligations, prohibitions, permissions, violations and sanctions), allowing reasoning about ideal states versus actual states of behaviour [18]. STIT logic is used to determine which agent should 'see to it that' a certain goal is achieved [10]. Since robot soccer is a highly regulated game with both time specific and role specific tasks and events, LCR is an appropriate language to describe it. In our domain we don't need the full specification of LCR because, for example, communication is limited in comparison to that in a human organization [33].

OperA's methodology contains several layers of design: the Organizational Model is by far the most elaborate layer, in which the characteristics of the domain are given in *social, interaction, normative* and *communicative structures*. The Social and Interaction Models are meant to instantiate specific agents and interactions for actual implementation, which is largely outside the scope of this work. OperA's methodology yields a formal model with descriptions of the roles, rules and interactions of the robot soccer system. This framework allows for extensions and adaptations of existing interactions.

3.1   Organizational Model

Roles and Dependencies. Roles are an important part of the framework. Two kinds of roles should be distinguished: facilitation roles and operational roles. Moreover, as our domain contains both humans and robots, a difference is made between roles that can be enacted by human or robot agents. Specification of domain concepts and entities is given in a domain ontology and in terms of *identifiers* respectively. The ontology is, to certain extent, developed in Protégé[1], similar to Opfer's approach for the Middle-Sized League [28]. It includes formulas describing (parts of) the field (e.g. $\forall x.isPartOf(x, OppArea) \leftrightarrow isPartOf$ $(x, OppHalf)$: all areas within the opponent's area are also within the opponent's half of the field). Identifiers are used as names for the sets of agents (both human and robot). Opponent robots are not included in the framework at this stage but can be added in future work.

The human roles of our domain are {head-referee, assistant-referee, Game Controller-operator, human-teammember} [7], whereas the robot roles are {goalkeeper, defender, attacker, coach}. The amounts of robots playing each role depends on specific team formation (except that there is always only one goalkeeper and one coach). At least one human should enact the human-teammember role, meaning that he/she is able to request for pick ups and time outs. Roles in the robot soccer domain are defined as tuples $role(r, Obj, Sbj, Rgt, Nor, tp)$ where $r \in Roles$ is the identifier of a role, $Obj \subseteq Act$ is the set of objectives of the role, $Sbj \subseteq Act$ is the set of sub-objectives sets of the role, $Rgt \subseteq Deon$ are the rights of the role and $Nor \subseteq Deon$ the norms of the role. $tp \in \{operational, institutional\}$ is the type of the role [10,33]. Institutional roles (e.g. referees) are typically enacted by impartial agents, ensuring global activity, while actors of operational roles aim to achieve their part of the society goals.

Based on these roles, agents have certain *(sub-)objectives* to achieve and *norms* to adhere to. For example, a coach should aim to send tactic messages to the players but is not allowed to leave its seated position beside the field. These norms and objectives are formalized in LCR to facilitate future implementation. Norms are based on the official RoboCup SPL rules and identified via a Norm Analysis method, yielding the responsible roles and triggers for each norm [10,33]. The coach role is given as an example in Table 1. The sub-objectives as defined in this table are merely suggestions for how to handle the plan recognition module in combination with a decisionmaking module yet to be developed. In Sect. 5 this will be discussed in more detail.

In order to achieve their objectives, enactors of roles depend on each other. These role dependencies determine the interactions that occur in the system. Role dependencies depend on the *power relations* between roles, for example, players can *request* things from one another while the coach's advice should perhaps have a higher priority. The robot soccer roles and their dependencies per objective are depicted in graph Fig. 1. For example for the dependencies written in red: the coach depends on the GameController-operator to send his messages to the players, and the head-referee depends on the assistant-referee(s) to apply his requests.

Interaction Scenes and Landmarks. The interactions, determined by role dependencies, are described as 'interaction scene scripts' and can be seen as the coordination of interactions among several roles. Scene scripts are tuples

---

[1] http://protege.stanford.edu.

Table 1. Role definition for the coach; t = window of observation, msg = message.

Role: Coach	
Role id	coach
Objectives	o1 := messaged-tactics
	o2 := followed-rules
Sub-objectives	Πo1 = ({∀p∈ Players: executed-plan-rec-module(p, role(p), t),
	got-plan(p, plan)), got-tactic-list(plan, formation, Tactics),
	decided-tactic(Tactics, tactic),got-msg(tactic, msg),
	message-sent(coach, GC-op, msg), wait(10s)}
	Πo1' =({∀p∈ Players: executed-plan-rec-module(p,t),
	got-role-map(plan(p), role(p))),
	got-formation-map(role(p), Formations),
	got-team-tactics(formation, TeamTactics),
	decided-tactic(TeamTactics, tactic), got-msg(tactic, msg),
	message-sent(coach, GC-op, msg), wait(10s)}
Rights	message-via-GC-op, decide-tactic(coach, (Team)Tactic)
Norms	PROHIBITED(coach, move(¬(head∧arms)))
	PROHIBITED(coach, communicate(coach, Robots, direct))
	PERMITTED(coach, have-clothes(anyColor, anyPattern))
	OBLIGED(coach, meet-msg-requirements(Msg, [Msg-Requirements]))
Type	operational

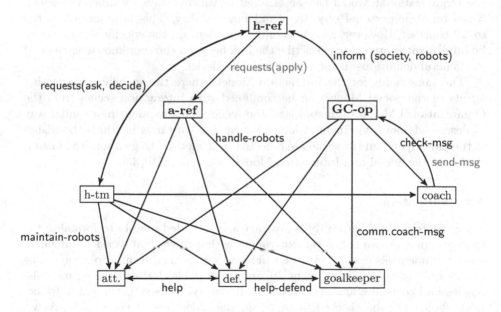

Fig. 1. Role dependency graph

scene(s,Rls,Res,Ptn,Nor) where **s** is the identifier of the scene, **Rls** is the set of identifiers of the roles enacting the scene, **Res** are the results of the scene in terms of *achievement expressions*, **Ptn** the interaction patterns (sub-achievements) and **Nor** the relevant norms of the agents in the scene. Achievement expressions are statements that describe the situation of a state after a certain goal is achieved (e.g. 'penalty-applied', 'goal-scored', 'ball-passed'). The notion of *landmarks* is used to represent such states. How exactly these landmarks have been reached is not defined at this level since it depends on the specifics of the participating agents.

In the Organizational Model, all roles, dependencies, norms and interactions of the 'robot soccer society' are formally defined. Since the entire model is quite elaborate, please see [33] for the complete framework.

3.2   Social and Interaction Models

Where the Organizational Model consists of the formal model of the society, the Social Model continues with the explicit representation of how an agent will enact such a role. At this level, the requirements and conditions of actual agents can be taken into account. When a specific agent is assigned a role, he becomes a *role-enacting agent (rea)*. The roles as described in the Organizational Model can be adjusted to the wishes of that specific agent about to enact it. For example, if an agent can only enact 'assistant-referee' for half a match, this can be adjusted in his *social contract*. That contract describes the role-enacting agent: his expected behaviour when he enacts that role. That is, a role as described in the Organizational Model can be enacted in various ways by different agents, based on their personal objectives and functionality. This is described in the social contract. However, as this is fully dependent on the specific agents yet to be implemented, we cannot describe the link between those unknown agents and the general definitions in the Organizational Model.

The same holds for the Interaction Model, where the specific role-enacting agents of the Social Model can be combined with interaction scenes from the Organizational Model to 'instantiate' the scenes. This happens in a similar way as described above, via instatiation of *interaction contracts* in which the wishes of the participants in the scene are reflected. The method to go about the instantiation of the Social and Interaction Models is given in [10,33].

3.3   Validation

Since our framework currently is abstract and intended mainly to formalise the specifications, we can not as yet experiment with it to verify it works as intended. OperA frameworks can however be validated and verified in multiple ways. The main requirements as presented in [10] are meant to check the model for inconsistencies and contradictions. Besides formally verifying its structures, the framework should also be checked to represent the objectives of the society. As we only developed the Organizational Model in detail, this verification step can only occur at the organizational level. On this level the roles, objectives and dependencies are confirmed to represent society purposes. Furthermore, we haven't found conflicts within or between the descriptions of roles, results, objectives, norms and scenes. Clearly, this has already been kept in mind in the development phase. When instantiating the Social and Interaction Models, one should make sure its contracts do not contradict the descriptions of the Organizational Model.

Besides a formal verification, a note on the robot soccer specific validation is in order. As this framework is based firmly on the official RoboCup SPL regulations and developed in collaboration with Edinferno's team (inspiration has been drawn from earlier implementations together with advice and ideas of current members), it is not only formally verified but also validated on content.

4   Plan Recognition

A necessary first step in implementing this framework is a plan recognition module for the coach. The idea here is that the coach should be able to identify a player's behaviour from observed actions only. We assume that behaviour in this sense is a set of goals and plans to achieve those goals, leading to characteristic *trajectories* on the field. The trajectories can be tracked by means of the player's self-localization module and collected in the form of a vector containing relative distances and angles to the possible goals. We choose to test on two behaviours representing the intention to go either towards the center of the 'own' goal or the 'opponent' goal. The problem is modelled as a parameterized Markov Chain $M : (S, A, T)$, where S is a finite set of states, A is a finite set of actions and T is the transition function: for any state and action $\mathbf{s} \in S$, $\mathbf{a} \in A$, the probability of each possible next state s' is $T(s' \mid \mathbf{s}, \mathbf{a})$. The state vector contains the distances and angles from the player, the action vector the distances travelled between the current state and a next state. For both behaviours, sets of states and actions are collected as training data, based on which probabilities are calculated: given a behaviour and a current state, what is the most likely next state for the player to be in. A Gaussian distribution is fitted to the training data to cover the entire soccer field for both behaviours.

In the test phase, for each observed state of the player, the most likely transition is calculated using the multivariate normal density function. Using these likelihoods, maximum log-likelihoods for entire observed trajectories are found. Behaviours are classified using Bayes' rule (Eq. 1): the behaviour (b) with the highest log-likelihood given the observations (O) is chosen as the correct one.

$$argmax_b \; p(b|O) = argmax \; p(O|b) \; p(b)$$

Decide b_1 if $p(b_1|O) > p(b_2|O)$; otherwise, decide b_2.   (1)

For the 'go to opponent goal' behaviour we collected 60 trajectories, for the 'go to own goal' behaviour 46. These were divided in a 80/20 ratio into training and validation sets, where the training sets yielded the probability distribution on the field given a certain behaviour. For 'go to opponent goal' we measured 91.6 % precision, for 'go to own goal' only 44.4 %. These numbers are the result of an initial implementation within the framework. More sophisticated models could have been used, for example Bayesian network representations [23]. Despite smoothing, some trajectories still did not represent their actual paths. That is, the logs showed jumps in positions or otherwise incorrect coordinations, mostly due to mirrorring of the field. Also, the amount of training data for the own goal behaviour is considerably less. A lot of the collected trajectories could not be used due to flaws with the player's self-localization. However, it should be noted that the differences between the maximum log-likelihoods that determined the classification decision were in most cases extremely small. A more sophisticated classifier appears necessary for stronger results in this direction.

5 Application

There are several roads to follow from here in implementing the coach. Two suggestions will be presented in this section. The first suggestion is to continue with the current plan recognition module and integrate it with a decision making and tactic adaptation module, for example in one of the following ways:

1. Tactics as independent sub-objectives
2. Tactics as dependent sub-objectives
3. Team tactics

For all these options, we assume that the coach has a library of plans and corresponding messages that he can send. The decision making module serves to compute the optimal plan given the observed current plan per single player (1 and 2) or for the entire team (3). In the first suggestion, we take the coach's advice to be obligatory for the player to perform and independent of the role of that player: the coach merely observes the game situation and decides tactic plans based on player position without including their role specific abilities. This could be a naive first test of such a decision module. The second option does consider the player's role: the coach decides optimal tactics like before, but sends messages to the role-enacting agent with the most suitable role and position to perform that tactic. The third option is to infer current roles and plans of the entire team before deciding an optimal team tactic. The players should re-arrange and divide the coach's tactic plans between themselves in a similar way as the current method of role switching (e.g. based on position relative to the ball). For example, if the coach sends a plan involving passing a ball, then the robot that is currently closest to the ball is in the best position to execute that plan.

However, should one want to connect the logical framework to a logical method of plan recognition, we suggest to continue along the lines of the work done by Sindlar et al. [35]. They propose a method for intention recognition via *mental state abduction* (MSA), were agents are assumed to be BDI agents

(Belief/Desires/Intentions [29]). Based on observations combined with knowledge of the rules and roles, it can be inferred why an agent performs a certain action.

MSA uses 'answer set programming' (ASP) for nonmonotonic, abductive reasoning in the agent programming language (2)APL [8]. In this language, a goal achievement rule of the form $n : \gamma \leftarrow \beta \mid \pi$, where n is the identifier of the rule, γ the goal to be achieved, β the beliefs that should be true in order to be allowed or able to perform plan π, where π generates observable action sequences leading to achieving that goal γ. That is, a plan can generate multiple observable sequences and also multiple *computation sequences*, meaning that various different routes can lead to the same goal. There is a subtle difference between *observable* and *seen* actions: an observable action is a possible observation (something that is possible according to the theory), while seen actions have actually been observed. Intuitively, if an action is seen, it should also be observable [35].

From these APL rules, a translation step is made to a logical theory and subsequently to a logical program in ASP style, making it directly implementable. We will give a short example for our domain to explain how MSA works. Consider the rule $R = \{$1: hold-ball $\leftarrow$ in-ownPA(g) and in-ownPA(b) | move(g,b) ; if B(opponent-near) then pickup(g,b) else skip$\}$, which would lead to the answer set program P_R:

```
2{g(hold-ball,0), b(conj(in-ownPA(g), in-ownPA(b)),0)}2 :- r(1,1)
2{g(hold-ball,0), b(conj(in-ownPA(g), in-ownPA(b)),0)}2 :- r(1,2)
                 2{o(move(g,b),1), o(pickup(g,b),2)}2 :- r(1,1)
                            1{o(move(g,b),1)}1 :- r(1,2)
                                 1{r(1,1), r(1,2)}1.
                                 :- s(A,T), not o(A,T),
```

where the last statement says that candidate answer sets that were seen, but not deemed observable at step T should be discarded. The reason that r(1,1) and r(1,2) are the same, is because the plan of this rule has two possible computation sequences, represented by the two possible observations given (depending on the belief of 'opponent-near'). The belief whether or not the opponent is near is a different kind of belief than β since it is a test case in the plan (π) part of rule R: B(opponent-near) does not need to be satisfied to execute rule R, while beliefs β (in-ownPA(g) and in-ownPA(b)) should.

Let us say we have seen the goalkeeper moving towards the ball: $P' = P \cup \{$s(move(g,b),1)$\}$. This can be explained by both r(1,1) and r(1,2). Next, we see him picking up the ball: $P'' = P' \cup \{$s(pickup(g,b), 2)$\}$; this can only be explained by r(1,1). We can now infer $P'' \models$ goal(hold-ball)$\wedge$bel(conj (in-ownPA(g), in-ownPA(b))) $\wedge$bel(opponent-near), revealing the mental state of our goalkeeper based on observed actions and known rules.

6   Discussion

The organizational MAS development method 'OperA' yielded a grounded formalization of the robot soccer society and a detailed description of its roles and coordination. However, a drawback of this method is that guidelines for actual implementation are not provided. Recent extensions like OMNI [11] and OperettA [27] explore possible means of implementation for such frameworks. The fact that agent designs are left out of the framework is presented as an advantage, as it gives the designer freedom to adjust the agents to his needs while they can still be used in the general yet formal framework [10]. This way, there are multiple options for future work, like the modular approach we started on or the BDI/abduction approach we suggested above.

This formalization of the coach role has been largely based on RoboCup's regulations [7]. Requirements for the coach's messages have been mentioned in [33], but the content of these messages has not been defined yet. This depends on the decisions made in the actual development of the coach: what would be most valuable for him to say in order to help the team. The format for the messages as described in the regulations has been included in the framework. All other specifics as mentioned in the regulations are included in this formalization [33]. Examples and additions to these specifics are inspired by (former) RoboCup members.

The plan recognition module as presented above is merely a naive first pass implementation that could be improved in several ways. For example, our initial implementation assumed that the coach has access to the player's self-localization logs, while ideally he should be able to use actual observations. Furthermore, we only tested on static goals, while ball behaviour would be more informative to decide tactics. However, from this first step we gained some important insights into the difficulties of the coach problem.

7   Conclusion and Further Research

In general, the OperA framework has shown to be very suitable to describe robot soccer in terms of roles and interaction structures. The coach role, as described by RoboCup's regulations, has been formalized within this framework. As it leaves agent designs undefined, the next step is to choose for example between the modular approach introduced in Sect. 4 and the logical approach based on BDI agents suggested in Sect. 5. As robot soccer is a highly regulated game, and all robots should have a knowledge base of the rules in any case, it seems intuitive to continue along the lines of mental state abduction. So far we only spoke of full observable sequences, but work on sequences with gaps (partially observable) already exists [34]. Besides implementation of a new and improved plan recognition module, the tactic decision making problem is also subject for further research.

References

1. Beetz, M., Bandouch, J., Gedikli, S.: Camera-based observation of football games for analyzing multi-agent activities. In: Proceedings of AAMAS 2006 (2006)

2. Bowling, M., McCracken, P.: Coordination and adaptation in impromptu teams. In: Proceedings of AAAI 2005, pp. 53–58 (2005)
3. Bui, H.H.: A general model for online probabilistic plan recognition. In: International Joint Conferences on AI, vol. 3, pp. 1309–1315 (2003)
4. Carberry, S.: Techniques for plan recognition. User Model. User-Adap. Inter. **11**, 31–48 (2001)
5. Carmel, D., Markovitch, S.: Model-based learning of interaction strategies in multi-agent systems. J. Exp. Theor. Artif. Intell. **10**(3), 309–332 (1998)
6. Charniak, E., Goldman, R.P.: A bayesian model of plan recognition. Artif. Intell. **64**, 53–79 (1993)
7. RoboCup Technical Committee. Robocup standard platform league (nao) rule book (2013)
8. Dastani, M.M.: 2apl: a practical agent programming language. Auton. Agent. Multi-Agent Syst. **16**(3), 214–248 (2008)
9. Desheng, X., Keijan, X.: Role assignment, non-communicative multi-agent coordination in dynamic environments based on the situation calculus. In: Proceedings of the WRI Global Congress on Intelligent Systems, vol. 1, pp. 89–93 (2009)
10. Dignum, V.: A model for organizational interaction: based on Agents, founded in Logic. Ph.D. Thesis, Utrecht University (2004)
11. Dignum, V., Vázquez-Salceda, J., Dignum, F.: OMNI: introducing social structure, norms and ontologies into agent organizations. In: Bordini, R.H., Dastani, M., Dix, J., El Fallah Seghrouchni, A. (eds.) PROMAS 2004. LNCS (LNAI), vol. 3346, pp. 181–198. Springer, Heidelberg (2005)
12. Esteva, M., Rodríguez-Aguilar, J.-A., Sierra, C., Garcia, P., Arcos, J.-L.: On the formal specification of electronic institutions. In: Sierra, C., Dignum, F.P.M. (eds.) AgentLink 2000. LNCS (LNAI), vol. 1991, pp. 126–147. Springer, Heidelberg (2001)
13. Ferber, J., Gutknecht, O.: A meta-model for the analysis and design of organizations in multi-agent systems. In: Proceedings of the 3rd International Conference on Multi-Agent Systems (ICMAS 1998) (1998)
14. Genter, K., Agmon, N., Stone, P.: Role-based ad hoc teamwork. In: Proceedings of PAIR-11 (Workshop at AAAI) (2011)
15. Goultiaeva, A., Lespérance, Y.: Incremental plan recognition in an agent programming framework. In: Proceedings of Plan, Activity and Intent Recognition (PAIR) (2007)
16. Grüninger, M., Fox, M.S.: Methodology for the design and evaluation of ontologies. In: Proceedings of the Workshop on Basic Ontological Issues in Knowledge Sharing, IJCAI 1995 (1995)
17. Isik, M., Stulp, F., Mayer, G., Utz, H.: Coordination without negotiation in teams of heterogeneous robots. In: Lakemeyer, G., Sklar, E., Sorrenti, D.G., Takahashi, T. (eds.) RoboCup 2006: Robot Soccer World Cup X. LNCS (LNAI), vol. 4434, pp. 355–362. Springer, Heidelberg (2007)
18. Dignum, F.P.M., Meyer, J.-J.C., Wieringa, R.J.: The role of deontic logic in the specification of information systems. In: Chomicki, J., Saake, G. (eds.) Logics for Databases and Information Systems, pp. 71–115. Kluwer Academics Publishers, Norwell (1996)
19. Johnston, B., Yang, F., Mendoza, R., Chen, X., Williams, M.-A.: Ontology based object categorization for robots. In: Yamaguchi, T. (ed.) PAKM 2008. LNCS (LNAI), vol. 5345, pp. 219–231. Springer, Heidelberg (2008)
20. Kautz, H.A., Allen, J.F.: Generalized plan recognition. In: Proceedings of AAAI 1986, vol. 86, pp. 32–37 (1986)

21. Kok, J.R., Spaan, M.T.J., Vlassis, N.: Non-communicative multi-robot coordination in dynamic environments. Robot. Auton. Syst. **50**(2–3), 99–114 (2005)
22. Levesque, H., Pirri, F., Reiter, R.: Foundations for the situation calculus. Comput. Inf. Sci. **3**(18) (1998)
23. Liao, L., Patterson, D.J., Fox, D., Kautz, H.: Learning and inferring transportation routines. Artif. Intell. **171**(5–6), 311–331 (2007)
24. MacAlpine, P., Barrera, F., Stone, P.: Positioning to win: a dynamic role assignment and formation positioning system. In: Chen, X., Stone, P., Sucar, L.E., van der Zant, T. (eds.) RoboCup 2012. LNCS, vol. 7500, pp. 190–201. Springer, Heidelberg (2013)
25. Mohr, M., Krustrup, P., Bangsbo, J.: Match performance of high-standard soccer players with special reference to development of fatigue. J. Sports Sci. **21**(7), 519–528 (2011)
26. Odell, J.J., van Dyke Parunak, H., Fleischer, M.: The role of roles in designing effective agent organizations. In: Garcia, A.F., de Lucena, C.J.P., Zambonelli, F., Omicini, A., Castro, J. (eds.) SELMAS 2002. LNCS, vol. 2603, pp. 27–38. Springer, Heidelberg (2003)
27. Okouya, D., Dignum, V.: Operetta: a prototype tool for the design, analysis and development of multi-agent organizations (demo paper). In: Proceedings of the 7th International Conference on Autonomous Agents and Multiagent Systems (2008)
28. Opfer, S.: Towards description logic reasoning support for ALICA. Master's thesis, Universität Kassel (2012)
29. Rao, A.S., Georgeff, M.P.: Modeling rational agents within a BDI-architecture. In: KR 1991, pp. 473–484 (1991)
30. Reis, L.P., Lau, N.: COACH UNILANG - a standard language for coaching a (Robo)Soccer team. In: Birk, A., Coradeschi, S., Tadokoro, S. (eds.) RoboCup 2001. LNCS (LNAI), vol. 2377, pp. 183–192. Springer, Heidelberg (2002)
31. Riley, P., Veloso, M.M.: Recognizing probabilistic opponent movement models. In: Birk, A., Coradeschi, S., Tadokoro, S. (eds.) RoboCup 2001. LNCS (LNAI), vol. 2377, pp. 453–458. Springer, Heidelberg (2002)
32. Saria, S., Mahadevan, S.: Probabilistic plan recognition in multiagent systems.In: Proceedings of ICAPS 2004, AAAI, pp. 287–296 (2004)
33. Schropp, G.Y.R.: Agent organization framework for coordinated multi-robot soccer. Master's thesis, Utrecht University (2014)
34. Sindlar, M.P., Dastani, M., Dignum, F.P.M., Meyer, J.-J.C.: Mental state abduction of BDI-based agents. In: Baldoni, M., Son, T.C., van Riemsdijk, M.B., Winikoff, M. (eds.) DALT 2008. LNCS (LNAI), vol. 5397, pp. 161–178. Springer, Heidelberg (2009)
35. Sindlar, M.P., Dastani, M.M., Meyer, J-J.Ch.: Programming mental state abduction.In: Proceedings of the 10th International Conference on Autonomous Agents and Multiagent Systems, vol. 1, pp. 301–308 (2011)
36. Stone, P., Veloso, M.: Task decomposition, dynamic role assignment, and low-bandwidth communication for real-time strategic teamwork. Artif. Intell. **110**(2), 241–273 (1999)
37. Visser, U., Drücker, C., Hübner, S., Schmidt, E., Weland, H.-G.: Recognizing formations in opponent teams. In: Stone, P., Balch, T., Kraetzschmar, G.K. (eds.) RoboCup 2000. LNCS (LNAI), vol. 2019, pp. 391–396. Springer, Heidelberg (2001)
38. Weigel, T., Rechert, K., Nebel, B.: Behavior recognition and opponent modeling for adaptive table soccer playing. In: Furbach, U. (ed.) KI 2005. LNCS (LNAI), vol. 3698, pp. 335–350. Springer, Heidelberg (2005)

39. Wooldridge, M., Jennings, N., Kinny, D.: The gaia methodology for agent-oriented analysis and design. J. Auton. Agent. Multi-Agent Syst. **3**(3), 285–312 (2000)
40. Zhu, G., Xu, C., Huang, Q., Gao, W.: Automatic multi-player detection and tracking in broadcast sports video using support vector machine and particle filter. In: Proceedings of the International Conference on Multimedia and Expo, pp. 1629–1632 (2006)

Is that Robot Allowed to Play in Human Versus Robot Soccer Games

Laws of the Game for Achieving the RoboCup Dream

Tomoichi Takahashi[1]($\boxtimes$) and Masaru Shimizu[2]

[1] Meijo University, Shiogamaguchi, Nagoya, Tempaku-ku 468-8502, Japan
ttaka@meijo-u.ac.jp
[2] Chukyo University, YagotoHonmachi, Nagoya, Showa-ku 466-8666, Japan
shimizu@sist.chukyo-u.ac.jp

Abstract. Many RoboCuppers share a dream: a team of fully autonomous humanoid robot soccer players that is capable of playing soccer games against human players by 2050. The demonstration of a human versus robot soccer game at RoboCup 2007 was an exciting display of the progress that has been in RoboCup community since 1997 in terms of the robot's autonomy, sensing ability, and physical features. This game also uncovered several new issues. For example, do human soccer players play games against robots in the same way that they would if they were playing against human players?

In this study, we investigate the features of RoboCup soccer leagues and examine whether these features are necessary and sufficient to realize the dream of RoboCuppers. We compare the current RoboCup soccer leagues to human soccer leagues and discuss metrics that indicate the similarity between robot soccer games and human soccer games. In other words, we discuss our progress toward realizing the dream. In addition, we propose amendments of laws for human and robot soccer that will deal with human-related issues and facilitate research toward achieving the dream.

1 Introduction

Many RoboCuppers share a dream: a team of fully autonomous humanoid robot players that is capable of winning a soccer game against the winner of the most recent World Cup while complying with the official FIFA rules by 2050 [15]. In RoboCup 2007, the first robot versus human soccer game was played between robots from the Middle Size League (MSL) and humans [19]. The game excited RoboCuppers but also brought about the fact that the RoboCup dream implicitly assumes that robots will be able to run as fast as humans and kick the ball as well as humans. The MSL uses the size 5 soccer ball, which is also used in human games. This was one of the factors that made the exhibition games possible. Other factors like the field size, floor and lighting conditions, and mechanical specifications of the robots, such as speed, size, and shape, are also important in realizing robot versus human soccer games.

© Springer International Publishing Switzerland 2015
R.A.C. Bianchi et al. (Eds.): RoboCup 2014, LNAI 8992, pp. 358–368, 2015.
DOI: 10.1007/978-3-319-18615-3_29

With each passing year, these factors have become increasingly similar to those that are used in human soccer games. The progress of the RoboCup soccer leagues were reported and issues to achieve the RoboCup dream have been reviewed [10,18]. Most of the issued that were reviewed were technical ones. On the other hand, safety to human players and abilities of robot players were questioned: Are they allowed to have omnidirectional visions? How much power is allowed to for kicks? However, these questions have not been discussed further, and the safety of human players in human versus robot soccer games remains unsettled.

Human players seem to be apprehensive about robots approaching at full speed as an iron block, and tend to be extra careful in order to avoid injuries due to the robots in the human versus robot soccer game. The robot's movements can potentially injure human players. This factor has prompted us to consider the themes presented in this paper. Player safety is important in soccer games and must be ranked as the highest priority in the event of robot-human contact. This leads to the following questions:

Q1: Can we realize our dream given the present league rules?
Q2: What challenges should we consider and promote in order to achieve our goals for robot versus human soccer games?

We believe that now is a good time for us to examine our approach and to clarify what must be done in order to realize the RoboCup dream by 2050. In this study, we compare the RoboCup robot soccer leagues to human soccer leagues, describe issues in the existing laws of the games, and propose new laws for human versus robot soccer games, which will enable us to achieve the RoboCup dream. Section 2 describes robots from other fields and the rules that were initially established for the use of those robots. Section 3 discusses differences in the perspectives of humans and robots with robot soccer. Section 4 discusses our proposals for amending the laws of the games in order to realize the dream and related challenge issues. Finally, Sect. 5 concludes the paper with a summary of our proposals.

2 Background and Related Works

We expect robots to eventually become a pervasive part of our lives. For example, in the automotive fields, Google has been presenting and testing driverless cars in many cities since the success of the DARPA Urban Challenge [1,5]. Driverless cars have been presented at exhibitions, while laws and guidelines for driverless cars have been drafted [6]. There are numerous issues that need to be resolved for us to use the driverless cars. For example, who is to blame if the driverless car is involved in an accident? Should the automaker that designed the technology, the car's owner, or a passenger accept responsibility?

For the standardization, terms such as "service robots" and "mobile robots" are being defined, and co-ordinate systems for mobile service robots are being discussed [12,13]. With regard to safety, protective measures and system requirements are being examined. This involves considering the possibilities of incorrect autonomous actions, contact with moving components, stopping and speed/force restriction. In ISO 8373 standards for robots and robotic devices, it is clearly stated that the robot shall either be designed to ensure either a maximum dynamic power of 80 W, or be designed to maintain a separation distance from the operator. The ISO statement is clear for designing robot soccer players, however human soccer players cannot sense how much force we apply.

At the beginning of the RoboCup initiative, Kitano et al. said that RoboCup poses significant long-term challenges that will take a few decades to meet. Due to the clarity of the final target, several sub-goals can be derived, which in tern lead to short-term and mid-term challenges [14]. The three short-term challenges involve synthetic agents, physical agents and infrastructure challenges. For the synthetic agent challenge, learning, teamwork, and opponent modeling challenges were proposed. For phase I of the physical agent challenge, the following three challenges were set [8]:

1. Moving the ball to a specified area with no, stationary, or moving obstacles,
2. Receiving the ball from an opponent or teammate, and
3. Passing the ball between two players.

The robot versus human soccer game at 2007 showed that these challenges have been achieved to some extent [7,16,19]. Stone et al. presented a paper on robot soccer followed by the 2007 human robot soccer game [17]. The topics covered a range of issues: the physical size of robots that are not bigger than human players, confidence of human players that they are not more likely to sustain an injury than were playing with people and funs who enjoy soccer game, rules that are suitable human and robot players and so on.

We also wondered whether the human players were apprehensive that robots would injure by playing with robots. This feeling applies to other leagues that have interactions with human and robots. In the RoboCup@Home league, robots showcase their abilities in homes, shopping malls, and public parks rather than in RoboCup competition venues. Robots soccer leagues are conducted in accordance with the official FIFA soccer rules. The robots are designed with considering not injuring human. Following Asimov's Three Laws of Robotics may express one of design guidance [9].

1. A robot must not harm a human. And it must not allow a human to be harmed.
2. A robot must obey a human's order, unless that order conflicts with the First Law.
3. A robot must protect itself, unless this protection conflicts with the First or Second Laws.

Soccer players compete with their opponents to gain control of the ball. Haddadin et al. discussed contact plays from physical human-robot interaction and studied joint elasticity to kick the ball for human-level soccer [11]. In addition to developing such mechanisms, we believe that rules similar to the Three Laws of Robotics are required to ensure the safety of human players.

3 Differences Between Human Versus Human and Human Versus Robots Soccer Games

3.1 Comparisons with Physical Aspects of Human Soccer Games

The human versus robot demonstration game involved five players per team. Table 1 lists the specifications of three different soccer games. The first category is for human soccer games. The table list data for leagues in the professional, junior high schools, and elementary categories. The second category are data for futsal that is played on a smaller field and mainly played indoor and blind soccer games. The third and fourth categories are data from RoboCup soccer games from real robot leagues and simulation leagues, respectively.

Parameters such as field size, running speed of players, and ball size indicate how close we are to achieving the RoboCup dream. The values in Table 1 are standard values or are taken from the rules and manuals of the leagues. For MSL and the Small Size League (SSL), the field sizes in 2003 and 2010 are listed to indicate the progress in each league. W, D of P_{size} denote the width and depth of players, respectively and H denotes the height of human players. The following metrics were set in order to normalize the game specifications.

Table 1. Comparison of parameters of human and robot soccer Games.

Leagues	Field Size (m × m)	Players	P_{size} (W × D(×H)cm)	P_{speed} (m/s)	P_{number} (m²/player)	P_{space}	P_{grid}	P_{time} (s)
Human full-size games [*1]								
Professional	105 × 68	11 vs. 11	50 × 30 × 180	9.09	325	210 × 227	2,164	12
JuniorHighS.	70 × 50	11 vs. 11	30 × 20 × 130	6.25	159	233 × 250	2,652	11
ElementaryS.	68 × 50 (halfsize)	8 vs. 8	30 × 20 × 100	4.50	213	227 × 250	3,542	15
Human mini-soccer games								
Futsal	40 × 20	5 vs. 5	50 × 30 × 180	9.09	80	80 × 67	533	4
Blid soccer [*2]				8	80			5
RoboCup Real Robot League								
Middle (2010)	18 × 12	5 vs. 5	30 × 40	3.00	22	36 × 30	108	6
(2003)	10 × 5		to 50 × 80	3.00	5	20 × 13	25	3
Small (2010)	6.05 × 4.05	5 vs. 5	18 × 14	3.00	2.5	34 × 23	76	2
(2003)	2.8 × 2.3			1.00	0.6	16 × 13	20	3
SPL/Humanoid	6 × 4	3 vs. 3	10 × 5 × 58	0.10	4.0	60 × 80	800	60
RoboCup Simulation League								
2D	105 × 68	11 vs. 11	60 × 60	1.20	325	175 × 113	902	88
3D	105 × 68	11 vs. 11	140 × 96 × 210	2.00	325	75 × 71	241	53

$P_{number} = \frac{Field\ Size}{Players}$, $P_{space} = \frac{Long\ Side\ of\ Field}{Player\ Width} \times \frac{Short\ side\ of\ Field}{Player\ Depth}$, $P_{grid} = \frac{P_{space}}{Players}$,
$P_{time} = \frac{Field\ long\ side}{P_{Speed}}$

[*1]: Students of Junior High and elementary school play soccer according to modified rules: for example cf http://www.usyouthsoccer.org/coaches/RulesSmallGames/
[*2]: There are three leagues: B1 and B2/3, for completely blind and fully sighted or partially sighted persons.

- P_{number} denotes the area size per player.
- P_{space} denotes the ratio of the field's long side to the player width and the ratio of the field's short side to the player depth. The numbers indicate the number of grids necessary to represent the movements of players.
- P_{grid} denotes the number of grids allocated to one player. This number corresponds to the size of area that agents use to monitor the changes in environments for playing soccer games.
- P_{time} denotes the time taken by a player to run from one end line to the other. This is related to the mobility and stamina of players.

P_{size} and P_{speed} are different for the three levels of human soccer leagues: professional, junior high school, and elementary school. The relative parameters, P_{space}, P_{time}, and P_{grid}, of the three levels of human leagues have a similar order of values. This implies that the rules are set such that the settings of games match the physical abilities and sizes of human players. The values for

Fig. 1. Pictures from local(left) and global(right) vision Systems. In a zigzag way from the left top screen shot to the right bottom one, they are snapshots of collision between two robots. The last picture was taken as the robot was falling backwards and most of the visual information is from the sky.

Table 2. Potential amendments of rules and time schedule to achieve the DREAM.

(a) Potential Rule Changes

FIFA LAWs	categories of rule amendments and others
1 The Field of Play	INP
2 The Ball	
3 The Number of Players	INP
4 The Players' Equipment	ESH
5 The Referee	
6 The Assistant Referees	CRT
7 The Duration of The Match	
8 The Start and Restart of Play	kick-in instead of throw-in in Futsal
9 The Ball In and Out of Play	
10 The Method of Scoring	
11 Offside	no offside rule in Futsal
12 Fouls and Misconduct	ESH
13 Free Kicks	
14 The Penalty Kick	
15 The Thrown-in	CRT
16 The Goal Kick	
17 The Corner Kick	

(b) Expected Schedule of Events to 2050

year	event
1997	start of RoboCUP
2007	1st Human vs. robot soccer games
2013	
2015	Draft of RoboCup Soccer Rules for human versus robot games <Transitional phase>
2020	Competition compliant with RoboCup Soccer Rules
2030, 2040, 2050	<Major updating of Rules per decade to Achieving Dream>

Futsal games rank between human soccer leagues and RoboCup soccer leagues. This provides a good example for determining the rules for RoboCup soccer games and working toward achieving the dream.

3.2 Local-Global Vision as an Example of Difference in Sensing Ability

In the early stages of MSL, all teams used video cameras. Now all teams use omnidirectional vision systems and laser range finder systems. In the SSL, two types of vision systems were used at the beginning. Some teams used video camera systems mounted on robots and other teams used a global vision system that provided a bird's eye view of the game. After the first a couple of competitions, all team used the global vision system. At present, vision cameras are set over the field by rules and the images from the cameras have been provided to teams.

Figure 1 shows images taken in a 3D simulation soccer games. In this league, individual robot players receive information around them from a server. The pictures on the left are pictures that contain the same information from the server. These pictures are obtained from a camera that is virtually attached to a player. The right pictures are images of global vision system that are displayed to spectators.

The local vision that human use is clearly inferior to the global or omnidirectional vision that robots use. In the case of Fig. 1, different commands are generated 46 times during 300 steps between using sensing data from global and local vision systems. This issue presents us with questions about extra sensing abilities that are available to robots and whether they should be permitted in human versus robot soccer games.

4   Proposal to Achieve Human Versus Robot Soccer Games

4.1   Rule Amendments and Time Schedule

The FIFA Laws of the Game include seventeen laws that address subjects ranging from field to plays to the corner kick. Burkhard et al. compared the FIFA rules and RoboCup rules and showed the main differences are the playing field, the number and skills of players, etc. They also pointed the difference of field size and the number of players would be used to measure the progress of robot soccer [10].

The FIFA Laws of the Game empower a referee to interpret the rules at his/her decision and the referee has full authority to control a match to which he/she is assigned [3]. For example, a direct free kick is awarded to the opposing team if a player commits one of seven offenses in a manner that the referee considered to be careless, reckless or using excessive force. One of the seven offenses is pushing an opponent. The interpretations of the FIFA Laws of the Game vary from one referee to another. It is interesting to note that the latter half of these laws includes a section about their interpretation and a section about guidelines for standardizing referees' decisions.

The left column of Table 2 (a) lists the seventeen laws. The right column of the tables lists the laws that we think will need to be amended in order to achieve the dream. We believe that laws 1, 3, 4, 6, 12 and 15 are potential ones to be amended and they are categorized to three groups:

INP(Introduction of New Parameters based on metrics) - Law 1 and 3 -:
It is a goal to play soccer games at fields fully compliant with the FIFA rules, however it takes a lot of space and money for teams. The metrics in Sect. 3.1 describe the features of human soccer games. Field size and the number of players need to be different from those specified by FIFA, but the metrics calculated from them should be similar to values for real soccer games, as listed in Table 1.

ESH(Examination from Safety of Human players) - Law 4 and 12 -:
We believe that there are two categories of issues related to safety. The first category deals with technical issues, which are concerned with how well the robots play. The second category deals with the mental approach of the

human players. These issues are concerned with the risks of injury and how those risks affect the way that human players play the game.

CRT(Challenge Research Themes) - Law 6 and 15 - :
In SSL league and simulation leagues, programmed computers have refereed games. For this discussion, we will assume that the referee is human and that one of the assistant referees is a robot that has the required software installed. The robot referee checks not only goals but also offside infraction in order to help human referees [2]. In humanoid leagues, throw-in-lay has been challenged [4].

Rule revisions influence research themes of participants and the operations of the RoboCup soccer leagues. Before anything else, RoboCuppers share what really needs to achieve the dream from the viewpoints of technically themes to safety of human players. Table 2 (b) shows our proposal how to proceed further discussions and rule changes.

- By 2015, a draft of RoboCup rules for human versus robot games will be presented. During the subsequent five years, every league will be required to change their rules to comply with the RoboCup 2015 rule set.
- From 2020, games will be played according to the 2015 rules. And the rules will be updated once per decade to achieve the dream.

4.2   What Should Be Considered in Order to Gain Acceptance from Human Teams

Soccer players compete with their opponents to take the ball. This implies real tackles, collisions and fouls between humans and robots. Law 12 (Fouls And Misconduct) prohibits players from making severe body contact that would hurt opponents or themselves. It does not seem feasible for robots to apply Asimov's Three Laws of Robotics to soccer games. We believe that laws similar to the Three Laws of Robotics are needed in order to provide human players with sense of ease to play games with robots. The following issues should be considered in examining the amendment of laws.

1. Equality of Physics:
   According to the action-reaction law in physics, the same force acts on both a human and robot when they collide with each other. This leads to the conclusion that the mass, speed, and acceleration of robots should be restricted to those comparable to human players. Skin and muscles make collisions between human players softer than those that would occur with robots having protective shells made of metal or plastics. Collision mitigation should be taken into considerations when setting the rules of robot soccer.
2. Ease of Playing:
   Players predict the next movement and action of opponents when they play games. Omnidirectional vision systems and movement mechanisms cause robots to move in ways that are distinctly different and also unpredictable. As a result, humans players often do not have a clue about what the robot's

next movement will be. This makes human players feel uneasy when playing with robots. It is important for human players to feel at ease while playing against robotic opponents.

3. Communication and Accountability:
In the unfortunate cases where human players get injured, the players are required to describe the situations to a referee or show that the play was not an intentional one. This includes not only communication to the referee but also explanation of which player did what kind of plays and where. The need for an account for one's own play and related plays requires robots to generate and record log data in text form. The data corresponds to memory of human players and other players recorded at consecutive steps are identified each other.

4.3 Challenges in Human Versus Robot Soccer

As we have described, human versus robot games and robot versus robot soccer games have different features. From a viewpoint of human versus robot soccer games, it is important to get acceptance form a human teams to play with robots. For the above second and three issues, followings will be agenda to be discussed:

1. Clue-giving Mechanism:
The direction of the gaze or body is an important factor that we use in the predictions of others' movements. A robot should have a mechanism to give clues about its future motions that will let human players predict the next motion of the robot. The direction of the face, body, or camera of the robot should be linked to the next motion of the robot for predictability and ease of playing.
2. Hostile (unfair) Play and Summarizing Ability:
Blocking an opponent player may be judged as a foul. A player with a captain's band can appeal a refereeing decision. An appeal should address the 5W1H questions(who, what, when, where, why and how). When a human player gets hurt, the robot player (or designer of the robot) explains that the plays are not intentional. This requires a textual summarization of the sequences of recorded plays in a natural language.

5 Discussion and Summary

The RoboCup community has been working toward achieving the dream of seeing robots playing soccer games against humans using the FIFA rules. The demonstration games between robots and humans that started at RoboCup 2007 and have continued showed that robots can play soccer games with humans at the level of an elementary school team. Honestly, one of authors was drawn strategically into what the robot teams of RoboCup 2014 champion would do next to human players as a spectator. The game uncovered several new issues regarding safety and the mental approach of the human players. Soccer is a contact sport and humans get injured sometimes during games. At robot-human soccer games, we expect robots to play competitively. Robots will be penalized by referees when they injure human players even if unintentionally.

This paper surveys issues on human-robot interaction from the view of human safety to realize our dream and raises the following questions:

Q1: Can we realize our dream given the present league rules?

Q2: What challenge should we consider and promote in order to achieve our goals for robot versus human soccer games?

This paper introduces various issues and themes that we must discuss in order to achieve the RoboCup dream in a concrete manner. We hope that this paper makes will serve a starting point for discussions about rules for human-robot soccer games and setting metrics that show how the leagues reach to the dream. And these outputs are assumed to support common functions of service robots that will coexist with us and will assist us in daily life.

References

1. http://archive.darpa.mil/grandchallenge, 25 April 2014
2. Goal line technology. http://www.fifa.com/aboutfifa/news/newsid=2310048/, 25 April 2014
3. Laws of the game. http://www.fifa.com/mm/document/footballdevelopment/refereeing/81/42/36/log2013en_neutral.pdf, 25 April 2014
4. Robocup soccer humanoid league rules and setup for the 2012 competition in mexico city. http://www.tzi.de/humanoid/pub/Website/Downloads/HumanoidLeagueRules2012_DRAFT_20111223.pdf, 25 April 2014
5. The self-driving car logs more miles on new wheels. http://googleblog.blogspot.jp/2012/08/the-self-driving-car-logs-more-miles-on.html, 25 April 2014
6. Liability issues create potholes on the road to driverless cars. Wall Street Journal Online, New York, 27 January 2013
7. Humans against Robots RoboCup 2009. http://www.youtube.com/watch?v=YI9xnZ9msn8, 25 April 2014
8. Asada, M., Kuniyoshi, Y., Drogoul, A., Asama, H., Mataric, M., Duhaut, D., Stone, P., Kitano, H.: The robocup physical agent challenge: Phase-I. Appl. Artif. Intell. **12**, 251–263 (1998)
9. Asimov, I.: I, Robot. BANTAM BOOKS
10. Burkhard, H.-D., Duhaut, D., Fujita, M., Lima, P., Murphy, R., Rojas, R.: The road to robocup 2050. Robot. Autom. Mag. **9**(2), 31–38 (2002)
11. Haddadin, S., Laue, T., Frese, U., Wolf, S., Albu-Schäffer, A., Hirzinger, G.: Kick it with elasticity: safety and performance in human-robot soccer. Robot. Auton. Syst. **57**(8), 761–775 (2009)
12. ISO:8373:2012. Robots and robotic devices - vocabulary
13. ISO:9787:1999. Manipulating industrial robots - coordinate systems and motion nomenclatures
14. Kitano, H., Tambe, M., Stone, P., Veloso, M., Coradeschi, S., Osawa, E., Matsubara, H., Noda, I., Asada, M.: The robocup synthetic agent challenge 97. In: International Joint Conference on Artificial Intelligence (IJCAI 1997) (1997)

15. RoboCup. http://www.robocup.org/02.html
16. RoboCup 2010: Man Vs Robot. http://www.youtube.com/watch?v=xxW78vVTIZ8, 25 April 2014
17. Stone, P., Quinlan, M., Hester, T.: Can robots play soccer? In: Richards, T. (ed.) Soccer and Philosophy: Beautiful Thoughts on the Beautiful Game (Popular Culture and Philosophy), vol. 51, pp. 75–88. Open Court Publishing Company, Chicago (2010)
18. Visser, U., Burkhard, H.-D.: RoboCup: 10 Years of Achievements and Future Challenges. AI Mag. **28**(2), 115–132 (2007). Summer
19. Humans vs Mid Size Robots at RoboCup '07 Atlanta. http://www.youtube.com/watch?v=ApspTluZO4Y, 25 April 2014

Towards Rapid Multi-robot Learning from Demonstration at the RoboCup Competition

David Freelan, Drew Wicke, Keith Sullivan, and Sean Luke[✉]

Department of Computer Science, George Mason University,
4400 University Drive MSN 4A5, Fairfax, VA 22030, USA
{dfreelan,dwicke,ksulliv2}@gmu.edu, sean@cs.gmu.edu

Abstract. We describe our previous and current efforts towards achieving an unusual personal RoboCup goal: to train a full team of robots directly through demonstration, on the field of play at the RoboCup venue, how to collaboratively play soccer, and then use this trained team in the competition itself. Using our method, HiTAB, we can train teams of collaborative agents via demonstration to perform nontrivial joint behaviors in the form of hierarchical finite-state automata. We discuss HiTAB, our previous efforts in using it in RoboCup 2011 and 2012, recent experimental work, and our current efforts for 2014, then suggest a new RoboCup Technical Challenge problem in learning from demonstration.

Imagine that you are at an unfamiliar disaster site with a team of robots, and are faced with a previously unseen task for them to do. The robots have only rudimentary but useful utility behaviors implemented. You are not a programmer. *Without coding them,* you have only a few hours to get your robots doing useful collaborative work in this new environment. How would you do this?

Our interest lies in rapid, real-time multi-robot training from demonstration. Here a single human trainer teaches a team of robots, via teleoperation, how to collectively perform tasks in previously unforeseen environments. This is difficult for two reasons. First, nontrivial behaviors can present a high-dimensional space to learn, yet one can only provide a few samples, as online training samples are costly to collect. This is a worst case for the so-called "curse of dimensionality". Second, when training multiple interactive robots, even if you can quantify the emergent *macro-level* group behavior you wish to achieve, in order to do learning, each agent needs to know the *micro-level* behavior he is being asked to do. One may have a micro→macro function (a simulator), but it is unlikely that one has the inverse macro→micro function, resulting in what we call the "multiagent inverse problem". These two challenges mean that real-time multi-robot learning from demonstration has proven very difficult and has a very sparse literature.

Over the past several years we have participated in the Kid-Size Humanoid League with a single objective: to successfully do a personal RoboCup-style technical challenge of our own invention, independent of those offered at RoboCup: **can we train multiple generic robots, through demonstration on the**

© Springer International Publishing Switzerland 2015
R.A.C. Bianchi et al. (Eds.): RoboCup 2014, LNAI 8992, pp. 369–382, 2015.
DOI: 10.1007/978-3-319-18615-3_30

field, how to play collaborative soccer at RoboCup solely within the preparatory time prior to the competition itself?

This is a very high bar: but over the past four years we have made major strides towards achieving it. In RoboCup 2011 we began by replacing a single hard-coded behavior in one attacker with a behavior trained on the field at the venue, and entered that robot into the competition. At RoboCup 2012 we expanded on this by training an attacker to perform *all* of its soccer behaviors (17 automata, Fig. 1), again at the venue. This trained attacker scored our winning goal against Osaka. This year we intend to train multiple robots, and ideally all four robots on the team, to perform collaborative behaviors.

Our approach, HiTAB, applies supervised learning to train multiple agents to perform behaviors in the form of decomposed hierarchical finite-state automata. HiTAB uses several tricks, notably task decomposition both per-agent and within a team, to break a complex joint behavior into smaller, very simple ones, and thus radically reduce its dimensionality. Sufficient domain knowledge is involved that HiTAB may fairly be thought of as a form of *programming by demonstration*.

This paper documents our past efforts at applying HiTAB on the field at RoboCup. We also discuss related penalty-kick experiments using the technique, and detail our success so far towards our 2014 goal. Finally, we propose a new RoboCup Technical Challenge in multiagent learning from demonstration.

1   Related Work

Learning from demonstration (or LfD) has been applied to a huge range of problems ranging from air hockey [2] to helicopter trajectory planning [14], but rarely to the multi-robot case [1]. Most of the multi-robot learning literature falls under *agent modeling*, where robots learn about one another rather than about a task provided by a demonstrator. The most common multi-robot LfD approach is to dismiss the macrophenomena entirely and issue separate micro-level training directives to each individual agent [11]. Another approach is to train individual robots only when they lack confidence about how to proceed [4].

1.1   Machine Learning at RoboCup

To put our "personal technical challenge problem" in context, it's worthwhile to survey how machine learning has been used at RoboCup in the past. Machine learning has been applied to RoboCup since its inception, coming to slightly less than 100 papers and demonstrations since 1997. We mention only a small number of the papers here.

The bulk of the machine learning RoboCup literature has involved single agents. This literature breaks down into three categories. First, learning algorithms have been applied about a dozen times to *sensor feature generation* tasks such as visual object recognition [13,31] and opponent behavior modeling and detection (for example [8,29]). Second, a equal amount of literature has applied machine learning to a robot's kinematics, dynamics, or structure. The lion's share of this work involves gait development (such as [18,19]), with some work on kicking [6,32], head actuation [5] and omnidirectional velocity control [17]. Third, about sixteen papers have concerned themselves with learning higher-level behaviors (for example [26,28]).

Cooperative Multiagent Learning. There have been approximately twenty five cooperative multiagent learning papers at RoboCup. The area breaks down into two categories. First, there is *team learning*, where a single learning algorithm is used to optimize the behaviors of an entire team. Some of this work has involved evolutionary computation methods to develop joint team behaviors (such as [10,15]); reinforcement learning papers have instead usually developed a single homogeneous behavior (for example [7,22]). In contrast the *concurrent learning* literature, where separate learners are applied per-agent, has largely applied multiagent reinforcement learning (such as [12,21]).

It is useful here to mention why this area is dominated by optimization methods (reinforcement learning, evolutionary computation): as mentioned before, multiagent learning presents a difficult inverse problem, and optimization is the primary way to solve such problems. However, optimization generally needs many iterations for even moderately high-dimensional spaces, meaning realistically such methods must employ a simulator, and so are not optimal for real-time training.

Training. Training differs from learning in that it involves a *trainer*, that is, a person who iteratively teaches behaviors, observes agent performance, and suggests corrections. This is a natural fit for soccer: but training is surprisingly rare at RoboCup. RoboCup has long sponsored a related topic, *coaching*, but the focus has more been on influencing players mid-game via a global view [27] than on training. One exception has used a coach to train action sequences as directed by human speech, then bind them to new speech directives [30]. This work resembles our own in that it iteratively trained behaviors as compositions of earlier ones. There is also work in *imitation learning*, whereby an agent learns by observing a (not necessarily) human performer [9,16], though without any trainer correction.

We know of two examples. besides our own, where training or related iterative learning was done *at* RoboCup. The Austin Villa has fed the previous night's results into an optimization procedure to improve behaviors for the next day [20]. Using corrective demonstration, the CMurfs coached a robot to select the correct features and behaviors from a hard-coded set in an obstacle avoidance task during the open technical challenge [3].

We also note that, like our own work, [27] does hierarchical decomposed development of stateless policies, albeit built automatically and for single agents.

2 HiTAB: Hierarchical Training of Agent Behaviors

HiTAB is a multiagent LfD system which trains behaviors in the form of hierarchical finite state automata (or HFA) represented as Moore machines. The system is only summarized here: for a fuller description see [23].

In the single-agent case, an automaton contains some number of *states* which are each mapped to a unique *behavior*, plus a distinguished *start* state whose behavior simply idles. A behavior may be *atomic*, that is, hard-coded, or it may be another finite-state automaton trained earlier. Some atomic behaviors trigger built-in features: for example, transitioning to the *done* (similarly *failed*) state immediately transitions to *start*, and further signals to the grandparent automaton that the parent HFA believes it is "done" with its task (or "failed"). Other built-in behaviors increment or clear counters. Every state has an accompanying *transition function* which tells HiTAB which state to transition to next time. Each iteration, HiTAB queries the current state's transition function, transitions as directed, then pulses the new state's behavior for an epsilon of time.

The trainer manually decomposes the desired task into a hierarchy of subtasks, then iteratively trains the subtasks bottom-up. In our experience, an experienced trainer need decompose only once. Training an automaton only involves learning its transition functions. In "training mode" the HFA transitions from state to state only when told to by the demonstrator. When the demonstrator transitions from state S to a new state $S' \neq S$, the automaton gathers the robot's current sensor feature vector $\vec{f}$, then stores a tuple $\langle S, \vec{f}, S' \rangle$ as a sample, and in many cases a "default sample" $\langle S', \vec{f}, S' \rangle$. A default sample says "as long as the world looks like $\vec{f}$, continue doing S' ", and is added only when transitioning to a continuous behavior (such as *walk*), as opposed to a one-shot behavior (like *kick*).

When training has concluded, the robot enters a "testing mode", at which point it builds an automaton from the samples. To do this, for each i the robot collects all tuples of the form $\langle S_i, \vec{f}, S' \rangle$, then reduces them to $\langle \vec{f}, S' \rangle$. These form data for a classifier $C_i(\vec{f}) \to S'$ which defines the transition function T_i accompanying state S_i. We use decision trees (C4.5) to learn these classifiers.

The trainer then observes the performance of the automaton. If he detects an incorrect behavior, he may correct it, adding a few new training samples, and then re-build the classifiers. HiTAB can also perform *unlearning*: use the corrective samples to determine which earlier samples had caused the erroneous behavior (either due to sensor noise or user error), then delete them [25]. Finally, the trainer can "undo" an incorrect sample he had just erroneously entered. When he is satisfied with the automaton, he can save it to the behavior library, at which point it becomes available as a behavior (and state) when training a later, higher-level automaton. A behavior saved to the behavior library can be revised in the future without retraining the entire HFA from scratch.

In HiTAB, both basic behaviors and sensor features may be parameterized: thus we may say "go to X" rather than "go to the ball"; and similarly "angle to X" rather than "angle to the nearest teammate". Use of parameterized behaviors or features in an automaton without binding them to ground values results in the automaton itself being parameterized as well. Of course, ultimately each parameter must be bound to a ground value somewhere in the hierarchy: the set of available ground values is, like basic behaviors, hard-coded by the experimenter.

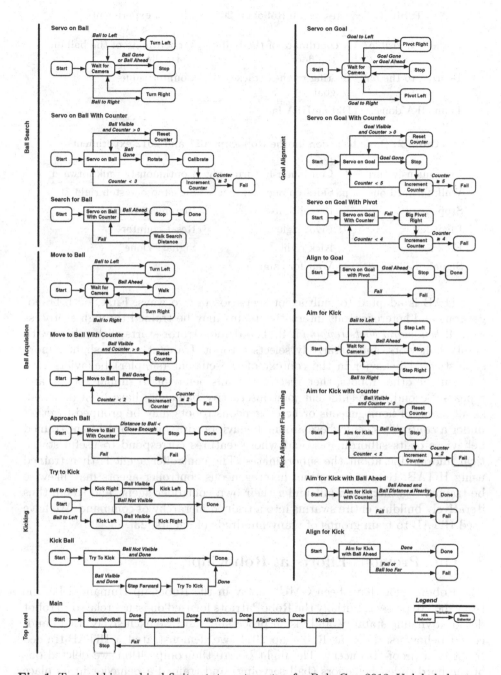

Fig. 1. Trained hierarchical finite-state automaton for RoboCup 2012. Unlabeled transitions are always executed. Note the significant repetition in pattern: part of this is simply behavior similarity, but part is because the 2012 HFA interpreter did not support parameterized behaviors or features (see Sect. 2).

Table 1. Features in the Robocup 2011 and 2012 experiments

Is the ball visible?	X coordinate of the ball on the floor	Y coordinate of the ball on the floor
Bearing to the ball	Bearing to the attacker goal	Counter value
Is an HFA done?	Did an HFA fail?	

Table 2. Basic behaviors in the Robocup 2011 and 2012 experiments

Continuously turn left	Continuously turn right	Continuously walk forward
Walk forward one step	Sidestep one step left	Sidestep one step right
Stop	Re-calibrate gyros	Increment counter
Pivot left	Pivot right	Reset counter
Kick left	Kick right	Signal "Done"
Signal "Failed"	Wait for camera	

HiTAB is adapted to multiagent scenarios in two ways. First, both homogeneous and heterogeneous interactive teams may be trained through a process we call *behavioral bootstrapping* [24]. The demonstrator starts with robots with empty behaviors, and iteratively selects a robot, trains it with a slightly more sophisticated behavior in the context of the current (simpler) behaviors running on the other robots, then distributes this behavior to similar robots, and repeats. Second, once sufficient joint interactive behaviors have been designed, small teams of homogeneous or heterogeneous robots may be grouped together under a *controller agent* whose atomic behaviors correspond to the joint trained behaviors of its subordinates, and whose features correspond to useful statistical information about the subordinates. The controller agent is then trained using HiTAB. Homogeneous and heterogeneous controller agents may likewise be trained together, then put under their own controller agent, and so on, thus iteratively building entire swarms into a trained hierarchy of command. We have used HiTAB to train groups of many hundreds of agents [23].

3   Our Previous Efforts at RoboCup

The RoboPatriots have been GMU's entry in the RoboCup Humanoid League from 2009 to present. Initially the RoboPatriots focused on issues related to robot design, dynamic stability, and vision processing, and we exclusively used hard-coded behaviors. Then at RoboCup 2011, we demonstrated a HiTAB-trained robot as a proof-of-concept. The night before the competition, we deleted one of the hard-coded behaviors (ball servoing) and trained a behavior in its place through direct tele-operation of the robot on the field of play. We then saved out the trained behavior, and during the competition, one attacker loaded this behavior from a file and used it in an interpreter alongside the remaining hard-coded behaviors. This trained behavior was simple and meant as a proof of concept, but it worked perfectly.

In 2012 we had a much more ambitious goal: to train the entire library of behaviors of a single robot on the field immediately prior to the competition. Our attacker robots in 2012 used a decomposition of 17 automata which collectively defined a simple "child soccer" style of behaviors without localization: search for the ball, approach the ball, align to the goal, align for kicking, kick, and repeat. Two days before the competition, we deleted the entire behavior set and proceeded to train an equivalent set of 17 automata in its place (Fig. 1), again through tele-operation of the robot on the competition field. The final HFA was saved to disk and run through an interpreter during game play.

The basic sensor features and robot behaviors we relied on to build these automata are given in Tables 1 and 2 respectively: these were essentially the same basic sensor features and behaviors used in the hard-coded version. Note that not all features and behaviors were used in every HFA. The *Wait for Camera* behavior ensured that we had new and complete vision information before transitioning (our vision system was slower than the HFA).

The top-level HFA behavior, *Main*, performed "child soccer" by calling the following second-level behaviors, which triggered additional hierarchical behaviors:

- *Search for Ball:* Using the bearing to the ball, the robot did visual servoing on the ball, with the additional constraint of performing a rotation if the ball was missing for several frames. If the robot had rotated several times, it then walked forward before resuming searching.
- *Approach Ball:* Using the bearing to the ball and distance to the ball, the robot moved towards the ball while performing course corrections en route.
- *Align to Goal:* Using the bearing to the goal, the robot oriented toward the goal while maintaining the ball near the robot's feet. The robot pivoted around the ball if it could not see the goal.
- *Align for Kick:* Using the $\langle X, Y \rangle$ position of the ball, the robot took small steps to get the ball in a box near its feet so a kick could be performed.
- *Kick Ball:* The robot kicked based on the X position of the ball. If after a kick the ball was still there, then the robot would kick with its other foot. If the ball was *still* there, the robot would take a step forward and repeat.

Issues such as referee box event response and recovery from falls were handled with hard-coded logic (in the second case, resetting to *Search for Ball*). The HFA included subroutines designed to handle high sensor noise: for example, *MoveToBallWithCounter* would robustly handle the ball disappearing due to a temporary camera error.

HiTAB can be used to rapidly retrain behaviors as needed. As an example, we had to train an additional HFA after the first day of competition. During our early matches, we observed that the *Aim for Kick* sub-behavior assumed that the ball would consistently be near the robot's feet. However, due to sensor noise the robot might enter *Align to Goal* when the ball was far away, and so when *Aim for Kick* was entered, it would take many, many baby steps towards the ball. We then trained a new version, *Aim for Kick With Ball Ahead* to also include a failure situation for when the ball was outside a box centered at the robot's feet. The new HFA was then used in our later matches.

Table 3. Number of data samples for each HFA trained at RoboCup 2012. *Provided Samples* are those directly provided by the user and do not include automatically inserted "default samples" for continuous sub-behaviors. The data for ServoOnBall-WithCounter was not saved, so the estimate is based on other HFAs which used a counter.

Behavior	Number of Samples	Number of Provided Samples
ServoOnBall	11	11
ServoOnBallWithCounter	(estimate) 10	(estimate) 9
SearchForBall	10	8
MoveToBall	9	9
MoveToBallWithCounter	10	9
ApproachBall	15	11
ServoOnGoal	9	9
ServoOnGoalWithCounter	12	11
ServoOnGoalWithPivot	9	7
AlignToGoal	12	9
AimForKick	9	9
AimForKickWithCounter	10	9
AlignForKickWithBallAhead	22	14
AlignForKick	42	35
TryToKick	10	10
KickBall	9	6
Main	34	19
Total	243	195

Table 3 shows the number of samples collected for all 17 trained HFAs. The first column includes automatically inserted default samples while the second column shows only the directly provided samples. Given the problem complexity, we were able to train on a remarkably small number of samples.

During our second match versus Team JEAP from Osaka University, **our trained robot scored the winning goal.** After discussion with colleagues at the competition, we believe that, to the best of our knowledge, this is the first time a competing robot at RoboCup has used a full behavior set trained in real time at the venue itself, much less scored a goal using those trained behaviors (Fig. 2).

Fig. 2. GMU's trained Johnny-5 (magenta #5) kicks the winning goal against Osaka.

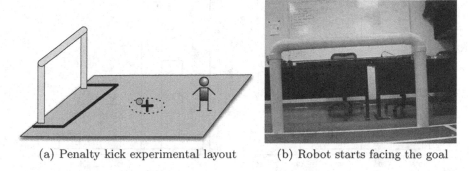

(a) Penalty kick experimental layout   (b) Robot starts facing the goal

Fig. 3. Penalty kick experimental layout. The robot cannot initially see the ball.

4   Penalty Kick Experiments

One claimed benefit of LfD is that the trained behaviors perform as well as hand-coded behaviors. After RoboCup 2012, we conducted experiments to verify this claim by comparing our trained soccer behavior with the hand-coded behavior deployed on our other attacker. The task was penalty kicks, similar to those used during the RoboCup competition.

The robot was placed 40 cm away from the penalty kick mark with a neutral head position and facing the goal. The ball was randomly placed within a 20 cm diameter circle centered on the penalty kick mark (see Fig. 3(a)). Initially, the robot could see the goal, but not the ball, as shown in Fig. 3(b). The metric was time to kick the ball, independent of whether a goal was scored. Both behaviors were run 30 times.

Figures 4(a)–(b) show histograms for the hard-coded and trained behaviors. For both behaviors, sensor noise caused one run to take significantly longer than the rest. The trained behavior had a mean execution time of 37.47 ± 5.51 sec. (95 % confidence interval), while the hardcoded behavior had a mean of 35.85 ± 3.08. The means were not statistically significantly different.

5   Set Plays: A Multiagent Training Proof of Concept

For RoboCup 2014 our goal is to train not just a single robot but a full team of humanoids to play interactive robot soccer. To that end we have begun with an experiment in multi-robot training on the soccer field: set plays.

Multi-robot training is notionally difficult because of the interaction among the robots and the challenges faced in coordinating them. To attack this problem at scale, HiTAB relies on manual decomposition of a swarm of agents under a hierarchy of trained "controller agents". However for small groups (two to four agents) we focus instead on developing joint behaviors among the agents. This is the case for the set-play scenario, which typically involves two agents.

How might one use HiTAB to train an interactive joint behavior among two robots without a controller agent coordinating them? We see three possibilities:

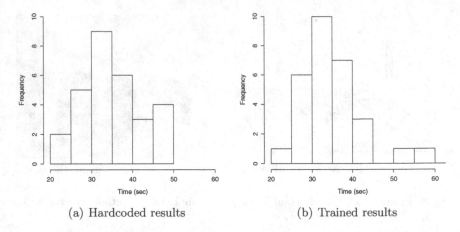

(a) Hardcoded results   (b) Trained results

Fig. 4. Penalty kick results. In both experiments, one run took longer than 60 seconds.

- *Train the Robots Independently.* We train one robot while tele-operating the other (the *dummy*), and vice versa. This is the simplest approach, but to us it does not intuitively feel like a match for multiagent training scenarios which involve a significant degree of interaction.
- *Bootstrap.* We train one robot to perform a rudimentary version of its behavior with the other robot doing nothing. We then train the second robot to do a slightly more sophisticated version of its own behavior while the first robot is performing its trained rudimentary behavior. This back-and-forth training continues until the robots have been fully trained.
- *Simultaneously Train.* We use two HiTAB sessions, one per robot, to train the robots at the same time while interacting with one another. This obviously requires much more effort on behalf of the demonstrator (or multiple demonstrators working together).

For 2014 we have new robots (Darwin-OP humanoids) and so have decided to base our system on a heavily modified version of the UPennalizers's open-sourced 2013 champion software. This code provides localization and helpful behaviors which we use as the foundation for basic behaviors and features in the set plays:

- *GotoPosition(P, L)* goes to location L on the field, facing the location of player or object P, then broadcasts a "Ready" signal for five seconds.
- *GotoBall* goes to the ball position.
- *AlignToTarget(R)* orients around the ball until the robot is facing player or object R.
- *KickBall* kicks the ball and broadcasts a "Kick" signal for five seconds.
- *TurnLeft* rotates to the left.

Each robot was also equipped with the robot sensor features *Kicked(P)* (did P raise the Kick signal?), *Ready(P)* (did P raise the Ready signal?), *Ball Lost*

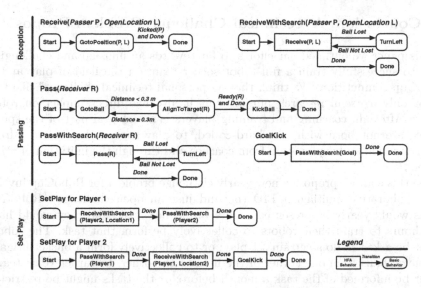

Fig. 5. Trained hierarchical finite-state automata for the 2014 set-play experiments. Passing and reception automata are shared among both robots, but each robot executes a different top-level set play automata.

(has the ball been lost for over three seconds?), and *Distance* (to ball). Note that the Goal, as a parameter, was considered to be always "Ready".

Clearly these behaviors and features are higher-level than those used in 2012, and the resulting automata are simple for a programmer to implement. We took (and are continuing to take) such baby-steps on purpose: real-time training of multirobot behaviors is notionally nontrivial, and previous examples for guidance are few and far between. Our goal is to show that such a thing is even feasible.

Using this foundation, we trained the robots independently via dummies to perform the joint set play behaviors shown in Fig. 5: Robot *A* would acquire the ball while *B* moved to a preset position. When both were ready, Robot *A* would then kick to *B* and move to a second preset position. Then Robot *B* would kick to *A*, which would then kick to the goal.

Though we had imagined that we would need to perform simultaneous training or bootstrapping, in fact we have been perfectly successful in training set plays separately using dummies. This surprising result is likely due to the small number (two) of robots involved, but it has nonetheless forced us to question the prevailing wisdom: does interaction *necessarily* complicate multi-robot learning?

Whether independent training will be sufficient for the remainder of the behaviors for 2014 remains to be seen; and ultimately we will need to train a virtual controller agent (likely residing on the goalie) to direct which behaviors and joint actions should be undertaken by the team at any given time.

6   Conclusion: A Technical Challenge Problem Proposal

In this paper we outlined our efforts so far towards an unusual and challenging goal: to successfully train a full robot soccer team on the field of play at the RoboCup competition. We think that a "personal technical challenge" like this is not only a useful research pursuit, but it also has direct impact on robot soccer. After all, coaching and training players is an integral part of the sport! People are not born with, nor hard-coded, to play soccer: they learn it from demonstration and explanation from coaches and through the imitation of other players.

To this end, we propose a new yearly challenge problem for RoboCup involving collaborative multiagent LfD (beyond just an open challenge). RoboCup teams would yearly be presented with a brand new task, and they would have four hours to train their robots to collectively perform that task. The robots might be asked to do a certain set play; or to collectively form a bucket brigade to convey balls from one corner of the field to the other. In earlier years teams might be informed of the task a month before; or the tasks might be restricted to single agents. But eventually the task should require multiple interacting agents and few clues provided beforehand except for the basic behaviors permitted. Differences in robot hardware or software architectures might constrain the available techniques, and so the challenge might need to be more a showcase than a judged competition.

Acknowledgments. Research in this paper was done under NSF grant 1317813.

References

1. Argall, B.D., Chernova, S., Veloso, M., Browning, B.: A survey of robot learning from demonstration. Robot. Auton. Syst. **57**(5), 469–483 (2009)
2. Bentivegna, D.C., et al.: Learning tasks from observation and practice. Robot. Auton. Syst. **47**(2–3), 163–169 (2004)
3. Meriçli, Ç., Veloso, M., Akin, H.L.: Multi-resolution corrective demonstration for efficient task execution and refinement. Int. J. Soc. Robot. **4**(4), 423–435 (2012)
4. Chernova, S.: Confidence-based robot policy learning from demonstration. Ph.D. thesis, Carnegie Mellon University (2009)
5. Fountain, J., Walker, J., Budden, D., Mendes, A., Chalup, S.K.: Motivated reinforcement learning for improved head actuation of humanoid robots. In: Behnke, S., Veloso, M., Visser, A., Xiong, R. (eds.) RoboCup 2013. LNCS, vol. 8371, pp. 268–279. Springer, Heidelberg (2014)
6. Hausknecht, M., Stone, P.: Learning powerful kicks on the Aibo ERS-7: the quest for a striker. In: Ruiz-del-Solar, J., Chown, E., Ploger, P.G. (eds.) RoboCup 2010. LNCS, vol. 6556, pp. 254–265. Springer, Heidelberg (2011)
7. Kalyanakrishnan, S., Liu, Y., Stone, P.: Half field offense in robocup soccer: a multiagent reinforcement learning case study. In: Lakemeyer, G., Sklar, E., Sorrenti, D.G., Takahashi, T. (eds.) RoboCup 2006. LNCS (LNAI), vol. 4434, pp. 72–85. Springer, Heidelberg (2007)

8. Kaminka, G.A., Fidanboylu, M., Chang, A., Veloso, M.M.: Learning the sequential coordinated behavior of teams from observations. In: Kaminka, G.A., Lima, P.U., Rojas, R. (eds.) RoboCup 2002. LNCS, pp. 111–125. Springer, Heidelberg (2002)
9. Latzke, T., Behnke, S., Bennewitz, M.: Imitative reinforcement learning for soccer playing robots. In: Lakemeyer, G., Sklar, E., Sorrenti, D.G., Takahashi, T. (eds.) RoboCup 2006. LNCS, pp. 47–58. Springer, Heidelberg (2007)
10. Luke, S., Hohn, C., Farris, J., Jackson, G., Hendler, J.: Co-evolving soccer softbot team coordination with genetic programming. In: Kitano, H. (ed.) RoboCup 1997. LNCS, pp. 398–411. Springer, Heidelberg (1998)
11. Martins, M.F., Demiris, Y.: Learning multirobot joint action plans from simultaneous task execution demonstrations. AAMAS, pp. 931–938 (2010)
12. Merke, A., Riedmiller, M.: Karlsruhe Brainstormers — a reinforcement learning approach to robotic soccer. In: Birk, A., Coradeschi, S., Tadokoro, S. (eds.) RoboCup 2001. LNCS, pp. 435–440. Springer, Heidelberg (2002)
13. Metzler, S., Nieuwenhuisen, M., Behnke, S.: Learning visual obstacle detection using color histogram features. In: Röfer, T., Mayer, N.M., Savage, J., Saranlı, U. (eds.) RoboCup 2011. LNCS, vol. 7416, pp. 149–161. Springer, Heidelberg (2012)
14. Nakanishi, J., et al.: Learning from demonstration and adaptation of biped locomotion. Robot. Auton. Syst. **47**(2–3), 79–91 (2004)
15. Nakashima, T., Takatani, M., Udo, M., Ishibuchi, H., Nii, M.: Performance evaluation of an evolutionary method for robocup soccer strategies. In: Bredenfeld, A., Jacoff, A., Noda, I., Takahashi, Y. (eds.) RoboCup 2005. LNCS (LNAI), vol. 4020, pp. 616–623. Springer, Heidelberg (2006)
16. Noda, I.: Hidden Markov modeling of team-play synchronization. In: Polani, D., Browning, B., Bonarini, A., Yoshida, K. (eds.) RoboCup 2003. LNCS (LNAI), vol. 3020, pp. 102–113. Springer, Heidelberg (2004)
17. Oubbati, M., Schanz, M., Buchheim, T., Levi, P.: Velocity control of an omnidirectional robocup player with recurrent neural networks. In: Bredenfeld, A., Jacoff, A., Noda, I., Takahashi, Y. (eds.) RoboCup 2005. LNCS (LNAI), vol. 4020, pp. 691–701. Springer, Heidelberg (2006)
18. Saggar, M., D'Silva, T., Kohl, N., Stone, P.: Autonomous learning of stable quadruped locomotion. In: Lakemeyer, G., Sklar, E., Sorrenti, D.G., Takahashi, T. (eds.) RoboCup 2006. LNCS (LNAI), vol. 4434, pp. 98–109. Springer, Heidelberg (2007)
19. Schwarz, M., Behnke, S.: Compliant robot behavior using servo actuator models identified by iterative learning control. In: Behnke, S., Veloso, M., Visser, A., Xiong, R. (eds.) RoboCup 2013. LNCS, vol. 8371, pp. 207–218. Springer, Heidelberg (2014)
20. Stone, P.: Personal conversation (2014)
21. Stone, P., Veloso, M.M.: Layered learning and flexible teamwork in robocup simulation agents. In: Veloso, M.M., Pagello, E., Kitano, H. (eds.) RoboCup 1999. LNCS (LNAI), vol. 1856, pp. 495–508. Springer, Heidelberg (2000)
22. Stone, P., Kuhlmann, G., Taylor, M.E., Liu, Y.: Keepaway soccer: from machine learning testbed to benchmark. In: Bredenfeld, A., Jacoff, A., Noda, I., Takahashi, Y. (eds.) RoboCup 2005. LNCS (LNAI), vol. 4020, pp. 93–105. Springer, Heidelberg (2006)
23. Sullivan, K., Luke, S.: Learning from demonstration with swarm hierarchies. AAMAS (2012)
24. Sullivan, K., Luke, S.: Real-time training of team soccer behaviors. In: Chen, X., Stone, P., Sucar, L.E., van der Zant, T. (eds.) RoboCup 2012. LNCS, vol. 7500, pp. 356–367. Springer, Heidelberg (2013)

25. Sullivan, K., et al.: Unlearning from demonstration. IJCAI (2013)
26. Takahashi, Y., Edazawa, K., Asada, M.: Behavior acquisition based on multi-module learning system in multi-agent environment. In: Kaminka, G.A., Lima, P.U., Rojas, R. (eds.) RoboCup 2002. LNCS (LNAI), vol. 2752, pp. 435–442. Springer, Heidelberg (2003)
27. Takahashi, Y., Hikita, K., Asada, M.: A hierarchical multi-module learning system based on self-interpretation of instructions by coach. In: Polani, D., Browning, B., Bonarini, A., Yoshida, K. (eds.) RoboCup 2003. LNCS (LNAI), vol. 3020, pp. 576–583. Springer, Heidelberg (2004)
28. Tuyls, K., Maes, S., Manderick, B.: Reinforcement learning in large state spaces. In: Kaminka, G.A., Lima, P.U., Rojas, R. (eds.) RoboCup 2002. LNCS (LNAI), vol. 2752, pp. 319–326. Springer, Heidelberg (2003)
29. Visser, U., Weland, H.-G.: Using online learning to analyze the opponent's behavior. In: Kaminka, G.A., Lima, P.U., Rojas, R. (eds.) RoboCup 2002. LNCS (LNAI), vol. 2752, pp. 78–93. Springer, Heidelberg (2003)
30. Weitzenfeld, A., Ramos, C., Dominey, P.F.: Coaching robots to play soccer via spoken-language. In: Iocchi, L., Matsubara, H., Weitzenfeld, A., Zhou, C. (eds.) RoboCup 2008. LNCS, vol. 5399, pp. 379–390. Springer, Heidelberg (2009)
31. Wilking, D., Röfer, T.: Realtime object recognition using decision tree learning. In: Nardi, D., Riedmiller, M., Sammut, C., Santos-Victor, J. (eds.) RoboCup 2004. LNCS (LNAI), vol. 3276, pp. 556–563. Springer, Heidelberg (2005)
32. Zagal, J.C., Ruiz-del-Solar, J.: Learning to kick the ball using back to reality. In: Nardi, D., Riedmiller, M., Sammut, C., Santos-Victor, J. (eds.) RoboCup 2004. LNCS (LNAI), vol. 3276, pp. 335–346. Springer, Heidelberg (2005)

RF-based Relative Position Estimation in Mobile Ad-Hoc Networks with Confidence Regions

Luis Oliveira[✉] and Luis Almeida

IT/Faculdade de Engenharia, Universidade do Porto, Porto, Portugal
loliveria@fe.up.pt

Abstract. Relative localisation of mobile robots can provide useful information to applications, from formation control, to joint exploration and inspection. One way to obtain relative localisation is to measure distances between the multiple robots. In this scope, distance estimates based on RF ranging data can be beneficial for small/inexpensive communicating robots that have no other means of measuring distances, or as disambiguation of multiple hypothesis in high accuracy localisation systems. In this work, we present a technique of estimating the relative positions of simple mobile robots in a small team using the distance information that can be captured by a wireless transceiver, only. Simulation results with a team of five mobile robots show that we can estimate their relative positions with an average accuracy of 1.3 m without any fixed reference and using RF information, only. The main contribution of our work is that we can provide consistent reliability information as the covariance of the obtained positions.

Keywords: Relative localisation · RF-ranging · Multidimensional scaling

1 Introduction

Accurate localisation is a key factor on most mobile robots applications which, due to the need of interaction with the world, has been mostly focused on absolute localisation. Particularly, substantial attention has been given to simultaneous localisation and mapping (SLAM) algorithms, using sensors as diverse as vision [1,15], lasers [19], or a combination of those [20], usually fused with dead reckoning. Such sensors provide high precision measurements, consequently they are used to provide high precision localisation. Despite that, lasers are bulky (for example to be carried by quad-rotors) and vision systems require high computational capacity. A viable alternative is to use external systems such as GPS or previously built infrastructures, however, the first may be undesirable since it is only available in outdoor environments and may be rendered unusable in forests and street canyons. Similarly, the latter are usually an undesired alternative, mainly because they are costly to deploy, and are useless if a catastrophic event changes the environment significantly.

© Springer International Publishing Switzerland 2015
R.A.C. Bianchi et al. (Eds.): RoboCup 2014, LNAI 8992, pp. 383–394, 2015.
DOI: 10.1007/978-3-319-18615-3_31

Alternatively it is possible to localise robots relatively to each other, and although this type of localisation may seem insufficient to support interactions with the environment, in multi-robot systems it is of key importance to support fusion of information from different robots, allowing robot cooperation/coordination as well as improvement of global localisation, e.g. using lasers to measure distance and angle to other robots and fusing that information with dead reckoning [21], or using relative positions to fuse SLAM maps [3].

In this work, we propose a relative position estimation technique that can be employed to localise small (approximately up to 10 elements) dynamic multi-robot teams. The sensing is performed by the wireless communications interface, which reports the Received Signal Strength Indicator (RSSI) and is able to measure the Round-Trip Time-of-Flight. No extra hardware is necessary. Despite the fact that RSSI readings are unreliable, they can still provide helpful information about the distance between robots and are unequivocal information since the robot can know who it is "talking" with. The main contributions of this work are:

- we provide a novel relative position tracking technique using RF-based ranging estimates, only
- we make no assumption on the dynamics of the robots, i.e. all robots are considered mobile with unknown velocity
- we provide positioning reliability information as the covariance of the positions.

2   Related Work and Proposal

Relative localisation of mobile robots (or more generically mobile nodes) can provide useful information to applications. Specifically, estimates of the topology of a network based on collected ranging data, can be beneficial to applications as diverse as coarse formation control, cooperative sensing and area coverage, and disambiguation of multiple hypothesis in high accuracy localisation systems.

A common approach to robot localisation is SLAM. A famous example is the FastSLAM algorithm [11], where a particle filter is employed to track several possible paths of the robot, and extended Kalman filters to estimate the positions of landmarks. Another example is Wifi-SLAM [8] where an automatic fingerprinting technique that exploits landmarks on the radio map is proposed. By fusing RSSI information with IMU data, it is possible to detect loop closure and to build the environment map and locate the user. Another interesting work is presented in [6], where authors propose a range only SLAM. The work in [6] proposes a technique using dead reckoning to track robot movement and ultrasound ranging equipment to measure the distance between itself and some beacons. The beacons have unknown positions but are able to measure the distance between themselves. MapCraft [24] assumes that a physical map containing the walls and doors is already available. Then, data from different sensors is fused and matched to the map to estimate positions. Despite all that, these techniques require the ability of measuring some static features, such as landmarks, walls, or the RSSI fingerprints of certain access-points. Conversely, in the work presented in [16] a method of estimating the positions of moving nodes in an anchor-less scenario is proposed. The authors use relative velocity, calculated based on dead-reckoning, together with RSSI measurements to provide a position estimate of a team of mobile nodes without resorting to anchor nodes.

Another very popular solution of calculating nodes position from ranging information is the MultiDimensional Scaling (MDS) algorithm [2,4] that minimises the dissimilarities of a connectivity matrix up to a rigid formation. In order to improve results under unknown line-of-sight/non-line-of-sight (LOS/NLOS) conditions and scarce ranging information, [5] uses another variant of MDS based on Weighted Least Squares algorithm, whose weights are assigned according to the reliability of the ranging measurements. The work of [17] proposes a method of estimating nodes positioning using a Maximum Likelihood Estimator (MLE). RSSI and ToA measurements of static nodes are collected using the location of four anchor nodes. In order to obtain distance measurements from RSSI values, the parameters of the path loss model are estimated using data collected prior to experiments. However, assuming known position of anchor nodes is undesirable, since they can be unavailable in many scenarios. Therefore, in [10,14], the authors propose a method for deriving the network topology from the RSSI data using MDS. The method presented is not a physical accurate localisation system, mainly because the work does not consider any propagation model. Other approaches involve iterative methods, such as the work in [25] that proposes solving an expectation-maximisation problem that jointly estimates the path loss model and the relative positions of the nodes, using MDS as the initial condition. Similarly, the authors in [22] propose to use a gradient descent algorithm to solve a minimisation problem that finds the topology that minimises the error of the distance, but this problem has non-linear constraints and many local minima. Finally, the technique of communication-based relative localisation has been also been extensively studied in the field of wireless sensor networks (WSN), such as [5,9,17]. However, the assumptions valid in most WSN scenarios can seldom be extended to mobile robot, particularly the absence of, or very limited, mobility that simplifies the problem.

Our paper differs from the previously referred works in several aspects. First of all, we do not use sensors other than the RF transceiver module to compute approximate relative positions between the nodes of a small team of mobile robots. Moreover, we make no assumption on the dynamics of the robots, and do not assume pre-installed anchor nodes or robots with known positions. Finally, in addition to the positions estimate, we provide an estimate of the covariance of the positions, allowing the user to define confidence regions around the estimate. To the best of the authors knowledge, this is one of the first works computing confidence regions associated to MDS-computed position estimates.

3   Estimating the Positions

In this section we explain the details of our proposal. First of all we explain the data that is collected from the robots, and how we use it to obtain a measurement of the positions. Then we explain how we filter that data and generate positions estimates.

3.1   Collecting Distance Information from Robots

The work presented in [13] proposes an RF-only, anchor-less technique that performs online estimation of the distance between mobile robots without previous knowledge. The authors use RSSI/Time-of-Flight measurements to perform online estimation of the path loss model. Then, the corresponding model is used to estimate the distance using the RSSI, or the RSSI and Time-of-Flight when available. Moreover, a Extended Kalman Filter (EKF) was used to perform the estimation. Consequently, the result is both the estimated distance between robots and the corresponding estimated variance. Despite that, the work in [13] approaches the problem we aim to solve up to the estimation of distances, only. The details of that work cannot be included here due to space constraints, but Fig. 1 explains the respective process.

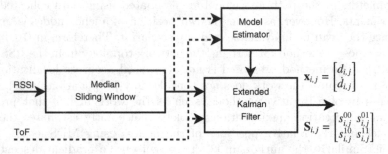

Fig. 1. RF-based ranging for each link: Dashed lines apply only when ToF data is available (adapted from [13]); In this figure, $\mathbf{x}_{i,j}$ is the estimated state (containing the distance and respective derivative) between robots i and j; $\mathbf{S}_{i,j}$ is the respective covariance matrix

Taking advantage of this work, we build and share amongst the robots two matrices based on the output of the EKF: the matrix $\mathbf{D}_{n \times n} = d_{ij}$, Eq. (1), where d_{ij} is the distance estimate between robots i and j as estimated by robot i, and the matrix $\mathbf{V}_{n \times n} = s_{ij}^{00}$, Eq. (2), where s_{ij}^{00} is the variance of the distance estimate between robots i and j, as estimated by robot i.

$$\mathbf{D}_{n \times n} = \begin{bmatrix} 0 & d_{12} & d_{13} & \cdots & d_{1n} \\ d_{21} & 0 & d_{23} & \cdots & d_{2n} \\ d_{31} & d_{32} & 0 & \cdots & d_{3n} \\ \vdots & \vdots & \vdots & \ddots & \vdots \\ d_{n1} & d_{n2} & d_{n3} & \cdots & 0 \end{bmatrix} \quad (1) \quad \mathbf{V}_{n \times n} = \begin{bmatrix} 0 & s_{12}^{00} & s_{13}^{00} & \cdots & s_{1n}^{00} \\ s_{21}^{00} & 0 & s_{23}^{00} & \cdots & s_{2n}^{00} \\ s_{31}^{00} & s_{32}^{00} & 0 & \cdots & s_{3n}^{00} \\ \vdots & \vdots & \vdots & \ddots & \vdots \\ s_{n1}^{00} & s_{n2}^{00} & s_{n3}^{00} & \cdots & 0 \end{bmatrix} \quad (2)$$

Finally, the algorithms we will use require these matrices to be symmetrical, which seldom happens due to different transmission power in different robots, non omnidirectional antennas, etc. That being said, in order to create and feed a symmetric distance matrix to the those algorithms, we define matrix $\mathbf{G}_{n \times n}$ according to Eq. 3, and matrix $\mathbf{W}_{n \times n}$ according to Eq. 4.

$$\mathbf{G}(i,j) = \begin{cases} \mathbf{D}(i,j), & \mathbf{V}(i,j) < \mathbf{V}(j,i) \\ \mathbf{D}(j,i), & \text{otherwise} \end{cases} \tag{3}$$

$$\mathbf{W}(i,j) = \begin{cases} \mathbf{V}(i,j), & \mathbf{V}(i,j) < \mathbf{V}(j,i) \\ \mathbf{V}(j,i), & \text{otherwise} \end{cases} \tag{4}$$

3.2   Estimating Positions from Distances

MultiDimensional Scaling (MDS) [2] is relatively simple to implement technique that can be used to compute the robots relative positions. MDS is a technique used in multivariate analysis that transfers a known $n \times n$ matrix of dissimilarities to n points of an b-dimensional Euclidean space in such a way that the pairwise distances between points are compatible with the dissimilarities matrix. Consequently, by limiting b to two, i.e. 2-dimensional positions, and by using the positive semi-definite matrix $\mathcal{G}_{n \times n}$ containing the pairwise distances between all robots, we can write Eq. (5). Where the MDS function returns the 2-dimensional $\mathcal{M}_{n \times 2}$ containing the positions of the robots.

$$\mathcal{M} = \text{MDS}(\mathcal{G}) \tag{5}$$

With respect to defining relative positions for a team of mobile nodes, MDS already sorts out certain ambiguities that are inherent to the relative localisation process, e.g. eigenvector switching. However, that has the consequence that a small perturbation in the distances matrix can bring totally different results for the coordinates, such as map flips. Since the nodes position is only recovered up to rigid motion, orientation of the team cannot be determined just with pairwise distances, neither can symmetry relationships. To obtain relative positions estimates that vary smoothly, we carry out the following adjustments of the coordinates provided by the MDS (considering only the result presented in 2D space, i.e. $b = 2$).

Let $\mathcal{M}_{n \times 2} = (\mathbf{m}_1; \mathbf{m}_2; \dots; \mathbf{m}_n)$ denote the coordinates determined with MDS, where $\mathbf{m}_i = (m_i^x, m_i^y)$ is the 2D position of robot i; $\mathcal{L}_{n \times 2} = (\mathbf{l}_1; \mathbf{l}_2; \dots; \mathbf{l}_n)$ denote a set of arbitrary reference positions, where $\mathbf{l}_i = (l_i^x, l_i^y)$ is the 2D position of robot i; and $\mathcal{F}_{n \times 2} = (\mathbf{f}_1; \mathbf{f}_2; \dots; \mathbf{f}_n)$ denote the final coordinates, where $\mathbf{f}_i = (f_i^x, f_i^y)$ is the 2D position of robot i.

We consider the robot making these calculations (herein referred by 1) as being in the origin. Then, because of the flip ambiguity, we generate $\mathcal{M}_{n \times 2}^I = (\mathbf{m}_1^I; \mathbf{m}_2^I; \dots; \mathbf{m}_n^I)$, Eq. (6), that represents the mirror image of the output of MDS along the y-axis.

$$\mathcal{M}^I = \mathcal{M} \times \begin{pmatrix} -1 & 0 \\ 0 & 1 \end{pmatrix} \tag{6}$$

In order to remove the rotation ambiguity, for each robot, we calculate the angle that would be required to align it with the reference $\phi = \text{atan2}(\mathcal{L}) - \text{atan2}(\mathcal{M})$. Where atan2 represents the four quadrant arctangent. Similarly, we calculate $\phi^I = \text{atan2}(\mathcal{L}) - \text{atan2}(\mathcal{M}^I)$. Using those two hypotheses we choose the best coordinate set according to Eq. (7). By selecting the set with the smallest standard deviation, we are selecting the topology with more similarities to the reference, i.e. the topology in which all robots require approximately the same rotation to match the reference. Finally, we remove the points whose residuals exceed one standard deviation (ϕ_e) from ϕ and ϕ^I, i.e. if $\text{abs}(\phi_e - \text{mean}(\phi)) > \text{std}(\phi)$ or if $\text{abs}(\phi_e - \text{mean}(\phi^I)) > \text{std}(\phi^I)$, and we calculate α, the clockwise rotation angle that minimises the square error of the angle between the estimate and the reference (Eq. (8)).

$$\mathcal{T} = \begin{cases} \mathcal{M}, & \text{std}(\phi) < \text{std}(\phi^I) \\ \mathcal{M}^I, & \text{otherwise} \end{cases} \tag{7}$$

$$\alpha = \begin{cases} \min\left[(\phi - \alpha) \cdot (\phi - \alpha)\right], & \text{std}(\phi) < \text{std}(\phi^I) \\ \min\left[(\phi^I - \alpha) \cdot (\phi^I - \alpha)\right], & \text{otherwise} \end{cases} \tag{8}$$

The last step to calculate the final coordinate $\mathcal{F}$, is rotating the selected topology as in Eq. (9). Note that ϕ or ϕ^I can contain angles similar in rotation but different in value, i.e., any values separated by 2π. Therefore, we analyse the residuals both between $[-\pi, \pi]$ and $[0, 2\pi]$, choosing the one with less standard deviation.

$$\mathcal{F} = \mathcal{T} \times \begin{pmatrix} \cos(\alpha) & \sin(\alpha) \\ -\sin(\alpha) & \cos(\alpha) \end{pmatrix} \tag{9}$$

3.3   Kalman Filter

In order to estimate the relative positions of a team of robots, we implemented a Kalman filter (KF) [23]. The state vector is given in Eq. (10), where $\mathcal{P}_{2n \times 1}$ is the state vector, and $\mathbf{p}_i = (p_i^x, p_i^y)$ is the estimated 2D position of robot i. The equations of the state space model are provided below.

$$\mathcal{P}_{2n \times 1} = \begin{bmatrix} \mathbf{p}_1 & \mathbf{p}_2 & \mathbf{p}_3 \cdots \mathbf{p}_n \end{bmatrix}' \tag{10}$$

The prediction equation is Eq. (11), where $\omega(k) \sim \mathcal{N}(0, \mathbf{R})$ is the process noise at instant k. Finally, we measure the state $\mathcal{P}$ directly by sampling the output of MDS, Eq. (12) where, $\overline{\mathcal{P}}(k)$ is the measurement and $\nu(k) \sim \mathcal{N}(0, \mathbf{Q})$ is the measurement noise.

$$\mathcal{P}(k) = \mathcal{P}(k - 1) + \omega(k) \tag{11}$$

$$\overline{\mathcal{P}}(k) = \mathcal{P}(k) + \nu(k) \tag{12}$$

In order to estimate the process noise ω, we apply an heuristic inspired on the technique used in [14] to estimate relative velocities. However, instead of using the speed between robots, we use the estimated variances of the distances between robots, Eq. (14), allowing us to use the uncertainty of the distance measurements as a measurement of the state progression. In detail, we use the estimated variances of the distances between robots ($\mathbf{W}$), and compute a unit vector $\mathbf{u}_{i,j}$, Eq. (13), pointing from the position of the robot i to the position of robot j. Then, the state variance of robot i ($\mathbf{r}_i$) is calculated as the sum of the absolute value of those vectors projected in the x and y axes, where $\mathbf{r}_i = (r_i^x, r_i^y)$. Finally, we divide the velocity by n, to remove the multiple inclusions of the same variance, obtaining the diagonal matrix $\mathbf{R}$, Eq. (15).

$$\mathbf{u}_{i,j} = (\mathbf{u}_{i,j}^x, \mathbf{u}_{i,j}^y) = \Big(\mathbf{p}_i(k-1) - \mathbf{p}_j(k-1)\Big) / \big|\mathbf{p}_i(k-1) - \mathbf{p}_j(k-1)\big| \qquad (13)$$

$$\mathbf{r}_i = \frac{1}{n} \times \sum_{j=1..n,\, j \neq i} \mathbf{W}(i,j) \cdot \left(\left|\mathbf{u}_{i,j}^x\right|, \left|\mathbf{u}_{i,j}^y\right|\right) \qquad (14)$$

$$\mathbf{R}_{n \times n} = \begin{bmatrix} r_1^x & 0 & 0 & \cdots & 0 & 0 \\ 0 & r_1^y & 0 & \cdots & 0 & 0 \\ \vdots & \vdots & \vdots & \ddots & \vdots & \vdots \\ 0 & 0 & 0 & \cdots & r_n^x & 0 \\ 0 & 0 & 0 & \cdots & 0 & r_n^y \end{bmatrix} \qquad (15)$$

3.4   Integrating Measurements

In order to facilitate the understanding of the remaining of this paper, we will now assume the output of MDS to be in the localisation system state format of the state, i.e. $\begin{bmatrix} x_1 \ y_1 \ x_2 \ y_2 \cdots x_n \ y_n \end{bmatrix}' = \text{MDS}(\mathbf{G})$.

Then, in order to compute the process noise we use a Monte Carlo (MC) approach, according to which we add random noise to the distance inputs, and repeatedly calculate the output, thus sampling the localisation function. With a sufficient number of runs we can determine the impact of the noise on the output distribution. Therefore, we execute MDS with the symmetrical matrix $\mathbf{G}$ obtaining the positions $\overline{\mathcal{P}}(k)$, and adjust the robots positions to minimise the error with relation to the current state estimate. Then, we execute the multidimensional scaling algorithm q times, Eq. (17), where q can be configured according to the precision required and the computational power available. For each of those executions we use as input $\mathbf{G} + a \times \mathbf{H}$, where $a \sim \mathcal{N}(0,1)$, and $\mathbf{H}$ is the matrix of standard deviations, obtained from the element-wise square root of $\mathbf{W}$. To obtain meaningful results after each execution, we adjust the robots positions to minimise the error with relation to $\overline{\mathcal{P}}(k)$. Finally we obtain the covariance matrix $\mathbf{Q} = \text{cov}(\mathbf{M}_z)$, where $z = [1..q]$.

$$\mathbf{H} = \begin{bmatrix} \sqrt{\mathbf{W}_{11}} & \sqrt{\mathbf{W}_{12}} & \sqrt{\mathbf{W}_{13}} & \cdots & \sqrt{\mathbf{W}_{1n}} \\ \sqrt{\mathbf{W}_{21}} & \sqrt{\mathbf{W}_{22}} & \sqrt{\mathbf{W}_{23}} & \cdots & \sqrt{\mathbf{W}_{2n}} \\ \vdots & \vdots & \vdots & \ddots & \vdots \\ \sqrt{\mathbf{W}_{n1}} & \sqrt{\mathbf{W}_{n2}} & \sqrt{\mathbf{W}_{n3}} & \cdots & \sqrt{\mathbf{W}_{nn}} \end{bmatrix} \qquad (16)$$

$$\mathbf{M}_z = \text{MDS}(\mathbf{G} + a \times \mathbf{H}) \qquad (17)$$

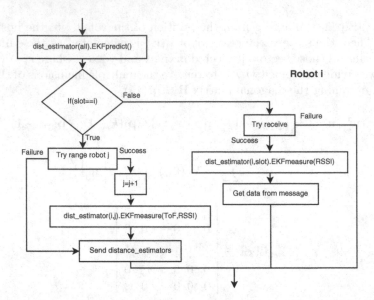

Fig. 2. Simulation loop for robot i – Predict all Kalman filters; If robot is a receiver, try to receive from sender and measure RSSI; If robot is the sender, try to range robot j and measure ToF and RSSI, finally estimate positions

4   Simulation Results

In this section, we first describe the simulation setup, Fig. 2, namely how we generate measurements and how we share the distance estimates between the robots. Then we present the results we obtained using our simulator on ground-truth (GT) collected from the CyberRescue@RTSS2009 competition [7].

4.1   Generating RSSI and ToF Measurements

In order to realistically simulate our proposal, we modelled the sensors measurements from the real experiments performed in [13]. Namely, we generate RSSI values taking into account the hardware 2dBm resolution, Eq. (18), where $\rho(d)$ is the medium propagation using the model in Eq. (19). For the RSSI the parameters are: $\sigma_\rho^2 = 20$, $\rho_0 = -39.6955$, $\alpha = 1.1558$, and $a_\rho \sim \mathcal{N}(0,1)$. In addition, we also generate ToF measurements from real distance according to Eq. (20). In this case $a_d \sim \mathcal{N}(0.3842,1)$ and $\sigma_d^2 = 0.4$.

$$\overline{\rho} = -2 \times \text{round}\left(0 \leqslant -\left(\rho(d) + a_\rho \times \sqrt{\sigma_\rho^2} + 35\right)/2 \leqslant 31\right) - 35 \qquad (18)$$

$$\rho(d) = \rho_0 - 10\alpha\log_{10}(d) \qquad (19)$$

$$\overline{d}(d) = d + a_d \times \sqrt{\sigma_d^2} \qquad (20)$$

4.2 Simulating the Communications

In order to simulate the delays of the information reaching a robot, we have devised a simple communication protocol. A TDMA schedule with slot size of 50 ms was created such that each robot i transmits in slot i, therefore, at each slot the simulator will run the loop in Fig. 2 for each robot. A round robin ranging schedule was implemented such that every i^{th} slot, robot i tries to range one of the other robots, then, in the next i^{th} slot it will try to range the next robot, and so on. Still in the same slot, robot i will broadcast a message containing its distance estimates and distance variance, which the other robots will try to receive. The communications were programmed with a probability of success of 94 % from measurements carried out by [12] in real conditions similar to our simulation, i.e. TDMA rounds with no external interference.

4.3 Results

In order to perform experimental evaluation of our proposal, we used logs collected in a simulated robots competition (CyberRescue@RTSS2009 [7]) using the Cyber-Physical Systems Simulator (CPSS) [18]. We used the GT positions collected in the 9 logs available on the website, as the path the robots travel through. Each log is composed of five robots moving in a 28 m by 14 m arena. For the purpose of this work, the walls were not considered for non-line of sight and reflection effects.

We ran the robots through our simulator, using the paths obtained from the logs to generate simulated measurements. The simulator output was used to run our positions estimator from the perspective of robot 1, and we set to 10 the number of MDS executions required for the measurement of the topology in the Monte Carlo approach. To be able to compare our generated positions to the log GT, we applied the same transformation techniques to the GT data using the positions estimate as the reference, placing them in the same reference frame as our data. Therefore, from this point on, when we mention GT, we are referring to the transformed GT.

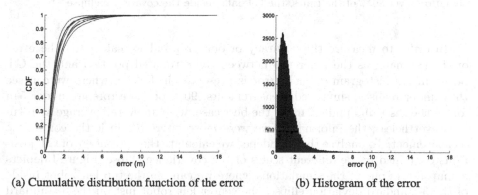

(a) Cumulative distribution function of the error   (b) Histogram of the error

Fig. 3. Error of the position estimates – 90 % of the errors are under 5 m for all cases and below 2 m in the best case, with an overall average of 1.3 m

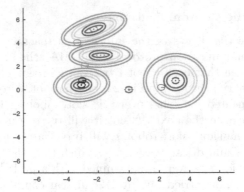

Fig. 4. A zoomed in snapshot of the simulation field – full line robots represent the ground-truth, dotted robots represent the estimations surrounded by the 1-standard deviation ellipse (red), by the 2-standard deviation ellipse (yellow), by the 3-standard deviation ellipse (green) (Color figure online)

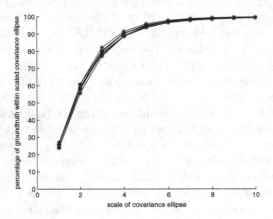

Fig. 5. Percentage of GT positions contained inside of the covariance ellipses at different scales, from 1 to 10 times the standard deviation – For 3 times the standard deviation, over 80 % of the times, the GT falls inside the covariance ellipse

In order to measure the accuracy of our proposal we calculated the error of the estimate, as the difference between the estimated position and the GT position. An histogram of such error is presented in Fig. 3a, where we can see that under realistic simulated measurements, 90 % of the errors are under 5 m for all the cases and under 2 m for the best case, with an overall average at 1.3 m

Nevertheless, the information we were more interested in is the confidence measurement. To analyse the confidence we calculate the percentage of GT positions contained inside different scales of the covariance ellipse. Figure 4 depicts a snapshot of one of the simulations where we can see the top left robot inside of the 3-standard deviation ellipse, the center left robot just on the 3-standard deviation ellipse, the bottom left robot inside of the 1-standard deviation ellipse, and the right robot inside of the 2-standard deviation ellipse. The total percentage of GT inside different scales of the covariance ellipse, from 1 to 10 times the standard deviation, is presented in Fig. 5, for all 9 simulations. We can see that for 3 times the standard deviation, over 80 % of the GT fall inside the covariance ellipse. For 5 times, this value raises to about 95 %.

5   Conclusions and Future Work

In this work we presented a novel technique that uses RF-based range estimates, only, to track the relative positions of a fully mobile team of robots providing reliability information as the covariance of the positions. We believe that this is one of the first works to propose confidence values to position estimates obtained through MDS, particularly for relative localization purposes without any anchors. Using RF-only information, our method could determine the relative localization of a team of 5 freely moving robots with an average error of $1.3\,m$ in a region of $28\times14\,m$. Results show that our approach can consistently provide similar performance across different experiments.

Currently we are implementing this approach on a real robots team, particularly a RoboCup MSL team, with which we plan to validate our proposal in a real scenario.

Acknowledgements. This work was partially supported by the Portuguese Government through FCT (Fun- dação para a Ciência e a Tecnologia) grant SFRH/BD/74292/ 2010 and by the Portuguese and Brazilian Governments through the CAPES-FCT bilateral project PICC. The authors would also like to acknowledge the valuable comments provided by Prof. Traian Abrudan from Oxford University.

References

1. Biswas, J., Veloso, M.: Depth camera based indoor mobile robot localization and navigation. In: 2012 IEEE International Conference on Robotics and Automation (ICRA), pp. 1697–1702, May 2012
2. Borg, I., Groenen, P.J.F.: Modern Multidimensional Scaling: Theory and Applications. Springer Series in Statistics, 2nd edn. Springer, New York (2005)
3. Burgard, W., Moors, M., Fox, D., Simmons, R., Thrun, S.: Collaborative multi-robot exploration. In: Proceedings of IEEE International Conference on Robotics and Automation, ICRA 2000, vol. 1, pp. 476–481. IEEE (2000)
4. Chen, Z.X., Wei, H.W., Wan, Q., Ye, S.F., Yang, W.L.: A supplement to multi-dimensional scaling framework for mobile location: a unified view. IEEE Trans. Signal Process. **57**(5), 2030–2034 (2009)
5. Destino, G., De Abreu, G.: Weighing strategy for network localization under scarce ranging information. IEEE Trans. Wirel. Commun. **8**(7), 3668–3678 (2009)
6. Djugash, J., Singh, S., Kantor, G., Zhang, W.: Range-only slam for robots operating cooperatively with sensor networks. In: 2006 Proceedings of IEEE International Conference on Robotics and Automation, ICRA 2006, pp. 2078–2084, May 2006
7. Facchinetti, T., Vedova, M.D.: Cyberrescue@rtss2009 (2009). http://robot.unipv. it/cyberrescue-RTSS09

8. Ferris, B., Fox, D., Lawrence, N.D.: WiFi-SLAM using Gaussian process latent variable models. In: IJCAI **7**, 2480–2485 (2007)
9. Jamaa, M.B., Koubaa, A., Kayani, Y.: Easyloc: RSS-based localization made easy. Procedia Comput. Sci. **10**, 1127–1133 (2012)
10. Li, H., Almeida, L., Wang, Z., Sun, Y.: Relative positions within small teams of mobile units. In: Zhang, H., Olariu, S., Cao, J., Johnson, D.B. (eds.) MSN 2007. LNCS, vol. 4864, pp. 657–671. Springer, Heidelberg (2007)
11. Montemerlo, M., Thrun, S., Koller, D., Wegbreit, B., et al.: FastSLAM: a factored solution to the simultaneous localization and mapping problem. In: AAAI/IAAI, pp. 593–598 (2002)
12. Oliveira, L., Almeida, L., Santos, F.: A loose synchronisation protocol for managing RF ranging in mobile Ad-Hoc networks. In: Röfer, T., Mayer, N.M., Savage, J., Saranlı, U. (eds.) RoboCup 2011. LNCS, vol. 7416, pp. 574–585. Springer, Heidelberg (2012)
13. Oliveira, L., Di Franco, C., Abrudan, T.E., Almeida, L.: Fusing time-of-flight and received signal strength for adaptive radio-frequency ranging. In: 2013 16th International Conference on Advanced Robotics (ICAR), pp. 1–6. IEEE (2013)
14. Oliveira, L., Li, H., Almeida, L., Abrudan, T.E.: RSSI-based relative localisation for mobile robots. Ad Hoc Netw. **13**(Part B), 321–335 (2014)
15. Olufs, S., Vincze, M.: Embedded vision-based Monte-Carlo robot localisation without additional sensors. In: AFRICON **2013**, 1–6 (2013)
16. Pan, T.W., Hou, T.C.: Localization of moving nodes in an anchor-less wireless sensor network. In: 2012 Wireless Communications and Networking Conference (WCNC), pp. 3112–3116. IEEE, (April 2012)
17. Patwari, N., Hero, A.O., Perkins, M., Correal, N., O'Dea, R.: Relative location estimation in wireless sensor networks. IEEE Trans. Signal Process. **51**(8), 2137–2148 (2003)
18. Pereira, A., Nuno Lau, T.F.: Cyber-physical systems simulator (2014). http://sourceforge.net/projects/cpss/
19. Pinto, M., Moreira, A.P., Matos, A., Sobreira, H., Santos, F.: Fast 3D map matching localisation algorithm. J. Autom. Control Eng. **1**(2), 110–114 (2013)
20. Pinto, M., Sobreira, H., Moreira, A.P., Mendon, H., Matos, A.: Self-localisation of indoor mobile robots using multi-hypotheses and a matching algorithm. Mechatronics **23**(6), 727–737 (2013)
21. Schneider, F.E., Wildermuth, D.: Influences of the robot group size on cooperative multi-robot localisation analysis and experimental validation. Robot. Auton. Syst. **60**(11), 1421–1428 (2012). towards Autonomous Robotic Systems 2011
22. Shioda, S., Shimamura, K.: Cooperative localization revisited: Error bound, scaling, and convergence. In: Proceedings of the 16th ACM International Conference on Modeling, Analysis Simulation of Wireless and Mobile Systems. MSWiM 2013, pp. 197–206. ACM, NY, USA (2013)
23. Thrun, S., Burgard, W., Fox, D.: Probabilistic Robotics. MIT press, Cambridge (2005)
24. Xiao, Z., Wen, H., Markham, A., Trigoni, N.: Lightweight map matching for indoor localisation using conditional random fields. In: Proceedings of the 13th International Symposium on Information Processing in Sensor Networks. IPSN 2014, pp. 131–142, IEEE Press, Piscataway, NJ, USA (2014)
25. Yin, F., Li, A., Zoubir, A., Fritsche, C., Gustafsson, F.: RSS-based sensor network localization in contaminated Gaussian measurement noise. In: 2013 IEEE 5th International Workshop on Computational Advances in Multi-Sensor Adaptive Processing (CAMSAP), pp. 121–124, December 2013

Learning Soccer Drills for the Small Size League of RoboCup

Carlos Quintero[1], Saith Rodríguez[1], Katherín Pérez[1], Jorge López[1],
Eyberth Rojas[1], and Juan Calderón[1,2(✉)]

[1] Universidad Santo Tomás, Bogotá, Colombia
{carlosquinterop,saithrodriguez,andrea.perez,jorgelopez,
eyberthrojas}@usantotomas.edu.co
[2] University of South Florida, Tampa, FL, USA
juancalderon@mail.usf.edu

Abstract. This paper shows the results of applying machine learning techniques to the problem of predicting soccer plays in the Small Size League of RoboCup. We have modeled the task as a multi-class classification problem by learning the plays of the STOx's team. For this, we have created a database of observations for this team's plays and obtained key features that describe the game state during a match. We have shown experimentally, that these features allow two learning classifiers to obtain high prediction accuracies and that most miss-classified observations are found early on the plays.

Keywords: Machine learning · RoboCup SSL · Learning soccer drills · Neural networks · Support vector machines

1 Introduction

One of the big challenges in robotics have been on how to provide the robots with the capability of efficiently using the available information to make decisions without human intervention [5,13]. In a sense, this process is related to the way humans do it. First, there is a step of raw data acquisition from the environment; then, these data are processed and finally a decision is made usually based on a specific reasoning structure or a previous similar experience.

The Small Size League (SSL) of the RoboCup international initiative has become a key scenario to push forward the state of the art in this matter as it provides an ideal framework where all the information of the current game state is available to a global controller that acts as a human coach with the additional capability of being able to transmit customized instructions to each player in the field [16]. Under this scenario, the ability of a team to predict or understand its opponent's strategies during a game seems crucial.

The approach that we propose in this paper resembles that of the human soccer in the following way: a group of experts (coaching group) of a given team is in charge of studying and learning, before the game, the strategies or plays that

© Springer International Publishing Switzerland 2015
R.A.C. Bianchi et al. (Eds.): RoboCup 2014, LNAI 8992, pp. 395–406, 2015.
DOI: 10.1007/978-3-319-18615-3_32

the opponent team has utilized against other teams in previous games. Based on this information, their task is to define specific strategies or actions to defend the team against such plays if these are used during the game. Furthermore, the team must have the ability of predicting which of the studied strategies will be implemented by the opponent team.

Other researchers have pointed out the importance of such ability mostly in the simulation league [12], where the conditions and gameplay are in general different than those found in the SSL. The SSL gameplay is usually fast due to the velocities of the robots and the ball and the relatively small size of the field. We have taken this into consideration in order to propose a learning framework in which the strategies to be learned from the opponent team correspond to soccer drills. These are plays that teams usually execute after a free kick command is issued from the referee box. In this work, we model the situation as a classification problem and use machine learning techniques to perform the task. We also evaluate the performance of the learning machines on unseen data and assess through simulated experiments whether the machines are able to correctly predict the drills or not.

The remainder of this paper is as follows: The following section shows the work that has been carried out by other researchers in similar environments. Then, we show the details of our approach by describing how we model the problem as a classification task, then, showing the set of experiments that were designed to obtained a dataset suitable for the problem and finally describing how we implemented the automatic learners. Section 4 shows the results of such experiments and finally we conclude and comment on future work.

2   Related Work

Researchers all over the globe have strived for providing their robotic soccer teams with the ability of autonomously recognize the opponent's strategies. It seems that all researchers agree in the fact that such information may be highly valuable in order to define the strategy that the team should carry out in a given game situation. However, these works highly vary in the concept that they try to learn and the modeling tools utilized to describe the robotic soccer challenge. This section briefly shows some of these works and highlight the similarities and differences between our work and theirs.

Certain authors have considered to use reinforcement learning as modeling tool to learn multi-agent strategies [1–3]. In these works, the idea is to model the soccer challenge as a state space for sequential decision making based on achieving specific goals. A notable work is shown in [4], where the authors propose learning how to play keepaway soccer using reinforcement learning. Finally, in [5], the authors propose a hierarchical learning methodology to learn skill actions, dynamic role assignment and action selection.

Other common approach is to use Case-Based Reasoning (CBR), a strategy that uses recorded behaviors to compare with the current situation based on a given similarity measure. CBR approaches have been used for action selection in the four-legged league of RoboCup [6], to recognize game state, team formations and positioning of the goalie [7], the behavior of agents in the RoboCup Simulation league [8], action selection [9] and other classification goals [10]. Improvements on these techniques have also been proposed in [9,11]. All these approaches share with our approach the use of historical data to predict unseen behaviors. However, their approach relies on finding a similarity between past and current situations instead of aiming at learning the underlying relationship in the data like ours.

Closest to our work we can find approaches where machine learning techniques such as decision trees, neural networks and SVMs are used as classifiers in order to predict adversary classes [13], opponent formations [14,15], strategies [16], game situations for each actor within the field [17] and actions [18]. The greatest similarity between our work and theirs is the use of features that describe the game state within the field. However, the criteria to be learned is different. To the best of our knowledge, there is no work that aims at learning drills for the RoboCup Small Size League in the way we propose here.

3 Our Approach

The Small Size League (SSL) is a highly traditional and widely known soccer league in the RoboCup initiative. In the league, a global vision system is in charge of acquiring the information of the current game state. This information is processed by a software framework whose main role is deciding in real time the actions that each robot should perform in the next period of time. Finally, this information is transmitted wirelessly to each team player who must be capable of successfully execute the commanded actions.

In the current state of the league each team is made of 6 players capable of moving up to 3 m/s into a 4 m×6 m field using a standard golf ball that reaches velocities of up to 8 m/s. These features provide the league with a highly dynamic and fast gameplay unique among the other leagues within the RoboCup initiative. This scenario has defined a very specific game style in which the ball remains within the field for short periods of time. A large proportion of the game relies on what is known as drills in human soccer: sets of plays that start with a direct kick, free kick or corner kick that human teams create during the training sessions that are used later in the official games and seek to surprise the opponent team to score a goal.

In the SSL, the idea remains the same: a set of plays previously defined by a programmer to perform a sequence of actions that may lead to score a goal. However, in the SSL they play a more significant role since there are more opportunities to perform a drill. Furthermore, a high amount of the goals scored are achieved as consequence of the execution of a drill.

Based on this observation we have proposed to build a learning machine capable of predicting the opponent's drills during a SSL game. For this, we have modeled the problem as a multiclass classification task where each class corresponds to one drill and each observation corresponds the state of the game in the current time frame represented as a set of features. In this framework, each time frame is labeled with the corresponding drill and the learner must assign drills to new unseen frames.

3.1   Learning Soccer Drills

As discussed above, we have modeled the problem of learning soccer drills in the SSL of RoboCup as a classification problem, initially aimed at learning some plays of the STOx's team. We have identified and selected six drills for this team's participation in RoboCup 2013 that we considered could demonstrate a proof of concepts for the problem of learning drills. In general, each drill is characterized by a sequence of movements performed by the robots after the referee issues the order. The identified drills of the STOx's team are be described next:

- **Direct Kick (DK)**: Two attackers move towards the opponent area and stand in front of the goalie. Two defenders remain in their own area. The attacker kicks directly towards the goalie.
- **Pass to Opposite Attacker (POA)**: Two attackers move towards the opponent area and stand in front of the goalie. Two defenders initially remain in their own area together with their goalie. However, one of them moves forward and stands in the opposite side of the ball. Next, the two attackers split up, the defender goes back and the kicker passes the ball to the opposite attacker. See Fig. 1.
- **Pass to Peer Attacker (PPA)**: Two attackers move towards the opponent area and stand in front of the goalie. Two defenders initially remain in their own area together with their goalie. However, one of them moves forward and stands in the opposite side of the ball. Next, the two attackers open up, the defender goes back and the kicker pass the ball to the attacker closest to it. Unlike the POA, when the two attackers open up, the peer attacker also moves backward in order to receive the pass properly.
- **Pass to Opposite Defender (POD)**: Two attackers move towards the opponent area and stand in front of the goalie. Two defenders initially remain in their own area together with their goalie. However, one of them moves forward and stands in the opposite side of the ball. Next, the kicker pass the ball to such defender.
- **Pass to Middle Defender (PMD)**: Two attackers move towards the opponent area and stand in front of the goalie. Two defenders initially remain in their own area together with their goalie. However, one of them moves forward and stands ahead the middle of the field. Next, the kicker pass the ball to such defender. See Fig. 2.
- **Five Attackers (FA)**: All the team members move forward to the opponent's area and get ready to receive the pass. The kicker pass the ball to the robot that is farther away.

Figures 1 and 2 show the most important steps of drills **POA** and **PMD** respectively as described above. The arrows show the movement of a robot or the ball

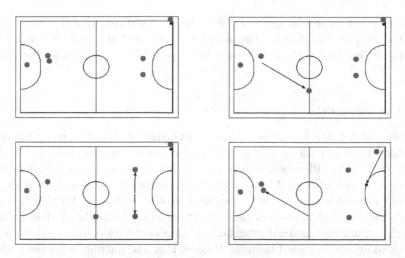

Fig. 1. Sequence of **Pass to Opposite Attacker** drill. The robots initially form in attack position. Then, one defender moves towards the center of the field. Next, the two attackers move apart of each other, the defender goes back and the kicker kicks the ball towards the opposite attacker.

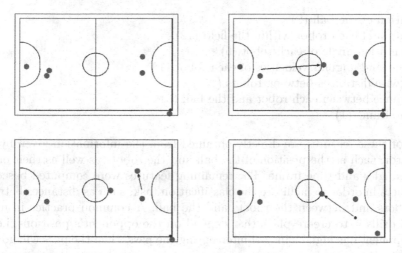

Fig. 2. Sequence of **Pass to Middle Defender** drill. The robots initially form in attack position. Then, one defender moves ahead the middle of the field. Next, the opposite attacker slightly moves and the kicker kicks the ball towards the defender in the middle.

from one figure to the next. This specific sequence of steps will allow the automatic classifier to predict the corresponding drill during a game. Notice that this approach requires at first human intervention in order to identify and label each frame's observation within the logs of the opponent team. Although this may seem burdensome, it mimics the behavior of human coaches before playing a game in real soccer. Furthermore, automatic identification and recognition of drills may be possible through the use of unsupervised learning techniques that find similarities in the data. However, such possibility requires further exploration.

3.2   The Dataset

We have built a dataset especially suited for the task at hand by using the grSim simulator. Although we do not show experiments in the real robots, we claim that there is little difference with respect to build the classifer with simulated data since the features used to perform the training are attributes such as positions and distances that are well emulated by the simulator. Furthermore, the simulator also considers the gaussian noise related to the artificial vision system.

Our program automatically begins the data acquisition immediately after a free kick command is issued from the Referee Box and outputs a log file with the information of each experiment. For each experiment, we have stored key features that describe the current game state and that provides enough information to allow the classifier perform its task. The features stored within the log files are mentioned below with the number of related features in brackets:

- Position of the ball (2)
- Position of each robot within the field (12)
- Orientation angle of each robot (6)
- Linear and angular velocities of each robot (18)
- Pairwise distances between robots (15)
- Distance between each robot and the ball (6)
- Time frame (1)

Some of these features are directly obtained from the information given by the simulator, such as the position of the ball and the robots as well as their orientation angles and time frame. The remaining features were computed based on the others in order to facilitate the classification task, such as distances between the robots and between the robots and the ball. A common practice in many team's drills is to create plays that depend on the opponent's positions, i.e., if one opponent has high chance of intercepting the pass, then the pass will go to a different player. We have performed experiments using the opponent's positions (not shown here) with a dynamic behavior and verified that these additional features make no difference in terms of the classifier's prediction accuracy. However, might be used to create dynamic defense strategies. For more complex plays sets, the opponent's positions could also be added to improve the learning capability of the classifier.

We have performed 24 repetitions of each drill with each defense strategy; 3 from the top of the field and 3 from bottom for a total of 144 runs. Each run records data at a rate of 60 frames per second.

3.3 Automatic Learners

Machine Learning (ML) is a discipline of computer science and statistics that seeks to build systems capable of improving their performance based on historical data. Usually, a learning machine is designed to perform tasks such as regression or classification by finding the underlying relationship within the data. Our approach consists on using the built dataset to create automatic learners capable of correctly classify observed frames into drills. For this, we have decided to perform experiments using two traditional ML techniques, namely Support Vector Machines (SVM) and Neural Networks (NN) to finally choose the one with highest prediction accuracy.

The artificial neural networks are well stablished learning machines that have been succesfully used in a variety of applications where historical data is available. They are systems made of connected nodes capable of approximating nonlinear functions of their weighted inputs. Training algorithms are usually designed to find suitable values for the connection weights according to a desired criteria. We have chosen a feed-forward multilayer perceptron architecture for the NN with one hidden layer and sigmoid activation function to perform the task. Additionally, the training process is performed using the Levenberg-Marquardt method, an optimization algorithm that solves nonlinear least square problems to fit a parameterized function to measured data. Its main advantage is that it combines two minimization methods: the gradient descent method and Gauss-Newton method.

The SVMs have become of paramount importance in the statistical learning community for the last two decades as well as in a variety of application areas where automatic classification, ranking, regression and pattern recognition tasks are required. The underlying concept behind SVMs is to find the hyperplane that separates instances of two different classes with maximum margin, i.e., the largest distance between the hyperplane and the closest training data point. More formally, the SVM algorithm solves the following optimization problem:

$$\arg\min_{w,\zeta,b} \frac{1}{2}\|w\|^2 + C\sum_{i=1}^{n} \zeta_i \tag{1}$$

subject to $y_i(wx_i - b) \geq 1 - \zeta_i$, where $\zeta_i > 0 \forall i = 1,\ldots n$ are slack variables that allow missclassification of data points for not linearly separable data, w is a vector that defines the slope of the separating hyperplane and C is a regularization parameter that defines a balance between large margin and the amount of missclassificated data.

Finally, the SVMs have been enhanced to perform nonlinear classification through the use of what is known as the kernel trick. The idea is to find a mapping between the original data space and a feature space, usually in a higher dimension with the idea that the hyperplane in the transformed space becomes a nonlinear surface in the original space. Such mapping is achieved by using kernel functions; a set of functions that define the new transformed space. Some common kernels used in general classification and regression problems include the polynomial kernel, the gaussian kernel and the hyperbolic tangent kernel. The gaussian kernel has the following relation:

$$k(x_i, x_j) = \exp(-\gamma \|x_i - x_j\|^2) \text{ for } \gamma > 0 \tag{2}$$

An important issue that needs to be solved in the classification learning framework is that of choosing the correct parameters for the automatic learners. For the case of the NN, the number of neurons in the hidden layer needs to be decided. For the SVMs, the regularization parameter C and kernel parameter γ require a tuning process. This process is carried out by using k-fold cross validation, a statistical procedure used to evaluate the generalization capability of a statistical analysis in a predictive model. The entire dataset is partitioned into k equal size subsamples and $k - 1$ of those subsamples are used for training. The remaining subsample is used to test the classifier's generalization capability. This process is repeated k times with each subsample used once as validation data and the idividual results are averaged. Cross validation avoids the possibility of overfitting the model which happens when the model complexity is too high compared to the complexity of the data, leading to poor generalization capabilities. The importance of the cross validation procedure is that the generalization error is calculated in data that have not been used during the training process.

4   Results

We begin by performing an unsupervised study of the features used to describe the game state as presented in Sect. 3.2 using the technique of Principal Component Analysis (PCA). In a PCA analysis the idea is to define a transformation from the original variables to a new set of variables with the constraint that they are orthogonal among them. We have applied the PCA algorithm on the entire dataset and found that 37 of the transformed variables are capable of explaining the 99.12 % of the variance within the data. This means that there is redundancy between the original variables and that a smaller subset of new features (37 out of 60) is enough to explain the information in the data. Notice that this is to be expected since certain features were computed based on others.

We have performed the training of the NN and the SVM with the built dataset of the STOx's drills using the Neural Network Toolbox of MATLAB and LIBSVM [19] respectively. We have used 10-fold cross validation in order to assess the generalization capability of each classifier together with a grid search on their respective parameters. For the NN, the parameter to be tuned is the number of neurons in the hidden layer. In the case of the SVM, the grid search is on the C and γ parameters and included the following values: $C = [2^{-3}, 2^{-1}, 2^1, 2^3, 2^5, 2^7, 2^9]$ and $\gamma = [2^{-7}, 2^{-5}, 2^{-3}, 2^{-1}, 2^1, 2^3, 2^5]$.

The training and testing procedures are performed for every combination of parameters for each classifier and the results with highest generalization accuracy are chosen in Table 1. On one hand, the training accuracy corresponds to the proportion of data points correctly classified on the data used to perform the training process. For the NN this value is related to the nonlinear function found after the Levenberg-Marquardt algorithm is executed and the quality of the optimal solution. In the case of the SVMs, this value is related to the balance between the parameters C and γ. On the other hand, the generalization accuracy is the proportion of data points correctly classified when using the test dataset (i.e., data points that were not used during the training phase). High values of the generalization accuracy ensure that there is no overfit on the constructed model.

Table 1. Results of the training and testing processes for the SVM and NN classifiers.

	SVM	NN
Algorithm parameters	$C = 512$	nNeurons $= 20$
	$\gamma = 0.0078$	MSE $= 0.2048$
	nSV $= 19637$	
Training accuracy	96.69 %	82.44 %
Generalization accuracy	91.45 %	81.96 %

It is noteworthy that the SVM attains a higher generalization accuracy by achieving 91.45 % compared to that of the NN which achieves 81.96 %. However, these results show that both classifiers are capable of correctly predict unseen frames.

We have also calculated the prediction accuracy of each classifier per class. In the Table 2 we can see that the frames of certain drills are easier to predict than others. For instance, it seems clear that the drill **FA** is significantly easier than all other drills. This makes sense since the configuration of the players is unique among the drills. There is no other drill in which all players move close to the opponent goal. On the contrary, drills **POA**, **PPA** and **POD** are more similar in terms of the robot's position at the beginning of the drill. In all of them, the initial position of the two attackers is very similar as well as the position of the defender that moves towards the opponent zone. This is reflected in their lower prediction accuracy for both classifiers.

Table 2. Prediction accuracy of each classifier per class.

Drill	SVM (%)	NN (%)
DK	94.75	76.81
POA	89.17	78.03
PPA	88.35	82.99
POD	86.01	73.60
PMD	93.79	82.78
FA	98.2	95.81

Finally, we have calculated the evolution of missclassified frames in one example of each drill by computing the accumulated number of errors per frame during the episodes in order to analyze the prediction pattern of each drill. These results are shown in Fig. 3.

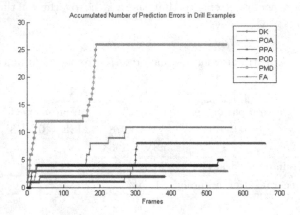

Fig. 3. Accumulated number of errors per frame in one example of each drill.

It is noteworthy that most classification errors occur mostly at the initial frames and no more after certain point of the episodes. It seems that there is a point where the drill becomes completely clear to the classifier which makes sense since all drills have similar frames at the beggining and become different after certain time. This information is highly valuable since it may be involved in the creation of dynamic defense strategies and it confirms that there is enough time for the team to defend against the executed drill.

A video showing episodes of each drill and the prediction performed by the SVM can be found in [20].

5   Conclusions and Future Work

We have demonstrated experimentally that it is possible to predict the play that certain team will perform in a SSL game of RoboCup by using historical data. We have done this by using machine learning techniques and modeling the task as a multi-class classification problem of learning the frames that are part of the drills. We have achieved high generalization accuracy with two learning algorithms namely support vector machines and neural networks using a thoroughly tuned process.

These results were achieved by using a dataset build from the strategies of the STOx's team in the Small Size League of RoboCup in 2013. However, we claim that the approach can be seamlessly extended to learn the drills of all teams that have participated in any RoboCup tournament for which data logs are available. Certain challenges still remain, such as managing the imbalance in the number of data points per class due to the fact that some drills may be more common than others.

At this point, the framework requires human intervention in order to know the number of drills of each team and label the data according to the logs. As future work, we plan to design a learning strategy capable of automatically obtained such information in order to assess the work. Also, the next step of this work consists on defining strategies to defend against the predicted drills and implement them during games in order to evaluate the impact of this approach on real games. Finally, a process to evaluate the features used to predict the drills is also of interest.

Acknowledgements. This work has been funded by "Octava Convocatoria Interna de Proyectos de Investigación FODEIN 2014 #048" at Universidad Santo Tomás Colombia, entitled "Desarrollo de un entorno para la validación de técnicas de robótica cooperativa".

References

1. Salustowicz, R.P., Wiering, M.A., Schmidhuber, J.: Learning team strategies: soccer case studies. Mach. Learn. **33**, 263–282 (1998)
2. Stone, P.: Layered Learning in Multiagent Systems: A Winning Approach to Robotic Soccer. MIT Press, United States (2000)
3. Bianchi, R. A., Ribeiro, C. H., Costa, A. H.: Heuristic selection of actions in multi-agent reinforcement learning. In: Proceedings of the 20th International Joint Conference on Artificial Intelligence, pp. 690–696. Morgan Kaufmann Publishers Inc., San Francisco (2007)
4. Stone, Peter, Kuhlmann, Gregory, Taylor, Matthew E., Liu, Yaxin: Keepaway soccer: from machine learning testbed to benchmark. In: Bredenfeld, Ansgar, Jacoff, Adam, Noda, Itsuki, Takahashi, Yasutake (eds.) RoboCup 2005. LNCS (LNAI), vol. 4020, pp. 93–105. Springer, Heidelberg (2006)
5. Duan, Y., Liu, Q., Xu, X.: Application of reinforcement learning in robot soccer. Eng. Appl. Artif. Intell. **20**, 936–950 (2007)
6. Karol, A., Nebel, B., Stanton, C., Williams, M.-A.: Case based game play in the RoboCup four-legged league. In: Polani, D., Browning, B., Bonarini, A., Yoshida, K. (eds.) RoboCup 2003. LNCS (LNAI), vol. 3020, pp. 739–747. Springer, Heidelberg (2004)
7. Marling, C., Tomko, M., Gillen, M., Alexander, D., Chelberg, D.: Case-based reasoning for planning and world modeling in the RoboCup small size league. In: IJCAI Workshop on Issues in Designing Physical Agents for Dynamic Real-time Environments, Acapulco (2003)
8. Wendler, J., Bach, J.: Recognizing and predicting agent behavior with case based reasoning. In: Polani, D., Browning, B., Bonarini, A., Yoshida, K. (eds.) RoboCup 2003. LNCS (LNAI), vol. 3020, pp. 729–738. Springer, Heidelberg (2004)
9. Ros, R., López de Màntaras, R., Arcos, J.-L., Veloso, M.M.: Team playing behavior in robot soccer: a case-based reasoning approach. In: Weber, R.O., Richter, M.M. (eds.) ICCBR 2007. LNCS (LNAI), vol. 4626, pp. 46–60. Springer, Heidelberg (2007)

10. Steffens, T.: Adapting similarity-measures to agent types in opponent modelling. In: Workshop on Modeling Other Agents from Observations at AAMA, pp. 125–128, New York (2004)
11. Ahmadi, M., Lamjiri, A.K., Nevisi, M.M., Habibi, J., Badie, K.: Using two-layered case-based reasoning for prediction in soccer coach. In: International Conference of Machine Learning. Models, Technologies and Applications, pp. 181–185. CSREA Press, Las Vegas (2003)
12. Lattner, A.D., Miene, A., Visser, U., Herzog, O.: Sequential pattern mining for situation and behavior prediction in simulated robotic soccer. In: Bredenfeld, A., Jacoff, A., Noda, I., Takahashi, Y. (eds.) RoboCup 2005. LNCS (LNAI), vol. 4020, pp. 118–129. Springer, Heidelberg (2006)
13. Riley, P., Veloso, M.: On behavior classification in adversarial environments. In: Parker, L.E., Bekey, G., Barhen, J. (eds.) Distributed Autonomous Robotic Systems, pp. 371–380. Springer, Tokyo (2000)
14. Visser, U., Drcker, C., Hbner, S., Schmidt, E., Weland, H.: Sequential pattern mining for situation and behavior prediction in simulated robotic soccer. In: Stone, P., Balch, T., Kraetzschmar, G.K. (eds.) RoboCup 2000. LNCS (LNAI), vol. 2019, pp. 391–396. Springer, Heidelberg (2001)
15. Faria, B.M., Reis, L.P., Lau, N., Castillo, G.: Machine learning algorithms applied to the classification of robotic soccer formations and opponent teams. In: 2010 IEEE Conference on Cybernetics and Intelligent Systems, pp. 344–349. IEEE Press, Singapore (2010)
16. Trevizan, F., Veloso, M.: Learning opponent's strategies in the RoboCup small size league. In: Proceedings of the AAMAS 2010 Workshop on Agents in Real-time and Dynamic Environments, pp. 45–52, Toronto (2010)
17. Konur, S., Ferrein, A., Lakemeyer, G.: Learning decision trees for action selection in soccer agents. In: Proceedings of the ECAI-04 Workshop on Agents in Dynamic and Real-time Environments. IOS Press, Valencia (2004)
18. Ledezma, A., Aler, R., Sanchís, A., Borrajo, D.: Predicting opponent actions by observation. In: Nardi, D., Riedmiller, M., Sammut, C., Santos-Victor, J. (eds.) RoboCup 2004. LNCS (LNAI), vol. 3276, pp. 286–296. Springer, Heidelberg (2005)
19. Chang, C.-C., Lin, C.-J.: LIBSVM: A library for support vector machines. ACM Trans. Intell. Syst. Tech. 2, 1–27 (2011). http://www.csie.ntu.edu.tw/~cjlin/libsvm
20. STOx's team webpage. http://www.stoxs.org/index.php/en/projects/robocup-ssl-2

Fast Path Planning Algorithm for the RoboCup Small Size League

Saith Rodríguez[1], Eyberth Rojas[1], Katherín Pérez[1], Jorge López[1],
Carlos Quintero[1], and Juan Calderón[1,2](✉)

[1] Universidad Santo Tomás, Bogotá, Colombia
{saithrodriguez,eyberthrojas,andrea.perez,jorgelopez,
carlosquinterop,juancalderon}@usantotomas.edu.co
[2] University of South Florida, Tampa, FL, USA
juancalderon@mail.usf.edu

Abstract. Plenty of work based on the Rapidly-exploring Random Trees (RRT) algorithm for path planning in real time has been developed recently. This is the most used algorithm by the top research teams in the Small Size League of RoboCup. Nevertheless, we have concluded that other simpler alternatives show better results under these highly dynamic environments. In this work, we propose a new path planning algorithm that meets all the robotic soccer challenges requirements, which has already been implemented in the STOx's team for the RoboCup competition in 2013. We have evaluated the algorithm's performance using metrics such as the smoothness of the paths, the traveled distance and the processing time and compared it with the RRT algorithm's. The results showed improved performance over RRT when combined measures are used.

Keywords: Path planning · Mobile robots · Real-time systems · RoboCup

1 Introduction

Collision-avoidance path planning has been a major challenge for robotics researchers. Different proposals have been made to address individual, but contrasting, requirements, such as following the shortest or smoothest trajectory or minimizing processing time. Our system has the goal to quickly find a smooth path with a low computational cost without ensuring that it is the shortest trajectory.

RoboCup has reveal a strong and rapid need to develop efficient path planning algorithms for complex environments. It is a global initiative in which researchers around the world present their best developments in the topics of robotics, artificial intelligence and related areas [1]. Based on the RoboCup initiative, every year many tournaments are held in different countries around the world, therefore teams participate in various disciplines [2].

© Springer International Publishing Switzerland 2015
R.A.C. Bianchi et al. (Eds.): RoboCup 2014, LNAI 8992, pp. 407–418, 2015.
DOI: 10.1007/978-3-319-18615-3_33

For the case of the Small Size League (SSL) [3], the challenge is a soccer contest in which full autonomous robots are able to cooperate to score and win a match. Its artificial vision system sends field images at a rate of 60 fps [4], so path planning and intelligence processing must be made within the span of 16 ms. Thus, the challenge involves a highly dynamic multi-agent environment which implies the need for obstacles avoidance and fast path planning algorithms.

Initially, we present a short review of related work, specifically describing the RRT [5] algorithm, some applications and results that this algorithm generates in situations with several obstacles. After that, we describe in detail the proposed algorithm and we suggest a set of benchmarking scenarios in the context of this league, including real game situations from the RoboCup 2013 competition.

Then, we show the results of our performance analysis for both algorithms over the proposed scenarios. We evaluated specific attributes such as path smoothness, distance traveled and the processing time. Finally, we include conclusions and future work sections.

2   Related Work

The problem of path planning under dynamic environments has been tackled by a variety of researchers. Most of them have focused on optimizing specific performance measures such as obtaining smooth paths in the less amount of time as possible. For example, Tsubouchi et al. analyze the behavior of a single robot within a multi-obstacle dynamic environment [6]. The navigation scheme in this work is based on a heuristic and assumes that obstacles move with a piecewise constant velocity.

On the Small Size League challenge, the robotic players have omnidirectional traction and the dynamic characteristic of the environment have to be carefully taken into account for a proper navigation. Han et al. studied the control of multiple non-holonomic robotic agents in which half the obstacles are non-controllable opponents whose dynamic patterns are unknown [7]. The algorithm in this work creates a set of halfway points between the initial robot position and its final destination. These points are calculated based on the evaluation of potential blockages in the route.

Kuffner et al. presented an algorithm based on Rapidly-exploring Random Trees (RRT) [8]. This algorithm is specifically suited to overcome the constraints that arise in dynamic environments. RRT creates the path that should be followed by an agent from its initial position to a target point by iteratively building search trees that quickly explore the environment. The general procedure is divided in 5 main processes as shown in Fig. 1. However, we have found that this approach can still be improved with regards to computational cost and path smoothness, in contexts like the RoboCup SSL environment.

The result of applying the RRT algorithm to an environment with a set of obstacles is shown in Fig. 2. Where the green region represents the obstacles, the white region is the obstacle-free configuration space, blue branches are the RRT, and the black line is the solution path between the starting configuration (blue) and the goal configuration (red).

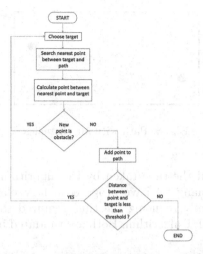

Fig. 1. Diagram RRT

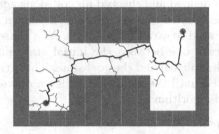

Fig. 2. Example RRT (Color figure online)

The RRT algorithm has been widely embraced in a variety of applications in different environments. This has been especially true for the teams that participate in the small size league of the RoboCup initiative. For instance, Bruce et al. showed a RRT-based planning system in simulation and its implementation on actual robots [9]. This algorithm is presented as ERRT and aims at reducing the cost of searching the nearest point to the path that is being build when compared to the original RRT. This feature increases the efficiency of the path planning procedure for real time applications even in dynamic and continuous environments.

Desaraju et al. presented a Decentralized Multi-Agent Rapidly-exploring Random Tree (DMA-RRT) algorithm [10]. This approach allows to perform an efficient planning by considering complex environments. It uses a coordination strategy to dynamically update the order in which the robots carry out their individual planning.

3 Proposal

After having reviewed path planning algorithms for dynamic environments, we have implemented RRT and validated it in simulated game situations. After

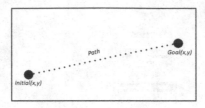

Fig. 3. Path without obstacles

that, we have measured the time taken by the algorithm to generate paths in complex situations. Finally we have proposed a new planning algorithm that is capable of dramatically reducing the time required to create the path and compared it with the RRT algorithm. Both are validated in real game situations of the RoboCup 2013.

The proposed algorithm is based on generating straight trajectories between an initial state and a goal state. To accomplish this, an initial straight trajectory between these points is defined and checked for collisions against all obstacles. If there is no obstacle, the selected route is returned, as shown in Fig. 3. Otherwise, a subgoal state is generated to avoid the obstacle. As a consequence, the original trajectory is split in two: one between the initial state and the subgoal, and another one between the subgoal and the goal state. Then, these new trajectories are recursively evaluated until the algorithm finds an obstacle-free path. Below we show the developed algorithm:

Pseudocode of the proposed path planning

```
function FastPathPlanning (environment,trajectory,depth)

  obstacle = trajectory.Collides(environment)

  if theres is an obstacle and depth < max_recursive then
  {
    subgoal=SearchPoint(trajectory,obstacle,environment);

    trajectory1=GenerateSegment(trajectory.start,subgoal);
    trajectory1=FastPathPlanning(environment,trajectory1,depth+1);
    trajectory2=GenerateSegment(subgoal,trajectory.goal);
    trajectory2=FastPathPlanning(environment,trajectory2,depth+1);
    trajectory=JoinSegments(trajectory1,trajectory2);
  }

  return trajectory;
```

The function *Collides* determines if there is any obstacle in the trajectory and if this is the case, the function returns the position of the closest obstacle in the trajectory.

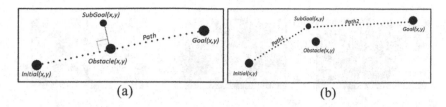

Fig. 4. (a) Subgoal selection. (b) New paths created with subgoal

The function *SearchPoint* assigns a new point (subgoal) at the side of this obstacle. This point is located from the obstacle to a distance equal to the robot diameter and 90 or −90 degrees related to the path between the initial point to the obstacle (the sign is a function parameter). See Fig. 4a. In the case the new point collides with an obstacle, the function keeps moving the point one robot diameter in the same direction, until an obstacle free point is found.

The function *GenerateSegment* generates a straight path between two points. It is used to create two new paths. The first one between the initial state and the subgoal and the second one between the subgoal and the goal state. These new paths will be analyzed recursively as shown in Fig. 4b. Figure 5 shows all the steps of execution until an obstacle free path is found.

Finally, the returned paths should be joined; this is performed by the function *JoinSegments*. It returns the path to reach the target point avoiding obstacles. Figure 6 shows an example of a game situation in which the algorithm performed recursion twice.

As we mentioned before, two possible subgoal points may be returned by the function *Collides*, one at 90 degrees and another at −90 degrees. This means that the obstacles will always be avoided in the same direction, either always in the 90 degree direction, or in the other one. In order to optimize the path length, both options are tested and the shortest path is chosen.

The Fig. 7a shows the two possible paths found by the algorithm in a scenario with multiple obstacles. Figure 7b shows a continuous line for the trajectory to be followed.

Finally, at Fig. 8 we present the solution found by the algorithm in a sample random scenario with multiple obstacles.

4   Results and Evaluation

The evaluation and comparison of the algorithms involved three different metrics. Namely, processing time required for the path generation, path smoothness and total path length. Additionally, a weighted sum evaluation (Eq. 1) of the trajectories was made according to Xiao's proposal [11] with some modifications, where W_t, W_s and W_d are the weights assigned to each criterion according to the desired relevance of processing time, smoothness and distance traveled respectively.

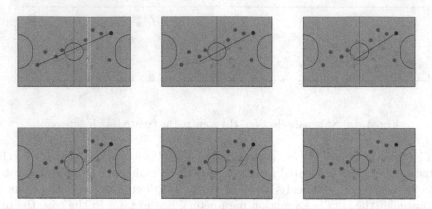

Fig. 5. The figures show the recursive steps of the algorithm. At each step, the black line shows the computed straight path and the magenta segments denote the alternative route to avoid the found obstacle. After computing an alternative route, the algorithm is applied over it, recursively (Color figure online).

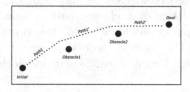

Fig. 6. Example of the proposed algorithm with recursion

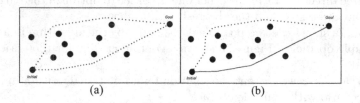

Fig. 7. (a) Possible paths (b) Path selection

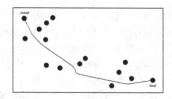

Fig. 8. Solution found by the algorithm in a random scenario with multiple obstacles

$$Eval(p) = W_t time(p) + W_d dist(p) + W_s smooth(p) \qquad (1)$$

We define *time*, *smooth* and *dist* as:

- *time(p)*, processing time
- $dist(p) = \sum_{i=1}^{n-1} d(m_i, m_{i+1})$, total path length, where $d(m_i, m_{i+1})$ is the distance between two adjacent nodes m_i and m_{i+1}.
- *smooth(p)* measures the smoothness of the path, defined as

$$smooth(p) = \frac{\sum \theta_i}{dist(p)}$$

where θ_i is the angle between adjacent segments of the route.

According to Eq. 1, smaller $Eval(p)$ values represent better performance measures. In addition, for the case of Small Size League, we assign more relevance to processing time $W_t = 0.5$, next in significance the smoothness $W_s = 0.3$ and finally the path length $W_d = 0.2$. As *time*, *smooth* and *dist* variables must be normalized, we normalized each variable using the higher value obtained for it from both RRT and the proposed algorithm.

The algorithm evaluation process was performed in two parts. The first one involves static random scenarios and second one includes real game dynamic environments taken from the RoboCup 2013 contest.

4.1   Part 1: Static Scenario

We consider two different game cases and for each one we generate 100 paths using the RRT algorithm. From these paths, we take average values of each one of the three criteria and then we compare with the trajectory generated by the proposed algorithm in the same case. We did not run our algorithm several times, since it would generate the same result every time, as it is not randomized.

Case 1. Figure 9a shows the trajectories generated by both algorithms in the shortest time. According to criterion 1 (processing time), the path generated the faster by RRT algorithm took less time than our proposal's. Figure 9b shows processing time for the 100 trajectories generated by RRT (green), their average (blue) and the time spent by the proposed algorithm (red). The obtained results are shown in Table 1.

Regarding the second criterion (smoothness of the path), Fig. 10 and Table 2 exhibit the results obtained for the same case. Last, in Fig. 11 and Table 3 we show the results according to the third criterion (path length).

The performance evaluation of this case is shown in Table 4. The path generated by the proposed algorithm is 3.4 times better than the average of those generated by RRT.

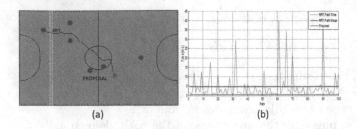

(a)   (b)

Fig. 9. (a) Proposal and RRT algorithm (Path that took less time to create). (b) Results Case 1, Criterion 1 (Process time) (Color figure online)

Table 1. Case 1 results: processing time analysis

Case 1, Criterion 1 (time)			
Algorithm	Minimum(ms)	Maximum(ms)	Mean(ms)
RRT	0.4223	40.3556	4.7193
PROPOSAL	0.5303	0.5303	0.5303

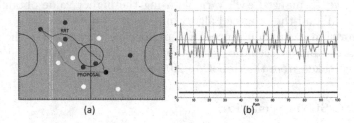

(a)   (b)

Fig. 10. (a) Proposal and RRT algorithm (smoother path). (b) Results Case 1, Criterion 2 (path smoothness)

Table 2. Case 1 results: smoothness analysis of path

Case 1, Criterion 2 (smoothness)			
Algorithm	Minimum(rad/m)	Maximum(rad/m)	Mean(rad/m)
RRT	2.378	5.148	3.668
PROPOSAL	0.3447	0.3447	0.3447

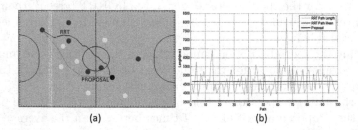

(a)   (b)

Fig. 11. (a) Proposal and RRT algorithm (shortest path). (b) Results Case 1, Criterion 3 (Path length)

Table 3. Case 1 results: path length analysis

Case 1, Criterion 3 (length)			
Algorithm	Minimum(mm)	Maximum(mm)	Mean(mm)
RRT	3780	8267	4667
PROPOSAL	3569	3569	3569

Table 4. Case 1 results: weighted performance evaluation

Case 1, Evaluation			
Algorithm	Time(p)	Smooth(p)	Dist(p)	Eval(p)
RRT	0.116	0.712	0.564	0.385
PROPOSAL	0.013	0.066	0.431	0.113

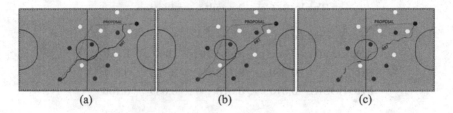

(a)   (b)   (c)

Fig. 12. Case 2. (a) Fastest generated path. (b) Smoother path. (c) Shortest path

Case 2. Figure 12 shows the second case, with (a) the fastest generated path by RRT, (b) the smoothest and (c) the shortest in length. All the results for this case are shown in Table 5.

The evaluation of this case is shown in Table 6. We observe that for this game situation, the proposed algorithm is 3 times better than RRT.

4.2   Part 2: Dynamic Scenario

For this case, we took data from a segment of one game of RoboCup 2013 (STOx's vs CMDragons). There are 281 continuous scenarios in total and for each one we generated trajectories using both RRT and the proposed algorithm. Full data of this match can be obtained from [12] and a video showing the segment can be found in [13].

The obtained results are shown in Fig. 13. This graphic presents (a) processing times for the 281 trajectories for both algorithms, (b) the smoothness and (c) distance traveled. Table 7, consolidates results obtained from both algorithms.

Finally, Table 8 shows the evaluation of both algorithms for the dynamic scenario and results indicate that the average of the proposed algorithm performance is 2.4 times better than RRT's.

Table 5. Case 2 results

Case 2				
Criterion	Algorithm	Minimum	Maximum	Mean
Time(ms)	RRT	1.710	68.3259	17.7726
	PROPOSAL	0.6523	0.6523	0.6523
Smooth(rad/m)	RRT	2.143	4.131	31.708
	PROPOSAL	0.3234	0.3234	0.3234
Dist(mm)	RRT	4842	7647	5776
	PROPOSAL	4883	4883	4883

Table 6. Case 2 results: evaluation

Case 2, Evaluation				
Algorithm	Time(p)	Smooth(p)	Dist(p)	Eval(p)
RRT	0.260	0.768	0.755	0.511
PROPOSAL	0.010	0.078	0.639	0.156

Table 7. Dynamic scenario results

Dynamic scenario				
Criterion	Algorithm	Minimum	Maximum	Mean
Time(ms)	RRT	1.16	35.595	7.891
	PROPOSAL	0.052	15.051	1.494
Smooth(rad/m)	RRT	1.398	6.178	2.546
	PROPOSAL	0.001	0.871	0.149
Dist(mm)	RRT	6062	9445	8310
	PROPOSAL	6344	7199	6728

Table 8. Dynamic scenario evaluation

Case 1, Evaluation				
Algorithm	Time(p)	Smooth(p)	Dist(p)	Eval(p)
RRT	0.222	0.412	0.880	0.410
PROPOSAL	0.042	0.024	0.712	0.171

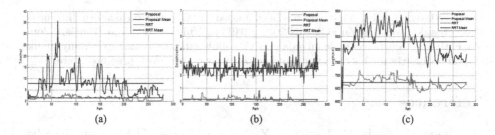

Fig. 13. Dynamic Scenario. (a) Time results. (b) Smooth results. (c) Dist results.

5   Conclusions

We consider RRT to be a robust and flexible algorithm when applied to path planning. However, we observe that its generality can turn into its own weakness, when aspects like processing time or path smoothness are critical. Thus, we have proposed an ad-hoc heuristic for path planning in non-cluttered dynamic environments. We have tested our approach over a set of artificial and real RoboCup Small Size League situations, and we have found it finds smoother and shorter paths faster in the average case.

It is clear that the good performance of the proposed algorithm is related with the constraints of the environment where we apply it. However, we claim that the characteristics of the SSL league with regards to path planning are similar to many other planning situations, e.g. autonomous car driving.

When testing the proposed heuristic, we have found that it outperforms the RRT implemented version by approximately 30 % on average, using the proposed metrics. Additionally, we have also observed that the proposal outperforms RRT in all three individual metrics, i.e. path length, smoothness and processing time.

After competition in RoboCup 2013, we can assert that the proposed algorithm generated paths avoiding obstacles and also processed all necessary data in real time, which was crucial to be among the top eight teams in the SSL.

6   Future Work

We propose to test the new algorithm on more complex environments to assess the validity of our claims regarding the applicability of our heuristic.

According with the outcomes of these tests, we will modify algorithm to make it more robust.

We also acknowledge the need to compare this algorithms with newer and improved versions of RRT-based path planners. Future work will include these comparisons.

Integrating the robots dynamics information into the planning process is also our priority. We believe that this is crucial to close the gap between the planned paths and the actual navigated ones, as well as to reach optimal plans.

Acknowledgements. This work has been funded by "Octava Convocatoria Interna de Proyectos de Investigación FODEIN 2014 #049" at Universidad Santo Tomás Colombia, entitled "Construcción de un enjambre de robots omnidireccionales". Many thanks to Martin Llofriu for his comments and suggestions on the work presented here.

References

1. Kitano, H., Asada, M., Kuniyoshi, Y., Noda, I., Osawa, E.: RoboCup: the robot world cup initiative. In: Proceedings of IJCAI-95 Workshop on Entertainment and AI/Alife, pp. 340–347. IEEE Press, Montreal (1995)
2. RoboCup Information. http://www.robocup2013.org/about-robocup/
3. Small Size Legue Information. http://robocupssl.cpe.ku.ac.th/rules:main
4. Ould-Khessal, N.: Botnia: a team of soccer plating robots. In: 2nd International Conference on Autonomous Robots and Agents, pp. 429–433. Palmerston North (2004)
5. LaValle, S.M.: Rapidly-exploring random trees: a new tool for path planning. Technical report No. 98–11 (1998)
6. Tsubouchi, T., Arimoto, S.: Behaviour of a mobile robot navigated by an iterated forecast and planning scheme in the presence of multiple moving obstacles. In: 1994 IEEE International Conference on Robotics and Automation, pp. 2470–2475. IEEE Press, San Diego (1994)
7. Han, K., Veloso, M.: Reactive visual control of multiple non-holonomic robotic agents. In: 1998 IEEE International Conference on Robotics and Automation, pp. 3510–3515. IEEE Press, Leuven (1998)
8. Kuffner, J.J., LaValle, S.M.: Rapidly-exploring random trees: progress and prospects. In: Donald, B.R., Lynch, K.M., Rus, D. (eds.) Algorithmic and Computational Robotics: New Directions, pp. 293–308. AK Peters, Massachusetts (2001)
9. Bruce, J., Veloso, M.M.: Real-time randomized motion planning for multiple domains. In: Lakemeyer, G., Sklar, E., Sorrenti, D.G., Takahashi, T. (eds.) RoboCup 2006: Robot Soccer World Cup X. LNCS (LNAI), vol. 4434, pp. 532–539. Springer, Heidelberg (2007)
10. Desaraju, V.R., How, J.P.: Decentralized path planning for multi-agent teams in complex environments using rapidly-exploring random trees. In: 2011 IEEE International Conference on Robotics and Automation, pp. 4956–4961. IEEE Press, Shanghai (2011)
11. Xiao, J., Michalewicz, Z., Zhang, L., Trojanowski, K.: Adaptive evolutionary planner/navigator for mobile robots. IEEE Transact. Evol. Comput. **1**, 18–28 (1997)
12. Log of Quaterfinal4 Small Size League RoboCup 2013. http://er-force.de/gamelogs/robocup2013/2013-06-29-165351_stoxs_cmdragons.log.gz
13. STOx's Team webpage. http://www.stoxs.org/index.php/en/projects/robocup-ssl-2/86-english-categories/stox-s-english/168-fast-path-planning-algorithm-en

AutoRef: Towards Real-Robot Soccer Complete Automated Refereeing

Danny Zhu, Joydeep Biswas$^{(\boxtimes)}$, and Manuela Veloso

School of Computer Science, Carnegie Mellon University,
Pittsburgh, PA 15213, USA
{dannyz,joydeepb,veloso}@cs.cmu.edu

Abstract. Preparing for robot soccer competitions by empirically evaluating different possible game strategies has been rather limited in leagues using real robots. Such limitation comes from factors related to the difficulty of extensively experimenting with games with real robots, such as their inevitable wear and tear and their usual limited number. RoboCup real robot teams have therefore developed simulation environments to enable experimentation. However, in order to run complete games in such simulation environments, an automated referee is needed. In this paper, we present AutoRef, as a contribution towards a complete automated referee for the RoboCup Small-Size League (SSL). We have developed and used AutoRef in an SSL simulation to run full games to evaluate different strategies, as we illustrate and show results. AutoRef is designed as a finite-state machine that transitions between the states of the game being either on or required to stop. AutoRef purposefully only uses the same visual and game information provided in SSL games with physical robots, which it uses to compute the features needed by the rules and to make decisions to transition between its states. Due to this real input to AutoRef, we have partially applied it to games of the physical robots. As AutoRef does not include all the rules of the real SSL games, we currently view it as an aid to human referees of SSL games, and discuss the challenges in automating several specific SSL game rules. AutoRef could be extended to other RoboCup real soccer leagues if a combined view of the game field, ball, and players is available.

1 Introduction

To date, the creation and evaluation of strategies in RoboCup SSL has tended to be rather *ad hoc*: teams work on their code throughout the year; when the tournament comes, they play a few games against other teams, and try to draw conclusions based on that. An issue with this is that teams are very limited by the small amount of time during which they can actually see the performance of their strategies. This is, in large part, due to the need for refereeing when games are played. A human must be present to set the game state and place the ball when it leaves the field, or else robots are unable to play games autonomously.

Apart from this need, the difficulty of maintaining a large number of real robots in the face of the normal wear and tear of the game is another significant

© Springer International Publishing Switzerland 2015
R.A.C. Bianchi et al. (Eds.): RoboCup 2014, LNAI 8992, pp. 419–430, 2015.
DOI: 10.1007/978-3-319-18615-3_34

limiting factor to adequate testing. Many teams have developed simulation environments to overcome this limitation, but such environments still require the full-time attention of a human to be fully useful.

In this paper, we focus on an automated referee, AutoRef, for the SSL. It is capable of working with either real robots using the standard SSL vision system [1], or with simulated robots using our team's simulation system. Using AutoRef, we have run many full games in simulation to demonstrate the evaluation of different strategies.

The core of AutoRef is a finite state machine that tracks the state of the game and determines whether play should be stopped or restarted at each moment. For the most part, it takes the same input whether it is running with real or simulated robots (the exception being the height of the ball, which is currently difficult to reliably obtain outside of simulation). Based on this input, AutoRef evaluates the rules of the game to determine how to advance at each timestep. Because of the great similarity between the real and simulated inputs to AutoRef, we are able to partially apply it to real games.

There are still two core challenges for an automated referee for a real robot league, preventing a full application. First, the ball needs to be positioned by the referee at specific positions after each stoppage of the game. Second, all of the events of the game, particularly those centered on ball motion, need to be checked for compliance with the rules based on real perception. For now, we do not address the first challenge, and rely on a human to position the ball on the field. (However, we do envision eventually having a mobile robot capable of handling the ball, as any other real robot soccer player, to act as the ball positioner.) In this work, we address the second challenge, but we do not yet implement the detection of all the rules of the game, as some are particularly challenging to identify from a perception point of view.

With either simulated or real robots, automated referees have several advantages over human ones. It can be difficult for a human referee to keep track of the fast-moving robots and ball, whereas this is not a problem for an automated referee. Similarly, an automated referee could enforce complicated rules such as the offside rule, which has not been adopted in the SSL precisely because of the burden it would place on the human referee.

2   Related Work

Automated online processing of RoboCup games has been addressed in several contexts. Several teams have discussed automated refereeing for simulations: Rocco [2], an early simulation commentator, uses a declarative event database to transform the continuous inputs from the game into the discrete events on which to comment. Röfer et al. [3] briefly discuss a simulator-based referee for the Standard Platform League (SPL), which behaves similarly to AutoRef in simulation, but the rule checking and state detection should be simpler due to the slower pace of the game.

In terms of observing real games, RFC Stuttgart [4] presented an automated cameramanthat could take a video of a complete game of the Middle Size League

(MSL). The cameraman uses shared information from a team to keep an onboard pan-tilt unit pointed toward the ball or other object of interest at all times. Merely following the ball involves limited interpretation of the game, while commentating involves some understanding of the elements of a game. Commentating is related to refereeing in terms of the real-time observation and analysis of games, though the decisions that need to be made based on the observations are different. Veloso et al. have previously created CMCast [5], a pair of humanoid robots (SONY QRIOs) that commentated a game of the RoboCup four-legged league with the SONY AIBO robots. The two CMCast robots are able to autonomously move along the field, localize themselves, and visually track objects and detect predefined events, which they then announced with a rich set of spoken utterancess. Tanaka-Ishii et al. presented Mike, a commentator for the simulation league [6], which they later joined with the torso of a humanoid robot to commentate on SSL games [7]. The commentators and referees face similar event detection challenges, but differ in the referee's physical embodiment and the commentator's desired interaction with spectators.

Two other projects to which this work is particularly comparable are the open-source `ssl-autonomous-refbox`[1], which provides rule checking based on perception during an SSL game, and an automated referee presented by Tech United Eindhoven for the MSL [8]. The work discussed here goes further than the former by broadcasting appropriate commands to the competing teams in response to its observations, and further than the latter, by incorporating more rules and, at least in simulation, being able to run a game from start to finish.

3   SSL: A CMDragons Perspective

The RoboCup SSL promotes research in multi-agent coordination in real-world adversarial domains. In this league, teams of six robots play soccer with a golf ball in a field of size 6.05 m × 4.05 m. The robots are constrained to each be contained entirely within a cylinder of diameter 180 mm and height 150 mm. The robots are controlled via radio commands sent by its team's offboard computer. Overhead cameras observe the field, one for each half, and the vision data is processed on a neutral computer by the SSL Vision System (SSL-Vision) [1] based on the early successful CMVision [9]. SSL-Vision broadcasts the detected locations and orientations of all the robots and the ball on the field to the team computers via a wired Ethernet network. With the global vision and centralized planning, the teams focus on the hardware development and coordination behaviors and high-level teamwork strategy. The SSL games are fast-paced, with the robots travelling up to 3 m/s and kicking the ball at speeds of up to 8 m/s (enforced by the rules). To illustrate the elements of an SSL team, we present our CMDragons [10], also built upon research used to create the previous CMUnited (1997–1999) and CMDragons teams (2001–2003, 2006–2010, 2013) and CMRoboDragons joint team (2004–2005).

---

[1] https://code.google.com/p/ssl-autonomous-refbox.

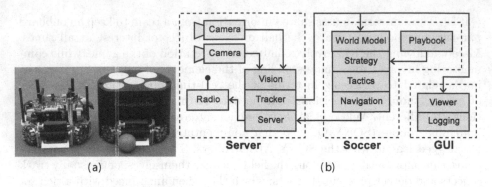

Fig. 1. (a) A CMDragons robot shown with and without a protective cover; (b) The general architecture of the CMDragons offboard control.

Robot Hardware. Our team consists of six homogeneous robots (designed and built by M. Licitra in 2006, and pursued and maintained by J. Biswas since 2010). Figure 1(a) shows an example robot with its omnidirectional drive with four custom-built wheels. The main kicker is a solenoid attached directly to a kicking plate; the chip-kicking device is implemented by a flat solenoid located under the main kicker, which strikes an angled wedge to kick the ball at an angle. Ball catching and handling are performed by a motorized rubber-coated dribbling bar [11].

Robot Behavior Architecture Overview. Figure 1(b) shows the behavior architecture for our offboard control. The major architectural components are a server system that performs vision and manages communication with the robots, and two client programs, namely (i) a soccer program, which implements the individual and team behavior strategy and robot navigation and control, and (ii) a graphical interface program for monitoring, replaying, and testing the team. The server acts as the central translation point between our team clients and the external systems, namely SSL-Vision and the human-operated referee box.

Simulator. We use a high-fidelity 3D soccer simulator [12] based on the NVIDIA PhysX engine[2]. Our simulator incorporates the exact geometric shapes of the robots for accurate collision modelling. Kicking and dribbling are simulated by directly imparting impulse linear or angular momentum to the ball. All The values of the impulse momenta are calibrated to match the speed of the kicked ball in the real world.

The simulator is a drop-in replacement for the soccer server. It accepts standard robot control messages from clients and sends vision messages back to them; however, instead of generating the vision messages based on actual cameras, it reports the positions of robots in the physical simulation [13], accounting for a simulated communications delay approximately matching that encountered when

---

[2] https://developer.nvidia.com/physx.

Fig. 2. A snapshot of our sophisticated graphical visualization program used for simulation, monitoring the real robots, and replaying logged game data.

using real robots. The positions of the robots are drawn with the extra predicted information. The recent speed or acceleration of the ball or any robot, and a text log from the soccer The viewer displays a very extensive set of information, which can be easily selected with multiple levels of detail (Fig. 2).

Over the several years that CMDragons has been competing, we have developed quite a complex set of algorithms for playing soccer and evaluating performance, and as a result it is important to be able to reduce the amount of effort and attention required to run test games with our robots; an automated referee, as we contributed in this work, provides a valuable step in this direction.

4   AutoRef

We describe the implementation details of AutoRef. This includes the means by which it communicates with the rest of the CMDragons codebase, the types of rules present in the game, and the internal details of how it enforces those rules.

4.1   AutoRef Interface with the Game

AutoRef executes as a standalone program; it operates based solely on vision information received from our main server, and sends commands to the server in the same format as commands from the normal human-operated referee box.

In particular, the design of our system means that, at least on the protocol level, AutoRef can be used nearly transparently with either the physical robots or the simulation, and indeed we have used it in this fashion. Currently, it can act as a sort of assistant to a human referee: It automatically signals a stoppage when the ball has gone off the field, indicating which team was the last to touch the ball. There are practical difficulties with using physical robots; for example, since SSL-Vision provides only locations of the ball and robots, not raw video feeds, it is impossible for teams to tell when the human referee has finished placing the ball and stepped off the field, so AutoRef waits for the human's signal to restart play. In simulation, it can simply signal to the simulator that the ball should be moved to a particular location; the simulator can then drag the ball smoothly to the desired location.

4.2 Types of Game Rules

The rules of the SSL fall into a few main categories, some of which are easier than others to handle with AutoRef.

Time-Related Rules. The conceptually simplest rules are those relating to game time. Each half of the game lasts ten minutes; AutoRef keeps track of the elapsed time in the game and ends the current half when appropriate. Time starts counting when the ready command is sent to teams at the beginning of a half, and each half is ten minutes long. At the end of the second half, AutoRef can either signal for a new game to begin or simply leave the robots halted. Timeouts are not currently handled.

Ball-Related Rules. The ball-related rules are primarily those that call for a game stoppage when the ball goes out of bounds.

- When the ball enters one of the goals, a goal is awarded to the appropriate team, the ball is placed in the center of the field, and a kickoff is signaled.
- When the ball leaves the field without entering a goal, a free kick is awarded to the team which touched the ball less recently. If the ball passes over a touch boundary, a throw-in is awarded. If the ball passes over a goal boundary, a corner kick is awarded if the defending team touched the ball more recently, or a goal kick otherwise. In each of these cases, the ball is reset to a particular position near the boundary of the field: throw-ins are taken from a point 100 mm from the point where the ball exited the field, while corner kicks and goal kicks are taken 100 mm from the nearest touch boundary and 100 mm or 500 mm, respectively, from the goal boundary that was crossed.

Either in simulation or with physical robots, AutoRef can track the ball position and signal a stop when the ball leaves the field. In simulation, it can also restart gameplay at an appropriate time (once all robots have stopped moving, with a fixed timeout to avoid waiting forever).

Other ball-related rules are that an indirect free kick is awarded to the opposing team if a robot either kicks the ball faster than $8\,m/s$ or kicks it into the air and directly into the goal. The former has not currently been implemented in AutoRef, but it would be easy to do so, given that the server provides filtered velocity estimates of the ball at each frame. The latter is more problematic, owing to the difficulty of implementing a robust chip kick detector.

Robot-Related Rules. Each robot must be at least 200 mm away from the opposing team's defense area, or an indirect free kick is awarded to the opposing team. AutoRef handles this rule.

Table 1. The values used as input by AutoRef. The time is given in seconds since the Unix epoch. All locations are given in millimeters, in a coordinate system with its origin at the center of the field, positive x-axis pointing toward a goal, and positive y-axis pointing 90° counterclockwise of the positive x-axis.

Name	Description
t	the current time
n	the number of robots on the field
$r_x^{(i)}, r_y^{(i)}$	x- and y-coordinates of the ith robot
b_x, b_y	x- and y-coordinates of the ball
c	the last team to touch the ball

A robot which touches the ball while partially or totally in its own team's defense area triggers a yellow card or penalty kick, respectively; this rule is not currently implemented but could be easily added.

A team may also be shown yellow or red cards for a variety of other offenses, such as if it "is guilty of unsporting behaviour," "is guilty of serious and violent contact," "persistently infringes the Laws of the Game," or several others [14]. These conditions are all somewhat subjective, so they are not handled.

4.3   AutoRef Input and State Machine

Once per camera frame, AutoRef receives as input the values in Table 1.

We devised a finite state machine with transitions governed by the input values, which allows the AutoRef to keep track of the current state of the game as described by the game rules. The machine updates once for each vision frame received from SSL-Vision. The two primary states are as follows.

- RUN: This is the state that occurs while normal gameplay is happening. In this state, AutoRef observes the state of the game until a condition is met that means that play should be stopped.
- WAIT_STOP: In this state, AutoRef waits for all robots to stop moving. This usually occurs after AutoRef has sent a "stop" or similar command to the server, to allow the robots to reach their final positions before another command is sent.

The other states, which occur less frequently, are as follows.

- INIT: This state occurs only once, immediately after execution begins. AutoRef simply saves the current positions of the robots and goes to WAIT_START.
  WAIT_START: This state occurs only near the beginning of execution, when waiting for clients to connect to the server. While in this state, AutoRef periodically compares the positions of the robots to the saved positions; once all robots have moved, it signals a kickoff and enters WAIT_STOP.
- DELAY_WAIT: This state only occurs for one frame at a time; it occurs when AutoRef needs to set the game state to a value which is only transmitted

briefly. This includes the command indicating that a goal has been scored, as well as the command to indicate that a half is starting.

See Fig. 3 for a graphical depiction of the state machine. Each transition is labeled with its preconditions (above the line) and the assignments that occur when it is taken (below the line). An empty condition indicates that the transition is always taken one frame after entering the source state. An assignmnent to `refstate` indicates the sending of a referee command to the server; `next` is an auxiliary variable within AutoRef. Not depicted is the handling of game time, which operates outside this diagram, since the end of a half can occur at any point relative to these states.

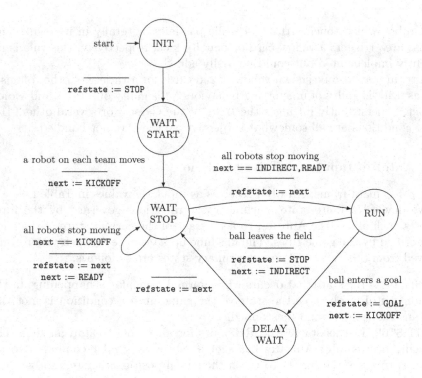

Fig. 3. The finite state machine describing AutoRef.

4.4 Calculations

AutoRef performs several kinds of numerical calculations in order to keep a game running. Primarily, it estimates when all robots on the field are ready for play to resume, by noting when they have all come more or less to a stop, and it computes not only when but where the ball exits the field, so that AutoRef can properly replace the ball afterward.

Robot Positions. When entering the `WAIT_STOP` state, AutoRef saves the current positions of the robots $\left\langle r_x^{(1)}, r_y^{(1)}, \cdots, r_x^{(n)}, r_y^{(n)} \right\rangle$. While remaining in that state, it periodically compares its saved values against the most recently received values $\left\langle r_y^{(1)'}, r_y^{(1)'}, \cdots, r_y^{(n)'}, r_y^{(n)'} \right\rangle$ to check whether all the robots have nearly stopped moving; that is, whether, for a small distance ϵ and every i,

$$\sqrt{\left(r_x^{(i)} - r_y^{(i)'} \right)^2 + \left(r_y^{(i)} - r_y^{(i)'} \right)^2} < \epsilon.$$

Currently, ϵ is set to $15\,\mathrm{mm}$, but that choice is somewhat arbitrary. If some i violates this condition, then AutoRef judges that the robots have not come to a stop yet and saves the new positions for later comparison. If all robots have come to a stop, AutoRef transitions out of `WAIT_STOP`.

AutoRef behaves similarly in `WAIT_START`, except that it the condition is inverted: it leaves the state when at least one robot on each team has moved a distance greater than ϵ.

Ball Position. When the ball goes outside the field, AutoRef needs to determine whether it has entered a goal, crossed a goal boundary without entering a goal, or crossed a touch boundary.

Define the following values:

- F_x: the shortest distance from the center of the field to the outside edge of a goal boundary
- F_y: the shortest distance from the center of the field to the outside edge of a touch boundary
- b_r: the radius of the ball
- G_w: half of the width of the opening of the goal
- G_h: the height of the opening of the goal
- b_x' and b_y': the previous coordinates of the ball

In order to distinguish the two boundaries when the ball is near a corner, it is necessary to compare the current and last positions of the ball to determine where it first was located entirely outside the field. This is the geometrical problem of finding the intersection between one line which is parallel to an axis and another line determined by two points. The intersection of the line between the points (b_x, b_y) and (b_x', b_y') with the line defined by $x = l_x$ is given by (l_x, c_y), where

$$c_y = \frac{b_y'(l_x - b_x) + b_y(b_x' - l_x)}{b_x' - b_x},$$

and a similar equation applies to a line which is defined by a constant y value. (This expression can be derived by finding the equation of the line defined by the two points and substituting $x = l_x$.)

AutoRef judges that the ball is out of the field if $b_x > b_r + F_x$ or $b_y > b_r + F_y$. In fact, this definition ignores a small region at each corner, but the difference is very small.

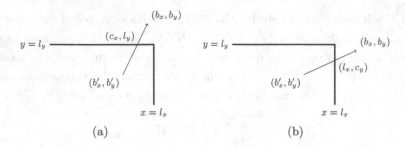

Fig. 4. Illustrations of the computation performed by AutoRef when the ball's trajectory intersects the field boundary. The thick lines depict the boundary of the field, expanded by a distance equal to the radius of the ball, and the arrow depicts the trajectory of the ball between two frames. In (a), since $|c_x| < |l_x|$, the intersection is on a touch boundary. In (b), that condition does not hold, so the intersection is on a goal boundary; the point (c_x, l_y) is farther to the right.

When the ball leaves the field, in order to determine how to respond, AutoRef computes the intersection points of the ball trajectory with the field boundaries as follows. Set $l_x = (F_x + b_r) \cdot \mathrm{sgn}(b_x)$ and $l_y = (F_y + b_r) \cdot \mathrm{sgn}(b_y)$; these represent the coordinates of the goal and touch boundaries closest to the ball, with the addition of b_r in order to capture the fact that the ball's coordinates describe the center of the ball, while the entirety of the ball must be past the lines to be counted as off the field.

Let c_x be the x-coordinate of the intersection of the trajectory with the line $y = l_y$, and let c_y be the y-coordinate of the intersection of the trajectory with the line $x = l_x$. If $|c_x| < |l_x|$, then the ball must have crossed a touch boundary, so a throw-in is awarded. See Fig. 4 for a graphical depiction of the process.

AutoRef judges that the ball is in a goal if the following conditions hold:

$$|b_x| > F_x + b_r,$$
$$|b_y| < G_w, \quad \text{and}$$
$$b_z < G_h.$$

These conditions simply specify that the ball is contained entirely inside the volume defined by the walls of the goal.

5 Demonstrations

To demonstrate the collection of statistics, we played repeated games between two instances of our own team code where one instance was handicapped in some way, either by having the acceleration and top speed of its robots reduced by some factor, or by having fewer robots on the field (Fig. 5). This resulted in a gap in scores that increased with the magnitude of the handicap. These results are unsurprising, but the collection of data across hundreds of realistically simulated full games is novel, as far as the authors are aware.

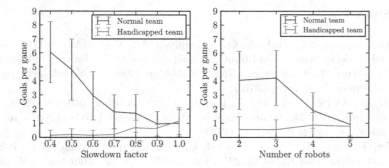

Fig. 5. The mean and standard deviation of the scores of games in which one team had its acceleration and top speed (left; 20 games per value) or number of robots (right; 60 games per value) reduced.

We also tested a limited version of AutoRef with the real robots. With a human referee present to position the ball and restart play after a stoppage, AutoRef observed the game and emitted referee commands as normal. It was able to immediately stop play when the ball left the field, as well as run a kickoff, which involves two transitions in the game state.

There are rules in SSL, which are important but not yet handled by AutoRef, including the conditions for fouls, as well as restrictions on goals scored by chip kicks. Some rules are difficult to implement due to an element of subjectivity by the referee, such as the condition of "deliberately" entering the referee walking area; some because positions alone may not give enough information to determine whether they have ben triggered, such as the rule against "serious and violent conduct"; and some because we lack a reliable chip kick detector. We aim at including all the rules in AutoRef and run it on real games, but until then we envision AutoRef acting in support of human referees, and including a dedicated ball-positioning robot referee.

6 Conclusion

We presented an autonomous referee system for reasoning about rule-related events in SSL games. It is capable of operating with the CMDragons soccer simulator, in which case it can run whole games automatically, and with real robots and vision, in which case it can support a human referee. This work may eventually be extended to a complete referee for real games; for now, we have created a framework for observing reasoning about high-level game state that may be used to fill either role. We also categorized the main rules of the game according to their impact on an automated referee, and demonstrated preliminary results from recording scores in automatically run games. We believe AutoRef can be extended to other RoboCup real soccer leagues if a combined view of the game field, ball, and players is available.

References

1. Zickler, S., Laue, T., Birbach, O., Wongphati, M., Veloso, M.: SSL-Vision: The shared vision system for the RoboCup small size league. In: Baltes, J., Lagoudakis, M.G., Naruse, T., Ghidary, S.S. (eds.) RoboCup 2009. LNCS, vol. 5949, pp. 425–436. Springer, Heidelberg (2010)
2. Voelz, D., André, E., Herzog, G., Rist, T.: Rocco: A robocup soccer commentator system. In: Asada, M., Kitano, H. (eds.) RoboCup 1998. LNCS (LNAI), vol. 1604, pp. 50–60. Springer, Heidelberg (1999)
3. Röfer, T., Laue, T., Müller, J., Graf, C., Böckmann, A., Münder, T.: B-human team description for RoboCup 2012. In: RoboCup 2012: Robot Soccer World Cup XVI Preproceedings, Mexico City, Mexico, RoboCup Federation (2012)
4. Käppeler, U., Zweigle, O., Rajaie, H., Häussermann, K., Tamke, A., Koch, A., Eckstein, B., Aichele, F., DiMarco, D., Berthelot, A., et al.: 1. rfc stuttgart team description 2010. RoboCup Team Description Papers (2010)
5. Veloso, M., Armstrong-Crews, N., Chernova, S., Crawford, E., McMillen, C., Roth, M., Vail, D., Zickler, S.: A team of humanoid game commenters. Int. J. Hum. Robot. 5, 457–480 (2008)
6. Tanaka, K., Nakashima, H., Noda, I., Hasida, K., Frank, I., Matsubara, H.: Mike: An automatic commentary system for soccer. In: Proceedings of the International Conference on Multi Agent Systems 1998, 285–292. IEEE (1998)
7. Frank, I., Tanaka-Ishii, K., Okuno, H.G., Akita, J., Nakagawa, Y., Maeda, K., Nakadai, K., Kitano, H.: And the fans are going wild! sig plus mike. In: Stone, P., Balch, T., Kraetzschmar, G.K. (eds.) RoboCup 2000. LNCS (LNAI), vol. 2019, pp. 139–148. Springer, Heidelberg (2001)
8. Schoenmakers, F., Koudijs, G., Martinez, C.L., Briegel, M., van Wesel, H., Groenen, J., Hendriks, O., Klooster, O., Soetens, R., van de Molengraft, M.: Tech united eindhoven team description 2013. In: Proceedings of the 17th RoboCup International Symposium, May 2013
9. Bruce, J., Balch, T., Veloso, M.: Fast and inexpensive color image segmentation for interactive robots. In: Proceedings of IROS-2000, Japan, October 2000
10. Biswas, J., Mendoza, J.P., Zhu, D., Etling, P.A., Klee, S., Choi, B., Licitra, M., Veloso, M.: CMDragons 2013 team description. In: Proceedings of the 17th RoboCup International Symposium, May 2013
11. Bruce, J., Zickler, S., Licitra, M., Veloso, M.: CMDragons 2007 Team Description. Technical report, Technical report CMU-CS-07-173, Carnegie Mellon University, School of Computer Science (2007)
12. Zickler, S.: Physics-based robot motion planning in dynamic multi-body environments. Ph.D. thesis, Carnegie Mellon University, Thesis Number: CMU-CS-10-115, May 2010
13. Zickler, S., Vail, D., Levi, G., Wasserman, P., Bruce, J., Licitra, M., Veloso, M.: CMDragons 2008 team description. In: Proceedings of RoboCup (2008)
14. Small Size League Technical Committee: Laws of the RoboCup Small Size League 2013 (2013)

Analytical Solution for Joint Coupling in NAO Humanoid Hips

Vincent Hugel[1](✉) and Nicolas Jouandeau[2]

[1] LISV, University of Versailles, Versailles, France
vincent.hugel@gmail.com
[2] LIASD, University of Paris 8, Saint Denis, France

Abstract. Usually the legs of humanoids capable of omnidirectional walking are not underactuated. In other words each one of the six degrees of freedom of the torso can be commanded independently from the leg joint angles. However the NAO humanoid robot has a coupled joint at the hips, which makes 11 degrees of freedom instead of 12 for the locomotor apparatus. As a consequence the trunk of the robot has only 5 independent degrees of freedom when the positions of both feet are fixed, and each leg cannot be commanded independently to execute walking steps. Up to now only bypass solutions have been proposed, where the coupled joint angles are not calculated exactly. This paper describes an analytical solution to determine the exact angle to be applied to the coupled joint. The method uses the positions of both foot ankles in the trunk reference frame and the angle between footprints as inputs, and calculates the yaw angle of the trunk. The solution was demonstrated in a dynamics simulator using the NAO model.

1 Introduction

The NAO humanoid robot [1] is widely used in the robotics community to design omnidirectional walking gaits [2,3], kicking moves [4], and stabilizers [5,6]. In the NAO robot the yaw joints at the hips are coupled and 45[deg] inclined with respect to the vertical [1] (Fig. 1). Therefore the hips of NAO are underactuated, with five independent degrees of freedom instead of six.

To deal with this underactuation, Graf et al. [3,7] introduced a fictive yaw joint at the swing foot, and applied a 6-DOF Inverse Geometric Model (IGM) based on the kinematic chain of this virtual leg, the yaw-pitch joint at the hip being fixed by the positioning of the support leg. However non-zero angle values for the fictive yaw joint constitute positioning errors. Nonetheless such errors have less impact than errors that would result from the addition of a virtual pitch or a virtual roll joint. Alcaraz-Jiménez [8] defined an iterative process to evaluate the torso yaw angle by using a proportional law that tends to minimize the yaw angle error of the swing foot. The torso yaw angle is then used by the inverse kinematics to update the joint angles. The process is iterated three sampling times to limit the control delay. Hugel et al. [9] used the torso longitudinal axis as a symmetry axis. They considered that the yaw angle between this axis and

© Springer International Publishing Switzerland 2015
R.A.C. Bianchi et al. (Eds.): RoboCup 2014, LNAI 8992, pp. 431–442, 2015.
DOI: 10.1007/978-3-319-18615-3_35

the right foot is the same as the yaw angle between this axis and the left foot. The symmetry property is only valid when both feet are on the ground, and is limited to the turning steps whose center is located on the torso axis. This method is still an approximation when the leg is lift off in this kind of turning steps.

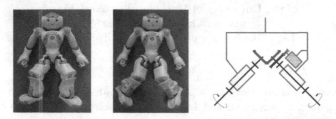

Fig. 1. Coupled hip-yaw rotary joints of the NAO pelvis.

All the methods mentioned above are approximation methods since they do not calculate the coupled joint angle exactly. This paper proposes an analytical solution that calculates the yaw angle of the trunk exactly given the positions of both feet in the Trunk Coordinate Frame (TCF) and the horizontal angle between the feet. This allows calculating the orientation of each foot within the TCF before applying the IGM to get all joint angles to command the legs to achieve the desired motion.

Section 2 presents the modeling convention used for the geometric calculations. Section 3 describes the leg model similar to the NAO model that is used throughout the paper. Section 4 details the analytical solution proposed to calculate the coupled joint angle exactly. Simulation results are presented in Sect. 5 followed by the discussion section.

2   Modeling Convention

Khalil and Kleinfinger [10] proposed a modified convention, named $DHKK$, for geometric modeling from the Denavit-Hartenberg convention [11].

There are four $DHKK$ parameters required to go from coordinate frame F_{i-1} to F_i, one for each transformation. Parameters are denoted by a_i, α_i, d_i and θ_i^*. a_i and d_i are distances, α_i and θ_i^* are angles. Figure 2 shows the four parameters at stage i. They involve the three axes z_{i-1}, z_i, and z_{i+1}.

- a_i is the algebraic distance $P_{i-1}P_{i-1}'$ along x_{i-1}.
- α_i is the rotation angle about x_{i-1} between z_{i-1} and z_i.
- d_i is the algebraic distance $P_{i-1}'P_i$ along z_i axis.
- θ_i^* is the rotation angle about z_i between x_{i-1} and x_i.

where x_{i-1} is along the segment that is orthogonal to z_{i-1} and z_i axes, from z_{i-1} to z_i. x_i is along the segment that is orthogonal to z_i and z_{i+1} axes, from

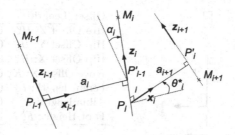

Fig. 2. Definition of $DHKK$ parameters for geometric modeling.

z_i to z_{i+1}. P_i is the intersection point of x_i with z_i. P'_i is the intersection point of x_i with z_{i+1}. M_i is a point that belongs to z_i axis.

The coordinate frame transformation T_i from F_i to F_{i-1}, with $F_i = (P_i, x_i, y_i, z_i)$, is written as follows, where R stands for rotation and D for translation:

$$T_i = D_{x_{i-1}}(a_i)R_{x_{i-1}}(\alpha_i)D_{z_i}(d_i)R_{z_i}(\theta_i^*) \tag{1}$$

$$T_i = \begin{pmatrix} \cos\theta_i^* & -\sin\theta_i^* & 0 & a_i \\ \cos\alpha_i\sin\theta_i^* & \cos\alpha_i\cos\theta_i^* & -\sin\alpha_i & -d_i\sin\alpha_i \\ \sin\alpha_i\sin\theta_i^* & \sin\alpha_i\cos\theta_i^* & \cos\alpha_i & d_i\cos\alpha_i \\ 0 & 0 & 0 & 1 \end{pmatrix} \tag{2}$$

In the following sections we will use the reduced notation: $T_x = \begin{pmatrix} R_x & D_x \\ 0 & 1 \end{pmatrix}$, where R_x represents the rotation matrix, and D_x the translation matrix to be applied.

3   Model of Legs

Figure 3 shows the skeleton of the humanoid trunk and legs in the reference position where legs are stretched vertically. The first joints at the hips are coupled, and their axes are inclined. The right-hand side of the figure shows the values of lengths and offsets related to two versions of NAO legs, of the 3D Soccer Simulation League (3D-SSL) and of the Standard Platform League (SPL). The inputs, namely points M_i and axes z_i in the reference position of Fig. 3, that are needed to calculate the $DHKK$ parameters automatically [12,13] are also given. Table 1 presents the different values of the $DHKK$ parameters that are related to the NAO leg ($\xi = 1$ for the right leg, and $\xi = -1$ for the left leg).

The geometric model for the NAO humanoid leg can be written as:

$$T_0 = T_sT_1T_2T_3T_4T_5T_6T_e \tag{3}$$

where

- T_s involves the first three z axes,
- T_e involves the last three z axes,
- T_i involves joint rotation of angle θ_i^*, and axes z_{i-1}, z_i, and z_{i+1},

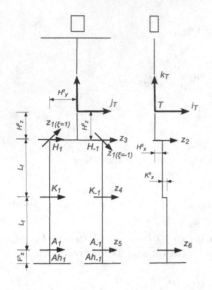

Offset/Length	3DSSL leg	SPL leg
Hip Offset Z: H_z^o	0.115	0.085
Hip Offset Y: H_y^o	0.055	0.050
Hip Offset X: H_x^o	0.010	0.000
Knee Offset X: K_x^o	0.005	0.000
Femur length: L_f	0.120	0.100
Tibia length: L_t	0.100	0.103
Foot Height: F_z^h	0.050	0.045

Point M_i	Joint axis z_i
$M_{s1} = T$	$z_{s1} = k_T = [0,0,1]^T$
$M_{s2} = T$	$z_{s2} = k_T = [0,0,1]^T$
$M_1 = H$	$z_1 = [0, \cos(\pi/4), \xi \sin(\pi/4)]^T$
$M_2 = H$	$z_2 = i_T = [1,0,0]^T$
$M_3 = H$	$z_3 = j_T = [0,1,0]^T$
$M_4 = K$	$z_4 = j_T = [0,1,0]^T$
$M_5 = A$	$z_5 = j_T = [0,1,0]^T$
$M_6 = A$	$z_6 = i_T = [1,0,0]^T$
$M_{e1} = Ah$	$z_{e1} = k_T = [0,0,1]^T$
$M_{e2} = Ah$	$z_{e2} = -j_T = [0,-1,0]^T$

Fig. 3. Humanoid skeleton with notations. Frontal view on the left-hand side and sagittal view on the right-hand side.

Table 1. Leg's DHKK parameters. [a] $r = (H_x^o - K_x^o)/L_f$. [b] $\delta = tan^{-1}(r)$.

i	$a_i[m]$	$\alpha_i[rad]$	$d_i[m]$	$\theta_i^*[rad]$	Joint
s	0	0	$-(H_z^o - H_y^o)$	π	dummy
1	H_x^o	$\pi/4(2-\xi)$	$-\xi.\sqrt{2}.H_y^o$	$\theta_1 - \pi/2$	hip yaw-pitch
2	0	$\pi/2$	0	$\theta_{2+} = \theta_2 + \xi.3\pi/4$	hip roll
3	0	$\pi/2$	0	$\theta_3 + \pi - \delta$ [b]	hip pitch
4	$L_f\sqrt{1+r^2}$ [a]	0	0	$\theta_4 + \delta$	knee pitch
5	L_t	0	0	θ_5	ankle pitch
6	0	$\pi/2$	0	$\theta_6 - \pi/2$	ankle roll
e	0	$\pi/2$	$-F_z^h$	$\pi/2$	dummy

- T_0 is the homogeneous matrix given by the user, that represents the wanted orientation R_0 of the foot's sole in the TCF, and the position D_0 of the projection of the ankle on the sole, named A_h, in the TCF. R_0 is the matrix whose columns contain the coordinates of the Sole Coordinate Frame (SCF) axes expressed in the TCF. R_0 can be interpreted as the transformation matrix to pass from the SCF into the TCF.

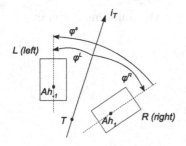

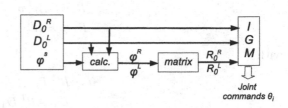

Fig. 4. Footprints in top view with right and left yaw angles φ^R and φ^L, and step angle φ^s.

Fig. 5. Scheme of the analytical solution. *IGM* stands for Inverse Geometric Model.

4   Analytical Solution

4.1   Objective

We assume that feet always remain parallel to the horizontal ground. This means that there is a rotation about the vertical of a footprint with respect to the other. The angle of this rotation is noted φ_s. The objective is to calculate the yaw angle φ^R between the robot's trunk longitudinal axis i_T and the right foot axis. The yaw angle with the left foot is $\varphi^L = \varphi^R + \varphi^s$ (Fig. 4).

The inputs for the analytical solution are the following:

– the 3D position of the right foot ankle projection within the TCF, D_0^R,
– the 3D position of the left foot ankle projection within the TCF, D_0^L,
– the angle between the right foot and the left foot, φ^s.

Figure 5 shows the scheme of the analytical solution that allows calculating all joint angles exactly taking into account the coupling between yaw-pitch joints at the hips. The next section details the *calc.* box of the scheme.

4.2   Analytical Expression of Tangent of Hip Yaw-pitch Joint Angle, θ_1

Equation 3 leads to[1]:

$$R_{123}(D_4 + R_4 D_5) = D' \tag{4}$$

$$R_{123}R_{456} = R' \tag{5}$$

with

$$R' = R_\varphi^T R_0 R_e^T \tag{6}$$

$$D' = -D_1 + R_s^T(D_0 - Ds - R_0 R_e^T D_e) \tag{7}$$

Squaring the first equation gives $(D_4 + R_4 D_5)^2 = D'^2$ that allows to solve for the knee angle θ_4.

---

[1] $R_{xy} = R_x R_y$.

By replacing R_{123} from Eq. (5) into Eq. (4), we get the following system:

$$R_6 u = R_5^T v \tag{8}$$

$$R_{123} = R'' \tag{9}$$

with

$$u = R'^T D'$$
$$v = D_5 + R_4^T D_4$$
$$R'' = R' R_6^T R_5^T R_4^T$$

u can also be expressed as:

$$u = -D_e + R_e R_0^T R_s \left[-D_1 + R_s^T (D_0 - D_s) \right] \tag{10}$$

By taking the last column of both matrices in Eq. (9), it comes[2]:

$$R_{123}.C_3 = R' R_6^T [0, 0, 1]^T$$

$$\begin{bmatrix} s_1 s_{2+} \\ \frac{\sqrt{2}}{2}(-\xi c_1 s_{2+} + c_{2+}) \\ \frac{\sqrt{2}}{2}(-c_1 s_{2+} - \xi c_{2+}) \end{bmatrix} = R' \begin{bmatrix} -c_6 \\ s_6 \\ 0 \end{bmatrix}$$

θ_2 varies inside $[-\pi/2, \epsilon]$ for the right leg and inside $[-\epsilon, \pi/2]$ for the left leg, where ϵ represents a positive value that is less than $\pi/4$. Hence θ_{2+} (see Table 1) varies inside $[\frac{\pi}{4}, \epsilon + \frac{3\pi}{4}]$ for the right leg and inside $[-\epsilon - \frac{3\pi}{4}, -\frac{\pi}{4}]$ for the left leg. Therefore s_{2+} is never equal to zero and we get[3]:

$$t_1 = \tan \theta_1 = -\frac{1}{\sqrt{2}} \frac{-R'_{11}c_6 + R'_{12}s_6}{-(\xi R'_{21} + R'_{31})c_6 + (\xi R'_{22} + R'_{32})s_6}$$

To determine θ_6 we use the last line of matrix Eq. (8):

$$-c_6 u_x + s_6 u_y = 0 \tag{11}$$

which gives $c_6 = \pm u_y/(u_x^2 + u_y^2)$ and $s_6 = \pm u_x/(u_x^2 + u_y^2)$. Therefore:

$$t_1 = \frac{1}{\sqrt{2}} \frac{R'_{11}u_y - R'_{12}u_x}{-(\xi R'_{21} + R'_{31})u_y + (\xi R'_{22} + R'_{32})u_x} \tag{12}$$

with:

$$u = -D_e + R'^T \overline{D_0} \tag{13}$$

$$\overline{D_0} = -D_1 + R_s^T (D_0 - D_s) \tag{14}$$

$$u_x = R'_{11} x_{\overline{D_0}} + R'_{21} y_{\overline{D_0}} + R'_{31} z_{\overline{D_0}} \tag{15}$$

$$u_y = -F_z^h + R'_{12} x_{\overline{D_0}} + R'_{22} y_{\overline{D_0}} + R'_{32} z_{\overline{D_0}} \tag{16}$$

---

[2] s_x and c_x stand respectively for $\sin \theta_x$ and $\cos \theta_x$.

[3] $R_{n_1 n_2}$ stands for the element at row n_1 and column n_2 of matrix R.

Then,

$$t_1 = \frac{(-F_z^h R'_{11} + y_{\overline{D_0}} d_1 + z_{\overline{D_0}} d_3)/\sqrt{2}}{(\xi R'_{21} + R'_{31})F_z^h + (\xi d_1 + d_3)x_{\overline{D_0}} + d_2(y_{\overline{D_0}} - \xi z_{\overline{D_0}})} \tag{17}$$

with:

$$d_1 = R'_{11}R'_{22} - R'_{12}R'_{21}, \quad d_2 = R'_{21}R'_{32} - R'_{22}R'_{31}, \quad d_3 = R'_{11}R'_{32} - R'_{12}R'_{31}$$

4.3   Determination of the Body Yaw Angle

Because of the coupling of both yaw-pitch joints at the hips, the following constraint of equality of right and left joint angles must be satisfied:

$$\tan \theta_1^R = \tan \theta_1^L \tag{18}$$

Given that the foot soles remain horizontal and that the trunk can be pitched forward or backward by an angle η in the sagittal plane during the walk, R_0 can be written as:

$$R_0 = \begin{bmatrix} c\eta & 0 & -s\eta \\ 0 & 1 & 0 \\ s\eta & 0 & c\eta \end{bmatrix} \begin{bmatrix} c & -s & 0 \\ s & c & 0 \\ 0 & 0 & 1 \end{bmatrix} \tag{19}$$

which is the matrix product of the rotation matrix of angle $(-\eta)$ about the y-axis and the rotation matrix of the walking step, with $c = \cos\varphi$ and $s = \sin\varphi$. $\varphi = \varphi^R$ for the right leg, and $\varphi = \varphi^L$ for the left leg.

Therefore R' becomes:

$$R' = \begin{bmatrix} -sc\eta & -s\eta & -cc\eta \\ c & 0 & -s \\ ss\eta & -c\eta & cs\eta \end{bmatrix} \tag{20}$$

$$d_1 = cs\eta, \quad d_2 = -cc\eta, \quad d_3 = s$$

and Eqs. (17) and (18) become:

$$t_1 = \frac{1}{\sqrt{2}} \frac{p_1 \tan\varphi + p_2}{p_3 \tan\varphi + p_4} \tag{21}$$

$$\frac{p_1^R \tan\varphi^R + p_2^R}{p_3^R \tan\varphi^R + p_4^R} = \frac{p_1^L \tan\varphi^L + p_2^L}{p_3^L \tan\varphi^L + p_4^L} \tag{22}$$

Replacing φ^L by $\varphi^R + \varphi^s$, and noting $t = \tan\varphi^R$, $t^s = \tan\varphi^s$ leads to the 2nd order equation:

$$t^2 + B.t + C = 0 \tag{23}$$

with

$$B = (p_1^R \sigma_1 + p_2^R \sigma_2 - p_3^R \sigma_3 - p_4^R \sigma_4)/A = B'/A$$
$$C = (p_2^R \sigma_1 - p_4^R \sigma_3)/A = C'/A$$

$$A = p_1^R \sigma_2 - p_3^R \sigma_4$$

$$p_1^{R,L} = F_z^h c\eta + z_{\overline{D_O^{R,L}}}, \quad p_2^{R,L} = y_{\overline{D_O^{R,L}}} s\eta$$

$$p_3^{R,L} = F_z^h s\eta + x_{\overline{D_O^{R,L}}}, \quad p_4^{R,L} = \xi(F_z^h + c\eta z_{\overline{D_O^{R,L}}} + s\eta x_{\overline{D_O^{R,L}}}) - c\eta y_{\overline{D_O^{R,L}}}$$

$$\sigma_1 = p_3^L t^s + p_4^L, \quad \sigma_2 = p_3^L - p_4^L t^s, \quad \sigma_3 = p_1^L t^s + p_2^L, \quad \sigma_4 = p_1^L - p_2^L t^s$$

The solution for φ^R is the one with the lowest magnitude:

$$\Rightarrow t = \frac{1}{2}(-B + sign(B).\sqrt{B^2 - 4C})$$

Then

$$\varphi^R = tan^{-1}(t) \tag{24}$$

The other one is the solution where the feet are pointing inwards with an approximately 90[deg] angle with respect to the first solution. If $(A, B') = (0, 0)$ there is no solution. If $B' = 0$, $t = -\sqrt{-C}$. If $A = 0$ then $t = -C'/B'$.

5   Simulation Results

The analytical solution was embedded in the locomotion code of NAO to be used in the 3D-SSL software (SimSpark application for the RoboCup 3D-SSL [14,15], namely *rcssserver3d*, based on the Open Dynamics Engine (ODE)). This code enables to calculate the joint command angles in real time to be sent to the server to make the NAO client walk on the simulator. Walking patterns were designed using the model of the 3D linear inverted pendulum and the Zero Moment Point technique [9].

The analytical solution was tested for three walking patterns. The first one is a forward walk of 0.1[m], the second one a sideways walk of 0.1[m], and the third one a turn-in-place of 60[deg]. In the forward walk the left leg makes a 0.05[m] step forward, then the right leg executes a 0.1[m] step forward, and finally the left leg makes a 0.05[m] step to come parallel with the right leg. In the sideways walk, the left leg executes a left sidestep of 0.05[m], then the right leg makes a left sidestep to comes parallel with the left leg. This sequence is reproduced once again to cover a total distance of 0.1[m]. In the turn-in-place walk the left leg executes a 30[deg] left turn, then the right leg makes also a 30[deg] left turn to get parallel with left leg. The sequence is reproduced once again to cover a total angle of 60[deg]. The center of rotation is located at 0.01[m] behind the middle of the ankles to give more space for the heels and prevent them from colliding into each other during the rotation motion of the leading leg. Each step lasts 0.24[s]. Before executing the first step of every waking pattern, the robot sways its hips outward, then inward to initiate the step. After the last step, there is also an outward-inward hip sway to stop the lateral oscillation of the torso. Figures 6, 7 and 8 display the variations over time of the yaw angle φ^R and of the coupled-yaw-joint angle θ_1 for the three walking patterns respectively, with

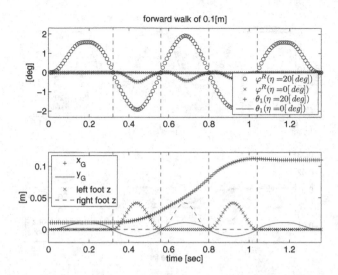

Fig. 6. Variations of the right yaw angle φ^R and the coupled-yaw joint angle θ_1 along a 0.1[m] forward walk. Angles values are given for the straight torso ($\eta = 0[\deg]$), and for a forward inclination of 20[deg] of the torso in the sagittal plane ($\eta = 20[\deg]$).

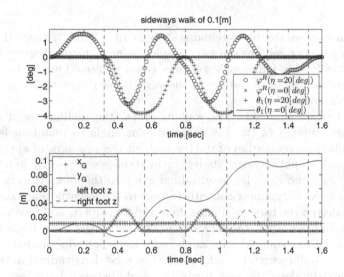

Fig. 7. Variations of the right yaw angle φ^R and the coupled-yaw joint angle θ_1 along a sideways walk 0.1[m]. Angles values are given for the straight torso ($\eta = 0[\deg]$), and for a forward inclination of 20[deg] of the torso in the sagittal plane ($\eta = 20[\deg]$).

the torso in straight position ($\eta = 0[\deg]$) and with the torso inclined forward ($\eta = 20[\deg]$). The bottom part of each of the figures gives the variation over time of the horizontal coordinates of the center of mass (x_G along the longitudinal axis, y_G along the lateral axis), and the variation with time of both foot heights.

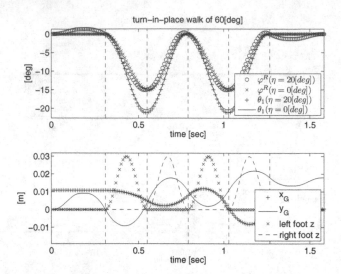

Fig. 8. Variations of the right yaw angle φ^R and the coupled-yaw joint angle θ_1 along a left turn-in-place of 60[deg]. Angles values are given for the straight torso ($\eta = 0$[deg]), and for a forward inclination of 20[deg] of the torso in the sagittal plane ($\eta = 20$[deg]).

In the case where the torso remains straight ($\eta = 0$[deg]), forward and side-ways walking do not require the yaw coupled joint θ_1 to participate in the leg motion. Actually this joint angle remains equal to zero. On the contrary the turn-in-place walk requires the use of the yaw coupled joint to design the turning trajectories of the legs.

During forward and sideways walking steps, the inclination of the torso requires the actuation of the yaw coupled-joint angle to maintain feet parallel. This leads to an oscillation of the torso about the yaw axis, of approx. 4[deg] of peak to peak amplitude for a 20[deg] inclination (see φ^R, angle between torso longitudinal axis and right leg longitudinal axis). In the forward walk the oscillation of the torso is symmetrical with respect to the longitudinal axis, whereas in the left sidestep the torso oscillates more right than left, and vice-versa. During the turn-in-place walk, the forward inclination of the torso leads to similar variations of the yaw coupled-joint angle, with some light fluctuations compared with the turn-in-place with straight torso. The torso longitudinal axis does not remain rigorously along the bisector of the angle between both feet, i.e. φ^L is not equal to φ^R except at the foot impacts.

6   Discussion

The analytical solving allows designing double support phases during the walk. For a fixed configuration of footprints the desired inputs to the inverse kinematics can be the center of mass of the robot and the torso bending angle in the sagittal plane, the roll angle being kept to 0[deg]. Due to the underactuation at the hips the yaw angle is constrained and can be calculated in real time to command the coupled joint angle. Compared with other approximate solutions [7–9], this analytical solving enables to calculate joint angles exactly and to prevent feet from sliding on the ground. In addition this contributes to reduce the stress on the joints.

The calculation presented here assumes that the feet soles are horizontal and that the torso remains vertical or is bent in the sagittal plane. This calculation could be especially useful for walking algorithms that embed closed loop control of torso inclination like the technique adopted by Gouaillier et al. [16] or the one used by Glaser et al. [17] for the NAO robot. However, if we take the same foot trajectory shape for the inclined torso as for the straight torso to design longitudinal and lateral walking patterns where feet remain parallel, the coupled yaw-pitch joint must also be controlled, this is because the roll axis at the hip is no more horizontal and therefore cannot move the foot in the frontal plane without actuating the coupled joint. This results in a yaw motion of the torso about the vertical, in the direction of the support leg when the other leg is lift off, and in the other direction when the other leg goes down for landing.

In the case of instantaneous double support phases, the swing trajectory of the leg can be tuned to ensure fixed torso heading by transferring the slight rotation of the torso to the swinging leg.

7   Conclusion

This paper has presented an analytical solution for the yaw coupled joint at the hips of the NAO humanoid robot. The calculation of the yaw angle can be made in real time, allowing the control of the coupled joint to enable the exact tracking of the foot Cartesian trajectories. From an inverse kinematics point of view the definition of the robot's center of mass position and the bending angle in the sagittal plane of the torso can be used as inputs to design walking moves. The study was conducted for torso bending in the sagittal plane but it can be extended to whatever bending including lateral bending. The solution proposed is especially useful for the command of the joints in the case of turning steps, or in all cases of walking patterns when the torso is inclined. In the case of instantaneous double support, it can be useful to design swing leg trajectories carefully in the trunk reference frame to avoid rotation of the torso about the vertical.

References

1. Gouaillier, D., Hugel, V., Blazevic, P., Kilner, C., Monceaux, J., Latourcade, P., Marnier, B., Serre, J., Maisonnier, B.: Mechatronic design of NAO humanoid. In: International Conference on Robotics and Automation, pp. 769–774 (2009)
2. Czarnetzki, S., Kerner, S., Urbann, O.: Applying dynamic walking control for biped robots. In: Baltes, J., Lagoudakis, M.G., Naruse, T., Ghidary, S.S. (eds.) RoboCup 2009. LNCS, vol. 5949, pp. 69–80. Springer, Heidelberg (2010)

3. Graf, C., Härtl, A., Röfer, T., Laue, T.: A robust closed-loop gait for the standard platform league humanoid. In: Proceedings of the 4th Workshop on Humanoid Soccer Robots, pp. 30–37 (2009)
4. Wenk, F., Röfer, T.: Online generated kick motions for the NAO balanced using inverse dynamics. In: Behnke, S., Veloso, M., Visser, A., Xiong, R. (eds.) RoboCup 2013. LNCS, vol. 8371, pp. 25–36. Springer, Heidelberg (2014)
5. Alcaraz-Jiménez, J.J., Missura, M., Martínez-Barberá, H., Behnke, S.: Lateral disturbance rejection for the Nao robot. In: Chen, X., Stone, P., Sucar, L.E., van der Zant, T. (eds.) RoboCup 2012. LNCS, vol. 7500, pp. 1–12. Springer, Heidelberg (2013)
6. Hengst, B., Lange, M., White, B.: Learning ankle-tilt and foot-placement control for flat-footed bipedal balancing and walking. In: IEEE-RAS International Conference on Humanoid Robots, pp. 288–293 (2011)
7. Röfer, T., Laue, T., Müller, J., Bösche, O., Burchardt, A., Damrose, E., Gillmann, K., Graf, C., de Haas, T.J., Härtl, A., Rieskamp, A., Schreck, A., Sieverdingbeck, I., Worch, J.-H.: B-Human Team Report and Code Release (2009). http://www.b-human.de/downloads/bhuman09_coderelease.pdf
8. Alcaraz-Jiménez, J.J.: Robust feedback control of omnidirectional biped gait on NAO humanoid robots. Ph.D. thesis, University of Murcia, Spain (2013)
9. Hugel, V., Jouandeau, N.: Walking patterns for real time path planning simulation of humanoids. In: IEEE RO-MAN, pp. 424–430 (2012)
10. Khalil, W., Kleinfinger, J-F.: A new geometric notation for open and closed-loop robots. In: IEEE International Conference on Robotics and Automation, pp. 1174–1180 (1986)
11. Denavit, J., Hartenberg, R.S.: A kinematic notation for lower-pair mechanisms based on matrices. Trans ASME J. Appl. Mech. **23**, 215–221 (1955)
12. Zorjan, M., Hugel, V., Blazevic, P.: Influence of hip joint axes change of orientation on power distribution in humanoid motion. In: IEEE International Conference on Automation, Robots and Applications, pp. 271–276 (2011)
13. Zorjan, M., Hugel, V.: Generalized humanoid leg inverse kinematics to deal with singularities. In: IEEE International Conference on Robotics and Automation, pp. 4791–4796 (2013)
14. Obst, O., Rollmann, M.: Spark – a generic simulator for physical multi-agent simulations. In: Lindemann, G., Denzinger, J., Timm, I.J., Unland, R. (eds.) MATES 2004. LNCS (LNAI), vol. 3187, pp. 243–257. Springer, Heidelberg (2004)
15. SimSpark, a generic physical multiagent simulator system for agents in three-dimensional environments. http://simspark.sourceforge.net/
16. Gouaillier, D., Collette, C., Kilner, C.: Omni-directional closed-loop walk for NAO. In: IEEE-RAS International Conference on Humanoid Robots, pp. 448–454 (2010)
17. Glaser, S., Dorer, K.: Trunk controlled motion framework. In: Proceedings of the 8th Workshop on Humanoid Soccer Robots (2013)

A Comparative Study and Development of a Passive Robot with Improved Stability

Hamid Reza Shafei[1], Soroush Sadeghnejad[2(✉)],
Mohsen Bahrami[1], and Jacky Baltes[3]

[1] Mechanical Engineering Department,
Amirkabir University of Technology (Tehran Polytechnic), Tehran, Iran
{hr.shafei,mbahrami}@aut.ac.ir
[2] Amirkabir Robotic Institute,
Amirkabir University of Technology (Tehran Polytechnic), Tehran, Iran
s.sadeghnejad@aut.ac.ir
[3] Autonomous Agents Laboratory, University of Manitoba, Winnipeg, Canada
jacky@cs.umanitoba.ca

Abstract. Passive walkers are robots that can produce a stable cyclical movement similar to walking on mildly inclined surfaces. In recent years, various investigations have been conducted on this subject. Since this is a new field of research, the effect of structural parameters on the movement of these walkers and the development of more human-like movement models can be further investigated. This paper compares three popular models for passive dynamic walkers: Garcia's mode, Wisse's model, and our own extended Wisse's model with arm. Our research shows that the extended models lead to more energy efficient and stable walking.

1 Introduction

Among the mobile robots, the bipedal robots have always been of interest, because of their adequate flexibility and movement speed; however, the high energy consumption of these robots has made their use impractical. Since the presentation of the first passive walker model by McGeer in 1990, a new window was opened to the investigation of the walking process in humans and to the designing of highly-efficient robots [4]. Through the passive walking mechanism using the two feet, humans are able to walk on inclined surfaces without needing any control, and by spending only a small amount of input energy. In Fig. 1, the energy utilization of most famous active and passive bipedal robots and of humans has been compared.

Walking appears as a very complicated process, which is controlled by the muscular and nervous systems. But, is this control system fully active, or can it operate passively in certain conditions? By definition, a mechanical system is said to have a passive movement, if, during the movement, the resultant of the input gravitational energy and the energy losses in it is zero and no outside energy and control are applied to the system. This notion was first introduced by McMohan et al. [3].

They maintained that the oscillatory motion of human feet can be convincingly modeled as a passive dual pendulum. They called this type of movement "Throw

© Springer International Publishing Switzerland 2015
R.A.C. Bianchi et al. (Eds.): RoboCup 2014, LNAI 8992, pp. 443–453, 2015.
DOI: 10.1007/978-3-319-18615-3_36

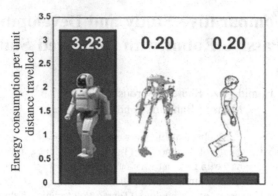

Fig. 1. Comparison of human energy, Honda's robot and semi-active biped robot at Cornell University [2]

Walking". Later on, in 1990, McGeer demonstrated that walking can be a completely autonomous [self-initiated] process. He stated: "*A set of bi-pedal machines exist for which, walking is a natural dynamic mode. The movement started on a mildly inclined surface by a machine belonging to this set leads to a stable movement comparable to natural walking, while it requires no input energy and control.*" [4] This stable movement by two feet is sustained by balancing the energy lost during the impact of feet with the ground and the other plausible energy losses during movement with the input energy due to gravity. McGeer demonstrated this fact theoretically as well as experimentally by building prototypes of bipedal walkers. This line of research was subsequently pursued by other investigators who presented different models. These models included the 3D models [5], models with upper bodies [6], and models with knees [7]. The most complex passive bipedal walker is a 3D model with knees and arms, which was built in the Cornell University under the supervision of Andy Ruina in 2001 (Fig. 2) [5]. In recent years, many universities around the world have conducted research works on passive and semi-active bipedal walkers [8–13]. In brief, the advantages of the passive movement scheme are as follows: (1) Saving in the consumption of energy: while the humanoid robot of the Honda Motors Company carries a 6 kg battery for 40 min of walking [1], Garcia et al. demonstrated that the energy utilization limit of a passive bipedal walker can be near zero, (2) the natural movement modes of passive walkers are very similar to human walking modes.

In 1988, Garcia designed the simplest bipedal robot model that could not be reduced further [1]. Garcia's model consisted of 2 massless rods, with 3 point masses at the ends of two feet and the thigh joint; and the whole model moved two-dimensionally (Fig. 1). He showed that the above mechanism can walk on mildly inclined surfaces (Fig. 3).

In 2004, the team of Wisse and Schwab of the Delft University analyzed a bipedal robot similar to Garcia's, but which had an upper body [6]. The upper body had a kinematic constraint that always kept it aligned with the bisector of the angle formed between the two feet (Fig. 4). The effect of the upper body on stability was analyzed and it was demonstrated that the existence of the upper body increases the walking speed and efficiency in the robot. Finally, it was shown that by adding a spring to the

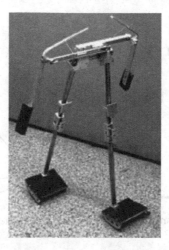

Fig. 2. 3D passive walking at Cornell University [7]

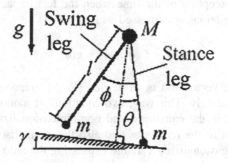

Fig. 3. Garcia's model [1]

above model, the stride length can be arbitrarily changed. In this paper, Wisse's robot
was modeled by placing a spring at the connection joint of the two feet and analyzed
the speed and robustness of the resulting walking gait. This robot is shown in Fig. 4.

In our research, we extended the previous models by Garcia and Wisse to create a
more realistic and accurate model with springs and arms for a walker robot and analyzed
its behavior with respect to walking speed and stability. The model is based on the same
prototype as presented by Wisse and Schwab, except that instead of using a kinematic
constraint to hold the upper body upright, a torsional spring is introduced (Fig. 5).

2   Modeling of Garcia's Robot

Due to the discretely continuous responses of the system, two sets of equations are
needed to describe the overall motion. The Euler-Lagrange relations and the Hamilton
principle are used to obtain the equations of motion governing the movement of the
bipedal robot. First, the potential and kinetic energy equations are formulated, and then

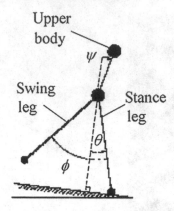

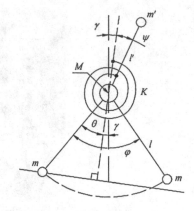

Fig. 4. Wisse's original model [12]   **Fig. 5.** Extended Wisse's model [9]

the equations of motions are derived by taking the derivative of the energy functions. The equations obtained at this stage, accurately describe the movement of the two feet at all times with the exception of the time when the feet impact on the ground. The general form of the equations is expressed as relation (1).

$$M(q)\ddot{q} + C(q,\dot{q})\dot{q} + g(q) = 0 \tag{1}$$

Here, q is the state vector and is written as $\{\theta, \ \phi\}^T$ representing the stance and swing leg angles respectively. This way, two equations of motion are obtained. Since many parameters exist in the equations, and non-dimensionalizing the equations can facilitate the evaluation of the responses and also the comparison with other models, the following non-dimensionalizing variable expressing the ratio of foot to hip mass is used:

$$\beta = \frac{m}{M} \tag{2}$$

Thus, by changing the above variable, the equations of motions are obtained as follows [1]:

$$\begin{cases} [1 + 2\beta(1 - \cos\varphi)]\ddot{\theta} - \beta(1 - \cos\varphi)\ddot{\varphi} \\ \quad - \beta(\dot{\varphi} - 2\dot{\theta})\dot{\varphi}\sin\varphi + \dfrac{\beta g}{l}[\sin(\theta - \varphi - \psi) - \sin(\theta - \psi)] \\ \quad - \dfrac{g}{l}\sin(\theta - \psi) = 0 \\ \ddot{\varphi} - (1 - \cos\varphi)\ddot{\theta} - \dot{\theta}^2\sin\varphi - \dfrac{g}{l}\sin(\theta - \varphi - \psi) = 0 \end{cases} \tag{3}$$

Now, by assuming $m < < M$, $\beta = 0$ (i.e., the mass of feet is ignored with respect to the mass of the waist), the equations take the following form [1]:

$$\begin{cases} \ddot{\theta} - \sin(\theta - \gamma)\dfrac{g}{l} = 0 \\ -\ddot{\varphi} + \ddot{\theta} + \dot{\theta}^2 \sin\varphi - \cos(\theta - \gamma) \sin\varphi = 0 \end{cases} \tag{4}$$

The impact condition for this robot is defined as follows:

$$\varphi = -2\theta \tag{5}$$

It should be mentioned that the above condition also applies when both angles are equal to zero; however, this situation is ignored in the analysis and the assumption is that the swing leg can swing freely through this condition.

By using the conservation of angular momentum about the stance point and impact point, the transfer matrix is obtained as:

$$\begin{bmatrix} \theta \\ \dot{\theta} \\ \varphi \\ \dot{\varphi} \end{bmatrix}^+ = \begin{bmatrix} -1 & 0 & 0 & 0 \\ 0 & \cos(2\theta) & 0 & 0 \\ -2 & 0 & 0 & 0 \\ 0 & \cos(2\theta)(1 - \cos(2\theta)) & 0 & 0 \end{bmatrix} \begin{bmatrix} \theta \\ \dot{\theta} \\ \varphi \\ \dot{\varphi} \end{bmatrix}^- \tag{6}$$

We simulated the resulting system using Matlab's builtin Runge Kutta solver for differential equations.

The initial conditions and the parameters are shown in Table 1.

Table 1. Parameters and Initial conditions for Garcia's model [1]

Initial conditions		Parameters	
θ	0.2073	γ	0.009
$\dot{\theta}$	−0.2061		
φ	0.4147		
$\dot{\varphi}$	−0.0175		

The results of the simulations are shown in Fig. 6.

As the above figure shows, the limit cycle is obviously stable. It should be mentioned that the above limit cycle has become stable at the slope angle less than 0.009.

3 Modeling of Wisse's Robot

The dynamic equations of Wisse's robot are obtained by employing the Lagrange formulation. The general form of these equations is expressed by relation (1), but the vector of state variables is extended by adding a third dimension – the hip angle

$$q = \theta, \varphi, \psi^T \tag{7}$$

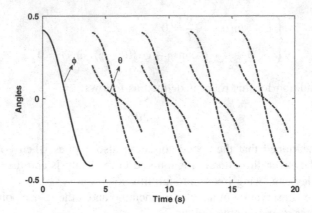

Fig. 6. Stable limit cycle, for slope angles 0.009 Rad

Consequently, three equations of motion are generated, and according to the previously mentioned explanations, these equations are converted to non-dimensionalized forms. So, the dimensionless parameters are defined as follows:

$$\alpha = \frac{m'}{M}, \quad \beta = \frac{m}{M}, \quad k = \frac{K}{M\lg}, \quad v = \frac{l'}{l} \tag{8}$$

Therefore, the following dimensionless equations are obtained [9]:

$$
\begin{aligned}
& \left[Ml^2 + 2ml^2(1 - \cos\varphi) + m'l^2\right]\ddot{\theta} \\
& - ml^2(1 - \cos\varphi)\ddot{\varphi} - m'l^2 v\cos(\psi + \theta)\ddot{\psi} \\
& - Ml^2\dot{\varphi}^2\sin\varphi + 2Ml^2\dot{\varphi}\dot{\theta}\sin\varphi + m'l^2 v\dot{\psi}^2\sin(\psi + \theta) \\
& - Mgl\sin(\theta - \gamma) - m'gl\sin(\theta - \gamma) \\
& - Mgl[\sin(\theta - \gamma) - \sin(\varphi - \theta - \gamma)] = 0
\end{aligned}
$$

$$\tag{9}$$

$$
\begin{aligned}
& ml^2(1 - \cos\varphi)\ddot{\theta} - ml^2\ddot{\varphi} + ml^2\dot{\theta}^2\sin\phi \\
& + mgl\sin(\theta - \varphi - \psi) = 0
\end{aligned}
$$

$$
\begin{aligned}
& - m'l^2 v\cos(\theta + \psi)\ddot{\theta} + m'l^2 v^2\ddot{\psi} \\
& + m'l^2 v\dot{\theta}^2\sin(\theta + \psi) - m'gl'\sin(\psi - \gamma) + k\psi = \tau
\end{aligned}
$$

Similarly, the impact condition for this robot is defined as:

$$\varphi = -2\theta \tag{10}$$

The transfer matrix for the Wisse's robot can be obtained by writing the conservation of angular momentum about the following three points:

About the hip, for the left foot.
About the hip, for the upper body.
About the flight foot, for the whole body.

$$
\begin{bmatrix} \theta \\ \dot{\theta} \\ \varphi \\ \dot{\varphi} \\ \psi \\ \dot{\psi} \end{bmatrix}^{+} = \begin{bmatrix} -1 & 0 & 0 & 0 & 0 & 0 \\ 0 & A & 0 & 0 & 0 & 0 \\ 2 & 0 & 0 & 0 & 0 & 0 \\ 0 & B & 0 & 0 & 0 & 0 \\ 0 & 0 & 0 & 0 & 1 & 0 \\ 0 & C & 0 & 0 & 0 & 1 \end{bmatrix} \begin{bmatrix} \theta \\ \dot{\theta} \\ \varphi \\ \dot{\varphi} \\ \psi \\ \dot{\psi} \end{bmatrix}^{-}
\tag{11}
$$

Where we have:

$$
\begin{aligned}
A &= \frac{\mu(\cos(2\theta) - \cos(2\psi)) + 2\cos(2\theta)}{2\beta \sin^2(2\theta) + 2\mu \sin^2(\theta - \psi) + 2} \\
B &= A(1 - \cos(2\theta)) \\
C &= \left[1 + \frac{\cos(\theta - \psi)}{v} \right](1 - A)
\end{aligned}
\tag{12}
$$

By performing the simulation, the limit cycle of the robot is obtained as follows (Table 2):

Table 2. Parameters and Initial conditions for Wisse's model [9]

Initial conditions		Parameters	
θ	0.2	γ	0.0045
$\dot{\theta}$	−0.2	β	0
φ	0.4	α	0.08
$\dot{\varphi}$	−0.02	v	0.13
ψ	−0.3659	k	0.0917
$\dot{\psi}$	1.7925		

And the results are as follows (Fig. 7):
The above figure clearly shows the stability of the limit cycle. The limit cycle is stable with a slope angle less than 0.0045.

4 Modeling of a Passive Robot with Arms

In this section, a 5-DOF robot consisting of two arms, two feet and also one degree of freedom in the trunk section is analyzed. The configuration of this robot has made it more human-like. This robot is shown in Fig. 8.

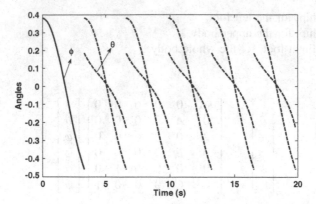

Fig. 7. Stable limit cycle, for slope angles 0.0045 Rad

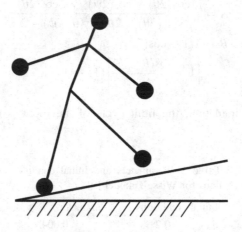

Fig. 8. 5 DOF model

Similar to the previous robot, Lagrange formulation is used to extract the dynamic equations of this system. The kinetic energy of this system is written as:

$$T_1 = \frac{1}{2} m_1 l_1^2 \dot{\theta}_1^2$$

$$T_2 = \frac{1}{2} m_2 l_1^2 \left(\dot{\theta}_1^2 + (\dot{\theta}_1 - \dot{\varphi})^2 - 2\dot{\theta}_1(\dot{\theta}_1 - \dot{\varphi}) \cos \varphi \right)$$

$$T_3 = \frac{1}{2} m_3 \left(l_1^2 \dot{\theta}_1^2 + l_2^2(\dot{\theta}_1 + \dot{\theta}_2)^2 - 2l_1 l_2 \dot{\theta}_1(\dot{\theta}_1 + \dot{\theta}_2) \cos(\theta_2 - \theta_1) \right) \quad (13)$$

$$T_4 = \frac{1}{2} m_4 \left(l_1^2 \dot{\theta}_1^2 + l_2^2(\dot{\theta}_1 - \dot{\theta}_3)^2 - 2l_1 l_2 \dot{\theta}_1(\dot{\theta}_1 - \dot{\theta}_3) \cos(\theta_1 - \theta_3) \right)$$

$$T_5 = \frac{1}{2} m_5 \left(l_1^2 \dot{\theta}_1^2 + (\dot{\theta}_1 - \dot{\psi})^2 l_3^2 + 2l_1 l_3 \dot{\theta}_1(\dot{\theta}_1 - \dot{\psi}) \cos(\theta_1 - \psi) \right)$$

And potential energy:

$$V_1 = m_1 g l_1 \cos(\theta_1 - \gamma)$$
$$V_2 = m_2 g l_1 (\cos(\theta_1 - \gamma) - \cos(\theta_1 - \varphi - \gamma))$$
$$V_3 = m_3 g (l_1 \cos(\theta_1 - \gamma) - l_2 \cos(\theta_2 - \gamma))$$
$$V_4 = m_4 g (l_1 \cos(\theta_1 - \gamma) - l_2 \cos(\theta_3 + \gamma))$$
$$V = m_5 g (l_1 \cos(\theta_1 - \gamma) + l_3 \cos(\psi + \gamma)) + \frac{1}{2} k \psi^2$$

(14)

By inserting the kinetic and potential energies into the Lagrange relation, the dynamic equations of the above system are obtained, employing the previously mentioned impact conditions and using the initial conditions and the dimensionless parameters which are defined below (Table 3):

Table 3. Parameters and Initial conditions for 5 DOF's model

Initial conditions		Parameters	
θ	0.2	γ	0.01
$\dot{\theta}$	−0.2	β	0.06
φ	0.38	α	0
$\dot{\varphi}$	−0.02	ν	0.2
ψ	−0.3659	k	0.0917
$\dot{\psi}$	1.7925	ξ	0.07
		σ	0.004

$$\alpha = \frac{m_2}{m_1}, \quad \beta = \frac{m_3 = m_4}{m_1}, \quad \sigma = \frac{m_5}{m_1}, \quad \frac{l_2}{l_1} = \nu, \quad \frac{l_3}{l_1} = \xi$$

(15)

By performing the simulations, the following results are obtained (Figs. 9 and 10):

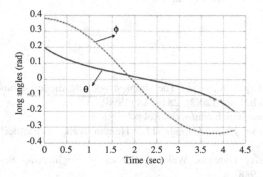

Fig. 9. Stable limit cycle, for slope angle 0.01 Rad

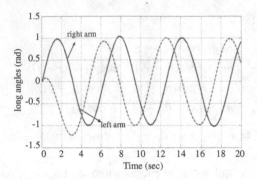

Fig. 10. The position of the robot's arm

The reason for changing the sign of angles θ and φ is the change in the measuring reference of these angles at the moment of impact. Because, as the support (fulcrum) foot suddenly switches, the two said angles find new definitions and a new measuring reference, and therefore, from a geometrical perspective, their signs will change. But since the measuring reference of angle ψ doesn't change during impact, this angle always remains the same; because, although the impact forces produce impulsive changes in acceleration and stepped changes in velocity, they are not able to change the angles themselves. So, if the measuring references of the angles are fixed, these angles will always remain constant. At this stage, the 5-DOF robot has attained a stable limit cycle at a slope less than 0.01.

5 Conclusion

In this paper, three models of passive robots were analyzed. For each system, a stable limit cycle was achieved by finding appropriate initial conditions and dimensionless parameters. In the Garcia's model, the robot attained a stable limit cycle at slope angle less than 0.009; whereas in the Wisse's model and the model with arms, the robots achieved stability at slopes angle less than 0.0045 and 0.01, respectively. Therefore, we can conclude that by extending the Wisse's robot model to a 5-DOF robot with arms, the new robot becomes more stable.

References

1. Garcia, M., Chatterjee, A., et al.: The simplest walking model stability, complexity, and scaling. J. Biomech. Eng. **120**(2), 281–288 (1998)
2. http://www-personal.umich.edu/~shc/Robot/
3. McMahon, T., Rose, J., and Gamble, J.G.: Human Walking, Biomechanical engineering, pp. 52–53, (1989)
4. McGeer, T: Passive Dynamic Walking, School of Engineering Science Simon Fraser University, May 1988

5. Kuo, A.D.: Stabilization of lateral motion in passive dynamic walking. Int. J. Robot. Res. **18** (9), 917–930 (1999)
6. Wisse, M., Schwab, A.L.: Passive dynamic walking model with upper body. Robotica **22**, 681–688 (2004). Cambridge University Press
7. Collins, S., Wisse, M., Ruina, A.: A three-dimensional passive dynamic walking robot with two legs and knees. Int. J. Robot. Res. **20**, 607–622 (2001)
8. Collins S., Ruina A., A bipedal walking robot with efficient and human-like gait. In: International Conference on Robotics and Automation - ICRA, pp.1983–1988, Barcelona, Spain (2005)
9. Farshimi, F., Naraghi, M.: A passive-biped model with multiple routes to chaos. J. Acta Mech. Sinica **27**(2), 277–284 (2011)
10. Safa, A.T., Saadat, M.G., Naraghi, M.: Passive dynamic of the simplest walking model replacing ramps with stairs. Mech. Mach. Theory **42**, 1314–1325 (2007). Elsevier Publication
11. Camp J.: Powered "Passive" dynamic walking, Masters of Engineering Project, The Sibley School of Mechanical and Aerospace Engineering, Cornell University, USA, pp. 1–8 (1997)
12. Wisse M.: Essentials of dynamic walking, analysis and design of two-legged robots, Delft University, Netherlands, pp. 33–50 (2004). ISBN 90-77595-82-1
13. Tedrake R., Zhang W., et al., Actuating a Simple3D passive dynamic walker. In: IEEE International Conference on Robotics and Automation, New Orleans, USA, pp. 1–6 (2004)

Object Motion Estimation Based on Hybrid Vision for Soccer Robots in 3D Space

Huimin Lu$^{(\boxtimes)}$, Qinghua Yu, Dan Xiong, Junhao Xiao,
and Zhiqiang Zheng

College of Mechatronics and Automation,
National University of Defense Technology, Changsha, China
{lhmnew,xiongdan,zqzheng}@nudt.edu.cn, yuqinghua@163.com,
junhao.xiao@ieee.org

Abstract. Effective object motion estimation is significant to improve the performance of soccer robots in RoboCup Middle Size League. In this paper, a hybrid vision system is constructed by combining omnidirectional vision and stereo vision for ball recognition and motion estimation in three-dimensional (3D) space. When the ball is located on the ground field, a novel algorithm based on RANSAC and Kalman filter is proposed to estimate the ball velocity using the omnidirectional vision. When the ball is kicked up into the air, an iterative and coarse-to-fine method is proposed to fit the moving trace of the ball with paraboic curve and predict the touchdown-point in 3D space using the stereo vision. Experimental results show that the robot can effectively estimate ball motion in 3D space using the hybrid vision system and the proposed algorithms, furthermore, the advantages of the 360° field of view of the omnidirectional vision and the high object localization accuracy of the stereo vision in 3D space can be combined.

1 Introduction

In the highly dynamic RoboCup Middle Size League (MSL) competition, accurate estimation of object motion states, such as the velocity and the shooting touchdown-point of the ball, is the basis of ball passing and intercepting for regular robots, and shoot defending for goalie robots. Furthermore, because the ball is often lifted by the robots' high kicks during MSL competition, the ball motion should be estimated in three-dimensional (3D) space to improve the performance of soccer robots.

In [1], a Kalman Filter was used to detect whether the ball is moving or stationary. In the corrector part of the Kalman Filter, a multilayer perceptron artificial neural network was integrated to reduce the affection of image noises caused by the motion vibration of the robot, so the robustness of the state detection could be improved. In [2], Lauer et al. assumed that, during a small piece of time, the motion of a ball rolling on the filed is a linear movement with a constant velocity. Therefore, the ball velocity estimation could be modelled as a standard linear regression problem, which could be solved by ridge regression — a least squares mathcing method (LSM). In [3], Kalman Filter was employed for

© Springer International Publishing Switzerland 2015
R.A.C. Bianchi et al. (Eds.): RoboCup 2014, LNAI 8992, pp. 454–465, 2015.
DOI: 10.1007/978-3-319-18615-3_37

ball position estimation; based on the estimated positions, an algorithm similar as that in [2] was utilized to evaluate the ball velocity, resulting improved ball velocity estimation accuracy.

However, the methods mentioned above can only be used when the ball is located on the ground field. In [4,5], Taiana et al. applied particle filter to track the ball using omnidirectional vision in 3D space, where the 3D shape of the ball was considered and the colour histograms of the inner and outer boundary on the panoramic image projected by the ball were used to construct the observation model in particle filter. The experimental results show that it can precisely track the ball and acquire the ball position in 3D space, but other motion states like velocity were not estimated.

In [6–8], a mixed stereo camera sensor was constructed based on omnidirectional vision and perspective camera, which was employed to recognize and localize the ball in 3D space. As a result, the advantages of the 360° field of view of the omnidirectional vision and the long field of view of the perspective camera can be combined. In [6,7], triangulation was used to calculate the ball position in 3D space. In [8], Käppeler et al. found that better results could be achieved when calculating the 3D ball position using the angle to the ball determined by the omnidirectional vision and the distance to the ball derived from the size in the image of the perspective camera in comparison with the triangulation method, but they did not discuss ball velocity estimation in 3D space. In [6], Voigtländer et al. extended the approach in [2] from 2D space to 3D space to estimate the ball velocity. In [7], Lauer et al. developed a maximum likelihood estimator based on the ECM approach and a Bayesian approach based on Gibbs sampling to estimate the ball position and velocity in 3D space. However, the field of view of this kind of mixed stereo camera sensor is quite limited for motion estimation in 3D space. Particularly, when the ball is kicked up higher than the omnidirectional vision, the omnidirectional vision can not work any more.

Tech United Eindhoven team used a Laser Range Finder (LRF) attached to the highest point of the goalie robot to detect lob balls, when the ball disappears out of the view of the omnidirectional vision [9]. After the ball is detected, the LRF is tilted further upwards with a servo to measure additional positions of the ball in the air. With these points, the complete path of the ball can be calculated using a parabolic fit. Recently, they use the RGB-D camera Kinect to detect and track the ball in 3D space [10]. The main drawbacks are limitations in resolution and field of view, and only balls within approximately six meters can be recognized, which is not enough for goalie robots to intercept lob balls successfully.

In this paper, we present a hybrid vision system combining omnidirectional vision and stereo vision, which is employed to recognize the ball and estimate the ball motion states in 3D space including the velocity and the shooting touchdown-point. The following sections are organized as follows: the system overview is introduced in Sect. 2; an algorithm to estimate the ball velocity based on RANSAC and Kalman filter is proposed in Sect. 3, using the omnidirectional vision to deal with the situation that the ball is located on the ground field; an object motion estimation algorithm including moving trace fitting and touchdown point prediction is presented in Sect. 4, which deals with the situation that the ball is kicked up into the air using the stereo vision; experimental results are presented in Sect. 5; Sect. 6 concludes this paper.

2   The System Overview

Omnidirectional vision can provide a 360° view of the robot's surrounding environment in a single image, thus is quite suitable for ball recognition and motion estimation when the ball is located on the ground field [11]. In the mixed stereo vision proposed in [6–8], the overlapped field of view of omnidirectional vision and perspective camera is quite small. Especially when the ball is higher than the robot, it is beyond the field of view of the stereo vision. Furthermore, since the imaging resolution of the omnidirectional vision is limited, the accuracy of ball localization can not be high in 3D space. Therefore, it is hard to deal with object motion estimation in 3D space well with such a mixed stereo vision. In this paper, we add Bumblebee2, a stereo vision developed by Point Grey Research, to construct the hybrid vision system for our soccer robots, as shown in Fig. 1. Typical images acquired by the omnidirectional vision and stereo vision, and the corresponding ball recognition results are also illustrated.

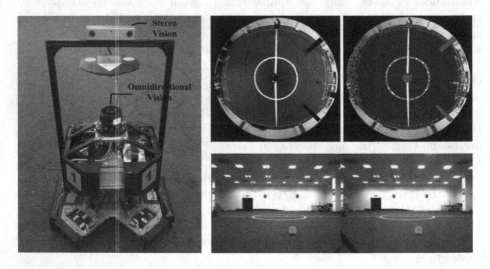

Fig. 1. The NuBot soccer robot equipped with the hybrid vision system constructed with omnidirectional vision and stereo vision. Typical images acquired by each vision and the ball recognition results are also shown.

The working architecture of the hybrid vision system is depicted in Fig. 2. Firstly, the omnidirectional vision searches the ball in the 360° view. If the ball is detected, the robot will turn to the ball. Afterwards, the stereo vision is employed

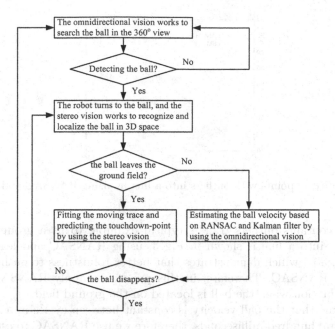

Fig. 2. The working architecture of the hybrid vision system to estimate the ball motion in 3D space.

to recognize and localize the ball in 3D space based on triangulation, by which the robot can determine whether the ball leaves the ground field. If the ball is located on the ground field, an algorithm based on RANSAC and Kalman filter is used to estimate the ball velocity with the omnidirectional vision. If the ball is kicked up into the air, a ball motion estimation algorithm is used to fit the moving trace and predict the touchdown-point where the robot should move to intercept the ball using the stereo vision. When the ball disappears from the view of the robot, the robot has to search the ball again using the omnidirectional vision. In such an architecture, the advantages of the 360° field of view of the omnidirectional vision and the high object localization accuracy of the stereo vision in 3D space can be combined.

3 Ball Velocity Estimation Based on RANSAC and Kalman Filter

When the ball is located on the ground field, omnidirectional vision is the best choice for ball recognition and motion estimation. In the highly dynamic RoboCup MSL competition, outliers and noises always exist in the information of ball positions due to the limited imaging resolution of omnidirectional vision, image noises, occlusions, motion blurs, etc., which will decrease the ball velocity estimation accuracy. RANdom SAmple Consensus (RANSAC) is an iterative method to estimate parameters of a mathematical model from a set of observed

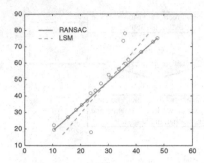

Fig. 3. Fitting points with outliers into a line by using RANSAC and LSM.

data which contains outliers [12]. A typical example to fit a group of points with outliers into a line is shown in Fig. 3, using RANSAC and least squares matching (LSM), which demonstrates that better robustness to outliers can be achieved by RANSAC. Therefore, in this paper, we apply RANSAC for ball velocity estimation when the ball is located on the ground field.

We assume that the ball velocity is constant between cycles in a short time such as several hundred milliseconds, therefore we use RANSAC to estimate the ball velocity with ball positions X_i and corresponding timestamp $t_i, i = 1, ..., n$. We can acquire the velocity value V_m as follows:

$$V_m = \frac{X_i - X_j}{t_i - t_j}, (i \neq j, i, j = 1, ..., n, m = 1, ..., n(n-1)/2) \qquad (1)$$

Then we can choose k values from $\{V_m, m = 1, ..., n(n-1)/2\}$ randomly as hypothetical inliers, and calculate the mean value $M_l = \sum_{i=1}^{k} V_i/k$ as a candidate model with a counter $C_l = 0$. All the other velocity values are tested against the model by comparing the distance between the value and the model. If the distance is less than a threshold, $C_l = C_l + 1$, and the value is added into hypothetical inliers as the consensus set. Once this test finishes, a better model M_l is updated by calculating the mean value from the consensus set. After performing this operation iteratively by L times, $\{M_l, C_l, l = 1, ..., L\}$ can be achieved. The M_l with the largest C_l is considered as the estimated ball velocity.

To further improve the robustness and accuracy, we also use Kalman filter to optimize the ball positions before estimating the ball velocity similar as that in [3]. As the ball velocity is assumed to be constant between cycles in a short time, we use the prediction model as follows:

$$\begin{pmatrix} X_{k+1} \\ V_{k+1} \end{pmatrix} = \begin{pmatrix} 1 & \Delta t \\ 0 & 1 \end{pmatrix} \begin{pmatrix} X_k \\ V_k \end{pmatrix} \qquad (2)$$

where X_k is the ball position, and V_k is the ball velocity. To model the measurement variance in the Kalman filter, we place the ball on the field with different distances to the robot, and then the robot recognizes and localizes the ball using the omnidirectional vision. As a result, a group of measurement variances is

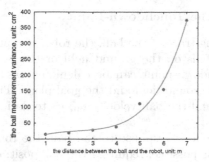

Fig. 4. The ball measurement variance when the ball is located on different distances to the robot.

acquired shown as the red points in Fig. 4. Afterwards, we fit these variances by using a four order polynomial, shown as the green curve in Fig. 4. The result is also useful in the simulation experiments in Sect. 5.1.

With Kalman filter, ball collision with other objects can also be detected, which happens frequently during the competition. We consider a collision happens when the distance between the ball measurement and the filtered result is larger than a threshold in five consecutive cycles. In that case, the Kalman filter will be restarted, and the ball measurements of the latest five cycles will be used as the initial data for ball velocity estimation.

4 Fitting the Moving Trace and Predicting of the Touchdown-Point in 3D Space

4.1 Fitting the Moving Trace in 3D Space

After obtaining the 3D position of the ball by triangulation using stereo vision, the robot can determine whether the ball leaves the ground field. If the 3D positions are higher than the ground field by 10 cm in consecutive three cycles, the ball is considered to be kicked up into the air. We use parabolic curve to fit the moving trace of the ball. The ball motion can be modelled as follows:

$$\begin{cases} x_i = a_0 * t_i + a_1 \\ y_i = a_2 * t_i + a_3 \\ z_i + g * t_i^2/2 = a_4 * t_i + a_5 \end{cases} \tag{3}$$

where x_i, y_i and z_i are 3D coordinates obtained by stereo vision in the timestamp t_i, and g is the gravity acceleration. The three equations are simple and linear. When a group of $\{x_i, y_i, z_i, t_i\}$ is obtained, $a_0, ..., a_5$ can be calculated by the least squares method in three coordinates respectively. Therefore, the moving trace of the ball can be acquired using the parabolic fit.

4.2   Predicting of the Touchdown-Point

After fitting the moving trace of the ball, the robot can predict the touchdown-point where the ball falls on the ground field or the ball passes through the goal plane. The touchdown-point can be calculated easily by intersecting the moving trace with the ground field and the goal plane. Then the robot can use the touchdown-point and the ball velocity a_0, a_2 to decide how to intercept or defend the ball.

To speed up the robot's response to the ball, the touchdown-point should be predicted as early as possible, requiring the ball positions used in fitting and predicting should be as few as possible. However, to improve the accuracy of ball intercepting or defending, more ball positions should be used in fitting and predicting to acquire more accurate prediction results. As a compromise, we use an iterative method to deal with this contradiction. Once the robot obtains five ball positions in the air, the first fitting and prediction is performed to acquire a coarse result of the predicted touchdown-point, thus the robot can respond to the ball very quickly. Then the robot will go on fitting and predicting with more data after obtaining new ball positions to update the prediction results. So the robot can achieve coarse-to-fine fitting and prediction results iteratively during ball intercepting or defending.

5   Experimental Results

In this section, we evaluate the two algorithms proposed in Sect. 3 and Sect. 4 respectively.

5.1   Ball Velocity Estimation Results by Omnidirectional Vision

Because no ground truth about ball positions and velocities can be provided in the actual experiments, we firstly perform simulation experiments to evaluate three algorithms: the proposed Kalman filter+RANSAC, Kalman filter+LSM, and Kalman filter. We add Gaussian noises into the simulated ball positions according to the variances in Fig. 4. Three different situations are considered including no collision, one collision, and multiple collisions of the ball. The ball positions and the estimated ball velocities in one experiment are shown in Fig. 5. We perform ten such experiments, and the statistics are shown in Table 1, where $\bar{E}$ is the mean estimation error, and $\bar{P}$ the mean ratio between the error and the real velocity. From Table 1, we see that when no collision happens, all the algorithms work well, and Kalman filter is the best algorithm to estimate the ball velocity, because the added noises are Gaussian in this simulation experiment. When collision happens, the motion direction of the ball changes, and better performance can be achieved by using Kalman filter+RANSAC and Kalman filter+LSM than Kalman filter. Furthermore, in comparison with Kalman filter+LSM, the estimation accuracy can be improved by 15 %~40 % when using Kalman filter+RANSAC.

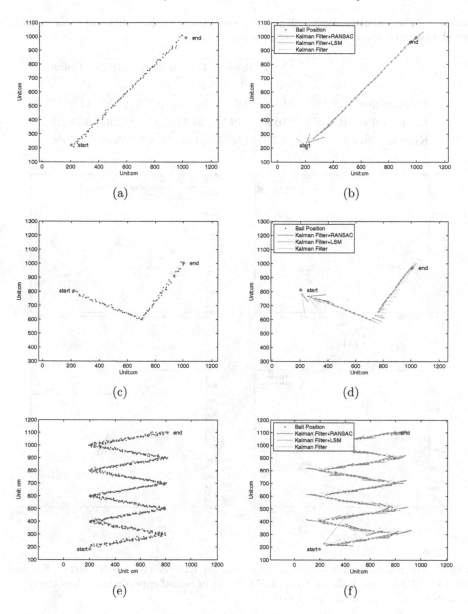

Fig. 5. The ball positions (a)(c)(e) and the ball velocities estimated by three different algorithms (b)(d)(f) in simulation experiments.

We also test Kalman filter+RANSAC and Kalman filter+LSM in the actual experiments using our NuBot soccer robot. The ball velocity estimation results are shown in Fig. 6 when the robot is stationary or moving. Because no ground truth can be provided, we only can say both algorithms work well. A video

Table 1. The statistics about ball velocity estimation by three different algorithms in simulation experiments. The unit of $\bar{E}$ is cm/s.

	No collision		One collision		Multi-collision	
	$\bar{E}$	$\bar{P}$	$\bar{E}$	$\bar{P}$	$\bar{E}$	$\bar{P}$
KalmanFilter+LSM	10.84	4.01 %	32.78	9.47 %	27.22	11.68 %
KalmanFilter+RANSAC	10.97	4.05 %	20.20	5.83 %	22.64	9.710 %
Kalman Filter	4.780	1.80 %	91.86	26.6 %	59.50	25.70 %

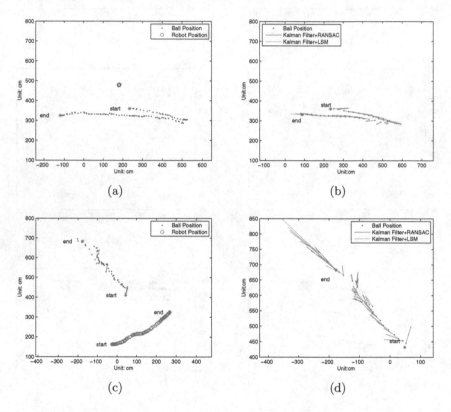

Fig. 6. The robot and ball positions (a)(c) and the ball velocities estimated by Kalman filter+RANSAC and Kalman filter+LSM (b)(d) in actual experiments when the robot is stationary (a)(b) or moving (c)(d).

showing that our robot estimating the ball velocity by using the proposed Kalman filter+RANSAC algorithm can be found on:

http://v.youku.com/v_show/id_XNzI2MDM5MzAw.html

Because the computation complexity of Kalman filter+RANSAC is quite low and it can be performed within several milliseconds, the discussion about the real-time performance is not necessary in this paper.

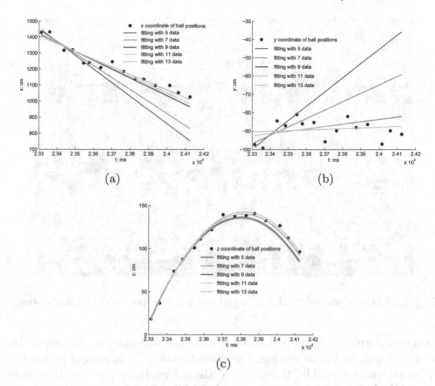

Fig. 7. The moving trace of the ball is fitted by the proposed algorithm in x (a), y (b), z (c) coordinate respectively when using different numbers of ball positions.

5.2   Ball Motion Estimation Results by Stereo Vision

The moving trace of the ball is fitted by the proposed algorithm in x, y, z coordinate respectively. A typical fitting process is shown in Fig. 7, where the first fitting with five ball positions, and the subsequent fitting with seven, nine, eleven, and thirteen ball positions are demonstrated. We see that the fitting results converge during the iterative process, which verifies that the fitting accuracy becomes higher as more data of ball positions are used. Although the accuracy of the first fitting is low, it is significant for the robot's quick response.

We use our goalie robot to test the fitting and prediction results by defending high shooting. During the experiment, a person simulates the high shooting by throwing the ball to the goal from different positions with different distances to the goal about 5~9 meters. The results show that the goalie robot can defend the high shooting with a successful rate of 80 %, which also verifies that the touchdown-point can be predicted effectively using the proposed algorithm. A typical successful defending is shown in Fig. 8. The video about this experiment can be found on: http://v.youku.com/v_show/id_XNzI2MDQwNDcy.html.

Again, the computation complexity of the proposed algorithm is quite low and it can be performed in less than one millisecond, so the discussion about the

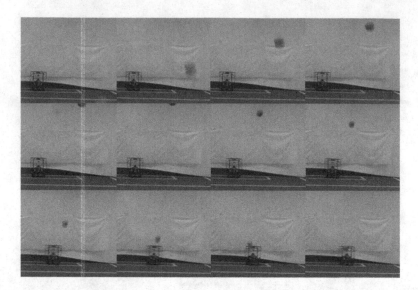

Fig. 8. A typical successful defending of our goalie robot to the high shooting.

real-time performance of the algorithm itself is not necessary in this paper. However, in our hybrid vision system, the omnidirectional vision and stereo vision often work simultaneously, therefore we should evaluate the real-time performance of the hybrid vision system as a whole. The robot's computer is equipped with a 1.66G CPU and 1.0G memory. When the robot works in the competition state, the omnidirectional vision can work on a frame rate of 25 fps, and the stereo vision can work on a frame rate of 20 fps, which meets the real-time requirement of RoboCup MSL competition.

6   Conclusion

In this paper, a hybrid vision system is constructed with omnidirectional vision and stereo vision to realize ball recognition and motion estimation in 3D space. To the best of our knowledge, it is the first time that this kind of hybrid vision is used in RoboCup MSL. When the ball is located on the ground field, a novel algorithm based on RANSAC and Kalman filter is proposed to estimate the ball velocity using the information from the omnidirectional vision. When the ball is kicked up into the air, an iterative and coarse-to-fine method is proposed to fit the moving trace of the ball with paraboic curve and predict the touchdown-point in 3D space using the information from the stereo vision. As a result, the quick response and accuracy can be met simultaneously for soccer robots to intercept or defend the ball. Experimental results show that the robot can realize effective estimation of ball motion in 3D space using the hybrid vision system and the proposed algorithms.

In the future, we would like to compare our ball motion estimation with other approaches existing in RoboCup MSL after trying to acquire the ground truth. We also want to realize better recognition and motion estimation of other objects like obstacles employing our hybrid vision system.

References

1. Taleghani, S., Aslani, S., Shiry, S.: Robust moving object detection from a moving video camera using neural network and Kalman filter. In: Iocchi, L., Matsubara, H., Weitzenfeld, A., Zhou, C. (eds.) RoboCup 2008. LNCS, vol. 5399, pp. 638–648. Springer, Heidelberg (2009)
2. Lauer, M., Lange, S., Riedmiller, M.: Modeling moving objects in a dynamically changing robot application. In: Furbach, U. (ed.) KI 2005. LNCS (LNAI), vol. 3698, pp. 291–303. Springer, Heidelberg (2005)
3. Silva, J., Lau, N., Rodrigues, J., Azevedo, J.L., Neves, A.J.R.: Sensor and information fusion applied to a robotic soccer team. In: Baltes, J., Lagoudakis, M.G., Naruse, T., Ghidary, S.S. (eds.) RoboCup 2009. LNCS, vol. 5949, pp. 366–377. Springer, Heidelberg (2010)
4. Taiana, M., Gaspar, J.A., Nascimento, J.C., Bernardino, A., Lima, P.: 3D Tracking by catadioptric vision based on particle filters. In: Visser, U., Ribeiro, F., Ohashi, T., Dellaert, F. (eds.) RoboCup 2007: Robot Soccer World Cup XI. LNCS (LNAI), vol. 5001, pp. 77–88. Springer, Heidelberg (2008)
5. Taiana, M., Santos, J., Gasper, J., et al.: Tracking objects with generic calibrated sensors: an algorithm based on color and 3D shape features. Robot. Auton. Syst. 58(6), 784–795 (2010)
6. Voigtländer, A., Lange, S., Lauer, M., et al.: Real-time 3D ball recognition using perspective and catadioptric cameras. In: Proceedings of 2007 European Conference on Mobile Robots (ECMR07) (2007)
7. Lauer, M., Schönbein, M., Lange, S., Welker, S.: 3D-objecttracking with a mixed omnidirectional stereo camera system. Mechatronics 21(2), 390–398 (2011)
8. Käppeler, U., Höferlin, M., Levi, P.: 3D object localization via stereo vision using an omnidirectional and a perspective camera. In: Proceedings of the 2nd Workshop on Omnidirectional Robot Vision, pp. 7–12 (2010)
9. Kanters, F.M.W., Hoogendijk, R., et al.: Tech united eindhoven team description 2011. In: RoboCup 2011 Istanbul, CD-ROM (2011)
10. Schoenmakers, F.B.F., Koudijs, G., et al.: Tech united Eindhoven team description 2013 middle size league. In: RoboCup 2013 Eindhoven, CD-ROM (2013)
11. Lu, H., Yang, S., Zhang, H., Zheng, Z.: A robust omnidirectional vision sensor for soccer robots. Mechatronics 21(2), 373–389 (2011)
12. Fischler, M.A., Bolles, R.C.: Random sample consensus: a paradigm for model fitting with applications to image analysis and automated cartography. Comm. ACM 24, 381–395 (1981)

Human Inspired Control of a Small Humanoid Robot in Highly Dynamic Environments or Jimmy Darwin Rocks the Bongo Board

Jacky Baltes, Chris Iverach-Brereton$^{(\boxtimes)}$, and John Anderson

University of Manitoba, Winnipeg, MB R3T2N2, Canada
chrisib@cs.umanitoba.ca
http://aalab.cs.umanitoba.ca/

Abstract. This paper describes three human-inspired approaches to balancing in highly dynamic environments. In this particular work, we focus on balancing on a bongo board - a common device used for human balance and coordination training - as an example of a highly dynamic environment. The three approaches were developed to overcome limitations in robot hardware. Starting with an approach based around a simple PD controller for the centre of gravity, we then move to a hybrid control mechanism that uses a predictive control scheme to overcome limitation in sensor sensitivity, noise, latency, and jitter. Our third control approach attempts to maintain a dynamically stable limit cycle rather than a static equilibrium point, in order to overcome limitations in the speed of the actuators. The humanoid robot Jimmy is now able to balance for several seconds and can compensate for external disturbances (e.g., the bongo board hitting the table). A video of the robot Jimmy balancing on the bongo board can be found at http://youtu.be/ia2ZYqqF-lw.

1 Introduction

This paper describes our research on active balancing reflexes for humanoid robots. Rapid progress in both hardware and software in recent years has led to impressive improvements in the performance of humanoid robots. For example, the soccer playing robots participating in the RoboCup competition [1] can walk and turn quickly, as well as stand up rapidly after falling. In the multi-event HuroCup competition [2], the world record in the sprint event (3 m walking forward followed by 3 m walking backward) has improved from 01:07.50 s. in 2009 to 00:25.50 s. in 2013. Similarly, the world record times in the marathon, which is traditionally held outdoors, improved from 37:30.00 over 42.195 m in 2007 to 13:24.39 over 120 m in 2013. Today, most humanoid robots have little difficulty traversing flat and even surfaces with sufficient friction.

The problem of traversing an irregular and potentially unstable surface, on the other hand, is still extremely difficult and remains without a general solution. Today's robots do not have sufficiently powerful actuators, nor enough sensors to be able to move over a rubble pile or similar environment.

© Springer International Publishing Switzerland 2015
R.A.C. Bianchi et al. (Eds.): RoboCup 2014, LNAI 8992, pp. 466–477, 2015.
DOI: 10.1007/978-3-319-18615-3_38

In recent years, we have therefore focused on balancing in challenging, yet achievable environments. Examples are our robot Tao-Pie-Pie [3], which active balanced over a uneven balance field, and our ice and inline skating humanoid robot Jennifer [4], which demonstrated gaits that were stable on moving wheels and on ice. In winning this year's FIRA Hurocup [5] in the kid-size division, our robots demonstrated a broad range of achievements in adaptive humanoid motion, scoring highly in weightlifting, climbing, and sprinting, as well as soccer, while using the same unaltered robot in these and other events.

Balancing skills are central to all of these, as well as most other humanoid movement. In this paper, we describe our work toward balancing on a bongo board using a small humanoid robot (Jimmy, a DARwin-OP robot made by Robotis). A bongo board is a device commonly used in human training for balance and coordination, and consists of a small board that is placed on top of a cylindrical fulcrum. Figure 1 shows our humanoid robot Jimmy on top of the bongo board used in this work. Jimmy is a Robotis DARwIn-OP robot [6], standing approximately 45 cm tall. He has 22 Degrees Of Freedom (DOFs) and has a six-axis gyroscope/accelerometer located in the torso. Jimmy has been modified from the manufacturer's stock configuration with two single-DOF hands and four-point force (FSR) sensors in the feet.

Fig. 1. Jimmy on the bongo board (left). The figure on the right shows how the bongo board corresponds to an inverted pendulum problem.

The bongo board's fulcrum can freely move left and right, forcing the robot to balance in those directions to keep the board from touching the ground on either side. Balancing on the bongo board is a non-trivial task even for humans. Moreover, because the fulcrum can move, shifting the centre of mass can allow the board to remain balanced and off the ground while shifting the fulcrum from side to side, and this and other tricks are used by human acrobats for entertainment purposes.

The remainder of this paper is organized as follows. Section 2 presents an analysis of the dynamics of the bongo board and shows the relationship to other inverted pendulum problems. Section 3 describes the design and implementation of our three control strategies for the bongo board. Section 3 also describes several challenges imposed by the robot hardware and how we overcame them. Section 4 provides a brief numerical evaluation of the three control strategies we developed. Additional discussion appears in with Sect. 5, which also provides directions for future work.

2   Analysis and Related Work

This section gives a brief introduction to the dynamics of an inverted pendulum [7].

2.1   Dynamics of the Inverted Pendulum Problem

The dynamics of the inverted pendulum problem are well-studied and well understood and form the basis of many motion control algorithms for bipedal humanoid walking robots [8].

There has been a lot of theoretical work in the area of highly dynamic balancing [9–11], but practical implementations are still lacking. Anderson et al. describe an adaptive torque based approach [12] that is able to balance a humanoid robot on a simple see saw. In simulation, their approach is also able to balance a humanoid robot on the more challenging bongo board.

A similar system is described by Hyon [13] is able to balance a robot on a see saw in the presence of unknown disturbances.

2.2   Dynamics of the Bongo Board

The problem of balancing on a bongo board is similar to the cart and rod problem, as can be seen in Fig. 1. The robot can be modelled as a single point mass balancing on top of the board, and the goal is for the robot and board to balance without touching the ground or the robot falling off the board. In other words, the inverted pendulum system formed by the robot and the bongo board should balance.

The difference between the bongo board and the card and rod problem is that when balancing on a bongo board, (a) the pivot point of the robot will rotate along the circumference of the wheel, and (b) the position of the pivot point cannot be controlled directly - only indirectly by controlling the motion of the humanoid robot balancing on the board. Figure 2 shows the how the robot manipulates its limbs in order to control the inclination of the board.

The robot uses its legs and hips to provide coarse control over the inclination, elevation, and lateral position of the CoM with respect to the deck of the bongo board. Moving the shoulders provides fine control over the lateral position of the CoM, as well as applying torque to the system which can be used to provide angular control over the system. Note that from the perspective of the rider d – the distance between the point of contact between the deck and the wheel and the lateral position of the rider's CoM – is unknown.

As with a more traditional inverted pendulum problem, the inclination and angular velocity of the mass (in this case the rider and the deck) must be controlled in order for the system to remain in a stable position.

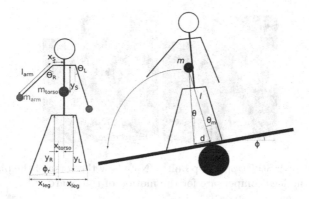

Fig. 2. A bongo board with the rider in an unstable position. The rider's CoM, m, is the weighted average of the CoMs of the torso and the arms. The robot's arms and legs can be adjusted independently to offset the torso laterally (x_{torso}) and vertically (y_L, y_R, ϕ_r). The angle between the point of contact between the bongo board's deck and wheel and m is given by θ_m. The distance between the point of contact and robot's CoM measured parallel with the deck is given by d. The CoM is located at height l above the deck.

3   Design and Implementation

Based on the analysis in the previous section, we began examining how people balance on a bongo board, and what considerations had to be made to adapt humanoid robot balancing to this task. Through the experimentation with a simple control regime, it became clear that significant complications arise with current robotic technology that are easily taken for granted in simple balancing tasks in humans. In particular, sensor noise, sensor latency, and actuator latency are major problems which required the development of more sophisticated control approaches. The three approaches that we moved through in our work are presented in the following subsections.

Previous research by Wang [14] has shown that Proportional-Derivative (PD) controllers are effective at controlling the angular velocity and inclination of an inverted pendulum. Because the bongo board is a similar problem to more traditional inverted pendulum problems we chose to use a PD controller as the basis of our approach.

3.1   Stiff-Upper-Lip Policy and Sensor Fusion

The Stiff-Upper-Lip control policy is directly inspired by the behaviour observed in humans balancing on a bongo board. The goal is to maintain the torso in an upright position and to compensate for the motions of the bongo board by moving the legs only and thus moving the centre of gravity. Figure 3 show stills from a video of a human using the stiff upper lip policy.

When using the Stiff-Upper-Lip policy a PD controller based on the robot's current inclination and angular velocity, as recorded by the robot's on-board

Fig. 3. Bongo Board: Stiff-Upper-Lip Policy. Note how the torso of the player is almost stationary and the legs compensate for the motion of the board.

sensors, is used to control the angular velocity and inclination of the torso, using the following control law:

$$\theta_{Torso} = K_p(\theta_{Torso}) + K_d(\dot{\theta}_{Torso})$$

The robot keeps its torso at a constant height relative to the deck and controls the inclination by extending and contracting its legs (y_L and y_R in Fig. 2).

The first problem with adapting this approach to current robot technology can be seen in Fig. 3 itself: it requires bending of the torso. The robot used in this work does not have the necessary DOF in the torso to execute this motion. Therefore, the necessary control can only be approximated by raising and lowering of the individual hip joints, as described above.

The second problem uncovered was that the gyroscope on the robot was not sensitive enough to detect any angular velocity until the robot was already moving at approximately 3°/s, as can be shown in Fig. 4. To compensate for this problem we use the robot's three-axis accelerometer to measure the inclination. The angular velocity is then estimated using the difference between the present and previous inclination divided by the time between readings. The time between readings is governed by the serial connection between the robot's main processor and its sub-controller and is approximately 8 ms.

3.2   Do-The-Shake Policy and Predictive Control

In spite of overcoming the sensitivity issue of the gyroscope, the other two problems still remain: (a) latency and (b) jitter in the control. To deal with these, we added a one time-step prediction for the PD controller. The result and the error of the prediction of the inclination angle and the angular velocity can be seen in Figs. 5 and 6. The average error in the prediction of the inclination was −0.002° with a standard deviation of 2.437. The average error in the prediction of the angular velocity was 0.0006°/s with a standard deviation of 2.931.

The prediction greatly improved the performance of Jimmy's balancing, but it was still limited by the slow speed of the actuators. Furthermore, the lack of a servo in the torso resulted in only a limited range of motion. However, shifting

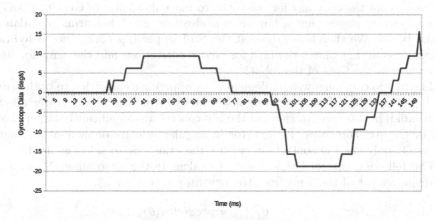

Fig. 4. Sensor Readings from the Y Plane Gyroscope Using the Stiff-Upper-Lip Policy. The gyroscope measures in approximately $3°/s$ increments, which was too coarse for balancing.

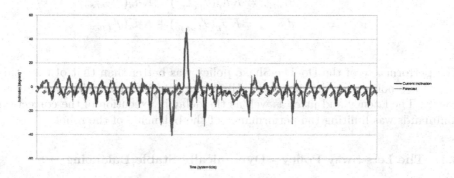

Fig. 5. Comparison between predicted and actual inclination angle

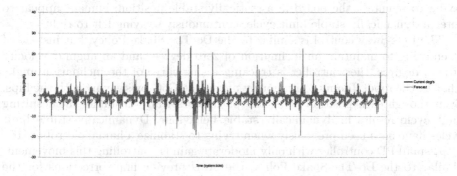

Fig. 6. Comparison between predictaed and actual angular velocity

the torso is not the only way for the robot to move its Center of Gravity (CoG). Figure 3 clearly shows that a human can also use his or her arms to balance on the board. We therefore extended the Stiff-Upper-Lip policy into a hybrid control scheme that moves the hips for coarse corrections and the arms for fine corrections to the CoG of the robot.

The hybrid controller was implemented by applying a correction to both arms and hips only when the error in angular velocity or inclination angle was above a threshold. In this case, the gain of the hip control was significantly larger than that of the arm controller. If the error in angular velocity or inclination angle was small, only the PD control for moving the arms side to side was used.

The following control law was used to calculate the torso angle θ_{Torso} and the displacement of the arms from the neutral position d_{Arms}.

$$\theta'_{Torso} = predicted(\theta_{Torso}, \dot{\theta}_{Torso})$$

Case 1 (Small inclination and angular velocity error):

$$d_{Arms} = Ka_p(\theta'_{Torso}) + Ka_d(\dot{\theta}'_{Torso})$$

Case 2 (Large inclination or angular velocity error):

$$\theta_{Torso} = Kh_p(\theta'_{Torso}) + Kh_d(\dot{\theta}'_{Torso})$$

$$d_{Arms} = Ka_p(\theta'_{Torso}) + Ka_d(\dot{\theta}'_{Torso})$$

The performance of the Do-The-Shake policy was better than that of the Stiff-Upper-Lip policy, but the robot was still not able to balance on its own continuously. The latency and jitter as well as the delay in execution of the correction commands was limiting the performance of the balancing of the robot.

3.3   The Lets-Sway Policy - Dynamically Stable Balancing

The latency in the system meant that it was impossible for Jimmy to correct for tilting of the bongo board quickly enough. By watching humans on the bongo board it became apparent that this is also a problem for humans. Instead of trying to maintain the board in a statically stable position, humans appear to enter a dynamically stable limit cycle, continuously swaying left to right.

The Lets-Sway control is similar to the Do-The-Shake Policy, but instead of attempting to maintain an inclination of zero degrees and an angular velocity of zero degress, the controller is tracking a sine curve of the inclination angle. That is, the robot Jimmy continuously moves the CoG by swaying with the hips. Even though each position along the path is statically unstable, the resulting limit cycle results in dynamically stable behaviour. Dynamically stable limit cycles have been used previously when trying to stabilize a humanoid robot [15].

A small PD controller with only moderate gain is controlling this movement. Similar to the Do-The-Shake Policy, the arms provide fine corrections for the centre of gravity.

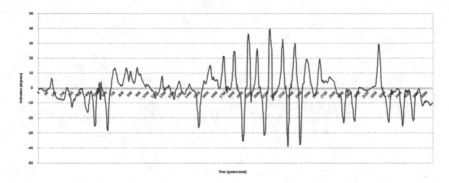

Fig. 7. Inclination Angle of the Do-The-Shake Policy. Using both arms and legs to control the CoG results in smoother balancing than the Stiff-Upper-Lip Policy.

$$\theta'_{Torso} = predicted(\theta_{Torso}, \dot{\theta}_{Torso})$$
$$\theta_{Desired} = sin(\omega t)$$
$$\theta_{Torso} = Kh_p(\theta'_{Board} - \theta_{Desired}) + Kh_d(\dot{\theta}_{Board} - \dot{\theta}'_{Desired})$$
$$d_{Arms} = Ka_p(\theta'_{Board} - \theta_{Desired}) + Ka_d(\dot{\theta}_{Board} - \dot{\theta}'_{Desired})$$

The Lets-Sway policy led to much better performance as can be seen when comparing the accelerometer data from Figs. 7 and 8. The resulting motion is more stable and regular as compared to that of the Do-The-Shake policy.

This was also apparent when watching the performance of the robot. The robot is able to balance for several cycles without help and can compensate if the board hits the table. A video of Jimmy rocking the bongo board using the Lets-Sway policy can be found on youtube (http://www.youtube.com/watch?v=ia2ZYqqF-lw).

4 Evaluation

To compare all three methods we developed we used the robot's average inclination and angular velocity. Under ideal circumstances the system should preserve an angular velocity and inclination near zero. Figures 9 and 10 show the inclination and angular velocity recorded when using each of the three control policies. Table 1 shows the average incliation and angular velocity as well as the standard deviations for each control policy.

From Fig. 9 we can see that the let's-sway policy exhibits cyclic spikes in its inclination; peaks and valleys occur at roughly regular intervals and are generally similar in magnitude. Comparitively the other two policies both exhibit much more extreme inclinations and show less regular cycles. Similarly, from Fig. 10 we

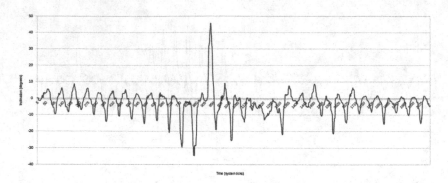

Fig. 8. Lets-Sway Policy. The robot attempts to maintain a dynamically stable limit cycle by moving its hips side to side. The arms are used for fine grained corrections.

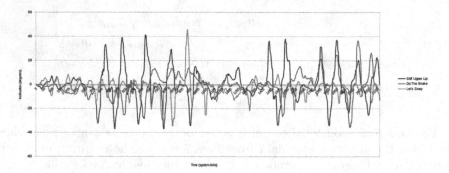

Fig. 9. The robot's recorded inclination when balancing using all three control policies.

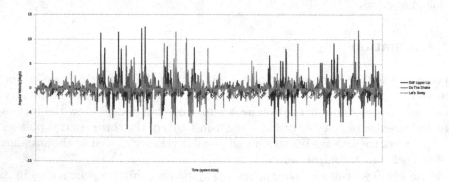

Fig. 10. The robot's recorded angular velocity when balancing using all three control policies.

Table 1. The average inclination and angular velocity recorded when balancing with each control policy

	Inclination	σ^2	Angular velocity	σ^2
Stiff Upper Lip	0.00415	13.9	−4.17E-18	1.57
Do The Shake	−0.178	9.45	0.00612	0.995
Let's Sway	−2.82	7.47	0.00244	1.23

can see that the let's-sway policy demonstrates fewer extreme peaks and valleys in its angular velocity than the other policies.

Of the three policies the stiff-upper-lip is the least stable; the robot's inclination oscillates wildly back and forth, frequently striking the ground on each side. The robot's angular velocity is frequently very high; more than $5°/s$.

The introduction of regular oscillations in the let's-sway policy appears to reduce the angular velocity recorded in the robot's torso, though this reduction is not statistically significant. When using the let's-sway policy the robot's inclination is maintained within a narrower range, suggesting that the introduction of a dynamically stable oscillation does improve the robot's ability to balance on the bongo board. This corresponds with our qualitative observations while testing the policies; the let's-sway policy appeared to be the most stable.

5   Conclusions and Future Work

The research described in this paper is still work in progress. The robot is currently able to balance for several seconds, but the board will often hit the table. This is due to the relatively small diameter of the supporting wheel, which means that the robot has very little time to correct and reverse the motion before the board hits the table. We are countering this by increasing the diameter of the supporting wheel by 1 cm.

We are currently in the process of evaluating the performance of our control approach to deal with unknown external disturbances. The experiments will include perturbation of the robot while balancing on the bongo board as well as sudden pushes to the robot while walking on a flat and even surfaces.

There are many possible directions for future research. We plan on adding visual feedback of the optical flow in the image to improve the robot's estimation of its inclination angle and angular velocity.

Furthermore, both the inverted pendulum and the cart and rod problem are textbook examples for applying machine learning techniques to solve control problems. In particular, reinforcement learning is able to solve these types of problem efficiently. We plan to apply reinforcement learning to the bongo board problem.

Another direction for future research is team balancing. The goal is for two robots to balance on a single bongo board, one robot to the right and one to the left of the wheel. Mathematical analysis shows that the combination of the two robots can be viewed as a single system with two separated actuators.

Finally, there are more complicated balancing devices than a bongo board on which these approaches could be adapted. The fulcrum of a bongo board is a cylinder, making banking motion the main focus for balancing, along with translation (sliding the board along the fulcrum). While it is still possible for the robot to fall forward or backward off the bongo board, the board itself is not intended to force movement in these dimension. A Wobble board, on the other hand, allows spherical motion across the fulcrum, making pitching and yawing motions just as important as the banking movements encompassed by a bongo board. On the other hand, a wobble board has a stationary base for its fulcrum, making it still somewhat restricted compared to a device with a free-moving spherical fulcrum.

References

1. Chen, X., Stone, P., Sucar, L.E., van der Zant, T. (eds.): RoboCup 2012. LNCS, vol. 7500. Springer, Heidelberg (2013)
2. Baltes, J., Tu, K.Y., Lip, S.L.: HuroCup Competition. FIRA, September 2013
3. McGrath, S., Anderson, J., Baltes, J.: Model-free active balancing for humanoid robots. In: Iocchi, L., Matsubara, H., Weitzenfeld, A., Zhou, C. (eds.) RoboCup 2008. LNCS, vol. 5399, pp. 544–555. Springer, Heidelberg (2009)
4. Iverach-Brereton, C., Winton, A., Baltes, J.: Ice skating humanoid robot. In: Herrmann, G., Studley, M., Pearson, M., Conn, A., Melhuish, C., Witkowski, M., Kim, J.-H., Vadakkepat, P. (eds.) TAROS-FIRA 2012. LNCS, vol. 7429, pp. 209–219. Springer, Heidelberg (2012)
5. Omar, K. (ed.): FIRA 2013. CCIS, vol. 376. Springer, Heidelberg (2013)
6. Robotis: Darwin-op, 09 2013. http://support.robotis.com/en/techsupport_eng.htm#product/darwin-op.htm
7. Pratt, J.E.: Exploiting Inherent Robustness and Natural Dynamics in the Control of Bipedal Walking Robots. Ph.D. thesis, MIT Computer Science (2000)
8. Sugihara, T., Nakamura, Y., Inoue, H.: Real-time humanoid motion generation through zmp manipulation based on inverted pendulum control. In: Proceedings of the IEEE International Conference on Robotics and Automation ICRA 2002, vol. 2, pp. 1404–1409 (2002)
9. Pratt, J., Carff, J., Drakunov, S., Goswami, A.: Capture point: A step toward humanoid push recovery. In: 2006 6th IEEE-RAS International Conference on Humanoid Robots, pp. 200–207 (2006)
10. Park, J., Haan, J., Park, F.: Convex optimization algorithms for active balancing of humanoid robots. IEEE Trans. Robot. **23**(4), 817–822 (2007)
11. Hyon, S., Cheng, G.: Gravity compensation and full-body balancing for humanoid robots. In: 2006 6th IEEE-RAS International Conference on Humanoid Robots, pp. 214–221 (2006)
12. Anderson, S., Hodgins, J.: Adaptive torque-based control of a humanoid robot on an unstable platform. In: 2010 10th IEEE-RAS International Conference on Humanoid Robots (Humanoids), pp. 511–517 (2010)
13. Hyon, S.: Compliant terrain adaptation for biped humanoids without measuring ground surface and contact forces. IEEE Trans. Robot. **25**(1), 171–178 (2009)

14. Wang, J.J.: Simulation studies of inverted pendulum based on PID controllers. Simul. Modell. Pract. Theor. **19**(1), 440–449 (2011). Modeling and Performance Analysis of Networking and Collaborative Systems
15. Goswami, A., Espiau, B., Keramane, A.: Limit cycles and their stability in a passive bipedal gait. In: Proceedings of the 1996 IEEE International Conference on Robotics and Automation, 1996. vol. 1, pp. 246–251. IEEE (1996)

Multi-robot Localization
by Observation Merging

Ahmet Erdem[✉] and H. Levent Akın

Department of Computer Engineering, Boğaziçi University, 34342 Istanbul, Turkey
ahmet.erdem1@boun.edu.tr

Abstract. In robot soccer, self-localization of robots may fail because of perception failure, falling down or by being pushed by another robot. In this study, our goal is to improve self localization of robots using the teammate robots' perceptions. Robots which have perceived more landmarks and have moved less can share their localization and observations with other robots to improve localization accuracy. Currently, in the RoboCup Standard Platform League it is not feasible to identify the jersey number of robots using vision. Therefore, we merge perceptions of all robots depending on their reliability, and then identify them in a probabilistic manner to increase localization performance and provide a common world model that can be used for planning. We show that our approach has a significant advantage with respect to single robot localization on estimating the poses of the robots.

Keywords: Multiagent localization · Map merging · Standard Platform League

1 Introduction

In the RoboCup Standard Platform League (SPL), robots use landmarks such as field lines and goal bars as observations for localization. However, these are symmetric with respect to the center line which makes localization harder. When a robot is near to the center of the field, its observations may be due to observations from either side of the field causing localization errors. The cases where the robots are kidnapped, even though not very frequent, also exist. When robots interfere with each others locomotion, this also results in serious localization errors. In addition, noise in perception and motion make it an even harder problem. In order to eliminate such noise and solve the kidnapping problem, we need our robots to work collaboratively. Robots merging their perceptions in a consistent way in order to reach a common world perception leading to better localization is a viable approach.

Collaborative localization methods can be implemented for either stationary or mobile robots. Localization of stationary robots are also known as *network localization* [1,2]. When it comes to mobile robots, most previous works assume that the identity of the robots can be perceived by the individual robots. But

© Springer International Publishing Switzerland 2015
R.A.C. Bianchi et al. (Eds.): RoboCup 2014, LNAI 8992, pp. 478–489, 2015.
DOI: 10.1007/978-3-319-18615-3_39

in SPL this is not the case. A similar problem with ours is solved in [3] where they studied absolute mutual localization with anonymous relative position measurements, with perception modules that do not provide identification of the robots. When designing a multirobot algorithm, we also need to consider time constraints, for during the game, the robots have to respond to changes quickly.

In this study, we introduce an efficient novel collaborative localization method without identification for robot soccer. The problem we address is to provide better localization for Nao [4] robots on a soccer field during an SPL game by merging their perceptions in a consistent way. Each robot is able to self-localize, and in our study we use a Monte Carlo Localization (MCL) algorithm [5] with a set of extensions as described in [6]. In MCL unique landmarks in the field are essential, however, there are no unique landmarks on the SPL field except the center circle. For this reason, we aim to use the team members as unique landmarks. There are, however, two major problems:

- There is no identification. In other words, the robots can perceive that there is a team mate in its field of view, but cannot perceive its jersey number. So, the robots can only observe that at position (x, y) there is a teammate instead of that player 3 is at position (x, y).
- Self-localization is noisy, because robots perceptions and motions are noisy. Therefore, we cannot rely on a single robot. If its localization knowledge is erroneous, the position of a robot seen by it is also wrong since perceptions are relative. For this reason, we need to estimate reliabilities of the robots before processing the information they send. In addition, a robot with good position estimate but bad orientation estimate may cause wrong position estimates of visible teammates.

In [7], a solution to this problem is proposed. They introduced a collaborative multi-robot localization (MRL) approach which improves performance of the self-localization. They make use of relative orientations of the robots. In this paper we extend their approach so that we make use of not only their relative orientations but also the relative distances. Another advantage of our approach is that it generates a common world model for the robots, which may be useful for solving planning problems.

The organization of the rest of the paper is as follows. In Sect. 2 we present the proposed approach. The experiments and the results are given in Sect. 3 and the Conclusions in Sect. 4.

2   Proposed Approach

In collaborative multi-robot localization the robots share some information that is used in the localization process. We describe below the details of the communication between the robots and the developed multi-robot localization approach.

2.1   Messaging

The Nao robots can communicate with each other via a wireless network. They send and receive broadcast messages including some important data such as the position of the ball. In order to realize our algorithm, we add a new message component which includes the following information:

- player number of the sender,
- orientation of the sender,
- mean and variance values of the sender's estimated position,
- number of robots seen by the sender and their relative positions.

2.2   Multi-robot Localization

The Circle Method. We represent the possible locations of the robots by circles with orientations. We use circles, because when a robot is observed, its orientation is unknown. If we consider a robot as a thick line, a known center and all possible orientations construct a circle. Each circle can be constructed by the coordinates of its center, an orientation, a reliability value between 0 and 1, and its type. Since they represent the robots, the circles can perceive each other. There are three types of circles:

- A Circle: This is the circle which a robot claims that it is on. Initially, the number of A circles must be equal to the number of robots.
- B Circle: This is the circle which a robot claims that it perceives a teammate on.
- O Circle: This is the circle which is occupied by a robot as its correct location. Initially, there is no O circle. The number of O circles eventually increases while the number of A and B circles decreases.

We assume that particles in the self-localization process (MCL) have a Gaussian distribution. We can calculate the variance and mean of the particles. Robots consider these means as their estimated positions. For an A circle, there is a negative correlation between the variance σ_A^2 and reliability ρ_A. We can say that over a certain variance threshold $maxVariance$, the robots are completely lost and have 0 percent reliability as in Eq. 1:

$$\rho_A = \max\left(0, 1 - \frac{\sigma_A^2}{maxVariance}\right) \tag{1}$$

The radius of a B circle is related to the distance between its location and its perceiver robots location. As the distance increases, the reliability decreases. Distance is normalized by the $visionDistanceLimit$ which is the maximum distance for perceiving a teammate visually as given in Eq. 2:

$$\rho_B = \rho_A \cdot \left(1 - \left(\frac{\text{distance}_{A,B}}{visionDistanceLimit}\right)^2\right) \tag{2}$$

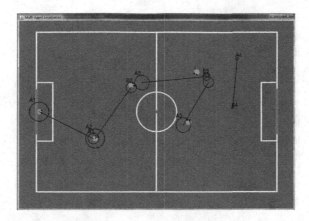

Fig. 1. Initial circle construction

The radius of a circle is proportional to the reliability value ρ. So, robots with low variance form larger A and B circles. In this way, the intersection possibility of reliable data increases. Initially, we construct A circles for each robots self location estimate and B circles for their perceived teammates as shown in Fig. 1.

Map Merging. After the construction of the circles, we may notice that some circles overlap as seen Figure in 1. We have to merge them in order to have one consistent world model. Binary merging of circles continues until there are no intersecting circles. In order to merge two circles, the following conditions have to be satisfied:

- They have to be perceived by different robots. This ensures that we do not see two robots, which are close, as a single robot.
- At most one of them may be an A circle. This ensures that we do not merge two different robot positions as one.

The Merging Process: The merging process goes through the following steps:

1. The new center is the center of mass of the two circles. The coordinates of the center is given in Eq. 3.

$$x_{\text{merged}} = \frac{(x_{c1} \cdot \text{area}_{c1} + x_{c2} \cdot \text{area}_{c2})}{\text{area}_{c1} + \text{area}_{c2}}$$

$$y_{\text{merged}} = \frac{(y_{c1} \cdot \text{area}_{c1} + y_{c2} \cdot \text{area}_{c2})}{\text{area}_{c1} + \text{area}_{c2}} \tag{3}$$

2. New reliability is calculated as in Eq. 4. It is the probability that at least one of the circles is in the correct position. Therefore, the merged circle is more reliable than its constructor circles.

$$\rho_{\text{merged}} = 1 - (1 - \rho_{c1}) \cdot (1 - \rho_{c2}) \tag{4}$$

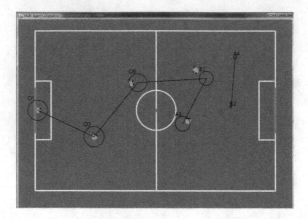

Fig. 2. Map merging

3. A circles which perceive these two circles start to perceive the new one instead of them.
4. If one of the merged circles is an A circle, the new circle perceives the circles which were perceived by the A circle before.
5. Constructor circles are deleted.
6. Finally, reliabilities of the circles which perceive a merged circle increase such that they have reliabilities greater than or equal to the merged circle's reliability. Reliabilities of the other B circles which do not participate in merging also increase when the reliabilities of A circles which perceive them increase. Consensuses between robots lead to more reliable robots.

After all intersected circles are merged; A circles with higher reliability than a certain threshold are converted to O circles as seen in Fig. 2. Map merging algorithm is given in Algorithm 1.

Algorithm 1. Map Merging

 1: **procedure** MERGEMAP(*circles*)
 2:     **while** haveIntersections(circles) **do**
 3:         mergeCircles(pickAnIntersectedPair(circles))
 4:     **end while**
 5:     **for all** $c_i \in circles$ **do**
 6:         **if** $c_i.type = a$ **and** $c_i.reliability > reliabilityThreshold$ **then**
 7:             $c_i.type \leftarrow o$
 8:         **end if**
 9:     **end for**
10: **end procedure**

Occupying Circles Based on the Bounty System. We want our players to occupy a circle which suits best for them. The best is selected based on the following three criteria. For each player which has not occupied any circle yet, we give bounty points to possible locations it can occupy. These locations are the A circle itself and B circles which are not perceived by this circle. Pairs $(player_i, circle_j)$ gain bounty points based on the reliability of the circle. We developed following metrics to assess reliability of a circle:

– *Bounty by Distance:* Pairs gain bounty points related to the distance between $circle_i$ and $circle_j$. Circles which are close to the current A circle gain more bounty points. In other words, there is a negative correlation between the distance to the A circle and the bounty points. This rewards the circles which are near to the location estimated by the robot itself. The A circle centered at the estimated location of the robot gets the highest bounty as in Eq. 5.

$$bounty_{i,j} = bounty_{i,j} + \text{BOUNTY}_{dist} \cdot \left(1 - \frac{\text{dist}(c_i, c_j)}{normalizer}\right) \qquad (5)$$

– *Bounty by Reliability:* Circles gain bounty points with respect to their reliability points. There is a positive correlation between the bounty points and reliabilities as in Eq. 6

$$bounty_{i,j} = bounty_{i,j} + \text{BOUNTY}_{reliability} \cdot reliability_j \qquad (6)$$

– *Bounty by Visibility:* We check candidate circles whether they are in the line-of-sight of occupied circles and they are not perceived by them. If this is the case, the circle loses bounty points. The magnitude of the loss depends on the visibility distribution which is calculated by Eq. 7. If the circle is in the line-of-sight of more than one occupied circle, losses are summed. (Actually, all circles gain visibility points, no one loses but they lose relative to the others' gains.) The visibility bounty point is calculated by Eq. 8.

$$visibility(k, j) = \left(1 - \frac{\text{dist}(c_k, c_j)}{visionDistanceLimit}\right) \cdot \left(1 - \frac{|c_k.headOrientation - slope(c_k, c_j)|}{visionDegreeLimit}\right)(7)$$

$$bounty_{i,j} = bounty_{i,j} + \text{BOUNTY}_{visibility} \cdot \left(1 - \sum_{k \in O} visibility(k, j)\right) \qquad (8)$$

After these operations, the pair with the biggest bounty is found. (This bounty should be above a certain limit.) If the circle is one of the B circles, the A circle which represents this player moves to this B circle. All the circles perceived by the A circle also move. The circle is converted to an O circle for this player. The process given in Algorithm 2 repeats until it cannot generate an O circle.

By this method, the robots select the best circles for themselves whenever they can. Being close to the robot's estimated position, high possibility of having a robot on it and low possibility of visibility contradictions makes a circle the best choice as seen in Fig. 3. One can tune the bounty coefficients to have better results. For example; if the bounty coefficient of distance is too small compared to others, it may cause "teleportation" situations.

Algorithm 2. Occupying Circles Based On Bounty System

1: **procedure** FINDHIGHESTBOUNTYANDOCCUPY($c1, c2$)
2:      $giveBountyByReliability(bounties)$
3:      $giveBountyByVisibility(bounties)$
4:      $giveBountyByDistance(bounties)$
5:      $[i, j] \leftarrow$ max(bounties)
6:      $occupy(i, j)$
7: **end procedure**

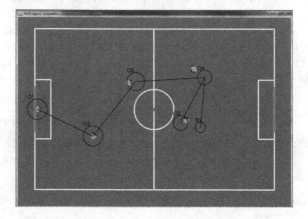

Fig. 3. Occupying circles based on bounty system

Rotation Correction. After these procedures, if there are some robots which perceive at least one B circle which is not perceived by the others, we can say that this robot may have incorrect knowledge of its orientation. Then we try to rotate it between $-\theta$ and $+\theta$ degrees. The orientation which causes the maximum number of intersections for the circles perceived by the robot becomes the suggested orientation for the robot.

After rotation correction, circles are again merged until there is no intersection as seen in Fig. 4. If there are still B circles which are perceived by only one O circle, this O circle is converted back to an A circle in order to prevent misleading localization information, because these circles may have true positions but faulty orientations.

Combining with Monte Carlo Localization. If a robot can occupy a circle at the end of the Multi-Robot Localization process, this circle is used in self-localization of the robot. Since the self-localization is a Monte Carlo Localization, we inject some particles around the center of the circle. These particles are constructed by adding 2D Quasi-Gaussian noise to the center. Quasi-Gaussian particles are preferred to Gaussian ones, because a previous work [8] has shown that quasi-random numbers have great advantage over actual random numbers

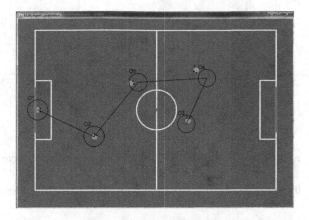

Fig. 4. Rotation correction

Algorithm 3. Rotation Correction

1: **procedure** CORRECTROTATION(*circles*)
2:     **for all** $c_i \in circles$ **do**
3:         **if** $c_i.type = o$ **then**
4:             $findAndSetBestRotation(c_i)$
5:         **end if**
6:     **end for**
7:     $mergeMap(circles)$
8:     **for all** $c_i \in circles$ **do**
9:         **if** $c_i.type = b$ (and) $sizeOf(c_i.perceivedByCircles) = 1$ **then**
10:             $c_i.perceivedByCircles[0].type \leftarrow a$
11:         **end if**
12:     **end for**
13: **end procedure**

Algorithm 4. Find And Set Best Rotation

1: **procedure** FINDANDSETBESTROTATION(*c*)
2:     $maxNumOfIntersections \leftarrow 0$
3:     **for** $i = -\theta$ **to** θ **do**
4:         $numOfIntersections \leftarrow c.rotateTo(c.orientation + i)$
5:         **if** $numOfIntersections >$ maxNumOfIntersections **then**
6:             $maxNumOfIntersections \leftarrow$ numOfIntersections
7:             $bestRotation \leftarrow c.orientation + i$
8:         **end if**
9:     **end for**
10:     $c.rotateTo(c.orientation + bestRotation)$
11: **end procedure**

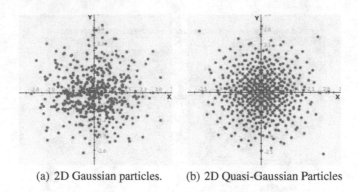

(a) 2D Gaussian particles.   (b) 2D Quasi-Gaussian Particles

Fig. 5. The difference between gaussian and Quasi-Gaussian random numbers.

in localization. Figure 5 illustrates the difference between Gaussian and Quasi-Gaussian distributions used in this study.

These particles are injected just after the resampling process. The number of injected particles is proportional to the reliability (i.e. radius) of the circle. Orientation of each injected particle is a Gaussian random number with the circle's orientation as a mean. Since we claim that we inject particles to the real position with the real orientation, these particles gain weights by observations. After resampling, more particles will be sampled from these particles due to their large weights. By this method, we expect to see particles converging to the real position of the robot after a few steps.

3 Experiments and Results

3.1 Simulation Experiments

In order to make the initial tests of the proposed approach, we have developed a 2D robot simulator since we want to control motion and perception noises. The other reason is that we can make a large number of experiments in a small amount of time without harming the actual robots. We can monitor the robot's real positions and estimated positions via the 2D simulation interface. The robots walk from one point to another point. We add noise to both motion and perception of the robot. Noise is proportional to the magnitude of actions. If a robot walks a large distance or perceives another robot which is far away, the noise gets larger.

We first tested the performance of our method with five robots using our simulator. In each simulation, we allow the algorithm to run for 100 steps. Different polygonal paths are planned for each robot and the robots try to follow these paths with respect to their estimated positions. In each step, the robots travel at most a 32 cm length path and rotate when necessary. Since the localization is noisy, the robots cannot follow the determined paths exactly. How much they deviate from the actual path depends on the quality of the localization process.

We have tested the algorithm with 1000 different runs and the average position error per player per step and average orientation error per player per step have been calculated as the measurement of performance. Setups with different number of stationary robots (robots which do not move and observe others) have also been tested. We have found that our method improves localization significantly in terms of position estimates. From Table 1 it can be observed that the performance improvement increases as the number of stationary robots increases. The improvement in orientation is not significant as shown in Table 2.

Table 1. Position error in the 2D simulation environment

Number of stationary robots	Error without MRL (mm)	Error with MRL (mm)	Improvement percentage
1	402	370	8.0
2	353	248	29.7
3	254	178	29.9
4	125	61	51.2

Table 2. Orientation error in the 2D simulation environment

Number of stationary robots	Error without MRL (degree)	Error with MRL (degree)	Improvement percentage
1	35	34	0.03
2	24	21	0.13
3	19	19	0.00
4	19	17	0.11

3.2 Experiments with Real Robots

After the initial simulator tests we tested our algorithm on Nao robots in the SPL field. In order to get the real position of the robot on the field we have modified the robot tracking system previously developed by Kavaklıoğlu [9] in his MSc thesis. In this setup, there are four webcams fitted to the ceiling of the field. They are calibrated for giving exact positions. We put a marker on top of the robot which we want to track. The marker consists of one blue and two brown blobs. These colors dont belong to any other objects. The blue one is for finding the orientation of the robot. The robot tracking system can easily detect these markers and send their positions on network. We receive these messages in order to draw the current position and the followed path and to calculate the localization errors.

We have designed a rectangular path for the localization test. We have tracked the position of the robot during the test by the cameras. MRL was disabled for the control group. In most of the cases in which MRL was disabled, robots left the field whenever they could not observe any landmarks. When we enabled MRL, robots completed the path with less localization errors. We show a sample from our experiments in Fig. 6.

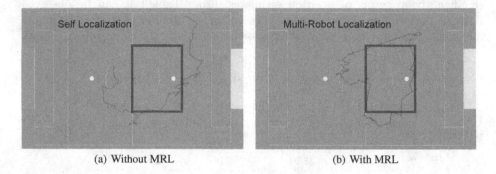

(a) Without MRL   (b) With MRL

Fig. 6. Effect of MRL on path following (Real Robots)

4   Conclusions

Collaborative localization is a challenging problem in multi-robot systems, when used efficiently it can lead to an overall improvement in localization. In the SPL, localization is a major issue since over the years unique landmarks have been removed one by one. In this study we introduce a novel method for MRL without player identification. We not only aim to improve self-localization performance of the robots, but also aim to provide a common world model for them to be used in planning. Since robot soccer is a real-time event, we have considered time constraints and proposed a time-efficient method.

The simulation experiments have shown that our approach has a significant advantage over single robot localization on estimating the position and the orientation, but we could not observe the same performance when we did the real world experiments. The main reason is that the real world may have more noise than we expected. In addition, we need better perception and motion modules in order to decrease these noises.

We have to mention that in the cases where the lost robot cannot see its teammates, we cannot give orientation information for that robot. It is a shortcoming of our method. It can be eliminated, if the ball is visible to that robot and at least one reliable robot. The position given by MRL and relative distance and orientation to the ball are sufficient for estimating the orientation.

The most challenging part of our method is determining which robots are reliable. Our criterion is the variance of particles in MCL, but sometimes particles may converge to a wrong position and the robots may not be aware that they are lost. As a further work, we intend to work on better estimating the reliability.

Acknowledgments. This work is supported by Boğaziçi University Research Fund through project 13A01P3.

References

1. Aspnes, J., Eren, T., Goldenberg, D.K., Morse, A.S., Whiteley, W., Yang, Y.R., Anderson, B.D.O., Belhumeur, P.N.: A theory of network localization. IEEE Trans. Mob. Comput. **5**(12), 1663–1678 (2006)
2. Aragues, R., Carlone, L., Calafiore, G., Sagues, C.: Multi-agent localization from noisy relative pose measurements. In: 2011 IEEE International Conference on Robotics and Automation (ICRA 2011), pp. 364–369, May 2011
3. Franchi, A., Oriolo, G., Stegagno, P.: Probabilistic mutual localization in multi-agent systems from anonymous position measures. In: 49th IEEE Conference on Decision and Control (CDC 2010), pp. 6534–6540. IEEE (2010)
4. Gouaillier, D., Hugel, V., Blazevic, P., Kilner, C., Monceaux, J., Lafourcade, P., Marnier, B., Serre, J., Maisonnier, B.: The nao humanoid: a combination of performance and affordability. CoRR abs/0807.3223 (2008)
5. Thrun, S., Fox, D., Burgard, W., Dellaert, F.: Robust monte carlo localization for mobile robots. Artif. Intell. **128**(1–2), 99–141 (2000)
6. Kaplan, K., Çelik, B., Meriçli, T., Meriçli, Ç., Akın, H.L.: Practical extensions to vision-based monte carlo localization methods for robot soccer domain. In: Bredenfeld, A., Jacoff, A., Noda, I., Takahashi, Y. (eds.) RoboCup 2005. LNCS (LNAI), vol. 4020, pp. 624–631. Springer, Heidelberg (2006)
7. Özkucur, N.E., Kurt, B., Akın, H.L.: A collaborative multi-robot localization method without robot identification. In: Iocchi, L., Matsubara, H., Weitzenfeld, A., Zhou, C. (eds.) RoboCup 2008. LNCS, vol. 5399, pp. 189–199. Springer, Heidelberg (2009)
8. Liu, B., Zheng, X., Wu, X., Liu, Y.: Quasi monte carlo localization for mobile robots. In: 12th International Conference on Control Automation Robotics & Vision (ICARCV 2012), pp. 620–625 (2012)
9. Kavaklıoğlu, C.: Developing a probabilistic post-perception module for mobile robotics. Master's thesis, Boğaziçi University (2009)

UAVision: A Modular Time-Constrained Vision Library for Soccer Robots

Alina Trifan$^{(\boxtimes)}$, António J.R. Neves, Bernardo Cunha, and José Luís Azevedo

IRIS Group, DETI / IEETA, University of Aveiro, 3810–193 Aveiro, Portugal
{alina.trifan,an,jla}@ua.pt, mbc@det.ua.pt

Abstract. The game of soccer is one of the main focuses of the RoboCup competitions, being a fun and entertaining research environment for the development of autonomous multi-agent cooperative systems. For an autonomous robot to be able to play soccer, first it has to perceive the surrounding world and extract only the relevant information in the game context. Therefore, the vision system of a robotic soccer player is probably the most important sensorial element, on which the acting of the robot is fully based. In this paper we present a new modular time-constrained vision library, named UAVision, that allows the use of video sensors up to a frame rate of 50 fps in full resolution and provides accurate results in terms of detection of the objects of interest for a robot playing soccer.

1 Introduction

The research area of robotic vision is greatly evolving by means of international competitions such as those promoted and organized once per year by the RoboCup Federation. The RoboCup initiative, through competitions like RoboCup Robot Soccer, RoboCup Rescue, RoboCup@Home and RoboCup Junior, is designed to meet the need of handling real world complexities, while maintaining an affordable problem size and research cost. It offers an integrated research task covering the broad areas of artificial intelligence, computer vision and robotics.

The soccer game in the RoboCup Middle Size League (MSL) is a standard real-world test for autonomous multi-robot systems. In this league, omnidirectional vision systems have become interesting in the last years, allowing a robot to see in all directions at the same time without moving itself or its camera [1]. The environment of this league is not as restricted as in the others and the pace of the game is faster than in any other league (currently with robots moving with a speed of 4 m/s or more and balls being kicked with a velocity of more than 10 m/s), requiring fast reactions from the robots. In terms of color coding, in the fully autonomous MSL the field is still green, the lines of the field and the goals are white and the robots are mainly black. The two teams competing are wearing cyan and magenta markers. For the ball color, the only rule applied is that the surface of the ball should be 80 % of a certain color, which is usually decided before a competition. The colors of the objects of interest are important hints

© Springer International Publishing Switzerland 2015
R.A.C. Bianchi et al. (Eds.): RoboCup 2014, LNAI 8992, pp. 490–501, 2015.
DOI: 10.1007/978-3-319-18615-3_40

for the object detection, relaxing thus the detection algorithms. Many teams are currently taking their first steps in 3D ball information retrieving [2,3]. There are also some teams moving their vision systems algorithms to VHDL based algorithms taking advantage of the FPGAs versatility [2]. Even so, for now, the great majority of the teams base their image analysis in color search using radial sensors [4–6].

In this paper we present a library for color-coded object detection, named UAVision, that is currently being used by the robots of the team CAMBADA, participating in the Middle Size League. The design of the library follows a modular approach as it can be stripped down into several independent modules (that will be presented in the following sections). Moreover, the architecture of our software is of the type "plug and play". This means that it offers support for different vision sensors technologies and that the software created using the library is easily exportable and can be shared between different types of vision sensors. These facts, on the other hand, make it appropriate for being used by robots in all other leagues. Another important aspect of our library is that it takes into consideration time constraints. All the algorithms behind this library have been implemented focusing on maintaining the processing time as low as possible. Realtime processing means to be able to complete all the vision dependant tasks within the limits of the frame rate.

The vision system for color-coded object detection within the RoboCup soccer games of Middle Size League that we have implemented using the UAVision library can work with frame rates up to 50 fps using a resolution of 1024 × 1024 pixels, both in Bayer, RGB or YUV color modes. Detailed processing time obtained will be presented in Sect. 3. Moreover, we provide experimental results showing the difference of working with the different frame rates in terms of the delay between the perception and the action. As far as we know, there is no previous published work that presents so detailed information about this issue.

The library that we are proposing comes as a natural development of the work already presented within the RoboCup community. After having implemented vision systems for robotic soccer players that perform both in the Standard Platform League [7] and Middle Size League [8–10], we are proposing this new cross-library that can be used by robots whose architecture might be different, but the goal remains the same: the game of soccer. We consider our work an important contribution for the RoboCup Soccer community since so far, there are no machine vision libraries used for the games of soccer that take into consideration time constraints. UAVision aims at being an open-source free library that can be used for robotic vision applications that have to deal with time constraints as are the RoboCup competitions. Moreover, we made publicly available the video sequences used in the experimental results of this paper, both in Bayer and RGB color modes, so that other researchers can reproduce our results and test their own algorithms.

This paper is structured in five sections, the first of them being this introduction. Section 2 describes the modules of the vision library. Section 3 presents the results that have been achieved using the library in the MSL robots. Section 4 concludes the paper and future lines of research are highlighted. Finally, in Acknowledgements section the institutions that have supported this work are acknowledged.

2   Library Description

The library that we are presenting is intended for the development of artificial vision systems for the detection of color-coded objects, being the robotic soccer the perfect application for its usage. The library contains software for image acquisition from video cameras supporting different technologies, for camera calibration and for blob formation, which stands at the basis of the object detection.

2.1   Image Acquisition

UAVision provides the necessary software for accessing and capturing images from three different camera interfaces, so far: USB cameras, Firewire cameras and Ethernet cameras. For this purpose, the Factory Design Pattern [11] has been used and a factory called "Camera" has been implemented. The user can choose from these three different types of cameras in the moment of the instantiation. An important aspect to be mentioned is that UAVision uses some of the basic structures from the core functionality of OpenCV library: the *Mat* structure as a container of the frames that are grabbed and the *Point* structure for the manipulation of points in 2D coordinates. Images can be acquired in the YUV, RGB or Bayer color format.

The module of Image Acquisition also provides methods to convert images between the most used color spaces: RGB to HSV, HSV to RGB, RGB to YUV, YUV to RGB, Bayer to RGB and RGB to Bayer.

2.2   Camera Calibration

The correct calibration of all the parameters related to the system is very important in any vision system. The module of camera calibration includes algorithms for calibration of the intrinsic and extrinsic camera parameters, the computation of the inverse distance map, the calibration of the colormetric camera parameters and the detection of the mirror, robot center and the definition of the regions of the image that do not have to be processed.

The result of the vision system calibration can be stored in a configuration file which contains four main blocks of information: camera settings, mask, map and color ranges. The mask is a binary image representing the areas of the image that do not have to be processed, since they contain only parts of the body of the robot, which are not relevant for the object detection. By ignoring these areas of the image, both the noise in the image and the processing time can be reduced. The map, as the name suggests, is a matrix that represents the mapping between pixel coordinates and real world coordinates.

The camera settings block is where the basic information is registered. Among others, these include the resolution of the image acquired, the Region of Interest regarding the CCD or CMOS of the camera and colormetric parameters, among others.

The color ranges block contains the color regions for each color of interest (at most 8 different colors as we will explain later) in a specific color space (ex. RGB, YUV, HSV, etc.). In practical means, it contains the lower and upper bounds of each one of the three color components for a specific color of interest.

The UAVision library contains algorithms for the self-calibration of most of the parameters described above, including some algorithms developed previously within our research group, namely the algorithm described in [8] for the automatic calibration of the colormetric parameters and the algorithms presented in [1,9] for calibration of the intrinsic and extrinsic parameters of catadioptric vision systems used to generate the inverse distance map. For the calibration of the intrinsic and extrinsic parameters of a perspective camera, we have used and implemented the algorithm for the "chessboard" calibration, presented in [12].

2.3   Color-Coded Object Detection

The color-coded object detection is composed by four sub-modules that are presented next.

• Look-Up Table

For fast color classification, color classes are defined through the use of a look-up table (LUT). A LUT represents a data structure, in this case an array, used for replacing a runtime computation by a basic array indexing operation.

This approach has been chosen in order to save significant processing time. The images can be acquired in the RGB, YUV or Bayer format and they are converted to an index image (image of labels) using an appropriate LUT for each one of the three possibilities.

The table consists of 16,777,216 entries (2^{24}, 8 bits for R, 8 bits for G and 8 bits for B) with one byte each. The table size is the same for the other two possibilities (YUV or Bayer), but the meaning of each of the components changes. Each bit in the table entries expresses if one of the colors of interest (white, green, blue, yellow, orange, red, blue sky, black, gray - no color) is within the corresponding class or not. A given color can be assigned to multiple classes at the same time. For classifying a pixel, first the value of the color of the pixel is read and then used as an index into the table. The 8-bit value then read from the table is called the "color mask" of the pixel. It is possible to perform image subsampling in this stage in systems with limited processing capabilities in order to reduce even more the processing time. The color classification is only applied to the valid pixels if a mask exists.

• Scanlines

To extract color information from the image we have implemented three types of search lines, which we also call scanlines: radial, linear (horizontal or vertical) and circular. They are constructed once, when the application starts, and saved in a structure in order to improve the access to these pixels in the color extraction module. This approach is extremely important for the reduction of processing time. In Fig. 1 the three different types of scanlines are illustrated.

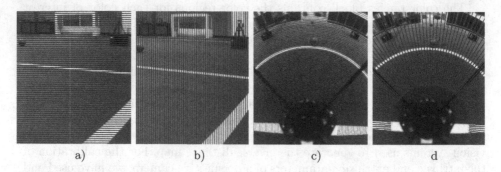

Fig. 1. Examples of different types of scanlines: (a) horizontal scanlines; (b) vertical scanlines; (c) circular scanlines; (d) radial scanlines.

• Run Length Encoding (RLE)

For each scanline, an algorithm of Run Length Encoding is applied in order to obtain information about the existence of a specific color of interest in that scanline. To do this, we iterate through its pixels to calculate the number of runs of a specific color and the position where they occur. Moreover, we extended this idea and it is optional to search, in a window before and after the occurrence of the desired color, for the occurrence of other colors. This allows the user to determine both color transitions and color occurrences using this approach.

When searching for run lengths, the user can specify the color of interest, the color before, the color after, the search window for these last two colors and three thresholds that can be used to determine the valid information.

As a result of this module, we obtain a list of positions in each scanline and, if needed, for all the scanlines, where a specific color occurs, as well as the amount of pixels in each occurrence (Fig. 2).

• Blob Formation

To detect objects with a specific color in a scene, we have to be able to detect regions in the image with that color, usually named blobs, and validate those blobs according to some parametric and morphological features, namely area, bounding box, solidity, skeleton, among others. In order to construct these regions, we use information about the position where a specific color occurs based on the Run Length module previously described (Fig. 2).

We iterate through all the run lengths of a specific color and we apply an algorithm of clustering based on the euclidean distance. The parameters of this clustering are application dependent. For example, in a catadioptric vision system, the distance in pixels to form blobs changes radially and non-linearly regarding the center of the image.

While the blob is being built, its descriptor is being updated. The description of the blobs currently calculated are, to name a few, center, area, width/height relation, solidity, etc.

- **Object Detection**

The last step of the vision pipeline is the decision regarding whether the colors segmented belong to an object of interest or not. In the vision system developed for the CAMBADA team using the proposed library, the white and black points that have been previously run-length encoded are passed directly to higher level processes, where localization based on the white points and obstacle avoidance based on the black points are performed.

For the ball detection, the blobs that are of the color of the ball have to meet the following validation criteria before being labelled as ball. First, a mapping function that has been experimentally designed is used for verifying a size-distance from the robot ratio of the blob (Fig. 2(c)). This is complemented by a solidity measure and a width-height ratio validation, taking into consideration that the ball has to be a round blob. The validation was made taking into consideration the detection of the ball even when it is partially occluded.

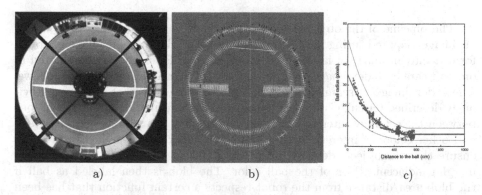

a)   b)   c)

Fig. 2. On the left, an image captured using the Camera Acquisition module of the UAVision library. In the center, the run length information annotated. On the right, illustration of the radius (in pixels) of the ball relative to the distance (in centimeters) from the robot at which it is found. The blue marks represent the measures obtained, the green line the fitted function and the cyan and red line the upper and lower bounds considered for validation (Color figure online).

3   Experimental Results

The UAVision library is currently used by the MSL team of robots CAMBADA team from University of Aveiro. These robots are completely autonomous, able to perform holonomic motion and are equipped, in terms of hardware, with a catadioptric vision system that allows them to have omnidirectional vision [10]. The architecture of the vision system is presented in Fig. 3.

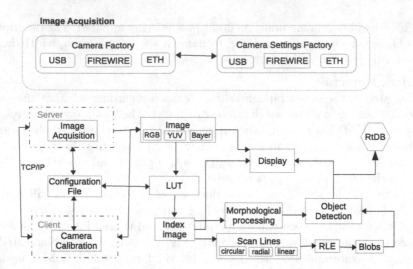

Fig. 3. Software architecture of the vision system developed based on the UAVision library.

The pipeline of the object detection procedure is the following: after having an image acquired, using a LUT previously built, the original image is transformed into an image of labels. This image of color labels, also denominated in our software by index image, will be the basis of all the processing that follows. The index image is scanned using one of the three types of scanlines previously described (circular, radial or linear) and the information about transitions between the colors of interest is run length encoded. Transitions between green and other colors of interest (white, ball color, black) are searched in order to ensure that the objects detected are inside the field area. Blobs are formed by merging adjacent RLEs of the ball color. The blob is then labeled as ball if the blob area/distance from the robot respects a certain function that has been experimentally determined (see Fig. 2(c)). Moreover, the width/height relation and solidity are also used for ball validation. If a given blob passes the validation criteria, its center coordinates will be passed to higher-level processes and shared on a Real-time Database (RtDB) [13]. For the obstacles and line detections, the coordinates of the detected points of interest are passed to higher-level processes through the RtDB.

A visual example of the detected objects in an image acquired by the vision system is presented in Fig. 4. As we can see, the objects of interest (balls, lines and obstacles) are correctly detected even when they are far from the robot. Moreover, the balls can correctly be detected up to 9 m (notice that the robot is in the middle line of the field and the further ball is over the goal line) even when they are partially occluded or engaged by another robot. No false positives in the detection are observed.

Several game scenarios have been tested using the CAMBADA autonomous mobile robots. In Fig. 5(a) we present a graphic with the result of the ball

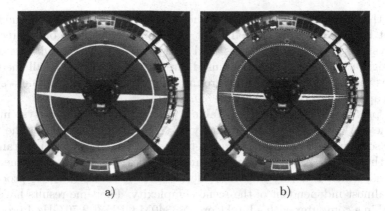

a)   b)

Fig. 4. On the left, an image acquired by the omnidirectional vision system. On the right, the result of the color-coded object detection. The blue circles mark the white lines, the white circles mark the black obstacles and the mangenta circles mark the orange blobs that passed the validation thresholds (Color figure online).

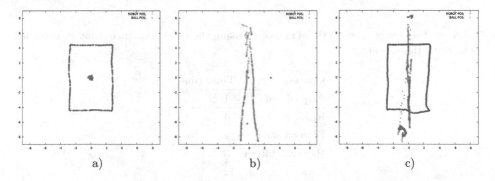

a)   b)   c)

Fig. 5. On the left, a graph showing the ball detection when the robot is moving in a tour around the soccer field. In the middle, ball detection results when the robot is stopped on the middle line on the right of the ball and the ball is sent across the field. On the right, ball detection results when both the robot and the ball are moving.

detection when the ball is stopped in a given position (the central point of the field, in this case) while the robot is moving. The graphic shows a consistent ball detection while the robot is moving in a tour around the field. The field lines are also properly detected, as it is proved by the correct localization of the robot in all the experiments. The second scenario that has been tested is illustrated in Fig. 5(b). The robot is stopped on the middle line and the ball is sent across the field. This graph shows that the ball detection is accurate even when the ball is found at a distance of 9 m away from the robot. Finally, in Fig. 5(c) both the robot and the ball are moving. The robot is making a tour around the soccer field, while the ball is being sent across the field. In all these experiments, no false positives were observed and the ball has been detected in more than

90 % of the frames. Most of the times the ball was not detected was due to the fact that it was hidden by the bars that hold the mirror of the omnidirectional vision system. The video sequences used for generating these results, as well as the configuration file that has been used, are available at [14]. In all the tested scenarios the ball is moving on the ground floor since the single camera system has no capability to track the ball in 3D.

The processing time shown in Table 1 proves that the vision system built using the UAVision library is extremely fast. The full execution of the vision pipeline software only takes on average a total of 12 ms, allowing thus a framerate greater than 80 fps. Moreover, the maximum processing time that we measured was 13 ms, which is a very important detail since it shows that the processing time is almost independent of the scene complexity. The time results have been obtained in a computer with a Intel Core i5-3340M CPU @ 2.70 GHz 4 processor, processing images with a resolution of 1024×1024 pixels (a Region Of Insterest centered in the CMOS of the camera used). In the implementation of this vision system we didn't use multi-threading. However, both image classification and the next steps can be parallelized if needed.

Table 1. Average processing times measured using the video sequences that we provide along with this paper.

Operation	Time (ms)
Acquisition	1
RLE	4
Blob creation	2
Blob validation	3
Total	12

The LUT is created once, when the vision process runs for the first time and it is saved in the cache file. If the information from the configuration file does not change during the following runs of the vision software, the LUT will be loaded from the cache file, reducing thus the processing time of this operation by approximately 25 times.

For the video sequences that we provide, the following number of scanlines have been built during the performance of the vision software:

– 720 radial scanlines for the ball detection.
– 98 circular scanlines for the ball detection.
– 170 radial scanlines for the lines and obstacle detection.
– 66 circular scanlines for the lines detection.

The cameras that have been used can provide 50 fps at full resolution (1280×1024 pixels) in RGB color space. However, some cameras available on the market can only provide 50 fps accessing directly to the CCD or CMOS data, usually a single channel image using the well known Bayer format. As described before, the LUT in the vision library can work with several color spaces, namely RGB, YUV and Bayer format. We repeated the three scenarios described above acquiring images directly in the Bayer format also at 50 fps and the experimental results show that the detection performance is not affected as expected, since the conversion between Bayer and RGB does not generate new information regarding the perception.

In addition to the good performance in the detection of objects, both in terms of number of times that an object is visible and detected and in terms of error in its position, the vision system must also perform well in minimizing the delay between the perception of the environment and the reaction of the robot. It is obvious that this delay depends on several factors, namely the type of the sensor used, the processing unit, the communication channels and the actuators, among others. To measure this delay in the CAMBADA robots, a setup was developed which is presented in Fig. 6. The setup consists of a led that is turned on by the motor controller board and the same board measures the time that the whole system takes to acquire and detect the LED flash, and send the respective reaction information back to the controller board. The vision system detects the led on and when it happens, the robotic agent sends a specific value of velocities to the hardware (via HWComm application). This is the normal working mode of the robots in game play.

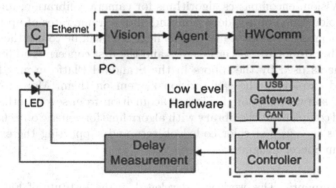

Fig. 6. The blocks used in our measurement setup. These blocks are used by the robots during game play.

As presented in Fig. 7, the delay time between perception and the reaction of the robot significantly decreases when working at higher frame rates. The average delay at 30 fps is 65 ms and at 50fps it is 53 ms, which corresponds to an improvement of 22 %. The jitter verified reflects the normal function of the several modules involved, mainly because there is no synchronism between the camera and the processes running on the computer.

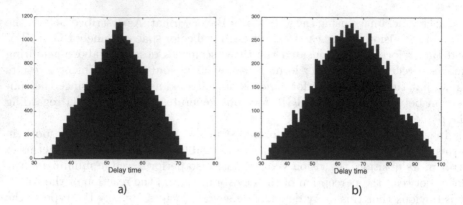

Fig. 7. Histograms showing the delay between perception and action on the CAMBADA robots. On the left, the camera is working at 50 fps (average = 53 ms, max = 74 ms, min = 32 ms). On the right, the camera working at 30 fps (average = 65 ms, max = 99 ms, min = 32 ms).

4 Conclusions and Future Work

In this paper we have presented a novel time-constrained computer vision library that has been successfully employed in the games of robotic soccer. The proposed library, UAVision, encompasses algorithms for camera calibration, image acquisition and color coded object detection and allows frame rates of up to 50 fps.

In what concerns the future work, the next step will be to use the developed library in other RoboCup Soccer Leagues and the first concern is adding support for the cameras used by the robots in the Standard Platform and Humanoid Leagues and employing the same vision system on them. Moreover, we aim at providing software support for image acquisition from several other types of cameras and complement the library with algorithms for generic object detection, relaxing thus the rules of color coded objects and supporting the evolution of the RoboCup Soccer Leagues.

Acknowledgements. This work was developed in the Institute of Electronic and Telematic Engineering of University of Aveiro and was partially supported by FEDER through the Operational Program Competitiveness Factors - COMPETE and by National Funds through FCT - Foundation for Science and Technology in a context of a PhD Grant (FCT reference SFRH/BD/85855/2012) and the project FCOMP-01-0124-FEDER-022682 (FCT reference PEst-C/EEI/UI0127/2011).

References

1. Neves, A.J.R., Pinho, A.J., Martins, D.A., Cunha, B.: An efficient omnidirectional vision system for soccer robots: from calibration to object detection. Mechatronics **21**(2), 399–410 (2011)

2. Kanters, F.M.W., Hoogendijk, R., Janssen, R.J.M., Meessen, K.J., Best, J.J.T.H., Bruijnen, D.J.H., Naus, G.J.L., Aangenent, W.H.T.M ., van der Berg, R.B.M., van de Loo, H.C.T., Heldes, G.M., Vugts, R.P.A., Harkema, G.A., van Brakel, P.E.J., Bukkums, B.H.M., Soetens, R.P.T., Merry, R.J.E., can de Molengraft, M.J.G.: Tech united eindhoven team description. In: RoboCup 2011, Istanbul, Turkey (2011)
3. Kappeler, U.P., Zweigle, O., Rajaie, H., Hausserman, K., Tamke, A., Koch, A., Eck-stein, B., Aichele, F., DiMarco, D., Berthelot, A., Walter, T., Levi, P.: RFC stuttgart team description. In: RoboCup 2011, Istanbul, Turkey (2011)
4. Huang, M., Ge, X., Hui, S., Wang, X., Chen, S., Xu, X., Zhang, W., Lu, Y., Liu, X., Zhao, L., Wang, M., Zhu, Z., Wang, C., Huang, B., Ma, L., Qin, B., Zhou, F., Wang, C.: Water team description. In: RoboCup 2011, Istanbul, Turkey (2011)
5. Lu, H., Zeng, Z., Dong, X., Xiong, D., Tang, S.: Nubot team description. In: RoboCup 2011, Istanbul, Turkey (2011)
6. Nassiraei, A.A.F, Ishida, S., Shinpuku, N., Hayashi, M., Hirao, N., Fujimoto, K., Fukuda, K., Takanaka, K., Godler, I., Ishii, K., Miyamoto, H.: Hibikino-musashi team description. In: RoboCup 2011, Istanbul, Turkey (2011)
7. Trifan, A., Neves, A.J.R., Cunha, B., Lau, N.: A modular real-time vision system for humanoid robots. In: Proceedings of SPIE IS&T Electronic Imaging 2012, January 2012
8. Neves, A.J.R., Trifan, A., Cunha, B.: Self-calibration of colormetric parameters in vision systems for autonomous soccer robots. In: Behnke, S., Veloso, M., Visser, A., Xiong, R. (eds.) RoboCup 2013. LNCS, vol. 8371, pp. 183–194. Springer, Heidelberg (2014)
9. Cunha, B., Azevedo, J., Lau, N., Almeida, L.: Obtaining the inverse distance map from a non-svp hyperbolic catadioptric robotic vision system. In: Visser, U., Ribeiro, F., Ohashi, T., Dellaert, F. (eds.) RoboCup 2007: Robot Soccer World Cup XI. LNCS (LNAI), vol. 5001, pp. 417–424. Springer, Heidelberg (2008)
10. Neves, A.J.R., Corrente, G.A., Pinho, A.J.: An omnidirectional vision system for soccer robots. In: Neves, J., Santos, M.F., Machado, J.M. (eds.) EPIA 2007. LNCS (LNAI), vol. 4874, pp. 499–507. Springer, Heidelberg (2007)
11. Gamma, E., Helm, R., Johnson, R., Vlissides, J.: Design Patterns: Elements of Reusable Object-oriented Software. Addison-Wesley Longman Publishing Co. Inc., Boston (1995)
12. Zhang, Z.: Flexible camera calibration by viewing a plane from unknown orientations. In: ICCV, pp. 666–673 (1999)
13. Neves, A.J.R., Azevedo, J.L., Cunha, B., Lau, N., Silva, J., Santos, F., Corrente, G., Martins, D.A., Figueiredo, N., Pereira, A., Almeida, L., Lopes, L.S., Pinho, A.J., Rodrigues, J., Pedreiras, P.: CAMBADA soccer team: from robot architecture to multiagent coordination. In: Papic, V. (ed.) Robot Soccer. ch. 2. I-Tech Education and Publishing, Vienna (2010)
14. http://sweet.ua.pt/an/uavision/. (Last visited March 2014)

Uncertainty Based Multi-Robot Cooperative Triangulation

Andre Dias[1]([⊠]), Jose Almeida[1], Pedro Lima[2], and Eduardo Silva[1]

[1] INESC TEC - INESC Technology and Science, ISEP/IPP - School of Engineering,
Porto, Portugal
{adias,jma,eaps}@lsa.isep.ipp.pt
[2] Institute for Systems and Robotics, Instituto Superior Técnico,
Universidade de Lisboa, Lisbon, Portugal
pal@isr.ist.utl.pt

Abstract. The paper presents a multi-robot cooperative framework to estimate the 3D position of dynamic targets, based on bearing-only vision measurements. The uncertainty of the observation provided by each robot equipped with a bearing-only vision system is effectively addressed for cooperative triangulation purposes by weighing the contribution of each monocular bearing ray in a probabilistic manner. The envisioned framework is evaluated in an outdoor scenario with a team of heterogeneous robots composed of an Unmanned Ground and Aerial Vehicle.

1 Introduction

The potential impact of cooperative perception within a group of autonomous vehicles is unquestionable in many application domains, such as rescue missions [1,2], forest fire monitoring [3], wildlife tracking, border control [4,5], surveillance [5,6] and reconnaissance [7], and battle damage assessment. One of the most common and versatile means of perception in multi-robot cooperative tasks is visual sensing with one or more cameras that are capable of acquiring visual information based on cooperative approaches. However, because technology miniaturization has improved significantly, there has been a tendency to decrease the vehicles' dimensions and payload [6]. This consequently brings new research challenges to the vision community with a natural transition to a bearing-only vision setup or to a smaller rigid baseline for stereo systems, which has an inherent impact in application scenarios where the goal is to estimate the position of targets whose depth distance exceeds the baseline. The result from a smaller stereo rigid baseline will be a 3D estimation position error increasing quadratically with depth [8,9]. Therefore, and in the context of an application scenario of rescue and border control missions, as depicted in Fig. 1, where the goal is to detect and estimate the position of a target in 3D, the main question is:

*How is it possible to produce 3D information based on **bearing-only** vision information using a team of robot observers?*

This paper proposes to address this issue with a novel multi-robot heterogeneous cooperative perception framework, defined as **U**ncertainty-based **M**ulti-**R**obot **Co**operative **T**riangulation (UCoT), capable of estimating the 3D dynamic

© Springer International Publishing Switzerland 2015
R.A.C. Bianchi et al. (Eds.): RoboCup 2014, LNAI 8992, pp. 502–513, 2015.
DOI: 10.1007/978-3-319-18615-3_41

Fig. 1. Motivation application scenarios. **Left**: Cooperative perception target tracking for the PCMMC FCT project. **Right**: Cooperative search and rescue task for the FP7 project ICARUS in La Spezia, Italy.

target position based on bearing-only measurements. The multi-robot cooperative triangulation method has been introduced in [10] and is extended in this paper in order to handle the uncertainty of the observation model provided by each robot by weighting the contribution of each monocular bearing ray in a probabilistic manner.

The motivation for the proposed framework emerged from open issues in the state-of-the-art on cooperative perception, to which the presented framework contributes:

- providing a multi-robot cooperative method to estimate 3D information based on bearing-only vision information;
- integrate all sources of uncertainty associated to the position, attitude and image pixel target position of each bearing-only sensor in a probabilistic manner, in order to weight the contribution of each visual sensor to the cooperative triangulation.
- improving a bearing-only vision technique applied to Unmanned Aerial Vehicles (UAVs) in order to estimate 3D information based on the flat earth assumption [11];

For the first point outlined, in the bearing-only vision setup it is intrinsically difficult to estimate depth and absolute scale [12]. Therefore, estimating 3D target position without a known target size is a research challenge [13]. Moreover, techniques such as monocular vision system Structure-from-Motion (SFM) or Visual Simultaneous Localization and Mapping (VSLAM) [14] have managed to achieve good results in depth estimation in indoor and even in outdoor map building scenarios [9]. However, they also present some constraints, such as high computational requirements, low camera dynamics, preferably with features available between frame for batch recursive process and large field of view. Therefore, and based on the motivation scenario, the available methods are not adequate to support scenarios where both robots and targets have dynamic behavior. Methods for the 3D target estimation with a bearing-only vision configuration based on the flat earth assumption [11], were developed for a particular case of aerial vehicles in order to estimate the 3D target position, with the depth information being provided by the vehicle's altitude without taking terrain morphology into consideration. The results of estimating the target position using this assumption are less accurate 3D information and the inability to estimate the position of targets that are not moving on the ground.

There are several techniques that can be used in the context of multi-robot cooperative 3D perception based on bearing-only vision information. An example of that is an offline collaborative VSLAM method, defined as CoSLAM [15]. The method is capable of estimate the pose between cameras with overlapped views by combining the conventional SFM method, Sukkarieh [16], in the ANSER project where the depth problem is solved with artificial landmarks of known size, and the Zhu [17] method, where two robots with an omnidirectional camera estimated the flexible baseline based on an object of known size.

The paper is organized as follows: Sect. 2 presents the proposed framework, the multi-robot architecture, as well as the mathematical formulation for the UCoT. The proposed framework is validated in Sect. 3 by presenting the results obtained in an outdoor scenario based on cooperative perception with a Micro Aerial Vehicle (MAV) and an Unmanned Ground Vehicle (UGV) tracking a target. Section 4 provides the conclusions and outlines future work topics.

2 Bearing-Only Multi-Robot Cooperative Triangulation

This section introduces our multi-robot cooperative triangulation framework to estimate 3D target position based on bearing-only measurements. The relative position and orientation provided by each robot are addressed by the framework, and based on the geometric constraints the 3D target position is estimated by establishing a flexible and dynamic geometric baseline for the cooperative triangulation. Therefore, two methods are formulated in this section: the **Mid-Point Multi-Robot Co**operative **T**riangulation(MidCoT) and the **U**ncertainty based Multi-Robot **Co**operative **T**riangulation (UCoT). In the first method, the framework selects the line that is perpendicular to the shortest segment for both rays, and assumes that both bearing-only cameras will contribute equally to the estimation of the target position, see Fig. 3. In the second method, all sources of uncertainty associated with the position, attitude and image plane pixel target position are addressed, and the covariance provided by the intersection rays is evaluated in order to weight, in a probabilistic manner, the contribution of each ray to the estimation of the target position, see Figs. 3 and 4.

The uncertainty of the observation model provided by each robot, including the robot uncertainty about its own pose, is described using the first order uncertainty propagation, with the assumption that all sources of uncertainty can be modeled as being uncorrelated Gaussian noise.

In the following sections, the notation ${}^{to}_{from}\boldsymbol{\xi}_n$ will be used to denote the transformation matrix *from* one coordinate *to* another coordinate frame. We call $\{B\}$ to the robot body frame and $\{W\}$ to the global frame expressed in the ECEF. The upper case notation in bold represents the matrix variables, while the lower case in bold represents the vectors, and finally the lower case represents the scalar variables.

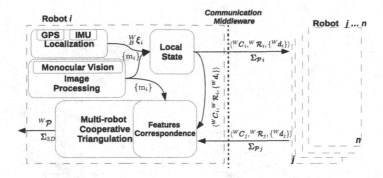

Fig. 2. Architecture framework for multi-robot cooperative triangulation.

2.1   Multi-Robot Architecture Framework

The envisioned multi-robot architecture framework, outlined in Fig. 2, describes the layers required to ensure a cooperative perception altruistic commitment that can share useful information between vehicles using a communication middleware. The architecture is independent from the position and attitude source of the information, as well as from the vision system.

The architecture framework is composed of the following components:

- **Localization**, responsible for providing the robot pose to the local state component, described by the following matrix $_B^W\xi$ related to the global frame. In the outlined architecture, this information is provided by an IMU as $\boldsymbol{u} \stackrel{\text{def.}}{=} [\phi\ \theta\ \psi]^T$, where (ϕ, θ, ψ) are respectively the roll, pitch and yaw angles, and a by GPS as $\boldsymbol{\varsigma} \stackrel{\text{def.}}{=} [\lambda\ \varphi\ h]^T$, where (λ, φ, h) are respectively the latitude, longitude, and altitude;
- **Local State**, provides an output 3 tuple $\langle ^W C, ^W \mathcal{R}, \{^W d\}\rangle$ related to the global frame composed by the camera position $^W C$, attitude $^W \mathcal{R}$ and ray vectors $\{^W d\}$, which represent the direction vector from the points detected by the monocular vision system $\{m\}$;
- **Feature Correspondence**, responsible for evaluating the tuples shared by other robots, related to the local state component. This evaluation is performed based on Euclidean distance between two points projected in the global frame from the intersection rays $\lambda_i^W d_i$ and $\lambda_j^W d_j$ with the perpendicular vector $^W d_\perp$[10], as depicted in Fig. 4;
- **Multi-Robot Cooperative Triangulation**, responsible for the 3D target estimation and covariance $\boldsymbol{\Sigma}_{Target}$ related to all sources of uncertainty, as described in Eq. (11) for the UCoT method.

2.2   Uncertainty Based Multi-Robot Cooperative Triangulation

The formulation proposed in this paper is an extension to the well-known stereo rigid baseline mid-point triangulation method proposed by Trucco [18], as

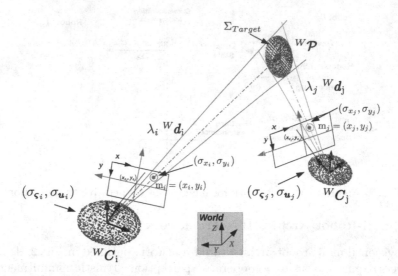

Fig. 3. Relative pose between robots i and j with a bearing-only vision system, estimating the 3D target position. The camera geometry change over time and provide a
flexible and dynamic baseline able to ensure cooperative triangulation.

depicted in Eq. (1), which is related to the reference frame of the camera on
the left. The proposed method removes the rigid baseline in order to support a
flexible and dynamic geometric baseline composed of the information provided
by each bearing-only vision system, camera position ^{W}C and attitude $^{W}\mathcal{R}$ relatively to the global frame, as depicted in Fig. 3.

$$\lambda_i^B d_i - \lambda_j \mathcal{R}^{TB} d_j + \alpha(^B d_i \wedge \mathcal{R}^{TB} d_j) = T \tag{1}$$

Therefore, based on a dynamic baseline approach, related to the body frame,
with a pair of bearing-only vision systems defined as i, j, and assuming that
each camera knows its own rotation and translation matrix $T_i, \mathcal{R}_i, T_j, \mathcal{R}_j$
relatively to their body reference frame, it is possible to obtain a $^B\mathcal{R} = \mathcal{R}_j^T \mathcal{R}_i$
and $^B T = \mathcal{R}_j^T(-T_j) - \mathcal{R}_i^T(-T_i)$. Therefore, replacing the $^B\mathcal{R}$ and $^B T$ in Eq. (1),
it is possible to obtain

$$\lambda_i^B d_i - \lambda_j^B d_j + \alpha(^B d_i \wedge^B d_j) = {}^B T_i - {}^B T_j \tag{2}$$

with $\lambda_i^B d_i$ and $\lambda_j^B d_j$ being the direction vectors related to the body frame, and
$^B d_\perp = {}^B d_i \wedge^B d_j$ the 3D intersection vector perpendicular to both i and j rays.
Solving the linear system from Eq. (2) to obtain the coefficients, $\lambda_i, \lambda_j, \alpha$, the
triangulated point $^B P$ will be over the midpoint of the line segments joining
$^B C_i + \lambda_i^B d_i$ and $^B C_j + \lambda_j^B d_j$

$$\begin{aligned}
^B P &= \Omega_i \, ^B P_i + \Omega_j \, ^B P_j \\
&= \Omega_i(^B C_i + \lambda_i^B d_i) + \Omega_j(^B C_j + \lambda_j^B d_j)
\end{aligned} \tag{3}$$

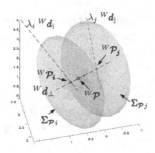

Fig. 4. Covariance 3D ellipse from the intersection rays Σ_{P_i} and Σ_{P_j}. The MidCoT method is represented by the blue dot and the UCoT by the blue cross. The purple line is the vector $^W\boldsymbol{d}_\perp$ perpendicular to $\lambda_i^W \boldsymbol{d}_i$ and $\lambda_j^W \boldsymbol{d}_j$.

with $^B\boldsymbol{C}_i$ and $^B\boldsymbol{C}_j$ being the camera position related to the body frame and Ω_i, Ω_j the weights to be derived. The 3D target position related to the global frame, as shown in Fig. 3, is derived from the transformation matrix $_B^W\boldsymbol{\xi}$ from body frame to world frame, as

$$^W\boldsymbol{P} = \Omega_i(^W\boldsymbol{C}_i + \lambda_i^W \boldsymbol{d}_i) + \Omega_j(^W\boldsymbol{C}_j + \lambda_j^W \boldsymbol{d}_j) \tag{4}$$

and for $\Omega_i = \Omega_j = \frac{1}{2}$ establishes a method that will be defined as MidCoT.

The geometric intersection in 3D between $\lambda_i^W \boldsymbol{d}_i$, $\lambda_j^W \boldsymbol{d}_j$ and the perpendicular vector $^W\boldsymbol{d}_\perp$ are shown in Figs. 3 and 4, as well as the mid-point derived from Eq. (4) between $^W\boldsymbol{P}_i$ and $^W\boldsymbol{P}_j$ (represented by the blue dot in the covariance ellipse, see Fig. 4). To ensure that all sources of uncertainty are taken into consideration when estimating the 3D target position, the Σ_{Target} will be estimated based on assumption that there is uncertainty in the input pixel localization $\sigma_{\boldsymbol{m}}$, in the cameras position $\sigma_{\boldsymbol{\varsigma}}$ and attitude $\sigma_{\boldsymbol{u}}$ relatively to the global frame, all of them modelled as uncorrelated zero-mean Gaussian random variables.

$$\sigma_{\boldsymbol{\varsigma}} = \begin{bmatrix} \sigma_\lambda & 0 & 0 \\ 0 & \sigma_\varphi & 0 \\ 0 & 0 & \sigma_h \end{bmatrix} \quad \sigma_{\boldsymbol{u}} = \begin{bmatrix} \sigma_\phi & 0 & 0 \\ 0 & \sigma_\theta & 0 \\ 0 & 0 & \sigma_\psi \end{bmatrix} \quad \sigma_{\boldsymbol{m}} = \begin{bmatrix} \sigma_{m_x} & 0 \\ 0 & \sigma_{m_y} \end{bmatrix} \tag{5}$$

Using the first-order uncertainty propagation, it is possible to approximate the distribution of the variables, defined in Sect. 2.1, as the input state vector $\boldsymbol{\nu}_{(i,j)} = [\boldsymbol{\varsigma}_i, \boldsymbol{u}_i, \boldsymbol{m}_i, \boldsymbol{\varsigma}_j, \boldsymbol{u}_j, \boldsymbol{m}_j]$, from Eq. (4), as multivariate Gaussians. The Σ_{Target} covariance matrix approximately models the uncertainty in the 3D target estimation, which is computed from the noisy measurements of the Multi-Robot Cooperative Triangulation, as follows:

$$\Sigma_{Target} = \boldsymbol{J_P}\Lambda_{i,j}\boldsymbol{J_P}^T \tag{6}$$

where $\boldsymbol{J_P}$ stands for the Jacobian matrix of $^W\boldsymbol{P}$ in Eq. (4) by

$$\boldsymbol{J_P}_{[3\times16]} = \nabla_{(\boldsymbol{\nu}_{(i,j)})}\,^W\boldsymbol{P}(\boldsymbol{\nu}_{(i,j)}) \tag{7}$$

with $\Lambda_{i,j}$ being the input covariance matrix represented by a diagonal line relatively to all sources of uncertainty present in Eq. (5) for each monocular vision system

$$\Lambda_{i,j[16\times16]} = \begin{bmatrix} \sigma_{\varsigma i[3\times3]} & \cdots & & & & \\ \cdots & \sigma_{ui[3\times3]} & \cdots & & & \\ & \cdots & \sigma_{mi[2\times2]} & \cdots & & \\ & & \cdots & \sigma_{\varsigma j[3\times3]} & \cdots & \\ & & & \cdots & \sigma_{uj[3\times3]} & \cdots \\ & & & & \cdots & \sigma_{mj[2\times2]} \end{bmatrix} \qquad (8)$$

As stated at the beginning of this paper, this work aims at addressing all sources of uncertainty provided by each intersection ray, using a probabilistic weight provided by the estimated uncertainty of each ray. Therefore, as previously described, using the first-order uncertainty propagation, it is possible to estimate the covariance $\Sigma_{\mathcal{P}i}$ and $\Sigma_{\mathcal{P}j}$ related to $^W\mathcal{P}_i$ and $^W\mathcal{P}_j$ as follows:

$$\Sigma_{\mathcal{P}i} = J_{\mathcal{P}i}\Lambda_{i,j}J_{\mathcal{P}i}^T \quad J_{\mathcal{P}i[3\times16]} = \nabla_{(\nu_{(i,j)})} {}^W\mathcal{P}_i$$
$$\Sigma_{\mathcal{P}j} = J_{\mathcal{P}j}\Lambda_{i,j}J_{\mathcal{P}j}^T \quad J_{\mathcal{P}j[3\times16]} = \nabla_{(\nu_{(i,j)})} {}^W\mathcal{P}_j \qquad (9)$$

where $J_{\mathcal{P}i}$ and $J_{\mathcal{P}j}$ are respectively the Jacobian matrix from $^W\mathcal{P}_i$ and $^W\mathcal{P}_j$, and $\Lambda_{i,j}$ is the input covariance matrix from Eq. (8). Therefore, with the uncertainty of each intersection ray $\Sigma_{\mathcal{P}i}$ and $\Sigma_{\mathcal{P}j}$, and the perpendicular vector $^W d_\perp$, the probabilistic weight of each ray is expressed as

$$\Omega_i = \frac{(^W d_\perp \, \Sigma_{\mathcal{P}j} \, ^W d_\perp{}^T)^2}{(^W d_\perp \, \Sigma_{\mathcal{P}i} \, ^W d_\perp{}^T)^2 + (^W d_\perp \, \Sigma_{\mathcal{P}j} \, ^W d_\perp{}^T)^2} \quad \Omega_j = \frac{(^W d_\perp \, \Sigma_{\mathcal{P}i} \, ^W d_\perp{}^T)^2}{(^W d_\perp \, \Sigma_{\mathcal{P}i} \, ^W d_\perp{}^T)^2 + (^W d_\perp \, \Sigma_{\mathcal{P}j} \, ^W d_\perp{}^T)^2}$$
$$(10)$$

Using these weight in Eq. (4) we effectively obtain the optimal solution over $^W d_\perp$, and the novel method becomes Uncertainty based Multi-Robot Cooperative Triangulation (UCoT). This cooperative triangulation method will ensure that the uncertainty of each bearing-only sensor will be weighted, using Ω_i, Ω_j, in a probabilistic manner comparatively to the 3D target estimation.

$$^W\mathcal{P} = \frac{(^W d_\perp \, \Sigma_{\mathcal{P}j} \, ^W d_\perp{}^T)^2}{(^W d_\perp \, \Sigma_{\mathcal{P}i} \, ^W d_\perp{}^T)^2 + (^W d_\perp \, \Sigma_{\mathcal{P}j} \, ^W d_\perp{}^T)^2}(^W C_i + \lambda_i^W d_i)$$

$$+ \frac{(^W d_\perp \, \Sigma_{\mathcal{P}i} \, ^W d_\perp{}^T)^2}{(^W d_\perp \, \Sigma_{\mathcal{P}i} \, ^W d_\perp{}^T)^2 + (^W d_\perp \, \Sigma_{\mathcal{P}j} \, ^W d_\perp{}^T)^2}(^W C_j + \lambda_j^W d_j) \qquad (11)$$

3   Implementation and Results

This section describes the outdoor scenario, the robot issues and the experimental results related to the implementation of single and multi-robot cooperative perception methods proposed in two experimental cases. The goal was to evaluate the 3D position estimation of a static and a dynamic target (Fig. 5).

Fig. 5. Left: UGV TIGRE. **Right**: Asctec Pelican MAV.

The outdoor scenario chosen to evaluate the proposed framework is a non-urban area with several landscape elements including vegetation and rocks, as depicted in Fig. 1. The heterogeneous robots used in both experimental cases were the UGV TIGRE [19] and the Asctec Pelican MAV. Both robots used the Linux operating system, with wireless communication. As far as the TIGRE's main hardware issues are concerned, there were two cameras with a resolution of 1278×958 in a stereo rigid baseline (~ 0.76 m), a Novatel GPS receiver and an IMU Microstrain. The Pelican MAV is a commercial platform to which was added a downward monocular camera with a resolution of 1280×1024.

The evaluation procedure in both experimental cases was composed of four methods used to estimate the 3D target position, including the proposed framework. The methods under evaluation are:

- **TIGRE - Stereo rigid baseline**, based on the Mid-Point Stereo Triangulation from Trucco [18], estimate 3D target position with the available stereo rigid baseline;
- **Pelican - Monocular 3D target estimation**, which follows the flat earth assumption method to obtain the 3D target position. The method was developed by Beard [11] and depth is provided by the MAV altitude;
- **Mid-Point Multi-Robot Cooperative Triangulation** (MidCoT), based on the assumption that each robot provides a bearing-only measurement, the framework will establish a dynamic baseline between the UGV and the MAV and with equally weight contribution, $\Omega_i = \Omega_j = \frac{1}{2}$, of each bearing-only measurement, as expressed in Eq. (4);
- **Uncertainty based Multi-Robot Cooperative Triangulation** (UCoT), based on the uncertainty of each bearing-only measurement $\Sigma_{\mathcal{P}i}$ and $\Sigma_{\mathcal{P}j}$, the method weights the contribution of each one of the bearing-only vision sensor, in a probabilistic manner, to estimate the 3D target position, as expressed in Eq. (11).

The experimental results from each method are represented in Figs. 6 and 7, depicted on the left side the trajectories of the UGV(blue triangle), of the MAV(magenta circle) and of the target(red star). The estimated position of the target related to the UCoT method is represented by the green circle. The 3D target estimation error for each method related to the Euclidean distance between the camera and the target are presented on the right side of both figures. The accuracy evaluation in both experimental cases was achieved based on the

Real-Time Kinematic (RTK) GPS attached to the target as an exogenous ground truth system. The 3D covariance ellipse from the intersection rays $\Sigma_{\mathcal{P}i}$ and $\Sigma_{\mathcal{P}j}$ are expressed in Fig. 8 for both experimental cases, and are related to three instances with different Euclidean distances between the camera and the target. The trajectory of both robots are represented by the blue triangle, in the case of the UGV and in magenta for the MAV.

In the static target experimental case, the UGV was positioned at a distance of ~ 35 m and the MAV hovering over the static target at a distance of ~ 15 m. Both robots were given a target location task: the MAV hovered over the target and the UGV performed an approximation maneuver relatively to the estimated target position. The target was moving at a velocity of ~ 0.8 m/s relatively to the dynamic target experimental case, and both robots had the same target location task as previously described, although now with a rule of safe distance between the robots and the target of ~ 2 m.

The results in both experimental cases related to the single perception methods are in accordance with the expected results, especially for the stereo rigid baseline, where the accuracy of the 3D target estimation is very low for targets where the depth distance exceeds the available baseline. In the case of the flat earth assumption applied to the MAV Pelican in order to estimate the 3D target, the results also reveal an error similar to the stereo baseline due to the terrain morphology of the outdoor scenario.

The MidCoT presented in Figs. 6 and 7 shows a 3D target estimation error independent from the distance to the target when compared to single perception methods. This behavior is reflected in the standard deviation improvement σ shown in Table 1. However, the overall mean μ result is less optimistic when compared to single perception methods. In this method, the 3-tuple rays $\langle {}^{W}C_{n,B}^{W}\xi n, \{ {}^{W}d_{n}\} \rangle$ are shared between robots, and will have the same weight Ω_{n} for the multi-robot cooperative triangulation. By assuming an equal contribution, the impact of the sensor uncertainty provided by each robot is disregarded. For the experimental case with a static target, the impact of both robots with the same weight $\{\Omega_{n}\}$ in the triangulation can be observed in Fig. 6 when the

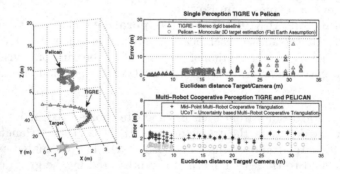

Fig. 6. Experimental case for a static target. 3D target estimation error for each method, relatively to the distance between the camera and the target.

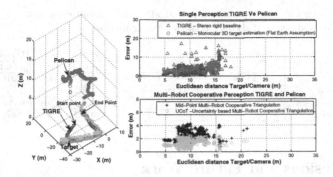

Fig. 7. Experimental case for a dynamic target. 3D target estimation error for each method, relatively to the distance between the camera and the target.

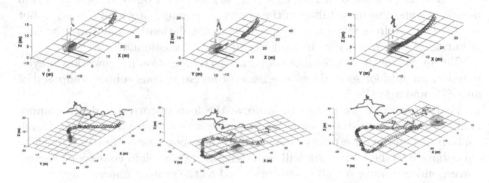

Fig. 8. Covariance $\Sigma_{\mathcal{P}i}$ and $\Sigma_{\mathcal{P}j}$ from the intersection rays related to three instances in a static (top) and dynamic target tracking (bottom) with the UCoT.

Euclidean distance between the UGV and the static target varies between 5 and 10 m, and the 3D target estimation error does not decrease as expected. Even though the UGV is closer to the target, its contribution will have the same impact as the MAV 3-tuple information, which is influenced by a high uncertainty in the estimated position($\sim$ 3 m). This is due to the low-cost GPS assembled, and also to the fact that the relative height to the target is stable at $\sim$ 15 m. This behavior can be also observed during the experimental case with a moving target (Fig. 7) because the distance between the moving target and the UGV varies between 5 and 15 m, and the resulting error in the 3D target estimation is dominated by the uncertainty of the MAV.

The UCoT presented in Figs. 6 and 7 illustrates an accuracy improvement in the 3D target estimation when compared to the three previous methods. The equal weight Ω_n contribution from each robot, described as a limitation in MidCoT, is overcome in UCoT because it assumes the camera position and attitude uncertainty of each ray in a probabilistic manner in the weight Ω_n. The overall improvement is depicted in Table 1 with a lower mean μ and standard deviation σ error in the 3D target estimation position.

Table 1. 3D target estimation mean μ and standard deviation σ error in meters for each method in a static and dynamic target experimental cases.

	Method	Static Target		Dynamic Target	
		μ (m)	σ (m)	μ (m)	σ (m)
Single Perception	TIGRE - Stereo rigid baseline	2.202	2.435	1.210	2.253
	Pelican - Monocular 3D target estimation	2.602	0.548	1.793	2.421
Multi-Robot Cooperative Perception	MidCoT	2.305	0.501	0.996	1.324
	UCoT	0.873	0.214	0.570	0.768

4   Conclusions and Future Work

This paper proposes a decentralized multi-robot cooperative framework to estimate the 3D position of highly dynamic targets based on bearing-only vision measurements. The uncertainty of the observation model provided by each robot is effectively address to achieve cooperative triangulation by weighting the contribution of each monocular bearing ray in a probabilistic manner.

The impact of the multi-robot cooperative triangulation framework was evaluated in an outdoor scenario with a team of heterogeneous robots composed of an UGV and a MAV.

For future work, we intend to improve the feature correspondence component through a multi-robot epipolar line with a dynamic narrow band search space based on the uncertainty associated to the robot pose and pixel error. The contribution to the framework will be a dynamic hyperboloid confidence narrow correspondence point search space correlated to the sensors uncertainty.

Acknowledgments. This work is co-financed by Project "NORTE-07-0124-FEDER-000060" by the North Portugal Regional Operational Programme (ON.2 O Novo Norte), under the National Strategic Reference Framework (NSRF) through the European Regional Development Fund (ERDF) and also by National Funds through the FCT within project PEst-OE/EEI/LA0009/2013.

References

1. Michael, N., Shen, S., Mohta, K., Mulgaonkar, Y., Kumar, V., Nagatani, K., Okada, Y., Kiribayashi, S., Otake, K., Yoshida, K., Ohno, K., Takeuchi, E., Tadokoro, S.: Collaborative mapping of an earthquake-damaged building via ground and aerial robots. J. Field Rob. **29**(5), 832–841 (2012)
2. Olson, E., Strom, J., Goeddel, R., Morton, R., Ranganathan, P., Richardson, A.: Exploration and mapping with autonomous robot teams. Commun. ACM **56**(3), 62–70 (2013)
3. Merino, L.: A cooperative perception system for multiple unmanned aerial vehicles. Application to the cooperative detection, localization and monitoring of forest fires. Ph.D. thesis, University of Seville (2007)
4. Marino, A., Caccavale, F., Parker, L., Antonelli, G.: Fuzzy behavioral control for multi-robot border patrol. In: 17th Mediterranean Conference on Control and Automation, MED 2009, pp. 246–251 (2009)

5. Xu, Z., Douillard, B., Morton, P., Vlaskine, V.: Towards collaborative multi-MAV-UGV teams for target tracking. In: 2012 Robotics: Science and Systems Workshop Integration of Perception with Control and Navigation for Resource-Limited, Highly Dynamic, Autonomous Systems (2012)
6. Kushleyev, A., Kumar, V., Mellinger, D.: Towards a swarm of agile micro quadrotors. In: Proceedings of Robotics: Science and Systems, Sydney, Australia (2012)
7. Zhang, F., Leonard, N.E.: Cooperative filters and control for cooperative exploration. IEEE Trans. Autom. Control **55**(3), 650–663 (2010). doi:10.1109/TAC.2009.2039240
8. Gallup, D., Frahm, J.M., Mordohai, P., Pollefeys, M.: Variable baseline/resolution stereo. In: 2008 IEEE Conference on Computer Vision and Pattern Recognition (2008)
9. Weiss, S.M.: Vision based navigation for micro helicopters. Ph.D. thesis (2012)
10. Dias, A., Almeida, J., Lima, P., Silva, E.: Multi-robot cooperative stereo for outdoor scenarios. In: 13th International Conference on Autonomous Robot Systems and Competition (2013)
11. Barber, D., Redding, J., McLain, T., Beard, R., Taylor, C.: Vision-based target geo-location using a fixed-wing miniature air vehicle. J. Intell. Rob. Syst. **47**(4), 361–382 (2006)
12. Achtelik, M.W., Weiss, S., Chli, M., Dellaert, F., Siegwart, R., Eth, Z.: Collaborative stereo. In: Proceedings of the IEEE/RSJ Conference on Intelligent Robots and Systems (IROS) (2011)
13. Ahmad, A., Nascimento, T., Conceicao, A., Moreira, A., Lima, P.: Perception-driven multi-robot formation control. In: 2013 IEEE International Conference on Robotics and Automation (ICRA), pp. 1851–1856 (2013)
14. Davison, A., Reid, I., Molton, N., Stasse, O.: Monoslam: real-time single camera slam. IEEE Trans. Pattern Anal. Mach. Intell. **29**(6), 1052–1067 (2007)
15. Zou, D., Tan, P.: Coslam: collaborative visual slam in dynamic environments. IEEE Trans. Pattern Anal. Mach. Intell. **35**(2), 354–366 (2013)
16. Sukkarieh, S., Nettleton, E., Kim, J.H., Ridley, M., Goktogan, A., Durrant-Whyte, H.: The ANSER project: data fusion across multiple uninhabited air vehicles. Int. J. Robot. Res. **22**(7–8), 505–539 (2003)
17. Zhu, Z., Karuppiah, D.R., Riseman, E.M., Hanson, A.R.: Dynamic mutual calibration and view planning for cooperative mobile robots with panoramic virtual stereo vision. Comput. Vis. Image Underst. **95**(3), 261–286 (2004)
18. Trucco, E., Verri, A.: Introductory Techniques for 3-D Computer Vision. Prentice Hall, Upper Saddle River (1998)
19. Martins, A., Amaral, G., Dias, A., Almeida, C., Almeida, J., Silva, E.: Tigre - an autonomous ground robot for outdoor exploration. In: 2013 13th International Conference on Autonomous Robot Systems (Robotica), pp. 1–6 (2013)

A Method to Estimate Ball's State of Spin by Image Processing for Improving Strategies in the RoboCup Small-Size-Robot League

Yuji Nunome[1]([✉]), Kazuhito Murakami[2], Masahide Ito[2],
Kunikazu Kobayashi[2], and Tadashi Naruse[2]

[1] Graduate School of Information Science and Technology,
Aichi Prefectural University,
Nagakute, Aichi 480-1198, Japan
im132007@cis.aichi-pu.ac.jp
[2] School of Information Science and Technology,
Aichi Prefectural University,
Nagakute, Aichi 480-1198, Japan
{murakami,masa-ito,kobayashi,naruse}@ist.aichi-pu.ac.jp

Abstract. This paper addresses an estimation problem of the ball's state of spin in RoboCup Small Size League (SSL). A spinning ball varies its speed after the ball bounces off the floor. This paper proposes an image-based estimation method of the ball's state of spin, in particular, by using inertia feature of co-occurrence matrix of the image sequence. The effectiveness of our proposed method is shown by some experiments.

1 Introduction

The RoboCup small size league (SSL) is one of the soccer leagues in the RoboCup [1,2]. The size of robot is limited less than 180 mm diameter and less than 150 mm height. Each team which consists of six robots plays the game on a field of size 8090 mm × 6050 mm. Four cameras are equipped above the field to overlook the entire field. The positions of the robots and the ball are recognized by the image. Based on the position information, each team decides an appropriate strategy.

Each robot can kick the ball in two different ways, straight-kick-type and loop-kick-type (called chip-kick), and also has a rolling mechanism which gives a strong spin to the ball. When the ball is chip-kicked with strong spin, the spinning ball varies its speed after the ball bounced off the floor. This makes it difficult to predict the locus of the ball. So, the estimation of the ball's state of spin is very important to dominate the game.

Few paper has been reported in the field of estimating ball's state of spin in the RoboCup SSL. In some ball games, several papers have been reported to estimate it by high-speed camera. Inoue et al. reported how to estimate the vector of spin axis and speed by tracing a seam pattern on the surface of the ball by using a high speed camera [3]. They obtained the rotation axis and spinning speed by comparing the input image with the image stored in database.

R.A.C. Bianchi et al. (Eds.): RoboCup 2014, LNAI 8992, pp. 514–524, 2015.
DOI: 10.1007/978-3-319-18615-3_42

Tamaki et al. reported a method to estimate spinning speed by image registration [4]. They utilized an information of the mark on the surface of the ball. Liu et al. proposed a method by using two high speed cameras to calculate both of ball's speed and spinning speed [5]. These methods could be applied if some kinds of mark patterns are attached or printed on the surface of the ball. In Robocup SSL, a golf ball is used, so it is difficult to apply these image processing methods straightly because there is no feature on the surface of the golf ball except maker logotype. And furthermore it is not so easy to set up and control expensive multiple high speed cameras.

Federico et al. reported how to estimate rotation of the ball without using high speed camera [6]. In their method, the vector of spin axis and speed are estimated by using the movement of the six marks that are placed on the surface of the soccer ball. However, their method could be used under the condition that the rotation speed of the soccer ball is 10 rps or less. In the RoboCup SSL, the rotation speed is very high (more than 30 rps), so it is difficult to apply this method.

When a ball is spinning, some kind of blur is observed in the region of the ball. One of the principal factors of blur is the dimples on the surface of the ball. The degree of blur changes according to the exposure time of the camera and the rotational speed of the ball. This paper proposes a new method to estimate the ball's state of spin by using inertia feature of co-occurrence matrix of the image sequence. This method doesn't need a special pattern attached on the surface of the ball.

In this paper, Sect. 2 describes our basic idea to estimate ball's state of spin and Sect. 3 explains concrete methods to calculate inertia feature and co-occurrence matrix and Sect. 4 expresses experimental results to show the effectiveness of the proposed method, and after that, Sect. 5 discusses the rotation speed of the ball and application to strategy.

2 Basic Idea

In the RoboCup SSL, an orange golf ball is officially used. There are many dimples on the surface of the ball. These dimples cause the changes of reflection and this reflection changes according to the degree of blur. Figure 1 shows the differences of the images. Figure 1(a) is a static ball's image and some highlight regions exist in it, on the other hand Fig. 1(b) is a spinning ball's image and it looks like a smoothed image. It will be possible to estimate the ball's state of spin by analyzing the distribution of reflection.

In order to observe the surface of the ball, let an additional camera (i.e. PTZ camera) be installed to the vision system as shown in Fig. 2. The data flow of our vision system is shown in Fig. 3. In the RoboCup SSL, the standard vision system called SSL-Vision [5] is utilized and it sends the information of the position and ID of each robot and the ball. In addition to these information, to send the ball's state of spin realizes a new strategy to cope with an irregular bounce of ball.

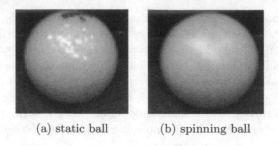

(a) static ball   (b) spinning ball

Fig. 1. Differences of the images

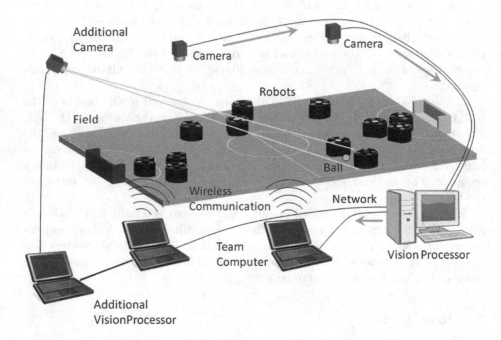

Fig. 2. Cameras' layout

3   Image Processing Method to Estimate Ball's State of Spin

The main flow of our image processing system is shown in Fig. 4. Our system is composed of 3 parts of image processing. First, the ball's region is extracted (*Step* 1), then co-occurrence matrix and inertia feature are calculated in the ball's region (*Step* 2), and finally the ball's state of spin is estimated (*Step* 3). Details of each step is as follows:

Step 1. **Extraction of ball's region**

   Let each frame of image sequences be inputted. An example of input image is

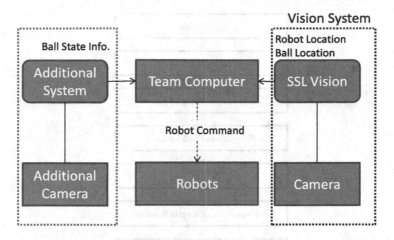

Fig. 3. Data flow of vision system

shown in Fig. 5. Then, make the grayscale image that is high value in the ball region (Fig. 6) and the grayscale image is threshold by discriminant analysis method. Threshold image is processed by expansion and contraction, then, threshold image is labelled. After the image is labelled, let the second largest region be that of the ball (Fig. 7). Then, extract the rectangle region that of the ball's region. In the rectangle region, covert to (R,G,B)=(0,0,0) except ball region. Then, let the rectangle region as shown in Fig. 8 be the processing region in the following processes.

Step 2. **Calculation of co-ocurrence matrix and inertia feature**
First, calculate co-ocurrence matrix $P_\delta(i,j)$ from the B-channel image of the extracted region in *step* 1. Here, Dx and Dy be the differences of x- and y-coordinates of two pixels, respectively, and $\delta = (D_X, D_Y)$ denotes the vector of them, and i and j are the values of these two pixels. The inertia feature Ine is calculated from $P_\delta(i,j)$ as

$$Ine = \sum_{i=0}^{n-1} \sum_{j=0}^{n-1} (i-j)^2 P_\delta(i,j) \tag{1}$$

Step 3. **Judgment of ball's state of spin**
For each frame, the intertia feature Ine is thresholded by an suitable value and the ball's state of spin is judged as follows:

$$BullSlule = \begin{cases} StaticState & (Ine > Threshold1) \\ UnknowSlule & (Threshold2 \leq Ine \leq Threshold1) \\ SpinningState & (Ine < Threshold2) \end{cases}$$

$Threshold1$ and $Threshold2$ are easily obtained, for example, by discriminant analysis, because the value of feature Ine are greatly different. $UnknowState$ is the transition between $StaticState$ and $SpinningState$.

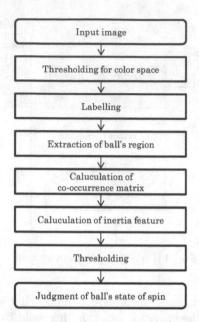

Fig. 4. Flowchart of image processing

Fig. 5. Original image

Fig. 6. Grayscale image

Fig. 7. Ball's region**** C ****

Fig. 8. Enlarged image to be processed

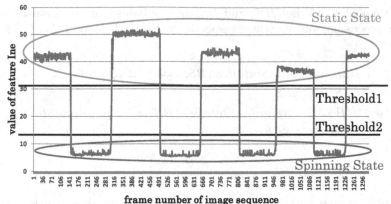

(a) The changes of the inertia feature

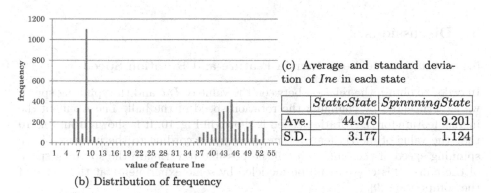

(c) Average and standard deviation of Ine in each state

	$StaticState$	$SpinningState$
Ave.	44.978	9.201
S.D.	3.177	1.124

(b) Distribution of frequency

Fig. 9. An example of the inertia feature for an image sequence (spin and stop is repeated)

4   Experiment

Figure 9(a) shows an example of the changes of the inertia feature for an image sequence in which the ball repeats spinning and stopping. It is known from the figure that Ine is changed largely between the ball's state is spinning and stopping. Figure 9(b) shows frequency of feature Ine and Fig. 9(c) shows average and standard deviation of Ine in $StaticState$ and $SpinningState$. Spinning and stopping images are taken by A601fc (Baslar make) [6] and the parameters of the camera are as follows:

- -frame size: 656×490 pixels
- -frame rate: 30 FPS
- -exposure time: 1/60 s

The performance of our proposed method is evaluated by calculating success rate. The ball's state was estimated by proposed method, where $Threshold1 = 35$

and $Threshold2 = 13$. Let BS_n and BS'_n be the ball's state in the n-th frame by image processing and its truth, respectively. $SUCEESS$ or $FAILURE$ was given by

$$if \quad ((BS_n == StaticState)\&\&(BS'_n == StaticState))$$
$$SUCCESS$$
$$else \ if((BS_n == SpinningState)\&\&(BS'_n == SpinninngState))$$
$$SUCCESS$$
$$else$$
$$FAILURE$$

6054 of 6100 frames succeeded, thus the success rate in this case was about 99.2 %.

5   Discussions

5.1   Relation Between Inertia Feature and Rotation Speed

In order to obtain the relation between the value of Ine and the spinning speed, we experimented by changing the rotation speed of the ball. The result of the inertia feature for the rotation speed is shown in Fig. 10. It is shown from Fig. 10 that the value of inertia feature Ine decreases according to the increase of the spinning speed of the ball, so it will be possible to estimate the spinning speed of the ball if this curve could be modeled by some experiments at the venue of the competition [9].

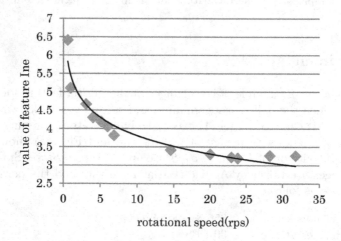

Fig. 10. An example of the changes of the inertia feature for an image sequence (spin and stop is repeated)

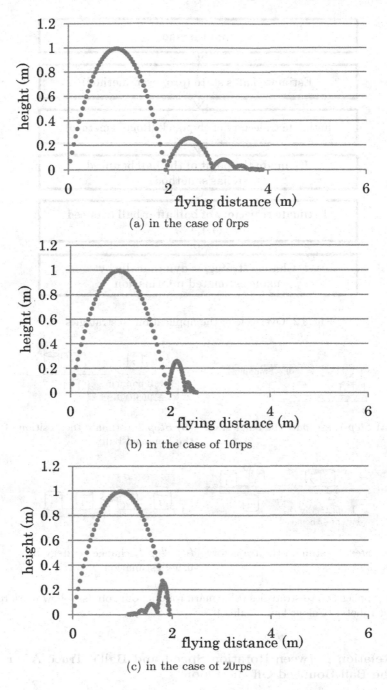

(a) in the case of 0rps

(b) in the case of 10rps

(c) in the case of 20rps

Fig. 11. Trajectory simulation with different rotational speed

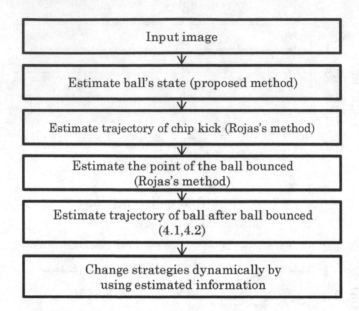

Fig. 12. Overview of the application to strategies

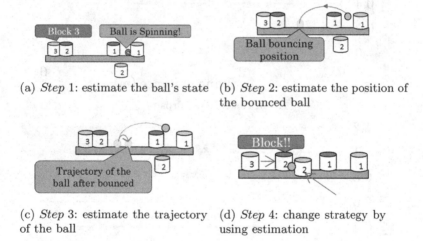

(a) *Step* 1: estimate the ball's state

(b) *Step* 2: estimate the position of the bounced ball

(c) *Step* 3: estimate the trajectory of the ball

(d) *Step* 4: change strategy by using estimation

Fig. 13. Application to strategies (blue mark robots : our robots; yellow mark robots : opponent robots) (Colour figure online)

5.2 Relation Between Rotation Speed and Ball's Trace After the Ball Bounced Off the Floor

It will be possible to predict the ball's movement by estimating the rotation speed of the ball. Figure 11 shows the trajectory simulations that the ball is kicked at 5 m/s and 60 degrees with different rotation speed. In the case of *StaticState*, the kicked ball is traced as shown in Fig. 11(a) and reaches to 4 m distance. In the case of 10 rps rotation speed, the kicked ball is moved as shown in Fig. 11(b). In the case of 20 rps rotation speed, the kicked ball is come back by strong spin as shown in Fig. 11(c).

The spinning ball varies its speed after the ball bound off the floor. Thus if the rotation speed of the ball can be estimated, the trajectory of the ball after the bound off the floor will be predicted more accurately.

5.3 Application to Strategies

The proposed method described can be applied to various strategies in the RoboCup SSL. Overview of the application to our strategies is shown in Fig. 12. It is possible to apply the strategies, if the trajectory of the ball can be predicted.

New defense strategies will be possible as shown in Fig. 13 (in this figure, the blue robots are ours, the yellow ones are opponent robots). In the defense, the following situation is expected:

Figure 13 shows an example of a defense strategy improved by the proposed method. We here suppose the following situation:

1. The yellow ID1 is going to kick the ball and the blue ID1 blocks the straight-kick pass.
2. The yellow ID3 is in the position where the ball will come after chip-kick by the yellow ID1,
3. And the blue ID2 is in the position where this robot can block the yellow ID3 as shown in Fig. 13(a).

In this situation, (*Step* 1) the ball's state of spin is estimated by the proposed method as shown in Fig. 13(a). Then, (*Step* 2) the point of the ball bounced off the floor is estimated by Rojas's method [10] as shown in Fig. 13(b). Next, (*Step* 3) the trajectory of the ball after bounced is estimated as shown in Fig. 13(c), and finally (*Step* 4) even if the yellow ID2 moves to the position of the ball after bounce, the blue ID2 robot is able to move to around the yellow ID2 robot for blocking shoot as shown in Fig. 13(d). Accordingly, we can prevent disadvantageous situation.

By estimating the rotation of the ball, it is possible to correspond to defense. Furthermore, by estimating the rotation speed, it is possible to improve the positions of robots for defense.

6 Conclusion

We have proposed a method to estimate the ball's state of spin by using inertia feature of co-occurrence matrix of the image sequence. The effectiveness of our proposed method was demonstrated some experiments. We also discuss the following things. (1) the possibility to estimate the spinning speed of the ball by the relation between the value of *Ine* and the spinning speed, and (2) an application to a defense strategy of the RoboCup SSL.

The proposed method works well in our prototype system. In future work, we need to confirm the robustness in various environment. It is also important to implement our method to a strategic learning system.

References

1. About Robocup. http://www.robocup.org/about-robocup/. Accessed 20 June 2014
2. Small Size Robot League - start. http://robocupssl.cpe.ku.ac.th/. Accessed 20 June 2014
3. Inoue, T., Uematsu, Y., Saito, H.: Stimation of rotational velocity of baseball using high-speed camera movie. Trans. Inst. Electr. Eng. Jpn. D, A publ. Ind. Appl. Soc. **131**(4), 608–615 (2011)
4. Tamaki, T., Sugino, S., Yamamoto, M.: Measuring ball spin by image registration. In: Proceedings of the 10th Korea-Japan Joint Workshop on Frontiers of Computer Vision : FCV 2004, pp. 269–274 (2004)
5. Liu, C., Hayakawa, Y., Nakasima, A.: An on-line algorithm for measuring the translational and rotational velocities of a table tennis ball. SICE J. Control Meas. Syst. Integr. **5**(4), 233–241 (2012)
6. Cristina, F., Dapoto, S.H., Russo, C.: A lightweight method for computing ball spin in real time. J. Comput. Sci. Technol. **7**(1), 34–38 (2007)
7. Small Size Robot League - sslvision. http://robocupssl.cpe.ku.ac.th/sslvision. Accessed 20 June 2014
8. Basler Industriekameras - A600 Series - A601fc. http://www.baslerweb.com/products/A600.html?model=311. Accessed 20 June 2014
9. Nunome, Y., Murakami, K., Kobayashi, K., Naruse, T.: A method to estimate ball's state of spin by image processing for strategic learning in RoboCup Small-Size-robot League. The Japanese Society for Artificial Intelligence. SIG-Challenge-B301-4, pp. 21–25
10. Rojas, R., Simon, M., Tenchio, O.: Parabolic flight reconstruction from multiple images from a single camera in general position. In: Lakemeyer, G., Sklar, E., Sorrenti, D.G., Takahashi, T. (eds.) RoboCup 2006: Robot Soccer World Cup X. LNCS (LNAI), vol. 4434, pp. 183–193. Springer, Heidelberg (2007)

Model-Instance Object Mapping

Joydeep Biswas[✉] and Manuela Veloso

School of Computer Science, Carnegie Mellon University, Pittsburgh, USA
{joydeepb,veloso}@ri.cmu.edu

Abstract. Robot localization and mapping algorithms commonly represent the world as a static map. In reality, human environments consist of many movable objects like doors, chairs and tables. Recognizing that such environment often have a large number of instances of a small number of types of objects, we propose an alternative approach, *Model-Instance Object Mapping* that reasons about the models of objects distinctly from their different instances. Observations classified as short-term features by Episodic non-Markov Localization are clustered to detect object instances. For each object instance, an occupancy grid is constructed, and compared to every other object instance to build a directed similarity graph. Common object models are discovered as strongly connected components of the graph, and their models as well as distribution of instances saved as the final Model-Instance Object Map. By keeping track of the poses of observed instances of object models, Model-Instance Object Maps learn the most probable locations for commonly observed object models. We present results of Model-Instance Object Mapping over the course of a month in our indoor office environment, and highlight the common object models thus learnt in an unsupervised manner.

1 Introduction

Robots deployed in human environments frequently encounter movable objects like tables and chairs. These movable objects pose a challenge for the long-term deployment of robots since they obstruct features of the map, and are difficult to map persistently. Areas where the majority of the observations of the robot are of unmapped objects are especially challenging, making it harder for a robot to accurately estimate its location globally.

There have been a number of long-term Simultaneous Localization and Mapping (SLAM) solutions proposed to tackle this problem by estimating the latest state of a changing world. While such approaches may be successful at tracking the changes of frequently observed environments over time, it may be infeasible for a robot to observe all the changes in all parts of a large deployment environment. In such an approach, if a robot were to infrequently visit a place with

---

This research was partially supported by the National Science Foundation award number NSF IIS-1012733, and by DARPA award number FA8750-12-2-0291. The views in this document are those of the authors only.

R.A.C. Bianchi et al. (Eds.): RoboCup 2014, LNAI 8992, pp. 525–536, 2015.
DOI: 10.1007/978-3-319-18615-3_43

many movable objects, it might fail to register the latest map to the older map due to large differences between them. Another approach is to maintain maps of the different observed states of the environment as local sub-maps. However, the set of possible states of an environment grows exponentially with the number of movable objects due to the exponential number of combinations of different poses of the movable objects.

In this paper, we propose an alternative approach to modelling a changing environment by separating the models of the movable objects from their distribution of poses in the environment. This proposed approach is borne of the realization that movable objects in human environments are often different instances of the same type of object. Furthermore, even though movable objects do move around, they still tend to be situated in roughly the same locations. For example, a work chair will most likely be observed in front of its accompanying work desk, even if the exact location of the chair might change over time. Our proposed approach, *Model-Instance Object Mapping*, models the objects independently of their instances in the environment. Each Model thus represents a unique *type* of object, while Instances represent the different *copies* of it that are observed. Observed short-term features, as classified by Episodic non-Markov Localization [1], are first clustered into separate point clouds for each object instance. An occupancy grid is built for each object instance, and are then compared in pairs to form a directed object similarity graph. Common object models are then detected by looking for strongly connected components in the graph.

2 Related Work

In a changing human environment, one common approach is to extend SLAM to maintain an up-to-date map of the environment when newer observations contradict the older map. Dynamic Pose Graph SLAM [2] is one such approach that extends pose graph SLAM. Dynamic Maps [3] extends SLAM and maintains estimates of the map over several timescales using recency weighted samples of the map at several timescales simultaneously.

An alternative to relying on an up-to-date map is to locally model the variations of a changing environment. The approach of "Temporary Maps" [4] models the effect of temporary objects by performing local SLAM, using the latest global map estimate as an initial estimate for the local map. Saarinen et al. [5] model a dynamic environment as an occupancy grid with associated independent Markov chains for every cell on the grid. Patch Maps [6] represents the different observed states of parts of the map and selects the one most similar to the robot's observation for localization.

There have been a few approaches to detecting movable objects in the environment. Recognizing that many objects in indoor human environments are of similar shapes, the approach of hierarchical object maps [7] assumes certain classes of shapes of objects that are matched to observed unmapped objects. The Robot Object Mapping Algorithm [8] detects moveable objects by detecting differences in the maps built by SLAM at different times. Detection and Tracking of

Moving Objects [9] is an approach that seeks to detect and track moving objects while performing SLAM. Relational Object Maps [10] reasons about spatial relationships between objects in a map. Bootstrap learning for object discovery [11] is similar to the "model" part of our work in that it builds models of unmapped objects, but does not reason about instances. Generalized Approach to Tracking Movable Objects (GATMO) [12] is an approach to tracking the movements of movable objects that is similar to our work in that it models movable objects. Our work however, further tracks every observed instance of the movable objects, thus allowing it to reason about the most likely poses of objects.

In contrast to the related work, our proposed approach, Model-Instance Object Mapping, is novel in its decoupling of the models of objects from the different instances of each model that are observed, and keeps track of every observed instance of the objects, to further reason about the most likely poses of objects.

3 Episodic non-Markov Localization

In a human environment, observations correspond to either immovable objects like walls, movable objects like chairs and tables, and moving objects like humans. Episodic non-Markov Localization (EnML) [1] classifies these different types of observations as Long-Term Features (LTFs), Short-Term Features (STFs) and Dynamic Features (DFs), respectively. LTFs correspond to features from a long-term map, while STFs are related to unmapped movable objects observed at different time steps. The correlations between STF observations across timesteps results in a non-Markovian algorithm. EnML represents the correlations between observations from different timesteps, the static map, and unmapped objects using a Varying Graphical Network (VGN). Figure 1 shows an example instance of a VGN.

EnML further limits the history of observations by partitioning the past observations into "episodes", where the belief of the robot's poses over the latest episode is independent of prior observations given the first pose of the robot in the episode. Episode boundaries commonly occur when the robot leaves one area for another, for example by going through a doorway, or turning a corner.

EnML thus estimates the robot's pose, as well as the pose of the unmapped STFs over each episode. EnML does not maintain a persistent history or database of the STFs that have been observed in the past. However, in a real human environment, the set of movable objects observable by a robot in a given area is finite, and the rest of this paper is devoted to building models of these STFs, and estimating their pose distributions in the world. EnML provides (aside from the poses of the robot) a set S of all the points corresponding to STFs observed in the world during the robot's deployment, and the corresponding poses P of the robot from where the points were observed. Figure 2 shows an example set S of STFs as detected by EnML.

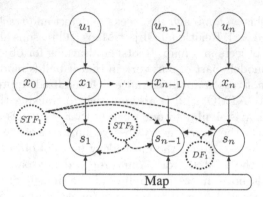

Fig. 1. An example instance of a Varying Graphical Network (VGN) for non-Markov localization. The non-varying nodes and edges are denoted with solid lines, and the varying nodes and edges with dashed lines. Due to the presence of short term features (STFs) and dynamic features (DFs), the structure is no longer periodic in nature. The exact structure of the graph will depend on the STFs and DFs present.

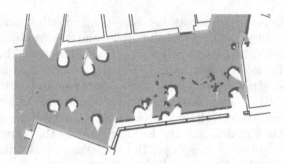

Fig. 2. Step 1 of Model-Instance Object Mapping: Observations S classified as Short-Term Features (purple) by Episodic non-Markov Localization are extracted (Colour figure online).

4 Finding Object Instances

The first step to building models of the objects is to cluster the observed points. The goal of this step is, given S (the set of points in global coordinates corresponding to STFs) and P (the set of poses of the robot from which points in S were observed), to form clusters $C = \{c_i\}_{i=1:n}$ where each cluster c_i is a non-overlapping subset of S, and every point in c_i is within a distance of ϵ of at least one other point in c_i. Here, ϵ is a configurable parameter that determines how close two objects can be, in order to be considered to be part of the same object. Note that each cluster c_i may contain points observed from different poses of the robot, as long as they are spatially separated by at most ϵ from other points in the cluster. We use the point cloud clustering Algorithm from [13] for this step, with a distance threshold of $\epsilon = 3\,\mathrm{cm}$. Figure 3 shows the clusters that were extracted from the example set S shown in Fig. 3.

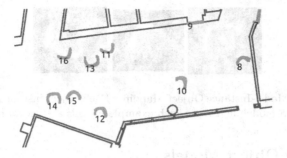

Fig. 3. Step 2 of Model-Instance Object Mapping: Observations S are clustered into set $C = \{c_i\}_{i=1:n}$. Each distinct cluster is denoted by its cluster index. Note that points in S that were sparsely distributed and not within distance ϵ of other points have been discarded.

For each cluster c_i, we build an occupancy grid [14] map m_i from the observed points to model the shape of the object. Algorithm 1 lists the algorithm to build an occupancy grid map m given cluster c, the associated set of poses x, and the width w and height h of the cluster. The grid map m and occupancy counts n are first initialized to zero matrices. For every pixel location l, for every observation that was made at that location, the occupancy value $m(l)$ and observation count $n(l)$ are both incremented. For every pixel location l that is observed to be vacant by observing a point beyond it, the observation count is incremented without incrementing the occupancy value. Finally, the occupancy value at every pixel location is normalized by the observation count for that pixel. Figure 4 shows three example occupancy grid maps constructed from the clusters in Fig. 3.

Algorithm 1. Build Object Model

1: **procedure** BUILDOBJECTMODEL(c, x, w, h)
2:      $m \leftarrow w \times h$ zero matrix
3:      $n \leftarrow w \times h$ zero matrix
4:      **for** each pixel l in m **do**
5:          **for** each point p in c, pose o in x **do**
6:              **if** p is observed at pixel l **then**
7:                  $m(l) \leftarrow m(l) + 1$
8:                  $n(l) \leftarrow n(l) + 1$
9:              **else if** l is between p and o **then**
10:                 $n(l) \leftarrow n(l) + 1$
11:             **end if**
12:         **end for**
13:     **end for**
14:     **for** each pixel l in m **do**
15:         $m(l) \leftarrow m(l)/n(l)$
16:     **end for**
17:     **return** m
18: **end procedure**

Fig. 4. Step 3 of Model-Instance Object Mapping: Each object instance is used to build an occupancy grid model, and three such example models are shown here.

5   Building Object Models

In human environments, and in particular in office environments, movable objects frequently occur as multiple instances of the same model. For example, in a common study area, there may be many chairs, but there will likely be either a single, or a small number of *types* of chairs. Hence, given the occupancy grid maps m_i for each object instance observed, there may be multiple instances of the same model of object, and the subject of the next section is on how to find these common object models.

As a robot encounters different instances of the same object model, it is likely to observe them at different locations and rotations. Furthermore, due to the presence of other objects, they may be partially occluded. Therefore, before comparing the models of different instances, we first align pairs of objects. For a given relative transform $T = \{\delta_x, \delta_y, \delta_\theta\}$ between the centroids of two object instances m_i and m_j, we define an alignment objective function F_{Align},

$$F_{\text{Align}}(m_i, m_j, T) = \sum_{l \in m_i} m_i(l) m_j(T \times l) \tag{1}$$

computed over all pixel locations l in m_i. Given this alignment object function, the optimal alignment T_{ij}^* between object instances m_i and m_j is defined as

$$T_{ij}^* = \arg \max_T F_{\text{Align}}(m_i, m_j, T). \tag{2}$$

This optimal alignment is found by exhaustive search over all possible relative transforms within a search window $\delta_x|| < \Delta_t, ||\delta_y|| < \Delta_t, ||\delta_\theta|| < \Delta_\theta$.

Given an optimal alignment T_{ij}^* between object instances m_i and m_j, we define a similarity metric F_{Similar} to compare the object instances:

$$F_{\text{Similar}}(m_i, m_j, T_{ij}^*) = \frac{\sum_{l \in m_i} I_\epsilon(m_i(l)) I_\epsilon(m_j(T_{ij}^* \times l))}{\sum_{l \in m_i} I_\epsilon(m_i(l))}. \tag{3}$$

Here, $I_\epsilon(\cdot)$ is the thresholded indicator function that returns 1 when the value passed to it is greater than the threshold ϵ. Note that F_{Similar} is not a symmetric metric : $F_{\text{Similar}}(m_i, m_j, T_{ij}^*) \neq F_{\text{Similar}}(m_j, m_i, T_{ji}^*)$. This is because the similarity metric is computed over all pixel locations l in the first object instance m_i, and the two object instances need not necessarily be of the same size. Furthermore, the similarity metric returns values in the range $0 \leq F_{\text{Similar}} \leq 1$.

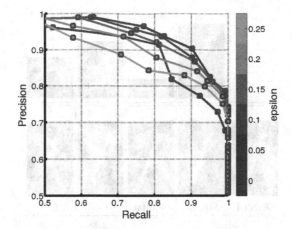

Fig. 5. Precision-recall curves for object similarity classification between pairs drawn from 25 objects, compared to hand-labeled classification.

Given two object instances m_i and m_j, and a similarity threshold Δ_{Similar}, m_i is classified as being similar to m_j if

$$F_{\text{Similar}}(m_i, m_j, T^*_{ij}) \geq \Delta_{\text{Similar}}. \qquad (4)$$

Similarity comparisons between pairs of objects thus depend on two parameters, ϵ and Δ_{Similar}. ϵ serves as an outlier rejector, ignoring low occupancy values in the object instance models, while Δ_{Similar} serves as a confidence threshold.

To determine the values of ϵ and Δ_{Similar} to be used, we hand-labelled similarity comparisons between 25 object instances. Next, for several values of ϵ, we varied Δ_{Similar} to evaluate the precision and recall of the similarity classification. Figure 5 shows the precision-recall curves for the different values of ϵ. The values $\epsilon = 0.05$ and $\Delta_{\text{Similar}} = 0.7$ yield a precision as well as recall of 90 %, and we use these values for the subsequent sections.

Figure 6 shows sample results of alignment and similarity comparison between four objects. Each object is classified as being similar to itself as expected, while different instances of the same model (instances 1 and 2, and instances 3 and 4) are correctly aligned and classified as being similar. Objects 1 and 2 were doors, and objects 3 and 4 were two instances of the same type of chair observed by the robot. Note that object instances 1 and 2, as shown by their occupancy grids, were observed in different poses, but still aligned correctly with respect to each other.

The similarity metric F_{Similar} is used to build a directed similarity graph $G_{\text{Similar}} = \langle V, E \rangle$ where vertices $v_i \in V$ denote the object instances the corresponding object instances m_i, and directed edges $e_k \in E$ indicate similarities between object instances. The presence of an edge e_k: $v_i \to v_j$ denotes that object instance m_i, corresponding to vertex v_i, is similar to object v_j, as evaluated by Eq. 4. Given this directed similarity graph G, we find sets of object

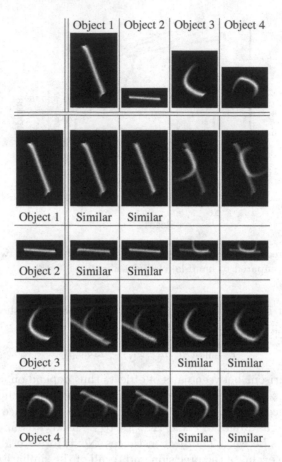

Fig. 6. Alignment and similarity comparison grid between four object instances. The top row and left-most column show the occupancy grids of the object instances. In each cell, the instance corresponding to the row is drawn in red, and the optimal alignment $T*$ of the instance corresponding to the column is drawn in green. Similarity matches are labelled in each cell (Colour figure online).

instances corresponding to the same object model by computing the strongly connected components of G. Let $S = \{s_i\}_{i=1:N_S}$ be the set of N_s strongly connected components extracted from graph G. Each strongly connected component s_i thus corresponds to a common model that is shared by the object instances represented by the vertices in s_i. Figure 7 shows an example similarity graph constructed by comparing the observed object instances from Fig. 3.

6   Updating the Model-Instance Map Across Deployments

So far we have focussed on how common models, detected as strongly connected components s_i, are extracted from the observations of STFs during a single

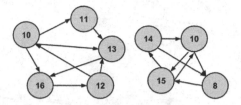

Fig. 7. Step 4 of Model-Instance Object Mapping: Every object instance is compared to every other object instance, and a similarity graph is constructed by connecting nodes (representing object instances) by directed edges to indicate similarity between them. Note that in this example, there are two strongly connected components, and hence the Model-Instance Object Mapping algorithm will find two common object models.

deployment log of the robot. However, as a lifelong learning algorithm, Model-Instance Object Mapping needs to reason about the persistence of object models and their instances across *all* deployments of the robot. In this section we present how models s_i from the latest deployment of the robot are merged with older persistent models p_j from all past deployments of the robot.

Persistent models p_j, just like models s_i, include a common occupancy grid model of the object, and a list of poses of the observed instances of the object. For the first deployment of the robot, the list of persistent objects is empty, and thus all models s_i are saved as unique persistent objects. For every subsequent deployment, every model s_i is aligned (Eq. 2) and compared to every persistent model p_j using the similarity metric F_{Similar} (Eq. 3). For each object s_i, every persistent object p_j that satisfies the similarity test of Eq. 4 bidirectionally is declared to be of the same model. All such sets of objects that are declared to be of the same model are then merged. Note that this merging step also attempts to merge persistent objects that were previously distinct, but which are determined to be bidirectionally similar to a newly observed object. This procedure for updating and merging persistent objects preserves the property that all instances of that object form a strongly connected component. Figure 8 shows the pose instances of an example persistent object, as extracted from the similarity graph of Fig. 7, which was constructed by comparing the object instances of Fig. 3.

7 Results

We have been deploying our robot, CoBot in an actual office environment since 2011, and have gathered extensive logs of the robot's deployments [15]. We processed a subset of the logs thus gathered between the January and April of 2014, in one of the floors in our building, through the Object Instance Mapping algorithm. There were 133 such deployments in total, over the course of which the robot traversed a total distance of more than 55 km. We present here the resultant Model-Instance map thus learnt.

There were a total of 2526 objects observed, out of which there were 221 unique object models discovered. Out of the 221 unique object models, 50 had

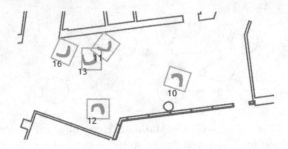

Fig. 8. Step 5 of Model-Instance Object Mapping: Poses for every instance of one example persistent objects are visualized here by duplicating the model of the persistent object at the poses of the instances on the map.

Table 1. Frequency counts of the unique object models discovered. The first row lists the range of the number of instances, and the second row lists the number of objects that were discovered within that range of instances.

Num Instances	1	2-5	6-10	11-20	21-40	41-80	81-160	161-1440
Num Objects	171	25	11	2	5	4	1	2

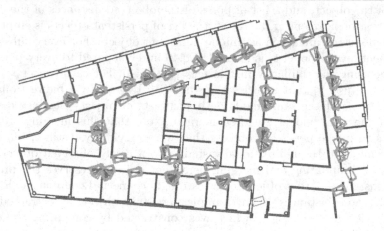

Fig. 9. The most commonly observed object, doors, as discovered by Model-Instance Object Mapping. Our algorithm correctly detected all the doors in the environment without any supervision.

2 or more observed instances. Table 1 lists the distribution of the frequency of the unique object models. The most commonly observed object model is that of doors, and the algorithm detected 1440 instances of them. Figure 9 shows the poses of the doors that were detected by Model-Instance Object Mapping without any supervision. The algorithm correctly detected all the doors in the environment, and also catalogued all the observed poses of the doors observed. From the figure, it is easy to discern the opening edges of the doors. Figure 10 shows some of the most commonly observed object models and their instances.

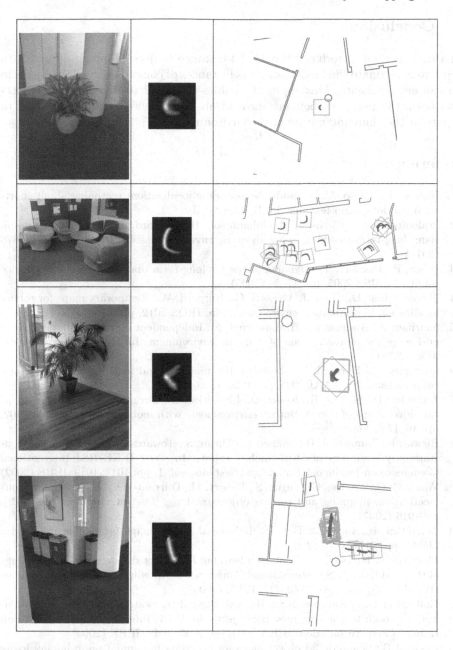

Fig. 10 Some examples of objects discovered by Model-Instance Object Mapping: (from top to bottom) a round planter, a commonly found chair, a square planter, and trashcans in the kitchen area. The first column shows a photograph of the object, the second column shows the occupancy grid model built, and the third column shows the instances that were observed for each object model, across all the deployments. Note that instances of these objects were detected at multiple locations, but only one location is shown for each object here.

8   Conclusion

In this paper we introduced the Model-Instance Object Mapping approach to separately estimate the model as well as instances of common recurring objects in human environments. This approach enables the robot to find common objects and track the poses of their instances in an unsupervised manner in order to represent the changing nature of the environment.

References

1. Biswas, J., Veloso, M.: Episodic Non-Markov localization: reasoning about short-term and long-term features. In: ICRA (2014)
2. Walcott-Bryant, A., Kaess, M., Johannsson, H., Leonard, J.: Dynamic pose graph slam: long-term mapping in low dynamic environments. In: IROS, pp. 1871–1878 (2012)
3. Biber, P., Duckett, T.: Dynamic maps for long-term operation of mobile service robots. In: RSS 2005, pp. 17–24 (2005)
4. Meyer-Delius, D., Hess, J., Grisetti, G., Burgard, W.: Temporary maps for robust localization in semi-static environments. In: IROS 2012, pp. 5750–5755 (2012)
5. Saarinen, J., Andreasson, H., Lilienthal, A.: Independent markov chain occupancy grid maps for representation of dynamic environment. In: IROS 2012, pp. 3489–3495 (2012)
6. Stachniss, C., Burgard, W.: Mobile robot mapping and localization in non-static environments. In: AAAI 2005, pp. 1324–1329 (2005)
7. Anguelov, D., Biswas, R., Koller, D., Limketkai, B., Thrun, S.: Learning hierarchical object maps of non-stationary environments with mobile robots. In: UAI 2002, pp. 10–17 (2002)
8. Biswas, R., Limketkai, B., Sanner, S., Thrun, S.: Towards object mapping in non-stationary environments with mobile robots. In: 2002 IEEE/RSJ International Conference on Intelligent Robots and Systems, vol. 1, pp. 1014–1019. IEEE (2002)
9. Wang, C.-C., Thorpe, C., Thrun, S., Hebert, M., Durrant-Whyte, H.: Simultaneous localization, mapping and moving object tracking. The Int. J. Rob. Res. **26**(9), 889–916 (2007)
10. Limketkai, B., Liao, L., Fox, D.: Relational object maps for mobile robots. In: IJCAI, pp. 1471–1476 (2005)
11. Modayil, J., Kuipers, B.: Bootstrap learning for object discovery. In: Proceedings of the 2004 IEEE/RSJ International Conference on Intelligent Robots and Systems (IROS 2004), vol. 1, pp. 742–747. IEEE (2004)
12. Gallagher, G., Srinivasa, S.S., Bagnell, J.A., Ferguson, D.: Gatmo: a generalized approach to tracking movable objects. In: IEEE International Conference on Robotics and Automation, ICRA 2009, pp. 2043–2048. IEEE (2009)
13. Rusu, R.B.: Semantic 3d object maps for everyday manipulation in human living environments. KI-Künstliche Intelligenz **24**(4), 345–348 (2010)
14. Elfes, A.: Using occupancy grids for mobile robot perception and navigation. Computer **22**(6), 46–57 (1989)
15. Biswas, J., Veloso, M.M.: Localization and navigation of the cobots over long-term deployments. The Int. J. Rob. Res. **32**(14), 1679–1694 (2013)

Detection of Aerial Balls Using a Kinect Sensor

Paulo Dias, João Silva, Rafael Castro, and António J.R. Neves$^{(\boxtimes)}$

IRIS Group, DETI / IEETA, University of Aveiro,
3810–193 Aveiro, Portugal
{paulo.dias,joao.m.silva,rafaelcastro,an}@ua.pt

Abstract. Detection of objects in the air is a difficult problem to tackle given the dynamics and speed of a flying object. The problem is even more difficult when considering a non-controlled environment where the predominance of a given color is not guaranteed, and/or when the vision system is located on a moving platform. As an example, most of the Middle Size League teams in RoboCup competition detect the objects in the environment using an omni directional camera that only detects the ball when in the ground, and losing any precise information of the ball position when in the air. In this paper we present a first approach towards the detection of a ball flying using a Kinect camera as sensor. The approach only uses 3D data and does not consider, at this time, any additional intensity information. The objective at this stage is to evaluate how useful is the use of 3D information in the Middle Size League context. A simple algorithm to detect a flying ball and evaluate its trajectory was implemented and preliminary results are presented.

1 Introduction

To our knowledge, most of the RoboCup teams of the Middle Size League (MSL) have limited vision systems regarding the detection of the ball when in it is in flight. Most teams use an omni directional camera on top of the robots that only detects correctly ball position when on the ground since they use projective geometry and a single camera.

Given that most of the robots shoot the ball through the air, the possibility to detect the ball when in flight is very relevant. Obvious solutions using more than a single camera (either using two additional cameras to provide stereo vision, or combining the information from the omni directional camera with additional cameras) can be considered as in [1–3]. However they present some limitations: first these additional cameras may point outside the field and cannot use background or color information to simplify ball segmentation (since a flying ball might be in the air with any possible backgrounds - tribunes, chairs, spectators, etc.) or might have limited field of view (most omni directional camera point downwards, meaning a maximum height of around 60 cm).

In a previous work [4], we developed algorithms based on color and shape detection using a single perspective camera. In this work, the above mentioned problems were detected. The work presented in this paper uses a different vision approach based on a depth sensor instead of an intensity sensor. This work

R.A.C. Bianchi et al. (Eds.): RoboCup 2014, LNAI 8992, pp. 537–548, 2015.
DOI: 10.1007/978-3-319-18615-3_44

presents several similarities with the work of Khandelwal et al. [5] that uses a Kinect sensors as a low cost ground truth detection system. As 3D sensor, a Kinect was chosen given its low price (making it possible to use in an aggressive environment such as RoboCup), its ability to directly provide 3D depth information and its refresh rate of 30 fps similar to the RGB camera used in the onmi direction vision system of the robots [6].

2   3D Data

Kinect is a motion sensing input device developed by Microsoft and launched on November 2010. The sensor includes an RGB camera, an infrared laser projector, a monochrome CMOS sensor, and other components less relevant for our application. The field of view is 57 degrees horizontally, and around 43 degrees vertically. Acquisition of the 3D data from the Kinect is done using a C++ wrapper for libfreenect that transforms depth images into an OpenCV matrix. For an initial stage of the application, ROS was used to access directly the 3D cloud of points. However, the time needed for processing and transmitting the 3D data was large and it was verified that direct access to the 2D depth data was faster. Besides, the CAMBADA team code structure is not ROS based, which would ultimately become an issue. The transformation from raw data to metric is done using a formula found in an online manual [7].

In its original configuration, Kinect normal working range is from 0.8 m to 4 m. However, the sensor provides distance measurements for longer distances while suffering from additional errors and loss of precision which causes a discretization effect to appears with distance increase. Figure 1 presents views of a typical 3D cloud of points with texture as provided by a Kinect. Two of the main limitations of the Kinect for this kind of application are clearly visible in this figure. For larger distances a discretization of space occurs turning the cloud of points into several parallel planes. Also, given the speed of the ball a trail of points appears around the flying objects as shadow points that do not exist and results from averaging between points in the object and in the rear wall. This effect is more significant for higher speeds of the objects and is well known in 3D data when jump edges occur resulting in a mixed distance between the foreground and the background object.

3   Ball Detection Algorithm

Despite the limitations inherent from the discretization of the space, more visible at higher distances, the use of 3D information still seems to be a good option.

3.1   Flying Objects

A first approach to ball detection using the Kinect cloud of points as a source would be to use geometry, for example fitting half spheres to the data and

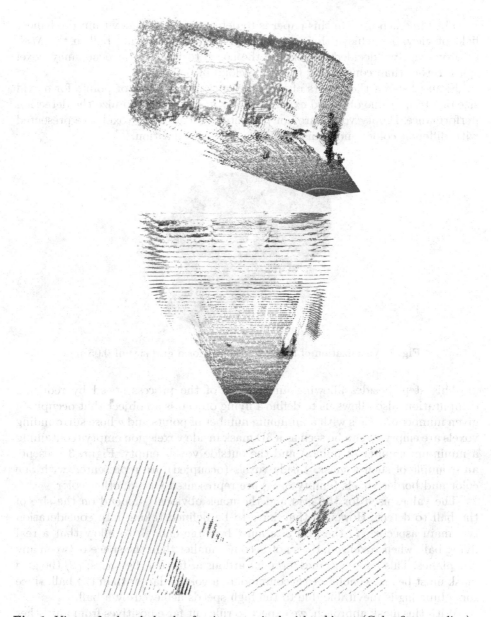

Fig. 1. Views of colored clouds of points acquired with a kinect (Color figure online).

trying to find areas of interest. Preliminary trials have been performed with this approach but, given the discretization effect, this fitting only appeared to provide reliable data when close to the sensor (within the typical Kinect working range, less than 4 m). For longer distance the spherical shape of the ball becomes difficult to detect.

The approach used in this paper is to detect flying objects within the Kinect field of view. To achieve this, given the properties of a flying ball in the MSL environment, we decided to voxelize the space to work in an occupancy voxel space rather than considering the while cloud of points.

Figure 2 shows the results of the voxelization of a cloud of points for a grid size of 0.05 m, value obtained experimentally in order to maximize the detection performance. Empty voxels are not presented and adjacent voxels are presented with different colors and transparency to ease visualization.

Fig. 2. Voxelization of a cloud of points for a grid size of 0.05 m.

This step, besides allowing an increase of the process speed by reducing computation, also allows us to define a flying object as an object that occupies a given number of voxels with a minimum number of points and whose surrounding voxels are empty. It can be seen as a 3D mask inside voxels non empty (containing a minimum number of points) and the outside voxels empty. Figure 3 presents an example of such a mask. Inside voxels (occupied) are represented with one color and border voxels (empty ones) are represented with another color.

The values used for the grid and the mask obviously depend on the size of the ball to detect. However, they have to be defined taking into consideration two main aspects: *(1)* the grid size must be large enough to allow that a real flying ball, when voxelized, does not become smaller than the space between any two planes. This issue becomes more hazardous at farther distances; *(2)* the grid mask must be large enough to accommodate a volume larger than the ball, since some blurring is inevitable due to the high speeds achieved by a ball.

With this mask approach, we expect to rule out false positives from any other object on the field of play, since all other artifacts during a game can only be a robot or a human. Since all of them have a clear "connection" with the ground, the mask will not allow a valid detection.

Also, any object that could effectively be identified by the mask as a ball, at this point, could be outside the field of play. Since this is a complementary vision system for our robots and since they know their position inside the field of play, further integration steps will be responsible for handling these possible false positives.

Fig. 3. Example of 3D mask used for flying object identification (Color figure online).

In our preliminary tests in a field with limited range and a wall at 7 m, we empirically set the following values for the grid and mask sizes: the grid size of the occupancy grid is 0.27 m, and the mask size (corresponding to the outside of the mask) is 5. As it is, and with our test scenario, the resulting mask size is around 1.35 m meaning that a flying object must be at least 0.27 m away from any other structure to be detected. These values were used to avoid wrong detection related to objects close to the wall at 7 m, where the discretization effect of the Kinect is significant. In Fig. 4 we present the ball correctly detected in three consecutive points corresponding to a kick away from the sensor.

3.2 Ground Objects

The algorithm presented in the previous section is only suitable to detect flying balls, but it can be easily adapted to detect ground balls by ignoring the bottom part of the 3D mask for balls lying on the ground. To apply this idea it is necessary to know where the ground is and apply a different 3D mask for voxels lying above the ground.

To allow for this additional detection, we align the grid with the ground by introducing an additional step in the first processed image. For the first acquired image, the ground plane is detected using Random Sample Consensus Algorithm (RANSAC) already available in the PCL package [8]. Given the transformation between the plane and the original coordinate reference, we compute the Rigid Body Transform between the original coordinate frame and the ground plane and apply this transformation to every point cloud ensuring that the XZ axis will be aligned with the ground plane. Computing the voxelization in this new coordinate frame ensure that the grid cells corresponding to $y = 0$ are containing the ground plane (Fig. 5).

For optimizing the process and since usually the height of the sensor will not change during acquisition, the rigid body transform between the original

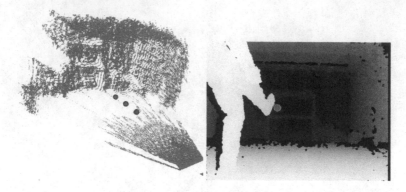

Fig. 4. The ball correctly detected in three different moments during a kick away from the sensor.

Fig. 5. 3D cloud of points in original Kinect coordinates (light blue), detected ground plane (red square), 3D cloud in new coordinate system, with ground plane aligned with xOz plane (Dark) (Color figure online).

coordinate frame and the ground is computed only once in the first image (since the RANSAC plane fitting is quite heavy) but any following point cloud is transformed to align the ground plane with the XY coordinate plane. The final algorithm will be exactly the same but will use a given mask for flying objects when $y > 1$ and the ground mask when $y = 1$.

Figure 6 shows the results of ground ball detection in 4 consecutive images corresponding to a kick away from the user with the ball on the field.

The general algorithm to detect ground and flying ball is the following:

Compute Ground Plane from first image
Compute transform to align ground with XZ plane
for *Each point cloud* **do**
    Align point cloud with XZ plane
    Voxelization of grid
    Detection of flying ball $y > 1$ *(use flying object mask)*
    if *no flying ball* **then**

Detect ground ball y=1 (use ground mask mask that ignores bottom part of flying mask)
  end if
end for

Fig. 6. Example of detection of ground ball in 4 consecutive 3D images.

4 Trajectory Estimation

Besides detecting the ball in the environment, it is important for the robot to estimate the ball trajectory in order to predict the best action to take. In robotic soccer, the algorithm for trajectory estimation presented in this section is useful so that the goalkeeper can move to a position in order to prevent a goal.

For flying objects, and considering that air resistance is negligible, the trajectory can be approximate by a simple ballistic trajectory. To perform this evaluation, we keep trace of the last ball positions. Currently, 10 previous position are kept since it is enough for most of the flying movements detected.

The trajectory estimation algorithm computes the 2D vector (x,y) between the actual ball and the initial point of the trajectory (the first ball detected that supports the actual trajectory). It then used this vector norms to estimate the point on the trajectory (according to the current estimated ballistic) at the same distance. If the Euclidean between the original ball position and the point on the trajectory is below a threshold (empirically set to 0.25 m in the current experiments) the position is considered as supporting the current trajectory.

Given a number of points supporting the trajectory, Singular Value Decomposition (SVD) is used to compute the parabolic equation that best fits the whole supporting coordinates. The algorithm used is the one implemented in the Eigen Library [9].

The general algorithm for trajectory estimation is as follow.

Update history
if *ball detected* **then**
  if *new position support previous trajectory* **then**
   Compute trajectory with all points

```
    else
        Compute new trajectory with last 3 points
    end if
    if If trajectory error below threshold then
        Update new trajectory
    end if
end if
if NOT(last 2 positions exist and support trajectory) then
    reset trajectory
end if
```

Figure 7 shows the results of the trajectory estimation for a flying ball. The history of balls used for the computation is presented as large red spheres and the parabolic trajectory estimated is represented by the small blue spheres.

With the trajectory estimated, the agent can use the current projection of the ball position on the ground and even the predicted touchdown point of the ball.

5   Kinect Position Calibration on the Robot

Since the objective is to use the 3D position detection and trajectory algorithm on board of a robotic soccer robot, calibration of the position of the Kinect relative to its position on a robot must be performed. The coordinates of the detected ball on the field coordinate frame can then be obtained easily by applying the transformation from Kinect to the robot and then from robot to world position (this transformation comes from robot positioning routines). To allow flexibility and avoid any calibration based on measurement, a calibration program was developed. The calibration program (see Fig. 8) acquires on demand an image from Kinect. It then allows to automatically detect a ball on the scene using the ground detection algorithm. Alternatively, the user can also pick the center of the ball directly in the 3D cloud of points. The process must be repeated for at least three ball positions.

Fig. 7. Example of trajectory estimation: small blue spheres represent the trajectory computed from the balls in the history (large red spheres) (Color figure online).

In the left table, the coordinate of the detected/selected points in the robot coordinates must be provided. Then the software evaluate the rigid body transform between the 2 coordinates system corresponding to the Kinect position and orientation relatively to the robot origin. This matrix is saved and must be read by the 3D detection program for a given configuration of the Kinect. A simple calibration process consist in positioning the robot in the center/origin of the field, and then acquire 3D images with the ball on well-defined landmarks of the field. A typical configuration of the calibration procedure is presented in Fig. 9.

6   Experimental Results

The work presented in this paper is still in an early stage of development. However, the first results obtained seems very promising. At this stage, the validation of

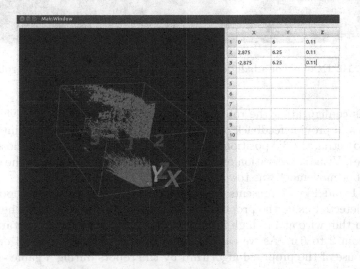

Fig. 8. Application for kinect position calibration with 3 reference points on the kinect cloud and the corresponding coordinate in robot coordinates.

Fig. 9. Example of kinect calibration set-up with a ball on a well known position.

the algorithm was performed by observation of the results using visual tools (see Figs. 4, 6 and 7). Based on these visual observations, the algorithm seems quite robust in open field (with no obstacles) and processing times are below 30 ms meaning that all the Kinect 30 images per second are processed. Some tests were performed with the Kinect on the top of a real robot. Figure 10 shows an example of the ball detection from the point of the view of the robot.

Fig. 10. Example of a ball detection from the point of view from a robot.

In this configuration the robot evaluated the ball position on several flying ball tests and provide feedback on the base station of the estimated intersection point (projecting the 3D position on the ground and computing the closest point to the line). Visual observation on the base station was according to the expected as the robot movement was toward the correct direction.

Table 1 and Fig. 11 presents experimental results regarding the position of the ball detected using the proposed algorithm. In this experiment, the ball was fixed by a thin wire at 1 m high in the front of the robot. Several distances were tested, from 2 to 6 m. As we can see, the error in position is small enough to allow the use of the proposed algorithm by the robots during a game.

Table 1. Ball position evaluation regarding the proposed vision system. The ball was fixed by a thin wire at 1 m high in the front of the robot ($x = 0$) at several distances (y).

Distance x/y (cm)	Average x/y (cm)	Std deviation x/y (cm)
0/200	-0.08/2.01	0.004/0.002
0/300	-0.11/2.96	0.007/0.029
0/400	-0.07/4.01	0.005/0.001
0/500	-0.06/5.00	0.169/0.718
0/600	0.01/5.97	0.002/0.003

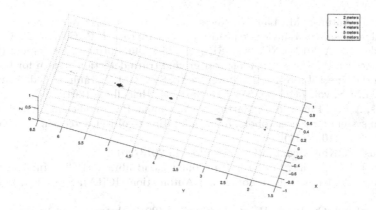

Fig. 11. Ball position detected by the proposed vision system with the kinect on the robot. The ball was fixed by a thin wire at 1 m high in the front of the robot ($x = 0$) at several distances (y).

7 Conclusion and Future Work

In this paper we present preliminary work toward the detection of 3D flying balls using only the depth information provided by a Kinect. The final objective is to use the algorithms presented here to extend the current vision of our MSL robotic agent to cope with 3D object positions. An algorithm based on an occupancy grid was developed. Preliminary results are encouraging since the algorithm can process the Kinect data in real time (processing at 30 fps) and the visual inspection of the detection is quite convincing.

Regarding future work, the first step is the fully integration of the Kinect on the platform to allow real experiments during a game, in the presence of other objects than the ball. Ball validation using the RGB image is also under development, in order to validate any false positive that can occur.

Aknowledgements. This work was developed in the Institute of Electronic and Informatic Engineering of University of Aveiro and was partially supported by FEDER through the Operational Program Competitiveness Factors - COMPETE and by National Funds through FCT - Foundation for Science and Technology in a context of a project FCOMP-01-0124-FEDER-022682 (FCT reference PEst-C/EEI/UI0127/2011).

References

1. Ahmad, A., Lima, P.: Multi-robot cooperative spherical-object tracking in 3D space based on particle filters. Robot. Auton. Syst. **61**(10), 1084–1093 (2013)
2. Voigtlnder, A., Lange, S., Lauer, M., Riedmiller, M.: Real-time 3D ball recognition using perspective and catadioptric cameras (2007)
3. Scaramuzza, D., Pagnottelli, S., Valigi, P.: Ball detection and predictive ball following based on a stereoscopic vision system. In: IEEE International Conference on Robotics and Automation, pp. 1561–1566 (2005)

4. Silva, J., Antunes, M., Lau, N., Neves, A.J.R., Lopes, L.S.: Aerial ball perception based on the use of a single perspective camera. In: Reis, L.P., Correia, L., Cascalho, J. (eds.) EPIA 2013. LNCS, vol. 8154, pp. 235–246. Springer, Heidelberg (2013)
5. Khandelwal, P., Stone, P.: A low cost ground truth detection system for RoboCup using the kinect. In: Röfer, T., Mayer, N.M., Savage, J., Saranlı, U. (eds.) RoboCup 2011. LNCS, vol. 7416, pp. 515–527. Springer, Heidelberg (2012)
6. Neves, A.J.R., Pinho, A.J., Martins, D.A., Cunha, B.: An efficient omnidirectional vision system for soccer robots: from calibration to object detection. Mechatronics **21**(2), 399–410 (2011)
7. Burrus, N.: Kinect calibration, consulted in 2013/2014
8. Rusu, R.B., Cousins, S.: 3D is here: point cloud library (PCL). In: IEEE International Conference on Robotics and Automation (ICRA), Shanghai, 9–13 May 2011
9. Guennebaud, G., Jacob, B., et al.: Eigen v3 (2010). http://eigen.tuxfamily.org

Ball Dribbling for Humanoid Biped Robots:
A Reinforcement Learning
and Fuzzy Control Approach

Leonardo Leottau$^{(\boxtimes)}$, Carlos Celemin, and Javier Ruiz-del-Solar

Department of Electrical Engineering and Advanced Mining Technology Center,
Universidad de Chile, Santiago, Chile
{dleottau, carlos.celemin, jruizd}@ing.uchile.cl

Abstract. In the context of the humanoid robotics soccer, ball dribbling is a complex and challenging behavior that requires a proper interaction of the robot with the ball and the floor. We propose a methodology for modeling this behavior by splitting it in two sub problems: alignment and ball pushing. Alignment is achieved using a fuzzy controller in conjunction with an automatic foot selector. Ball-pushing is achieved using a reinforcement-learning based controller, which learns how to keep the robot near the ball, while controlling its speed when approaching and pushing the ball. Four different models for the reinforcement learning of the ball-pushing behavior are proposed and compared. The entire dribbling engine is tested using a 3D simulator and real NAO robots. Performance indices for evaluating the dribbling speed and ball-control are defined and measured. The obtained results validate the usefulness of the proposed methodology, showing asymptotic convergence in around fifty training episodes, and similar performance between simulated and real robots.

Keywords: Reinforcement learning · TSK fuzzy controller · Soccer robotics · Biped robot · NAO · Behavior · Dribbling

1 Introduction

In the context of soccer robotics, ball dribbling is a complex behavior where a robot player attempts to maneuver the ball in a very controlled way, while moving towards a desired target. In case of humanoid biped robots, the complexity of this task is very high, because it must take into account the physical interaction between the ball, the robot's feet, and the ground, which is highly dynamic, non-linear, and influenced by several sources of uncertainty.

Very few works have addressed the dribbling behavior with biped humanoid robots; [1] presents an approach to incorporate the ball dribbling as part of a closed loop gait, combining a footstep and foot trajectory planners for integrating kicks in the walking engine. Since this work is more focused to the theoretical models and controllers of the gait, there is not included a dribbling engine final performance evaluation. On the other hand, [2] presents an approach that uses imitative reinforcement learning for dribbling the ball from different positions into the empty goal, meanwhile [3] proposes an approach that uses corrective human demonstration for augmenting a

© Springer International Publishing Switzerland 2015
R.A.C. Bianchi et al. (Eds.): RoboCup 2014, LNAI 8992, pp. 549–561, 2015.
DOI: 10.1007/978-3-319-18615-3_45

hand-coded ball dribbling task performed against stationary defender robots. Since these two works are not addressing explicitly the dribbling behavior, not many details about the specific dribbling modeling, or performance evaluations for the ball-control or accuracy to the desired target are mentioned. Some teams that compete in humanoid soccer leagues, such as [4, 5], have implemented successful dribbling behaviors, but to the best of our knowledge, no publications directly related about their dribbling's methods have been reported for comparison. It is not clear whether in these cases a hand-coded or a learning-based approach has been used.

The dribbling problem has been addressed more extensively for the wheeled robots case, approaches based on the use of automatic control and Machine Learning (ML) has been proposed; [6, 7] apply Reinforcement Learning (RL), [8, 9] use neural networks and evolutionary computation, [10] applies a PD control with linearized kinematic models, [11] uses non-linear predictive control, and [12–14] apply heuristic methods. However, these approaches are not directly applicable to the biped humanoid case, due to its much higher complexity.

Although several strategies can be used to tackle the dribbling problem, we classify these in three main groups: (i) based on human experience and/or hand-code [2, 3], (ii) based on identification of the system dynamics and/or kinematics and mathematical models [1, 10, 11], and (iii) based on the on-line learning of the system dynamics [6–9]. In order to develop the dribbling behavior, each of these alternatives has advantages and disadvantages: (i) is initially faster to implement but vulnerable to errors and difficult to debug and re-tune when parameters change or while the system complexity increases; (ii) could be solved completely off-line by analytical or heuristic methods since robot and ball kinematics are known, but to identify the interaction between the robot's foot while it is walking, with a dynamic ball and the floor, could be anfractuous; in this way those strategies from (iii) which are capable to learn about that robot-ball-floor interaction, while find an optimal policy for the ball-pushing behavior, as RL, is a promise and attractive approach.

The main goal of this paper is to propose a methodology to learn the ball-dribbling behavior in biped humanoid robots, reducing as many as possible the on-line training time. In this way, the aforementioned alternatives (ii) are considered for reducing the complexity of behaviors learned with (iii). The proposed methodology models the ball-dribbling problem by splitting it in two sub problems, *alignment* and *ball-pushing*. The *alignment* problem consists of controlling the pose of the robot in order to obtain a proper alignment with the final ball's target. The *ball-pushing* problem consist of controlling the robot's speed in order to obtain, at the same time, a high speed of the ball but a low relative distance between the ball and the robot, that means controllability and efficiency. These ideas are implemented by three modules: (i) a fuzzy logic controller (FLC) for aligning the robot when approaching the ball (off-line designed), (ii) a foot-selector, and (iii) a reinforcement-learning (RL) based controller for controlling the robot's speed when approaching and pushing the ball (on-line learned).

Performance indices for evaluating the dribbling's speed and ball-control are measured. In the experiments the training is performed using a 3D simulator, but the validation is done using real NAO robots. The article is organized as follows: Sect. 2 describes the proposed methodology. Section 3 presents the experimental setup and obtained results. Finally, conclusions and future work are drawn in Sect. 4.

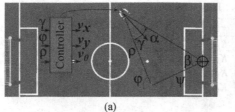
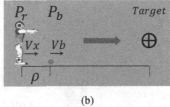

(a)   (b)

Fig. 1. Variables definition for: (a) the full dribbling modeling, (b) the *ball-pushing* behavior reduced to a 1-Dimensional problem.

2   A Methodology for Learning the Ball-Dribbling Behavior

2.1   Proposed Modeling

As mentioned in the former section, the proposed methodology splits the dribbling problem in two different behaviors: *alignment* and *ball-pushing*. Under this modeling, *ball-pushing* is treated as a one dimensional (1D) problem due to the ball must be pushed over the ball-target line when the robot is aligned; the *alignment* behavior is responsible to enforce this assumption, correcting every time the robot desired direction of movement.

The description of the defined behaviors will use the following variables: v_x, v_y, v_θ, the robot's linear and angular speeds; α, the robot-target angle; γ, the robot-ball angle; ρ, the robot-ball distance; ψ, the ball-target distance; β, the robot-target-ball angle; and, φ, the robot-ball-target complementary angle. These variables are shown in Fig. 1(a). where the desired target ($\oplus$) is located in the middle of the opponent goal, and with x axis pointing always forwards, measured in a robot's centered reference system. The behaviors are described as follows:

i. *Alignment:* in order to maintain the 1D assumption, it is proposed to implement a FLC which keeps the robot aligned to the ball-target line ($\varphi = 0$, $\gamma - 0$) while approaching the ball. The control actions of this subsystem are applied all the time over v_θ and v_y, and partially applied over v_x, only when the constraints of the 1D assumption are not fulfilled. Also, this behavior uses the foot selector for setting the foot that pushes the ball, in order to improve the ball's direction. Due to the nature of this sub-behavior, kinematics for the robot and ball can be modeled individually. Thus, we propose the off-line design and tuning of this task.

ii. *Ball-pushing:* following the 1D assumption, the objective is that the robot walks as fast as possible and hits the ball in order to change its speed, but without losing the ball possession. That means that the ball must be kept near the robot. The modeling of the robot's feet–ball–floor dynamics is complex and inaccurate because kicking the ball could generate several unexpected transitions, due to uncertainty on the foot shape and speed when it kicks the ball (note that the foot's speed is different to the robot's speed v_x). Therefore it is proposed to model this behavior as a Markov Decision Process (MDP), in order to solve and to on-line learn it using a RL scheme. The behavior is applied only when the constraints of 1D assumption are fulfilled, i.e. when the robot's alignment is achieved. Figure 1(b) shows the variables used in this behavior.

Table 1. The v_x rule base.

$\gamma\backslash\varphi$	$-H$	$-L$	$+L$	$+H$
$-H$	$K_{Lx\rho}$	$K_{Lx\rho}$	$K_{Lx\rho}$	$K_{Hx\rho}$
$-L$	$K_{Hx\rho}$	$K_{Hx\rho}$	$K_{Hx\rho}$	$K_{Hx\rho}$
$+L$	$K_{Lx\rho}$	$K_{Hx\rho}$	$K_{Hx\rho}$	$K_{Hx\rho}$
$+H$	$K_{Lx\rho}$	$K_{Lx\rho}$	$K_{Lx\rho}$	$K_{Lx\rho}$

2.2 Alignment Behavior

The *alignment* behavior is compound of two modules: (a) the FLC which sets the robot speeds for aligning it to the ball-target line; and (b) the foot selector which depending on the ball position and robot pose decides which foot must kick the ball.

(a) Fuzzy Controller. The FLC is inspired in a linear controller that tries to reduce φ and γ angles for being aligned to ball and target, while reduces ρ distance for approaching to the ball:

$$[v_x \quad v_y \quad v_\theta]^T = [(k_{x\rho}\cdot\rho) \quad (k_{y\varphi}\cdot\varphi) \quad (k_{\theta\gamma}\cdot\gamma - k_{\theta\varphi}\cdot\varphi)]^T \tag{1}$$

In order to perform better control actions for different operation points, constant gains of the three linear controllers can be replaced by adaptive gains. Thus, three Takagi-Sugeno-Kang Fuzzy Logic Controllers (TSK-FLCs) are proposed, which maintain the same linear controller structure for their polynomial consequents.

This linear controller and its non-linear counterpart based on FLC are proposed and compared in [15], please refers to that work for details about the proposed FLC. Table 1 depicts the rule base for the v_x FLC. Its consequent has only the gain $k_{x\rho}$, however the antecedent has the angle γ and φ. The rules basically set a very low gain if the robot is very misaligned to ball ($|\gamma| \gg 0$); else the robot goes straight and fast.

The v_y FLC's rule base is described in Table 2(a). Based on 1, the control action is proportional to φ. The FLC makes adaptive the gain for φ; e.g., $k_{y\varphi}$ tends to zero where the ball is away, avoiding lateral movements which speed-up the gait.

Table 2. (a) The v_y rule base. (b) The v_θ rule base.

v_y Controller rules	v_θ Controller rules
If φ is Low & ρ is Low, then $k_{y\varphi}$ is Low	If ρ is Low, then $k_{\theta\gamma}$ is High, $k_{\theta\varphi}$ is Low
If φ is Low & ρ is High, then $k_{y\varphi}$ is Zero	If ρ is Med., then $k_{\theta\gamma}$ is Low, $k_{\theta\varphi}$ is High
If φ is High & ρ is High, then $k_{y\varphi}$ is Zero	If ρ is High, then $k_{y\varphi}$ is High, is Low
If φ is High & ρ is Low, then $k_{y\varphi}$ is High	

The v_θ FLC's rule base is described in Table 2(b). The control action is proportional to $(\gamma - \varphi)$, the FLC adapts the gain $k_{\theta\varphi}$, setting it close to zero when ρ is low or high, in that cases the robot is aligned to the ball minimizing γ; but when ρ is medium, the robot tries to approach the ball aligned to the target minimizing φ.

The fuzzy sets' parameters of each TSK-FLC are tuned by using the Differential Evolution (DE) algorithm [16]. It searches for solutions that minimize the fitness function F expressed in 2, whose performance indices is the time (t_f) used by the robot for achieving the ball being aligned to the target.

$$F = \frac{1}{S}\sum_{i=1}^{S} \frac{t_{fi}}{(\rho_i)},\tag{2}$$

where S is the total number of trained scenes with different initial robot and ball positions, whereas ρ_i is the $i-th$ initial Robot-Ball distance. Please refers to [15].

(b) **Foot Selector.** Since the proposed FLC is designed to align the robot with the ball-target line without using footstep planning, it cannot control which foot (right or left) is going to kick the ball. The FLC is designed to get the center of the ball aligned with the midpoint of the robot's footprints. This could generate undesired ball trajectories, because the NAO robot's foot-shape is rounded (see Fig. 2b). This could be improved if the ball is hit with the front side of the foot, therefore the 1D assumption is more enforceable. Thus, it is proposed to align the robot to a point beside to the ball with an offset, a new biased position, called virtual ball (V). So, it is required to include a module that computes V and modifies the input variables depicted in the Fig. 1(a).

Figure 2 depicts the required variables for computing the position of the virtual ball, where $\oplus = [x_T y_T]'$, $B = [x_b y_b]'$ are target and ball positions referenced to the local coordinates system of the robot, σ is the angle of the target-ball vector, and Ω is a unit vector with angle σ whis is calculated as:

$$\Omega = \begin{bmatrix} cos(\sigma) \\ sin(\sigma) \end{bmatrix} = \begin{bmatrix} x_b - x_T \\ y_b - y_T \end{bmatrix} \|B - \oplus\|^{-1}\tag{3}$$

Due to the 1D assumption that robot achieves the ball aligned to the target (i.e. $\varphi = 0°$ $\sigma = 180°$), depending of the selected foot for pushing the ball, robot should be displaced over its y axis towards left if the right foot is selected and vice versa. This sideward shifting is applied with orthogonal direction regarding Ω direction, it means to direction of translation vector T_L or T_R if left or right foot has been selected respectively. These vectors are shown in Fig. 2(a). The 90° added to σ are positive for selecting left foot and negative in other case. The sideward shifting ($S_{sideward}$) expressed in 4 has an amplitude (S_{offset}) that depends of the physical structure of robot, it is the distance from the middle point between feet to the flattest edge as is shown in Fig. 2(b).

$$S_{sideward} = \begin{bmatrix} cos(\sigma + f[k]90°) \\ sin(\sigma + f[k]90°) \end{bmatrix} S_{offset}\tag{4}$$

The proposed criteria for selecting the foot is based on where the robot comes from, particularly depends of the sign of φ, e.g. in the case of Fig. 2(a), the robot would have to select left foot for walking a shorter path towards a pose aligned to ball and target.

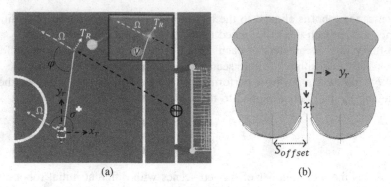

(a)   (b)

Fig. 2. a) Angles and Vectors taken into account for computing the Virtual Ball (V). b) Footprint of NAO robot.

Therefore, the sign of the angle added is opposite to the sign of φ. Equation (5) describes the rule for selecting the foot, where $f[k] = 1$ indicates left foot selected and $f[k] = -1$ the right one. In this rule some constraints are proposed for avoiding undesirable changes of the selected foot related to noisy perceptions: φ_h is a hysteresis for those cases when the φ angle is oscillating around zero, then the foot selected is changed only for considerable magnitude changes of φ; ρ_{th} is a threshold that avoids foot changes when robot is closer to the ball.

$$f[k] = \begin{cases} -sgn(\varphi), \rho > \rho_{th} \& \varphi > \varphi_h \\ f[k-1], otherwise \end{cases} \tag{5}$$

The position of the virtual ball regarding the robot reference system, is given as $V = B + S_{sideward}$.

2.3 Reinforcement Learning for Ball-Pushing Behavior

Since no footstep planners or specific kicks are performed, the ball is propelled by the robot's feet while it is walking, and the distance travelled by the ball depends on the robot's feet speed just before to hit it. Moreover, since our variables to be controlled are the robot speed relative to its center of mass, and not directly the speed of the feet, the robot's feet–ball dynamics turn complex and inaccurate. In this way, the RL of the *ball-pushing* behavior is proposed.

Speed Based Modeling (*M1*). Our expected policy is walking fast while keeping the ball possession. That means to minimize ρ, and at the same time to maximize v_x. So, a first modeling for learning the speed v_x depending on the observed state of ρ is detailed in Table 3. Only one feature composes the state space, ρ, which is discretized with intervals of *50 mm* (approximately the diameter of the ball in the SPL[1]). On the other

---

[1] http://www.informatik.uni-bremen.de/spl/bin/view/Website/Downloads.

Table 3. States and Actions description for *M1*

States space: $s = [\rho]$, a total of *11* states.			
	Min	Max	Discretization
Feature $\quad \rho$	*0* mm	*500* mm	*50* mm
Actions space: $a = [v_x]$, a total of *5* actions.			
	Min	Max	Discretization
Action $\quad v_x$	*20* mm/s	*100* mm/s	*20* mm/s
There are *55* state-action pairs.			

hand, the robot speed v_x composes the actions space, so the agent has to learn about its self foot' dynamics handling v_x.

Acceleration Based Modeling (*M2*). In this case, it is proposed to use speed increments or decrements as the action space for avoiding unreachable changes of speed in a short time and keeping a more stable gait. In this way, the expected policy is to accelerate for reaching the ball faster and to decelerate for pushing the ball soft enough. This modeling considers two additional features regarding *M1*: dV_{br}, the difference between ball (v_b) and robot speed (v_x), in order to track the ball; and v_x, in order to avoid ambiguous observed states and learn about the robot walking engine capabilities. It is expected that agent will be able to learn more about the dynamics between its walking speed, v_b, and ρ, regarding the selected foot to hit the ball. Since v_x is already a feature, it is possible to reduce the number of actions by using just three acceleration levels in the action space. The modeling for learning the acceleration acc_x is detailed in Table 4.

Reward Function 1 (*R1*). The proposed reward $r_1(s, a)$ introduced in 6 is a continuous function which punishes the agent every step along the training episode. Since ρ is a distance and v_x a speed, its resulting quotient can be seen as the predicted time to achieve the ball assuming constant speed. So, the agent is punished according this time; it would be more negative if ρ is high and/or v_x is very low (not desired), otherwise $r(s, a)$ tends to zero (good reward). If the training environment is episodic and defines a terminal state, the cumulative reward would be better (less negative) if the agent ends the episode as fast as possible, always being near the ball.

$$r_1(s, a) = -\rho/v_x \tag{6}$$

Reward Function 2 (*R2*). The proposed reward $r_2(s, a)$ introduced in 7 is an interval and parametric function that punishes the agent when it loses the ball possession or when it walks slowly, and rewards the agent when it walks fast without losing the ball. *R2* could be more intuitive and flexible than *R1* because it includes threshold parameters to define an acceptable interval for ρ defined by ρ_{max}, and a desired minimum speed defined by V_{th}. Moreover, it is possible to increase the punishment or the reward according a specific desired performance. For example, to increase the punishment when $\rho > \rho_{max}$ in order to prioritize the ball control over the speed.

$$r_2(s,a) = \begin{cases} -1 & , \text{if } \rho > \rho_{max} \\ -1 & , \text{else if } v_x < V_{th} \\ +1 & , \text{oterwise} \end{cases} \tag{7}$$

The SARSA(λ) Algorithm. The implemented algorithm for the *ball-pushing* behavior is the tabular SARSA(λ) with the replacing traces modification [17]. Based on previous work and after several trials, the SARSA(λ) parameters have been chosen prioritizing fastest convergences. In this way, the following parameters are selected: learning rate $\alpha = 0.1$, discount factor $\gamma = 0.99$, eligibility traces decay $\lambda = 0.9$, and epsilon greedy $\varepsilon = 0.2$ with exponential decay along the trained episodes.

3 Results

3.1 The Ball-Pushing Behavior

The goal of this experiment is the reinforcement learning of the *ball-pushing* behavior. Figure 3 depicts the initial robot positions (Pr_i) and the initial ball positions (Pb_i) for experiment *i*. In this case, the initial positions are set for $i = 1$ in order to configure a training environment similar to Fig. 1(b). The terminal state is fulfilled where the robot crosses the goal line, then the learning environment is reset and a new episode of learning is started.

Table 4. States and Actions description for *M2*

States space: $[\rho \; dv_{br} \; v_x]$, total of *110* states.				
		Min	Max	Discretization
Feature	ρ	*0* mm	*500* mm	*50* mm
Feature	dv_{br}	−*100*	*100*	*Negative or positive*
Feature	v_x	*20* mm/s	*100* mm/s	*20* mm/s
Actions space: $a = [acc_x]$, a total of 3 actions.				
		Min	Max	Discretization
Action	acc_x	−*20* mm/s^2	*20* mm/s^2	*Negative, zero, and positive*
There are *330* state-action pairs.				

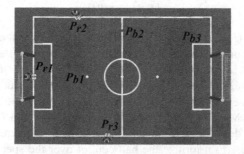

Fig. 3. Robot and ball initial positions for the tested Dribbling scenes. The field has 6 × 4 m.

In Sect. 2.3, two modeling and two reward functions have been proposed. The four possible combinations of them (M1-R1, M1-R2, M2-R1 and M2-R2) are used for learning the *ball-pushing* behavior. They are tested and compared by using the following performance indices:

- the episode time t_f, i.e. how long the agent takes to push the ball up to the target,
- the % *cumulated time of faults*: the cumulated time t_{faults} when the robot loses the ball possession, that means $\rho > \rho_{max}$, then:

$$\% \; cumulatedtimeoffaults = t_{faults}/t_f$$

- a global fitness function expressed as:

$$F = \frac{BTD + \psi_{tf}}{\psi_0} \int_{t=0}^{tf} \rho(t)/V_{rx}(t) \cdot dt,$$

where BTD is the total distance traveled by the ball and ψ_{tf} is the ball-target distance when the episode is finished. In the ideal case BTD = ψ_0 and ψ_{tf} = 0.

The results of the experimental procedure for learning the *ball-pushing* behavior are presented in the Fig. 4. As it can be observed, there is a trade-off between convergence speed and performance. It is possible to see in Fig. 4(a) those modeling using *M2* have the best performance while those modeling using *M1* achieve the fastest convergence. Also, from Fig. 4(b) it can be noticed that *M2* prioritizes the dribbling speed meanwhile modeling with *R1* cares the ball possession as shown Fig. 4(c). The best performance is achieved with *M2* and reward function *R1*, but its convergence is the worst. The fastest convergence is obtained with *M1*, independently from the reward function being used. Although when using *M2-R1* the final performance is almost 10 % better than the obtained with *M1-R1*, Fig. 4(a) shows a time reduction in learning convergence of 75 % with *M1* respect to *M2-R1*. This means that *M1* is more convenient for learning with a physical robot.

Figure 4(b) shows that *M1* carries out the dribbling about 15–20 % slower than *M2*. However, *M1-R1* cares more for the ball possession (Fig. 4(c)), which implies walking with lower speed and off course more time is taken. Notes that *M1-R1* gets about a half of the time of faults compared with the other three modeling as the Fig. 4(c) depicts.

Modeling *M1* is simpler, less state-action pairs, so, it learns faster. Modeling *M2* has three features and more state-action pairs, so, it learns slower than *M1* but improves its performance. On the other hand reward *R1* is simpler, for both modeling obtains better performance, but, since it is less explicit for a detailed task, it convergence time is slower.

After to carry out several training episodes, testing different types of rewards and learning parameters for the proposed *ball-pushing* problem, we have concluded that the use of parameterized and interval rewards as *R2* is a very sensitive problem to small parameter changes. For example, a right selection of the magnitude of each interval reward, ρ_{max}, and V_{th} for handling the tradeoff between speed and ball-control, in

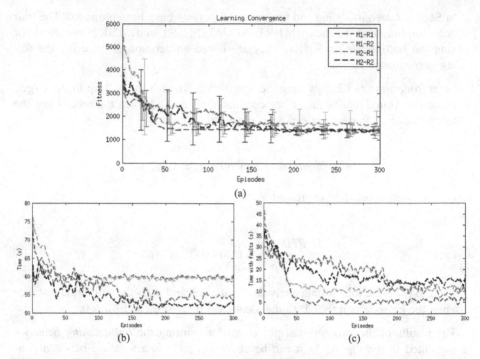

Fig. 4. Learning convergence through episodes by the modeling *M1* and *M2* using *R1* or *R2*, (a) global fitness average and intervals of confidence, (b) average of Time used for dribbling, (c) average of percentage of time with faults during dribbling.

Table 5. Validation results of the dribbling engine with three different scenes.

	Physical NAO			Simulated NAO	
	Dribbling time (s)	Time increased (%)	Time increased St.Dev.	Time increased (%)	Time increased (%)
Scene 1	53.71	38.56	11.48	31.91	31.91
Scene 2	49.57	49.57	11.79	37.88	37.88
Scene 3	44.38	32.39	10.06	9.98	9.98

addition to other learning parameters such as α, γ, λ and the exploration type, could dramatically modifies the learning performance.

3.2   Validation: The Full Dribbling Behavior

For the final validation of the dribbling behavior, the policy learning with modelling *M1* and reward function *R1 (M1-R1)* has been selected because its best tradeoff between performance and convergence speed. Since the resulting policy is expressed as a Q-table with 55 state-action pairs, a linear interpolation is implemented in order to make a continuous input-output function. The FLC (*alignment*) and the RL based controller (*ball-pushing*) switches for handling v_x if the robot is or not into the *ball-pushing* zone. Finally, both controllers are transferred to the physical NAO robot, a hand parameter adjusts in the foot selector and FLC is carried out in order to compensate the so-called *reality gap* with the simulator.

The entire dribbling engine is validated using the 3D simulator SimRobot [4] and physical NAO robots, with the three different experiments/scenes described in Fig. 3, For each scene, the robot have to dribble the ball up to the target, a scene is finished when the ball cross the goal line. 50 runs are carried out in order to statistically significant results. For these tests, the performance indices are:

- the average of t_f, the time that robot takes for finishing the dribbling scene,
- the average of % *time increased*, if t_{walk} is the time that robot takes to finish the path of the dribbling scene without dribbling the ball and walking at maximum speed, then: % *time increased* $= (t_f - t_{walk})/t_f$
- the standard deviation of the % *time increased*.

Table 5 shows the validation results of the entire dribbling engine with the simulator and the real robot. As it is usual, performances are better on simulation. This is more noticeable in scene 3, the most challenging for the motion and perception modules of the real robot. The dribbling time in scene 1 validates the asymptotic convergence values shown in Fig. 4(b), in addition, for this particular case, the % *time increased* indicates that physical NAO spend 38 % less time, if it walks without dribble the ball (i.e., $\approx$33 s). The final performance of the designed and implemented dribbling engine can be watched in [18].

Figure 5 shows the learned policy for the robot speed (v_x). Assuming the minimum in v_x as the learned speed for pushing the ball ($v_{x\_push} \approx 40$ mm/s), it can be noticed that the agent learns to request v_{x_push} to the walking engine when ρ is biased from zero, $\rho_{biased} \approx 170$ mm. It can be interpreted as the agent learns about its own walking-request delays. Moreover, when the agent observes ρ less than ρ_{biased}, it learns to increase the speed, meanwhile the mentioned delay is over and the ball is pushed.

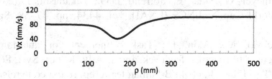

Fig. 5. Policy from modeling *M1*; v_x speed dependent on distance to the ball (ρ).

4   Conclusions and Future Work

This paper has presented a methodology for modeling the ball-dribbling problem in the context of humanoid soccer robotics, reducing as many as possible the on-line training time in order to make achievable futures implementations for learning the ball-dribbling behavior with physical robots.

The proposed approach is splitted in two sub problems: the *alignment* behavior, which has been carried out by using a TSK-FLC; and, the *ball-pushing* behavior, which has been learned by using a tabular SARSA(λ) scheme, a well-known, widely used and computationally inexpensive TD-RL method.

The ball-pushing learning results have shown asymptotic convergence in 50 to 150 training episodes depending on the state-action model used, which clarifies the feasibility of future implementations with physical robots. Unfortunately, according to the best of our knowledge, no previous similar dribbling engine implementations have been reported, in order to compare our final performance.

From the video [18], it can be noticed some inaccuracies with the alignment after pushing the ball. This could be related to the exclusion of the ball and target angles from the state space and reward function because the 1D assumption. Thus, as future work it is proposed to extend the methodology in order to learn the whole dribbling behavior avoiding switching between the RL and FLC. In this way, we plan to transfer the FLC policy of the pre-designed *alignment* behavior and refine it using RL in order to learn the *ball-pushing*. For that porpouses, transfer learning for RL is a promising approach. In addition, since the current state space is continuous and it will increase with the proposed improvements, RL methods with function approximation and actor critic will be considered.

Acknowledgments. This work was partially funded by FONDECYT under Project Number 1130153 and the Doctoral program in Electrical Engineering at the *Universidad de Chile*. D.L. Leottau was funded by CONICYT, under grant: CONICYT-PCHA/Doctorado Nacional/2013-63130183.

References

1. Alcaraz, J., Herrero, D., Mart, H.: A closed-loop dribbling gait for the standard platform league. In: Workshop on Humanoid Soccer Robots of the IEEE-RAS International Conference on Humanoid Robots (Humanoids), Bled, Slovenia (2011)
2. Latzke, T., Behnke, S., Bennewitz, M.: Imitative reinforcement learning for soccer playing robots. In: Lakemeyer, G., Sklar, E., Sorrenti, D.G., Takahashi, T. (eds.) RoboCup 2006: Robot Soccer World Cup X. LNCS (LNAI), vol. 4434, pp. 47–58. Springer, Heidelberg (2007)
3. Meriçli, Ç., Veloso, M., Akin, H.: Task refinement for autonomous robots using complementary corrective human feedback. Int. J. Adv. Robot. Syst. **8**(2), 68–79 (2011)
4. Röfer, T., Laue, T., Müller, J., et al.: B-human team report and code release 2012. In: Chen, X., Stone, P., Sucar, L.E., Van der Zant, T. (eds.) RoboCup-2012: Robot Soccer World Cup {XVI}. Springer, Heidelberg (2012)
5. HTWK-NAO-Team: Team Description Paper 2013. In: RoboCup 2013: Robot Soccer World Cup XVII Preproceedings. Eindhoven, RoboCup Federation, The Netherlands (2013)
6. Carvalho, A., Oliveira, R.. Reinforcement learning for the soccer dribbling task. In: 2011 IEEE Conference on Computational Intelligence and Games (CIG), Seoul, Korea (2011)
7. Riedmiller, M., Hafner, R., Lange, S., Lauer, M.: Learning to dribble on a real robot by success and failure. In: 2008 IEEE International Conference on Robotics and Automation (ICRA). IEEE, Pasadena, California (2008)
8. Ciesielski, V., Lai, S.Y.S.Y.: Developing a dribble-and-score behaviour for robot soccer using neuro evolution. Work. Intell. Evol. Syst. **2001**, 70–78 (2013)

9. Nakashima, T., Ishibuchi, H.: Mimicking dribble trajectories by neural networks for RoboCup soccer simulation. In: IEEE 22nd International Symposium on Intelligent Control, ISIC 2007 (2007)
10. Li, X., Wang, M., Zell, A.: Dribbling control of omnidirectional soccer robots. In: Proceedings 2007 IEEE International Conference on Robotics and Automation (2007)
11. Zell, A.: Nonlinear predictive control of an omnidirectional robot dribbling a rolling ball. In: 2008 IEEE International Conference on Robot Automation (2008)
12. Emery, R., Balch, T.: Behavior-based control of a non-holonomic robot in pushing tasks. In: Proceedings 2001 ICRA. IEEE International Conference on Robotics and Automation (Cat. No.01CH37164), vol. 3 (2001)
13. Damas, B.D., Lima, P.U., Custódio, L.M.: A modified potential fields method for robot navigation applied to dribbling in robotic soccer. In: Kaminka, G.A., Lima, P.U., Rojas, R. (eds.) RoboCup 2002: Robot Soccer World Cup VI. LNCS, vol. 2752, pp. 65–77. Springer, Heidelberg (2003)
14. Tang, L., Liu, Y., Qiu, Y., Gu, G., Feng, X.: The strategy of dribbling based on artificial potential field. In: 2010 3rd International Conference on Advanced Computer Theory and Engineering (ICACTE), vol. 2 (2010)
15. Celemin, C., Leottau, L.: Learning to dribble the ball in humanoid robotics soccer (2014). https://drive.google.com/folderview?id=0B9cesO4NvjiqdUpWaWFyLVQ3anM&usp=sharing
16. Storn, R., Price, K.: Differential Evolution - A simple and efficient adaptive scheme for global optimization over continuous spaces (1995)
17. Sutton, R., Barto, A.: Reinforcement Learning: An Introduction. MIT Press, Cambridge (1998)
18. Leottau, L., Celemin, C.: UCH-Dribbling-Videos. https://www.youtube.com/watch?v=HP8pRh4ic8w. Accessed 28 April 2014

Offensive Positioning Based on Maximum Weighted Bipartite Matching and Voronoi Diagram

Mohammadhossein Malmir(✉), Shahin Boluki, and Saeed Shiry Ghidary

Amirkabir University of Technology (Tehran Polytechnic), Hafez Ave., Tehran, Iran
{mhmalmir,sh.boluki,shiry}@aut.ac.ir

Abstract. In this paper we propose a modification to the well known Delaunay Triangulation based positioning in the attacking situation positioning of the agents in 2D Soccer Simulation environment. Due to advanced defensive skills such as marking skill, the attacker agents should have a dynamic positioning with respect to the rival team defenders. The proposed method employs the vertices of the Voronoi Diagram of the defending team agents as potential positions for the attacker team agents, since these positions are dynamic and change with the movement of the defending team agents and are always safe positions regarding the distance to the defending team agents, and also have a good coverage of the field. So the attacking agents can increase the chance of receiving pass by the ball owner agent and the scoring chance by taking positions on these vertices. This proposed method then applies Maximum Weighted Bipartite Matching to match these vertices to the attacking agents. This algorithm can be applied by each agent individually, but in order to reduce the possible decision conflicts in this matching which is the result of the limitation in the incoming information of the field from the agents' sensors, this algorithm can be performed by one agent and then this agent should inform the other attacking team agents of the result by communication skills like "say ability" in 2D Soccer Simulation (SS) environment. This method shows better performance in offensive situation than the conventional Delaunay Triangulation based positioning. It is tested in 2D SS environment as a highly dynamic multi-agent environment but its application is not restricted to the 2D SS League.

Keywords: Delaunay Triangulation · Voronoi diagram · Maximum weighted bipartite matching · Offensive positioning

1 Introduction

Positioning involves finding the best target position for agents who do not possess the ball regarding the field situation and team strategy [1]. Positioning is of great importance in a soccer match, either in defensive situation or in offensive situation. Soccer league of Robocup is no exception. In offensive positioning which is the case when one agent of us owns the ball, all the other agents of us

© Springer International Publishing Switzerland 2015
R.A.C. Bianchi et al. (Eds.): RoboCup 2014, LNAI 8992, pp. 562–570, 2015.
DOI: 10.1007/978-3-319-18615-3_46

should modify their position in field in order to increase the chance of receiving pass from the ball owner and increase the scoring chance. We propose and test our algorithm for offensive positioning on 2D Soccer Simulation environment due to the fact that in 2D Soccer Simulation we have a multi-agent system that our agents do not have the limitations of real robots in their movement, nevertheless the maximum acceleration and speed of movement of agents, and the similar energy factor of humans are simulated by the stamina factor and other parameters in this simulated environment. Therefore, considering the stamina and smooth movement in positioning is a crucial task. The problem of positioning within the Robocup context has been widely investigated in the past years and the methods vary a lot and it seems that there is no standard approach [2]. Some differences are surely influenced by the special specifications of each league of Robocup. A survey of approaches is out of the scope of this paper, but a nearly complete list of important approaches can be found in [1,2]. In general the solutions within the simulation leagues are more elaborated and computationally complex in comparison to solutions within hardware leagues due to the more reliable world model and less motion and low level control considerations within the simulation leagues [2]. In this paper we focus our attention to a widely used algorithm of positioning in Robocup Soccer Simulation League, Delaunay Triangulation based positioning depending on the ball's position in the field, and we propose a modification to improve the offensive situation positioning of agents. In 2008 Hidehisa Akiyama and Itsuki Noda used Delaunay Triangulation of the field depending on the ball position to determine the agents' positions [3–5]. In this method a representative set of potential ball positions are the vertices of these triangles and in each potential ball position we can set the positions of all of our agents. During the match the positions of the agents are determined by an interpolation method between the vertices of the triangle in which the current position of the ball lies. This method has shown a great performance and was adopted by most of the teams participating in Soccer Simulation League of Robocup Competitions in recent years. One of the benefits of this method is its simplicity in implementation and initialization by the developer, since it can be done by a visual software namely *fedit* provided by Hidehisa Akiyama [6] and all the considerations of smooth movement of players between positions can be handled by the human intuition operation in a visual manner. In Fig. 1 the Delaunay Triangulation of the field based on the representative set of ball positions in the GUI of *fedit* is shown. In each numbered potential position of the ball, the human developer determines the positions of all the agents. But this method has a major defect that it does not consider the opponent agents and is static with respect to opponent players. Due to recent defensive skill advancements and strong marking skill of Soccer Simulation Teams, receiving passes in the opponent's penalty area and making scoring chances requires a dynamic positioning regarding the opponent's agents to escape their marking skills. In this paper we propose an algorithm to achieve this goal in offensive positioning by applying Voronoi Diagram of opponent's agents and Maximum Weighted Bipartite Matching [7–10]. To be more precise, the proposed method

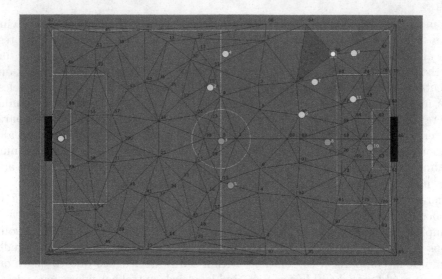

Fig. 1. Agents' positioning based on Delaunay triangulation.

considers the vertices of the defending team agents' Voronoi Diagram as potential positioning points for attacking team agents, since they have a good and safe distance from all the defending players and positioning in these points increase the chance of receiving passes by teammate agents. This algorithm then makes a bipartite graph, assuming these points on one side and attacking team agents on the other side, so our problem reduces to an assignment problem to determine which agent should choose which vertex to take position on. Besides, Soccer Simulation environment is a multi-agent system in which we have an increased chance of decision conflicts when each agent applies the algorithm and reaches to a conclusion individually, due to the limitations and uncertainties in each agent's information of the field situation gathered by its sensors. For this reason in the implementation of this algorithm one agent which has the best and most confident information of the field applies the algorithm and informs the other teammate agents of the result by the "say ability". In the proceeding we first give a short description of Maximum Weighted Bipartite Matching and Voronoi Diagram, and then we describe our algorithm in detail. In Sect. 5 we define and employ performance measures to compare our algorithm with the conventional Delaunay Triangulation employed currently by most of the soccer simulation teams regarding offensive positioning in opponent's half-field and near goal situations. At the end a conclusion is presented in Sect. 6.

2   Maximum Weighted Bipartite Matching

Graph $G = (V, E)$ in which V is the set of vertices and E is the set of edges is called Bipartite if the set V can be divided into two parts A and B such that,

$$A \cap B = \emptyset \tag{1}$$

$$A \cup B = V, \tag{2}$$

and also there does not exist any edge in E that connects two vertices in the same set [7]. A subset M of the set E is a matching (collection of edges) when each vertex of V is at most incident to one edge of M. Without loss of generality we can assume our graph complete by adding dummy vertices and edges of weight zero [7,8].

If each edge of the graph is assigned a weight we have a weighted bipartite graph [11]. If the sum of the weights of the edges in a matching (M_i) is called the weight of that match $W(M_i)$,

$$W(M_i) = \sum_{e \in M_i} w(e), \tag{3}$$

a maximum weighted matching M is a match in such a way that every other matching has lower weight than the weight of M [12]. In the proposed algorithm we employ *Hungarian* method to solve our assignment problem. This Maximum Weighted Bipartite Matching has been used in defensive skill in [11,12] as well.

3   Voronoi Diagram

The Voronoi diagram is a versatile geometric structure and has many applications in social geography, physics, astronomy, robotics, and many more fields [13,14]. The following description and the rest of this section is a summary of the definition originally presented in [15].

Let S denote a set of n sites (e.g., defending team's agent positions) in the two-dimensional plane. For two distinct sites $p, q \in S$, the dominance of p over q is defined as the subset of the plane being at least as close to p as to q, that is,

$$dom(p, q) = \{x \in \mathbb{R}^2 \mid d(x, p) \leq d(x, q)\} \tag{4}$$

where $d(x, p)$ is the Euclidean distance between two points x and p. Let $\overline{pq}$ denote the line segment between p and q. The perpendicular bisector of $\overline{pq}$ divides the plane into two halves. This perpendicular bisector, denoted by $\overline{pq}^{\perp}$, is called the separator of p and q. We denote the open half-plane that contains p by $h_p(q)$. Any point on the separator $\overline{pq}^{\perp}$ is equidistant to p and q. Any point within a half plane $h_p(q)$ has the distance to p smaller than the distance to q. Therefore, $dom(p, q)$ is the perpendicular bisector line $\overline{pq}^{\perp}$ plus the half of the plane $h_p(q)$. The region of a site $p \in S$ is the portion of the plane lying in all the dominance of p over the remaining sites in S, that is,

$$reg(p) = \cap_{q \in S - \{p\}} dom(p, q) \tag{5}$$

The region $reg(p)$ comes from intersecting $n-1$ half planes, and it is a convex polygon. Furthermore, the boundary of $reg(p)$ consists of at most $n - 1$ edges (maximal open straight-line segments) and vertices (their endpoints). Each point on an edge is equidistant from exactly two sites, and each vertex is equidistant from at least three sites. As a consequence, the edges and vertices of all regions form a polygonal division of the whole plane. This partition is called the Voronoi diagram, $Vor(S)$. These edges are called Voronoi edges and the vertices are called Voronoi vertices. A region $reg(p)$ is called a Voronoi cell (or Voronoi polygon).

4   Proposed Method

In the proposed approach, in general the conventional Delaunay Triangulation depending on the ball position is used as the method for positioning but in the attacking and near opponent's penalty area and cross situations, due to increasing development of marking abilities and defense power of 2D SS teams, scoring goal has become a hard and challenging task. Attackers should escape the marking of the opponent's defenders in order to create space for them to have the chance to score.

For this we apply the "Voronoi Diagram" idea and "Maximum Weighted Bipartite Matching". The algorithm is as follows. First the Voronoi diagram of opponent's defenders positions is created and the vertices of the diagram are found. Considering factors such as offside line and the fact that the vertex lies on the field (not out of field or near out of field) some vertices from the possible choices are omitted. The remaining vertices of the opponent's agents Voronoi Diagram are considered as potential positioning points for attacking team agents, since the vertices of the Voronoi diagram are good positions for the attackers because they have safe distance from opponent's defenders and positioning in

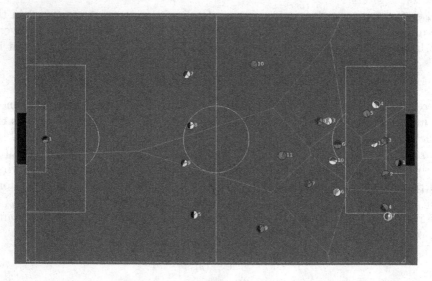

Fig. 2. Potential positioning places based on Voronoi diagram of the defending team's agents (red agents) (Color figure online).

these points increases the chance of receiving passes from teammate agent ball owner, as can be seen in Fig. 2. In Fig. 2 the Voronoi Diagram of the defending team agents (red agents) is depicted, and it can be seen that the vertices of this diagram are potential safe and good positions for the attacking team agents (yellow agents) to take position on. For assigning the attackers to vertices positions without any conflict we apply "MWBM" to our problem employing *Hungarian* algorithm for solving it. The attackers are assumed on one side and the vertices positions on the other side as the graph's nodes. Then a weight $w(i, j)$ is given to each edge respecting some features. The weight in fact indicates the importance and priority of choosing the point j by the attacker i for positioning.

In fact, if F_k is the value of feature k and R_k is the related coefficient and N is the number of the features the weight of each edge is the sum of the features value multiplied by their related coefficients as shown in (6).

$$W_{(i,j)} = \sum_{k=1}^{N} R_k \times F_k(i, j) \qquad (6)$$

Some of the factors that should be taken into account are the ones that have a great effect on stamina consumption of the agents. Factors like distance from current position or distance from formation position. In fact by considering these factors another goal is achieved in which the least deviation from the pre determined formation positions exists.

Some of the important elements in the weight calculation are shown in the Table 1.

In order to prevent the possible conflicts in applying the algorithm by each agent individually, e.g. choosing the same vertex for positioning by two or more agents due to different final matches concluded by each agent individually which is the result of the limitations of information of the current field situation, one of the agents who have the best view in attacking situations executes the matching and informs the attacker agents of the result by using "say ability".

5   Tests and Results

For comparing our algorithm in attacking situations with the conventional Delaunay Triangulation depending just on ball position, we employ the agent 2D 3.1.1 source code (team) which is released by Helios Soccer Simulation Team and is the base code of most of the participant teams in 2D Soccer Simulation League of Robocup. We have conducted two tests: one with its conventional and default method of positioning and the other with our modification in attacking situation. In both cases the formation file and settings are the same and the default version of agent 2d 3.1.1 formation designed in *fedit* software is used. We tested both scenarios against the same binary files of famous teams of around the world participating in World Robocup Competitions. In these tests parameters such as

Table 1. Important factors of our positioning decision

Factor	Feature	Description	Value
F1	Opponent mean distance	The opponent's agents mean distance to the position	The higher the distance is, the higher value is allocated to F1
F2	Home position distance	The distance between the formation position of the attacker and the vertex position	The lower the distance is, the higher value is allocated to F2
F3	Formation changing	The amount of the mess of our formation with that positioning	The less the amount is, the higher value is allocated to F3
F4	Distance from goal	The distance between the vertex and opponent's goal	The lower the distance is, the higher value is allocated to F4
F5	Current position distance	The distance between the vertex and the current position of attacker	The lower the distance is, the higher value is allocated to F5
F6	Distance to ball	The distance between the position and ball position	In a specified range the longer the distance is, the higher value is allocated to F6; and in other cases the shorter the distance is, the higher value is allocated
F7	Distance to teammate	The position's distance from our nearest teammate to that position	In a specified range the longer the distance is, the higher value is allocated to F7; and in other cases the shorter the distance is, the higher value is allocated

the number of completed passes in opponent's half-field, number of goals scored and number of cycles ball was owned that are suitable features for this comparison were measured. Each match was run three times. The measured mean values are shown in Table 2. The resultant values in tests against a small selection of teams participated in 2013 Robocup competitions are also shown in Table 3. In addition, the values measured against the agent 2D 3.1.1 base code itself is also reported in Table 3. These tables show that the number of correct passes and goals scored are increased by employing the proposed method. The number of cycles which the team owns the ball, however, have been decreased. Although it might seem unacceptable, due to rather more stamina consumption caused

Table 2. Total measured values and test results

Factor	Measure	Description	Value(D:Delaunay/ V:Voronoi)
M1	Successful pass	The average number of complete passes in the opponent half-field in each match	D: 26.33, V: 38.46
M2	Goals scored	The average number of goals scored in each match	D: 1.33, V: 1.56
M3	Possession cycles	The average number of cycles the team owns the ball in each match	D: 2184.5, V: 1995.7

Table 3. Test results against a selection of teams

Rival team	Successful pass (D:Delaunay/ V:Voronoi)	Goals scored (D:Delaunay/ V:Voronoi)	Possession cycles (D:Delaunay/ V:Voronoi)
AUT	D: 41, V: 63.66	D: 1.33 V: 1.66	D: 3180.33, V: 2850.33
Helios	D: 22.33, V: 31.33	D: 0.0, V: 0.0	D: 1650 , V: 1400.33
Yushan	D: 31.66, V: 37.33	D: 0.0, V: 0.33	D: 2440.33 , V: 2750.33
agent 2D	D: 22, V: 26.33	D: 1.66, V: 4.33	D: 2790.66, V: 2410

by the method's sensitiveness to the opponent agents, this occurrence is obvious. This algorithm has also been used in our AUT Soccer Simulation team since 2012 and the great number of goals scored by this team in different competitions such as Robocup 2012, Iranopen 2013 and Robocup 2013 is another indicator of this algorithm acceptable performance. The coefficients R_k mentioned in Sect. 4 that are used for scoring method regarding the features can be determined by the developer as a human observer and expert or can be optimized by defining optimization scenarios and using advanced AI methods to improve the performance of the method.

6 Conclusion

We proposed a method for attacking situation positioning of the agents in 2D Soccer Simulation, but its application is not restricted to the 2D SS League. The proposed method employs the vertices of the Voronoi Diagram of the defending team agents as potential positions for the attacker team agents. It then applies Maximum Weighted Bipartite Matching to match these vertices to the attacking agents. As shown in the previous section this method shows better performance in offensive situation than the conventional Delaunay Triangulation based positioning. But as mentioned in Sect. 5, this method is more sensitive to opponent agents movements, and might consume more stamina by causing some redundant move actions of the agents. Another fact is that the defensive skills of soccer teams are more effective near their goal, and escaping from the advanced marking skills of opponent team agents is very important in cross situations and near goal situations in order to score more goals. Therefore, our opinion is that the proposed method should be used mostly in cross situations and near goal and penalty area situations, and for the attacking moments that the ball is far from the defending team penalty area a simple free run behavior can be performed by the attacking agents in order to save more stamina and increase the chance of receiving through passes by them. This is the same combined method employed in our AUT Soccer Simulation team, which has shown good performance in different Robocup competitions.

References

1. Dashti, H.T., Kamali, S., Aghaeepour, N.: Positioning in robots soccer. In: Lima, P. (ed.) Robotic Soccer, pp. 29–44. I-Tech Education and Publishing, Austria (2007)
2. Kaden, S., Mellmann, H., Scheunemann, M., Burkhard, H.D.: Voronoi based strategic positioning for robot soccer. In: Proceedings of the 22nd International Workshop on Concurrency, Specification, and Programming (CS&P), Warsaw, pp. 283–293 (2013)
3. Akiyama, H., Noda, I.: Multi-agent positioning mechanism in the dynamic environment. In: Visser, U., Ribeiro, F., Ohashi, T., Dellaert, F. (eds.) RoboCup 2007: Robot Soccer World Cup XI. LNCS (LNAI), vol. 5001, pp. 377–384. Springer, Heidelberg (2008)
4. Akiyama, H., Noda, I., Shimora, H.: Helios 2008 team description paper. In: RoboCup (2008)
5. Akiyama, H.: Helios 2007 team description paper. In: RoboCup (2007)
6. Part of Helios team released materials. http://sourceforge.jp/projects/rctools/
7. Lecture notes from Michael Goemans class on Combinatorial Optimization. http://math.mit.edu/~goemans/18433S09/matching-notes.pdf (2009)
8. Paul, M.: Algorithmen fur das maximum weight matching problem in bipartiten graphen. Master's thesis, Fachbereich Informatik, Universität des Saarlandes, Saarbrücken (1989)
9. Kreyszig, E.: Advanced Engineering Mathematics, 10th edn. Wiley, Hoboken (2010)
10. West, D.B.: Introduction to Graph Theory, 2nd edn. Prentice Hall, Englewood Cliffs (1999)
11. Norouzitallab, M., Javari, A., Noroozi, A., Salehizadeh, S.M.A., Meshgi, K.: Nemesis team description paper. In: RoboCup (2010)
12. Malmir, M., Simchi, M., Boluki, S.: AUT Team Description Paper 2012. In: RoboCup (2012)
13. Aurenhammer, F.: Voronoi diagrams - a survey of a fundamental geometric data structure. ACM Comput. Surv. **23**(4), 345–406 (1991)
14. Berg, M., Cheong, O., Kreveld, M., Overmars, M.: Computational Geometry: Algorithms and Applications. Springer, Berlin (2008)
15. Wang, B.: Coverage Control in Sensor Networks. Springer, New York (2010)

Keyframe Sampling, Optimization, and Behavior Integration: Towards Long-Distance Kicking in the RoboCup 3D Simulation League

Mike Depinet$^{(\boxtimes)}$, Patrick MacAlpine, and Peter Stone

Department of Computer Science, The University of Texas at Austin,
Austin, USA
{msd775,patmac,pstone}@cs.utexas.edu

Abstract. Even with improvements in machine learning enabling robots to quickly optimize and perfect their skills, developing a seed skill from which to begin an optimization remains a necessary challenge for large action spaces. This paper proposes a method for creating and using such a seed by (i) observing the effects of the actions of another robot, (ii) further optimizing the skill starting from this seed, and (iii) embedding the optimized skill in a full behavior. Called KSOBI, this method is fully implemented and tested in the complex RoboCup 3D simulation domain. To the best of our knowledge, the resulting skill kicks the ball farther in this simulator than has been previously documented.

1 Introduction

Every optimization needs a starting point. If the starting point is not in a region of the search space with a meaningful gradient, optimization is unlikely to be fruitful. For example, if trying to maximize the speed of a robot's walk, the robot has a much greater chance of success if given a stable walk to begin with. We refer to the starting point of an optimization for skill learning as a *seed skill*. Even with improvements in optimization processes, developing seed skills remains a challenge.

Currently most seed skills are written by hand and then tuned by a human until they resemble the desired skill enough to begin an optimization. Some seeds can also be acquired by having a robot mimic a human in a motion capture suit [1,2]. We propose a third way of creating a seed, called *keyframe sampling*, which uses learning by observation. In this case, a robot observes the effects of actions of another object, and does its best to reproduce those effects. In our work robots observe another robot with the same model, although in principle this methodology could be applied to robots with different models or to humans using transfer learning ([3]) and different body mappings as described in [1,2].

This paper considers a 3-step methodology of keyframe sampling, optimization, and behavior integration (*KSOBI*), which guides the development of a skill from watching a teacher to using the skill as part of an existing behavior. First, we describe KSOBI in Sect. 2, focusing on the keyframe sampling (KS) step.

© Springer International Publishing Switzerland 2015
R.A.C. Bianchi et al. (Eds.): RoboCup 2014, LNAI 8992, pp. 571–582, 2015.
DOI: 10.1007/978-3-319-18615-3_47

We introduce the robot soccer domain in Sect. 3, and apply keyframe sampling and optimization (O) to kicking in robot soccer in Sect. 4. The robot soccer domain has the added complication that a skill is only useful if it can be incorporated into the robot's existing behavior. We describe behavior integration (BI) for our application, in Sects. 4.5 and 4.6, and conclude with a summary and future work in Sect. 6.

2   KSOBI Overview and Keyframe Sampling

The goal of KSOBI's KS step is to use observations of another robot to quickly create an imitation skill. This imitation will later be used as the seed for an optimization, the O step, to create an optimized skill, which hopefully matches or improves upon the observed skill. Finally, that skill will be incorporated into the robot's existing behavior during the BI step. KSOBI is outlined in Fig. 1.

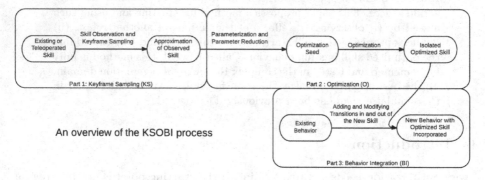

Fig. 1. An outline of KSOBI

Keyframe sampling assumes that the actions of the observed robot have observable effects. In the case of a physical robot, the robot's actions could be the torque applied to each motor while the effects are the change in rotation of joints. From the observed effects, a *keyframe skill* can be created directly.

A *keyframe* is defined to be a complete description of joint angles, either in absolute values or relative to the previous keyframe, with a *scale* for each joint indicating the percentage of the motor's maximum torque allowed to be used to reach the target angle. (The torque applied at any point in time is determined by a controller - often a PID controller, as in this work - but is multiplied by this value to affect how quickly a target angle is achieved.) A keyframe $k \in \mathcal{K} := \mathbb{R}^n \times \mathbb{R}^n \times \{0, 1\}$ where n is the number of joints, 0 indicates absolute angles, and 1 indicates relative angles.[1] The first n-vector gives target angles for each joint,

---

[1] Note that in many robotic domains, including robot soccer, the distinction between relative and absolute joint positions is unnecessary since the joints have a specified non-overlapping range of possible values. This fact simplifies the above process since all keyframes may be unambiguously absolute.

while the second n-vector gives their scales. For example, the 3 joint keyframe $k_1 = ((0, 0, 0), (0.5, 0.5, 0.5), 0)$ indicates all joints should be set to 0° using half maximum torque, while $k_2 = ((180, 180, 0), (1, 1, 1), 1)$ indicates that the first and second motors should be rotated 180° with maximum torque.

A *keyframe skill* (or *skill* unless otherwise noted) is defined as a list of keyframe-time pairs, where the time indicates how long to hold the paired keyframe. A skill $s \in (\mathcal{K} \times \mathbb{R})^m$ where m is the number of keyframes in the skill, and each of the m (k, t) pairs indicates that keyframe k should be the target for the next t seconds. For example, using k_1 and k_2 as defined above, the skill $s1 = ((k_1, 1.0), (k_2, 1.0))$ would indicate that the robot should take 1 second to get all its joints to 0° (using at most half their torque) if possible, remaining there until 1 second has expired if time remains, then take another second to rotate joints 1 and 2 by 180° as quickly as possible.

If the joint angles of an observed robot are directly observable, then a skill can be generated by recording each joint angle at specified time steps. This idea is the heart of keyframe sampling, which is made rigorous in the pseudocode below, where n is the number of joints in the robot model and T is the total time required by the skill divided by the time step:

```
define angle θ_{j,t} for j ∈ [1, n] ∩ ℤ and t ∈ [0, T) ∩ ℤ
define keyframe k_t for the same values of t
skill observeSkill(robot teacher, duration timeStep):
    int t = 0
    repeat:
        sampleKeyframe(teacher, t++)
        wait(timeStep)
    until teacher.skill.isDone()
    skill s = ((k_0,timeStep), (k_1,timeStep), ...(k_{T−timeStep},timeStep))
    return skill
void sampleKeyframe(robot teacher, int t):
    for joint j in teacher:
        θ_{j,t} = j.angle
    if t == 0:
        k_t = ((θ_{1,t}, θ_{2,t}, ..., θ_{n,t}), (1, 1, ..., 1), 0)
    else:
        k_t = ((θ_{1,t} − θ_{1,t−1}, θ_{2,t} − θ_{2,t−1}, ..., θ_{n,t} − θ_{n,t−1}), (1, 1, ..., 1), 1)
```

Using this method, the skill s will assume the observed starting position and, at each time step, attempt to assume the next set of observed joint angles as quickly as possible, imitating the observed object.

The generated skill s likely will not replicate the observed skill exactly, since it is just a sampling of several points in a presumably continuous motion, as described in more detail in [4]. However, as will be seen in Sect. 4, it may be close enough to use as the seed for an optimization. The hope is that the optimization will overcome the discontinuities in the seed skill s and create a skill which replicates or improves upon the observed motion.

Prior to optimization, it is necessary to parametrize the generated skill, allowing each value set by keyframe sampling to be varied by the optimization. Often it will then be necessary to freeze a subset of the parameters, preventing them from changing during the optimization and reducing the dimension of the parameter space. Parameter reduction is addressed in Sect. 4.2. Having chosen which values may vary, a fitness function should be chosen and the optimization may begin. The optimization process is described in Sect. 4.3. Finally, once the new skill has been optimized in isolation, it must be incorporated into existing behavior. We illustrate the behavior integration (BI) process as applied to our application in Sects. 4.5 and 4.6.

3   Application Domain: RoboCup 3D Simulation League

The target application for this work is the 3D Simulation RoboCup domain. In the 3D Simulation League simulated Nao robot agents play 11 vs 11 soccer in a physically realistic environment. The field is a 30 m x 20 m scale model of a full sized human field, and the robots are about 0.5 m tall. Every 0.02 s, agents respond to information given from a game server with the torque to apply to each of their 22 motors.

In recent years, the RoboCup 3D Simulation League has been won primarily by creating fast and robust walks ([5,6]). However, teams are now developing their own kicks ([7–9]). Kicking is difficult for three main reasons. First, robust kicking requires a smooth transition from walking and most walks involve some noise in reaching a target point. Second, kicking requires high precision in that a difference of a couple degrees on any joint in any keyframe will likely result in a failed kick. Third, there are many joints involved in a long distance kick and there are many keyframes between planting the foot and kicking the ball. This complexity results in a large search space for optimal kicks. Existing machine learning techniques help alleviate some of these problems, but there remains a need for finding reasonable starting seeds to guide the search through such a large parameter space, as well as a methodology for incorporating the resulting optimized skill into a full behavior, as is provided by KSOBI.

4   Learning to Kick from a Fixed Point

We begin learning a kick under the assumption of a chosen starting location. In the RoboCup domain, an agent can expect this situation for its own kickoff. To learn a kick skill for a fixed starting location, we observe the previously furthest documented kick, belonging to FC Portugal [10] (see Fig. 2). Using keyframe sampling, we create an approximation of this kick, which we use as a seed for optimization. The result of this optimization is the new longest known kick in the RoboCup 3D simulation environment.

Fig. 2. The observed kick. Video available at: http://www.cs.utexas.edu/~Austin Villa/sim/3dsimulation/AustinVilla3DSimulationFiles/2014/videos/FCPKick.ogv

4.1 Observing a Seed: Keyframe Sampling

The RoboCup server currently only provides the location of the head, torso, each leg, and each arm of robots to observers. This is not enough information to mimic another robot since there are multiple sets of joint angles that give the same locations for each body part. To solve this problem, we modify the server such that observers receive all of the joint angles of the observed robot, as required for keyframe sampling. The joint angles could reasonably be estimated by a real robot watching another real robot, so it seems like a reasonable level of detail to request. With this added information, we apply keyframe sampling at 16.67 Hz (every 3 server cycles). Higher sampling rates give a seed skill more similar to the observed skill, but also result in more parameters to optimize. The 16.67 Hz rate gave a sufficiently similar skill while keeping the parameter space reasonable in this case. As expected, the result is not an exact match of the observed skill. The imitation skill results in the robot kicking the ground behind the ball and falling over (Fig. 3). However, the imitation is close enough to use as a seed for the optimization.

Fig. 3. The seed after observation. Video available at: http://www.cs.utexas. edu/~AustinVilla/sim/3dsimulation/AustinVilla3DSimulationFiles/2014/videos/ InitialKick.ogv

4.2 Fixed Point Training

The observed seed in Sect. 4.1 results in a skill with 89 key frames, each with every joint included, giving a total of 1958 parameters to train. Although this search space is prohibitively large, many of the parameters can be safely ignored. In fact, there is a need for parameter reduction before optimization in general when the seed is created by keyframe sampling. The goal in parameter reduction is to freeze parameters whose values will not affect the skill, removing them from the optimization and reducing the dimension of the search space. In addition to domain heuristics, seeds from keyframe sampling offer some general heuristics.

First, any joint that does not change significantly between two keyframes can be fused between frames. Second, beginning and ending keyframes can sometimes be removed entirely. In this case, it is important to be careful not to disrupt the transition in and out of the skill (e.g. you would not want to remove keyframes responsible for setting the plant foot from a kicking skill).

In the case of our kicking seed, removing the head joints (a domain heuristic), any joint that does not change by more than $0.5°$ between two frames, and the keyframes before the plant foot is set limits the skill to only 59 parameters. Adding 3 parameters for the starting location (x, y, and angle), results in 62 parameters to optimize. This is still a large state space, but it is manageable.

With the optimization parameters chosen, the next step is to define a fitness function. We use the distance traveled by the ball

$$fitness_{initial} = \begin{cases} -1 & : \text{Failure} \\ finalBallLocation.x & : \text{Otherwise} \end{cases}$$

where a "Failure" is any run in which the robot falls over, kicks backward, or runs into the ball before kicking it.

4.3 Optimizing with CMA-ES

Optimizing a set of parameters is the same as finding the global maximum of a fitness function $g(\mathbf{x}) : \mathbb{R}^n \rightarrow \mathbb{R}$ where n is the number of parameters being tuned. For optimization we use Covariance Matrix Adaptation Evolutionary Strategy (CMA-ES) [11] as we have had previous success using CMA-ES for optimizing a walk, as documented in [12].

4.4 Fixed Point Results

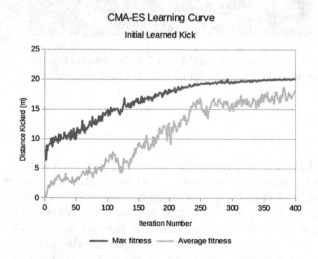

After 400 iterations of CMA-ES with a population size of 200, the resulting skill is able to kick the ball 20 meters on average (see Fig. 4). In addition to solving the problem of the robot kicking the ground behind the ball and helplessly falling over, the optimization produced a kick which exceeded the length of the original observed kick by more than 5 m (Fig. 5)!

Fig. 4. CMA-ES Learning Curve for Initial Kick

Fig. 5. The first learned kick. Video available at: http://www.cs.utexas.edu/~Austin Villa/sim/3dsimulation/AustinVilla3DSimulationFiles/2014/videos/LearnedKick.ogv

We also consider two other fitness functions, producing slightly different resulting kicks. The first is a fitness function centered around accuracy. This function uses the same ball distance fitness as before except with a Gaussian penalty for the difference between the desired and actual angles. Optimization with this function gives the powerful and predictable kick seen in Fig. 6.

$$f_{accuracy} = \begin{cases} -1 & : \text{Failure} \\ finalBallLoc.x * e^{-angleOffset^2/180} & : \text{Otherwise} \end{cases}$$

Fig. 6. Improved accuracy. Video available at: http://www.cs.utexas.edu/~Austin Villa/sim/3dsimulation/AustinVilla3DSimulationFiles/2014/videos/AccuracyKick.ogv

The second is a fitness function centered around distance in the air. As the idea is to kick the ball long distances above opponents' heads, the fitness function heavily rewards the distance traveled by the ball before descending to 0.5 m above the ground. It also rewards total distance and heavily penalizes missing the goal (ignoring any other tests of accuracy). Optimization with this function results in a noisy kick, but one that travels over 11 m in the air (see Fig. 7).

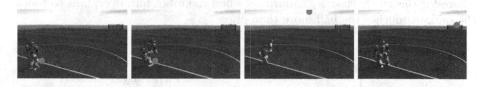

Fig. 7. Increased air distance. Video available at: http://www.cs.utexas.edu/~Austin Villa/sim/3dsimulation/AustinVilla3DSimulationFiles/2014/videos/AirDistKick.ogv

$$f_{air} = \begin{cases} -1 & : \text{Failure} \\ 0 & : \text{Missed goal} \\ 100 + finalBallLoc.x + 2 * airDist & : \text{Otherwise} \end{cases}$$

These results are summarized in Table 1. All kicks are executed by the original NAO model used in the RoboCup 3D simulation for ease of comparison. In addition to being the longest documented kicks to our knowledge, these kicks are also the first ones able to score from any point in the offensive half of the field.

Table 1. Kick distances

Kick	Avg Distance (m)	Notes
Observed seed	About 15	
FCPortugal	About 17	Based on empirical data and verbal confirmation
Learned Kick	20.0($\pm$0.12)	
Accuracy Kick	18.8($\pm$0.29)	With placement 1.3°($\pm$1.78°) from target angle
Air Distance Kick	19.2($\pm$0.38)	With 11.4 m($\pm$0.25 m) higher than 0.5 m

4.5 Multi-Agent Training

With the kick optimized in isolation, we continue now to the behavior integration (BI) step. As the kick was optimized from a fixed point, integration into a legal kickoff is a natural first integration.

Unfortunately, scoring from the kickoff is illegal unless someone else touches the ball first in soccer. To rectify this, we introduce another agent with another skill which moves the ball as little as possible then gets out of the way. After optimizing this skill alone, using the server's play mode and the distance of the ball's movement to determine fitness, we optimize the touch and kick together, using the same fitness function as used for the kicker with an added penalty for either agent missing the ball or the kicker hitting the ball before the toucher.

$$f_{touch} = \begin{cases} -1 & : \text{Failure} \\ 10 - finalBallLoc.magnitude & : \text{Otherwise} \end{cases}$$

where for this single case, a "Failure" is when the robot falls over, fails to touch the ball, or touches the ball more than once.

$$f_{kickoff} = \begin{cases} -1 & : \text{Failure} \\ -1 & : \text{Wrong touch order} \\ -1 & : \text{Either agent missed} \\ 100 + finalBallLoc.x + 2*airDist & : \text{Otherwise} \end{cases}$$

4.6 Multi-Agent Results

After a successful 400 iteration optimization with population size of 150 (see Fig. 8), two agents are able to legally and reliably score within 8 s of the game

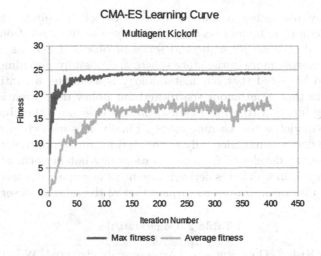

Fig. 8. CMA-ES Learning curve for multiagent kickoff

Fig. 9. Multiagent Kickoff. Video available at: http://www.cs.utexas.edu/~AustinVilla /sim/3dsimulation/AustinVilla3DSimulationFiles/2014/videos/Kickoff.ogv

starting and within 3 s of the ball first being touched (see Fig. 9). Adding this kickoff behavior and changing nothing else dramatically improves the team's overall performance (see Table 2). If in addition we alter the agent to improve the accuracy resulting from the server's beam command before kickoffs (and re-run the multi-agent optimization with this improvement), we get the scoring percentages presented in Table 3.

The percentage of kickoffs which score varies with the opponent team. Most kickoffs which fail to score are a result of opponent player formation. We have not found a kick that makes it all the way to the goal in the air, so a player located where the ball bounces on its path to the goal effectively stops kickoff goals. That said, the location at which the ball bounces is not always the same. In the future, the kicking agent could have multiple kickoff kicks and trajectory information for each and could use that information at run-time to choose a kick that misses opponents.

5   Related Works

Other forms of learning from observation have been explored, and are described in [13]. In the context of that review, keyframe sampling has several qualities.

First, we may use either a human or a robot for teaching, whereas most approaches require a human teacher. Secondly, the data set from which we learn is limited to a single sequential series of observed actions, rather than a larger set covering more initial state spaces and possibly providing repetition. It should also be noted that our methodology is assuming a continuous state space (discrete time, continuous space) and our policy derivation is completed by a mapping function (the identity map in this case since the observed robot has the same model as the learning robot). Finally, it is important to note that the policy we derive is intended only as the seed for further optimization. While initially the policy developed from observation may not perform as well as the observed policy from which it is derived, in general we expect the learned policy's performance after optimization to surpass that of the original observed policy.

Table 2. Game statistics

Using Kickoff	Opponent	Average goal differential	W-L-T
No	FCPortugal	0.335 (+/-0.023)	368-85-547
Yes	FCPortugal	1.017 (+/-0.029)	775-6-219
Yes	WithoutKickoff	0.385 (+/-0.022)	416-44-540

Table 3. Percentage of scored kickoffs against the top 4 finishers from RoboCup 2013

Opponent	Beginning of Half	During Half
FCPortugal	92.19 %	77.15 %
UTAustinVilla	76.70 %	54.15 %
SeuJolly	77.00 %	77.66 %
Apollo3D	89.30 %	65.60 %

Another approach which uses learning by observation to begin an optimization is described in [2]. In that work, a human's motions are captured via a Microsoft Kinect, converted to a robot model, and optimized to achieve balance despite differences in joints and mass distributions. This is similar to the KS and O steps of KSOBI, with some important differences in the sampling rate and the parameter reduction steps. In this work, the authors sample at 50 Hz then model the motion of each joint as a function in order to hopefully reduce the search space of the optimization. In our application of KSOBI, we use a smaller sampling rate (5-17 Hz) and use several heuristics to further reduce the parameter search space. Additionally, this work requires a human teacher, whereas KSOBI can use either a robot or a human teacher.

In [14], the authors describe two methods of having humans teach robots by demonstration. One way is to record the robots motion as a human moves it (trajectory-based). The other (keyframe-based) involves inferring the trajectory between two points set by a human. The authors try to make the interaction as easy as possible for humans by allowing a hybrid approach, in which either trajectory-based or keyframe-based methods may be recorded. This methodology is similar to our keyframe-sampling approach, except that our approach does not require a human teacher, and instead records keyframes at a certain rate instead of at important points. Depending on the sampling rate, keyframe-sampling could be more similar to the trajectory-based approach described in this article. Moreover, this article expects to create a new behavior by learning from demonstration alone, while in KSOBI, demonstration is only the first step.

6   Summary and Future Work

This paper introduced the KSOBI process, guiding the development of a skill from watching another robot (keyframe sampling - KS), to optimizing the resulting sampled skill (optimization - O), to integrating the optimized skill into an existing behavior (behavior integration - BI). The full KSOBI process was applied in a complex simulated domain. Along the way, we showed the success of this method with a set of new kicks which raise the bar for how far agents can kick in the RoboCup 3D simulation league.

Future work includes another desirable behavior integration, which requires being able to approach the ball before kicking it. Such an integration would allow the kick to be used for shooting and passing during regular gameplay. Both the approach and the kick would need to be quick for the kick to be useful during a game. Although we've made progress on this front ([4]), it remains an important area for future work.

It may be possible to make more robust kicks by defining them using trajectories relative to the ball instead of fixed joint angles. The UTAustinVilla codebase already has a set of kicks parametrized by trajectories relative to the ball [9], however they seldom exceed a distance of 5 m. Future work may also include finding a happy medium between the flexibility of those kicks and the distance achieved by fixed joint angle kicks.

References

1. Setapen, A., Quinlan, M., Stone, P.: Marionet: motion acquisition for robots through iterative online evaluative training. In: Ninth International Conference on Autonomous Agents and Multiagent Systems - Agents Learning Interactively from Human Teachers Workshop (AAMAS - ALIHT) (2010)
2. Seekircher, A., Stoecker, J., Abeyruwan, S., Visser, U.: Motion capture and contemporary optimization algorithms for robust and stable motions on simulated biped robots. In: Chen, X., Stone, P., Sucar, L.E., van der Zant, T. (eds.) RoboCup 2012. LNCS, vol. 7500, pp. 213–224. Springer, Heidelberg (2013)
3. Taylor, M.E., Stone, P.: Transfer learning for reinforcement learning domains: a survey. J. Mach. Learn. Res. 10, 1633–1685 (2009)
4. Depinet, M.: Keyframe sampling, optimization, and behavior integration: a new longest kick in the robocup 3D simulation league. Master's thesis, University of Texas at Austin, Undergraduate Thesis (2014)

5. Bai, A., Chen, X., MacAlpine, P., Urieli, D., Barrett, S., Stone, P.: WrightEagle and UT austin villa: robocup 2011 simulation league champions. In: Röfer, T., Mayer, N.M., Savage, J., Saranlı, U. (eds.) RoboCup 2011. LNCS (LNAI), vol. 7416, pp. 1–12. Springer, Heidelberg (2012)
6. MacAlpine, P., Collins, N., Lopez-Mobilia, A., Stone, P.: UT austin villa: robocup 2012 3D simulation league champion. In: Chen, X., Stone, P., Sucar, L.E., van der Zant, T. (eds.) RoboCup 2012. LNCS (LNAI), vol. 7500, pp. 77–88. Springer, Heidelberg (2013)
7. Ferreira, R., Reis, L.P., Moreira, A.P., Lau, N.: Development of an omnidirectional kick for a NAO humanoid robot. In: Pavón, J., Duque-Méndez, N.D., Fuentes-Fernández, R. (eds.) IBERAMIA 2012. LNCS, vol. 7637, pp. 571–580. Springer, Heidelberg (2012)
8. Cruz, L., Reis, L.P., Lau, N., Sousa, A.: Optimization approach for the development of humanoid robots' behaviors. In: Pavón, J., Duque-Méndez, N.D., Fuentes-Fernández, R. (eds.) IBERAMIA 2012. LNCS, vol. 7637, pp. 491–500. Springer, Heidelberg (2012)
9. MacAlpine, P., Urieli, D., Barrett, S., Kalyanakrishnan, S., Barrera, F., Lopez-Mobilia, A., Ştiurcă, N., Vu, V., Stone, P.: UT austin villa 2011: a champion agent in the RoboCup 3D soccer simulation competition. In: Proceedings of 11th International Conference on Autonomous Agents and Multiagent Systems (AAMAS) (2012)
10. Lau, N., Reis, L.P., Shafii, N., Ferreira, R.: Fc portugal 3D simulation team: team description paper. In: RoboCup 2013 (2013)
11. Hansen, N.: The CMA evolution strategy: a tutorial (2009). http://www.lri.fr/~hansen/cmatutorial.pdf
12. MacAlpine, P., Barrett, S., Urieli, D., Vu, V., Stone, P.: Design and optimization of an omnidirectional humanoid walk: a winning approach at the RoboCup 2011 3D simulation competition. In: Proceedings of the Twenty-Sixth AAAI Conference on Artificial Intelligence (AAAI) (2012)
13. Argall, B.D., Chernova, S., Veloso, M., Browning, B.: A survey of robot learning from demonstration. Rob. Auton. Syst. **57**, 469–483 (2009)
14. Akgun, B., Cakmak, M., Jiang, K., Thomaz, A.: Keyframe-based learning from demonstration. Int. J. Soc. Robot. **4**, 343–355 (2012)

Generalized Learning to Create an Energy Efficient ZMP-Based Walking

Nima Shafii[1,2,4(✉)], Nuno Lau[3,4], and Luis Paulo Reis[1,5]

[1] LIACC - Artificial Intelligence and Computer Science Laboratory,
Porto, Portugal
nima.shafii@fe.up.pt, lpreis@dsi.uminho.pt
[2] Department of Informatics Engineering, Faculty of Engineering,
University of Porto, Porto, Portugal
[3] Department of Electronics, Telecommunications and Informatics,
University of Aveiro, Aveiro, Portugal
nunolau@ua.pt
[4] IEETA - Institute of Electronics and Telematics Engineering of Aveiro,
Aveiro, Portugal
[5] Department of Information Systems, School of Engineering,
University of Minho, Guimaraes, Portugal

Abstract. In biped locomotion, the energy minimization problem is a challenging topic. This problem cannot be solved analytically since modeling the whole robot dynamics is intractable. Using the inverted pendulum model, researchers have defined the Zero Moment Point (ZMP) target trajectory and derived the corresponding Center of Mass (CoM) motion trajectory, which enables a robot to walk stably. A changing vertical CoM position has proved to be crucial factor in reducing mechanical energy costs and generating an energy efficient walk [1]. The use of Covariance Matrix Adaptation Evolution Strategy (CMA-ES) on a Fourier basis representation, which models the vertical CoM trajectory, is investigated in this paper to achieve energy efficient walk with specific step length and period. The results show that different step lengths and step periods lead to different learned energy efficient vertical CoM trajectories. For the first time, a generalization approach is used to generalize the learned results, by using a programmable Central Pattern Generator (CPG) on the learned results. Online modulation of the trajectory is performed while the robot changes its walking speed using the CPG dynamics. This approach is implemented and evaluated on the simulated and real NAO robot.

Keywords: Humanoid walking · Energy efficiency · Zero Moment Point · Central pattern generators

1 Introduction

In competitive, non-deterministic environments like RoboCup soccer humanoid robot leagues, stable omnidirectional biped walking is one of the keys to win a match. Such movements must accomplish different requirements. For example, in case the target is

© Springer International Publishing Switzerland 2015
R.A.C. Bianchi et al. (Eds.): RoboCup 2014, LNAI 8992, pp. 583–595, 2015.
DOI: 10.1007/978-3-319-18615-3_48

far away, the humanoid robot must be capable walking with minimum energy usage, since the energy resources of a humanoid robots is limited.

Biped walking can be defined as the modeling of the predefined Zero Moment Point (ZMP) references to the possible body swing or horizontal CoM trajectory. For approximating the ZMP dynamics, simple physical models have been used, such as Cart-on-a-table [2] and classical Inverted Pendulum Model [3]. Kajita has utilized an Cart-on-a-table model as an approach to generate stable CoM trajectory [4]. The major drawback of the Cart-on-a-table model is its simplification, It models the walking by considering the CoM height is a fixed position, but biomechanical studies show that the CoM height is variant during walking [5]. The vertical CoM movement is also important for energy consumption [1].

A ZMP based walking with variable height can be generated by using inverted pendulum model [3]. Recently, Kormushev et al. have presented an approach for generating an energy efficient ZMP based walking [6]. They have used a vertical CoM trajectory generator, modeled by a Spline representation, and a policy search reinforcement learning. This approach was applied to minimize the energy consumption of a walk with only a specific and predefined step length and step period; however, the shape of an energy efficient vertical CoM trajectory is different for each walking characteristics, including step length and step period.

In this paper, a novel learning scenario is presented to achieve energy efficient ZMP-based walking with variable step lengths and step periods. Fourier basis functions are used as a policy representation for modeling the vertical CoM trajectory. In order to generate the horizontal CoM trajectory, first the position of the foot during a walking is planned. Then, the ZMP trajectory is generated based on the support foot polygon. In the next step, the position of the horizontal CoM trajectory is calculated by using the approach presented in [11], which is able to solve the differential equations of the inverted pendulum model numerically with respect to the input predefined ZMP and vertical CoM trajectories. The Covariance Matrix Adaptation Evolution Strategy (CMA-ES) [7] is used as a black-box optimizer to find an optimized and energy efficient vertical CoM trajectory policy. For different input walk speeds, including step lengths and step periods, different hip height trajectories are learned. The generated trajectories based on Fourier series are the input to the programmable Central Pattern Generators (CPGs) based on Hopf oscillator [8]. CPGs prepare online modulation of trajectories, in the face of changing the walk speed. The system overview of the proposed approach is provided in Fig. 1, which illustrates the role of each component.

The paper structure is as follows. First, A ZMP-based humanoid walking approach to control the walking stability is introduced. Then a CoM vertical trajectory generator based on Fourier basis functions is presented which can produce the periodic vertical hip motion. This trajectory is the input signals of a CPGs approach, which is also introduced in Sect. 3. The learning scenarios with using the CMA-ES are explained in Sect. 4. At the end of the article, the results of learning scenarios are presented and the efficiency of the method to generate energy efficient walking is shown by experiments.

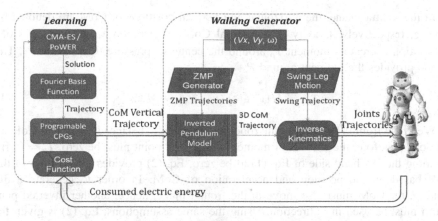

Fig. 1. The interaction between each component of the proposed approach

2   Biped Locomotion Control Approach

A humanoid robot contains many degrees of freedom. It is difficult to maintain the balance of humanoid robot walking. The Zero Moment Point (ZMP) criterion [9] is widely used as a stability measurement in the literature. For a given set of walking trajectories, if the ZMP trajectory keeps firmly inside the area covered by the support foot or the polygon containing the support legs, the given biped locomotion will be physically balanced.

ZMP based biped walking is assumed as a problem of balancing an inverted pendulum model, since in the single supported phase human walking can be represented as an inverted pendulum a predefined vertical CoM trajectory is the input of this model. Kagami et al. proposed an approach to generate the horizontal CoM trajectory by solving numerically inverted pendulum equations [10]. Figure 2-a shows the NAO humanoid robot and the inverted pendulum model. Figure 2-b shows a schematic view of the inverted pendulum model in *XZ* plane or in the sagittal plane. Two sets of inverted pendulum are used to model a 3D walking. One is for movements in the frontal plane; another is for movements in the sagittal plane.

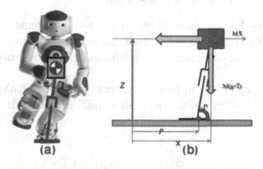

Fig. 2. (a) Frontal view of the NAO robot and the inverted pendulum model (b) A schematic view of the inverted pendulum model

In the sagittal plane, the horizontal and vertical positions of CoM are denoted by x and z, respectively. Gravity g, horizontal CoM acceleration $\ddot{x}$, and vertical CoM acceleration $\ddot{z}$, create a moment T_p around the center of pressure (CoP) point P_x. The Eq. (1) provides the moment around P.

$$T_p = M\left(g + \ddot{z}\right)\left(x - P_x\right) - M\ddot{x}z \tag{1}$$

We know from [9] that when the robot is dynamically balanced, ZMP and CoP are identical, therefore, the amount of moment in the CoP point must be zero, $T_p = 0$. By assuming the left hand side of Eq. (1) to be zero, Eq. (2) provides the position of the ZMP based on the position and acceleration of CoM. In order to generate a 3D walking, the CoM must also move in the frontal plane; hence, another inverted pendulum must be used in y direction. Using the same assumptions, Eq. (2) is given for movements in the frontal plane denoted by y.

$$P_x = x - \frac{z}{g + \ddot{z}}\ddot{x} \qquad P_y = y - \frac{z}{g + \ddot{z}}\ddot{y} \tag{2}$$

In order to apply the inverted model in a biped walking problem, first the positions of the support foot during a walk must be determined. In a forward walk, the support foot positions are calculated based on the desired input step length. Then, the ZMP trajectory is designed based on support foot positions and the input step period. The vertical CoM position and acceleration trajectory must also be determined as the input of the inverted pendulum model, our approach to generate vertical CoM trajectories is explained in Sect. 3. In the final step, the horizontal position of the CoM is calculated by solving the differential equations (2). The main issue of using the inverted pendulum is how to solve these differential equations. The solution is explained in Sect. 2.1. Finally, an inverse kinematics method is used to find the angular trajectories of each joint based on the planned position of the feet and generated CoM position. We used and developed an inverse kinematic approach, which was applied on the NAO humanoid soccer robot, see details in [11].

2.1   Horizontal CoM Trajectory Generation

Kagami et al. proposed an approach to generate walking patterns by solving the ZMP equations numerically [10]. Kajita et al. used this numerical approach and the inverted pendulum model in order to generate the horizontal CoM trajectory of a ZMP based biped running [3].

In this numerical approach, in order to generate horizontal CoM, first the position and acceleration of CoM are discretized with a small time step Δt.

$$x(i\Delta t) \rightarrow x(i) \tag{3}$$

$$\ddot{x}(i\Delta t) \rightarrow \frac{x(i-1) - 2x(i) + x(i+1)}{\Delta t^2}$$

Then, a tridiagonal system for the Eq. (2) is written as:

$$P_x = a_i x(i-1) + b_i x(i) + a_i x(i+1) \tag{4}$$

Where,

$$a_i = -\frac{1}{\Delta t^2}\left(\frac{z(i\Delta t)}{g + \ddot{z}(i\Delta t)}\right) \qquad b_i = 1 + \frac{2}{\Delta t^2}\left(\frac{z(i\Delta t)}{g + \ddot{z}(i\Delta t)}\right) \tag{5}$$

For generating CoM trajectory, the linear system is obtained, which is presented in Eq. (6). In order to solve this tridiagonal system Thomas algorithm can be applied. The solution can be obtained in $O(n)$ operations. Here, $n = T_s/\Delta t$, in this study Δt is assumed to be 0.005 s, and T_s is the total time in which CoM is calculated.

$$
\overbrace{
\begin{bmatrix}
P_{(1)} \\
P_{(2)} \\
\\
\vdots \\
\\
P_{(n)}
\end{bmatrix}
}^{ZMP_x}
=
\overbrace{
\begin{bmatrix}
b_1 & a_1 & & & \cdots & & 0 \\
a_2 & b_2 & a_2 & & & & \\
& a_3 & b_3 & a_3 & & & \\
& & & \ddots & \ddots & & \vdots \\
& & & a_{n-1} & b_{n-1} & a_{n-1} \\
0 & & & \cdots & & a_n & b_n
\end{bmatrix}
}^{k}
\overbrace{
\begin{bmatrix}
x_{(1)} \\
x_{(2)} \\
x_{(3)} \\
\vdots \\
\\
x_{(n)}
\end{bmatrix}
}^{CoM_x}
\tag{6}
$$

Since n must be a finite value, therefore, boundary conditions must be used. Kagami et al. [10] assumed T_s to be a given time period of the walking step, and he presented boundary conditions for the beginning and the end of the walking step. The walk was assumed to be statically stable in the beginning and end of each walking step, and $a_1 = a_n = 0$. In this algorithm, the initial and final positions of CoM during a walk step need to be defined before the robot actually starts to walk. These are drawbacks of this approach, since the beginning of each walking step is not always statically balanced, i.e. the initial acceleration and velocity of the CoM have not been considered. In addition, before the robot starts to walk, it is not possible to define the exact position of CoM in the beginning and final of the walking step.

In order to remedy the aforementioned problems, we use the presented method by assuming $x_{(1)}$ equal to the middle of the feet at the beginning of the previous two walk steps, and $x_{(n)}$ is also equal to the middle of the feet in the predicted position of the following five steps. Consequently, the ZMP trajectory is given to the algorithm for a walk, which has been starting in the previous two steps and lasting in the future five steps. The CoM trajectory of the current walking step is extracted from the calculated CoM trajectory. At the beginning of each walking step, the presented procedure is repeated.

3 Vertical CoM Trajectory Model

We consider the height trajectory as the periodic movement. In this study, vertical CoM trajectory is represented by the first five terms of the Fourier basis functions. Therefore, the equation of our vertical CoM trajectory generator is given in (7).

$$F(t) = \mathbb{C} + \sum_{i=1}^{i=2}\left(\beta_i \cos\left(\frac{i\pi t}{L}\right) + \alpha_i \sin\left(\frac{(i-1)\pi t}{L}\right)\right) \tag{7}$$

The parameter L is equal to the step period, therefore the generator has five parameters such as $\mathbb{C}$, β_1, β_2, α_1 and α_2. A black-box optimization approach can be applied in order to find the optimized hip height trajectory generator with respect to energy efficient walking, in the Sect. 4 we describe our optimization scenario. The generated trajectory by Fourier basis functions is the input to the programmable CPGs.

In the CPGs implementation studies, nonlinear oscillators i.e. Hopf are interesting because of their synchronization properties when they are coupled with other oscillators or with an external drive signal. Most CPGs use phase-locking behavior for their coupling method [12]. If intrinsic frequency of the oscillator is close to frequency component of the periodic input, phase-lock behavior will appear, and synchronization will be done perfectly. In 2006, Righeti et al. designed an adaptive oscillator based on Hopf oscillator which was able to learn CPGs frequency from the frequency of periodic input signals [8]. They called their adaptive mechanism dynamic Hebbian learning because it shared similarities with correlation-based learning found in neural networks [13]. The structure of the network of adaptive Hopf oscillators is shown in Fig. 3.

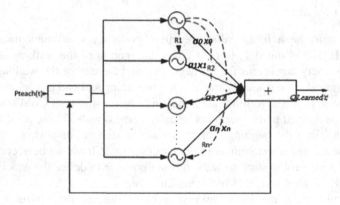

Fig. 3. A Schematic view of network of adaptive Hopf oscillators as a CPG block

In this study the external drive signal, or teaching trajectory, is the output trajectory of the presented Fourier based generator. Each oscillator is responsible for learning one frequency component of the signal. The network can be designed by four oscillators and each oscillator is denoted by i. The output of the system is the weighted sum of the output of the oscillators $Q_{\text{learned}}(t) = \sum_i a_i x_i$, here a_i is assumed the amplitude of each learned frequency. By using negative feedback loop, the already learned frequencies will be subtracted from the teaching signal $F(t) = P_{\text{teach}}(t) - Q_{\text{learned}}(t)$. It leads the system to adapt to remaining frequencies component which have not yet converged. According to the fact that each oscillator has its own phase shift, a variable encoding phase difference between the oscillator and the first oscillator of the network is associated with each of them. In order to reproduce any phase relationship between the oscillators, the Kuramoto coupling scheme [14] is used. The Equations describing the Total CPG's learning and dynamics are given as follow.

$$\dot{x}_i = \gamma(\mu - r_i^2)x_i - \omega_i y_i + \in F(t) + \tau \sin(\theta_i - \emptyset_i) \tag{8}$$

$$\dot{y}_i = \gamma(\mu - r_i^2)y_i - \omega_i x_i \tag{9}$$

$$\dot{\omega}_i = \in F(t)\frac{y_i}{r_i} \tag{10}$$

$$\dot{\alpha}_i = \beta x_i F(t) \tag{11}$$

$$\dot{\emptyset}_i = \sin(\frac{\omega_i}{\omega_0}\theta_0 - \theta_i - \emptyset_i) \tag{12}$$

$$\theta_i = sgn(x_i)cos^{-1}(-\frac{y_i}{r_i}) \tag{13}$$

Equations 8, 9 and 10 are representing Hopf oscillator and its frequency learning, where γ controls the speed of recovery after perturbation. In Eq. 8, the Kuramoto coupling method is represented by $\tau \sin(\theta_i - \emptyset_i)$ in order to achieve phase synchronization between oscillators. Each adaptive oscillator is coupled with oscillator 0, with strength τ to keep correct phase relationships between oscillators. $\emptyset_i$ is the phase difference between oscillator i and 0.

Equations 12 and 13 shows how $\emptyset_i$ can converge to the phase difference between the instantaneous phase of oscillator 0, θ_0 scaled at frequency ω_i and the instantaneous phase of oscillator i, θ_i Learning rule for updating a_i is presented by Eq. 11, where β is learning rate. Learning rule shows how correlation between x_i and $F(t)$ will be maximized. The correlation will be positive on average and will stop increasing when frequency component ω_i disappears from $F(t)$ because of the negative feedback loop. The negative feedback is working like amount of the error, and learning rule is working like the perceptron rule and since the input signal is linearly separable, the above online algorithm will converge.

As conclusion, applying learning rules given as differential equations, parameters such as intrinsic frequencies, amplitudes, and weights of phase coupling can be automatically adapted to a teaching signal. One of its interesting aspects is that the learning is completely embedded into the dynamical system, and does not require external optimization algorithms.

Using Fourier basis functions together with presented CPG concept the proposed trajectory generator model, or policy representation, has the following advantages:

Smoothness: Using the CPGs increase basin of stability of walking. CPGs generate smooth and continuous trajectories without sudden accelerations, which enable the robot not fall and also reduce its energy consumption. By changing the step length and period during the walk, the robot may change its energy efficient vertical CoM trajectory; CPGs make able this change to be smoothly. CPGs also have the ability of frequency adaptation when walking step period and CoM trajectory period changed.

Periodicity: The Fourier basis function is easily able to represent periodic or cyclic movement. A biped walk often consists of periodic movements.

Convergence: the frequency of a walk is equal to the frequency of its vertical CoM trajectory. Using Fourier basis function the frequency parameters is eliminated, therefore the robot converges to the energy efficient walk faster compared to the approaches uses Spline basis function [6].

4 Learning Scenario

In this study, the vertical CoM trajectory is represented by the first five terms of the Fourier basis functions. The optimized energy efficient vertical CoM motion must be achieved for different step periods and step lengths. Since the step length and step period are continuous variables, they are discretized with a proper resolution. The boundaries of the step lengths and step period resolutions used in this work are the following:

Step Length = [0.06.....0.18] *m*; *Resolution* = 0.04.
Step Period = [0.4.... 0.8] *s*; *Resolution* = 0.2.

There are 12 possible combinations of the step periods and step lengths, and for each of them the energy optimization is performed. The optimized values of the Fourier basis functions terms must be found, with respect to minimization of actuator electrical power consumption. Bipedal walking is known as a complicated motion since many factors affect walking style and stability, such as robot's kinematics and dynamics, collision between feet and the ground. In such a complex motion, relation between gait trajectory and walking characteristic, e.g. energy consumption, is nonlinear. Stochastic optimization algorithms can be applied to find the optimized parameter values of the CoM vertical trajectory generator with respect to generate an energy efficient walk.

In this paper, Covariance Matrix Adaptation Evolution Strategy (CMA-ES) is used as a stochastic optimization algorithm for our gait optimization scenario. CMA-ES is a population-based stochastic, derivative-free method, which can be used in black-box optimization problems or direct policy search reinforcement learning. It has been successfully applied previously on gait optimization scenarios [15, 16]. It is also reported that CMA-ES could achieve better results and faster convergence compared to other famous stochastic optimization techniques such as particle swarm optimization (PSO) and Genetic Algorithm (GA) [16].

CMA-ES generates a set of candidates, as the population, sampled from a multi-variate Gaussian distribution. After generating the population, CMA-ES evaluates each candidate with respect to a fitness measure. After evaluating all the candidates in the population, the mean of the multivariate Gaussian distribution is recalculated as a weighted average of the candidates with the highest fitness. The covariance matrix of the distribution is also updated to bias the generation of the next set of candidates toward directions of previously successful search steps. In this study the population size is assumed to be 8.

For minimization of electrical power and energy consumption, the electrical power must be measured. The electrical power for a motor can be given in a simple form by

$P_m = I^2 R$, where I is the current, R the resistance. The motor stall torque is calculated by $\tau = K_t I$, where K_t is the motor torque constant. By combining these expressions, the electrical power can be rewritten as $P_e = \frac{r}{K_t^2}\tau^2$. Therefore, in this study, the cost metric is measured by the sum of the joint-torques squared.

5 Results and Discussions

In this study, a simulated NAO robot is used in order to test and verify the approach. The NAO model is a kid size humanoid robot that is 58 cm high and 21 degrees of freedom (DoF). The simulation is carried out using the RoboCup soccer simulator, rcsssever 3d, which is the official 3D simulator released by the RoboCup community, in order to simulate humanoids soccer match. The simulator is based on Open Dynamic Engine (ODE). The ODE can report the produced torque of each joint in each simulation step time, therefore the sum of the joint-torques squared can be calculated as the cost function.

Using the CMA-ES, after 30 iterations and 240 trails, the robot could reduce its energy usage by 25 percent, on average in all learning scenarios. The optimization is performed for 10 s walking with all the step lengths and step periods, which were presented in Sect. 4. Figure 4, shows the convergence of the cost function of the learning scenario for the walk with step length 0.1 m and step period 0.4 s. The optimized vertical CoM trajectory for this walk is also shown in the Fig. 5.

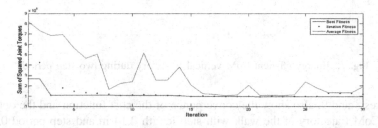

Fig. 4. CMA_ES convergence for walk with step length 0.10 m and step period 0.4 s

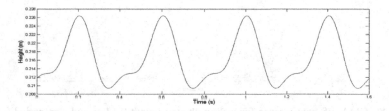

Fig. 5. CoM vertical trajectory for a walk during 1.6 s

The walking achieved by the above learning scenario presented in Fig. 4 is compared to the fixed height walking with the height assumed to be the offset of learned trajectory. The results show the sum of squared torques of the walk with variable height is reduced by 25 percent compared to the walk with fixed height. The same walking parameters with minor tuning are tested on a real NAO robot and the robot achieved to a more energy efficient walking. Please refer to https://www.dropbox.com/s/ft1en4 blotnwcrx/Real_Robot_Experiment.wmv to watch the video of the fixed height walking versus the variable height walking. Figures 6 and 7, show the convergence of the cost function and the energy efficient vertical CoM vertical trajectory for the walk with step length 0.1 m and step period 0.8 s.

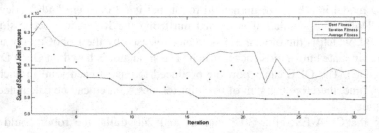

Fig. 6. Learning convergence for walking with step length 0.10 m and period 0.8 s

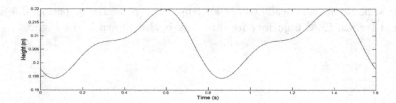

Fig. 7. Energy efficient CoM vertical trajectory during two step periods

Figures 8 and 9, also show the convergence of the cost function and the optimized vertical CoM trajectory of the walk with step length 0.14 m and step period 0.4 s.

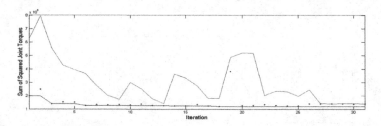

Fig. 8. CMA_ES convergence for walk with step length 0.14 m and step period 0.4 s

As shown in Figs. 5, 7 and 9, the optimized CoM vertical trajectory for different walking characteristics are different. By using programmable CPGs, the robot can change its walking speed, and modulation of the CoM trajectories can be done

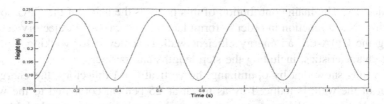

Fig. 9. Optimized vertical CoM vertical trajectory for the above walking scenario

autonomously. Figure 10 shows the modulation of CoM trajectory of a walk when the robot, changes its walking characteristics from walk with step length 0.10 m and step period 0.8 s to the walk with step length 0.14 m and step period 0.4. This change is happening after the four seconds from the starting of the walk.

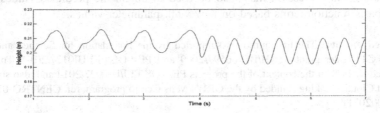

Fig. 10. Modulation of the vertical CoM trajectory by using the CPGs

For the same walking scenario shown in the Fig. 10, we test the change of the vertical CoM trajectories, this time with only use the Fourier basis function generator. Figure 11, shows this experiment. This figure also illustrates that, at the time when the change is happening, the CoM vertical trajectory has the sudden acceleration. In our experiment the robot in this scenario fell four times in 10 tests, this happens because of the explained sudden acceleration in vertical CoM trajectory. Nevertheless, by using CPGs for the same walking scenario the robot did not fall in 10 times tests, because of the smooth change in vertical CoM trajectory, as it is shown in Fig. 10.

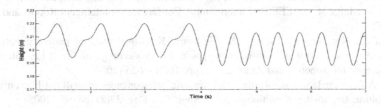

Fig. 11. Changing the vertical CoM trajectory by the Fourier based generator

6 Conclusions

This paper presented an approach to create an energy efficient walk. The walking controller approach is a ZMP based approach, which the ZMP dynamics is modeled by an inverted Pendulum model. A numerical approach is used to generate the horizontal

CoM trajectory. The main contribution of this paper is the using of the CPG approach with Fourier based function in order to formulate the vertical CoM trajectory generator. By using the CMA-ES, an energy efficient walk is achieved for walking types with different characteristics, including the step lengths and periods.

The results show that by optimizing the vertical CoM trajectory, the energy consumption of the walk is reduced by as much as 25 percent compared to the walking with fixed height. For different step lengths and periods, the optimized CoM vertical trajectory is different in shape and characteristics. By using CPGs the online modulation and change of the vertical CoM trajectories is done smoothly and without jerk.

Since a ZMP-based approach is used and CPGs can generate smooth trajectories, the generated walking is stable, and the risk of hardware damage during the gait learning procedure is low. Therefore, Future work will be concerned with performing the gait learning directly on a real NAO robot. For improving the learning generalization, the linear regression may also be used to obtain the predicted values of the Fourier basis function terms based on new walk parameter values.

Acknowledgements. The first author is supported by the Foundation for Science and Technology (FCT) under grant SFRH/BD/66597/2009 and PEst-OE/EEI/UI0127/2014. This work was funded by FCT in the context of the projects PEst-OE/EEI/UI0027/2014 and also supported by project Cloud Thinking (funded by the QREN Mais Centro program, ref. CENTRO-07-ST24-FEDER-002031).

References

1. Gordon, K.E., Ferris, D.P., Kuo, A.D.: Metabolic and mechanical energy costs of reducing vertical center of mass movement during gait. Arch. Phys. Med. Rehabil. **90**, 136–144 (2009)
2. Kajita, S., Kanehiro, F., Kaneko, K., Yokoi, K., Hirukawa, H.: The 3D linear inverted pendulum mode: a simple modeling for a biped walking pattern generation. In: Proceedings of the IEEE/RSJ International Conference on Intelligent Robots and Systems, pp. 239–246 (2001)
3. Kajita, B.Y.S., Nagasaki, T., Kaneko, K., Hirukawa, H.: ZMP-based biped running control. In: Proceedings of the IEEE/RSJ International Conference on Intelligent Robots and System (2007)
4. Kajita, S., Kanehiro, F., Kaneko, K., Fujiwara, K.: Biped walking pattern generation by using preview control of zero-moment point. In: Proceedings of the IEEE International Conference on Robotics and Automation, pp. 1620–1626 (2003)
5. Kuo, A.D., Donelan, J.M., Ruina, A.: Energetic consequences of walking like an inverted pendulum: step-to-step transitions. Exerc. Sport Sci. Rev. **33**(2), 88–97 (2005)
6. Kormushev, P., Ugurlu, B., Calinon, S., Tsagarakis, N.G., Caldwell, D.G.: Bipedal walking energy minimization by reinforcement learning with evolving policy parameterization. In: Proceedings of the IEEE/RSJ International Conference on Intelligent Robots and Systems, pp. 318–324 (2011)
7. Hansen, N.: The CMA evolution strategy: a tutorial. Technical report, TU Berlin, ETH Zurich (2005)

8. Righetti, L. Ijspeert, A.J.: Programmable central pattern generators: an application to biped locomotion control. In: Proceedings of the IEEE International Conference on Robotics and Automation, pp. 1585–1590 (2006)
9. Vukobratović, M., Juricić, D.: Contribution to the synthesis of biped gait. IEEE Trans. Biomed. Eng. **16**(1), 1–6 (1969)
10. Kagami, S., Nishivaki, K., Inaba, M., Inoue, H.: A fast dynamically equilibrated walking trajectory generation method of humanoid robots. Auton. Robots **12**(1), 71–82 (2002)
11. Domingues, E., Lau, N., Pimentel, B., Shafii, N., Reis, L.P., Neves, A.J.: Humanoid behaviors: from simulation to a real robot. In: Antunes, L., Pinto, H. (eds.) EPIA 2011. LNCS, vol. 7026, pp. 352–364. Springer, Heidelberg (2011)
12. Pikovsky, A., Rosenblum, M., Kurths, J.: Synchronization: A Universal Concept in Nonlinear Sciences. Cambridge University Press, New York (2003)
13. Righetti, L., Buchli, J., Ijspeert, A.: Dynamic Hebbian learning in adaptive frequency oscillators. Phys. D Nonlinear Phenom. **216**(2), 269–281 (2006)
14. Acebrón, J., Bonilla, L., Pérez Vicente, C., Ritort, F., Spigler, R.: The Kuramoto model: a simple paradigm for synchronization phenomena. Rev. Mod. Phys. **77**(1), 137–185 (2005)
15. MacAlpine, P., Barrett, S., Urieli, D., Vu, V., Stone, P.: Design and optimization of an omnidirectional humanoid walk: a winning approach at the RoboCup 2011 3D simulation competition. In: Proceedings of the Twenty-Sixth AAAI Conference on Artificial Intelligence, pp. 1047–1053 (2012)
16. Farchy, A., Barrett, S., MacAlpine, P., Stone, P.: Humanoid robots learning to walk faster: from the real world to simulation and back. In: International Conference on Autonomous Agents and Multi-Agent Systems, pp. 39–46 (2013)

Special Track on the Advancement
of the RoboCup Leagues

On the Progress of Soccer Simulation Leagues

Hidehisa Akiyama[1], Klaus Dorer[2]([✉]), and Nuno Lau[3]

[1] Fukuoka University, Fukuoka, Japan
akym@fukuoka-u.ac.jp
[2] Hochschule Offenburg, Offenburg, Germany
klaus.dorer@hs-offenburg.de
[3] DETI/IEETA, Aveiro University, Aveiro, Portugal
nunolau@ua.pt

Abstract. Soccer simulation league is one of the founding leagues of RoboCup. In this paper we discuss the past, present and planned future achievements and changes. Also we summarize the connections and inter-league achievements of this league and provide an overview of the community contributions that made this league successful.

1 Introduction

The soccer simulation league is one of the founding leagues of RoboCup [11]. From the first year of RoboCup it was played with teams of 11 versus 11 players. It is therefore the league that sets the standards for collaboration, team play and opponent modelling. It is also the league in which learning algorithms play a key role since the simulated robots are not subject to wear and tear.

After many discussions, in 2004 the league was split into the 2D and 3D Simulation branch. The goal has been to still have a league with 11 player teams, but add another league that keeps the strengths of simulation leagues, yet closes the gap to real robot leagues.

The rest of the paper is organized as follows: Sects. 2 and 3 describe the development and future of 2D and 3D soccer simulation. In Sect. 4 we point out the achievements and connections to other RoboCup soccer leagues and real soccer. Section 5 summarizes the wealth of contributions that made this league possible.

2 2D Simulation

2.1 History

The RoboCup Soccer Simulator (RCSoccerSim) is the official simulator for the 2D Simulation league since 1996. The official RoboCup competition was started in 1997, but the first 2D simulation competition was held in pre-RoboCup 96 in conjunction with IROS-96. RCSoccerSim was designed and developed by Itsuki Noda [19] and later developed and maintained by the RoboCup Soccer Simulator Maintenance Committee as an Open Source project. All software for 2D Simulation is freely available from the official site[1].

---

[1] http://sf.net/projects/sserver.

© Springer International Publishing Switzerland 2015
R.A.C. Bianchi et al. (Eds.): RoboCup 2014, LNAI 8992, pp. 599–610, 2015.
DOI: 10.1007/978-3-319-18615-3_49

Fig. 1. A screenshot of 2D soccer simulator.

The 2D Simulation league is designed as a soccer competition which is played on a virtual 2D plane field (Fig. 1). All objects, such as the ball and players, are modeled as circle on a 2D plane, therefore players never jump and kick the ball into the air. The players' actuators are also simplified and very much different from real robots. However, almost all soccer rules are implemented and the simulator provides a completely distributed multiagent system which realizes full 11 vs 11 soccer games. The aim of these simplifications is to encourage RoboCup teams to concentrate on the research of teamwork.

With progress in a gameplay level, various rule changes have been made to push the progress by the league:

1998 Two important features were implemented. (1) A goal keeper was introduced. This special player possesses an additional action namely 'catch ball'. (2) An offside rule was implemented in the automatic referee. By introducing these features, the human soccer rules are almost fully implemented. Moreover, a stamina model was introduced. It imposes a resource model for the players. To perform actions, stamina needs to be invested and the model implements short-term expenditures with long-term recovery phases. This encourages the development of strategic resource management.

1999 The `turn_neck` command was introduced. This command enables players to change their head direction. As players' visible area is restricted, this command enables players to gather more environment information without interfering with body actions. In order to encourage teams to consider online adaptation, an online coach was introduced. The online coach can observe the whole field during a game and can send advice messages to players.

2000 The ball kick power was increased. This change accelerated the game pace. As a result, the competitions became more attractive.

2001 Heterogeneous players and player substitution were introduced. Teams still can use the default (homogeneous) player, but can select a heterogeneous player if necessary. Only the coach can substitute players during non-play on period. Because the physical abilities of heterogeneous player are randomly generated for each game, online coaches have to deliberate the player assignment according to their team strategy.

A new contest was added: the Coach Competition. In this competition the coach agent advises a team trying to improve its performance. To allow for universality of communication, a standardized coach language was introduced. The rules of this competition have changed frequently and in 2005 the coach task became that of identifying strategies (opponent modelling) instead of providing efficient strategies for the own team.

2002 In order to introduce new challenges related to communication, the length of auditory communication messages among players was shortened. Compensating this restriction, two new commands, `attentionto` and `pointto`, were introduced. The `attentionto` command can be used to focus players' attention on a particular player's auditory message. If `attentionto` is off, the player will hear one auditory message from each team selected randomly. The `pointto` command enables players to point to a spot on the field using a virtual arm and other players can see the arm direction. Players can send a location, and indirectly their intention to other players as a visual message. As the improvements of the dribbling skill progressed further, it became too difficult for defenders to block a smart dribbler. The new `tackle` command enables players to kick the ball in a range wider than the `kick` command succeeding with a probability based on the ball position relative to the tackling player yet causes immobility for a few cycles.

2003 An automatic penalty mode was added to the simulator. If the game ends in a draw, the automatic penalty mode may be started. As a RoboCup-based testbed for machine learning, a simplified scenario was implemented, the so-called Keepaway problem. In this game, the one team (the "keepers") attempt to keep the ball away from the other team (the "takers"). Details of the Keepaway task as well as some experiments can be found in [26].

2008 The number of heterogeneous player types was increased from 7 to 14. Assigning the default player type was forbidden except for the goal keeper, and assigning the same heterogeneous player type to several players was forbidden. Under the new rule, each player has to be assigned different player type. This means the player type assignment became more difficult. The range of the goal keeper's catch action was reduced to encourage teams to use more effective positioning of the goal keeper and defensive players.

2009 The stamina capacity model was introduced. This model restricts the total amount of stamina value for each player during the game. Until 2008, players could always recover their stamina. If the stamina capacity is exhausted, players never recover their stamina. As a result, they cannot run by their maximum power. In order to avoid this situation, players have to manage their stamina more carefully and online coaches have to consider the timing of player substitution according to the game situation and the players' tiredness. The dash model was extended to enable players to accelerate their body in 4 directions. Players can adjust their position more flexibly and quickly by using the extended model, but it requires more complicated planning for all move actions. In soccer games, players need to adjust their body direction in order to start the next behavior quickly. Under the previous dash model, players have to turn and dash to the target position, then turn their body to the desired direction. The new dash model enables players to adjust their position without turn action. However, there are tradeoffs, for example, between time and stamina cost.

2010 The tackle command was extended to introduce an intentional foul action. When the intentional foul action is performed, the success probability of action is increased but the player may be penalized. The foul detect probability is defined for each heterogeneous player type. In the current implementation, the foul detection depends only on the probability. If an intentional and dangerous foul is detected by the automatic referee, the referee penalizes the player by giving the yellow or red card. A new heterogeneous parameter that affects the goal keeper's catch range was introduced. Teams are still allowed to assign the default type player to the goal keeper in order to keep the Compatibility with previous team binaries. The dash model parameter was changed. Players are now able to accelerate their body in 8 directions.

2.2   Present and Future

Although the community held numerous discussions for introducing new features, no major changes were introduced since 2010. This does not mean the 2D Simulation league stops its progress. Gabel et al. evaluated the performance improvement of teams quantitatively [8]. They compared the performance of teams that participated in the competitions from 2003 to 2007 which has been possible, because no major changes were applied in that period. The results showed that newer champion teams would always overcome older champions.

Of course, new challenges that discover new research are always required. Although an online coach is available since 1999, online game analysis and online adaptation are still important research topics in the 2D simulation league. In the past few years, several teams started to prepare more than one strategy and switch them according to the opponent team. The role of online coach will become more important in next few years. With the improvement of online coach, the standard coach language will also be updated.

Another important topic is the collaboration with real human soccer. Because the 2D Simulation league focuses on the research of teamwork, 2D Simulation has begun to apply decision making techniques and game analysis techniques not only to 2D league but also to human soccer like in [1].

3   3D Simulation

In 2004 the 3D soccer simulation competition was born. Its main goals are those of the 2D soccer simulation, i.e., to keep the focus on multiagent system coordination research and to use the new 3D simulator to conduct research that cannot be performed using real robots, either because of time, money or hardware constraints. It adds the third dimension to the game seeking to make it more realistic and uses more realistic robot models and environment dynamics.

3.1   History

2004 The first RoboCup 3D Simulation Competition took place in 2004. This competition used the Simspark generical physical simulator platform [20] to build a 11 against 11 soccer game simulator called rcssserver3d. The simulation includes several innovations such as (obviously) the 3D model of the environment, but also the use of the ODE physics engine library to model and update the dynamics of simulated objects and a middleware for agent simulations, SPADES [23], that manages the distributed simulation, and ensures results do not depend on network or system load, and a new timing model, where agent thinking times are taken into account.

To keep the 3D environment in line with 11 vs 11 games, the 3D robot model used in 2004 was quite simple and its shape was a sphere, (Fig. 2). All actions were performed by applying forces either to the agent or to the ball.

2005 The 2005 rules and models were identical to the ones used in 2004. However, there were several changes inside the simulator to make it more efficient and to remove bugs. Big effort was also spent in the documentation of the new simulator as in its initial year, the documentation was quite scarce.

2006 The robot vision perceptor was changed to restrict agents vision to a limited field of view. To control the looking direction a PanTilt effector was added to the robot. A limited bandwidth broadcast communication model was also introduced in this year. Soccer rules were also changed to approximate FIFA rules, and an offside rule was added to the simulator. Optimizations made to the simulator turned it into a much more efficient application.

2007 In the 2007 competition, the robot model was changed from the simple sphere model towards the targeted humanoid robot model. The humanoid model used in 2007 was based on real humanoid HOAP2 from Fujitsu [3], which can be seen in Fig. 2. Each of the joints could be controlled setting the desired angular velocity. The new model introduced many new research challenges to the Simulation League, enabling this league to conduct research on humanoid robotics. The trade-off was that the number of players per team had to be reduced from 11 to 2, as the simulator could not cope with more than 4 humanoid agents in total, and the restricted vision was also changed to a global vision perceptor. The development of efficient humanoid skills became the most important point for the success of the team, making coordination research less important in this year. Nevertheless, this change was necessary to bring Simulation League research closer to the real robots and, as predicted, the number of players per team was increased in the following years making coordination essential again.

2008 With the introduction of Aldebaran's Nao Robot in the Standard Platform League, also 3D soccer simulation switched to the simulation of Nao robots. In 2008 games were played 3 versus 3. With this step, the 3D soccer simulation league again made a step towards narrowing the gap between soccer simulation leagues and real robot leagues.

2009 Saw an increase of the number of robots per team to 4 with the goal in mind to reach 11 vs 11 players eventually. The robots' omnidirectional view was replaced by a restricted view with a cone similar to real Nao robots.

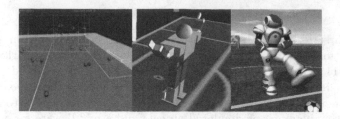

Fig. 2. Robot models used. Spheres (2003-2006), Hoap robot model [3] (2007) and NAO (from 2008).

2010 The major change for 2010 in Singapore has been the increase of teams of 6 simulated Nao players. Better hardware and software optimizations made it possible to increase the number of players and the field size accordingly. Also, to boost the development of simulation server and infrastructure, a 3D development competition was held.

2011 The last intermediate step to 11 versus 11 was taken in 2011 with teams of 9 versus 9 NAOs. Again the field size has been increased to keep the space per robot roughly the same. The visualization was done using a new visualizer RoboViz [25]. With a special 3D beamer and glasses the games could be watched in 3D in Istanbul.

2012 A major milestone has been achieved in 2012 simulating games of 11 vs 11 NAO robots for the first time. The soccer field size has been increased to represent the size of a real soccer field with respect to the size of the robots. As of 2012, both soccer simulation leagues were the only leagues with the full final team size played in 2050.

2013 A major strength of simulation leagues was first exploited in 2013 with the introduction of heterogeneous robots. In no other league it is so cheap to change robot models and use different variations of a robot model in one and the same game. The rational behind introduction variations of the NAO robot has been to shift from programming good behavior skills for a very specific robot to creating good algorithms for the behaviors to any similar robot, especially for walking. The exact robot models have only been published some days before the competition. Only teams that are able to adjust their algorithms and have the modified robots learning the skills succeed.

Also a drop-in player challenge was held in conjunction with several other leagues. Mixed teams with two players from each participant team played against other mixed teams put together from other participating teams. The challenge is to play with other players without a common strategy agreed upon. A couple of games with a changed mix of teams determined the winner.

3.2   Present and Future

For 2014, a new goal has been introduced in 3D soccer simulation: the league committed itself to work towards having the first running robots. In a running robots challenge, robots are evaluated for their run speed, but also on how much

Fig. 3. 11 vs 11 Nao robot game.

time both feet are off the ground. This is possible in 3D soccer simulation since the simulated Nao robots have slightly stronger motors than real Nao robots. Nao models with toes were introduced for this challenge to allow more human-like walking and running. The idea behind is to explore in a cheap way what hardware setups are required to make running with biped robots possible. While in 2013 the usage of heterogeneous robot models was optional, in 2014 teams are forced to use heterogenous robots. At most 7 robots may use the standard NAO type. The remaining 4 robots have to use 3 different variations of the NAO.

The main challenge for the future will remain to find a good balance between having a multiagent simulation for intelligent robotics research with ideally 11 vs 11 games and having a more realistic simulation in terms of real hardware. Having the 11 vs 11 goal achieved, concrete suggestions have been made for the later: a model for energy consumption and motor warming has been implemented and suggested, stiffness of motors, or more realistic noise models for sensors. Also a transition to new, more realistic simulators like gazebo is an option.

3D Simulation should also play a role in research for new robot models and hardware. This has started with the introduction of heterogeneous robot models including a model with toes. It is now continued with the running challenge in which teams for the first time can suggest their own robot model variations. And it should be further developed in the future to, for example, use new sensors and actuators if these sensors are biologically plausible and likely to be developed on real robots. Suggestions include linear actuators, touch suites and many more (Fig. 3).

4   Inter-League Achievements

Technologies that result from research in the Simulation League are often used in other RoboCup leagues. In some cases, the same institution creates new teams to compete in new leagues using the knowledge from a previous Simulation League team or starts its participation at several leagues simultaneously, this is the case, for example, of CMUnited (2D, Small, MSL, 4legged, SPL), Brainstormers (2D, MSL, 3D), FC Portugal (2D, 3D, Mixed Reality, 4legged, SPL, MSL), UT Austin Villa (2D, 3D, SPL), WrightEagle (2D, 3D, 4legged,SPL), Bold Hearts

(3D, Humanoid) or magmaOffenburg (3D, Humanoid). In other cases the technologies are adopted by completely independent teams from different leagues.

One of the major challenges in soccer is the positioning of the team in the field. Several positioning systems have been proposed in the Simulation League [2,4,14,21,27] and have been used in other leagues [10,12,27].

Simulation is a very adequate environment for the automatic development of behaviors. Several teams have developed machine learning to enhance their 2D and 3D teams. A very significant example is the research developed by the Brainstormers team, that has been applied in the Simulation League and also in the Middle-Size League [7,22]. Using humanoid models at the 3D Simulation league fostered automatic humanoid behavior generation research. The techniques that have been developed in the simulation league are mostly based on optimization and machine learning, either using model-free or model-based approaches and have had a strong impact on real robot leagues [5,6]. It is very interesting to see that, as referred to before, several 3D Simulation Teams have created teams that participate in the Standard Platform League or at the Humanoid League.

Teams that maintain a strong link to Simulation League research and developments have been very successful in other leagues. There are several examples of this kind of teams that became RoboCup champions in SSL (CMUnited, CMDragons), 4Legged (CMPack), MSL (Brainstormers Tribots, CAMBADA) and SPL (UT Austin Villa) (Fig. 4).

5   Community Contributions

More than in most leagues, soccer simulation leagues' progress is heavily influenced by community contributions. The main workhorses for 2D and 3D Simulation are the soccer simulators. Both have been fully developed and maintained by the community specifically to ensure independence from external vendors and so that the league retains the right to fully use and modify them as the requirements of the league progress. Around the simulators, a wealth of other software was created to visualize, analyze and comment games as well as run games or complete tournaments.

Simulator Development. The original 2D soccer simulator was proposed and developed by Noda et al. in 1997 [19]. Many people contributed to the development under the lead of Itsuki Noda, Tom Howard and later and until now by Hidehisa Akiyama. The simulator runs on Linux and has reached version 15[2]. A more detailed history of changes is provided in Sect. 2.1.

The 3D Simulator[3] was initiated by work from Markus Rollmann, Oliver Obst [20], Jan Murray and Joschka Boedecker who provided the Spark simulator as a generic simulator for 3D. It is based on ODE physics engine[4] to simulate physically realistic objects in 3D. The actual soccer simulation is a separate module based on top of Spark. It provides the robot models, soccer rules, automated

---

[2] http://sourceforge.net/projects/sserver/.

[3] http://sourceforge.net/projects/simspark/.

[4] http://ode.org/ode-latest-userguide.html.

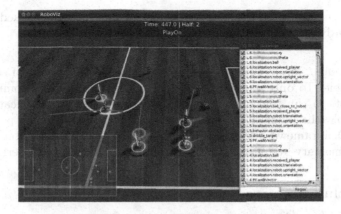

Fig. 4. Debugging features of RoboViz [25].

referee and more. It proved its flexibility many times, for example whenever new robot types were used in the simulation. Today it provides the base for having heterogeneous robots types available in one and the same game.

Again many people from the community contributed to the server development under the lead of the aforementioned and later lead by Hedayat Vatankah.

Visualizers. To the public, the quality of the simulations is mainly associated with the quality of visualizing it. A couple of 2D visualizers are available for free for Linux and Windows, many of them adding debugging features and logfile replaying. 3D Simulation offers two visualizers, rcssmonitor3D and RoboViz [25][5]. The later is now used as main visualizer in competitions and by many teams for team-specific graphical visualization of real time debugging information.

Source Code Releases. It is a condition of participation that binaries of all teams are released after a RoboCup. This allows all teams to test against old teams and to perform scientific research including other team's binaries.

Source code release, however, is voluntary. Nevertheless, more than 15 teams have released source code libraries or fully functional agent code in 3D. Many teams in 2D based their code on source code bases from CMUnited 99, FCPortugal, UVATrilearn and Helios. The code is available for C++, Java, C#, Clojure, C, Prolog and Javascript. This way, new teams do not have to redevelop low level communication or geometrical transformations, but can focus on high level skills and decision making[6].

Benchmarks. Keepaway (see Sect. 2.1), a subtask of 2D Simulation has been a well established benchmark domain in the reinforcement community for many years [26]. The magmaRunChallenge is a currently developed benchmark tool

---

[5] https://sites.google.com/site/umroboviz/.
[6] http://wiki.robocup.org/wiki/Soccer_Simulation_League.

for the 3D running robot challenge in which robots are benchmarked for speed and the relative amount of time both feet are off the ground.

Miscellaneous. Many more contributions added to the success of soccer simulation league. To only mention a few: the league managers, a comprehensive set of scripts to run tournament rounds automatedly, a communication proxy that decouples the simulation from any network or simulation server performance issues in 3D, a framework to host a 2D tournament on a remote system, a couple of live commentary systems [9, 18, 28] or the ssil, an automated internet league.

6   Conclusion

In this paper we have summarized the past, present and future of the 2D and 3D RoboCup Soccer Simulation leagues as an excellent domain for multi-agent, machine learning and humanoid robot research. Due to its low cost, it is a good entry point for new teams into the RoboCup community. Many teams with roots in soccer simulation now participate in hardware leagues with some of them even winning other league competitions. The success of the soccer simulation league is due to the inpour of community contributions and the perseverance of the simulator maintainers which is an essential tool for mapping the path towards the Grand Challenge of 2050.

Appendix

(See Table 1)

Table 1. Table of Champions in 2D and 3D Simulation.

Year	2D champion	Country	3D champion	Country
1997	AT Humboldt	Germany	-	-
1998	CMUnited	USA	-	-
1999	CMUnited	USA	-	-
2000	FC Portugal	Portugal	-	-
2001	TsinghuAeolus	China	-	-
2002	TsinghuAeolus	China	-	-
2003	UvA Trilearn	Netherlands	-	-
2004	STEP	Russia	Aria [24]	Iran
2005	Brainstormers	Germany	AriaKavir	Iran
2006	WrightEagle	China	FC Portugal [13, 17]	Portugal
2007	Brainstormers	Germany	WrightEagle	China
2008	Brainstormers	Germany	SEU-RedSun [29]	China
2009	WrightEagle	China	SEU-RedSun	China
2010	HELIOS	Japan	Apollo3D	China
2011	WrightEagle	China	UT Austin Villa [14, 16]	USA
2012	HELIOS	Japan	UT Austin Villa [15]	USA
2013	WrightEagle	China	Apollo3D	China

References

1. Abreu, P., Moreira, J., Costa, I.: Human versus virtual robotics soccer: a technical analysis. Eur. J. Sport Sci. **12**(1), 26–35 (2012)
2. Akiyama, H., Noda, I.: Multi-agent positioning mechanism in the dynamic environment. In: Visser, U., Ribeiro, F., Ohashi, T., Dellaert, F. (eds.) RoboCup 2007: Robot Soccer World Cup XI. LNCS (LNAI), vol. 5001, pp. 377–384. Springer, Heidelberg (2008)
3. Boedecker, J., Dorer, K., Rollmann, M., Xu, Y., Xue, F., Buchta, M., Vatankhah, H.: Simspark users manual. Version 1, 17–18 (2008)
4. Dashti, H.A.T., Aghaeepour, N., Asadi, S., Bastani, M., Delafkar, Z., Disfani, F.M., Ghaderi, S.M., Kamali, S., Pashami, S., Siahpirani, A.F.: Dynamic positioning based on voronoi cells (DPVC). In: Bredenfeld, A., Jacoff, A., Noda, I., Takahashi, Y. (eds.) RoboCup 2005. LNCS (LNAI), vol. 4020, pp. 219–229. Springer, Heidelberg (2006)
5. Domingues, E., Lau, N., Pimentel, B., Shafii, N., Reis, L.P., Neves, A.J.R.: Humanoid behaviors: from simulation to a real robot. In: Antunes, L., Pinto, H.S. (eds.) EPIA 2011. LNCS, vol. 7026, pp. 352–364. Springer, Heidelberg (2011)
6. Farchy, A., Barrett, S., MacAlpine, P., Stone, P.: Humanoid robots learning to walk faster: from the real world to simulation and back. In: Proceedings of the 12th International Conference on Autonomous Agents and Multiagent Systems (AAMAS), May 2013
7. Gabel, T., Hafner, R., Lange, S., Lauer, M., Riedmiller, M.: Bridging the gap: learning in the RoboCup simulation and midsize league. In: Proceedings of the 7th Portuguese Conference on Automatic Control (Controlo 2006). Portuguese Society of Automatic Control, Porto (2006)
8. Gabel, T., Riedmiller, M.: On progress in RoboCup: the simulation league showcase. In: Ruiz-del-Solar, J. (ed.) RoboCup 2010. LNCS, vol. 6556, pp. 36–47. Springer, Heidelberg (2010)
9. Jung, B., Oesker, M., Hecht, H.: Virtual RoboCup: real-time 3D visualization of 2D soccer games. In: Veloso, M.M., Pagello, E., Kitano, H. (eds.) RoboCup 1999. LNCS (LNAI), vol. 1856, pp. 331–344. Springer, Heidelberg (2000)
10. Kaden, S., Mellmann, H., Scheunemann, M., Burkhard, H.D.: Voronoi based strategic positioning for robot soccer. In: Proceedings of the 22nd International Workshop on Concurrency, Specification and Programming (CS&P), vol. 1032, pp. 271–282. CEUR-WS.org (2013)
11. Kitano, H., Asada, M., Kuniyoshi, Y., Noda, I., Osawa, E.: Robocup: The robot world cup initiative. In: IJCAI-1995 workshop on entertainment and AI/Alife. Montreal, Quebec, August 1995
12. Lau, N., Lopes, L., Corrente, G., Filipe, N., Sequeira, R.: Robot team coordination using dynamic role and positioning assignment and role based setplays. Mechatronics **21**(2), 445–454 (2011). Cited by (since 1996) 6
13. Lau, N., Reis, L.P.: FC Portugal - high-level coordination methodologies in soccer robotics. In: Robotic Soccer. I-Tech Education and Publishing (2007)
14. MacAlpine, P., Barrett, S., Urieli, D., Vu, V., Stone, P.: Design and optimization of an omnidirectional humanoid walk: a winning approach at the RoboCup 2011 3D simulation competition. In: Proceedings of the Twenty-Sixth AAAI Conference on Artificial Intelligence (AAAI), July 2012

15. MacAlpine, P., Collins, N., Lopez-Mobilia, A., Stone, P.: UT Austin Villa: RoboCup 2012 3D Simulation League Champion. In: Chen, X., Stone, P., Sucar, L.E., van der Zant, T. (eds.) RoboCup 2012. LNCS (LNAI), vol. 7500, pp. 77–88. Springer, Heidelberg (2013)

16. MacAlpine, P., Urieli, D., Barrett, S., Kalyanakrishnan, S., Barrera, F., Lopez-Mobilia, A., Ştiurcă, N., Vu, V., Stone, P.: UT Austin Villa 2011: A champion agent in the RoboCup 3D soccer simulation competition. In: Proceedings of the 11th International Conference on Autonomous Agents and Multiagent Systems (AAMAS), June 2012

17. Marques, H., Lau, N., Reis, L.P.: Architecture and basic skills of the FC Portugal 3D simulation team. Revista do Departamento de Electrnica e Telecomunicaes da Universidade de Aveiro 4(4), 478–485 (2005)

18. Matsubara, H., Frank, I., Tanaka-Ishii, K., Noda, I., Nakashima, H., Hashida, K.: Automatic soccer commentary and RoboCup. In: Asada, M., Kitano, H. (eds.) RoboCup 1998. LNCS (LNAI), vol. 1604, pp. 34–49. Springer, Heidelberg (1999)

19. Noda, I., Matsubara, H., Hiraki, K., Frank, I.: Soccer server: a tool for research on multiagent systems. Appl. Artif. Intell 12(2–3), 233–250 (1998)

20. Obst, O., Rollmann, M.: Spark – a generic simulator for physical multi-agent simulations. In: Lindemann, G., Denzinger, J., Timm, I.J., Unland, R. (eds.) MATES 2004. LNCS (LNAI), vol. 3187, pp. 243–257. Springer, Heidelberg (2004)

21. Reis, L.P., Lau, N., Oliveira, E.C.: Situation based strategic positioning for coordinating a team of homogeneous agents. In: Hannebauer, M., Wendler, J., Pagello, E. (eds.) ECAI-WS 2000. LNCS (LNAI), vol. 2103, pp. 175–197. Springer, Heidelberg (2001)

22. Riedmiller, M., Gabel, T., Hafner, R., Lange, S.: Reinforcement learning for robot soccer. Auton. Robots 27(1), 55–74 (2009)

23. Riley, P., Riley, G.: SPADES - a distributed agent simulation environment with software-in-the-loop execution. In: Winter Simulation Conference Proceedings, vol. 1, pp. 817–825 (2003)

24. Solgi, M., Dezfouli, S., Baghi, H., Ghaderi, A., Mola, O., Kazempour, V., Akhondian, S., Montazeri, H., Nickabadi, A., Moradi, S.: Aria 2005 3D soccer simulation team description. In: Proceedings of the RoboCup Symposium 2005, Osaka, Japan (2005)

25. Stoecker, J., Visser, U.: RoboViz: programmable visualization for simulated soccer. In: Röfer, T., Mayer, N.M., Savage, J., Saranlı, U. (eds.) RoboCup 2011. LNCS, vol. 7416, pp. 282–293. Springer, Heidelberg (2012)

26. Stone, P., Kuhlmann, G., Taylor, M.E., Liu, Y.: Keepaway Soccer: from machine learning testbed to benchmark. In: Bredenfeld, A., Jacoff, A., Noda, I., Takahashi, Y. (eds.) RoboCup 2005. LNCS (LNAI), vol. 4020, pp. 93–105. Springer, Heidelberg (2006)

27. Veloso, M., Stone, P., Bowling, M.: Anticipation as a key for collaboration in a team of agents: a case study in robotic soccer. In: Proceedings of SPIE Sensor Fusion and Decentralized Control in Robotic Systems II. vol. 3839, pp. 134–143, SPIE, Bellingham, WA, September 1999

28. Voelz, D., André, E., Herzog, G., Rist, T.: Rocco: a RoboCup soccer commentator system. In: Asada, M., Kitano, H. (eds.) RoboCup 1998. LNCS (LNAI), vol. 1604, pp. 50–60. Springer, Heidelberg (1999)

29. Yuan, X., Yingzi, T.: Layered omnidirectional walking controller for the humanoid soccer robot. In: IEEE International Conference on Robotics and Biomimetics, ROBIO 2007, pp. 757–762, December 2007

RoboCup Small-Size League: Past, Present and Future

Alfredo Weitzenfeld[1(✉)], Joydeep Biswas[2], Mehmet Akar[3],
and Kanjanapan Sukvichai[4]

[1] University of South Florida, Tampa, FL, USA
aweitzenfeld@usf.edu
[2] Carnegie Mellon University, Pittsburgh, PA, USA
joydeepb@ri.cmu.edu
[3] Boğaziçi University, Istanbul, Turkey
mehmet.akar@boun.edu.tr
[4] Kasetsart University, Bangkok, Thailand
sukvichai@gmail.com

Abstract. The Small Size Robot League (SSL) was among the founding RoboCup leagues in the 1997 competition held during IJCAI'97 in Nagoya, Japan. Since then, the league has experienced various advances in terms of robot design, number of robots, field size, software algorithms and other infrastructure used during the games, among these the recent standardization of the vision system shared by all teams. The SSL league has been one of the fastest paced leagues in RoboCup where teamwork, coordination, high-level strategies and artificial intelligence have played a critical role in the league development. As robots speeds have greatly increased in the past years, the league has witnessed the development of advanced control and cooperative algorithms. In parallel, shared open software, in particular the shared vision system has made it easier for new teams to join the league. In this paper we discuss the past, present and future of the Small Size League in its path towards the goal of achieving robot vs. human soccer in 2050.

Keywords: Small-size league · Shared vision · Omnidirectional control · Artificial intelligence · Wheeled robots

1 Introduction

Since the original robot soccer proposal by Mackworth [1] and the following RoboCup initiative [2], the Small Size League (SSL) has been a unique and pioneering league within the RoboCup initiative consisting of an off-board vision system to perceive all robots in the field [3]. The global vision system simplifies the task of robot localization and mapping problems, enabling teams to focus more on the software algorithms, hardware and control engineering.

Since its foundation a SSL soccer game takes place between two teams of five and just recently six – robots, where each robot must conform to the F180 dimension rule specifying that individual robots must fit within an 180 mm diameter circle and must be no higher than 15 cm. An orange golf ball has been used since SSL foundation, with

© Springer International Publishing Switzerland 2015
R.A.C. Bianchi et al. (Eds.): RoboCup 2014, LNAI 8992, pp. 611–623, 2015.
DOI: 10.1007/978-3-319-18615-3_50

the robots playing soccer on a green carpeted field with its size having increased throughout the years [4], as shown in Fig. 1.

The complete game configuration is shown in Fig. 2. All objects on the field – robots and ball - are tracked by a global vision system that processes the data provided by two cameras that are attached to a camera bar located 4 m above the playing surface. The shared vision system, i.e. SSL-Vision [5], is an open source project maintained by the league's community. The shared vision perceptions are processed by off-field computers belonging to each of the playing teams to provide wireless control of team robots, typically using a dedicated commercial FM transmitter/receiver unit. The off-field computers also receive communication from a referee box or game controller providing state of the game. Typically, these computers also perform most, if not all, of the processing required for coordination and control of the robots.

In general, building a successful team requires clever design, implementation and integration of many hardware and software sub-components into a robustly functioning whole making Small Size robot soccer a very interesting and challenging domain for research and education. In the rest of the paper we will briefly highlight the past, present and future of the league.

Fig. 1. Small size league robots facing against each other at the start of a game with orange golf ball at the center of green carpeted field (Color figure online).

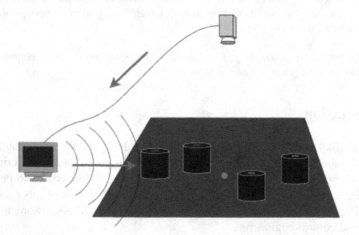

Fig. 2. Diagram illustrates the small size league robot control configuration with global vision system feeding off-field computer with ball position and robot positions and orientations. The off-field computer provides wireless control to the blue and yellow labeled team robots (Color figure online).

2 Past

The Small Size Robot League (SSL) rules constraints have evolved over the years since 1997. The SSL robots in 1997 were limited to have a maximum linear dimension of 18 cm and a maximum projected area of 180 cm^2. Today, the rules limit the robots to have a maximum diameter of 180 mm and maximum height of 150 mm. In 1997, SSL teams were limited to a maximum of 5 robots, and this limit was only recently increased to 6 robots in 2012. The global vision system was initially processed through a single overhead camera, but this was later increased to two cameras, and with the introduction of the double-size field in 2014, to four cameras. The most important changes in time involved primarily the increase in field size and the robot design, starting as two wheeled differential control to the current four wheeled omnidirectional drive. Many of the initial challenges in SSL are still current as described by the CMUnited-97 team that won the first SSL competition [6].

2.1 Vision

One of the primary challenges of the SSL since 1997 had been to devise fast and accurate vision processing algorithms that could process camera images of the field an estimate the locations of all the robots and the ball on the field. The constraints of the vision system are numerous, and the vision system must:

1. Be capable of processing all the images in real time at full frame rates, which today consist of images of size 780 × 580 pixels captured at a rate of 60 Hz,
2. Have minimal latency so as not to adversely affect the motion control of the robots,
3. Be capable of simultaneously tracking multiple robots and the ball on the field,
4. Be robust to robots touching each other without losing track of each individual robot, and
5. Correctly disambiguate between different robots in different orientations on the field, even when touching each other.

Due to these limitations, novel image processing algorithms, including CMVision [7] had to be developed to efficiently and robustly track all the robots and the ball by color segmentation of the images. CMVision has since been adopted as the de facto vision processing algorithm not only by the SSL, but also the other leagues of RoboCup, including the Standard Platform League.

2.2 Robot Design

The first soccer-playing robots in the SSL were differential-drive robots with two actuated wheels per robot and one or more passive castors. Since then, the drive systems of the SSL robots have evolved to use omnidirectional wheels, in particular the "Swedish Wheel" or "Meccanum Wheel" designs [9]. Initially three-wheel designs were most common among teams. However, the three-wheel designs suffered from significantly varying maximum acceleration and velocity profiles as a function of the directions that the robots drove along. To combat this problem, teams later adopted four-wheel designs, which provided more even distribution of the maximum acceleration and velocity as a function of the drive direction. The motors used in initial designs were brushed DC motors, whereas the SSL robots today use 3-phase brushless motors which are more efficient, provide higher torque, and more durable. New control strategies had to be developed in order to control the omnidirectional robots, including a real-time near-optimal minimum time motion controller, introduced by the SSL team Cornell Big Red [10].

SSL robots in 1997 had no special actuators to manipulate the ball, and instead bumped into the ball to kick it around the field. The first dedicated kicking mechanisms used rack-and-pinion linear actuators to propel the ball forward. Today, custom solenoid based kickers are used to efficiently and powerfully kick the ball at speeds of up to 15 m/s, although they are limited by software to comply with the kicking speed limits imposed by the rules. In addition to the main kicker, most teams have an additional solenoid kicker called the "chip-kicker" that can be used to propel the ball into the air to pass over opponent robots. A third manipulation mechanism involves a dribbling horizontal roller that is spun up at high speeds to impart back-spin to the ball on contact, thus allowing robots to hold on to the ball for short periods of time.

3 Present

There have been gradual changes to the Small-Size League rules since 1997 until now, among these, the addition of 1 robot for a total of 6 robots in each team, the introduction of a shared global vision system currently processing two overhead cameras, a significant increase in the field size, and most important the drastic advancement in the design of each robot from the original two wheels to the current four wheels. The ball is a standard orange golf ball, approximately 46 g in mass and 43 mm in diameter. In 2014, the organizing committee is offering teams to compete in either the single-sized field of 6050 mm × 4050 mm (see Fig. 3) or the optional double-size field of 8090 mm × 6050 mm. The double sized field will become the standard field for SSL in 2015. For either field, the playing surface is green felt mat or carpet and the floor under the carpet is level, flat, and hard. The field surface includes an additional 675 mm surface beyond the boundary lines on all sides. The outer 425 mm of this runoff area are used as a designated referee walking area. At the edge of the field surface, a 100 mm tall wall should prevent the ball and robots from running off the edge. All lines are 10 mm wide and painted white. The field of play is divided into two halves by a halfway line. The center mark is indicated at the midpoint of the halfway line. A circle with a diameter of 1000 mm is marked around it (Fig. 4).

A defense area is defined at each end of the field as follows for the two field sizes. In the single-size field, two quarter-circles of radius of 800 mm are drawn on the field of play. A line of length 350 mm parallel to the goal line connects these quarter circles. In the double-size field, two quarter-circles of radius of 1000 mm are drawn on the field of play. A line of length 500 mm parallel to the goal line connects these quarter-circles. The area bounded by this arc and the goal line is the defense area.

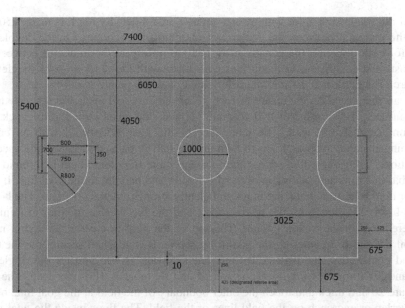

Fig. 3. The field dimensions of the single-size field. The dimensions include boundary lines. Dimensions of the field, goals, and special field areas are in millimeters.

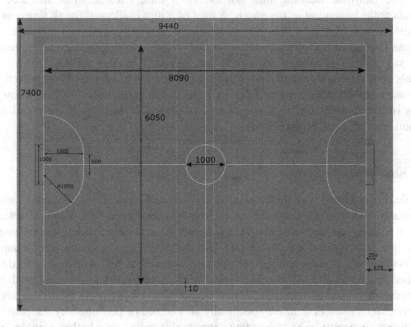

Fig. 4. The field dimensions of the single-size field. The dimensions include boundary lines. Dimensions of the field, goals, and special field areas are in millimeters.

On the single-size field, within each defense area a penalty mark is made 750 mm from the midpoint between the goalposts and equidistant to them. On the double-size field, for each field half the penalty mark is 1000 mm from the midpoint between the goalposts and equidistant to them, thus coinciding with the outer edge of the defense area arc. The mark is a 10 mm diameter circle of white paint.

Goals are placed on the center of each goal boundary and anchored securely to the field surface. They consist of two 160 mm vertical side walls joined at the back by a 160 mm vertical rear wall. The inner face of the goal is covered with an energy absorbing material such as foam to help absorb ball impacts and lessen the speed of deflections. The goal walls, edges, and tops are white in color. There is a round steel cross bar that runs across the top of the goalmouth and parallel to the goal line. It is no thicker than 10 mm in diameter, but is sufficiently strong to deflect the ball. The bottom of the bar is 155 mm from the field surface, and the bar is dark in color to minimize interference with the vision system. The top of the goal is covered in a thin net to prevent the ball from entering the goal from above. It is attached securely to the cross bar and goal walls. The distance between the sidewalls is 700 mm for the single-size field and 1000 mm for the double-size field and the goal is 180 mm deep. The goal walls are 20 mm thick and touch the outer boundary of the field at the goal line, but do not overlap or encroach on the field lines or the field. The floor inside the goal is the same as the rest of the playing surface.

A game lasts two equal periods of 10 min. Teams are entitled to an interval at half time. The half-time interval must not exceed 5 min. Each team is allocated four timeouts at the beginning of the match. A total of 5 min is allowed for all timeouts. For example, a team may take three timeouts of one-minute duration and thereafter have only one timeout of up to two minutes duration. Timeouts may only be taken during a game stoppage. The time is monitored and recorded by the assistant referee who controls the a referee signaling device supplied during the game to convert the referee's commands into Ethernet communication signals that are transmitted to both teams. For games on the double-size field, the number of allocated timeouts is increased to six timeouts and the total time to 7.5 min.

3.1   Shared Vision System

Each field is provided with a shared central vision server and a set of shared cameras. This shared vision equipment uses the community-maintained SSL-Vision software to provide localization data to teams via Ethernet in a packet format. Teams need to ensure that their systems are compatible with the shared vision system output and that their systems are able to handle the typical properties of real-world sensory data as provided by the shared vision system (including noise, latency, or occasional failed detections and misclassifications). Teams are not allowed to mount their own cameras or other external sensors, unless specifically announced or permitted by the respective competition organizers. The two or four cameras depending on whether a single-size or double-size field is used are mounted across bars provided 4 m above the field. The bar runs above the filed midline across goals in the case of the single-size field, while additional bars and cameras are added in the double-size field configuration.

Additionally, all teams must adhere to the operating requirements of the shared vision system having a flat surface on their top containing a unique "butterfly" color pattern [8] as shown in Fig. 5. The color of the robot top must be black or dark grey and have a matte (non-shiny) finish to reduce glare. Before a game, each of the two teams has a color assigned, namely yellow or blue. All teams must be able to be either yellow or blue color. The assigned team color is used as the center marker color for all of the team's robots. No two robots are allowed to use the same color assignment.

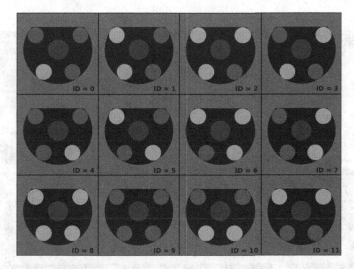

Fig. 5. The standard color assignments for use with the shared vision system.

3.2 Robot Design

A robot must fit inside a 180 mm diameter cylinder and have a height of 150 mm or less as shown in Fig. 6 (left). Robot wheels (or other surfaces that contact the playing surface) must be made of a material that does not harm the playing surface. Robots can use wireless communication to computers or networks located off the field. The robotic equipment is to be fully autonomous. Human operators are not permitted to enter any information into the equipment during a match, except at half time or during a time-out.

Dribbling devices may be included in the design of the robot as shown in Fig. 6 (right) as long as they actively exert backspin on the ball while the spin exerted on the ball must be perpendicular to the plane of the field. Vertical or partially vertical dribbling bars, also known as side dribblers, are not permitted. The use of dribbling devices has certain restrictions of usage during a game.

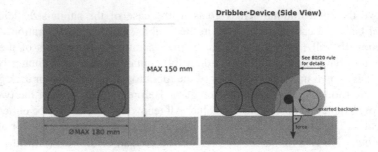

Fig. 6. (Left) The maximum robot dimensions; (Right) How a dribbler may work.

Fig. 7. Robocup small size league 2013 participants.

3.3   Artificial Intelligence and Team Coordination

One aspect that makes the Small Size League so attractive and unique among all RoboCup soccer leagues involving physical robots is the advanced pace of the game. The multi-robot aspect of soccer has been one of the most challenging areas of research in the SSL right from its inception. The first teams performed mostly as individual units of robots even on the same team, with no planned passes among teammates. The Skills-Tactics-Plays (STP) architecture [11] was among the first formal approaches to plan for coordination among teammates, and is still used by most of the teams in the SSL. More recently, there has been an increasing interest in modeling opponents in order to learn their plays [12], robot ball passing and shooting [13], and dynamic selection, planning and learning of behaviors and strategies [14,15]. Robot deception, in the form of the "Coerce And Attack Planner" [16], was introduced in the SSL in RoboCup 2013 to coerce opponents away from strategic locations, allowing goals to be scored by exploiting the strategic openings.

4 Future

There are several aspects that the league has recently discussed as part of its future roadmap. We discuss immediate and future league changes. The small size league is planning to move by 2015 to the double-sized field. The main reason for the increase is having more space for robot game playing. The increase also includes a larger defense area and corresponding rule changes to reduce the number of penalty shots due to the defense trying to cope with the increasingly aggressive offense. An aspect that is currently being discussed is what should be the ratio of goal width to player size as it is currently far smaller than in real soccer. To make the goal area more realistic, we are considering length scales and rules more in line with the FIFA laws [17].

In addition to field increase goal width and defense area should be increased accordingly. In particular, the goal width should be increased such that even an entire team cannot block it. The goal width to player ratio of real soccer (7.32x) is appropriate for this, and this would lead to a goal width of 1.3 m. There is also a need to amend the rules to balance offense and defense with the larger goal width and goal area. Note that this increase in field increase considers maintaining the same sized F180 robot as specified since the beginning of SSL.

4.1 Shared Vision System

The increase in field size implies technical challenges to the existing shared vision system and possibly its conversion to a totally different system. We are currently analyzing different alternatives that should be further evaluated in the future:

- Include additional field markings for calibration with 4 cameras with the existing SSL-Vision.
- Use tripod - mounted cameras with wide-angle and telephoto lenses to cover the field.
- Switch to a completely different vision system, like the Vicon motion capture system.
- Add a non-vision system such as RFID tags under the carpet or other radio based system.

4.2 Robot Design

At this moment we do not have any plans of changing the size of the robot in order to keep compatibility with existing robots and many of the existing league benefits such as easiness for transportation, inexpensive components, and established designs shared throughout the league. As other aspects of SSL are expected to change, such as field size, shared vision system, communication etc., we expect the robots to accommodate these aspects.

Considering the great success of the shared vision system, we propose to include a shared communication system or "standard radio communications server" that all teams will use. This will avoid direct communication links from teams to their robots and would be comparable to other leagues using communication servers. A standardized communication server would also permit "pickup" games where robots from different teams may be exchanged and enable joint teams from different institutions.

An aspect directly related to the robot designs is the ball specification. Considering the need to restrict the kicking strength to limit ball speeds, we would like to consider aspects such as ball bounce and dynamics as compared to human soccer. Alternatives to the current golf ball include hockey and squash balls.

Given the increased capabilities of microcontrollers and the proposed inclusion of the standardized radio server, we will like to evaluate fully onboard computing without a PC controlling the robots. Instead, the PC would be used only to monitor the game and make offline changes to the robots.

4.3 Artificial Intelligence and Team Coordination

In terms of the advanced game playing, there are several proposals that have been discussed. Among these, the most important is the inclusion of an automatic refereeing system that is currently being tested. Other changes affect primarily the game itself including rules being applied.

We would like to discourage "unintelligent" gameplay by modifying rules involving far away kicks that directly score goals, such as chip kicks that travel more than half the width of the field, analogous to the "icing" rule in ice hockey.

We would like to increase the penalization for robot collisions, where for example, if two robots from opposing teams collide, then the team with the slower team gets an indirect free kick, and after 3 such indirect free kicks, the opposing team will get a yellow card. Implementing this rule will require an automated referee, and a formal definition of "collision" that the referee should be able to detect with a low false positive rate.

There is a need also to keep updating the game rules such as to balance the defense and offense, abolish the rule prohibiting multiple defenders in the defense area, and add the offside rule. This change might need additional checks like limiting the number of defenders in the defense area to the number of offense players plus one.

In terms of technical challenges, we would like to introduce compulsory technical challenges for teams, for example to proceed past the quarter finals, by making technical challenges more relevant to the game and geared to address technical shortcomings in the league, such as:

- Intercepting chip kicks (from both teammates as well as opponents).
- Obstacle avoidance with moving opponents.
- Score goals against opponents with fixed (but unknown) handicaps to promote AI, opponent modeling, learning, etc.
- Successfully steal a ball from an opponent robot.
- Advanced plays involving collaboration among multiple robots.

Finally, we would like to evolve the league to have further interaction with humans as in real soccer games by introducing the role of the coach, though voice or written commands, so teams may modify their strategies during the game,

5   Timeline

We propose the following timeline:

- 2015: Play under double-size field.
- 2016: Introduction of standardized radio communication server.
- 2017: Introduction of automatic referee system and increase in number of robots (mixed teams).
- 2018: Introduction of fully onboard robot computing (elimination of controlling PC).
- 2019: Increase in field size to enable 11×11 robots and extension of shared vision system.
- 2020: Introduction of automated coaching during game.

6   Conclusions and Discussion

We have summarized in this paper the past, present and future of the small size league as currently envisioned by the executive committee based on prior discussion with the technical committee and team leaders. We summarized immediate challenges such as the recent increase to double the field size with corresponding changes such as the support of up to 4 cameras that will be tested in 2014 as a technical challenge and will become official in 2015. A critical aspect we would like to keep in the league is the size of the robots. As technology advances and component prices are reduced, in particular in relation to brushless motors, robots have become much more powerful while costing less. We would like to avoid a robot "power race" and instead see an advance in the "power algorithms" to control and coordinate among the robots to take advantage of the increased robot power by producing improved game play and collaborative agent strategies. We would also like to further encourage sharing of designs among teams, such as the introduction a few years back of mandatory ETDPs (Extended Team Description Papers) for top teams, and increase the use of shared systems in the league, such as the shared vision system (SSL-) that has encouraged new teams to enter the league. Finally, we would like to encourage Junior teams to take advantage of the advances of this league by developing a lower cost using established robotic kits or by exploiting lower cost components to develop a RoboCup Junior Small Size League analogous to that proposed several years ago [18].

Acknowledgements. The RoboCup Small Size League Executive Committee acknowledges the contribution of the members of the current and prior technical committees in specifying the rules of the league, the organizing committees in bringing these rules into actual competitions, the previous exec committees in guiding this league throughout the years, and the RCF trustees for supporting the SSL league. We also acknowledge all teams that have participated in SSL in the past years. Figure 7 shows a picture of RoboCup SSL 2013 participants.

References

1. Mackworth, A.K.: On seeing robots. In: Basu, A., Li, X. (eds.) Computer Vision: Systems, Theory, and Applications, pp. 1–13. World Scientific Press, Singapore (1993)
2. Kitano, H., Tambe, M., Stone, P., Veloso, M., Coradeschi, S., Osawa, E., Matsubara, H., Noda, I., Asada, M.: The RoboCup synthetic agent challenge 97. In: Proceedings of the Fifteenth International Joint Conference on Artificial Intelligence. Morgan Kaufman, San Francisco (1997)
3. Sahota, M.K., Mackworth, A.K., Barman, R.A., Kingdon, S.J.: Real-time control of soccer-playing robots using off-board vision: the dynamite testbed. In: IEEE International Conference on Systems, Man, and Cybernetics, pp. 3690–3663 (1995)
4. Small Size Robot League Wiki (2014). http://robocupssl.cpe.ku.ac.th/
5. Zickler, S., Laue, T., Birbach, O., Wongphati, M., Veloso, M.: SSL-vision: the shared vision system for the RoboCup small size league. In: Baltes, J., Lagoudakis, M.G., Naruse, T., Ghidary, S.S. (eds.) RoboCup 2009. LNCS, vol. 5949, pp. 425–436. Springer, Heidelberg (2010). http://code.google.com/p/ssl-vision/
6. Veloso, M., Stone, P., Han, K., Achim, S.: The CMUnited-97 small robot team. In: Kitano, H. (ed.) RoboCup 1997. LNCS, vol. 1395. Springer, Heidelberg (1998)
7. Bruce, J., Balch, T., Veloso, M.: Fast and inexpensive color image segmentation for interactive robots. In: Proceeding of the Intelligent Robots and Systems, pp. 2061–2066 (2000)
8. Bruce, J., Veloso, M.: Fast and accurate vision-based pattern detection and identification. In: Proceedings of ICRA2003: 2003 IEEE International Conference on Robotics and Automation, Taipei, Taiwan, pp. 1277–1282, May 2003
9. Diegel, O., Badve, A., Bright, G., Potgieter, J., Tlale, S.: Improved mecanum wheel design for omni-directional robots. In: Proceedings of the 2002 Australasian Conference on Robotics and Automation, Auckland, pp. 117–121, November 2002
10. Kalmár-Nagy, T., D'Andrea, R., Ganguly, P.: Near-optimal dynamic trajectory generation and control of an omnidirectional vehicle. Robot. Auton. Syst. **46**(1), 47–64 (2004)
11. Browning, B., Bruce, J., Bowling, M., Veloso, M.: STP: skills, tactics, and plays for multi-robot control in adversarial environments. J. Syst. Control Eng. **219**(1), 33–52 (2005)
12. Erdogan, C., Veloso, M.: Action selection via learning behavior patterns in multi-robot domains. In: Proceedings of the International Joint Conference on Artificial Intelligence, pp. 192–197 (2011)
13. Nakanishi, R., Bruce, J., Murakami, K., Naruse, T., Veloso, M.M.: Cooperative 3-robot passing and shooting in the RoboCup small size league. In: Lakemeyer, G., Sklar, E., Sorrenti, D.G., Takahashi, T. (eds.) RoboCup 2006: Robot Soccer World Cup X. LNCS (LNAI), vol. 4434, pp. 418–425. Springer, Heidelberg (2007)
14. Trevizan, F.W., Veloso, M.M.: Learning opponent's strategies in the RoboCup small size league. In: 9th International Conference on Autonomous Agents and Multi-Agent Systems, Springer (2010)
15. Yasui, K., Kobayashi, K., Murakami, K., Naruse, T.: Analyzing and learning an opponent's strategies in the RoboCup small size league. In: Behnke, S., Veloso, M., Visser, A., Xiong, R. (eds.) RoboCup 2013. LNCS, vol. 8371, pp. 159–170. Springer, Heidelberg (2014)

16. Biswas, J., Mendoza, J.P., Zhu, D., Choi, B., Klee, S., Veloso, M.: Opponent-driven planning and execution for pass, attack, and defense in a multi-robot soccer team. In: Proceeding of Autonomous Agents and Multi-Agent Systems, pp. 493–500 (2014)
17. FIFA, Laws of the Game (2013/2014) http://www.fifa.com/mm/document/footballdevelopment/refereeing/81/42/36/log2013en_neutral.pdf
18. Baltes, J., Sklar, E., Anderon, J.: Teaching with RoboCup. In: Greenwald, L., Dodds, Z., Howard, A., Tejada, S., Weinberg, J. (eds.) Accessible Hands-on AI and Robotics Education. AAAI Spring Symposium Series, vol. 1, pp. 146–152 (2004)

RoboCup MSL - History, Accomplishments, Current Status and Challenges Ahead

Robin Soetens[1]($\boxtimes$), René van de Molengraft[1], and Bernardo Cunha[2]

[1] Eindhoven University of Technology, Eindhoven, The Netherlands
robinsoetens@gmail.com
[2] IRIS Lab/IEETA/DETI, University of Aveiro, Aveiro, Portugal

Abstract. The RoboCup Middle-Size League (MSL) is one of the founding leagues of the annual RoboCup competition. Ever since its birth it has been a league where development of hard- and software happens simultaneously in a real-world decentralized multi-robot soccer setting. Over the years the MSL achieved scientific results in robust design of mechatronic systems, sensor-fusion, tracking, world modelling and distributed multi-agent coordination. Because of recent rule changes which actively stimulate passing, matches in RoboCup MSL have become increasingly appealing to a general audience. Approximately five thousand spectators were present during last years final match. In this paper we present our plan to build on this momentum to further boost scientific progress and to attract new teams to the league. We also give a historical overview and discuss the current state of the MSL competition in terms of strengths, weaknesses, opportunities and threats.

Keywords: RoboCup · Middle-size league · Roadmap · History · Academic competition · Open source · Survey · Outlook

1 Introduction

Picture two teams of five robotic players each, playing with an official FIFA ball on an indoor field. The team in possession of the ball coordinates passes and dribbles to bring the defending robots out of position, eventually they find a weak spot, score, and win. Perception is done by on-board sensors only, there is no centralized vision system. Robots communicate and build a world model based on information coming from itself and from team mates. No human intervention is allowed, except for the referee and a dedicated human coach who is allowed to provide high-level strategic instructions (but only during dead-time and only by showing signs, making gestures, or shouting).

Our perception, as spectators, brings us into the thrill of watching a real football game. The heart beats faster and we are anticipating the next best move, or we are plunging into a feeling of frustration by the missed opportunity or the incomprehensible action of a specific player. The sensations follow up and down at the rhythm and dynamics of the game. You find yourself yelling "*shoot*" or "*pass*," while mapping human emotions to your favourite robot football team.

© Springer International Publishing Switzerland 2015
R.A.C. Bianchi et al. (Eds.): RoboCup 2014, LNAI 8992, pp. 624–635, 2015.
DOI: 10.1007/978-3-319-18615-3_51

This is what a RoboCup MSL game unveils to those who watch it live. Players move around on wheels, are short and gross while lacking arms and legs. They are 80 cm tall, reach speeds of 60 % to 70 % of an average human top speed and weight up to 40 kg each. Those who watch them for the first time suddenly realize that "*Yes, they are fully autonomous, they play soccer all by their own.*" The same feeling also arises for those who, on the other side of the chain, are working towards a solution for one of hundreds of challenges still to be tackled.

These are not yet, for sure, robots that have fulfilled the dream stated by the RoboCup Federation as "*By 2050, a team of fully autonomous humanoid robot soccer players shall win a soccer game, complying with the official FIFA rules, against the winner of the most recent world cup.*" But, at this time, in the path towards that goal, they rise the imagination of a spectator to a level where a real soccer game can be seen beyond the science and technology hidden inside.

1.1   Relevance to Society

Since its first tournament in 1997, the RoboCup competition boosts both research and development of technology in artificial intelligence and robotics. At the same time the appealing nature of doing science by playing soccer also inspires young people and promotes science and technology among a general audience. It has potential to attract thousands of spectators to watch and cheer for a team of robots (Fig. 1). For participants, the element of competition brings up additional energy. A RoboCup tournament provides an open atmosphere, which facilitates knowledge exchange. It allows to acquire hands-on experience with real-world systems, while working under high pressure of time in a team with experts in different domains. RoboCup experience is a tremendous addition to the training of an engineer.

Boosting research, promoting science and training future engineers are three pillars of societal relevance that justify the allocation of resources required to organize a RoboCup competition. MSL contributes to each of these pillars. It boosts research by providing a real-world multi-robot benchmark involving real hardware and the disturbances and constraints that come with a real game of soccer. It promotes science and technology because its format is easy to grasp and explain to a general audience and to media. And since teams typically consist of ten to fifteen people, with a mix of experienced and less experienced team members, MSL is an ideal cradle for education and training of future engineers.

1.2   Organization

The remainder of this paper is organized in three sections: First we provide an historical perspective on evolution and revolution we have seen in MSL. Next we present the current status of the league and elaborate on how challenges are being addressed. Lastly, in the final section, we look at the future and put efforts being made today in the perspective of a long-term goal. Short conclusions and a link to a reference digest conclude the article.

(a) Over 5000 spectators during the final match of RoboCup 2013 in Eindhoven.

(b) Team Water from the Beijing Information Science and Technology University after winning RoboCup 2013.

(c) Team Tech United from Eindhoven University of Technology after losing the final match of RoboCup 2013.

Fig. 1. Public engagement and team emotions in RoboCup MSL. Photographers: Bart van Overbeeke (a), Albert van Breemen (b) and (c)

2   The Past

Although neither of the authors of this paper participated in the league from the beginning, from a scientific and a technical point of view it is possible to identify a number of well-constrained epochs, each representing a period of time where development was established through evolution towards a certain challenge. Epoch boundaries, on the other hand, represent instants where development was driven by a leap step, aiming new, more daring, scientific targets.

Fig. 2. From left to right: Sharif CE (Iran) vs. Azzurra Robot Team (Italy) - 1999; Golem Team (Mexico) vs. COPS (Germany) - 2000; CS-Freiburg (Germany) vs. RoboSix (France) - 2000 (Color figure online)

Throughout the years, this evolution versus revolution approach was balanced based on the ability of the community to evolve and share the produced knowledge. Powerful tools to achieve this balance include the annual updates to the MSL rulebook and the technical challenge competition, held at every RoboCup final tournament. A short overview of each of the epochs is presented next.

First Epoch: 1997 Through 2001. This was the starting up period. Original field dimensions were 9×5 m, limited by surrounding walls that kept the ball always inside the field (Fig. 2). Goals were color coded (yellow and blue) and the ball was bright orange. The rules were very simple, although already FIFA based. Teams played with up to four robots, and start and stop commands could be issued through a wireless connection by each team.

Illumination was artificial and uniform with variations below 300 lux. Teams were allowed to reposition the robots manually throughout the game. Research was centred on basic navigation and vision, mainly focussing on color-based classification and detection of objects. Speed was typically below 1 m/s. Differential traction solutions and simple electro-mechanical kickers were the focus at the mechatronics level, where solenoid-based kickers are starting to be used (e.g., at the Minho team). The first omnidirectional solutions, both at vision and traction level, start appearing around 2000.

Second Epoch: 2002 Through 2006. In the first year of this epoch the outside walls of the field were removed. Instead, color coded posts were placed on the four corners of the field. Manual repositioning of the robots was no longer allowed, except for a kick-off. Also, to simplify the game control, robots were supposed to stop if the ball was removed from the ground. Base colors were kept untouched as well as uniform artificial illumination over the field. Starting in 2004, the size of the field was increased to 12×8 m, now including both a penalty area and a goal area. In 2005, the first version of the referee box was introduced to allow a team-independent control of the game flux. Team description papers became mandatory in 2004. Common themes were real-time adaptive color segmentation, stronger and more precise kicking devices based on pneumatic or solenoid actuation, solutions for catadioptric vision systems, efficient omnidirectional driving, open loop dribbling devices, early sensor-fusion techniques, self-localization and first solutions for team coordination.

Third Epoch: 2007. This was a transition year where the size of the field got increased to 18×12 m, which is still the current MSL official size. Artificial uniform illumination was no longer required and the first attempt to move into a more cooperative game was introduced by imposing that a goal could only be validated on the first ten seconds after a restart if touched by a second player of the team playing the restart. Posts and goals kept their original color coding but the changes boosted the necessity to increase research on efficient dynamically adjustable vision systems together with real-time coordination of the teams. Concepts as dynamic role-changes, dynamic team-formations, force-adjustable kicking systems for both lob and floor-level shots, world modelling including opponent recognition, efficient ball-tracking on the floor, path planning and distributed real-time databases for information sharing among team mates are some examples of the research tackled in this period.

Fourth Epoch: 2008 to 2011. In the beginning of this epoch the colored corner posts were finally removed and goals became white. For the first time, in a field that looks like a real soccer field, teams had to find solutions for self localization without external visual aids while also having to disambiguate the own and opponent side of the field. Also, in 2009, maximum bandwidth used by the team was limited to 20 % of the IEEE 802.11b standard bandwidth. In 2010, the ball no longer had to be orange, being pre-selected by the LOC. Manual repositioning of the robots was forbidden. To enforce the players ability to perform passes efficiently, all restarts imposed a minimum distance between the ball, team mates and opponents. Own players at 1 m, opponents at 2 m (since 2009) and own players at 2 m, opponents at 3 m (since 2011). Two additional rules were introduced in 2011, changing dramatically the dynamics of the game. On one side, goals could only be validly scored when the shot was taken within the opponent side of the field, thus reducing the run towards stronger and stronger kickers and increasing the necessity to improve game planning. At the same time, it was also imposed that while disputing the ball, only two robots, one for each team, could be in direct contact with the ball.

The new rules brought the league to a higher level of artificial intelligence and multi-agent coordination. Real-time efficient communication among robots got improved, as well as world modelling and role assignment. Other topics addressed in MSL within this period were novel solutions for active ball handling mechanisms, multi-robot coordination using set-plays, dynamic strategy changing during the game, efficient simulation tools for highly dynamic cooperation, arbitrary ball recognition, 3D ball tracking in real-time, high density data-fusion, cooperative control through objective achievement and the use of utility fields for passing.

Fig. 3. Left to right: Tech United (Netherlands) vs. CAMBADA (Portugal) - 2012, Tech United (Netherlands) vs. Water (China) - 2013.

Fifth Epoch: 2012 Until Today. By 2011 it was clear that extra efforts should be put into changing dramatically the way cooperation and game play was going. Robots, at this time, were able to perform fast dribbles (between 3 and 4 m/s) in a controlled way, using dynamic path planning. Without the ball speeds above 4 m/s were reached. This left the efforts of teams who were trying to exploit more cooperative solutions unrewarded, since speed became more important than strategic passing. A radical rule was introduced in 2012: Robots could no longer dribble the ball over the mid-line when progressing from their side to the opponent side. Furthermore, robots should actually make a pass to a team mate on the opponent side that could then either dribble or shoot towards the opponent goal. This counter-intuitive rule paid out. Robot average speed reduced dramatically, pushing offences became sparse, actual in-game ball passing became a reality while new strategies including man-to-man cover, zone-cover or mixes of both started to appear. Active ball-interception, effective use of utility fields and passes into open zones were also thriving (Fig. 3, left).

3 The Present

In a contemporary RoboCup MSL match two teams of up to five robots and a base station pc play on a soccer field with an adapted version of the FIFA rules. Robots are completely autonomous (i.e., all sensors and computing equipment is on-board). They must fit in a $50 \times 50 \times 80$ cm box and weigh no more than 40 kg. The field is similar to a human soccer field, scaled down to 18×12 m and with white lines over a green carpet. Goals are also scaled down.

A regular size official FIFA soccer ball is used in the tournaments. The only required human intervention comes from the referee, through a dedicated application called "referee box," which is controlled during the match by an assistant referee. Wireless networking is allowed between robots and between robots and the base station using unicast or multicast (no ad hoc or broadcast permitted). Throughout the match, consisting of two halves of 15 min, the base station cannot use any form of sensing and cannot be touched. No permanent external active or passive beacons are allowed. Most teams adopt holonomic motion, based on a three or four wheel configuration, as well as omnidirectional vision, sometimes fused with panoramic and 3D vision. Ball handling can be done by active or passive devices as long as no more than a third of the ball is covered. Teams can use heterogeneous robots, where the goalkeeper, by itself, is a exceptional case, since it can have extensible parts that can momentarily extend the robot to the right, left or above, by at most 10 cm.

For 2014, the rule prohibiting a dribble over the mid-line was dropped and replaced with a more general and natural rule: A valid goal can only be scored after the ball has been received or touched by a team mate within the opponent side of the field after rolling freely for at least one meter. Furthermore, to reduce long dribbles, a robot is now limited to dribble the ball continuously for a maximum radius of three meters with respect to the point it received the ball. The ball has to be completely released before crossing this threshold. Both of these rules enforce further cooperative and dynamic passing while the latter also encourages more dexterity in ball handling. High level human coaching by means of a QR-code is now also allowed but only during off time (e.g., when robots are positioning themselves for a free kick). In order to make progress towards playing on regular soccer fields, RoboCup 2014 will have a challenge where teams have to show basic soccer playing skills on an artificial grass field.

3.1 Participants

The number of participating teams in RoboCup MSL grew steadily and peaked in 2004, 2005 and 2006. For each of these tournaments the technical committee had to select a maximum of 24 teams that could participate, although more teams applied. After RoboCup Bremen in 2006, a downturn has been experienced, with only 11 teams participating in 2007, 13 in 2008 and eventually a minimum of 6 in Mexico in 2012. An average of 10 teams have participated regularly in the RoboCup main event since 2007, while nearly 30 teams stayed active and participated in regional tournaments throughout these years.[1]

The main reasons for the downturn after RoboCup Bremen were the high overall costs compared to other leagues, sub-optimal knowledge sharing and an international focus swing from multi-robot research to research involving a general purpose service robot. Currently, the number of participants is stable, with new teams steadily appearing and taking over the place of those who fade away. From a geographical point of view, the league is mainly represented in Europe, Asia and the Middle-East. Although research budgets in some parts of the world are a barrier, the international profile of the league would profit from new teams coming from the Americas and Africa. MSL also consistently attracts participants from industry (e.g., Philips and more recently VDL and ASML).[2]

---

[1] Google Maps, MSL activity over the past two years, https://goo.gl/maps/AI9aW.

[2] VDL, a conglomerate of manufacturing companies founded an MSL team "VDL Robot Sports" in 2012: http://www.robotsports.nl/. ASML, a lithography equipment manufacturer founded an MSL team "ASML Robo Team" in 2013.

3.2   League Profile

Among the spectrum of different leagues, the MSL takes its own well-defined approach in its contribution to solving the RoboCup grand challenge. This approach is rooted in a number of fundamental propositions:

- **Hard- and Software:** In robotics, ingenious and robust design of hardware is equally important as clever software. There is as much science in developing state-of-the-art mechatronic systems as there is in advancing coordination and action. Hence, in MSL workshops and challenges we actively stimulate knowledge sharing with respect to both hard- and software design.
- **Real Disturbances and Constraints:** Proper benchmarking involves a real-world test where algorithms are exposed to disturbances and constraints that come with real hardware in a realistic environment. Hence MSL uses an official ball, plays on a large field, and is now focussing on playing on artificial grass instead of a smooth carpet.
- **Open Mind:** Progress is frequently hampered by what we know. Often we construct a robot arm that looks like a human arm because that is all we know, not because it is optimal for the task at hand. Hence, robot limitations in MSL by no means prescribe what a robot should look like. This stimulates creativity and allows taking the shortest route to FIFA compliant soccer versus a team of humans, though not necessarily with humanoid robots.
- **Multi-Robot:** Sharing knowledge between robots is one of the key challenges in robotics. Any general purpose service robot will have to be cooperate with other robots, or at least interact with other machinery. Hence, in MSL, we actively stimulate robot to robot cooperation and standardization of interfaces with, e.g., enforced passing and mixed team challenges.
- **Focus and Integration:** Robotics research groups often focus on specific topics like computer vision, motion control or machine learning. While focus is important, there is also science in integrating methods from all these fields into a single robotic system. Both approaches are possible in MSL. Teams can design and build an entire robot, but can also acquire a standardized platform (Sect. 3.3) and focus on, e.g., redesigning the vision unit.
- **Semi-closed World:** Deploying robots in an unstructured world still is a huge challenge. Therefore the MSL scenario is carefully placed halfway between the closed world of factories and the open world of human environments. White lines on a field of a known dimension, a preselected ball and a fixed number of ten mandatory black players provide enough structure to facilitate recognition and localisation. This allows teams to do research beyond perception, while at the same time roadmap elements like, e.g., a free choice of team shirts or playing with an arbitrary ball gradually shift the MSL scenario towards a more open world.

With the propositions above we defined MSL's focus and approach with respect to the RoboCup Grand Challenge, and with respect to other leagues. This competition profile comes with specific league properties most of which can be explained both as a strength and a weakness. In the upcoming subsections we list these and we distinguish opportunities and threats that arise from them.

Fig. 4. Human vs robots games are played yearly since 2007 (left). MSL robot outper-forms humans in shooting accuracy (right).

Strengths

- Developing hard- and software simultaneously, while facing real-world distur-bances and limitations. Therefore the league is a proper benchmark test.
- High-speed game flow and easy to understand, which makes the league appeal-ing to the general audience, to future researchers and to media.
- League with a long history, which implies lots of knowledge has been accumu-lated already and a large community exists.
- Highly dynamic environment with "aggressive" opponents resulted in robust hardware design and efficient real-time algorithms.
- Realistic field and ball size make the competition suitable for large numbers of spectators and allow human versus robot challenges. MSL facilitated the first ever human versus robot match in RoboCup history in 2007 and in shooting accuracy an MSL robot already beats humans (Fig. 4).
- Strong focus on robot to robot cooperation, which is important in many appli-cation domains of robotics.

Weaknesses

- Compared to other leagues, high start-up costs and time for new teams.
- High costs of participation, because teams are typically large and multiple robots have to be transported.
- League with a long history, therefore difficult for new teams to catch-up.
- Multiple robots and "aggressive" opponents implies lots of maintenance.
- Currently not all continents are represented in the MSL community.

Opportunities

- League has a long history so there is lots to document and valorize. When properly done new teams and the outside world can profit from that.
- Media and spectator friendly nature of the league makes it attractive to spon-sors. Also direct participation of companies in the league can be further stim-ulated. In the recruiting battle for high-tech talent, companies can profit from visibility at RoboCup.

– Real soccer dimensions, human versus robot possibilities and highly dynamic gameplay make MSL robots very useful for demoing purposes, e.g., at technology fairs or in-house days for prospect students. When properly executed, this can be an additional source of budget for teams.

Threats

– Teams quitting because of high costs.
– Difficult to keep balance between scientific output and management and maintenance of large teams of robots and people.
– Failure in balancing the evolution versus revolution threshold over time. Too much forward pressure could lead to teams quitting because they cannot keep up, too little forward pressure could lead to high-ranking teams quitting because the competition is not challenging enough.
– League getting inaccessible for new teams because of high startup effort.

3.3   Strategy

The necessarily short analysis presented in the previous subsection can be summarized in a single fundamental question: How are we going to keep making technical and scientific achievements, while also attracting more teams and reducing costs? The answer, in our view, is threefold:

– Increase of **knowledge sharing** by encouraging scientific publications, release of well established and tested middle-ware and sharing of hardware solutions. A significant effort is already going on in all these areas. Examples include the three day RoboCup MSL workshop held at the University of Kassel last year and the launch of ROP (Robotic Open Platform).[3] With a wiki, a Q&A section and a CAD file repository, ROP facilitates the release of hardware designs of robots and modules under an open hardware license. MSL robots of team Tech United have been fully released already. A release of the CAMBADA robot, among others, is currently being prepared.
– Design and production of an **affordable robot platform for MSL**, able to provide a starting point for any new team. A robot that plays basic soccer out of the box and allows new teams to literally build on knowledge that is already accumulated in the league. With this goal in mind, the TURTLE-5k consortium was initiated shortly after RoboCup Mexico.[4] At the time of writing a first team of robots has been sold and completed.
– Discussion and annual **evaluation of the MSL roadmap**, targeting objectives for the short, medium and long run and using the Technical Challenge to prepare for upcoming rule or environment changes.

---

[3] ROP, An online platform for Open Hardware releases, initiated by Eindhoven University of Technology, http://www.roboticopenplatform.org/.
[4] TURTLE-5k Consortium: ACE, VEDS, Frencken and Eindhoven University of Technology, http://www.turtle5k.org/.

4   The Future

MSL owns an important piece in solving the RoboCup challenge. It continuously gets fed by the simulation, standard platform and small-size leagues and, in turn, knowledge accumulated at MSL finds its way to other leagues.[5] Eventually, the RoboCup Challenge is formulated in terms of a *humanoid* team. Therefore, as soon as MSL is able to build a team of non-humanoid robots able to beat a human team, while playing by the FIFA rules on an official turf, close cooperation with the humanoid leagues should be started. It is not unthinkable, by many it is even expected, that around 2025 the MSL has evolved into a legged league as well,[6] although this will not be enforced by the rules.

In the short run we are focussing on other things. Further improving and maintaining the 5k platform is crucial for the league, as well as adding more robot designs to ROP. In order to better present ourselves to universities and companies potentially interested in joining, we are working on a remake of the MSL website. The new site will have search functionality and tagging to browse trough a list of publications related to MSL.

The artificial grass challenge during RoboCup Brazil has the potential to be revolutional. Once teams manage to drive, dribble, pass and shoot on artificial grass, we can speak of a new epoch in MSL. The different characteristic of artificial grass in comparison to carpet will have their effect on base movement and base stability. Probably it will also influence robot vision, which has to deal with much more tilt and vibrations, while active ball handling systems will have to be redesigned to deal with the reduced friction. One of the teams has announced to introduce a fully suspended soccer robot during the technical challenge in Brazil, which would be unique within MSL.

Also new in Brazil will be standardized logging of game data. Via the refbox pc, at a fixed sample rate, teams will fill in a struct with parameters like, e.g., the estimated location of the ball and a list of candidate obstacles. Once further extended, publishing of such a standardized struct can ease up mixed team soccer playing. Data will also be put online, enabling post game analysis or application of machine learning techniques to robot soccer.

The decision to log data and other plans presented in this paper were made during the 2013 RoboCup MSL workshop in Kassel. With this years workshop planned to take place in Eindhoven, and the one next year already planned in Aveiro it will be an annual event where proposed rule changes are discussed and teams obtain hands-on experience in working with each others development environment and toolchain.

Lastly, although continuously under discussion, we present snippets of some important subjects on our roadmap (Table 1).

---

[5] The robots of RoboCup@Home teams CAMBADA and Tech United Eindhoven are largely based on the soccer robots made by those teams.

[6] Already in 2010 a first prototype of a hexapod robot for MSL was presented during the scientific challenge. This project is still ongoing.

Table 1. Part of the proposed roadmap for in-game changes and technical challenges (TC). Subject to yearly review.

	2014	2015	2016	-
Ground	Artificial grass (TC)	Artificial grass (demo match)	Artificial grass (competition)	Real grass outdoors (TC)
Passing	One pass before scoring	One pass before scoring, ex. rest.	Two passes before scoring	Stimulate in-game lob pass
Dribbling	Circle max 3 m radius	Circle max 2 m radius	Circle max 1 m radius	Limit ball-holding time
Team shirts	Cyan/Magenta on black	Saturated colors from predef. set	Random color opponent (TC)	Free saturated colors
Coaching	QR and color code allowed	Call robot with voice/gest. (TC)	Only voice and gestures allowed	Directly respond also to referee
Mixed team	Standardized data logging	Common simulator	Demo match	Heterogeneous robot teams

5 Conclusions

RoboCup MSL takes the shortest route to FIFA compliant, but not necessarily humanoid, autonomous robot soccer playing. The annual match of the MSL winner against the trustees is an exposure of the worldwide state-of-the art in human versus robot soccer playing. Epochs of evolution and revolution observed over the past 18 years show that MSL contributed significantly to solving the RoboCup Grand Challenge. With an ambitious roadmap and multiple initiatives to allow new teams to catch up quickly, substantial contributions can also be expected in the years to come.

Acknowledgments. The authors express their deepest gratitude to every team or individual that, since 1997, either at the regional or worldwide level, has contributed to the technical and scientific progression of RoboCup MSL.

References
More than 450 technical and scientific documents have been produced in relation to RoboCup MSL. On our wiki a pdf and bibtex list is maintained.[7]

---

[7] http://wiki.robocup.org/wiki/Middle_Size_League.

The Standard Platform League

Eric Chown[1]([⊠]) and Michail G. Lagoudakis[2]

[1] Bowdoin College, Brunswick, ME, USA
echown@bowdoin.edu
[2] Technical University of Crete, Chania, Crete, Greece
lagoudakis@ece.tuc.gr

Abstract. The Standard Platform League is unique among RoboCup soccer leagues for its focus on software. Since all teams compete using the same hardware (a standard robotic platform), success is predicated on software quality, and the shared hardware makes quality judgments simpler and more objective. Growing out a league based on the Sony AIBO quadruped robots, the league has constantly evolved while moving ever closer to playing by human rules, and currently features the Aldebaran NAO humanoid robots. The hallmark of the league has been a focus on individual agents' skills, such as perception, localization, and motion, at the expense of more team-oriented skills, such as positioning and passing. The league has begun to address this deficiency with the creation of the Drop-in Challenge, where robots from multiple teams will work together. This new focus should force teams to work on multi-agent coordination in more abstract and general terms and promises to create fruitful new lines of research.

Keywords: RoboCup · SPL · NAO · Multi-agent cooperation · Machine learning

1 Introduction

The Standard Platform League (SPL) www.tzi.de/spl was referred to as the "Sony Quadruped Robot Football League" in the original rules published in 1998 [9] and was later renamed "Four-Legged League" (4LL). The original league adopted a standard hardware platform, the Sony AIBO, as a way to make the focus of the competition *software*, rather than hardware. The AIBO was a robust robot platform, which afforded teams the ability to do a significant amount of development and testing. AIBO was progressively replaced by the NAO robot between 2007 and 2009 and the league coined its current name. The latest edition of the league featured teams of five robots each, wearing red and blue jerseys, playing soccer autonomously on a 6×9 m field with yellow goals on both sides, using an orange street-hockey ball (Fig. 1).

The advantages of a standard platform are many. There are prominent teams in the league, which could not produce their own robots, but wish to use robots

© Springer International Publishing Switzerland 2015
R.A.C. Bianchi et al. (Eds.): RoboCup 2014, LNAI 8992, pp. 636–648, 2015.
DOI: 10.1007/978-3-319-18615-3_52

Fig. 1. SPL final game at RoboCup 2013 in Eindhoven, The Netherlands. [10] (Color figure online)

and to work on software development. Further, the standardization of the hardware puts teams on theoretically equal footing, making the competition ultimately about the quality of the developed software. Since the teams all use the same robots, differences in the speed of motion, the accuracy of perception, the quality of localization, etc. can be directly attributed to software, rather than having to be disentangled from differences in hardware. This required focus on software has arguably pushed software development further in the SPL than in any other RoboCup humanoid league. This can be seen, for example, by the performance of Team DARWin which has won the KidSize class competition of the Humanoid League three years in a row using a code base developed for the SPL. That code base has since been widely cloned in the KidSize class. The focus on software can also be seen in the large number of research papers presented at the RoboCup Symposium that originate in the SPL.

2   History

2.1   The AIBO Years

The AIBO was an ideal platform for the original league, as the robots were relatively robust, and having four legs meant that their motion was relatively simple, as compared to two-legged robots, to control. Nevertheless, getting a small robot to play soccer presented a number of difficult problems in the early stages, especially with regard to vision and localization.

To help mitigate these problems the league used brightly-colored solid goals, in addition to a number of multi-colored beacons that ringed the field. While these unique landmarks helped to make localization much simpler, ironically they made vision something of a larger problem, because they increased the number of landmarks to be recognized, as well as the number of colors that a vision system needed to discern. In particular, AIBOs had trouble with the colors blue/skyblue, used in goals and beacons, and red/orange, used in balls and uniforms. Blue/skyblue colors were a problem, because they were close to so many colors that appear around a typical AIBO game (e.g. blue jeans worn by spectators), but also because the AIBO camera had a peculiar feature, called "chromatic distortion", that shifted parts of the image towards the blue end of the color spectrum. Red and orange colors were also a problem, because the balls used were close in color to the uniforms that the robots wore. And, since the robots were so low to the ground, it was easy to confuse their uniforms for balls.

The standard approach to vision was to use a color look-up table (LUT). Teams would take images from the competition venue and use them to "paint" a color table. The process was slow and tedious and prone to error. Because of this, attempts were made to improve the process with machine learning [1,8] Once a color table was built, most teams used a version of run-length encoding to extract blobs of the various colors that made up the landmarks around the field. Blobs were then examined and various sanity checks were run to see, if, for example, an orange blob was truly a ball and not part of a uniform. Notably, there was very little use of modern and general machine vision techniques, such as Hough and other transforms, mainly because of a desire to run at the highest frame rate possible. In addition, dedicated, bright, invariant, uniform lighting of the field was a necessity in those years for successful visual object recognition.

The fixed landmarks of the field (goals, beacons, lines) provided the cues for robot localization. Nearly every team ran some instantiation of the generic probabilistic Markov localization approach based on the idea of Bayesian filtering for state estimation. To update the state (position and orientation) of each robot, some kind of Particle Filter (Monte-Carlo Localization) or Kalman Filter (Gaussian Localization) or a combination of both was used. The advantage of Kalman Filters relied again on computational efficiency, as it was difficult to use enough particles to make Particle Filters work adequately on the limited AIBO processor. Similar techniques were also used for tracking the (moving) ball on the field. Over the years, the localization problem became harder, by altering the number and the quality of landmarks; the initially six multi-colored beacons around the field, became four, and finally just two, while the goals were simplified towards colored goalposts. In the last two editions of the Four-Legged League (2007 and 2008), only blue and yellow were used as landmark colors. Lines were not utilized much as landmarks due to their inherent perceptual ambiguity in contrast to unique landmarks and the additional complexity required in properly recognizing them, given that the robots themselves were also white-colored.

In terms of behaviors, the Four-Legged League was marked by a succession of improvements in individual skills and the development of basic behavioral frameworks, but very little obvious teamwork was evident. On an individual level, the AIBOs went from walking on their "paws", like normal dogs, to walking on what amounts to the forearms of their front legs. This made them much more stable and ultimately much faster and more agile. The dogs would kick the ball by putting their front legs up at a 90° angle and then forcefully bringing them down on the ball. This led to two important behavioral developments: the "grab" and the "dodge." Robots would grab the ball by throwing their chin on top of it and putting their front legs around it. This would stabilize the ball and put it under their control. It then naturally led to a behavior, where the robot could actually move from side to side, while holding the ball, to get to a better shooting angle, to the effect that attackers could actually dodge goalies and score goals by walking into the goal with the ball. The combination of these two strategies meant that the best teams could easily score multiple goals during a game. What really made this interesting, however, was that these behaviors could be acquired and optimized using machine learning.

The last few years of the Four-Legged League were marked by a significant rise in machine learning [2]. Teams discovered that their walks could easily be optimized by any of a variety of machine learning techniques [3, 7]. Furthermore, individual behaviors, such as the grab [5], could be learned, as well as various aspects of color learning [8].

2.2   The NAO Years

After Sony ceased production on the AIBOs in 2006, the league began transitioning to a new robot platform. Following an open call for proposals presented at RoboCup 2007, the new standard platform was chosen to be the two-legged Aldebaran NAO robot. In 2008, SPL ran two competitions (AIBO and NAO) and in 2009 it transitioned completely to NAO. Moving to the biped NAO robot solved a number of problems that AIBOs had, but the change also created significant new problems.

The NAOs were able to benefit from ten years of improvements in robot hardware. For example, the NAOs has significantly better cameras than the AIBOs and faster processors. Among other things, this enabled the league to move to goal posts that were true posts. In addition beacons were no longer needed because the NAO's increased height enabled it to see more of the field than the AIBOs could. As a result, vision became simpler and ultimately localization improved, because the NAO's increased field of view made exploiting field lines possible. Eventually, these developments allowed to color both sets of goals posts uniformly yellow. While the new symmetric field created interesting new localization challenges, it also made vision even simpler by eliminating the blue color from the landmarks.

The reduction in colors (e.g. no more pinks on beacons, blues on goals or beacons) and landmarks (only one type of goal, no beacons, simpler uniforms), as well as better cameras and faster processors, has made more experimentation in vision possible. Teams, such as NAO Team HTWK and NAO Devils, have experimented with approaches to vision that do not use LUTs at all and instead infer the colors of the field of the objects by a combination of knowledge (e.g. most of the floor is green) and changes in luminance. Other teams have begun to move towards modern and general machine vision approaches, such as using Hough transforms to identify lines. Such approaches would not have been possible on the AIBO. What has not happened though is any sort of counterbalance to make vision a hard problem again. Ideas in this direction can easily be found in real soccer; the balls, for example, vary in color and patterns from game to game, as do team uniforms. Meanwhile, even though field lighting is not as carefully controlled as it was during the AIBO years, NAO games are still played indoors with fairly invariant ceiling, but no dedicated, lighting.

On the other hand, the league did work hard to counterbalance the improvements the NAOs brought to localization. The biggest change was moving to perfectly symmetric fields with regard to colors. No longer are there any unique landmarks that will tell a robot which direction it is facing. The current environment represents a significant contrast that to the early years, when there were two uniquely-colored goals, as well as six uniquely-colored beacons. Although the problem has gotten much harder, the underlying algorithms in localization have not changed significantly, besides incorporating field lines as landmarks and the need to rely on correct initialization. Probabilistic approaches based on Particle and Kalman Filters continue to dominate, but have been refined and extended with the addition of techniques, such as shared ball models and global (team) localization approaches, which in turn can help robots to disambiguate their location between the symmetric positions on opposite sides of the field.

By far, however, the biggest area of change and research, since the switch to the NAOs, is work on motion. Four-legged robots do not fall over and are relatively agile. Two-legged robots can fall and thus far are still far from being agile. In RoboCup 2009, the 2nd year of the NAO competition, the top three finishers were marked by the fact that each had developed their own walk engine (B-Human [6], Northern Bites [11], and NAO Devils [4]). Since then, there have been other teams who have developed very successful walks, e.g. rUNSWift and NAO Team HTWK, but the league has begun to converge on a single walk engine. Numerous teams now use some version of the B-Human engine. The reasons for this are pretty simple: a good motion system is still the dominant factor in robot soccer at the present. The best vision, localization, and behavior systems in the world cannot compete against a team that is simply faster to get to the ball and falls down less often.

While the motion systems used in the SPL now are vastly improved from what they were five years ago, the transition has cost the league in other areas. Robots that are slow, not agile, and prone to falling down do not make reliable teammates for coordinated activities, such as passing. Meanwhile, the robots are capable of kicking the ball more than the full-field length. This has led to a situation similar to the AIBO days, where there has been more emphasis on individual skills, such as directed kicking, than on coordination and passing. However, recent developments in the league point to a change towards this direction. One reason for this has been the development of "motion kicks", where the robots can kick the ball without first coming to a stop. Such kicks are not as powerful, but are much more useful against good teams, because they move the ball quickly and so it is less likely to be stolen by an opponent. Since motion kicks are less powerful, robots using them are not a threat to score from literally anywhere on the field and so moving the ball down the field in a series of kicks has become more important. As more teams use the B-Human, or similarly reliable, walk engine, walk speed will be cease to be the dominant differentiation factor between teams and team behaviors will become increasingly important. With precise kicking increasingly more prominent in the SPL, positioning of players and ball passing will start to become the primary factors that separate teams.

3   The Present

In an effort to spur development of better team play, the league has begun to experiment with "*drop-in*" games. Such games consist of teams of robots drawn randomly from the population of competing teams. To emphasize the importance of this competition, it is required that all qualified teams compete and some teams are invited to compete, even if they did not qualify for the main soccer competition. From the perspective of the league, there are two major challenges in running such a competition. The biggest challenge is *how to rank* the teams. Since one of the major goals of the competition is to increase interest in passing and teamwork, a blind, peer-review scoring system has been created with metrics that reward such teamwork activities. Nevertheless, judging a robot's quality in a given game is, at best, a difficult proposition. A poor decision may be the result of bad information provided by the teammates, but also a glimpse in the robot's own behavior system. The hope is that such variables will tend to even out over the course of a number of games. It is also clear that judging metrics will necessarily evolve over time to better meet the needs of the league.

A second challenge is *communication*. Drop-in games are not even possible, unless robots use a common, agreed upon communication protocol, so that they can share basic state information. To facilitate these games, the league has moved to a standard message packet format that each team is required to use in all games. The packet contains a number of standardized data fields for information, such as the robot's location and the location of the ball, and it also contains data fields that can be customized for any team's own needs, when playing normal games. The construction of the packet is not trivial, as the decisions have consequences for how well drop-in teams might perform. Early versions of the packet, for example, only contained basic information about the id of the robot sending the packet, where the robot was, and where the ball was. Such packets give no sense whatsoever of a robot's intentions. Is it going to the ball? Is it playing a particular position? And, so on. These things would need to be inferred by the robot's teammates. Ultimately, the league has chosen to include more information in the packet, including the robot's intended destination, where it is shooting to (or, is trying to shoot to), and a description of its intentions in terms of taking roles/positions, such as defender or keeper. Of course, the quality and even the truth of the information conveyed through these messages has to be considered by any robots receiving it.

There are many challenges involved in making such drop-in games work, and drop-in games could open rich areas of research for the league. For any given robot, for example, it is necessary to figure out which of its teammates it can trust. Some will be faster than others or have better localization systems. Some will communicate a ball location that isn't correct, etc. One change that this system may foster is the development of a more human-like position system. A theoretical advantage that robots have over humans is that they can all run the same code and have the same hardware, meaning that players can swap positions at any time. A single player might move from defense to striker and back to defense again, all in the space of a single point. If this is done seamlessly, it affords many advantages. In an ad hoc game, however, the amount of coordination to pull such a system off may not be achievable. Furthermore, in such scenarios robots might really have different capabilities. Therefore, it makes more sense for robots to adopt a human-like strategy, where players take a role in the beginning and stick with it. That way some of the communication that is lost by using a simple common protocol, can be implicitly gained back by knowing the basic roles of teammates. A human midfielder need not communicate explicitly with all of her teammates to know generally where they are and what they are doing, if she understands the basic principles of team soccer.

This year the league is also experimenting with a *coaching robot*. The idea comes from human soccer, where coaches provide general strategy for teams from the sideline. So, this year teams will be allowed to place a coaching robot on the sideline, where it can observe the game and communicate with the field players. The danger of such an addition, of course, is that the coach, who will have an fairly good view of the field, could provide solve problems for the individual players, such as which side of the field that they are on. To combat this, the league is experimenting with limiting the number and the style of communiques that the coach can provide, as well as putting a time delay on coach-player communications. The idea is that the coach should merely be communicating overarching strategies, not real-time state information. These communication limits are mirrored somewhat by a large drop in the amount of packets that field players are allowed to send and the requirement to use the new standard message format for communicating.

Meanwhile, solving the localization problems involved in a symmetric and fairly populated field remains an area of ongoing research. To this point no best practice has yet emerged, rather most teams seem to use a variety of heuristics. The most common has probably been to rely on teammates. For example a goalie can inform teammates when the ball is on its side. Other solutions have ranged from having a team's goalie make distinctive sounds to reveal the location of the home goal to using a variant of the SURF algorithm to try to build an on-the-fly map of the field's surroundings.

In terms of the mechanics of soccer, the league has once again progressed back to the point where the games were at the end of the AIBO era; there are five players on each side and the field has grown to be 9×6 m. Typically, the league sticks with a given field size and number of players for approximately two to three years, so it is likely that the league will look to use larger fields with more players in the next year. While this growth represents important progress towards the ideal of playing on human fields, it brings its own set of challenges. Some of these are purely practical—not many labs are large enough to support a full-sized field and, of course, more robots require commensurate amounts of additional money. There is a further complication as well having to do with how much teams can run practice games.

In principle, one of the advantages SPL has over all other RoboCup soccer leagues is the common standard robot platform. Among other things, this means that it should be possible to run regular practice games between players of the same team and even to play practice games against other teams, if their code (or their executables) are available. This actually happened a fair amount of times in the last few AIBO years. For example, the Northern Bites were able to compete remotely in the RoboCup German Open competition in 2008. In turn, the German Team provided the Northern Bites with executable versions of their code so that the Northern Bites could play practice games against the German Team in their lab. The advantages of such arrangements are myriad, but boil down to the fact that the way to get better at anything is to practice. In RoboCup, practice games provide essential data and opportunities for debugging, considering the variation offered by playing against other teams. In some ways there are more opportunities for these kinds of practice games now than ever, as several teams release their code on a yearly basis. However, as the league rules, including field dimensions and number of players, change, the amount of work necessary to get one of these releases running, obeying the updated rules, quickly becomes prohibitive. Further, even with the current settings, running a full practice game requires a total of 10 healthy robots, a number that very few teams can afford.

4   The Future

The move to two-legged robots was a case of the league needing to take a number of steps backwards in terms of quality of play in order to make some very large steps forward possible. The difficulty of walking stably and quickly was clearly the dominant focus of the SPL for several years after the switch. At this point, the low-hanging fruit of what the NAOs can do has more or less been reached and many of the remaining limitations can now be attributed to the hardware. On the one hand, such limitations are a roadblock to further development in terms of motion, but, on the other, it means that the standings of the league will not be so dependent on who has the best walk engine, which ought to force developments in other areas. The biggest area with the most potential for future developement is almost certainly team play, such as team formations, role assignment, player positioning, and ball passing. In order for a team to develop a successful passing strategy, there must be a combination of ability and need. On the ability side, robots must know where their teammates, as well as their opponents, are and be able to kick the ball with a fair amount of precision. Receiving robots must be agile enough to stop and control the ball. On the need side, it must be the case that it is better for a robot to pass the ball than to simply make a long kick towards the opponent goal or try to dribble the ball upfield on its own.

In terms of ability, many of the pieces are in place. Robots in the league localize well enough that most robots very reliably know where their teammates are with a relatively small amount of error. While opponent localization is a much harder problem, many teams can now reliably identify opposing robots in their field of view and are actively working on incorporating them into a shared model. Kicking the ball accurately remains difficult, because of the shapes of the robots' feet, limitations on the movements of their joints, and differences in field conditions. Nevertheless, robots are able to kick the ball with some accuracy over short distances. The biggest limitation seems to lay in the receiving robots. Even the best humans do not always make perfect passes, but human receivers do quickly adjust to imperfect passes. Humans can quickly shift their feet, their motion path, or whatever is necessary in order to receive and control a pass. Robots in the SPL are simply not capable of doing much of that yet. Simple passing, not nearly as fluid as it is for humans, still remains to be seen.

In terms of need, a major reason there is so little passing seems to be that there is not enough payoff to try it very often. A pass should provide the passing team with a clear advantage over not passing. In RoboCup, a player near the ball has three basic options: shoot, dribble, or pass; right now, passing is the third choice for most teams. This mainly has to do with the fact that shooting is such a high-reward, low-risk option. If a player is aligned toward the goal, there is little reason not to shoot. Players can easily kick the ball hard enough to score from anywhere on the field. Even if the goalie stops such a shot, the ball has gone to a place where the shooting team is at an offensive advantage, particularly since there is no offside rule and many teams position players near the opposing goal. The only disadvantage of such long shots is the possibility that the ball goes out of bounds; a rule introduced in recent years to discourage random shooting dictates that, in such a case, the ball is replaced in the field one meter behind the kicking robot. On top of that, robots that take the time to stop and set their feet to kick over long distances risk having the ball stolen by opponents. Given the lack of precise kicking, the next best option is to dribble, possibly using motion kicks, which are less powerful, but much quicker and almost akin to dribbling. Some addition to the rules may be required to discourage continuous dribbling and encourage work on the next option (passing). However, this may naturally occur, as the number of players and the field dimensions change.

One long-term goal of the SPL has been to be the first humanoid robot league to play 11 versus 11 players' games. With more players on the field necessarily comes an increased premium on teamwork. In the interim the league will continue to push towards games that are more realistic. Changes expected to occur in the near term include changing the goal colors to white and letting teams wear custom jerseys. Both will present new challenges for league vision systems, and both are necessary to the evolution of the league, but neither represents the kind of challenge that will significantly advance the quality of play in the league.

The very fact that RoboCup is a competition encourages teams to stick to "best practices" at the expense of breaking ground on radically new approaches. In recent years this has been reinforced, as the best teams have once again begun to regularly release their code and numerous other teams have used that code as the basis for their systems. On the one hand, this is desirable, as it pushes the overall quality of play higher, allows new teams to get a working system up and running, and forces the best teams to keep improving. On the other hand, many teams make little or no effort to truly push the boundaries of robot soccer.

Going forward the league requires balancing several things. The field needs to get bigger and more robots need to play at a time. On the other hand, if the trend is pushed too far/fast it will eliminate the ability of some teams to practice or even compete. In the past the league has discussed facing the issue by forcing the merger of teams into larger regional teams. The current move towards playing drop-in games, however, seems to be a more pragmatic solution and one that has payoffs in other dimensions. With drop-in games established, each research team no longer necessarily needs to have more than a few robots. The downside is that practice games are practically eliminated. In the long term, it may raise the need for more local competitions. Currently these local competitions are mainly seen as a tune-up for the main RoboCup event, but perhaps with more incentives in place they could be more useful and could even include remote participation. Ultimately, robots are so useful as a tool of science, precisely because they act in the real world. The more data that we can get from such actions the better.

The Standard Platform League needs to work to ensure that it leverages the qualities that make it unique among RoboCup leagues as much as possible. The Drop-in challenge is a strong, positive step in that direction. With the increased emphasis on this portion of the competition teams should begin to think beyond how to engineer the best performance for their team and more about how to create flexible agents capable of thriving under a wider variety of circumstances and with changing teammates. In turn, this could lead the league to explore research areas that it would not have considered otherwise.

To provide a specific example of how such research exploration can be realized, we note that any kind of communication between players in the SPL currently relies solely on the availability of a reliable wireless network. This is, however, an artificial communication channel, missing in real soccer games, whereby human players exchange and share information using audiovisual signals they generate. In addition, experience from the most recent RoboCup competitions indicates that even the current state-of-the-art technology in wireless communication is insufficient to support reliably the needs of SPL games. In the recent 2012 and 2013 SPL games many robots could not even connect to the network, players could not receive game state from the game controller, data packets were constantly missing, teammates could not communicate to coordinate, and eventually most efforts for successful coordination among team players were gone. Since communication is obviously necessary to advance teamwork, the key question that the league may need to address in the future is the following: How could robots exchange and share information, if there was no wireless or any other

Fig. 2. SPL participants at RoboCup 2013 (picture credit: Brad Hall).

kind of network in the SPL field? Could they use the physical world? Could they generate visual and/or auditory cues that could be perceived by other robots and thus convey information in a more natural and human-like way?

As a concrete example towards audiovisual communication, consider the problem of generating an intuitive visual signal for sharing information about the most important element in a soccer game: the current position of the ball in the field A robot having direct visual observation of the ball can assist its teammates by generating a visual signal that reveals the current position of the ball. The most natural and intuitive visual signal one could imagine is to fully extend one of the two arms, so that it points directly to the ball. Another robot could visually perceive this signal through its camera and use the vector defined by the first robot's arm to pinpoint the position of the ball. Note that this kind of information sharing about the position of the ball does not necessarily require any kind of localization of the two robots in the field or relative to each other. The perceiving robot could simply "follow" visually the line indicated by the pointing robot to locate the ball. Furthermore, such a visual signal could be used for directing attention to any desired position or object in the field; a lead player could direct a teammate towards a certain position in the field and a future robot referee could easily point out the player that was penalized or the place where the ball went out of bounds! Considering also the new addition of a robot coach on the sideline, images of real human coaches would come to mind, if one watched the robot coach pointing to positions and players, possibly "shouting" instructions at the same time.

The elimination of the (artificial) wireless communication network will open up opportunities for researching several human-like behaviors, which may not be strictly restricted to soccer. Teams will have to develop their own (or adopt existing) sign languages to convey information between their players, possibly taking into account the fact that opponents are also "watching" and information may have to be encoded. Robot perception will not be limited anymore on just correct object recognition (ball, goals, lines, players, etc.), but will also have to succeed in correctly recognizing postures, signals, gestures, sounds, even complete "oral" sentences in a human or even in an (artificial) robot language. Needless to say that all competing players will have to obey referee instructions, signaled with the sounds of a whistle, possibly complimented with details provided "orally". Under the described setting, not only will human spectators be naturally engaged into the game play strategies of each team, but also the technology developed could be further exploited in numerous human-robot and robot-robot interaction situations beyond robot soccer.

5   Conclusion

The SPL remains one of the most vibrant leagues in RoboCup (Fig. 2). It has always been a leader in advancing humanoid robot soccer in increasingly realistic directions in terms of the field dimensions, the number of robots on the field, and the setup of the field. In turn, the emphasis on software has led the league to develop and refine techniques in motion, vision, localization, communication, behavior, and coordination, but also to experiment with applying machine learning and optimization techniques to a range of tasks. In the present the league is continuing to lead as it experiments with the drop-in challenge and with coaching robots. In the future, it will continue to push the boundaries of what robots can do on a soccer field.

References

1. Bruce, J., Balch, T., Veloso, M.: Fast and inexpensive color image segmentation for interactive robots. In: Proceedings of IEEE/RSJ International Conference on Intelligent Robots and Systems (IROS), vol. 3, pp. 2061–2066 (2000)
2. Chalup, S., Much, C., Quinlan, M.: Machine learning with AIBO robots in the four-legged league of RoboCup. IEEE Trans. Syst. Man Cybern. Part C Appl. Rev. **37**(3), 297–310 (2007)
3. Chernova, S., Veloso, M.: An evolutionary approach to gait learning for four-legged robots. In: Proceedings of the IEEE/RSJ International Conference on Intelligent Robots and Systems (IROS), vol. 3, pp. 2562–2567 (2004)
4. Czarnetzki, S., Kerner, S., Urbann, O.: Observer-based dynamic walking control for biped robots. Robots Auton. Syst. **57**(8), 839–845 (2009)
5. Fidelman, P., Stone, P.: The chin pinch: a case study in skill learning on a legged robot. In: Lakemeyer, G., Sklar, E., Sorrenti, D.G., Takahashi, T. (eds.) RoboCup 2006: Robot Soccer World Cup X. LNCS (LNAI), vol. 4434, pp. 59–71. Springer, Heidelberg (2007)
6. Graf, C., Härtl, A., Röfer, T., Laue, T.: A robust closed-loop gait for the standard platform league humanoid. In: Proceedings of the 4th Workshop on Humanoid Soccer Robots at the IEEE-RAS International Conference on Humanoid Robots, pp. 30–37 (2009)
7. Kohl, N., Stone, P.: Policy gradient reinforcement learning for fast quadrupedal locomotion. In: Proceedings of the IEEE International Conference on Robotics and Automation (ICRA). vol. 3, pp. 2619–2624 (2004)

8. Quinlan, M., Chalup, S., Middleton, R.: Application of SVMs in colour classification and collision detection with AIBO robots. In: Advances in Neural Information Processing Systems (NIPS) (2004)

9. RoboCup: Sony Quadruped Robot Football League Rule Book (1998). www.tzi. de/spl/pub/Website/History/Rules1998.pdf

10. Snapshot from BotSport.TV. www.youtube.com/user/BotSportTV (2013)

11. Strom, J., Slavov, G., Chown, E.: Omnidirectional walking using ZMP and preview control for the NAO humanoid robot. In: Baltes, J., Lagoudakis, M.G., Naruse, T., Ghidary, S.S. (eds.) RoboCup 2009. LNCS, vol. 5949, pp. 378–389. Springer, Heidelberg (2010)

RoboCup Humanoid League Rule Developments 2002–2014 and Future Perspectives

Jacky Baltes[1]([✉]), Soroush Sadeghnejad[2], Daniel Seifert[3],
and Sven Behnke[4]

[1] University of Manitoba, Winnipeg, MB R3T 2N2, Canada
jacky@cs.umanitoba.ca
http://www.cs.umanitoba.ca/~jacky
[2] Amirkabir University of Technology, No. 424, Hafez Ave., Tehran, Iran
s.sadeghnejad@aut.ac.ir
http://autman.aut.ac.ir
[3] International Systems and Robotics, Freie Universität Berlin,
Arnimallee 7, Berlin, Germany
daniel.seifert@fu-berlin.de
http://www.fumanoids.de
[4] Autonomous Intelligent Systems, University of Bonn, 53113 Bonn, Germany
behnke@cs.uni-bonn.de
http://www.ais.uni-bonn.de/behnke

Abstract. This paper describes the major achievements in the history of the RoboCup Humanoid League from its start in 2002 to 2014. We provide a perspective for the future of the league with a strong push towards larger robots and FIFA-like playing fields. We also discuss some risks associated with these intended changes.

1 Introduction

RoboCup is an international initiative to promote science and technology through the organization of robot competitions and scientific meetings. The stated ultimate goal of RoboCup is: "By the middle of the 21st century, a team of fully autonomous humanoid robot soccer players shall win a soccer game, complying with the official rules of FIFA, against the winner of the most recent World Cup." [1] Hence, many of the competitions focus on soccer as a challenge problem for artificial intelligence and robotics. However, RoboCup also includes competitions for domestic service robots, rescue robots, and industry-inspired mobile manipulators.

This paper describes the history and future perspective of the RoboCup Humanoid League. The history of the RoboCup Humanoid League can be broken up into several periods. The early years, where simple walking and kicking were formidable challenges, are introduced in Sect. 2. As described in Sect. 3, rapid improvements in mechanics, electronics, perception, and control algorithms resulted in much more capable human-like robots that were able to play soccer games, starting in 2005 for KidSize and 2010 for TeenSize robots. The third period

© Springer International Publishing Switzerland 2015
R.A.C. Bianchi et al. (Eds.): RoboCup 2014, LNAI 8992, pp. 649–660, 2015.
DOI: 10.1007/978-3-319-18615-3_53

of humanoid robot development, discussed in Sect. 4 resulted in robots playing as a team, where winning and loosing was more determined by the perception of the game situation, localization on the field, and team coordination than by individual robot skills. Section 5 describes how commercially available platforms provided teams with an opportunity to speed up their development and the impact of those commercial platforms on the competition. Section 6 gives an introduction to the major changes for the RoboCup 2014 competition in João Pessoa, Brazil. The future evolution of the RoboCup Humanoid League is characterized by a strong push towards larger robots, bigger teams, and more FIFA-like soccer rules and environments as shown in Sect. 7. Some concerns and issues with the current road map are discussed in Sect. 8. Section 9 describes the publications and workshops provided by the league. The paper draws conclusions in Sect. 10.

2   The Early Years (2002–2004)

The Humanoid League is one of the youngest soccer playing leagues in the RoboCup competition. Its inaugural event took place at RoboCup 2002 in Fukuoka, Japan.

Building a fully autonomous humanoid robot that is able to play soccer games is still a challenging problem today, but it was clearly a blue sky—extremely ambitious with a high chance of failure—project in 2002.

At that time, some impressive humanoid robots developed by the Japanese industry like Honda Asimo and Sony Qrio existed. However, these robots were not available to the Humanoid League teams, but only showed demonstrations in controlled environments. Even if they would have been available for purchase, their price tag would have been prohibitive. One of the authors remembers that in 2003, a representative from Sony introduced the Sony Qrio - the successor to the widely successful Sony Aibo robot. In his speech, he was asked about the cost of the Qrio and stated that the cost would be about that of a car. Many researches were extremely excited, since the cost seemed very reasonable for such an advanced platform. When many members of the audience said that they would like to order one immediately, the Sony representative corrected himself by saying: "No. You don't understand. I mean a Ferrari".

In spite of the very ambitious goal, a hodge-podge of about a dozen teams entered the inaugural RoboCup Humanoid League competition in 2002 in

Tao-Pie-Pie (2nd 2002/3) Nagara (1st 2002) HITS (1st 2003) VisiON (1st 2004)

Fig. 1. Early RoboCup Humanoid League Competitors

Fukuoka, Japan (see Fig. 1. The robot designs varied from 20 cm to 180 cm tall robots. There were also many other differences between the robot designs. Another platform that was commercially available during that time was the Fuji HOAP series of robots. Their cost was about \$150,000 USD, but they were not able to act fully autonomously because they did not have sufficient processing power available on-board. So all vision processing etc. had to be done off-board on a PC. Another difference was that several other teams were unable to move autonomously under battery power and had to be powered externally. Some teams were even completely unable to act autonomously and used remote control to move the robot towards the ball.

Due to these constraints, the first RoboCup Humanoid League competition consisted of three events: balancing on one leg, free style demonstration (a panel of judges graded a short free style demonstration by the team), and penalty kicks. To allow these robots with very different capabilities to compete in the same event, the RoboCup Humanoid League Technical Committee (TC) introduced performance factors in order to level the playing field. For example, the performance factor for remote controlled operation was 100 %, so a goal scored with remote controlled robot counted 50 % of a goal scored by a fully autonomous robot.

Apart from team Joitech that used the Fujitsu HOAP, all competitors developed their own hardware for the RoboCup Humanoid League and similar competitions (e.g., Japan Robo-One fighting robots). Since the Humanoid League TC realized that building your own robot was a significant challenge at that time and since it wanted to encourage teams to explore design ideas and build their own robots, commercial platforms were also penalized by a 20 % performance factor.

The RoboCup Humanoid League TC was acutely aware of the fact that building larger humanoid robots was even harder than building smaller humanoid robots, but that to achieve the goal of 2050 large humanoid robots are of strategical importance. Therefore, the RoboCup Humanoid League TC separated the league into three size classes: small (<60 cm), medium (60 cm to 80 cm) and large (>80 cm) robots.

The constraints of these early years influenced rule development in successive years and still influence the culture of the Humanoid League.

The performance of the Humanoid League robots developed quickly. By 2004, all robots acted fully autonomously and all processing was handled on-board. Therefore, the need for the performance factors vanished and they were removed from the rule book. The rules evolved to provide fair and entertaining competitions that could still act as benchmark problems for our research into developing capable fully autonomous soccer robots. The main tournament was now played as penalty shoot out. Standing on one leg was replaced by a Humanoid Walk competition, where robots had to footrace around a pole. Each year, a new technical challenge was introduced, in order to encourage development of new skills that were not yet applicable in the main tournament. In 2004, the technical challenges consisted of an obstacle walk, a passing task, and balancing across a sloped ramp.

Fig. 2. Some Humanoid League Finals: (a) 2004 Penalty Kick: Team Osaka vs. Robo-Erectus; (b) 2005 2 vs. 2 Soccer: NimbRo vs. Team Osaka; (c) 2009 TeenSize Penalty Kick: CIT-Brains vs. NimbRo.

The results of the individual competitions are aggregated into a Best Humanoid ranking. Since robots from the different size classes cannot be compared directly, the overall Best Humanoid Award, the Louis Vuitton Cup, is determined by voting of the team leaders, based on robustness, walking ability, ball handling, and soccer skills.

3   From Penalty Kicks to Soccer Games (2005–2010)

As teams improved the robustness and walking ability of their robots, it became possible to start 2 vs. 2 soccer matches for the small size class. After demonstration events during RoboCup 2003 in Padua, Italy, and RoboCup 2004 in Lisbon, Portugal, soccer matches were introduced as main KidSize (<60 cm) tournament in 2005.

The larger TeenSize robots (initially >65 cm, later the minimal size was increased to 100 cm) continued to play penalty kick, which was developed in 2007 to the Dribble and Kick competition. Dribble and Kick is played between two robots - a striker and the goal keeper. The striker robot starts in the center of the field and the ball is placed randomly on the striker's goal box. The task of the striker is to move back to approach the ball, dribble the ball across the center line and then kick the ball into the opposing goal.

In order to remove subjective judgment from the competition as much as possible, quantitative measures (e.g., goals scored) were seen as much more desirable by the teams. Therefore, the free demonstration event was removed from the competition.

Team VStone from Osaka, Japan, set a new bar for all competitors with their small robot platform. The robots were able to move quickly and stably across the playing field. Furthermore, the robots used an omnivision system in the head of the robot, which allowed the robots a 360-degree view of the playing field. The VStone robots were extremely successful in the soccer competitions and won them two times in succession ([2], Fig. 2).

During this time, many formal and informal discussions were held among the technical committee and the participants. After several years, it became apparent that most participants felt that humanoid robots should be limited to human-like kinematics and human-like sensors. As a result of these discussions, the use of omni-vision was disallowed. Furthermore, active sensors (e.g., LIDAR, ultrasound, IR distance sensors) were also forbidden. This was not a major restriction, since from the beginning, color cameras were the most dominant sensor for perceiving the environment.

During these years, the performance of the larger robots also improved significantly. Partially driven by the availability of affordable high powered servos, the performance of the smaller TeenSize robots (80 cm to 120 cm and less than 10 kg) had improved to the point where 2 vs. 2 soccer matches became possible. There was a strong push from those teams to introduce soccer games for larger robots. But the largest (>120 cm) and heaviest robots were still too fragile to survive a fall. Furthermore, since some of the robots weighed more than 40 kg, they posed a real danger to other robots or participants should they fall. As a consequence, the larger robots were split again into two size classes: The smaller TeenSize robots (100–120 cm) started to play 2 vs. 2 soccer games in 2010, while the AdultSize robots (>130 cm) continued with Dribble and Kick competitions.

The rapid improvements in the robots' capabilities also led to an increase in the complexity and diversity of the technical challenges. The technical challenges introduced during this time includes: walking over uneven terrain, dribbling the ball around multiple poles, dribbling the ball trough randomly placed obstacles, and double passing.

4   From Individual Skills to Team Play (2008–2012)

In 2008, the number of players in the KidSize soccer matches was increased from 2 to 3 players per team. Furthermore, most teams had successfully solved the problem of locomotion and were able to walk stably over flat even surfaces such as hardwood floors or carpets. For these two reasons, the localization (where is the robot?) and the perception of the game situation (where are the ball and the other players?) became more and more important. Whereas individual robot skills (fast walking and strong kicking) were key to success in previous years, now team play and coordination became more important.

This was reflected in the rules, e.g. by placement disadvantages for robots which could not autonomously walk to their kickoff positions. Two teams from Germany (Team NimbRo [3], University of Bonn, and Darmstadt Dribblers [4], TU Darmstadt) were powerhouses during that period and won the competition several times.

The next wave of major rule changes aimed at making visual perception and localization more realistic. Additional landmarks in the corners of the field and later on the side lines were removed. In 2010, extra lighting on the field was removed, which resulted in much larger variability in brightness due to the influences of environmental lighting. The size of the playing field was extended, and detection of the goals was made more difficult: First the colored goal back walls were removed, leaving only the goal posts as landmarks, and in 2013 both goals were colored yellow [5].

The technical challenges now included throw-ins and high kicks. In the soccer matches, throw-ins are replaced by the referee putting the ball back into play without stopping the match, since a throw-in is an often occurring event that is a time consuming task for a humanoid robot. The throw-in challenge encouraged teams to use throw-ins by the goalie in actual games. The high-kick challenge also encourages the use of the third dimension in the games.

5   The DARwIn and NimbRo-OP Platforms (2012–now)

From its start in 2002, the number of participating teams had increased drastically over the years. So the number of participating teams in the KidSize class had to be limited to 24 and qualification for the RoboCup competition had become competitive. For qualification, teams had to submit a team description paper (TDP) and a video of their robot playing soccer. In that video, the robot needed to demonstrate the ability to perceive and approach a ball, line up with the goal, and to kick the ball into the goal. It also needed to demonstrate the ability

Fig. 3. Darwin-OP (left) and igus Humanoid Open Platform (right)

to stand up after a fall from various positions (i.e., falling forward and falling backwards).

In 2011, the Korean company Robotis introduced the DARwIn-OP robot, which they had developed in conjunction with Dennis Hong from Virginia Tech [6]. A year later, a similar collaboration between Robotis and Sven Behnke from Bonn University resulted in the development of NimbRo-OP [7], a teen sized humanoid robot, which is now further developed together with igus GmbH (Fig. 3). The introduction of these platforms had a big impact on the Humanoid League. Instead of designing and building their robots from scratch, teams could now simply purchase a robot platform that was able to walk and kick a ball and recover from a fall out of the box. So it made qualification and entry into the league much easier for new teams. E.g., the DARwIn-OP had a large impact on the kid size league. Figure 4 shows the Humanoid League teams participating in RoboCup 2013. In 2014, 50 % of the KidSize teams that submitted qualification material used the DARwIn-OP platform or based their robot on it.

Many teams that built their robots from scratch felt that it was unfair that in spite of the fact that they had spent much hard work, time, and money on building their own robots, other teams could just purchase a robot and qualify for the RoboCup competition with much less effort. Other teams felt that only the performance of the robot should be the determining factor in qualification. The RoboCup Humanoid League TC discussed the issue and decided on a compromise. The stated policy for qualification is that teams that purchase a robot had to clearly highlight what advancements and improvements they had made to the out of the box system for qualification.

Fig. 4. Teams of the Humanoid League at RoboCup 2013 in Eindhoven, NL.

6 Brazil Ole Ola (2014)

At the end of the 2013 RoboCup competition, the RoboCup Board of Trustees issued a challenge to all leagues as they felt that progress in the leagues had been limited to incremental improvements rather than radical breakthroughs.

After discussions with the team leaders, one major change in the Humanoid League was an aggressive push towards larger robots. The 2014 competition [8] introduced radical changes in the sizes of the Kid, Teen, and AdultSize classes. For example, the maximum height of the robots in the KidSize was raised by 50 % to 90 cm. Furthermore, the height limits of the Kid and Teen and the Teen and AdultSize classes were chosen with some overlap on the upper and lower limits. This allows teams to more easily transition into larger size classes, since they do not need to build a completely new robot. For example, a team could build an 85 cm tall robot and compete in the KidSize class in the first year and use the same robot in the TeenSize class the next year. Figure 5 shows some of the new generation of robots planning to participate in 2014.

Consequently, the field area for KidSize was increased by 125 % to 6 × 9 m, and the size of the goals, and the size and weight of the ball were adjusted to accommodate the larger robots. To enhance team play, the number of players for the soccer matches was increased to four players per team.

The complexities of the challenges also increased. The dribble and kick competition for AdultSize robots will introduce two obstacles (representing opposing players) that must be avoided by the striker robot.

7 The Future of the Humanoid League (2015–2050)

The RoboCup Humanoid League continues its rapid advance towards smarter and more capable robots. The goal is to move quickly towards more realistic soccer matches. There are several extremely difficult problems that need to be overcome with respect to the environment, the players, and the matches.

Fig. 5. Humanoid Soccer Robots preparing for RoboCup 2014 in Brasil.

One future direction of the RoboCup Humanoid League is to move to more human-like playing fields and environments. For example, in 2015 the RoboCup Humanoid League TC plans to reduce the colors in the environment even further and to move towards requiring shape or texture-based segmentation. The plan is to remove the yellow goal posts and to replace them with white goal posts. Furthermore, the orange ball might be replaced by a small version of a real soccer ball, that is a ball that is mostly white or gray with some texture to it.

Also, the playing surface might be changed from a carpeted floor to astro turf and eventually a real grass playing field. This means that active balancing and uneven terrain walking will become more important for the robots. As the number and speed of the robots increases, collisions between players are more likely to occur. The RoboCup Humanoid League TC might introduce push recovery challenges to test the ability of the robots to compensate for pushes from various directions.

In addition, corner flags similar to human soccer may be introduced. These changes combined will make the playing fields in the RoboCup Humanoid League scaled down versions of the FIFA soccer playing fields.

The progress in the TeenSize class has shown that it is now possible to have soccer matches for 80 cm to 120 cm tall humanoid robots. But the ambitious goal of the RoboCup Humanoid League TC is to introduce rules that move towards robots capable of playing against human players. To this end, the minimum height of the robots might be increased in stages from the current 40 cm to 140 cm.

As the capabilities of the robots increases, the RoboCup Humanoid League will play with larger and larger robots that become more and more similar to human players in their kinematics, dynamics, and sensing.

One technical challenge that could foster the use of the third dimension would be a header challenge, where a robot would need to score by using its head.

Lastly, the rules of the game have to be adjusted to match exactly the FIFA rules for human soccer. This requires that robots must be able to act fully autonomously during all aspects of the game (including kick-offs, substitutions, and free kicks). Furthermore, the games will include throw-ins and direct and indirect free kicks. It is interesting to note that this rule progression towards more human-like soccer is not always linear. For example, free kicks were included in the rules from 2004 to 2007, but slowed down the game significantly. Therefore, all free kicks were replaced by 30 second removal penalties similar to ice hockey rather than soccer. This made games much more entertaining and made the RoboCup Humanoid League one of the most exciting leagues in RoboCup. However, as teams improve the skills and capabilities of their robots, free kicks can be re-introduced while still resulting in exciting matches.

The number of players will be increased further in the following years. We start playing with 4 vs. 4 players in 2014 and will eventually reach 11 vs. 11 players in 2050. The RoboCup Humanoid League TC realizes that few teams will be able to afford 11 players and has also started to build the necessary technical infrastructure as well as amendments to the rules to encourage joint or mixed teams.

So to encourage more team collaboration, there have been several efforts directed at creating suitable communication protocols and infrastructure that will allow players from different teams to play soccer effectively. A good example for this approach is the RoboCup Standard Platform League. In the Standard Platform League (SPL) many teams have based their software on the yearly code-release of team B-Human ([9]). This resource has a great impact on the development of the Standard Platform League, since all teams must use an unmodified NAO robot and therefore the software can be used directly by other teams. Similarly, several Humanoid League teams have released their robots' source code and hardware designs ([10]). However, the benefit of those contributions is much less immediate. Firstly, teams use often different hardware designs, so inverse kinematics, walking gaits, device drivers, low level controllers need to be adjusted. Furthermore, even higher level functionality in the software (e.g., localization, vision, and behaviour coordination) are implemented using different and often custom middle-ware. There are now several initiatives to implement soccer robot middle-ware for important modules such as vision, localization, walking engine, and communication. The Robot Operating System (ROS) is a popular candidate to simplify Inter-operability by software developed by different teams.

The other issue with the robots in the RoboCup Humanoid League are their robustness and energy efficiency. The use of compliance in control and construction of the actuators and links as well as soft materials on the outer shells will be necessary for improved soccer capabilities, such as running, high-speed kicks, robustness to falls, and safe robot-robot and human-robot physical interaction [11]. To test the robustness of robots drop test may be introduced.

The energy efficiency of the robots also needs to be greatly improved. Currently the soccer games in the RoboCup Humanoid League last only 20 min since robots cannot operate for much longer, due to the limited capacity of the batteries, the relatively poor power to weight ratio of the servo motors. Furthermore, few of the robots are able to use the inherent dynamics of the motion (e.g., the swing leg needs to be actively driven rather than swinging freely, because of the friction in the gear box) or are able to store energy in springs or other mechanics. To encourage teams to develop more energy efficient robots, the duration of the games will be increased in the coming years to ultimately match that of human soccer matches – 90 min per game.

8 Risks and Issues

This paper would be incomplete if it were not to include some words of warning for the future development of the league. The initiatives described in Sect. 7 are far reaching and ambitious. As such, it is clear that they entail a certain amount of risk.

The first issue is that moving to larger robots will greatly increase the costs associated with RoboCup participation for all teams. Larger robots require much more torque and thus much more expensive motors. Furthermore, large robots cannot be brought in check-in luggage and often must be shipped as cargo.

Combined with the ever increasing registration fees of RoboCup, this may lead to teams deciding not to participate in RoboCup. For example, several of the teams that participated in the RoboCup Humanoid League for many years (e.g., Tamkung University, Damshui, Taiwan, NCKU, Tainan, Taiwan, and FUmanoids, Berlin, Germany) will not participate at RoboCup 2014 for financial reasons.

For example, the number of participants in the TeenSize and AdultSize classes remained low with four to six teams each. The low number of participants was in spite of the best efforts of the RoboCup Humanoid League and the supportive attitude of the RoboCup Federation in general to promote large humanoid robots. One possible way to make soccer with larger robots easier would be the introduction of an affordable AdultSize platform. For example, Robotis Inc. is currently developing a large humanoid robot THOR-OP.

Another problem is that the move to larger robots will make entry for new teams much harder; the current road map does not provide a path for new teams. One suggestion discussed by the RoboCup Humanoid League TC was to keep the current KidSize league, but to require the best teams to move to a larger size class after two years. Another idea is to promote mentor teams where more experienced teams form joint teams with less experienced teams and provide them with technical support. A third possible way would be to include small humanoid robot competitions in the RoboCup Junior leagues. The issue may also be mitigated in the future by the general commercial availability of better and cheaper robot platforms that are more suitable to robotic soccer.

9 Humanoid Soccer Workshops, Schools and Publications

The Humanoid League does not only foster development through the organization of competitions, but has also a strong focus on advancing research via publications, workshops, and schools.

The team descriptions required for qualification are archived together with the other qualification material and the competition rules at the Humanoid League web page [10].

Members of the league submitted many high-quality contributions to the annual RoboCup International Symposium and major robotics conferences (e.g. IROS, ICRA), some of which were honored with Best Paper Awards. There is also a large number of journal publications originating from the league and two special issues of leading journals (Robotics and Autonomous Systems and International Journal on Humanoid Robots) appeared.

In addition, members of the league contribute heavily to the organization of and the submission to the annual Humanoid Soccer Workshop, which is organized since 2006 at the IEEE-RAS International Conference on Humanoid Robots, the flagship conference for humanoid robotics research.

Finally, since 2012, member of the Humanoid league have organized week-long humanoid soccer schools (see Fig. 6). These schools provide unique opportunities for about 40 researchers and hobbyists alike to learn from some of the leading experts in the field. But in contrast to scientific conferences and workshops, the humanoid soccer schools include practical components. A lot of time is made available to students to complete exercises and/or test their own ideas on real systems. The humanoid soccer schools also include a series of social events making it easier for researchers to socialize. The hope is that this will lead to closer collaboration between the teams in the future.

Amirkabir UT., Tehran, Iran, 2014

Drachenfels, Germany, 2013

All these scientific activities ensure that (a) the research developed as part of the RoboCup initiative is widely disseminated to other researchers, and (b) that researchers participating at RoboCup learn about the latest

Fig. 6. Participants of the humanoid soccer schools

research results from other humanoid robotics researchers. This means that especially new teams do not have to start from scratch and can learn from leading teams.

10   Conclusions

The paper describes the history of the RoboCup Humanoid League from its humble beginnings in 2002. It describes the historical evolution of its competitions and rules to provide the reader with an insight into the culture and traditions of the RoboCup Humanoid League. This will make it easier to understand the current state as well as future plans for the development of the RoboCup Humanoid League toward the 2050 goal.

The authors would like to thank previous and existing members of the RoboCup Humanoid League community for their input during many years of rule discussions and development. In particular, we would like to thank the other members of the RoboCup Humanoid League Technical and Organizing committees (Reinhard Gerndt, Luis F. Lupian, Marcell Missura) and RoboCup trustee Oskar von Stryk.

References

1. Kitano, H., Asada, M.: The RoboCup humanoid challenge as the millennium challenge for advanced robotics. Adv. Rob. **13**(8), 723–737 (2000)
2. Matsumura, R., Ishiguro, H.: Development of a high-performance humanoid soccer robot. Int. J. Humanoid Rob. **5**(03), 353–373 (2008)
3. Behnke, S., Stückler, J.: Hierarchical reactive control for humanoid soccer robots. Int. J. Humanoid Rob. **5**(03), 375–396 (2008)
4. Friedmann, M., Kiener, J., Petters, S., Thomas, D., Stryk, O.V., Sakamoto, H.: Versatile, high-quality motions and behavior control of a humanoid soccer robot. Int. J. Humanoid Rob. **5**(03), 417–436 (2008)
5. RoboCup Humanoid League Technical Committee. RoboCup Soccer Humanoid League rules and setup for the 2013 competition in Eindhoven (2013)
6. Ha, I., Tamura, Y., Asama, H., Han, J., Hong, D.W.: Development of open humanoid platform DARwIn-OP. In: Proceedings of SICE Annual Conference (SICE), pp. 2178–2181. IEEE (2011)
7. Schwarz, M., Pastrana, J., Allgeuer, P., Schreiber, M., Schueller, S., Missura, M., Behnke, S.: Humanoid teensize open platform NimbRo-OP. In: Behnke, S., Veloso, M., Visser, A., Xiong, R. (eds.) RoboCup 2013. LNCS, vol. 8371, pp. 568–575. Springer, Heidelberg (2014)
8. RoboCup Humanoid League Technical Committee. RoboCup Soccer Humanoid League rules and setup for the 2014 competition in Joao Pessoa (2014)
9. Röfer, T., Laue, T., Müller, J., Bartsch, M., Batram, M.J., Böckmann, A., Böschen, M., Kroker, M., Maaß, F., Münder, T., Steinbeck, M., Stolpmann, A., Taddiken, S., Tsogias, A., Wenk, F.: B-human team report and code release (2013). http://www.b-human.de/downloads/publications/2013/CodeRelease2013.pdf
10. RoboCup Humanoid League Webpage. http://www.informatik.uni-bremen.de/humanoid
11. Haddadin, S., Laue, T., Frese, U., Wolf, S., Albu-Schäffer, A., Hirzinger, G.: Kick it with elasticity: safety and performance in human–robot soccer. Rob. Auton. Syst. **57**(8), 761–775 (2009)

RoboCup Rescue Simulation Innovation Strategy

Arnoud Visser[1]([⊠]), Nobuhiro Ito[2], and Alexander Kleiner[3]

[1] Universiteit van Amsterdam,
Science Park 904, Amsterdam, The Netherlands
A.Visser@uva.nl
[2] Aichi Institute of Technology,
1247 Yachigusa, Yakusa Cho, Toyota City, Japan
[3] Linköpings Universitet, SE-581 83 Linköping, Sweden

Abstract. The RoboCup rescue simulation competitions have been held since 2001. The experience gained during these competitions has supported the development of multi-agent and robotics based solution for disaster mitigation. The league consists of three distinct competitions. These competitions are the agent competition, the virtual robots competition, and the infrastructure competition. The main goal of the infrastructure competition is to increase every year the challenge and to drive the innovation of the league, while the agent and virtual robot competition are focused on developing intelligent agents and robot control systems that can cope with those challenges. This paper provides an overview on the current state-of-the-art in the league and developments and innovations planned for the future.

1 Introduction

Robots are designed to do work in dull, dirty and dangerous environments. A disaster can be classified as dirty and dangerous. In several occasions rescue robots were applied after a disaster, as described by Murphy [1]. An example is the application of the Quince robot at the Fukushima Daiichi Nuclear Plant. The prototype of the Quince robot was developed at the RoboCup [2]. Yet, the typical application of a robot in this circumstance is the teleoperation of a single robot, while the scale of the disaster could benefit from a multi-agent approach.

The rescue simulation league (RSL) aims to develop realistic simulation environments for benchmarking intelligent software agents and robots which are expected to make rational decisions autonomously in a disaster response scenario. The RSL has two major competitions, namely the *agent* and the *virtual robot* competition. The agent competition consists of a simulation platform that resembles a city after an earthquake. In that environment, intelligent agents can actively mitigate the impact of the disaster and influence the cause of events after the disaster occurred. The agents have the role of police forces, fire brigades, and ambulance teams, and are mainly in charge to remove debris from the roads, extinguish fires, and to rescue civilians. The virtual robot competition has as its goal the study of how a team of robots can work together to get a situation assessment of a devastated area as fast as possible. This will allow first responders to be well informed when they enter the danger zone.

R.A.C. Bianchi et al. (Eds.): RoboCup 2014, LNAI 8992, pp. 661–672, 2015.
DOI: 10.1007/978-3-319-18615-3_54

This paper builds upon the leagues' status report from eight years ago [3]. In Sect. 2 developments and advancements of the rescue simulation league during the last decade are described. Section 3 provides an overview on the current state-of-the-art of the league and Sect. 4 outlines developments and directions in the future.

2   Past 10 Years: Rescue Simulation League

The RoboCup Rescue Simulation League (RSL) started in 2001. The RSL aims to develop simulation environments which benchmark the intelligence of software agents and robots with the capabilities for making the right decisions autonomously in a disaster response scenario. Both the two major competitions of the RoboCup Simulation League, namely the *Agent* and the *Virtual Robot* competitions are described in the subsequent sections.

2.1   Agent Competition

The rescue agent competition aims to simulate large scale disasters and to explore new ways of autonomous coordination of rescue teams as an approach of disaster relief after real world incidents [4,5]. The competition consists of a simulation platform which resembles a city after an earthquake. For example, Fig. 1 depicts the simulation of fires and building collapse on a model of a virtual city. Teams of fire brigade, police and ambulance team agents try to extinguish fires and rescue victims in the collapsed buildings. Scoring is based on the number of victim saved on time and the number of buildings with fire damage. The problem of disaster mitigation requires the coordinated action of several heterogeneous and decentralized units. Due to the variation of different potential disaster scenarios, typically strongly diverse teams of actors are required rather than a single agent type. RSL fosters the development of algorithms for coordinating heterogeneous agents. Different types of intelligent agents can be spawned into the simulation environment for mitigating the effects of virtual threats, such as building collapse and fire. To this end, the agents may take on different roles such as *police force*, *fire brigade*, and *ambulance teams* each having different capabilities.

The overall goal is to develop robust software systems that are capable of efficiently coordinating large agent teams for Urban Search and Rescue (USAR). This goal raises several research challenges, such as the *exploration* of large scale environments in order to search for survivors, the *scheduling and planning* of time-critical rescue missions, *coalition formation* among agents, and the *assignment* of agents and coalitions to dynamically changing tasks, also referred to as *extreme teams* [6]. In the rescue domain, this issue is even more challenging due to restricted communication bandwidth. Moreover, the simulated environment is highly dynamic and only partially observable by a single agent. Under these circumstances, the agents have to plan and decide their actions asynchronously in real-time while taking into account the long-term effects of their actions.

Fig. 1. A typical RoboCup rescue simulation scenario.

Several authors tried to solve this challenge with a formal approach [7–9]. Their approaches are generally applicable, but still have difficulties to deal with real-time constraints when compared to heuristic approaches of specialized champion teams. To overcome this problem, the simulation league has initiated a new type of challenge [10]. The idea is to extract from the entire problem addressed by the agents certain aspects such as task allocation, team formation, and route planning, and to present these sub-problems in an isolated manner as stand-alone problem scenarios with an abstract interface. As a consequence, participating teams can focus on their research on an aspect of the game, without having to solve all low-level issues. This challenge is more detailed described in Sect. 3.2.

Over the years the winning entries in the competition showed a strong focus on highly optimized computations for multi-agent planning and model-based prediction of the outcome of the ongoing incidents. Several techniques for multi-agent strategy planning and team coordination in dynamic domains have also been developed based on the rescue simulator. Nairo *et al.* [11] first described the task allocation problems inherently found in this domain. Ferreira *et al.* [8] evaluated solutions developed for distributed constraint optimization problems (DCOPs) using problems generated by the simulation. Ramchurn *et al.* [12] modeled the problem as coalition formation with spatial and temporal constraints (CFST) and also adopted state-of-the-art DCOP algorithms for solving CFSTs. Kleiner [13] identified and described the problem of scheduling rescue missions and also introduced a real-time executable solution based on genetic algorithms.

Furthermore, there has been substantial work on building information infrastructure and decision support systems for enabling incident commanders to efficiently coordinate rescue teams in the field. For example, Schurr *et al.* introduced a system based on software developed in the rescue competitions for the training and support of incident commanders in Los Angeles [14].

Urban Disaster Relief Simulator (Early Releases). The first rescue simulation package (version 0) has been released in 1999. This package was further improved and used until 2009. The server was mainly coded in the C/C++ language on FreeBSD. The structure, as shown in Fig. 2, is similar to the package of the soccer simulation league: A central kernel intermediates between the actions performed by the agents (shown within the dashed box) and the state updates computed by the simulators (shown as boxes around the kernel). The initial state and structure of the city model is provided by the GIS component, Communication between all modules is implemented by TCP/IP based message passing.

The purpose of this release was twofold. Firstly, to provide a challenging testbed to the robotics and multi-agent communities. Second, to develop through several competitions intelligent and efficient disaster relief strategies that can make contributions to mitigation solutions developed in the real world.

Urban Disaster Relief Simulator (Advanced Releases). The second simulation package (version 1) is written in Java and thus made the code much more accessible to participants during the last years. The new simulator uses area-models with polygonal representation of the road network for computing

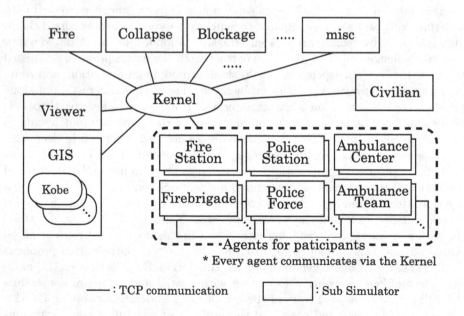

Fig. 2. The structure of rescue simulation package

the traffic simulation. In contrary, previous traffic simulation adopted a simple graph structure to represent roads and intersections, which made the moving of agents rather simple. The new traffic simulation provides much more realistic situations, such as congestions, when multiple agents act simultaneously on the map. In order to simplify the implementation of path planning for participants, a SDK for navigation on the map is provided with the release.

During the last years, several variations of the basic competition setup were introduced. These variations were, for example, maps of constantly increasing size, an increasing of the number of agents that have to be controlled, an increasing number of fire ignition points, changing communication conditions between agents, also including no communication at all, and adding gas stations and water points. Also the simulators within each release were constantly improved. For example, teams from the league developed a more realistic fire simulator, a 3d viewer, and several other simulators and tools. Most of these improvements were essentially suggested through the Infrastructure competition.

In agent competition has organized itself with a maintenance committee. The committee maintains and implements new features for the simulator. Therefore, this maintenance committee also takes part in the rule discussion with technical committees.

2.2 Infrastructure Competition

The infrastructure competition has started in 2004 with the purpose of promoting the development of new simulators and tools to continuously improving the rescue simulator. The simulation of various disaster situations turns out to be complicated and difficult to validate. Therefore, the infrastructure competition has been launched for supporting the maintenance and development of the simulator. For example, the fire simulator [15] and 3D viewer (Fig. 1) were both developed by the winner of the infrastructure competition in 2004.

The Aladdin project[1], funded by the Engineering and Physical Sciences Research Council, has strongly stimulated the conversion towards version 1 of the simulator. The original version of the simulator was chosen as the testbed for ALADDIN technologies and was further improved and extended as part of the project. The end-result is a well-engineered simulator with realistic traffic simulation (developed in collaboration with Meijo University) and GIS map conversion (in collaboration with Freiburg University) that allows disasters to be simulated in selected parts of any given map (OpenStreetMap or GIS formats).

Another simulator component proposed in the infrastructure competition, which is actively used, is the flood simulator [16], illustrated in Fig. 3. Recently, also an extension towards flying robots has been proposed, both in the Virtual Robot [17] and Agent competition [18].

The winner of the infrastructure competition is expected to join the maintenance committees.

---

[1] http://www.aladdinproject.org/robocuprescue/.

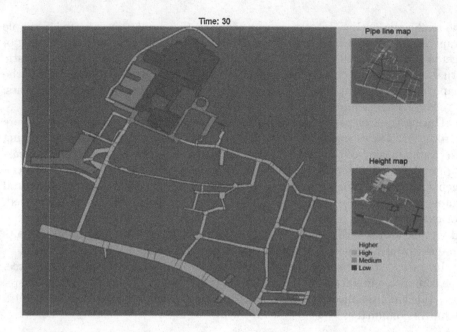

Fig. 3. The flooding of the city based on a height map. Courtesy [16]

2.3   Virtual Robot Competition

RoboCup Virtual Robot competitions are being held since 2006. The competition attracts mainly academic teams from universities often with experience in the RoboCup Rescue Robot League. The important research issues related to the competition are:

Utility-based mapping - autonomous generation of maps from the fused data of multiple vehicles to be used by both robots and humans for exploration and marking victims.

Victim detection - automatic detection of victims from fused sensor data (image processing, acoustics, etc.).

Advanced mobility - robust control algorithms capable of autonomous navigation in small spaces on non-flat flooring.

Multi-robot control - the ability to control multiple platforms with a single operator in realistic environments means that the robots have to be semi-autonomous.

Originally a single scoring formula was used to evaluate the solutions associated with these issues. Simplified challenges were later introduced in 2009 to create more objective measures of performance. Each challenge was about a particular sub-problem with only one measure of performance and a corresponding automatic scoring tool. These challenges, namely mapping, navigation and coordination over a mesh network, were used for qualification for the semifinals. Since 2011 the challenges have been combined again into a single mission. The goal is now defined in terms of an entirely automated scoring procedure. The scoring program is expected to allow for head-to-head competition between two competing teams. This also allows permanent installations of servers, each with its own world, which can be used for testing in preparation of the RoboCup. In this way, the teams can test their approach prior to the competition hence providing a lower barrier of entry for new teams. Furthermore, the scoring procedure takes into account the individual sub-problems solved in the comprehensive challenge so that the teams can assess their performance in all of the domains independently.

The main challenge for the teams is the control of a large team of robots (typically eight) by a single operator. This is state-of-the art; the only real comparison is the champion of the Magic competition [19], where 14 robots were controlled by two operators. The single operator has to use high level commands (such as the areas to be searched, routes to be followed, etc.) but is also needed to verify observations whether or not one of the robots has detected a victim (based on color and/or shape). Due to poor lighting and the number of occlusions, the conditions are generally not favorable for automatic victim detection, and manual conformation is always needed. The approach to a victim is quite critical (the robot should come within the communication range (<1 m), but is not allowed to touch the body or any of the limbs). This means that the workload for the operator is quite high, providing an advantage for the teams which are able to automate the decision making within the robot team as far as possible, and only involve the operator when needed.

The shared map generated during the competition has a central role in the coordination of such large robot teams. This is where the sensor information selected to be broadcasted via often unreliable communication links is collected and registered. The registration process is asynchronous; some information may arrive at the base-station even minutes after the actual observation. There is no guarantee that the operator has time to look at this information directly, which implies that the map within the user interface has to be interactive and should allow the operator to call back observations that were made at any point of interest (independent of when the observation was made and by which robot). At the same time the registration process should keep the map clean (no false positives or wrong associations), because it is the area where the coordination of the team behaviors is done.

3   Current Rescue Simulation League

3.1   Overview

The main purpose of current simulator (version 1) is the benchmark for decision making for multi-agent systems [7,8,20]. By providing the benchmark framework of disaster relief simulator, it is possible to compare the effectiveness of the different approaches.

The initial purpose was mainly to stimulate contributions which could be directly applied in the real world. Yet, realistic planning problems include hundreds of agents, which stimulated the major version update from the earlier releases to the advanced releases. The current competitions include the challenge to plan optimal decisions for large teams of agents.

Changes of the number of pre-registered teams are shown in Table 1. In the Table 1, we can see the number of teams has become about half from 2010, and the number of teams has not been recovered yet. Besides, the rate of newcomers among the participants is rather low. This implies that our current approach is not attractive enough to extend our community.

Table 1. The numbers of teams participating in the agent competition

2001	2002	2003	2004	2005	2006	2007	2008	2009	2010	2011	2012	2013	2014
7	8	30	34	40	42	31	35	22	33	21	22	16	24

In the following section we describe approaches the technical committee has undertaking in order to make the competition more attractive to a larger community.

3.2 Multi-agent Challenge

Our first approach is to provide the useful benchmark framework, *RMasBench*. The RMasBench is a new type of challenge that has been introduced in 2011 [10]. The idea is to extract from the entire problem addressed by the agents certain research relevant aspects such as task allocation, team formation, and route planning, and to present these sub-problems in an isolated manner as stand-alone problem scenarios with an abstract interface. Consequently, the participating teams are able to focus more on topics relevant to their own research rather than dealing with all the low-level issues. At the current stage, *RMasBench* introduces a generic API for distributed constraint optimization problem (DCOP) algorithms and reference implementations of state-of-the-art DCOP solvers, such as *DSA* [21] and *MaxSum* [22].

3.3 Communication Library

The second approach is the development of a library that implements communication protocols for agens. The rescue simulation league releases every year the source codes from the top-three teams and the specific scenarios that were used during the competitions. However, it turned out that the code released by teams in the past is hard to be re-used by other researches. To solve this problem, we are developing the communication library, of which the first version has already been released. When teams use this communication library, it is easy to share their source code, since team approaches are based on the same communication protocol. This library might also allow to create something comparable to the drop-in challenge of the Soccer competition; the cooperation of fire-agents from one team with police-agents from another team.

3.4   Virtual Evacuation with 3D Viewer

The current 2D viewer is not very attractive for the audience. To solve this problem, the 3D viewer has been released for the earlier simulator (version 0). But, the viewer is not ported for the advanced simulator (version 1). Therefore, we also provide the challenge to develop new 3D viewer. Through this new 3D viewer, the spectator is embedded inside the disaster experience, which could be well used in the training of rescue forces (Fig. 4).

Fig. 4. 3D viewer for the agent competition

3.5   Evacuation Simulator

By expanding the current simulator, an evacuation simulator has been developed [23]. It enables to simulate whether the evacuation instructions of commanders are effective or not. Besides, the situations are also directly visualized through the simulations. Also this extention could be used in the training of rescue force commanders.

3.6   Virtual Robot Competition

The number of participating teams is relatively small (Table 2). The small number of teams is partly due to difficulty of the challenge, and partly due to the strict qualification rules of this competition. Yet, the potential for a large participation in this competition could be quite large, as is demonstrated with the 100 teams which have registered in the simulation track of the DARPA Robotics Challenge [24,25]. Unfortunately, the RoboCup cannot offer the same prize money as the DARPA organization, so we should attract teams with our inspiring academic climate.

Table 2. The numbers of teams participating in the virtual robot competition

2006	2007	2008	2009	2010	2011	2012	2013	2014
8	8	10	11	5	6	4	4	5

The challenge has the tendency to become harder ever year. Semi-autonomy was always a pre-requisite to be able to control a large team of robots with a single operator. In 2013 a hands-off period of k minutes was introduced in the rules; a period where the robots have to explore the environment fully autonomously. In 2014 this period is already increased to $k = 10$ min: 50 % of the duration of a mission.

In addition, the scale of the simulation has increased. Outdoor maps with a size of 1 km by 1 km have been created, which is close to the size of the maps used in the agent competition. Yet, the game engine is not (yet) able to simulate 100 robots, including their sensory payload, in this environment.

4   Future Directions

In the short term the league could step-by-step by further developed towards the diverse situations which could be encountered during an disaster. An example is the introduction of a smoke simulator in the agent competition, comparable with the smoke simulator introduced in the virtual robot competition [26].

The RoboCup federation formulates the goal of the Rescue Simulation League as follows: *The purpose of the RoboCup Rescue Simulation league is twofold. First, it aims to develop simulators that form the infrastructure of the simulation system and emulate realistic phenomena predominant in disasters. Second, it aims to develop intelligent agents and robots that are given the capabilities of the main actors in a disaster response scenario.* On the one hand side, because the virtual robot competition simulates the real robot competiton, it is in line with this purpose. But, the competition requires advanced skills in perception, mapping and exploration while multi-agent researchers want to concentrate on decision making at the exploration level. On the other hand side, the agent competition has brought the planning and high-level decision making to another level. The competition has been proven to be a stable benchmark for a select group of researchers. Still, it remains one of our goals to integrate the agent competiton with the virtual robot competition, and aim to improve our simulator so that it can contribute the results in the real world.

5   Conclusion

In RoboCup, the rescue simulation league allows many academic teams without the resources to travel with several rescue robots over the world to contribute to the developments in this field. Many researchers have contributed to this league in the past. Therefore, we have to decide the direction of our community so as to be sufficient for both new and established researchers. To differentiate from the soccer simulation, our purposes should be directed to the contributions for real world through our simulations. The competition was initiated to show that the robotics community could contribute with a social relevant application. The league will continue to contribute to RoboCup and the real world by keeping innovating.

References

1. Murphy, R.: Disaster Robotics. Intelligent robotics and autonomous agents series. MIT Press, Cambridge (2014)
2. Nagatani, K., Kiribayashi, S., Okada, Y., Otake, K., Yoshida, K., Tadokoro, S., Nishimura, T., Yoshida, T., Koyanagi, E., Fukushima, M., et al.: Emergency response to the nuclear accident at the Fukushima Daiichi nuclear power plants using mobile rescue robots. J. Field Robot. **30**, 44–63 (2013)
3. Skinner, C., Barley, M.: RoboCup rescue simulation competition: status report. In: Bredenfeld, A., Jacoff, A., Noda, I., Takahashi, Y. (eds.) RoboCup 2005. LNCS (LNAI), vol. 4020, pp. 632–639. Springer, Heidelberg (2006)
4. Kitano, H., Tadokoro, S., Noda, I., Matsubara, H., Takahashi, T., Shinjou, A., Shimada, S.: RoboCup rescue: search and rescue in large-scale disasters as a domain for autonomous agents research. In: IEEE Conference on Man, Systems, and Cybernetics (SMC 1999) (1999)
5. Tadokoro, S., Kitano, H., Takahashi, T., Noda, I., Matsubara, H., Shinjou, A., Koto, T., Takeuchi, I., Takahashi, H., Matsuno, F., Hatayama, M., Nobe, J., Shimada, S.: The RoboCup-rescue project: a robotic approach to the disaster mitigation problem. In: Proceedings of the IEEE International Conference on Robotics and Automation (2000)
6. Scerri, P., Farinelli, A., Okamoto, S., Tambe, M.: Allocating tasks in extreme teams. In: Proceedings of the Fourth International Joint Conference on Autonomous Agents and Multiagent Systems, pp. 727–734. ACM (2005)
7. Chapman, A., Micillo, R.A., Kota, R., Jennings, N.: Decentralised dynamic task allocation using overlapping potential games. Comput. J. **53**, 1462–1477 (2010)
8. Ferreira, P., Dos Santos, F., Bazzan, A., Epstein, D., Waskow, S.: Robocup rescue as multiagent task allocation among teams: experiments with task interdependencies. Auton. Agent. Multi-Agent Syst. **20**, 421–443 (2010)
9. Dos Santos, F., Bazzan, A.L.: Towards efficient multiagent task allocation in the robocup rescue: a biologically-inspired approach. Auton. Agent. Multi-Agent Syst. **22**, 465–486 (2011)
10. Kleiner, A., Farinelli, A., Ramchurn, S., Shi, B., Maffioletti, F., Reffato, R.: RMAS-Bench: benchmarking dynamic multi-agent coordination in urban search and rescue. In: Proceedings of the Twelfth International Conference on Autonomous Agents and Multiagent Systems (AAMAS2013), pp. 1195–1196 (2013)
11. Nair, R., Ito, T., Tambe, M., Marsella, S.: Task allocation in RoboCup rescue simulation domain. In: Proceedings of the International Symposium on RoboCup (2002)
12. Ramchurn, S., Farinelli, A., Macarthur, K., Jennings, N.: Decentralized coordination in RoboCup rescue. Comput. J. **53**, 1447–1461 (2010)

13. Kleiner, A., Brenner, M., Bräuer, T., Dornhege, C., Göbelbecker, M., Luber, M., Prediger, J., Stückler, J., Nebel, B.: Successful search and rescue in simulated disaster areas. In: Bredenfeld, A., Jacoff, A., Noda, I., Takahashi, Y. (eds.) RoboCup 2005. LNCS (LNAI), vol. 4020, pp. 323–334. Springer, Heidelberg (2006)
14. Schurr, N., Tambe, M.: Using multi-agent teams to improve the training of incident commanders. In: Defence Industry Applications of Autonomous Agents and Multi-agent Systems, pp. 151–166 (2008)
15. Nüssle, T.A., Kleiner, A., Brenner, M.: Approaching urban disaster reality: the ResQ firesimulator. In: Nardi, D., Riedmiller, M., Sammut, C., Santos-Victor, J. (eds.) RoboCup 2004. LNCS (LNAI), vol. 3276, pp. 474–482. Springer, Heidelberg (2005)
16. Shahbazi, H., Abdolmaleki, A., Salehi, S., Shahsavari, M., Movahedi, M.: Robocup rescue 2010 – rescue simulation league team description paper - brave circles - infra-structure competition. In: Proceedings CD of the 14th RoboCup International Symposium (2010)
17. Dijkshoorn, N., Visser, A.: An elevation map from a micro aerial vehicle for urban search and rescue. In: Proceedings CD of the 16th RoboCup International Symposium (2012)
18. Gohardani, P.D., Ardestani, P., Mehrabi, S., Yousefi, M.A.: Flying Agent: An Improvement to Urban Disaster Mitigation in Robocup Rescue Simulation System. Mechatronics Research Laboratory, Qazvin, Iran (2013)
19. Olson, E., Strom, J., Morton, R., Richardson, A., Ranganathan, P., Goeddel, R., Bulic, M., Crossman, J., Marinier, B.: Progress towards multi-robot reconnaissance and the MAGIC 2010 competition. J. Field Robot. **29**(2012), 762–792 (2010)
20. Oliehoek, F.A., Visser, A.: A decision-theoretic approach to collaboration: principal description methods and efficient heuristic approximations. In: Babuška, R., Groen, F.C.A. (eds.) Interactive Collaborative Information Systems. SCI, vol. 281, pp. 87–124. Springer, Heidelberg (2010)
21. Fitzpatrick, S., Meetrens, L.: Distributed coordination through anarchic optimization. In: Distributed Sensor Networks A Multiagent Perspective, pp. 257–293. Kluwer Academic (2003)
22. Farinelli, A., Rogers, A., Petcu, A., Jennings, N.R.: Decentralised coordination of low-power embedded devices using the max-sum algorithm. In: Proceedings of the Seventh International Conference on Autonomous Agents and Multiagent Systems (AAMAS 2008), pp. 639–646 (2008)
23. Okaya, M., Takahashi, T.: Proposal for everywhere evacuation simulation system. In: Röfer, T., Mayer, N.M., Savage, J., Saranlı, U. (eds.) RoboCup 2011. LNCS, vol. 7416, pp. 246–257. Springer, Heidelberg (2012)
24. Ackerman, E.: Darpa robotics challenge trials: what you should (and shouldn't) expect to see. IEEE Spectrum, vol. 19 (2013)
25. Luo, J., Zhang, Y., Hauser, K., Park, H.A., Paldhe, M., Lee, C.S.G., Grey, M., Stilman, M., Oh, J.H., Lee, J., Kim, I., Oh, P.: Robust ladder-climbing with a humanoid robot with application to the darpa robotics challenge. In: IEEE International Conference on Robotics and Automation (2014)
26. Formsma, O., Dijkshoorn, N., van Noort, S., Visser, A.: Realistic simulation of laser range finder behavior in a smoky environment. In: Ruiz-del-Solar, J. (ed.) RoboCup 2010. LNCS, vol. 6556, pp. 336–349. Springer, Heidelberg (2010)

RoboCup Rescue Robot League

Johannes Pellenz[1]([✉]), Adam Jacoff[2], Tetsuya Kimura[3], Ehsan Mihankhah[4],
Raymond Sheh[2,5], and Jackrit Suthakorn[6]

[1] BAAINBw, Koblenz, Germany
pellenz@uni-koblenz.de
[2] National Institute of Standards and Technology, Gaithersburg, MD, USA
[3] Department of System Safety, Nagaoka University of Technology, Niigata, Japan
[4] School of Electrical and Electronic Engineering,
Nanyang Technological University, Singapore, Singapore
[5] Department of Computing, Curtin University, Bentley, Australia
[6] Faculty of Engineering, Mahidol University, Nakorn Pathom, Thailand

Abstract. The RoboCup Rescue Robot League (RRL) aims to foster
the development of rescue robots that can be used after disasters such
as earthquakes. These robots help to discover victims in the collapsed
structure without endanger the rescue personnel. The RRL has been
held since 2000. The experience gained during these competitions has
increased the level of maturity of the field, which allowed to deploy robots
after real disasters, e.g. at the Fukushima Daiichi nuclear disaster. This
article provides an overview on the competition and its history. It also
highlights the current state of the art, the current challenges and the
way ahead.

1 Competition Overview

After a natural or man-made disaster, the first task for the responders is to find
and rescue survivors in the collapsed buildings. Even though technologies such as
microphone arrays or search cameras help the responders, it is still a dangerous
task: the stricken building might collapse completely and bury the rescue team
in case of an aftershock. Search and Rescue robots can help to reduce this risk
during these Urban Search and Rescue (USAR) missions. The idea is to send
these robots into the rubble pile, and let them search for victims using their
sensors such as cameras or heat sensors and report to the rescuers once they
have found a person.

The RoboCup Rescue Robot League (RRL) competition aims to foster development in this field [1]. The arenas in this competition resemble partly collapsed
buildings, with obstacles consisting of standardised and prototypical apparatuses from the DHS[1]-NIST[2]-ASTM[3] International Standard Test Methods for
Response Robots [2]. The robots perform 20 min missions in the test arena,

---

[1] US Department of Homeland Security.
[2] National Institute of Standards and Technology, US Department of Commerce.
[3] Formerly the American Society for Testing and Materials.

© Springer International Publishing Switzerland 2015
R.A.C. Bianchi et al. (Eds.): RoboCup 2014, LNAI 8992, pp. 673–685, 2015.
DOI: 10.1007/978-3-319-18615-3_55

in which they try to find as many simulated victims as possible. The complexity of the terrain increases in the different parts of the arena (see Fig. 1): In the yellow part of the arena (which has to be cleared by autonomous robots which do not rely on radio communication) continuous ramps challenge the sensors of the robot, but they do not challenge its mobility. In the orange part, the ramps are not continuous any more and the robot has to go over smaller unstructured objects such as sand and gravel. Finally, in the red part obstacles such as partly blocked stairs and pipe steps (which simulate extremely slippery curb stone edges) have to be climbed to approach a simulated victim. This part also includes the Symmetric Stepfield [3], a standardised pattern of wooden blocks that resembles the structure of a rubble pile. The easier yellow part of the arena must be cleared by autonomous robots, while in the more difficult parts mainly remote controlled robots are used. Here the challenges are mobility, manipulation, user interfaces for good situational awareness, reliable wireless communications and autonomous assistive features.

Fig. 1. Overview of a RoboCup rescue arena, composed of standardized test elements, organized in a maze. This image depicts a typical International Championship arena (in this case from the 2012 competition in Mexico City)(Color figure online).

The robot scores points for each victim it can find. The simulated victims can be detected by different signs of life, which are all simulated: the visual appearance (shape, color, and movement), the body heat, the sound and the CO_2 emission. The more information can be discovered for a victim, the more points the robot gets. Another important information is where the victim is located. This is not easy to determine, since in an indoor building classical localization approaches such as GPS do not work. A detailed map of the building along with a precise victim location yields extra points.

Beyond the detection of victims, rescue robots can be used to supply items to the victims, such as water bottles, two-way radios or medicine. If the robot is able to supply objects to the victim, it is awarded extra points. Usually, a robot arm (also called a manipulator) is used for the delivery. This manipulator can also be used to open doors by pressing the door handle or to clear the path of the robot from obstacles.

In contrast to the other RoboCup related activities, such as RoboCup soccer, scenarios in rescue do not include adversarial agents but instead inherently demand solutions for unpredictable, unstructured, complex and unknown environments. The goal of the RoboCup Rescue competitions is to compare the performance of different solutions for coordinating and controlling single or multiple robots performing disaster mitigation in a simulated but realistic environment. Since the circumstances during real USAR missions are challenging and always changing [4], benchmarks based on challenges such as the RoboCup Rescue competition can be used for assessing the capabilities of the robots. The competitions are held regularly as a part of the yearly RoboCup World Championship. Regional competitions are held in RoboCup Opens in the United States, Germany, Japan, Iran, Thailand and China. The RoboCup Opens are organized by the regional committees, whereas the international competitions are organized by league organizing committees. The rules are determined by the technical committee whose members are elected annually.

The objective of the RRL is to promote the development of intelligent, highly mobile, dexterous robots that can improve the safety and effectiveness of emergency responders performing hazardous tasks. As a league, we demonstrate and compare advances in robot mobility, perception, localization and mapping, mobile manipulation, practical operator interfaces, and assistive autonomous behaviors. We use the annual RoboCup competitions and subsequent field exercises with emergency responders to accomplish the following:

- Increase awareness about the challenges involved in deploying robots for emergency response applications.
- Provide objective performance evaluations based on DHS-NIST-ASTM International Standard Test Methods for Response Robots.
- Introduce Best-In-Class implementations to emergency responders within their own training facilities.
- Support ASTM International Standards Committee on Homeland Security Applications (E54.08.01) [2].

The teams that performed best during the preliminaries and the finals are awarded the first, second and third place. To highlight the achievements in the different sections special awards are given for *Best-in-Class Autonomy*, *Best-in-Class Mobility*, *Best-in-Class Manipulation* and *Best-in-Class Small Unmanned Aerial System*. The Best-in-Class Autonomy award is given to the robot that found the highest number of simulated victims during the competition and that was able to clear the largest part of the arena and produced the most accurate map during a separate Best-in-Class Autonomy mission. The Best-in-Class Mobility award is assigned to the robot that found the highest number of victims in the red arena and was able to handle the highest number of difficult obstacles (such as stairs, incline planes, and pipe steps) in a certain time. The Best-in-Class Manipulation award is given to the robot with the highest number of objects successfully placed into victim boxes and points gained during a special mission where the robot has to move wooden blocks and to handle different types of switches or valves. The Best-in-Class Small Unmanned Aerial System award is assigned to the flying robot that found the most victims flying and showed autonomous behavior such as station keeping during a separate mission.

Beyond the RRL, the RoboCup competition hosts Rescue Simulation Leagues. An overview of all the rescue leagues can be found at [5].

In important aspect of the RRL is that the teams do not compete against each other, but they fight the simulated disaster to finally help first responders to save lives.

2　History of the Rescue Robot League

After the Great Hanshin Earthquake 1996 in Kobe, the Japanese government decided to promote research related to the problems encountered during large scale urban disasters. One of the outcomes of this initiative was the RoboCup Rescue competition. The competition started in 2000 as a part of the AAAI Mobile Robot Competition and Exhibition. In 2001, the rescue robot competition was integrated into RoboCup competition. The USAR scenario offers a great potential for inspiring and driving research in robot systems that work in unstructured environments.

In 2013, more than 20 teams from 12 nations (Austria, Australia, China, Germany, Greece, Iran, Japan, Mexico, Portugal, Thailand, United Kingdom, United States of America) qualified for the RoboCup Rescue world championship held in Eindhoven (Netherlands). In addition, more than 100 teams compete in RoboCup regional competitions that are offered in Thailand, Japan, Iran and Germany. A good performance at the regional competitions qualifies for the world championship. An overview of the historic competitions can be found on the NIST website[4].

2.1　Towards Standardization of the Arena

During the first years of the competition, the arena showed a typical apartment or house interior, strewn with random debris. This was already a mobility challenge for some of the robots which mostly came from tidy research labs. However, it was not easy to reproduce the arena at other locations and get comparable test results. To get more repeatable results, the National Institute of Standards and Technology (NIST) defined a set of standard test elements for response robots. Starting in 2007, a collection of these elements were used to build the RoboCup Rescue Arena. Using these test elements, the difficulty and complexity of the competition arenas could be defined. The test elements are now standardized under the ASTM (American Society for Testing and Materials) body and

---

[4] http://robotarenas.nist.gov/competitions.htm.

further developed by the ASTM International Standards Committee on Homeland Security Applications, Operational Equipment and Robots (E54.08.01). The standardized test elements are also used to evaluate and compare commercial robots. They also help the customers to define the requirements they have.

To keep up with the progress of the robots, the test elements are constantly modified. For example, the ramps and stairs became steeper over the last couple of years. The RoboCup Rescue competition is the place where prototypes of these new test element are tested. If they matured, their standardization is discussed at the ASTM committee. Beyond the standardized arena, regional open arenas can be adapted to the typical regional construction methods. For example, the Japan Open features an arena inspired by typical debris present after earthquakes in Japan which involve finer wooden structures.

2.2 Increased Mobility Challenges: From Flat Ground to Blocked Stairs

One of the constant changes over the years is to make the terrain more and more difficult. In indoor environments, the pose of the robot may be measured using wheel encoders. This approach fails if the ground is slippery, e.g. due to water or mud. To simulate this challenge, in 2005 loose paper was added to the arena. To cope with this, teams had to apply sensor fusion in their localization solutions. Wheel encoders are still a common approach for a rough local pose estimation, refined by matching consecutive laser scans or by matching the laser scan against an obstacle map. This approach has matured over time, and many teams do not use wheel based odometry for the pose estimation any more. Localization and mapping solutions can be used today independent from a specific robot – even stand-alone as a hand-held solution.

For years, the ground in the yellow arena was just flat and easy to handle for the autonomous robots. Starting in 2007, small 10° ramps were introduced. This was not so much a challenge for the mobility, but the rigidly mounted sensors suddenly sensed fake obstacles: when the robot was tilted due to a ramp, the laser range finder reported a spurious obstacle in front of the robot, which was actually only the floor. These fake obstacles can make autonomous navigation impossible. This forced the teams with autonomy to deal with this problem either by scanning the environment in 3D or by using an active sensing approach, controlling the orientation of the sensors depending of the attitude sensor. Today, the ramps in this easiest part of the arena have a 15° slope, and most of the autonomous robots can easily handle them.

Also the terrain in the orange and red part of the arena become more and more difficult. A major step forward was the introduction of Stepfield Terrains [3]. Initially a psuedo-random pattern of wooden blocks that simulate a rubble pile, these were refined into the Symmetric Stepfield terrain and test method over several years of competition. When the stepfields were introduced in 2005, none of the robots could overcome this obstacle reliably. But quickly, the robots evolved, and the stepfields become more difficult over the years. The current version of the wooden stepfield involves obstacles from 10 cm up to 50 cm height.

(a)   (b)

Fig. 2. New mobility apparatuses at the RoboCup 2013 in Eindhoven, Netherlands: (a) Stepfield made of concrete blocks. (b) Occluded stair and ramp apparatuses.

An alternative version of the stepfield – used at the 2013 world championship – is built of concrete blocks. The elements are askew and there are gaps between the bricks (see Fig. 2). While originally approached as a challenge to teleoperated mobility, there has also been work on autonomously and semi-autonomously traversing these terrains. Approaches include the use of machine learning and behavioural cloning [6] for mechanically simple platforms and assistive features for controlling high degree of freedom mobility platforms [7]. One lesson from the Fukushima Daiichi nuclear power plant disaster was that stairways cluttered with debris were a major problem to overcome for the robots [8]. This experience was introduced in the stair and ramp test elements, in the form of the Occluded Stair and Ramp. During the final runs in the competition, the rescue robots have to deal with wooden bars blocking these elements as shown in Fig. 2. The robots can either remove the bars with their manipulator or climb over them.

2.3   Advances in Mapping and Autonomy

During the early years, the autonomous robots could apply easy reactive techniques such as wall following to reach and score the victims over flat terrain. With the increased size of the arena, simple navigation techniques were not sufficient any more, and sophisticated exploration strategies become more and more common. Since the arena is unknown to the robot in the beginning, the robot has to build an obstacle map to plan its path efficiently. The quality of these maps improved steadily over the years, see Fig. 3 for two examples. Not only improved mapping algorithms help to get better maps. Also the availability of new sensors helped, such as small laser range finders with larger ranges (e.g. 30 m range instead of 4 m). New, affordable 3D sensors enabled the robots to acquire data from complex environments. These new capabilities have inspired new tests for 3D perception, including 3D versions of the nested Landolt-C visual acuity test artefacts as shown in Fig. 4.

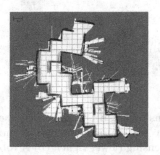
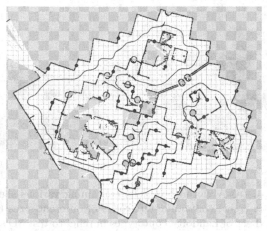

Fig. 3. Improvement of maps generated by autonomous robots during RoboCup Rescue finals over 5 years. Left: 2007 - map of the autonomy winner team Resko from University of Koblenz-Landau (Germany). Right: 2012 - map of the autonomy winner team Hector from TU Darmstadt (Germany). Note that both maps are shown to roughly the same scale.

2.4   Semi-automated Map Scoring

In the early years, the teams could turn in maps of the explored area in arbitrary formats. The Technical Committee (TC) then scored these maps be looking at the area explored by the robot and the accuracy of the map. Points were deducted if the map showed blurry walls or even double walls, which is a result of localization inaccuracies. With an increasing number of teams that produced maps and constantly growing arena sizes, the manual grading become more and more difficult. In 2012, unique visual tags, QR codes, were introduced at the walls of the arena. The robots have to detect and map the tags. For map grading, the number of detected tags conforms with the area covered by the robot. The precision of the localization equals the accuracy of the map. In 2013, different sizes of QR codes were introduced, testing also the accuracy of the sensors. By using these QR codes, map grading became much easier and more reliable. An example for such an QR code can be seen in Fig. 4.

2.5   Manipulation: Placing Objects and Opening Doors

In the initial years of the competition, the goal of the robots was information gathering by passive observation. In recent years, this goal has been extended to that of manipulation of objects in the environment. The initial task was to take wooden blocks and deliver them to the victims. This was extended to delivering operationally significant objects such as water bottles and radios. Most recently, additional challenges have been added including that of dexterously inspecting complex objects, opening doors (including those requiring the robot

(a)   (b)

Fig. 4. (a) A victim box (left) containing various signs of life, a QR Code test sheet (upper right) containing QR codes of different sizes to test autonomous vision and a nested set of Landolt-C optotypes in relief to test 3D sensing (lower right). (b) A robot stacking blocks in the box crib shore configuration.

to simultaneously rotate and pull a door handle) and stacking blocks in a box crib shore configuration as shown in Fig. 4.

2.6 Aerial Robots

Aerial robots are becoming increasingly popular among both the responder community and the general public with a plethora of highly capable, small, low cost vehicles. The league saw aerial robots as early as 2004 with a robotic airship. Powered, heavier-than-air robots such as quadcopters have been incorporated into the competition since 2009. This initially took the form of an aerial arena, equipped with prototypical test methods for aerial response robots and separated from the main arena by a safety net. In more recent competitions, the main arena as a whole has been covered with a safety net that aims to both shield spectators from the aerial robots as well as to introduce overhead obstacles that the aerial robots must avoid. Aerial robots are treated similarly to other robots in the competition and have demonstrated their ability to build maps and identify victims. While they do not have to tackle the terrain challenges that face ground robots, they must tackle other issues such as low overhead and side clearances, complex structures in the immediate vicinity of victims and air currents.

3 Today's Competition Structure and Challenges

Because of the standardized test elements, the progress of the league can be seen over the years. When the random stepfields were introduced in 2005, none of the robots was able to cross them reliably. Today, driving through the stepfields is state of the art for the remote controlled robots. During the same time, the fully autonomous robots had to learn how to handle uneven terrain, because their section of the arena is not flat any more, but filled with continuous or crossed ramps. This challenging floor does increase the complexity of the path planning and requires an active sensing approach with stabilized distance and victim sensors. Nevertheless the quality of the generated maps increased over the years, and mapping the whole arena is not state art. An example of a very precise map is given in Fig. 3. The newer map on the right not only shows the obstacles in the arena, but also contains the victim locations and other landmarks (QR codes) found autonomously by the robots' vision system automatically. The maps are scored by looking at the completeness of the exploration (using the number of QR codes found) and at the accuracy of the recorded landmark locations. The scientific output of the league includes solutions for online SLAM solutions [9, 10] and solutions for efficient exploration strategies [11]. The performance of these approaches was demonstrated and tested in practice during several RoboCup competitions.

3.1   Re-usable Software Modules

Due to the complexity of the problems addressed in the competition, the barrier for the entry of new teams is high. Therefore, one of the current challenges of the league is to make the successful software modules available for new teams. To this end, the league has established an open source standard software solution based on ROS for the RRL. This standard software solution makes excellent solutions broadly reusable thus enables new teams to reach quickly to the world class performance level. This way, each team only needs to focus on their own topics of interest, such as mapping, exploration or victim detection[5]. Code sharing in the league is done on voluntary basis. However, sharing good contributions is often honored with special awards such as the innovation award or the HARTING Open-Source Award at the German Open, which were both awarded in the past to the team Hector Darmstadt (Germany) for their major contribution to the software repository.

3.2   Rescue Robots in the Field

The obstacles that the robots must overcome in the competition consist of test method apparatuses. These are in turn developed by taking real world response robot operational scenarios and distilling them into elemental tests. Thus robots that perform well in the competition also answer critical real world needs. This was graphically demonstrated during the immediate aftermath of the Great East Japan Earthquake. Visibility into the reactor buildings at the Fukushima Daiichi Nuclear Power Station was critical and while several commercial robots were able to gain entry, the stairs proved too steep to climb. The robot that succeeded in climbing to the upper floors of the reactor buildings was a modified Quince robot, shown in Fig. 5. This robot was developed and refined over successive years, within the RoboCup RRL, by the International Rescue System Institute

---

[5] The current status of the common software framework for RoboCup Rescue initiative is documented in the ROS wiki at http://www.ros.org/wiki/robocup_rescue.

(Japan), a consortium of Japanese universities lead by Tohoku University [12]. Developments from the RRL are frequently demonstrated to first responders and robot manufacturers during a Response Robot Evaluation Exercises. These presentations help to show the state of the art in rescue robotics, and inspired new products, such as mapping capabilities for commercial robots.

3.3   Academic Summer Schools

The goal of the league is to advance the state of technologies for search and rescue. While the competitions are a vital tool for fostering collaboration between teams, actual collaboration during the competition can be difficult. The League runs academic summer schools in the second half of the year, where Best-in-Class performers from the league and wider research community can share their capabilities with the rest of the league and representatives of responder organisations. These events started in 2004, hosted by the University of Rome "La Sapienza" and held at the Istituto Superiore Antincendi (firefighter training academy) in Rome. Since then they have been hosted by Mahidol University in Thailand, the University of Koblenz-Landau in Germany, the Technical University of Graz in Austria, Linköping University in Sweden, Robolit LLC in Turkey and Curtin University in Australia [13].

Fig. 5. The Quince robot in competition in Singapore in 2010 (left) and equipped to sample radiation and water levels in the Fukushima Daiichi Nuclear Power Station (right, courtesy of CIT, Tohoku University, IRS and NEDO).

3.4   The DARPA Robotics Challenge

The DARPA[6] Robotics Challenge (DRC)[7] is a related competition that shares many of the same goals and principles as the RRL. NIST also plays a leading

---

[6] US Defence Advanced Research Projects Agency.

[7] http://www.theroboticschallenge.org/.

role in developing the challenges for the DRC and there are competitors in the DRC who also participated actively in the RRL. The DRC and RRL are complimentary competitions that differ in scope and emphasis. The DRC has much harder tasks that demand significantly greater levels of autonomy, mobility, dexterity, reliability and system integration. The RRL, in contrast, focuses on more incremental improvements that allow all teams to demonstrate their capabilities, including those who only specialise in a particular capability. The cost of entry in terms of personnel and resources is also significantly lower in the RRL. The relationship between the competitions allows for synergies between the two competitions. For example, the process of developing the challenges for the DRC involved evaluating different options within the RRL. Similarly, it is anticipated that many of the capabilities and tools developed for the DRC, including the DRC Simulator, will be adopted by teams within the RRL.

4   Future Goals of the League

The RRL continues to evolve in order to guide teams towards answering gaps in the capabilities of deployed systems. Developed hand-in-hand with the DHS-NIST-ASTM International Standard Test Methods for Response Robots, the challenges of the competition continue to be grounded in the needs of emergency responders across a wide variety of applications. Particular areas of focus for future attention include increasing assistive and full autonomy throughout the arena, improved human-system interaction especially for mobility, navigation and manipulation, the introduction of environmental effects such as fog and smoke, co-operation between robots and fully automated scoring and evaluation.

4.1   Autonomy

Advances in 3D depth perception enable the robots to capture the environment in greater detail. However, most autonomous robots use this knowledge so far to just stay out of difficult terrain. In the future, the robots will be required to deal with more difficult terrain autonomously, by increasing the complexity of the yellow arena.

4.2   Human-System Interaction

One of the greatest challenges in response robotics is in the interface between the robotic system and the human operator. For several years now the League has presented a technical award for Innovative User Interfaces. This has in the past been awarded for such innovations as wearable user interfaces and interfaces that support semi-autonomous operation. In the future, the League seeks to encourage the further development of improved user interfaces for advanced mobility, navigation and manipulation. Examples of such innovations include automatic movement of high degree of freedom mobility bases, such as those equipped with flippers, to address different mobility obstacles [7], improved ways of presenting real time map data to the operator in order to maintain situational awareness and innovative ways of allowing operators to control high degree of freedom manipulators to perform dexterous inspection and manipulation in confined spaces.

4.3   Fog and Smoke

A common characteristic of some response environments is the presence of fog and smoke. The competition is in the early stages of introducing these characteristics, with simulated smoke, in the form of clouds of water vapour, denoting potential incipient fires that are of interest to the robots. In the future, this may be extended to include fog and smoke that are intended to also interfere with the robots' sensing and requiring additional sensors, such as infrared and sonar, to overcome.

4.4   Co-operation Between Robots

The competition allows already the use of multiple robots during the mission, as long as they are controlled by a single operator. A common practice of the teams is to send in two robots: one autonomous and one remote-controlled. To search the arena more efficiently, the number of autonomous robots – working together as a team – should be increased. From a scientific point of view this would push the research on robot teams, working together in unstructured environments.

4.5   Automated Scoring

So far, the scoring of the robot's performance relies on a judge who keeps track of the result on a paper sheet. In the future, this process will be further automated. Ideally, the robot sends the discovered data directly to a judge box, which compares the data with the ground truth data and assigns points automatically.

The next competition will be in July 2014 in Brazil, followed in 2015 by a competition in Thailand. More information is available on the RoboCup website[8].

References

1. Sheh, R., Jacoff, A., Virts, A.M., Kimura, T., Pellenz, J., Schwertfeger, S., Suthakorn, J.: Advancing the state of urban search and rescue robotics through the robocuprescue robot league competition. In: Proceedings International Conference on Field and Service Robotics (2012)
2. Jacoff, A., Sheh, R., Virts, A.M., Kimura, T., Pellenz, J., Schwertfeger, S., Suthakorn, J.: Using competitions to advance the development of standard test methods for response robots. In: Proceedings of the Workshop on Performance Metrics for Intelligent Systems (2012)
3. Jacoff, A., Downs, A., Virts, A., Messina, E.: Stepfield pallets: repeatable terrain for evaluating robot mobility. In: Proceedings on Performance Metrics for Intelligent Systems Workshop. NIST Special Pub. 1090 (2008)

---

[8] http://wiki.robocup.org/wiki/Robot_League.

4. Murphy, R.R., Tadokoro, S., Nardi, D., Jacoff, A., Fiorini, P., Choset, H., Erkmen, A.M.: Search and rescue robotics. In: Siciliano, B., Khatib, O. (eds.) Springer Handbook of Robotics, pp. 1151–1173. Springer, Heidelberg (2008)
5. Akin, H.L., Ito, N., Jacoff, A., Kleiner, A., Pellenz, J., Visser, A.: Robocup rescue robot and simulation leagues. AI Mag. 34(1), 78–86 (2013)
6. Sheh, R., Hengst, B., Sammut, C.: Behavioural cloning for driving robots over rough terrain. In: Proceedings of IEEE International Conference on Intelligent Robots and Systems (2011)
7. Ohno, K., Morimura, S., Tadokoro, S., Koyanagi, E., Yoshida, T.: Semi-autonomous control system of rescue crawler robot having flippers for getting over unknown-steps. In: Proceedings of IEEE/RSJ International Conerence on Intelligent Robots and Systems, pp. 3012–3018 (2007)
8. Yoshida, T., Nagatani, K., Tadokoro, S., Nishimura, T., Koyanagi, E.: Improvements to the rescue robot quince toward future indoor surveillance missions in the fukushima daiichi nuclear power plant. In: Proceedings International Conference on Field and Service Robotics (2012)
9. Kohlbrecher, S., Meyer, J., von Stryk, O., Klingauf, U.: A flexible and scalable slam system with full 3d motion estimation. In: Proceedings IEEE International Symposium Safety, Security and Rescue Robotics, Kyoto, Japan. IEEE, 1–5 November 2011
10. Pellenz, J., Paulus, D.: Stable mapping using a hyper particle filter. In: Baltes, J., Lagoudakis, M.G., Naruse, T., Ghidary, S.S. (eds.) RoboCup 2009. LNCS, vol. 5949, pp. 252–263. Springer, Heidelberg (2010)
11. Wirth, S., Pellenz, J.: Exploration transform: a stable exploring algorithm for robots in rescue environments. In: Procedings IEEE International Workshop Safety, Security, and Rescue Robotics, pp. 1–5 (2007)
12. Nagatani, K., Kiribayashi, S., Okada, Y., Tadokoro, S., Nishimura, T., Yoshida, T., Koyanagi, E., Hada, Y.: Redesign of rescue mobile robot quince - toward emergency response to the nuclear accident at fukushima daiichi nuclear power station on March 2011. In : Proceedings IEEE International Symposium Safety, Security, and Rescue Robotics, pp. 13–18 (2011)
13. Sheh, R., Collidge, B., Lazarescu, M., Komsuoglu, H., Jacoff, A.: The response robotics summer school 2013: bringing responders and researchers together to advance response robotics. In: Proceedings IEEE/RSJ International Conference on Intelligent Robots and Systems (2014, to appear)

On RoboCup@Home – Past, Present and Future of a Scientific Competition for Service Robots

Dirk Holz[1]([⊠]), Javier Ruiz-del-Solar[2], Komei Sugiura[3], and Sven Wachsmuth[4]

[1] Autonomous Intelligent Systems Group, University of Bonn, Bonn, Germany
holz@ais.uni-bonn.de
[2] Department of Electrical Engineering, AMTC,
Universidad de Chile, Santiago, Chile
[3] National Institute of Information and Communications Technology, Kyoto, Japan
[4] Center of Excellence Cognitive Interaction Technology,
Bielefeld University, Bielefeld, Germany

Abstract. RoboCup@Home is an application-oriented league within the annual RoboCup events. It focuses on domestic service robots and mobile manipulators interacting with human users. Participating robots need to solve tasks ranging from following and guiding human users to delivering objects, e.g., in a supermarket.

In this paper, we present the @Home league and how it evolved over the last seven years since its existence. We place particular emphasis on how we evaluate the teams' performances over the years and how we use the obtained statistics to drive the development of the league. This process is shown in detail on two examples—following human guides, and finding and manipulating objects. Finally, we will outline possible future directions and developments.

1 Introduction

Scientific competitions are becoming more and more common in many research areas of Robotics and Artificial Intelligence (AI). They provide a common test-bed for comparing different solutions and enable exchange of research results. Furthermore, they are interesting for the general audience and industries. Particularly interesting in many of these competitions, is the opportunity of defining standard benchmarks for solving specific problems, comparing different solutions, and making the best solutions available to the community. Moreover, the tasks and the results of the competitions are often used in scientific papers to compare new approaches with respect to existing ones.

One of the classic competitions in AI and robotics research is RoboCup. RoboCup started in 1997 with the ambitious goal of bringing forth a team *"of fully autonomous humanoid robot soccer players"* that *"shall win a soccer game, complying with the official rules of FIFA, against the winner of the most recent World Cup"*. It incorporates many interesting problems from both AI and robotics research. The central point in the first years (up to now), was to have highly sophisticated gameplay in abstracted environments. That is, from year to year

© Springer International Publishing Switzerland 2015
R.A.C. Bianchi et al. (Eds.): RoboCup 2014, LNAI 8992, pp. 686–697, 2015.
DOI: 10.1007/978-3-319-18615-3_56

the complexity is increased by increasing the complexity of environment (e.g., color-coded goals etc.) and setup (e.g., no shot on goal without previous passing).

RoboCup@Home follows a different (and, in fact, inverted) approach: it started with very simple tasks in a complex (real) environment as opposed to the highly complex task of playing soccer in a simplified environment. RoboCup-@Home is a competition where domestic service robots are performing several tasks (called *tests*) in a home environment, interacting with people and with the environment in a natural way. Each task requires a combination of different skills (like navigation, object perception and manipulation, person detection and tracking, etc.) and the score is related to the accomplishment of the task. RoboCup@Home started in 2006 and has the main characteristic of changing tasks every year while maintaining the same basic skills. By changing the difficulty and the combinations of the skills to be integrated, we aim at pushing the teams to develop general and robust solutions. Indeed, with this setting of the competitions, it is too difficult to implement many specific systems to solve each of the tasks instead of one general solution for all tasks.

This paper presents the RoboCup@Home league in detail discussing its history and present rules as well as the system used to drive changes and possible future developments.

2   History of the RoboCup@Home League

The first ideas on RoboCup@Home have been proposed in 2005 by Tijn van der Zant and Thomas Wisspeintner [1]. The competition was held as a demonstration at RoboCup 2006 in Bremen, Germany. In 2007, RoboCup@Home became an official league and was featured in both local RoboCup competitions and the main RoboCup competition in Atlanta. Eleven teams participated in this first competition. At that time, the competition consisted of simple tasks testing only single basic skills such as navigation, following a person or finding an object. Every test was conducted in two phases. In the first phase, teams could customize (and simplify) the test setup, for example, by using own objects, artificial markers or own team members who knew how to interact with the robot. In the second phase, tasks were conducted as they had been planned by the Technical Committee (TC): the robot was operated by an independent referee, an official set of objects was used (that was not known to the teams before the competition), and no simplifications to the environment were allowed. Obviously, teams who had a general solution could score in both phases with the same approach.

After these first two years of RoboCup@Home competitions, we discussed how to evaluate the performance of the competition over the years in terms of improvement demonstrated by the teams. Three problems were identified: (1) improvement is difficult to measure because the tasks change every year, (2) performance is difficult to evaluate if situations differ, and (3) the scoring system, based on Boolean scores (either success or failure of the entire test), was not adequate for this analysis. While the first problem is inherent in a dynamic

Fig. 1. Teams and robots at the 2013 RoboCup@Home competition in Eindhoven.

yearly competition, the second and third could be addressed. Since 2008, the scoring system was changed to have scores not per task but per sub-task. Moreover, since most teams in 2007 came up with general solutions (as expected and intended), the first phase was removed and, nowadays, all tasks are run as defined by the Technical Committee and in the very same fashion (involved objects, people, situations, etc.) for all teams. The latter aspect provides a better testbed for benchmarking the presented approaches. Since 2008 up to the recent RoboCup 2014 in João Pessoa, this new scoring system was used together with a systematic analysis carried out every year in order to find out how the rules should be changed (e.g., keeping or removing complete tasks, making (sub-)tasks more complex, removing or adding sub-tasks etc.).

With becoming a regular league and with the improved scoring system, the number of participating teams quickly grew. Since 2009, RoboCup@Home has a stable average number of participating teams around 19 to 20 teams from all over the world, with a peak attendance of 24 teams (maximum number of teams allowed) in Singapore, 2010. Figure 1 shows the teams and robots participating in the 2013 competition in Eindhoven.

The development of the league can best be seen in the evolution of tests conducted within the competition. In 2007, tests have been as simple as navigating in a static apartment (with enough time to build static maps beforehand), following a human guide gently walking in front of the robot without further disturbances, or finding an object in the apartment. Nowadays, the league features, amongst other skills, following a human guide through a crowded public place with various disturbances like people blocking the path and sight between robot and guide, navigating and manipulating objects in previously unknown public places like supermarkets, or performing any task being asked for on demand where the task is only restricted to belong to the expected skills over all tests. The development can also be seen in details such as the complexity of object recognition:

Fig. 2. Typical RoboCup@Home arena (left) and objects (right).

over the years not only the number of objects throughout the tests has been steadily increased (set of objects known to the teams), but we also introduced unknown objects which pose particular problems on object perception.

Naturally, developing a robotic system integrating all the needed capabilities is very challenging. For example, finding and manipulating an object is not only complex (especially in uncontrolled environments) but was also solved by only very few teams in the beginning. However, in the past years we could observe an increase in the performance in both individual capabilities and integrated systems being capable of all skills and able to score in all tasks.

3   The Current Competition—Tasks and Sub-tasks

The RoboCup@Home competition runs in a realistic setting where an apartment with different functional rooms and typical furniture and objects is realized (see Fig. 2). Since it is not completely specified beforehand, it may differ in its implementation and teams do not know any information (such as number of rooms, dimensions, material of the floor, colors of the wall, kinds of furniture and objects, etc.) before arriving at the competition venue. This ensures the development of general solutions as well as easy and efficient setup procedures. Moreover, the arena is subject to constant changes: especially minor changes such as moved furniture or objects lying on the ground may happen right before or even during tests. Furthermore, some tasks are conducted outside this area in a public space such as a restaurant or a shopping mall (that is not known to the teams beforehand).

The competition is formed by about eight predefined tasks (called *tests*), two open demonstrations, a technical challenge and the finals. A stage system is used: Stage I is performed by all participating teams. The (better) half of them—or at least the best ten teams—advance to Stage II, and finally only the best five teams reach the finals. Previously achieved scores and a jury evaluation in the finals determine the winner of the competition.

The main skills required in the tests are the following (chosen as to reflect common capabilities a general purpose service robot in a domestic environment should possess): navigation and mapping, person recognition and tracking, object recognition and manipulation, and speech and gesture recognition. In addition, integration of the skills and higher level cognitive skills are of particular interest.

Each test requires a combination of some of these skills. For example, *"Follow me"* is a test in which the robot has to follow a person in a crowded area of the venue of the competition, enter an elevator with the person in order reach a location far away and on a different floor with respect to the starting position. The guide is not known in advance by the robot and a quick automatic calibration procedure must be done at the beginning of the test, when the person appears in front of the robot. During the test, other persons are allowed to pass between the robot and the guide and at some point the guide hides away from the view of the robot, that must be able to reacquire his/her position. Finally, entering and exiting the elevator is guided by speech or gestures. This test integrates navigation, person tracking, person recognition and speech/gesture recognition.

Another test is *"Cocktail party"*, in which the robot has to welcome unknown guests in the apartment. Five persons (unknown to the robot) are in a room of the apartment either sitting or standing. When the robot enters the room, three of these people (one after the other) call the robot by waving and order a drink using a speech command when the robot gets close to them. The robot has to go to the kitchen, grab the drink ordered by the person and bring it back to him/her. This test integrates navigation, person and speech recognition, object detection and manipulation.

The *"Restaurant"* test is executed in a real restaurant (in previous years in a real supermarket). The robot is guided by a user (a team member) through the environment (unknown to the robot) and some locations, e.g., tables and shelves with drinks and food, are described by the user to the robot during this visit. After this introduction phase, the robot receives an order to bring specific food or drink items to some of the locations previously visited and the robot is expected to reach the shelves, grasp the correct items and bring them to the correct locations. This test integrates navigation, mapping, person tracking, speech recognition, object detection and manipulation.

Finally, the test *"General Purpose Service Robot"* focuses on the ability of the robot to understand its goals and reason on them. The task is not specified beforehand, but generated by a random command generator and it is given to the robot through a spoken command. The robot has to understand the desired goal and to accordingly plan actions to execute the behavior requested. For example, a user request may be "bring me a drink" and the robot has to acquire possibly missing information such as the drink to get, and then plan a sequence of actions to go to the location at which drinks are usually stored, grab the ordered drink and bring it to the user. In this test, all the skills may be required (the task is unknown and each possible task actually requires different skills), but in particular the cognition-related skills must demonstrate the ability of the robot to understand the current situation as well as the user request, and to perform a complex task not specified beforehand. Moreover, the user requests may refer to missing information (e.g., underspecified aspects that the robot has to ask for), erroneous information (e.g., wrong aspects the robot has to cope with), and sequences of tasks (i.e., giving a sequence of three commands).

The tests in the competition are of two kinds: *standard tests*, in which the skills and their combination are decided by the Technical Committee, and *open tests*, for which each team can decide which skills to show. Standard tests are evaluated by a partial scoring system described in the next section, while open tests are evaluated in a peer-to-peer fashion with different juries (jury of team leaders, Technical Committee jury, Executive Committee jury, or external jury).

It is important to notice that the tests are improved (or replaced) every year, by making them more difficult and with unpredicted and difficult situations occurring. This evolution is important in order to prevent the development of local optima solutions that specialize too much on a particular instance of the problem without general applicability.

4   Scoring System and Analysis

Tests in the RoboCup@Home competition are divided into multiple phases (or sub-goals) and each phase, when accomplished, provides the team with a score. The total score of the test is thus given by the sum of the scores of all the accomplished phases. If all the phases are correctly performed a full score is gained, otherwise only a partial score is collected. Each phase in a test is evaluated in a Boolean manner: it is either fully accomplished or not accomplished at all.

This definition of the score is used to compare and rank the teams and thus to provide the final results of the competition.

However, in order to analyze more in detail the results of the teams during the competition and to compare results over the years, we need a method to measure the performance of the teams in the tests with respect to the desired skills. This further analysis does not affect the final results of the competitions, but it is used to evaluate the performance of the entire competition, as discussed in the next section.

To this end, we associate each phase of a standard test with a set of skills that are required to accomplish the (sub-)task. When a phase is successfully accomplished, we can state that the skills associated with it have been successfully implemented. On the other hand, if the phase is not accomplished, we can state that at least one of the associated skills was not successful, but we cannot say exactly which one, since we do not have any access to the internal state of the system under test.

In order to relate phase scores with the skills, we also define a *weight* for each skill in each phase of a test. This weight is a value in $[0, 1]$ and the weights of all the skills associated to a phase sum to 1. These weights can be intuitively explained as the percentage of contribution of a given skill to achieve a phase of the test, or, in other terms, the probability that it has been the cause of failure, if the phase is not accomplished.

Obviously, determining the weights is not a straightforward task. In practice, we are using an estimation based on our experience and discussed within the league's Technical Committee. Although approximated, we believe that the results obtained in this way are useful to evaluate the overall progress of the competition based on the average performance of the teams.

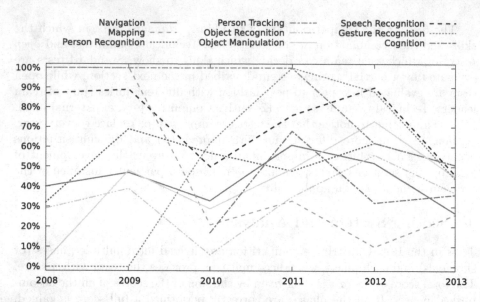

Fig. 3. Part of the statistics carried out after every competition. Looking at the scores achieved by the best teams (per skill) gives a good impression on what was solved well and there still problems exist.

The scoring system allows analyzing the results of the teams in the different skills; moreover, it can be used to evaluate the progress of the entire competition. The results are important for the organizers and Technical Committees of a competition, since they show how the competition is progressing and possibly how to modify the rules in order to drive the development of the competitions. The main goal of this analysis and of the consequent measures is to keep a reasonable level of difficulty over the years and to balance the development on all the skills. Referring to Fig. 3, a skill that has a high score for many years is an indicator that the problem addressed is too simple and that specialized solutions for the particular setting have been found. Moreover, in development of integrated research in AI and robotics, a good balance of all the skills is required. The low standard deviation obtained over the years in RoboCup@Home demonstrate our efforts in developing systems that properly integrate many skills. Some general rules that have been applied in RoboCup@Home are:

- if a skill has a high score (e.g., mapping in 2009), then increase its difficulty in the next year;
- if a skill has a low score (e.g., object recognition in 2010), keep the difficulty unchanged;
- if a skill is not developed at all (e.g., gesture recognition in 2008 and 2009), make it mandatory and increase the value and the phases in which it is used.

For more details on the conducted statistics and methods to derive changes, the interested reader is referred to [2]. In the following, we will give two particular examples for how tests evolved over the years and how both complexity and overall performance increased.

4.1 Following a Human Guide

In addition to (direct) human-robot interaction and mobile manipulation, we consider following (and guiding) human operators an important capability of a domestic service robot. In RoboCup@Home, we began testing human detection, tracking and following from the beginning in the so-called *"Follow me"*-test.

2007: The first implementation provided a proof-of-concept in a simplified setup, e.g., in the first phase special markers on the guide have been allowed. Many teams succeeded (or in case of failure could at least show that the robot possesses the ability of detecting and following a human guide). However, some teams also managed to present a general solution that already worked for arbitrary operators (and did not require customization of the test setup). Following was considered an important aspect, but had to be increased in complexity.

2008: In order to stabilize in performance and form a common test bed, artificial markers were no longer allowed. Instead, the teams could use a previously known person (e.g., a team member) and calibrate on this very guide, e.g., by tracking shirt color etc. Furthermore, the test was made competitive in order to integrate a measure of time. In this first implementation, two teams were competing at the same time, and the team finishing the track faster received a bonus. In addition, a demo run outside the arena was conducted where guides and robots walked through the venue (not part of the competition).

2009: Instead of having guides known beforehand, independent referees were used, but the test stayed competitive (two teams in parallel). The same guide(s) have been used for all teams. At the beginning of the test, every robot got one minute for calibration on the guide right after being commanded to follow that person. To integrate new aspects like collision avoidance, the tracks of the robots were chosen to cross uch that the robots had to avoid each other while still following their respective guides.

2010: The robustness of the approaches (not only for person following) in the arena considerably improved. In 2010, we introduced two tests that take place outside of the known RoboCup@Home arena and in (possibly) crowded public places—one of them being *"Follow me"*. That is, the environment is (1) no longer controllable and (2) the environment is not known beforehand and, consequently, no map is available.

2011: After following (even in previously unknown environments) has matured and enough teams succeeded, in 2011, we introduced pre-defined interferences to increase the complexity of detecting, tracking and following even further. The track was split into waypoints. Whereas at one waypoint people passed in between guide and robot), at another the guide disappeared. When coming back to the robot, the guide was accompanied by another person effectively requiring that the robot recognized its guide and did not continue following the wrong person.

Since 2012: These interferences have been made more complex to incorporate a person directly passing and stopping between robot and guide and a situation in which the guide sneaks through a crowd so that the robot has to circumvent the crowd to continue following. In addition, we introduced an elevator. The elevator required that guide and robot leave in opposite order of entering. It also requires guide and robot to interact in order to coordinate. In this setup, the test is conducted since 2012 since no team has so far been able to complete the track without failures.

Possible Extensions: The split into sub-tasks in this test allows for easily removing solved aspects of tracking and following while adding new more complex problems. More general extensions that are planned are (1) endurance and (2) guidance.

(1) Right now, the maximum time for this test is 10 min and the overall track length does not exceed 100 m. In order to foster the endurance aspect and to get closer to a real-world application, a possible future development is to extend the duration of the test and to let robots follow their guides through crowded public places for a longer period of time, e.g., 30 min.

(2) Another possible future development is guidance, e.g., for one part of the track the robot takes over the role of the guide and brings the operator to a certain position while keeping track of the operator (again, with interference as in following).

Another possible future test could include person search in public, crowded environments such as restaurants or supermarkets.

4.2   Finding and Manipulating Objects

Object search and manipulation were not limited to one recurring test, but evolved over a tree of tests over the years.

2007: A first implementation was *"Lost'n'found"* in the 2007 rulebook where robots had to find an object somewhere in the apartment. Manipulation was not involved. In another test (called *"Manipulate"*), an object should be grasped but the location of that object was known.

2008-2009: In order to obtain a more realistic setup, the procedure was changed in the *"Fetch'n'carry"*-test where teams could pick one object out of five (defined by the TC) together with five possible approximate locations of that object (e.g., "on the couch table"). The TC chose and placed the object. The robots were given commands for retrieving the object from the chosen location, but had to find the exact location and grasp the object themselves.

2010-2011: Since the approximate location of the objects was known, real search was not involved. Also, since there was only one object at the approximate location, solely grasping any object without recognition led to success because of the simplified setup. In 2010, these aspects were emphasized in the *"Go get it"*-test by distributing four objects to grasp and four objects to ignore in the whole apartment (all from the set of known objects). Two robots were operating simultaneously and the first robot to find, grasp and deliver one of the objects got a bonus. In the same year, we also introduced the *"Shopping mall"* test in which objects had to be found and manipulated in previously unknown public areas.

Since 2012: Up to 2011, manipulation only considered grasping but not placing objects. In 2012, the *"Clean up"*-test was introduced in which the robots had to find objects, recognize them and bring them back where they belong to (e.g., beverages on the kitchen table). We also introduced unknown objects that had to be brought to the trash bin.

Possible Extensions: Over the years, the number of objects that the robots needed to know and recognize was steadily increasing (to now 25 known objects). It is planned to further increase this number every year simply to increase recognition complexity. Also, objects are so far easy to manipulate (e.g., easily graspable). It is planned to introduce more complex objects, e.g., requiring for two-handed manipulation, being large or heavy, or even fragile. Another interesting aspect is dealing with unknown objects, e.g., categorizing them, retrieving missing information from the Internet etc. A possible future test may include object search in complex setups, such as the ones found in a kitchen or supermarket shelves or corridors.

5   Possible Future Directions

Just as the RoboCup soccer leagues, RoboCup@Home has a long-term goal: robots that find their way in the real world and cope with everyday problems in the real world. A possible application scenario are robots assisting people depending on help while they are traveling to the RoboCup competition. The robots would need to take over many responsibilities from planning the trip over packing everything needed to guiding and assisting their operators on the road, e.g., using public transport to the airport, finding check-in and gate, boarding the airplane, and at the destination find their way to the competition. Obviously, such systems are far from ready to be implemented, but there are many sub-problems on the way that can already be addressed, e.g.,

- **Interaction:** intuitively interacting with people (not used to robots), e.g., asking for the way and taking care of necessary communications (check-in etc.) possibly in different languages.
- **Finding the Way:** obtaining information on the Internet, reading signs, maps and pictograms or interpreting gestures or hand-drawn maps.
- **Long-Term Operation:** being able to run for longer periods of time.

Some problems on the long run towards *real* real-world applicability are already integrated into the RoboCup@Home competition or planned for the next years.

Endurance and Long-Term Operation: In contrast to soccer games that have a defined duration, real-world service robot tasks are open-ended. Moreover, tasks being coped with may take considerably longer than what is currently doable with a single charge of the batteries. That said, to enforce endurance, teams need to work on both hardware and software side for making robots longer (reliably) operable and at the same time address the problems arising in long-term operation (e.g., knowledge management). Already implemented in RoboCup@Home is the *"Enhanced General Purpose Service*

Robot"-test in which robots need to operate for 30 minutes (in contrast to the usual test duration of 10 min) and solve tasks on demand. It is planned, to have more and more parts of the competition running for a longer period of time possibly with several robots being tested simultaneously. We believe that this would also make the competition more interesting for spectators as there is always something happening.

Real-World Application: In contrast to the RoboCup@Home arena in which most tests take place, the real-world is unpredictable, may be very crowded, and may require the robot to interact with people not used to see, hear, or operate robots. Right now, the league features two tests that take place in public areas. However, these areas are still modified by controlling direct access of the audience or simplifying task and environment, e.g., avoid arising problems that—from experience of the Technical Committee—are unsolvable right now. However, it is planned to (1) have more and more tests taking place in the real world and (2) to perform fewer customizations and simplifications.

Semantic Perception and Mapping: Up to now, semantic perception and mapping capabilities are not explicitly being tested in RoboCup@Home. It is planned to foster the development of such capabilities by introducing new tests that also aim for cognitive skills in general. A possible test could be searching for a place in unknown environments like looking for the kitchen or the restroom in a house or restaurant.

Intuitive (multimodal) Interaction: When the league started, the only allowed way of interacting with the robot was by natural speech (i.e., spoken commands). While this was the most obvious and natural way of interacting, it is not always the most convenient or successful way. For example, in places where it is really loud like, for example, the RoboCup venue during a soccer match, giving speech commands is condemned to failure. Instead, we are aiming at multimodal intuitive interaction by having, for example, a combination of speech and gestures, buttons/displays on the robot for direct cooperation, and intuitive touch pad interfaces allowing to remotely command and operate the robots.

Shortcuts in Test Implementations: Many tasks in RoboCup@Home build upon one another or depend on certain user inputs or other events. For the first time in 2014, we integrate more and more shortcuts and workarounds that can be used to continue a test in case of failures, e.g., command being misunderstood or not understood at all, referees and operators acting wrong or giving wrong commands etc. If successful, we are going to foster these workarounds as they (1) allow for evaluating components otherwise inaccessible in the test and (2) make the competition more attractive for the audience.

In addition, we are going to consider the following aspects to be incorporated in the upcoming rulebooks:

– Fostering the benchmarking character: in order to better assess the performances of teams and establish RoboCup@Home as a widely accepted benchmark for domestic service robots, we aim at improving its benchmarking character, amongst others, by having

- more tests per capability and team for better statistics,
- tests that are easy to set up and reproduce, and
- outcomes that are easy to evaluate and compare.

This will, for sure, be accompanied by the introduction of new tools, e.g., for system monitoring or automated task assignment and evaluation.

- Semantic interpretation and categorization of previously unknown objects and complex, unknown environments, such as houses never seen before.
- Improving the cognitive and social skills of robots by integrating adequate tests. This may include language skills, but also social behaviors.
- Improving safety and security aspects, especially for interaction and cooperation with non-experts.
- Human-robot cooperation and inter-team robot-robot cooperation.
- Development and provision of a standard platform for RoboCup@Home in order to be more attractive for teams not willing to develop and maintain their own hardware.

6 Conclusion

In this paper we have presented the RoboCup@Home league—a league for domestic service robots—which addresses (in the long run) everyday problems in the real world. In particular, the competition is implemented as an annual benchmark for different robot capabilities ranging from intuitive human-robot interaction to mobile manipulation. Part of this benchmark is evaluating and keeping track of the teams' performances over the years and to use this information to drive the league's development. We could show that the implemented changes in the last years considerably increased complexity while keeping the overall performance nearly stable. That is, the participating teams not only robustified existing capabilities but also improved on them and implemented new ones. We detailed some of the capabilities within the competition and gave an outlook on future developments that—in the long run—will hopefully allow domestic service robots to find their way in the real world and cope with everyday problems in the real world.

References

1. van der Zant, T., Wisspeintner, T.: RoboCup X: a proposal for a new league where RoboCup goes real world. In: Bredenfeld, A., Jacoff, A., Noda, I., Takahashi, Y. (eds.) RoboCup 2005. LNCS (LNAI), vol. 4020, pp. 166–172. Springer, Heidelberg (2006)
2. Holz, D., Iocchi, L., van der Zant, T.: Benchmarking intelligent service robots through scientific competitions: the RoboCup@Home approach. In: Proceedings of the AAAI Spring Symposium Designing Intelligent Robots: Reintegrating AI II (2013)

Special Track on Open-Source Developments

Design of a Modular Series Elastic Upgrade to a Robotics Actuator

Leandro Tomé Martins[1,2], Roberta de Mendonça Pretto[1,2], Reinhard Gerndt[2], and Rodrigo da Silva Guerra[1(✉)]

[1] Centro de Tecnologia, Univ. Federal de Santa Maria,
Av. Roraima, Santa Maria, RS 1000, Brazil
{leandromartins,robertapretto}@mail.ufsm.br, rodrigo.guerra@ufsm.br
[2] Department of Computer Sciences, Ostfalia University of Applied Sciences,
Am Exer 2, 38302 Wolfenbüttel, Germany
r.gerndt@ostfalia.de

Abstract. In this article we present a compact and modular device designed to allow a conventional stiff servo actuator to be easily upgraded into a series elastic actuator (SEA). This is a low cost, open source and open hardware solution including mechanical CAD drawings, circuit schematics, board designs and firmware code. We present a complete overview of the project as well as a case study where we show the device being employed as an upgrade to add compliance to the knee joints of an existing humanoid robot design.

Keywords: Series elastic actuator · Passive compliance

1 Introduction

Traditional robot manipulators, such as the ones designed for use in controlled industrial settings, typically use very stiff joints, heavy and rigid structures and powerful actuators. This is a practical way to isolate the influence of reaction forces caused by the load being manipulated. Nowadays, however, there has been an increasing interest in the design of humanoid robots with compliant joints, capable of sharing their workspace with people. These joints allow for much safer and smoother human/robot interaction. Compliance also allows these joints to absorb the energy of possible impacts, preventing gear damage. This energy can also be released back into the environment in a controlled way, allowing for efficient dynamic walking or even jumping or running.

Most solutions for adding compliance into the design of robot joints can be divided in two groups: (1) active (or simulated) compliance and (2) passive (or real) compliance. Simulated compliance is achieved through software, by continuously controlling the impedance of back-drivable electric motors (see for instance [5]). Real or passive compliance is achieved through the employment of real elastic elements, typically mechanical springs, in the design of the joints (see for instance [3]). For a while there has been some debate on the advantages

© Springer International Publishing Switzerland 2015
R.A.C. Bianchi et al. (Eds.): RoboCup 2014, LNAI 8992, pp. 701–708, 2015.
DOI: 10.1007/978-3-319-18615-3_57

and disadvantages of choosing active versus choosing passive compliance [10]. However, with regard to human/robot interactions, there is consensus that passive compliance ensures higher levels of safety. In a recent review [2] pointed to the importance of real compliance when building robotic arms for assistive technology.

This paper describes the design of an open-hardware/open-software Series Elastic Actuator (SEA). All project files are made available at our group's website [9]. A SEA is a type of actuator of passive compliance, where a spring is placed in series with the rigid output thus granting elasticity to the system. Our designed device consists of software, firmware, electronics and a mechanical accessory that can be easily attached to the popular Dynamixel MX-28 series servo actuator, manufactured by Robotis, transforming it into a SEA. This servo actuator was chosen as the base due to its wide popularity within the RoboCup community, however the general idea could be easily adapted to fit most servo actuators of similar "RC-servo-style" design. The distinctive features of our design are its modularity and its versatility. Adding to these features the low manufacturing cost, we believe this device has great potential for application within the RoboCup context and elsewhere.

The remainder of this work is organized as follows: Sect. 2 explains the main details regarding the design as well as the modelling of the SEA. Section 3 shows some data regarding the actual construction of the device and a robot upgrade case. Section 4 presents the closing remarks and outlines potential future work.

2 Methodology

This section is divided in three subsections: Subsect. 2.1 presents the mechanical design of the SEA. Subsection 2.2 shows the electronics and firmware design. Subsection 2.3 gives a general idea of how the system is modelled as a whole.

2.1 Mechanics

The presented SEA follows a compact, modular, low cost mechanical design similar to that presented by Meyer et al. [8]. The device was designed such that it could be easily adapted to existing robot projects, requiring as little change as possible in the mechanics and electronics. It consists mainly of two parts that can rotate relative to each other, a set of two springs and a lid. See exploded view of Fig. 1. The bottom part consists of a disc with a wedge fixed to it. The middle part is a solid cylinder with a C-shaped cut. Both bottom and middle parts are designed to fit on top of each other with the bottom wedge inside the C-shaped cut, forming two arc-shaped chambers, where helical springs are installed. Finally a lid is designed to fit on top of the cylinder, in order to keep the springs enclosed inside their respective chambers. Figure 2 shows the assembled device (without electronics).

When a torque is applied, the bottom wedge slides through the C-shaped cut, expanding the spring of one chamber while compressing the spring in the other

Fig. 1. CAD exploded view   **Fig. 2.** Assembled SEA

chamber. In order to simplify their fixture design, both springs are designed to work always under compression. Notice that the angular range of motion in this elastic element does not need to be large because the servo actuator can dynamically extend this range, in a closed feedback loop. Focusing on their specific application Meyer et al. [8] used elastomer based, non-linear springs in an asymmetric design (compliant in one direction, stiff in the other). We focused instead in a lower cost, more general design, so we used linear helical springs in a symmetrical setup allowing compliance in both directions.

2.2   Electronics

In order to read the displacement of the springs we designed a magnetometer-based contact-less encoder circuit. A disc shaped, radial rare-earth magnet (typically made of neodymium) is placed in the center of the assembly, right below the lid, but attached to the bottom part through a hole in the middle. The sensor board is placed on the lid above it, allowing the magnetometer to measure the direction of the generated magnetic field.

The chosen chip was the AS5043 manufactured by AMS, the same used inside the Dynamixel MX-28R. This SIC offers a 10 bit DAC interface resulting in a resolution of $360 \deg/2^{10}$ steps $= 0.3515625 \deg$/step. There also is an analog output which could be combined with a custom external DAC to allow for even higher resolution, but for this version we used the 10 bit digital interface. The circuit was separated in two boards: (1) a small one just for the magnetometer, placed on top of the lid, and (2) an interface board to read the raw data from the sensor and communicate through RS485 protocol. Figures 3 and 4 show the schematics of the magnetometer and interface boards, respectively. To simplify the development of the firmware, the interface board was made compatible with the widely used Arduino standard [7].

The interface's firmware was programmed to communicate using Dynamixel's RS485 protocol. Each device can be programmed to receive a distinct id thus allowing them to communicate through the same bus as the original servo actuators, using the same protocol. This means no change is required in the electronics

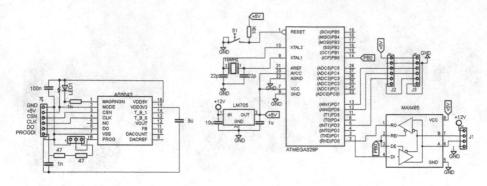

Fig. 3. Magnetometer circuit   **Fig. 4.** Magnetometer's interface circuit

of existing robot projects, except for the extra wires to include the additional devices to the existing communication bus. Torque measuring and control is achieved via software, as explained in the section below.

2.3 Modelling

The spring converts angular deformation into torque and vice-versa thus allowing us to transform a position control problem into a torque control problem. In the case of a linear spring this relation is given by Hooke's law, which here takes the form

$$\tau = -k \cdot \alpha \tag{1}$$

where τ is the torque, k is the linear spring stiffness constant and α is the angular displacement of the device.

See the block diagram of Fig. 5 for a more complete overview of the model. The dotted box at the top represents the computer responsible for the mid-level control of the joints[1]. In the case of an upgrade we suppose the same original device used for communicating instructions and data with the servo actuators can be used. The dotted box below represents the original Dynamixel servo actuator, without any firmware changes.

As usual, the rigid servo actuators are programmed for position control. These try to minimize an error $e = \beta^* - \beta$, where β^* and β denote respectively the desired and the current angular position. When there is no external load ($\tau = 0$) the SEA assembly rotates as a whole and $\alpha = 0$. However when there is some load ($\tau \neq 0$) then a corresponding angular displacement of the spring α will be read by the controller. The total angle of the joint as a whole can be easily obtained by summing $\beta + \alpha$ and the result can be used normally (e.g. to calculate direct kinematics).

---

[1] Tipically a higher level fully featured computer is used for things like planning, vision and data fusion while another mid-level computer takes care of controlling all the joints in real time.

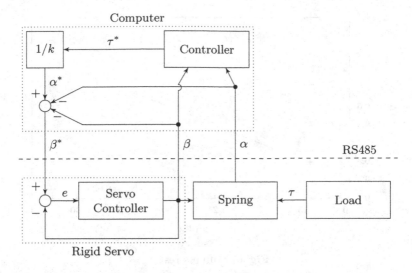

Fig. 5. Block diagram of the SEA system

With Eq. (1), the angular displacement of the spring α is used by the controller to calculate the torque τ being applied by the load into the actuator. At this point, depending on the final application, either or both τ and $\beta + \alpha$ can be used combined with other tools such as robot's inverse kinematics, Jacobian and dynamic models in order to decide a desired torque τ^* to be applied to the joint. Given this desired torque τ^*, Eq. (1) can be used again, this time to calculate a desired spring displacement α^*, and consequently the desired angular position to be sent to the servo actuator is calculated as $\beta^* = \alpha^* - \alpha - \beta$.

3   Results

The parts were machined in aluminium using CNC code generated from the CAD drawings. To measure the stiffness of the assembled SEA we attached a lever to one of its sides while the other side was fixed to a bench using a vise. Then the lever was placed in the horizontal position and known weights were hung to it, while the resulting angular displacement was measured. To calculate the applied moment we used the projected length of the lever on the horizontal plane and the applied weight combined with the weight of the lever itself. Figure 6 shows the laboratory test results. Notice that larger displacements move the spring outside its range of linear behaviour. Considering only the linear range, the spring stiffness constant was estimated to be approximately $k = 0.02 \, \mathrm{deg/Nm}$. Given that the original actuators are rated at 2.35 Nm, the resulting torque range of up to 0.7 Nm found in this prototype is restrictive for many applications, however this can be easily adjusted by choosing a different set of springs.

In order to confirm our SEA's potential for upgrading existing designs we adapted the device to the knees of an existing humanoid robot designed for the

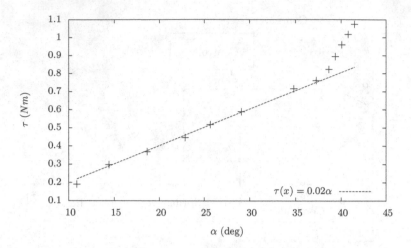

Fig. 6. Stiffness test

RoboCup Humanoid KidSize Soccer Tournament [1]. This robot is based on the DARwIn-OP platform [4]. All we needed to do was to partially change the design of a single part to accommodate the larger joint width so that the spring could be attached to the existing servo actuator. Figures 7 and 8 show respectively the exploded CAD view and the picture of the upgrade.

4 Discussion and Future Work

This paper presented the implementation of a modular and low cost SEA to be used for adding compliance and torque control to existing rigid robot designs. This is the first version of an open software and open hardware system which we hope can be copied and improved upon by other roboticists. We are currently focusing on improving the quality of the design and working on the dynamical model. We have also started testing the design using an alternative rubber-like material instead of helicoidal springs.

Upon completion of field tests with current setup with small servos and robots, the authors plan to apply the findings to larger robots with stronger servo drives. We consider compliance of the joints an increasingly crucial property for the robustness of larger robots, especially with larger weights and falling heights.

On another case study, we have recently started working on a two-link planar robot inspired on the robot-aided neuro-rehabilitation technology developed by Krebs and Volpe [6]. In this device we installed two of our SEAs, one in each joint transforming it into a compliant position and force controlling system. We used an ordinary pen as the end-effector. Pen movements are monitored allowing the application of corrective forces to user movements. This general framework allows not only for game-based neuro-rehabilitation applications but also for computer based dexterity enhancement, where the inference of user intent can be used to improve upon his or her actions (Fig. 9).

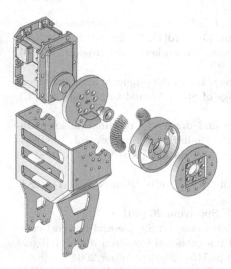

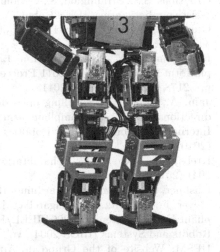

Fig. 7. Exploded view of the knee upgrade showing internal parts of the SEA

Fig. 8. Humanoid robot based on the DARwIn-OP platform, adapted to use the SEA in its knees

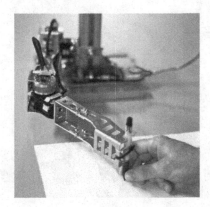

Fig. 9. Prototype of a two-link planar robot for assisting on hand movements

The same technology allows for a large variety of human-robot interactions, including exoskeletons, teaching by demonstration, telepresence with force-feedback, and much more.

References

1. Gerndt, R., Krupop, S., Ciesielski, S., Carstensen, J., Gillich, F., Bolze, T.: WF Wolves KidSize team description RoboCup 2011. In: Workshop RoboCup Singapore (2011)

L.T. Martins et al.

2. Groothuis, S.S., Stramigioli, S., Carloni, R.: Lending a helping hand: toward novel assistive robotic arms. IEEE Robot. Autom. Mag. **20**(1), 20–29 (2013)
3. Guizzo, E., Ackerman, E.: The rise of the robot worker. IEEE Spectr. **49**(10), 34–41 (2012)
4. Ha, I., Tamura, Y., Asama, H., Han, J., Hong, D.: Development of open humanoid platform DARwIn-OP. In: 2011 Proceedings of SICE Annual Conference (SICE), pp. 2178–2181, September 2011
5. Jain, A., Kemp, C.C.: Pulling open doors and drawers: Coordinating an omnidirectional base and a compliant arm with equilibrium point control. In: IEEE International Conference on Robotics and Automation (ICRA), pp. 1807–1814 (2010)
6. Krebs, H.I., Volpe, B.T.: Rehabilitation robotics. Handb. Clin. Neurol. 110(283–294) (2013)
7. Kushner, D.: The making of arduino. IEEE Spectrum 26 (2011)
8. Meyer, F., Sprowitz, A., Lungarella, M., Berthouze, L.: Simple and low-cost compliant leg-foot system. In: 2004 IEEE/RSJ International Conference on Intelligent Robots and Systems (IROS 2004), vol. 1, pp. 515–520, September 2004
9. UFSM: Website of the Grupo de Automação e Robótica Aplicada (GARRA) (2014). http://garra.ufsm.br/
10. Wang, W., Loh, R.N.K., Gu, E.Y.: Passive compliance versus active compliance in robot-based automated assembly systems. Ind. Robot **25**(1), 48–57 (1998)

Collaborative Behavior in Soccer: The Setplay Free Software Framework

Luís Mota[1], João A. Fabro[2]($\boxtimes$), Luis Paulo Reis[3,4], and Nuno Lau[5,6]

[1] ISCTE-IUL-Universitary Institute of Lisbon, Lisbon, Portugal
luis.mota@iscte.pt
[2] UTFPR-Federal University of Technology - Paraná, Curitiba, Brazil
fabro@utfpr.edu.br
[3] LIACC - Artificial Intelligence and Computer Science Laboratory, Porto, Portugal
[4] Department of Information Systems, School of Engineering,
University of Minho, Braga, Portugal
lpreis@dsi.uminho.pt
[5] Deptartment of Electronics, Telecommunications and Informatics,
University of Aveiro, Aveiro, Portugal
[6] IEETA - Institute of Electronics and Telematics Engineering of Aveiro,
University of Aveiro, Aveiro, Portugal
nunolau@ua.pt

Abstract. The Setplay Framework (available from SourceForge as free software) is composed of a C++ library (Project name: *fcportugalsetplays*), a fully functional RoboCup Simulation 2D demonstration team (*fcportugalsetplaysagent2d*), and a complete graphical tool (*SPlanner*), that can be used to design and plan the collaborative behavior between the soccer player agents. In order to demonstrate the usage of the Setplay library, a complete 2D simulation team, based on Agent2D, was developed. This example team uses the framework to execute previously planned collaborative behavior. This framework can be used both within simulated environments, such as the Robocup Soccer Simulation leagues, and with real soccer playing robots. This paper presents the free software Setplay Framework, and provides the necessary information for any team to use the framework with the goal of providing collaborative behavior to a team of soccer playing robots.

Keywords: Robocup · Collaborative robotics · Setplay

1 Introduction

Over the last couple of years, a complete framework for the specification, execution and graphic design of Setplays was designed and developed [1–5]. Setplays are sequences of actions and player strategic positions that should be executed cooperatively by a set of players in order to achieve an objective during the match, more commonly following a stoppage, butpossibly also in play-on situations. Setplays are frequently used for corner kicks and direct free kicks.

© Springer International Publishing Switzerland 2015
R.A.C. Bianchi et al. (Eds.): RoboCup 2014, LNAI 8992, pp. 709–716, 2015.
DOI: 10.1007/978-3-319-18615-3_58

The Setplay Framework evolved to a complete set of concepts, with representation and execution mechanisms, to achieve effective collaborative behavior [1]. The concepts of the framework were developed and used in implementations of the FCPortugal 2D Simulation team since 2009 [6,7]. In order to simplify the specification and development of the collaborative behavior, a graphical tool was developed [8], which later evolved to a complete environment for collaboration design [9], allowing the graphical definition of complex setplays.

The remainder of the paper presents the Setplay Library (Sect. 2), the SPlanner tool (Sect. 3), and a complete example of a team based on Agent2D (Sect. 4) that uses the library to execute setplays. Finally, in Sect. 5, some conclusions and possibilities of future work are presented.

2   The Setplay Library

Setplays (or set pieces) are pre-planned interactions among team members, used in specific situations by many teams, in collective sports such as football, rugby and soccer[1]. In the scope of the Setplay Framework, setplays are understood as predefined collaborative plays, used most commonly after stoppage situations (corner kicks, throw-ins, etc.). A setplay's goal is to surprise the opponent team with favorable sequences of passes and movements, ideally leading to a clear chance to shoot at goal.

An initial version of the Setplay Library [1] was developed and evaluated specifically for the FCPortugal 2D Simulation Team. Besides this, the library has also been successfully applied to the FCPortugal3D team (3D Simulation League) and the Middle Size League team CAMBADA [10]. The Setplay Framework is built upon a standardized, league independent, S-Expression based specification language, that defines setplays in enough detail for direct interpretation and execution in any soccer league, possibly including human soccer teams, at run-time.

The setplay is defined by a set of basic parameters, such as its name, identification number, the region of the field where it is planned to start, players involved and the game mode (that can be any situation in the game, e.g. Kick-off, Throw-in, Corner-kick). Figure 1 presents a diagram of the entities present in the definition of a setplay. A setplay's definition is, roughly speaking, a sequence of *Steps*. Each *Step* has a waiting time for action execution and a time-out to abort the setplay, as well as a set of *Participants*. Entering into a *Step* also depends on a *Condition*, that must be satisfied beforehand. *Steps* have one or more possible *Transitions* that indicate a set of actions that must be executed in order to reach another *Step*. In Fig. 2 the possible actions are graphically depicted.

The complete library that implements this specification was recently made available as free software[2]. Together with the library itself, a complete example

---

[1] See http://www.professionalsoccercoaching.com/free-kicks/soccerfreekicks2 for a complete example.

[2] Available for download at http://sourceforge.net/projects/fcportugalsetplays.

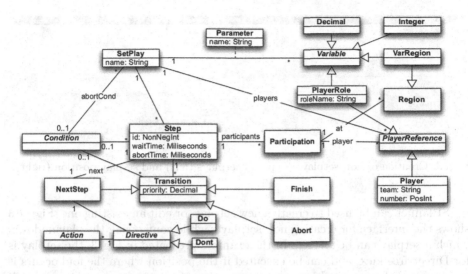

Fig. 1. Setplay concepts and their relationships

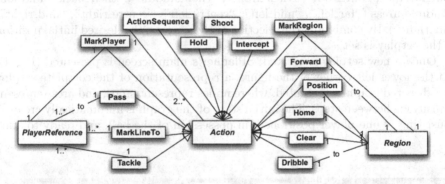

Fig. 2. Action base class and its derived classes

of its integration in a fully functional 2D team, Agent2D version 3.1.1 [11], that can use all the functionalities of the library, was also made public[3].

3 The SPlanner Graphic Specification Tool

In order to make it easier to specify complex setplays, a graphic tool (Playmaker [8]) was developed to allow the graphic design of setplays. An enhanced version of this tool, resulting from a complete re design and integration with both the 2D Soccer Simulator and LogPlayer was called SPlanner [9], and has also been made publicly available[4].

---

[3] http://sourceforge.net/projects/fcportugalsetplaysagent2d.
[4] http://sourceforge.net/projects/fcportugalsplanner.

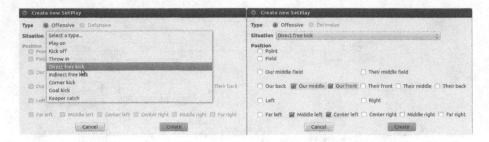

Fig. 3. Creation of new setplay - Types of setplays (left) and setplay position (right).

SPlanner can be used to create a new setplay, or edit an existing one. Figure 3 shows the interface for creating new setplays, while configuring the playmodes in which a setplay can be started. In the example presented in Fig. 3, the setplay is for Direct free kick, and can be executed if the position where the foul occurs is inside one of the accepted areas, an intersection of transverse areas ("our_back", "our_middle", "our_front", "their_front", "their_middle", "their_back") and longitudinal areas ("far_left", "mid_left", "centre_left", "centre_right", "mid_right", "far_right"). By combining the selection of these areas, the desired initial position of the setplay is set.

Once a new setplay is created, SPlanner's main screen is presented (Fig. 4). On the lower left corner of the figure, a representation of the complete setplay as a directed graph is presented, where nodes represent steps, and arcs represent transitions between steps. The initial state of the graph is automatically created, consisting of one player positioned, in possession of the ball, and ready to start

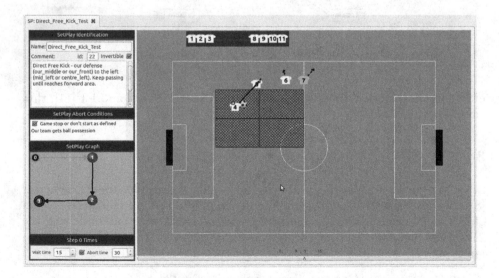

Fig. 4. Example of setplay definition inside the SPlanner graphic tool

the setplay. After the initial positioning the players in the field (realized by simple "drag-and-drop" operations), actions can be assigned to each player by clicking over them, in a context menu.

Figure 5 presents all the possible actions that can be assigned to a player. The context menus only allow possible actions to be assigned to each player. As such, the player with the ball can execute the actions direct pass(a), forward pass(b), dribble(c), hold ball(d) or shot at goal(e). In any step of the setplay execution, the following positioning actions can be associated to each player not in possession of the ball: hold(f), run(g), or run to the offside line(h). If no action is specified, the player will remain in its initial position.

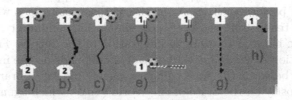

Fig. 5. Graphic design of the actions a player can execute during a setplay: (a) direct pass, (b) forward pass, (c) dribble, (d) hold ball, (e) shoot at goal, (f) hold, (g) run, (h) run to the offside line

In Fig. 6, the steps of a setplay following a corner kick are detailed. The setplay starts on the left side of the field, with player 6 taking the kick and passing the ball to player 7, while player 8 runs to a point inside the goal area (a). In the next step, player 7, after receiving the ball, forwards the ball to the free space in front of player 8 (b). Finally, player 8 tries to shoot at goal(c). If the setplay is labeled as invertible, it can also be executed on corner kicks on the right side of the field: all player positions and actions will simply be mirrored about the x-axis.

By assigning actions to different players at each step, it is possible to create complex collaborative setplays for each situation. The SPlanner tool allows for

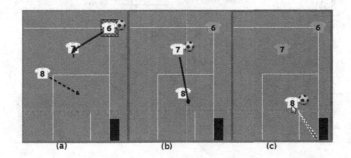

Fig. 6. Complete example of setplay in SPlanner

the graphic specification of such complex behavior, with an user-friendly interface. To evaluate and test the execution of setplays, it is possible to export the setplay to a configuration file, and even start the simulation directly from the application.

4   Using the Setplay Library to Obtain Coordinated Colaborative Soccer Playing Behavior

In order to enhance a robot-soccer playing team with the ability to execute a coordinate collaborative behavior, as the one described in SPlanner and made possible by the Setplay Library, it is necessary to apply the framework to a team with good performance. As Agent2D's [11] source code is well known and used by various teams from the 2D Soccer simulation community, this code was chosen to be the basis of an example integration of the Setplay Framework.

The application of the Framework to a team depends on the implementation of an interface between the soccer-playing agent and the library. This interface depends mostly on the State-of-the-World (SoW: which agent has the ball, where each agent is positioned on the field, what is the current play-mode, etc.). The Framework defines an abstract class (*Context*) to model the SoW, whose functions have to be implemented. These functions are mostly lookup functions for predicates (ball possession, ability to shoot at goal, etc.). Other methods allow the inspection of important information: current time of play, ball position, player position, the play-mode and field dimensions.

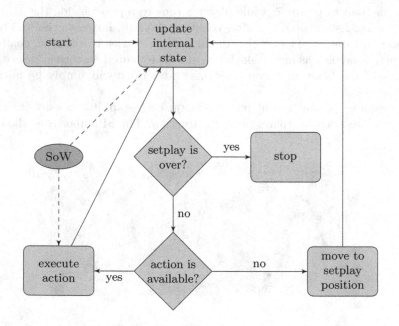

Fig. 7. Execution flow while executing a setplay

As the Framework manages setplay execution, it will need to have some actions executed. It is thus required that an abstract class (*Action::Executor*) is implemented, which has methods to execute all the *Actions* defined by the Framework, such as forwarding or passing the ball, dribble, shoot or move to a position on the field.

Each agent can at anytime use the Framework to determine if some setplay is available for execution, i.e., that setplay's pre-conditions are satisfied. If this happens to be the case, the Framework can also choose the players in the most favorable positions to participate in the setplay.

After the setplay has been started, the control flow will be as presented in Fig. 7: repeatedly, the agent will update the Framework's internal state and check if the setplay has ended. If this is the case, it will exit setplay execution and resume normal behavior. If the setplay is running, it will check if some action should be executed. If this is not the case, due to, e.g., the wait time not having elapsed, it will move to the setplay position.

5 Conclusions

This paper presents the Setplay Framework, composed of the C++ Setplay Library, the SPlanner graphic specification tool, and a complete example of the application of the framework to a team, based on the Agent2D code. Since Agent2D and its related libraries are used by a large portion of the teams participating in the 2D Simulation Robocup league, this release of the framework as free software will certainly be of interest to this community. It allows easy integration of complex collaborative setplays with minimal programming effort. Such collaboration scenarios can easily be planned, designed and tuned using a dedicated graphic tool.

The main contributions in this work to the Robocup community are thus: development and release, as free software, of the Setplay Library, the SPlanner graphical specification tool, and the complete functional example of its use based on the Agent2D source code. The framework is not limited to application to the 2D simulation league, as its generality allows the use by any Robocup soccer team, both in simulated and real robots leagues. Experiments using the setplay library, with both FCPortugal3D and CAMBADA Middle Size League team [10], have proved the generality of the proposed approach. The availability of the complete framework as free software can allow teams from every league to execute complex collaborative behavior, making robotic soccer matches more realistic and exciting to program and watch.

In terms of future development, it is planned to apply the Setplay Framework to teams from other leagues (such as Standard Platform), which will provide new feedback about possible future improvements. The Framework will also be used to bring together players from different teams, by making them commonly execute setplays.

Acknowledgements. The authors would like to thank for the support received from RoboCup Federation, under the scope of the "Call for Project Proposals for Promoting

RoboCup", that allowed the development of this work. The second author would like to acknowledge the scholarship provided by CAPES, Process N°. 9292-13-6, that allowed his participation in this project. The authors would also like to acknowledge Helios Team [11] for their contribution to the Robocup community.

References

1. Mota, L., Reis, L.P.: Setplays: achieving coordination by the appropriate use of arbitrary pre-defined flexible plans and inter-robot communication. In: Winfield, A.F.T., Redi, J. (eds.) First International Conference on Robot Communication and Coordination (ROBOCOMM 2007). ACM International Conference Proceeding Series, Vol. 318, p. 13. IEEE (2007)
2. Mota, L., Reis, L.P., Lau, N.: Co-ordination in Robocup's 2D simulation league: setplays as flexible, multi-robot plans. In: IEEE International Conference on Robotics, Automation and Mechatronics (RAM 2010), Singapore (2010)
3. Mota, L., Reis, L.P., Lau, N.: Multi-robot coordination using setplays in the middle-size and simulation leagues. Mechatron. **21**(2), 434–444 (2011)
4. Mota, L.: Multi-robot coordination using flexible setplays: applications in robocup's simulation and middle-size leagues. Ph.D. thesis, Faculdade de Engenharia da Universidade do Porto (2012)
5. Reis, L.P., Almeida, F., Mota, L., Lau, N.: Coordination in multi-robot systems: applications in robotic soccer. In: Filipe, J., Fred, A. (eds.) ICAART 2012. CCIS, vol. 358, pp. 3–21. Springer, Heidelberg (2013)
6. Reis, L.P., Lau, N., Mota, L.: FC portugal 2009–2D simulation team description paper. In: Baltes, J., Lagoudakis, M.G., Naruse, T., Shiry, S. (eds.) RoboCup 2009: Robot Soccer World Cup XIII, CD Proceedings (2009)
7. Reis, L.P., Lau, N., Mota, L.: FC portugal 2D simulation: team description paper. In: del Solar, J.R., Chown, E., Ploeger, P. (eds.) RoboCup 2010: Robot Soccer World Cup XIV, CD Proceedings (2010)
8. Lopes, R., Mota, L., Reis, L.P., Lau, N.: Playmaker: graphical definition of formations and setplays. In: Workshop em Sistemas Inteligentes e Aplicações - 5. Conferência Ibérica de Sistemas e Tecnologias de Informação (CISTI 2010) (2010)
9. Cravo, J.G.B.: SPlanner: uma aplicação gráfica de definição flexível de jogadas estudadas no RoboCup. Master's thesis, Faculdade de Engenharia da Universidade do Porto (2011)
10. Lau, N., Lopes, L.S., Corrente, G., Filipe, N.: Multi-robot team coordination through roles, positionings and coordinated procedures. In: 2009 IEEE/RSJ International Conference on Intelligent Robots and Systems - IROS 2009, St. Louis, USA, October 2009
11. Akiyama, H.: Rctools webpage. http://rctools.sourceforge.jp/pukiwiki/

Author Index

Printed in the United States
By Bookmasters